HEAT AND MASS TRANSFER IN MATERIALS PROCESSING

HEAT AND MASS TRANSFER IN MATERIALS PROCESSING

Edited by

I. Tanasawa
Institute of Industrial Science
University of Tokyo, Japan

N. Lior
Department of Mechanical Engineering and Applied Mechanics
University of Pennsylvania, Philadelphia, USA

HEMISPHERE PUBLISHING CORPORATION
A member of the Taylor & Francis Group

New York Washington Philadelphia London

HEAT AND MASS TRANSFER IN MATERIALS PROCESSING

1 2 3 4 5 6 7 8 9 0 EBEB 9 8 7 6 5 4 3 2 1

A CIP catalog record for this book is available from the British Library.

Library of Congress Cataloging-in-Publication Data

Heat and mass transfer in materials processing / edited by N. Lior, I. Tanasawa.
p. cm.
"Proceedings of the 1990 Oji International Seminar on 'Advanced Heat Transfer in Manufacturing and Processing of New Materials', held on 28 to 31 October 1990 at Tomakomai, Hokkaido, Japan"—Pref.
Includes bibliographical references and index.

1. Manufacturing processes—Congresses. 2. Heat—Transmission—Congresses. 3. Mass transfer—Congresses. I. Lior, Noam. II. Tanasawa, I. (Ichiro), date. III. Oji International Seminar on "Advanced Heat Transfer in Manufacturing and Processing of New Materials" (1990 : Tomakomai-shi, Japan)
TS183.H43 99
670.42—dc20 91-27259
ISBN 1-56032-192-X CIP

Contents

GENERAL TOPICS (1)

GENERAL TOPICS (2)

STEEL PROCESSING

CASTING AND MOLDING (1)

CASTING AND MOLDING (2)

PAPERMAKING

PHYSICAL PROPERTIES

Preface

This book has been published as the proceedings of the 1990 Oji International Seminar on "Advanced Heat Transfer in Manufacturing and Processing of New Materials," held on 28 to 31 October, 1990, at Tomakomai, Hokkaido, Japan. This book contains forty papers presented at the seminar. The papers are representative of the seminar's scope, and include plasma spraying, laser and electron beam processing, crystal growth, solidification, steel processing, casting and molding, and papermaking, as well as fundamental heat transfer issues and physical properties underlying all of the above. The seminar emphasized thorough discussion of the presentations and of the subfields. Brief summaries of the discussions are presented in the rapporteurs' reports.

The importance of the field of heat and mass transfer in materials processing seems to be two-fold.

In the first place, the progress of this field is important from the materials science point of view. Human civilization progressed in unison with the materials that people used. This is well symbolized by the way that we date the history of human civilization as the Stone Age, the Bronze Age and the Iron Age. Human beings have passed through the Iron Age long ago and are now in the middle of the New Materials Age. This New Materials Age started with new metals and alloys, and was then followed by plastic materials. At present we are surrounded by an abundant variety of different materials including composite materials, semiconductors, crystals, amorphous metals, ceramics, and biomaterials, all supporting the rapid development of current technology. Any further industrial development is vitally dependent on concomitant developments in materials science and engineering.

In the second place, the progress of this field is also important from purely a heat transfer point of view. The diagram on next page illustrates the relationship between heat transfer research and engineering and technology. The items in the rectangular boxes, namely, energy, materials, and information are considered to be the three major areas supporting technology. Among these three, energy is the most closely connected with heat transfer research: heat is just one form of energy. Probably nearly 70 to 80% of heat transfer research has had something to do with energy technology, as signified by the large arrow leading from the central circle towards "energy." This close relationship is very important and effective especially in Japan, which does not have indigenous resources of energy.

The heat transfer research community also has interest in the other two areas depicted in the diagram, namely materials and information, and has the

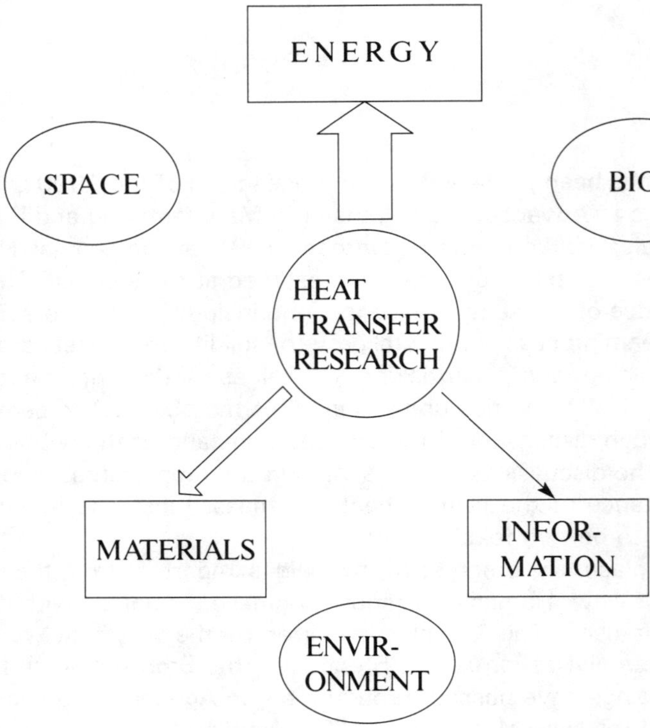

fundamental knowledge and experience to meet the challenges of manufacturing and processing of new materials. It is with this in mind that the editors organized this seminar.

This international seminar is the third in what has become a series of meetings, occurring at five year intervals, to present the state of the art and determine future research and development needs in current areas of key interest for the heat and mass transfer community. The first of these seminars, organized by Professors Tokuro Mizushina and Wen-Jei Yang, was held in 1980 in Tokyo [cf., T. Mizushina and W.-J. Yang (eds.): Heat Transfer in Energy Problems, Hemisphere, Washington, D.C. (1981)] and the second, organized by Professors Wen-Jei Yang and Yasuo Mori, was held in San Diego in 1985 [cf., W.-J. Yang and Y. Mori (eds.): Heat Transfer in High Technology and Power Engineering, Hemisphere, Washington, D.C. (1986)].

The focus this time is on heat and mass transfer in manufacturing and processing of new materials, whereas the last two seminars were energy-oriented. Even in the case of traditional materials, such as steel and paper, the processes described result in products with new or improved properties. Heat and mass transfer are of critical importance in materials manufacturing and processing, and improved and more detailed modeling is therefore needed.

This is especially true when new materials are manufactured, when new properties are sought in traditional materials, and when more economical and energy-conservative processes are desired.

In addition to creating a forum for the presentation of the present frontiers of the state of the art, for recommendations for improvement, and for the identification of priority areas for additional research and development, the seminar clearly identified the benefits of interdisciplinary and international collaboration in this important area.

The support for this seminar of the Japan Society for Promotion of Science and the Fujihara Foundation of Science is gratefully acknowledged.

Ichiro Tanasawa
Noam Lior

PLASMA SPRAY

Heat Transfer in Plasma Spraying

KUNIO HIJIKATA and KENZI MITUI
Department of Mechanical Engineering Science
Tokyo Institute of Technology
2-12-1 Ohokayama, Meguroku, Tokyo 152, Japan

ABSTRACT

A Bi_2Te_3 film was directly coated by a plasma spraying and its heat transfer process was experimentally investigated. A new thermal probe for measuring the temperature field was developed and its accuracy was checked from a structure of coated film. The Seebeck coefficients of Bi_2Te_3 films made under different ambient conditions were compared, and it was determined that the cooling condition during film deposition had a great effect on the thermoelectric performance of the film, especially of Bi_2Te_3 films. It was also shown that a thick thermoelectric film is able to be directly coated on the heat transfer pipe, which may bring about a large improvement in the conversion efficiency caused by the contact resistance between the thermoelectric elements and a heat source.

INTRODUCTION

Plasma spray coatings have already been used to make hard films on materials [1]-[3]. The most important characteristic of a plasma splay coating is the ability to make thick films. A plasma CVD or sputtering can make for very homogeneous and high quality films, but their thickness is limited. From the stand point of energy-related equipment, like a thermoelectric power generator, the thin film is unusable because of its low power density [4].

A thermoelectric power generator has the advantages of no moving parts and the ability to obtain electric power from low quality low temperature energy sources. The new material, iron silicide, discovered in Japan [5] has a very high Seebeck coefficient. However, its thickness is too thin (below 100 μ) to use for electric power and its working temperature is about 300°C, which is too high to utilize low temperature energy sources. The main reason for the high output power of $FeSi_2$ is its amorphous state. Therefore, in this study, a method of making a thick thermoelectric film by using a plasma spraying is investigated from a heat transfer aspect by changing the cooling speed of a coated film. The compound Bi_2Te_3 was used as a test material because it has the highest coefficient of thermoelectric performance in the low temperature region.

The other benefits of using a plasma spraying is the possibility of a direct coating of a thermoelectric film on the pipe wall of a heat source. Even if thicker films can be made by other methods like CVD, those films can not be coated directly on the pipe wall. For the utilization of low quality energy sources by a thermoelectric system, a contact resistance between the thermoelectric element and the heat transfer pipe is important, especially when the film is relatively thin [4]. Therefore, the direct coating of the thermoelectric element on the pipe wall is extremely important to get a high thermal-to-electric conversion efficiency.

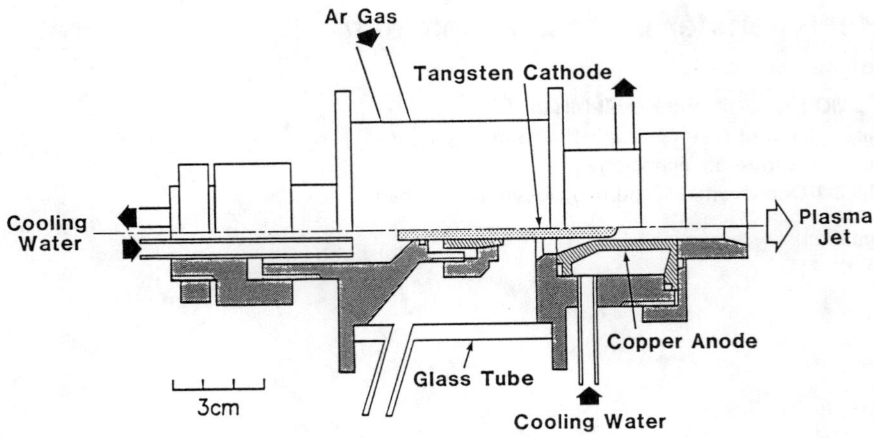

Fig.1 Schematic diagram of plasma gun

In a plasma spraying process, since the temperature histories of sprayed particles is important [6]-[8], an accurate determination of the temperature field is the first goal of this investigation. Methods for measuring the thermal field of plasma exist[9], most of which are electrical and optical in nature. The electron temperature can be easily obtained from a Langmuir probe, but the gas temperature is lower than that due to the non-equilibrium state between the gas molecule and the electron. For the gas temperature, an emission or an absorption process from /by the plasma is used. However, calibration is needed for an absolute determination of temperature, and that is very difficult. In this paper, therefore a newly developed thermal probe was used to determine the gas temperature, but the gas velocity in the plasma must also be known for determination of the heat transfer coefficient.

EXPERIMENTAL APPARATUS

A hand made plasma gun is shown in Fig. 1. The exit diameter of the plasma nozzle gun is 8 mm and the flow rate is about 0.0005 m^3/s. The diameter of the tungsten cathode, having 3% cerium, is 6 mm [10]. To get a stable plasma, adjustment of the gap between the electrodes is very important; a special jig is used to adjust it. An anode is made of copper for better cooling and the heat flux to the cooling water is in the film boiling region. Since the nozzle melted many times (a small pin hole was made first), large fins are attached to the outside of the nozzle to establish nucleate boiling at the fin tips. The cathode is also cooled by water, but its cooling is not enough to establish a long, stable operating time.

Argon was used as the working gas and sprayed particles were entrained by it. To get a stable plasma, a vortex motion was introduced in the inlet chamber of the nozzle. However, the centrifugal force due to this vortex motion moves the sprayed particles to the outer side of the nozzle flow. To see the motion of introduced particles, the inlet chamber wall was made of glass. The power supply was rated for 500 A and 20 kW , and was originally used for Ar welding. The trigger for the plasma generation was a high-frequency electric wave. Since the power supply was used to supply a constant current, the flow rate had to be increased when the plasma temperature decreased for some reason. Since it was our first time to make a plasma gun, the stability of the plasma was not very good and the temperature range of the plasma jet was limited. A very stable plasma, however has been obtained by small improvements in the apparatus.

To change the temperature history of sprayed particles, they were also injected into the plasma jet at the exit of the nozzle. A nozzle length of was 50 mm, and the distance from the nozzle to the coating plate was about 20 mm to 50 mm. Therefore, the residential time of the particle was reduced by 50% when the particle was introduced at the nozzle exit. A coating plate was mounted on a special turn table whose axis was located 20 mm from the plasma jet axis. The turntable was made of copper, and was cooled from the other side. Due to the rotation of the coating plate by the turntable, the substratum of coating did not attached a high temperature. To investigate the effect of the substratum temperature, nitrogen gas evaporated from nitrogen liquid was also used for cooling. The chamber containing the plasma gun and the turntable was evacuated to change the ambient condition for the coating.

In order to feed a very few amount of the particles, many types of particle feeders were tested. Finally, a vibrating type feeder was adopted, in which a part of the carrier gas was injected from the bottom to levitate the particles, and the remaining gas entrained the particles to the nozzle. A vibration motor was used to sustain a steady floating particle stream. The materials used for coatings were nickel, bismuth, and tin, whose melting temperatures are 1728°C, 544°C and 505°C. The evaporation temperatures are 3004°C, 1833°C and 2543°C, respectively. The melting temperature of tin is lower than that of bismuth, but the evaporation temperature of tin is much higher. An alloy of bismuth telluride was also used for making a thermoelectric element, whose thermophysical properties are almost the same as those of bismuth. All of the materials were screened by a 200 mesh screen, and the maximum size of the particles was 100 μ.

MEASURING METHOD OF TEMPERATURE AND VELOCITY

Optical Method

In a thermal plasma, local thermodynamic equilibrium is normally established, and the difference between an electron's and an ionized molecule's temperature is not so large. Under this condition, light emission from the ionized molecule or atom is given as follows:

$$\varepsilon_{ul} = \frac{1}{4\pi} A_{ul} h \nu_{ul} \frac{n_i}{z_i} g_{i,u} \exp\left(-\frac{E_{i,u}}{k\,T_{ex}}\right) \quad (1)$$

where ε_{ul} is the volumetric emission coefficient, A_{ul} is the corresponding transition probability, h is Planck's constant, ν_{ul} is the frequency of the photons emitted due to transition from the upper excited level u to the lower level l, T_{ex} is a temperature, Z_i is the partition function, $g_{i,u}$ is the statistical weight of the upper excited state, $E_{i,u}$ is the energy of the upper excited state, and n_i is the concentration of the chemical species irespectively. If the ratio of the two volumetric emission coefficients of different bands is used, the temperature can be obtained from the following equation:

$$\frac{\varepsilon_1}{\varepsilon_2} = \frac{A_1 \nu_1 g_1}{A_2 \nu_2 g_2} \exp\left(-\frac{E_1 - E_2}{kT_{ex}}\right) \quad (2)$$

When the emission from many bands are measurable, the temperature is obtained from a logarithm plot of (ε/Ang) against E [11],[12].

One typical example measured by this method is shown in Fig.2. The slope of the line in the figure corresponds to the reciprocal of the temperature. However, the measurement accuracy of this method is not so high for the following reasons. First, this theory is applicable only to the simultaneous measurement of multi-band emission. Secondly, the slope of the line connecting the data points becomes lower for higher temperatures.

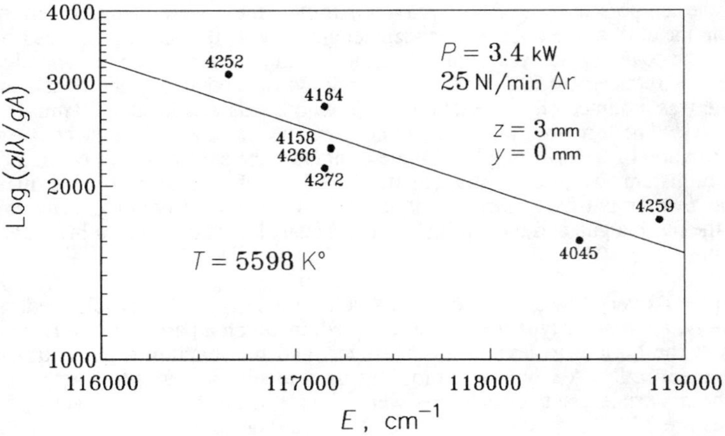

Fig. 2 Emission intensity against the energy of excited level

Langmuir probe

If an electrical probe is inserted into a plasma and an electrical potential is applied to it, the current is a function of the supplied voltage. The current to the probe consists of the thermal motion of the electrons and the drift velocity due to the supplied voltage. Therefore, the magnitude of the thermal motion can be estimated from the current to voltage characteristics. If the Maxwell velocity profile is assumed for the electron motion, the following equation can be obtained:

$$I_p = n_e\, e\, S \left(\frac{kT_{ex}}{2\pi m_e}\right)^{\frac{1}{2}} \exp\left[\frac{-e\,(V_p - V_s)}{k\,T_{ex}}\right] \qquad (3)$$

where n_e is the number density of electrons, S is the surface area of the probe, e and m_e are the electrical charge and the mass of an electron, and k is the Boltzmann constant. Since V_p and V_s are the potentials of the probe and of the plasma, the electron temperature T_{ex} can be obtained from the relation between the probe current I_p and the supplied probe potential V_p. Normally, the electron temperature is higher than that of the gas or ionized molecules, because of the high mobility of electrons. Its difference, however, is not so large for atmospheric pressures, and gives an approximate value of the gas temperature under these conditions.

Thermal probe

The most typical method for measuring the gas temperature is to use an enthalpy probe, in which the plasma gas is sampled and cooled by a coolant whose enthalpy change is measured. If the dissociation and ionization energy of the plasma are negligible, the temperature can be estimated from the enthalpy change. However, the size of this probe is too large to use in our experiment. A now thermal probe was thus developed in our study.

The structure of the probe is very simple. It is a stainless steel pipe of 1 mm O.D. and is cooled by water from the inside. Only the temperature difference between the inlet and the outlet of cooling water, and the cooling water flow rate, are measured. Since the total transfered heat is obtained from the experiment, the temperature difference between the probe and the plasma can be obtained from the theoretical value of the heat transfer coefficient of the small cylinder. The velocity profile of the plasma used in the calculation of the heat transfer coefficient also needs to be measured. Here, it was measured by a cooled Pitot tube. When A_q is an aqueous pressure head, the velocity u is given as follows:

$$u = \left(2 \frac{\rho_l}{\rho} g A_q \right)^{\frac{1}{2}} \tag{4}$$

where g is gravitational acceleration, ρ_l is the liquid density, and ρ is the reciprocal mean density between the wall and the main flow. The Nusselt number is assumed to be given as follows:

$$Nu = 0.52\, F(T_w, T_\infty) Re^{0.5} Pr^{0.37}, \quad Re > 40 \tag{5}$$

where F is the correction factor for variable physical properties. All of the nondimensional parameters are defined at the main flow temperature. If the temperature in the main flow is known, we can get the velocity from Eq.(4) and the Re number is thus known. From Eq.(5), we can get Nu, and the temperature difference is then calculated by using the heat flux obtained from the experiment. Iteration is continued until the obtained main flow temperature is equal to the assumed one.

Since the measured heat flux is an averaged value along the thermal probe, the Abel transformation is necessary to obtain the local temperature. In the Abel transform, the derivate of the heat flux with respect to the transverse coordinate is needed, which means that the results are very sensitive to the measurement accuracy. Therefore, in our calculation, a Gaussian distribution of heat flux was assumed. The comparison the experimental data and the assumed distribution is shown in Fig. 3.

The local heat flux $q(r)$ is obtained from the derivative of $Q(y)$ shown in Fig. 3 by the following equation:

$$q(r) = -\frac{1}{\pi^2 d} \int_r^{r_0} \frac{Q'(y)\, dy}{\sqrt{y^2 - r^2}} \tag{6}$$

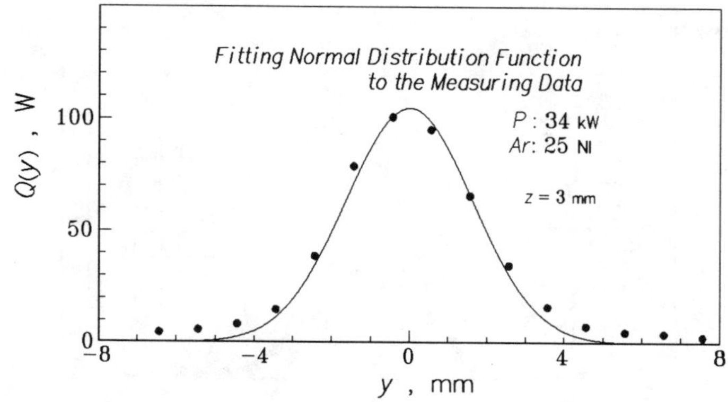

Fig. 3 Measured and assumed heat distribution of thermal probe.

The correction factor for variable physical properties was estimated from the theory of the two-dimensional stagnation flow, because the main heat transfer in the low Re number region is at the front surface. The equations of motion and energy can be expressed as follows:

$$f'^2 - ff'' - \frac{1}{\rho} = (\mu \rho f'')' \tag{7}$$

$$-f\theta' = \frac{1}{Pr_\infty}(\lambda \rho \theta')' \tag{8}$$

where f and θ are the nondimensional stream function and temperature. The quantities ρ, μ and λ are the nondimensional density, viscosity and thermal conductivity, respectively. All variables are nondimensionalized by the main flow values.

In the calculation, it is assumed that μ and λ are proportional to the 0.7 power of the absolute temperature, and ρ is inversely proportional to the absolute temperature. From the calculation, it was clarified that the correction factor for argon plasma is nearly equal to unity for a wide range of main flow temperature. Therefore, f is assumed to be 1 in the calculation for obtaining the main flow temperature.

TEMPERATURE PROFILE IN A PLASMA JET

In Fig. 4, temperature profiles along an argon plasma jet are shown. Most of the data were measured by the thermal probe. Results obtained by the Langmuir probe and the spectroscopic method are also shown. The results obtained by the thermal probe are consistent with the electron temperature, but the optical measurement gives a very low temperature. The reliability of the thermal probe is discussed later. The temperature at the center of the jet decreases rapidly, but the width of the jet does not expand rapidly. In the plasma jet, the change of density is very large. A reduction of the temperature causes an increase in the density, which brings about a reduction in the volume flow rate along the flow direction. As shown in Fig. 5, this is easily understood from the velocity profiles. The inlet flow rate is larger than that of Fig.4, but the shape of the profile is independent of the inlet flow rate. The velocity at the center of the jet decreasing monotonically, but the width of the jet does not change. From a comparison between Figs. 4 and 5, the half-width of the velocity is wider than that of the temperature.

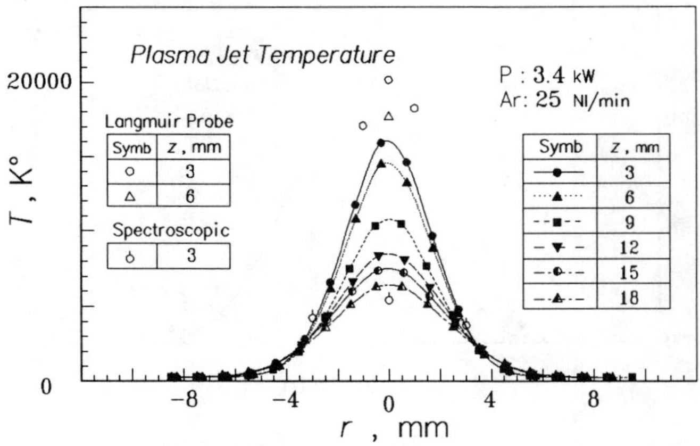

Fig. 4 Temperature distribution in the cross section perpendicular to the jet.

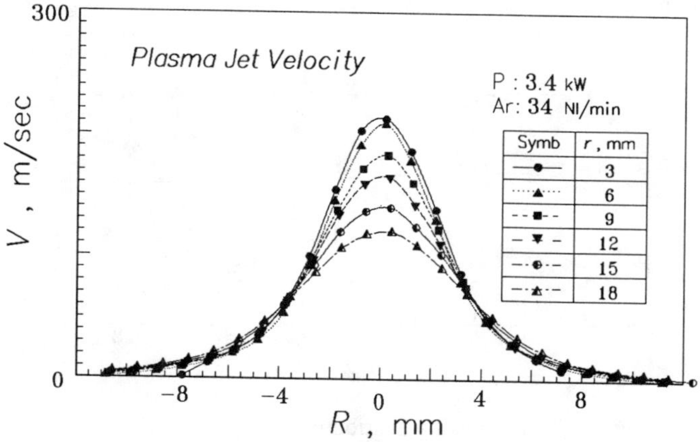

Fig. 5 Velocity distribution in the cross section perpendicular to the jet.

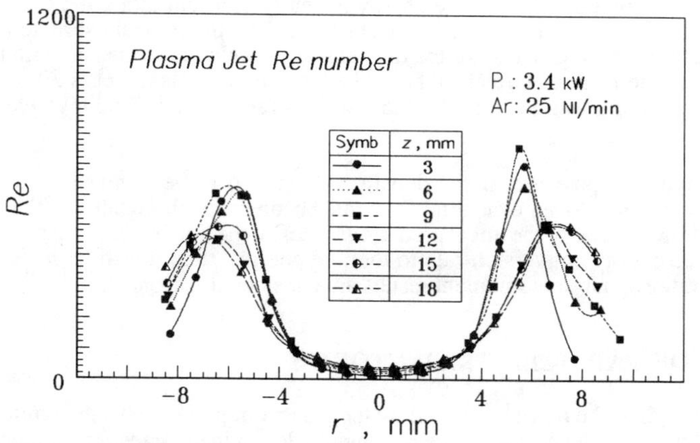

Fig. 6 *Re* number distribution in the cross section perpendicular to the jet.

The maximum temperature in the stable condition increases with a decrease in flow rate and with an increase of the input power. It was very difficult, however, to change the flow rate and input power continuously. Therefore, in this study, only two cases of the input power and two cases of the flow rate were examined. From these results, we can check the of measurement accuracy of the velocity and the temperature. The most realistic ways to evaluate the measurement accuracy are by checking the flow rate and the total enthalpy in the jet. The flow rate in the jet increased along the flow direction and reached twice or more of the inlet flow rate due to entrainment. Near the center of the plasma jet, the velocity and the viscosity were high, but the density was low. The density of the entrained gas was high compared with that of the plasma, and the increase in flow rate in the plasma jet is much higher than that in the low temperature jet flow. From these characteristics, the local *Re* number for the thermal probe exhibits a very strange behavior as shown in Fig. 6. The local *Re* number at the center of the jet is below 100, even though the velocity is over 250 m/s. The *Re* number reaches a minimum at the maximum velocity point. The *Re* number just outside the jet reaches a maximum. This behavior clearly shows that the entrainment has an important effect on the plasma jet.

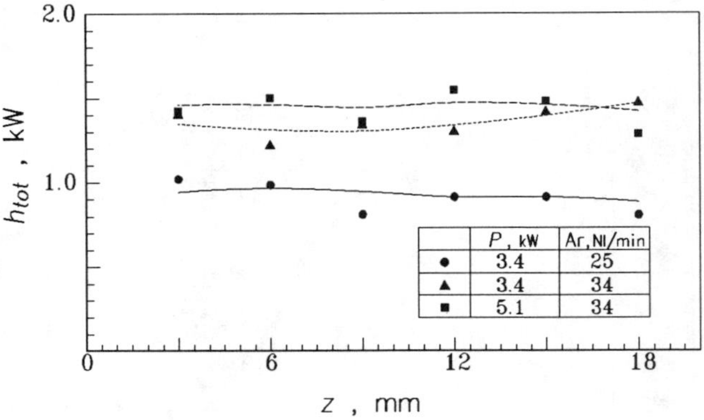

Fig. 7 Total enthalpy distribution along the jet flow.

From the temperature profile, we can determine the variation of the total enthalpy along the jet. In Fig.7, the total enthalpy in the cross section perpendicular to the flow is plotted against the flow direction. The total enthalpy is nearly constant along the flow direction, which means that heat loss by radiation is not large. The theoretical estimate [13] of the radiation heat loss is less than 10% of the total enthalpy, which is consistent with this result. The conversion rate from electrical to thermal energy in the plasma jet is 30 to 50%. The heat loss to the cooling water is high for the low flow rate of the jet and high electric input power.

For a given dynamic pressure, if the density is assumed to be higher heavier (i.e., the temperature is assumed lower), the total flow rate becomes much higher and the difference of the inlet flow rate and the measured flow rate in the jet becomes larger. For this assumption, the energy conversion rate to thermal energy should also be reduced. From these considerations, our measurement results are considered reasonable.

ANALYSIS OF PARTICLE TRAJECTORIES

In a plasma spraying, the particles heated by the plasma impinge on a substratum plate and make a thin film on it. The heating process is mainly determined by the temperature histories of the particles. Since the velocity and the temperature fields are already determined, as stated in the previous section, the particle temperatures can be calculated by assuming heat transfer process from the plasma to the particles [14][15]. The heat transfer resistance in the particle is much lower than that in the plasma, and thus only heat transfer from the plasma to the particle needs to be considered. For small particles, the drag force and the transferred heat are given as follows:

$$F_D = 6\pi\mu(v_g - v_p)D, \qquad \dot{q} = \frac{2\lambda}{D}(T_g - T_p) \tag{9}$$

where D is the diameter, v is the particle velocity and the subscripts g and p denote the gas and the particle, respectively. The initial condition of the particle velocity is zero at the inlet of the nozzle and the gas velocity profile in the nozzle is assumed to be parabolic. When the electrical power input is low, the particles are introduced at the nozzle exit.

The temperature changes of bismuth particles along the flow direction are shown in Fig. 8(a). A particle passing through the center of the jet is heated faster, which melts in a few mm traveling and becomes vapor in the nozzle. However, if the particle is introduced from

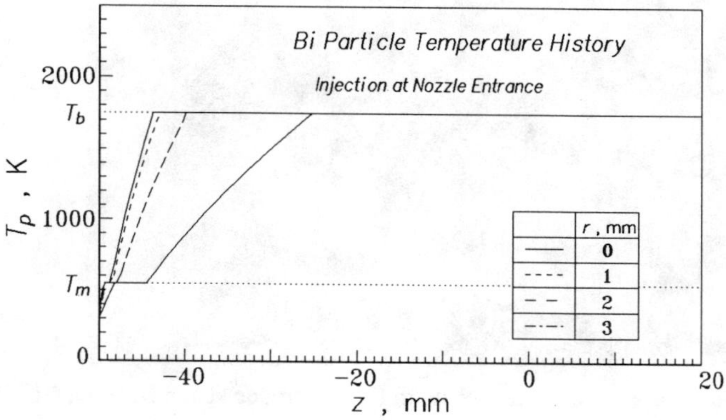

Fig. 8(a) Temperature histories of bismuth particles in the jet.

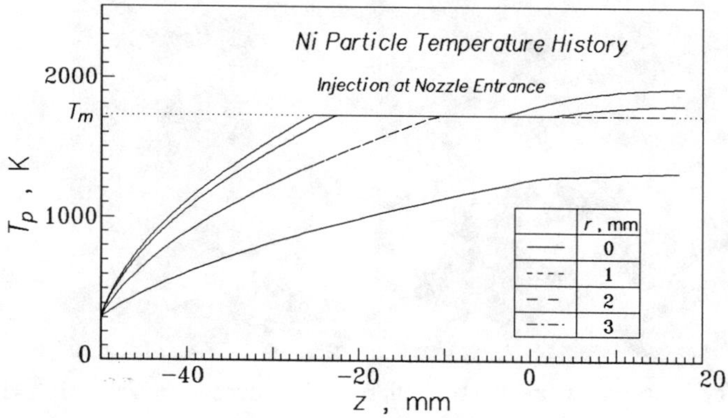

Fig. 8(b) Temperature histories of bismuth particles in the jet.

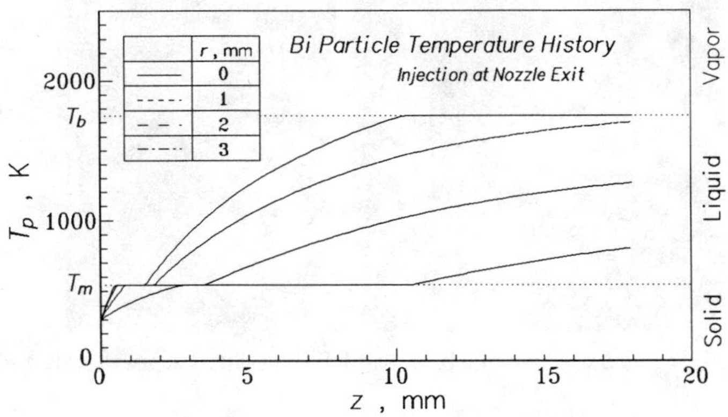

Figs. 9 Temperature history of nickel particle in the jet.

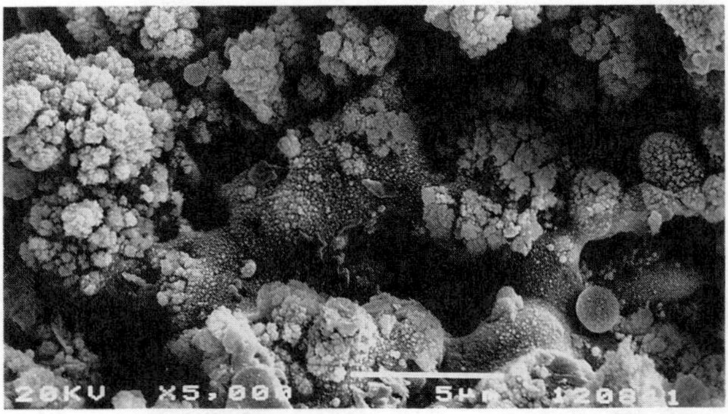

Fig. 10(a) The surface structure of a bismuth film (introduced at nozzle inlet, z=15mm)

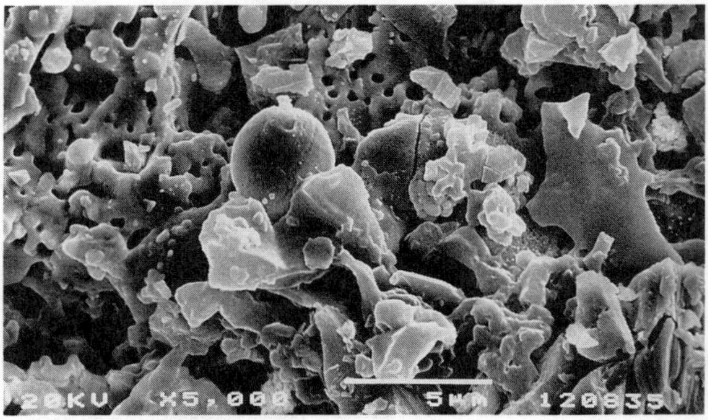

Fig. 10(b) The surface structure of a bismuth film (introduced at nozzle exit, z=15mm)

Fig. 10(c) The surface structure of a bismuth film (introduced at nozzle exit, z=45mm)

the nozzle exit, some particles reach the coating plate without evaporating, as shown in Fig. 8(b).

On the other hand, if nickel particles are used for the coating, the particles never evaporate in the plasma jet, and some particles reach the plate in the solid state without melting, even if they are introduced at the nozzle inlet as shown in Fig. 9. When nickel particles are introduced from the nozzle exit, the particles never reach their melting temperature. In the case of tin particles, they evaporate if they are introduced at the nozzle inlet, but they only melt, and do not evaporate, if they are introduced at the nozzle exit.

Since the measurement of a flying particle's temperature is very difficult, the accuracy of the previous analysis must be determined through indirect means. The surface structure of a bismuth film determined electron microscopy, (EM) is shown in Fig. 10(a). This film was deposited 15 mm downstream from the nozzle, and the particles were introduced at the nozzle inlet. Many fine particles whose sizes are below 1 μ, can be seen. All of these fine particles are thought to have originated from the vapor phase, because the smallest initial particle is larger than 10 μ. For the particles introduced at the nozzle exit, these fine particles are never detected, as shown in Fig. 10(b) . However, when the coating substratum is set far from the nozzle (45 mm), the same fine particles are observed, as shown in Fig. 10(c). These results are quite consistent with the previous calculation.

For nickel particles, the shapes of coated particles are not round, and have many corners, even if they were introduced at the nozzle. This indicated that the nickel particles melted but that the temperature was not much greater than the melting temperature. In the case of a tin coating, fine particles generated from the gas phase were also detected. These observations from EM photos led creditability to the appropriateness of the theory.

The growing rate of the film also indicates from what phase the film is made. As shown in Fig. 11, the growing rate of the film is very high for bismuth, when introduced at the nozzle exit, and for nickel when introduced at the nozzle inlet. In both cases, the calculation shows that the particles should impinge on the plate in the melted state. However, for bismuth particles introduced at the nozzle inlet and for nickel particles introduced at the nozzle exit, the film growth rates are very small. In the former case, the particles are evaporated, and in the latter case, the calculation indicates that the particles are still in solid phase.

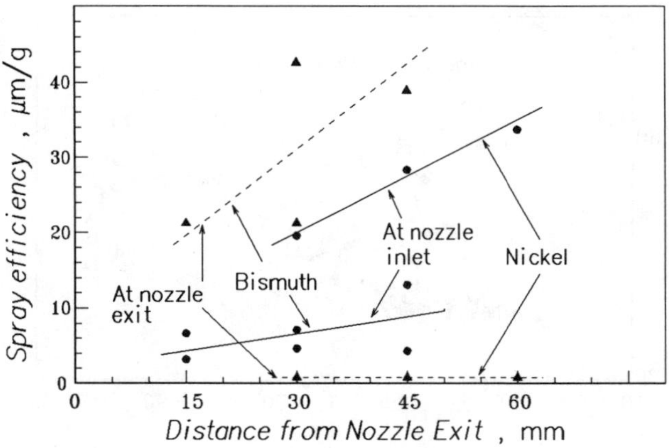

Fig. 11 Film growing rates for different spraying conditions

Production of Bi_2Te_3 Film

Since the thermophysical properties of Bi_2Te_3 are quite similar to those of bismuth, a Bi_2Te_3 film was made by introducing the particles at the nozzle exit. An alloy of Bismuth-Tellrium has high efficiency of thermal to electric energy conversion, and it is always p type, but for obtaining yet higher efficiencies, Sn, Sb, Pb, As or Bi are doped for p type as an accepter, and AgI, CuTe and CuBr are doped for n type as a donor [16]. In this study, bismuth is used as an accepter, and $Bi_{45}Te_{55}$ is made. The substratum is made of copper, and the thermoelectric film is coated on it. The film thickness is about 1 mm.

In order to measure the Seebeck coefficient, the copper electrode was also coated by the plasma spraying. In the measurement of the Seebeck coefficient, the electrode temperatures were measured by fine copper-constantan thermocouples. Films were produced under different ambient conditions, which were 1) the substratum was cooled by water and the ambient gas was atmospheric air; 2) the substratum was cooled by water but the ambient gas was argon; and 3) the substratum was cooled by liquid nitrogen and the ambient gas was argon.

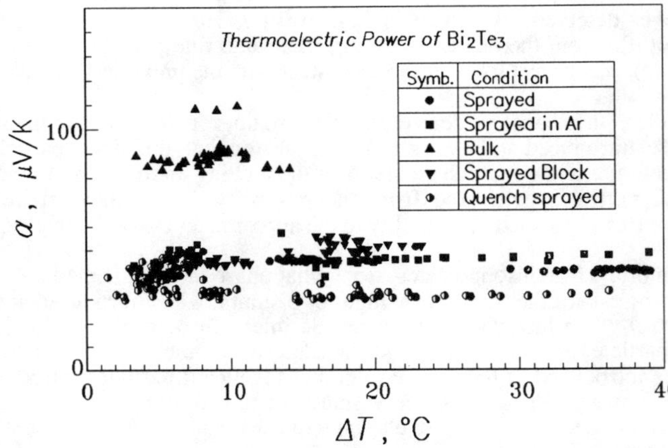

Fig. 12 The Seebeck coefficients of Bi_2Te_3 made by different cooling condition.

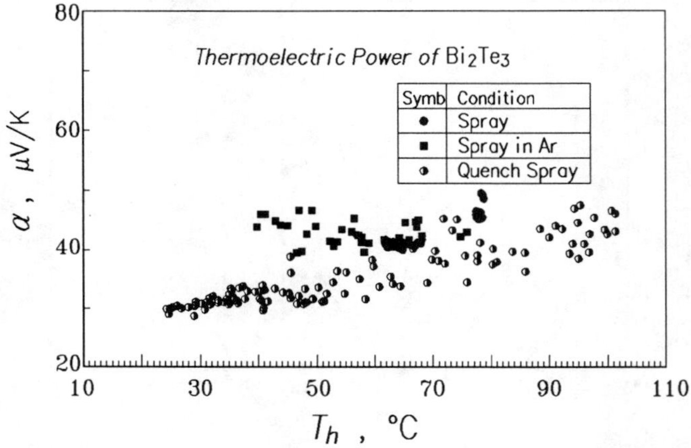

Fig. 13 The Seebeck coefficients of Bi_2Te_3 against the working temperature.

The Seebeck coefficients of these films are shown in Fig. 12 as a function of the temperature difference between electrodes. All of the Seebeck coefficients of the materials made by the plasma spray are less than 50% of the balk material of bulk value of Bi_2Te_3. The coefficient is minimized in the case of liquid nitrogen cooling. These results show that the plasma spraying is not an adequate technique for making a thermoelectric element of Bi_2Te_3. It is known that Bi_2Te_3 is a nonisotropic crystal, and that the Seebeck coefficient along one axis of Bi_2Te_3 crystal is higher than the other. Therefore, a bulk alloy made without considering the crystal orientation has a lower Seebeck coefficient than that of a well-aligned specimen, and a film made by rapid cooling, especially by liquid nitrogen cooling, might have a lower performance compared with that made by melting in a crucible.

However, a film made by rapid cooling, using liquid nitrogen, demonstrates the different behavior, as shown in Fig. 13. The Seebeck coefficient of the film cooled by water is independent of the working temperature, but the Seebeck coefficient of the film cooled by liquid nitrogen gradually increases with the working temperature. These different trends are considered to be caused by the difference in the crystal structure. Since a clear difference in the crystal structure is not detected from an X-ray analysis, EM is used to clearly observed and prove that the surface structures are different. Therefore, it is thought that the crystal structure is still affected by the cooling speed.

CONCLUSIONS

Heat transfer in a plasma spraying process has been investigated and thermoelectric elements were made by this process. From this investigation, the following results have been obtained.

1. A new thermal probe to measure the argon plasma temperature has been developed.
2. The temperature and velocity fields measured by this probe give appropriate results.
3. The temperature histories of coating particles were calculated from the measured velocity and the temperature, and they accurately describe the state of particles impinged on the substratum plate.
4. The Seebeck coefficient of Bi_2Te_3 largely depends on a manufacturing and cooling procedures.
5. By using a plasma spraying, a thermoelectric element can be deposited directly on a heat transfer pipe.

REFERENCES

[1] D.R.Mash, N.E. Weare and D.L. Walker, "Process Variables in Plasma-Jet Spraying", *Journal of Metals*, July 1961, pp.473-478

[2] N.N.Rykalin and V.V. Kudinov, "Plasma Spraying", *Pure and Applied Chemistry*, Vol. 48, 1976, pp229-239

[3] M.R Jackson and etal., "Production of Metallugical Structures by Rapid Solidification Plasma Deposition", *Journal of Metals*, November, 1981, pp23-27

[4] K.Hijikata and etal., "Study on Heat Transfer Aspects of a Thermoelectric Power Generation System Operating under Small and Medium Temperature Difference Conditions", *Heat transfer Japanese Research,* 1988, pp.48-64

[5] K.Matsubara and T.Koyanagi, "Development of Amorphous Transition Metal Silicides as Highly Efficient Thermoelectric Materials", *Research on Effective Utilization and Densification of Electric Energy*, SPEY 24, 1987, pp.87-94

[6] I.Groma and B.Veto, "Melting of Power grains in Plasma Splaying", *Journal of Heat and Mass Transfer*, Vol.29. No. 4., 1986, pp.549-554

[7] A.Vardelle and eial., "Influence of Velocity and Surface Temperature of Alumina Particles on the Properties of Plasma Splayed Coating", *Plasma Chemistry and Plasma Processing*, Vol2, No.3, 1982, pp.255-291

[8] P. Fauchais and etal. "Correlation of the Physical Properties of Sprayed Ceramic Coating to the Temperature and Velocity of the Particles Travelling in Atmospheric Plasma Jets", *Thin Solid Films*, 121 1984, pp.303-316

[9] O.Auciello and D.L.Flamm, "Plasma Diagonostics", Academic Press, 1989

[10] F.Matsuda, M.Ushio and T. Kumagai, "Study on Gas-Tungsten-Arc Electrode", *Transaction of JWRI*, Vol.15, No.1, 1986, pp. 13-19

[11] H.N.Olsen, "Measurement of Argon Transition Probabilities using the Thermal Arc Plasma as a Radiation Source", *Journal of Quantum Spectroscopic Radiation Transfer*, Vol.3, 1963, pp.59-76

[12] H.H.Li and C.J.HUmphreys, "New interferometric observations of Ar 1 in the Photographic region", *Journal of OSA*, Vol.64, No.8, 1974, pp. 1072-1082

[13] X. Chen and E. Pfender, "Unsteady Heating and Radiation Effects of Small Particles in a Thermal Plasma", *Plasma Chemistry and Plasma Processing*, Vol.2 No.3, 1982, pp. 293-316

[14] E.Bourdin, P.Fauchais and M.Boulos, "Transient Heat Conduction under Plasma Conditions", *Int. J. Heat and Mass Transfer*, Vol.26, No.4, 1983, pp. 567-582

[15] J.Fiszdon, "Melting of Powder Grains in a Plasma Flame", *Int. J. Heat Mass Transfer*, Vol.22, 1979, pp.749-761

[16] S.Scherrer and H.Scherrer, "Thermoelectric Optimization of Bismuth Telluride by Stoichiometric Deviation", *SAE,* 1985, pp.3.596-3.573

Structure of Ultra-short Pulse Plasma for CVD Processing

KEN OKAZAKI
Department of Energy Engineering
Toyohashi University of Technology
Hibariga-oka, Tempaku-cho
Toyohashi 441, Japan

YUKIO YAMAGUCHI
Thin Film Laboratory, Research Center
Mitsubishi Kasei Corporation
1000 Kamoshida-cho, Midori-ku
Yokohama 227, Japan

AKIRA MIZUNO
Department of Electrical and
Electronic Engineering
Toyohashi University of Technology
Hibariga-oka, Tempaku-cho
Toyohashi 441, Japan

ABSTRACT

High-voltage and ultra-short DC pulse discharge plasma is one of the promising technologies to improve the deposition rate and the quality of thin films in CVD processing because of its very high yield rate of radical species and high controllability of radical reactions. This study intends to clarify a peculiar structure of ultra-short pulse plasma by investigating current-voltage characteristics, time lags of current rise from that of voltage and emission intensity profiles between cathode and anode. In the ultra-short pulse discharge, very high voltages up to several kilovolts can be successfully applied without any plasma non-uniformity on the electrode surface. High emission intensities of CH and H radicals near the anode have been observed in the high voltage pulse plasma, while the usual DC plasma has an opposite well-known structure where the largest emission intensity appears in the negative glow region near the cathode. The mechanism to realize the newly found structure in the pulse plasma has been well predicted by a self-consistent one-dimensional theoretical model consisting of conservations of electrons, positive and negative ions, neutrals, radicals and charges. With the increase of applied voltage, the gamma coefficient of secondary electron emission at the cathode increases by the bonbardment of high-energy ions and excited molecules and the ionization becomes rather active in the anode sheath, which results in the transition of luminescent zone from the cathode to the anode.

1. INTRODUCTION

Plasma CVD (Chemical Vapor Deposition) process is widely applied in the manufacture of various thin films, such as amorphous silicon solar cells, diamond coating on cutting tools and so on. The development of various new reactor systems or their improvements have been made with the practical point of view of improving the film quality and the deposition rate or, to attain a uniform deposition on the large area of the substrates. However, the plasma CVD process is a complicated process, as shown in Fig.1 [1], consisting of:

(1) a mechanism to sustain a stable plasma,
(2) formation of radicals and ions by collisions of high-energy non-equilibrium electrons with parent gas molecules,
(3) secondary reactions of radicals and ions,

(4) transport processes of radicals and ions to the substrate,
(5) surface reactions including initial nucleation, surface migration,lattice formation, etc.

The actual deposition process is strongly affected by all of these coupled steps and, the mechanism of plasma CVD is still far from a comprehensive understanding or model, in spite of the large amount of fundamental studies [2-5], recently. Most of the previous improvements or optimizations of plasma CVD reactors have been made only by changing the combination of the bulk parameters, such as gas conditions (pressure, temperature, composition, flow rate and flow pattern), plasma formation conditions (input power, frequency and electrode configuration), substrate conditions (temperature, material, shape and surface structure) or reactor conditions (structure and size). Hence, it is considered that the next step in optimizing the plasma CVD process should be the direct control of the plasma structure itself, in space and time, to realize the optimized reaction field, considering the reaction selectivity or the life-time of radicals and ions [6].

For that purpose of active control of the plasma structure, we have adopted a high-voltage and ultra-short DC pulse discharge plasma, where the voltage is up to several kilovolts and the pulse-width is much smaller than one micro-second. Such instantaneous high-voltages and hence high power inputs could be possible only by the ultra-short pulse discharge. This plasma could have a very unique advantage, for the use of controlling the CVD processing, in that the voltage, the pulse width and the pulse frequency can be controlled independently. However, no previous research on the fundamental structure of the high-voltage, ultra-short pulse discharge plasma nor its application to CVD processing has been reported before, except in the cases of rather large pulse-widths of several tens of micro-seconds [7].

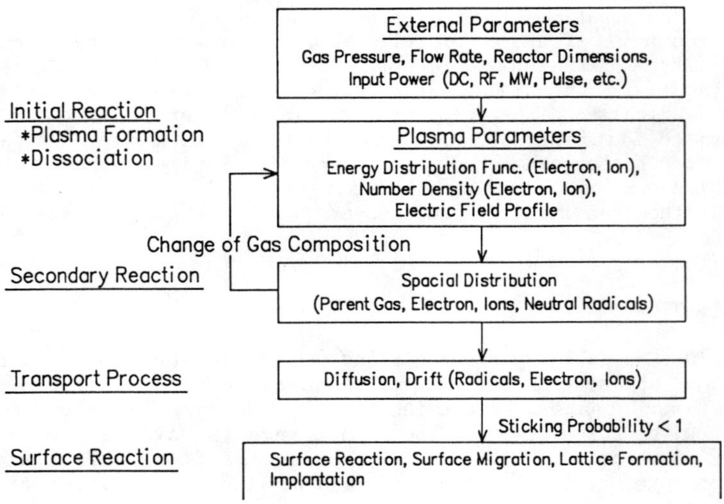

FIGURE 1 Schematic diagram of plasma CVD process [1].

This study intends to clarify a new and peculiar structure of high-voltage, ultra-short pulse discharge plasma by experimentally studying the current-voltage characteristics, time lags of current rise and profiles of radical emission intensities between cathode and anode. The mechanism to realize the actual plasma structure obtained in the experiments will also be discussed with the help of a theoretical model.

2. EXPERIMENTAL SYSTEM AND PROCEDURE

The experimental system used in the present study consists of a gas feeding system, an electric power supply system, a plasma CVD reactor and the measurement systems for elecrical characteristics and optical emission spectroscopy. Although the gas feeding system has many different gas lines, and is so designed as to be used not only for plasma CVD processing but also, for various kinds of plasma chemistry experiments, only the results obtained by using methane-hidrogen system will be presented here.

The configurations of the reactor and the optical measurement system are shown schematically in Fig.2. The reactor chamber is made of

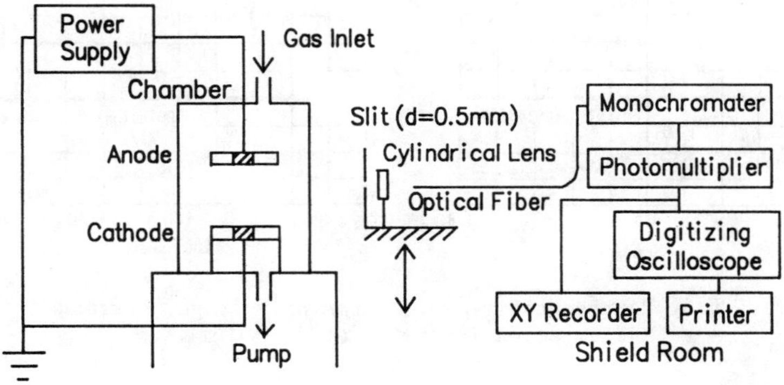

FIGURE 2 Experimental apparatus and measuring system.

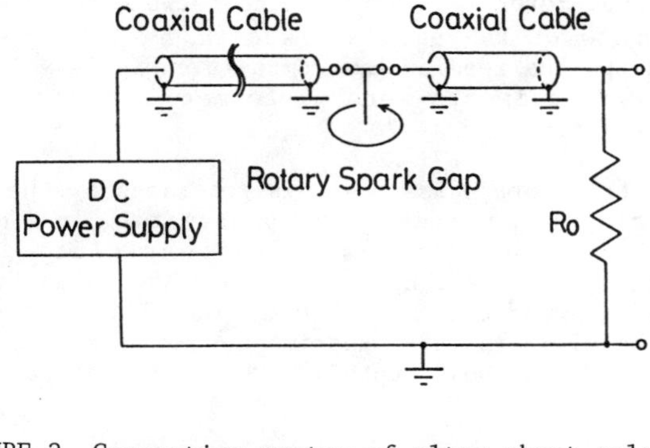

FIGURE 3 Generation system of ultra-short pulse power.

TABLE 1 Experimental conditions

Gas	Flow Rate (ccm)	Pressure (Torr)	Pulse Width (ns)	Voltage (kV)	Electrode	
					Anode	Cathode
CH_4(5%) H_2(95%)	20	4	(DC)	0.30–0.91	Graphite	Stainless Steel
			50	1.50–2.10		
			150	1.10–2.40		
			400	0.90–2.70		
			400	1.10–2.80	Copper	Copper
		10	400	1.80–2.40		
		20	400	2.70–3.20		

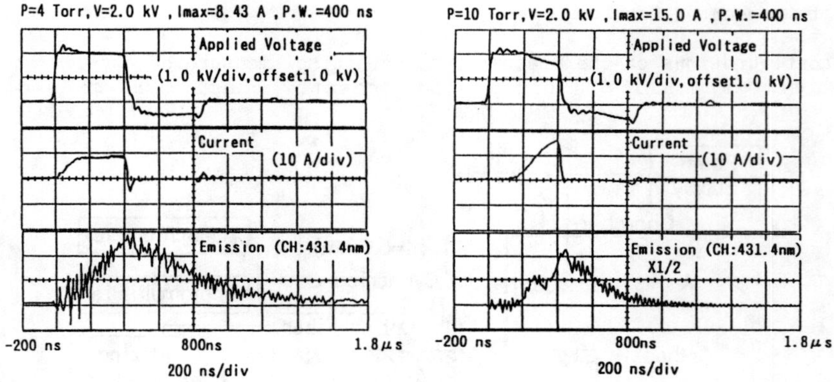

FIGURE 4 Representative wave forms of voltage, current and emission intensity at x = 20 mm.

cylindrical quartz glass of 150 mm diameter and 350 mm length, so that the discharge phenomena and the plasma structure could be easily observed from the outside. Two parallel, flat-plate, circular electrodes are set inside the reactor as anode and cathode. Each electrode is 10 mm in diameter and the distance between the electrodes is 40 mm. The flow rate of the gas is fixed to be 20 ccm (cubic centimeter per minute) and the feeding gas composition is 5% vol. of CH_4 in dilute gas of H_2. Experiments were performed for the various conditions as listed in Table 1.

The high-voltage and ultra-short pulse power is generated by the system shown in Fig.3, by using the voltage wave propagating in a co-axial cable of fixed length. Pulse voltages of various widths can be generated by changing the cable length. Although the pulse frequency can also be changed by changing the rate of rotation of the rotor, the frequency is fixed to be 45 Hz in this experiment. By this power supply system, very high instantaneous power upto 100 kW can be introduced into the plasma reactor.

Optical emission signals from radicals in the cross-section of the plasma column are collected through the cylindrical lens to an optical fiber, as shown in Fig.2, and then sent to the monochromater. The signals are amplified by the photomultiplier and collected by the fast digitizing oscilloscope with a time resolution of 2.5 ns. By this system, very fast change of optical emission from the radical species in the plasma can be obtained with corresponding voltage and current wave forms. Only the emissions from CH (431.4 nm) and H_α (656.3 nm) radicals were measured in this experiment because of their significant strength in the emission spectrum in CH_4- H_2 systems. All of the data acquisition systems are set inside an electrically shielded room to prevent the strong noises formed by the pulse power supply. Some of the representative time-resolved wave forms for the applied voltage, current and optical emmision intensity of CH radicals measured at the center between anode and cathode are shown in Fig.4.

3. EXPERIMENTAL RESULTS AND DISCUSSIONS

The current-voltage characteristics at a pressure of 4 Torr, for both the cases of ususal DC discharge and pulse discharge, are shown in Fig.5, with the pulse width as a parameter. In the usual DC discharge, the transition from normal glow (a) to abnormal glow (b) is clearly seen, while, in the region above the curve, the discharge mode will tend to an arcing one with rapid increase in current density and decrease of voltage, and heating-up the cathode surface, where the thermionic emission of electrons becomes dominant.

On the other hand, in the pulse discharge, very stable diffuse mode discharge can be easily achieved even in the high voltage region, up to several kilovolts, and the current density becomes two or three orders higher than that in the usual DC discharge, even though the values of current density and voltage are the instantaneous peak ones. The effect of pulse width on the current-voltage characteristics is not so significant and the data points seem to fall on the extension of the curve for the case of usual DC discharge, except near the lower end of the voltage for each pulse width. For a given pulse width, a stable plasma cannot be sustained for applied voltages below some critical value.

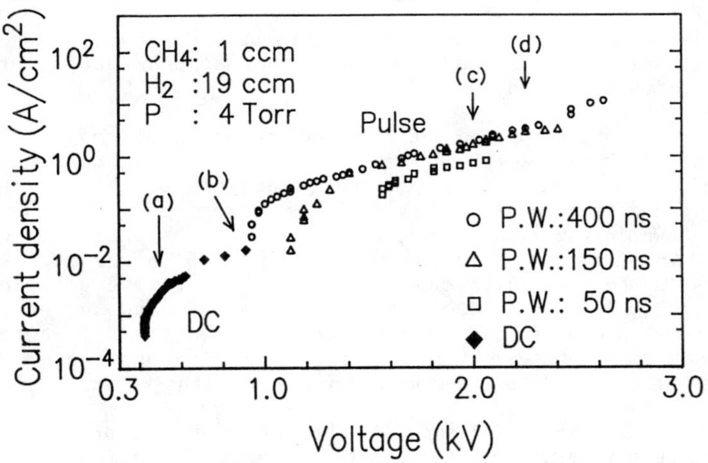

FIGURE 5 Current-Voltage characteristics of usual DC and high-voltage, ultra-short pulse discharge, and effects of pulse width.

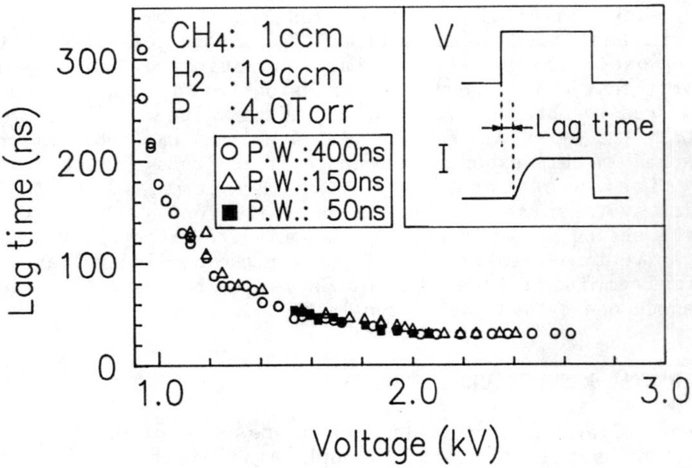

FIGURE 6 Lag times of current rise from the voltage input.

The lower limit of applied voltage to sustain a plasma can be well explained by the time lags of current rise from the input of pulse voltage, as shown in Fig.6. The lag time of current rise increases with the decrease of applied voltage and the data for each pulse width coincide almost on the one line. When the lag time becomes larger than the pulse width, the initiation of plasma becomes impossible. For the case of 150 ns pulse width, the lag time reaches the value of pulse width at the applied voltage of about 1.1 kV and no stable plasma can be sustained for the lower values of applied voltage. On the other hand, with the increase of applied voltage, the lag time decreases and tends to be almost constant at about 30 ns. This minimum lag time is not considered to be due to the time lags in the voltage and current measurements, because those values are estimated to be almost the same value of 18 ns and 15 ns, respectively, in the present measuring system. The time lags of current rise and effects of various factors on them will give us valuable informations about the basic mechanism of the initiation of pulse plasma.

The structures of high-voltage and ultra-short pulse plasma should be much different from that of the usual DC plasma. Hence, corresponding to the conditions of (a), (b), (c) and (d) in Fig.5, photographs of plasma field and emission intensity profiles between the cathode and the anode are shown in Fig.7 and Fig.8, respectively. The usual DC normal glow discharge, as shown in (a) of Figs.7 and 8, has a well-known structure with the most luminescent zone in the negative glow region near the cathode. With the increase of voltage and current, the discharge mode changes to the abnormal glow as shown in (b), where the luminescent zone spreads into the core region and the emission intensities of CH and Hα becomes much larger than that of (a).

In the pulse discharge of (c) and (d), the plasma shows a drastic change in its structure from that of the usual DC discharge. At the very high power region of (d), the luminescent zone has moved to the anode side and the instantaneous emission intensities of CH and Hα are three orders

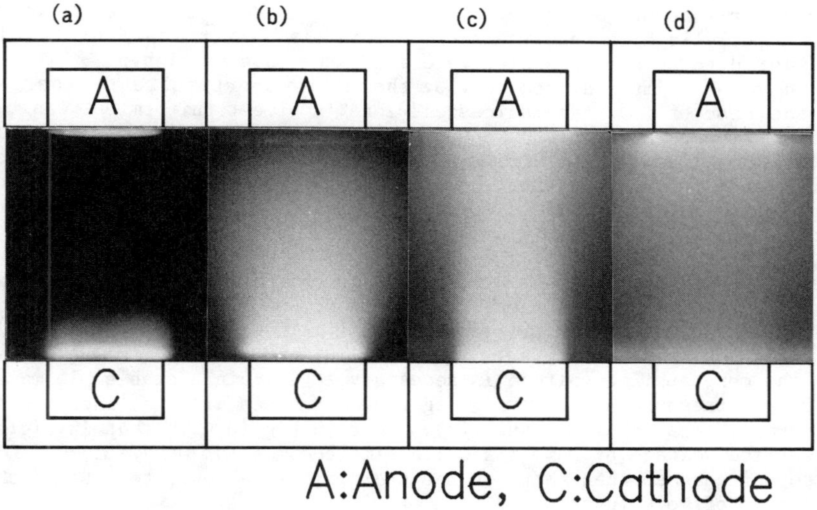

FIGURE 7 Photographs of various kinds of plasmas. (a), (b), (c) and (d) correspond to the conditions in Fig.5.

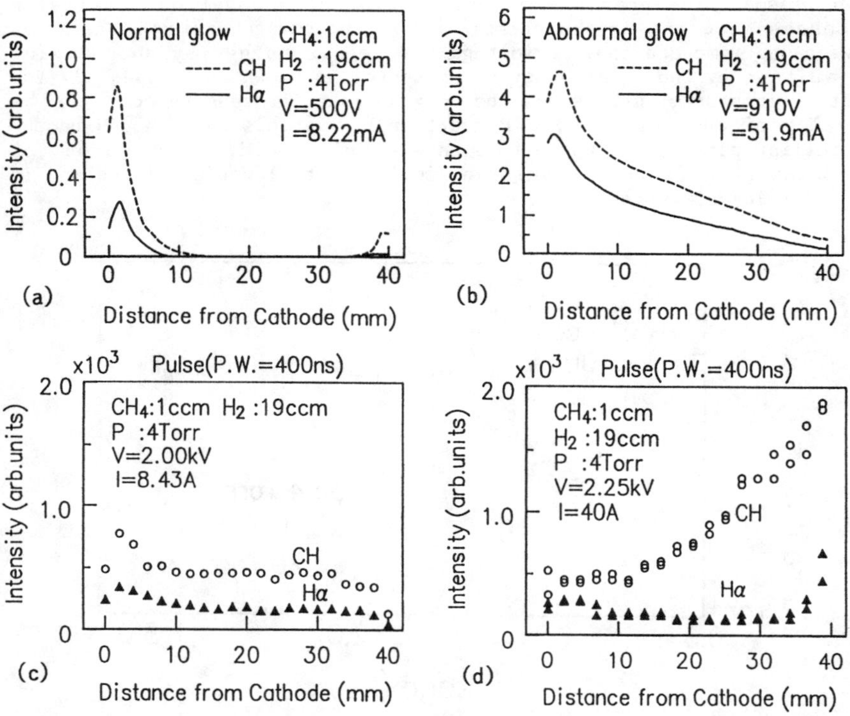

FIGURE 8 Profiles of Emission intensities of CH and H_α between anode and cathode. (a), (b), (c) and (d) correspond to the points in Fig.5.

higher than that of (a) or (b). It should be noted that the emission of CH is especially high near the anode. This plasma structure of (d) is considered to be a representative one for the case of high power region and can be actually realized only by the ultra-short pulse discharge. For the case of (c), the luminescent zone is almost uniform between the anode and the cathode, although the emission intensities are still much higher than that of the usual DC discharge. Mechanisms to establish such a newly found plasma structure by the high-voltage, ultra-short pulse power will be clarified by the theoretical model in the later part of this paper.

Effects of pressure on the current-voltage characteristics and the time lags of current rise have been also examined, where copper is used for both anode and cathode as listed in Table 1. Those results are shown in Fig.9 and 10, with the pressure as a parameter. It is clearly shown that the more applied voltage is necessary to sustain a stable plasma at the higher pressure and that the lag time increases with the increase of pressure. Comparing the result for 4 Torr in Fig.10 with that in Fig.6, the applied voltage at some fixed lag time becomes slightly higher for the copper cathode than that for the stainless steel one, because of the difference of work functions.

The increase of lag time with the increase of pressure as shown in Fig.10 is considered to be due to the decrease of mobilities of ions in the direction to the cathode, if the gammma effect of secondary electron emission on the cathode is the dominant mechanisn to initiate a high power plasma. The pressure effect should include various collision processes occuring in the plasma, and so the comprehensive correlation between lag time and applied voltage has been obtained by normalizing the voltage by the square root of pressure, as shown in Fig.11. If the drift of ions or electrons are the only dominant process to control the time lag of current rise, E/P and then V/P in this case should be the controlling parameters, where E is an electric field. The reason for the somewhat different result in Fig.11 is still not clarified and is now under investigation.

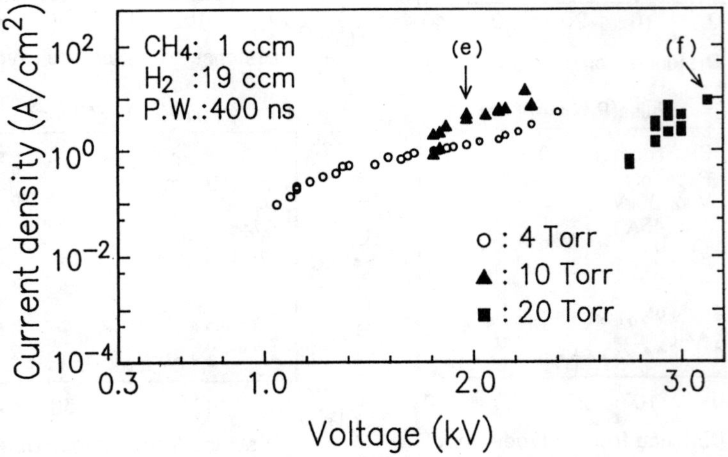

FIGURE 9 Current-voltage characteristics of pulse discharge at different pressures.

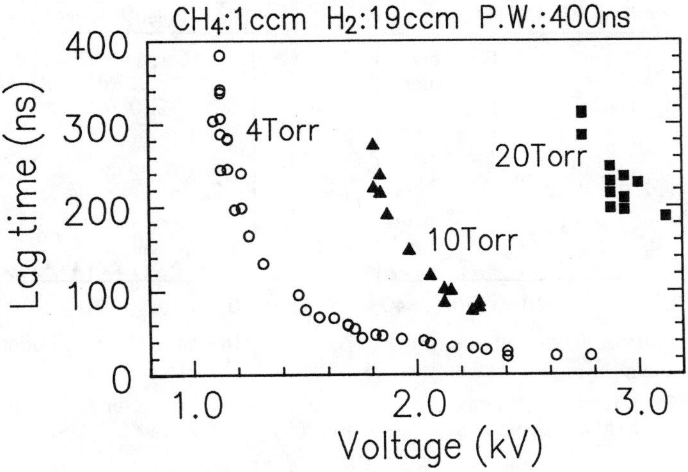

FIGURE 10 Effects of pressure on the lag times of current rise.

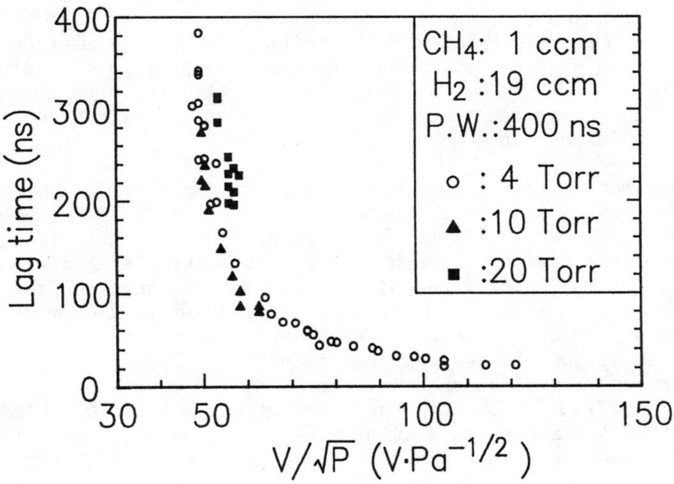

FIGURE 11 Conprehensive correlation between lag time and applied voltage normalized by square root of pressure.

Profiles of emission intensities of CH and Hα between anode and cathode for the conditions corresponding to (e) and (f) in Fig.9 are shown in Fig.12. The emission profiles in (e) and (f) show the intermidiate ones between (b) and (d) in Fig.8, although the intensity itself in (e) is stronger than that of (c) in Fig.8 because of the increase of radical density with the pressure rise. However, it is interesting to note that the emission intensity in (f), especially that of Hα, becomes smaller than that of (e) in spite of the increase of pressure.

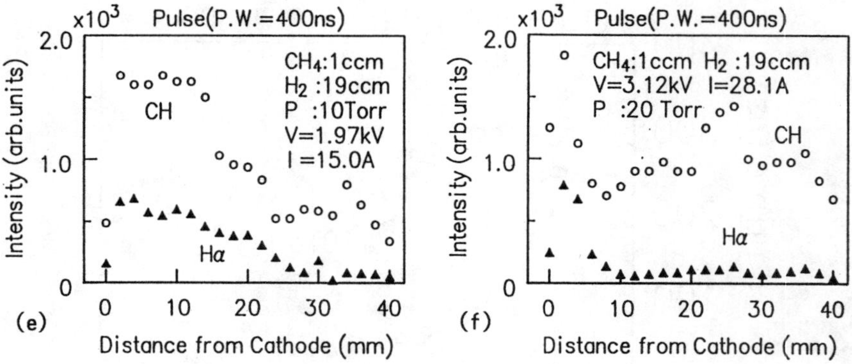

FIGURE 12 Emission intensity profiles between anode and cathode at higher pressures. (e) and (f) correspond to the points in Fig.9.

4. THEORETICAL MODELING OF HIGH POWER DC PLASMA

A few theoretical studies based on a physical model have been carried out for DC and RF systems, and have succeeded in determining both electric fields and density distributions of species [8-12]. The deposition rates are also calculated from the profile of radical density with the assumption of diffusion-controlled transfer to a substrate. However, there are few theoretical studies about V-I characteristics in the high power region.

The experimental apparatus corresponding to this analysis consists of two co-axial cylinders of 6 and 11 cm radii. The height of the cylinders is 50 cm. SiH_4 gas is introduced between two electrodes from the top of the reactor through a mass flow controller and exhausted from the bottom. The inner grounded cylinder is a cathode and the outer one is an anode. a-Si:H films are deposited on the surface of the electrodes made of aluminium. The temperature of the cathode surface is kept at about 300 C when in deposition [13]. Pressure employed here is 1.1 Torr, and SiH_4 flow rate is from 80 to 150 sccm(cubic centimeter per minute at STP). V-I characteristics measured upto 803 volts are shown in Fig.13. Applied voltage is a value between electrodes. The current density depends on the gas flow rate. As the gas flow rate increases, the current density increases.

A one-dimensional self-consistent description is executed for the system of a DC glow discharge. The one-dimensional model is accurate in itself to describe the discharge in a cylindrical system. The result is also applicable to a plate-plate system such as the case of the experiments in the former part of this paper. Though the model used here is under a steady state condition, the results could be useful to understand the structure of a high-voltage and ultra-short DC pulse plasma. The transient process of initiating a plasma will be analyzed elsewhere. Since swarm parameters available are only for the represented species, the fragmentation of both ions and radicals cannot be discussed here. The local field approximation [9,12] is employed. Then, swarm parameters are functions ,F, of only the local reduced-field E/P, where E is the electric field.

The conservation equations of charged species are:
electron continuity:

$$(1/r)d(rn_e v_{de})/dr = (\alpha - \eta)n_e v_{de} N/N_0, \qquad (1)$$

positive ion continuity:

$$(1/r)d(rn_+ v_{d+})/dr = \alpha n_e v_{d+} N/N_0, \qquad (2)$$

negative ion continuity:

$$(1/r)d(rn_- v_{d-})/dr = \eta n_e v_{d-} N/N_0, \qquad (3)$$

where, n_e, n_+ and n_- and v_{de}, v_{d+} and v_{d-} are the number densities and the drift velocities of electrons, positive and negative ions, respectively. α and η employed here [14] are interpolated using a spline function. N_0 is the number density of neutrals corresponding to the pressure before the discharge, and N is the residual density of neutrals during the discharge. The electric field can be determined from Poisson's equation:

$$(1/r)dE/dr = e(n_+ - n_e - n_-)/\epsilon_0, \qquad (4)$$

where, ϵ_0 is the permittivity of free-space. The electric field E is defined as:

$$E = -dV/dr. \qquad (5)$$

The conservation of neutrals N and radicals n^* produced from electron-neutral collisions are represented as:

$$v(N_0 - N)/H = D(1/r)d(rdN/dr)/dr - R_d n_e N, \qquad (6)$$

$$-v^* n^*/H = D(1/r)d(rdn^*/dr)/dr + R_d n_e N, \qquad (7)$$

where, v, v^*, D and D^* are the linear velocities calculated from F and the diffusion coefficients of neutrals and radicals of SiH_4, respectively. R_d is the total dissociation rate. The swarm parameters employed here are the same as in the previous works [13,14]. The decrease of N during discharge can be neglected under low electric power conditions, but it must be taken into account under high electric power conditions. The consumption of N makes a discharge to reach a saturation value for a given condition. In the sense, the equations from (1) to (5) describing a discharge are coupled with the equations of (6) and (7) associated with deposition. The effect of flow rates is considered in equation(6).

The boundary conditions are set at the cathode and the anode, respectively, as,
at the cathode:

$$n_e v_{de} = \gamma n_+ v_{d+}, \quad n_- = 0, \quad V = 0, \quad dN/dr = 0, \quad n^* = 0, \qquad (8)$$

at the anode:

$$n_+ = 0, \quad V = V_0, \quad dN/dr = 0, \quad n^* = 0. \qquad (9)$$

It is assumed that the surface reaction of radicals is extremely high, and the sticking coefficient of radicals is unity. In general, alpha and gamma effects are hardly separated because, gamma has the possibilty to be changed due to etching and deposition during measurements. In our case, since radicals and ions continue to deposit on the aluminium substrate during V-I measurements, its surface conditions seem to be kept constant. Two effects are easily distinguished in the calculation.

The set of equations is discretized using the finite difference method. The non-linear algebraic equation set is solved using the Newton-Raphson method. The step size is taken to be 0.125 cm throughout in the computation. The computation error defined as the root-mean-square is less than 0.1%. When V_o, γ, P and F are given, the current density can be calculated by solving the equation set. When the current density, instead of gamma, is given, gamma can be calculated and then, it is found that gamma increases with the increase of the applied voltage, as shown in Fig.13. The points B,C and D are obtained not by measurements, but by calculations. As the applied voltage increases, the number densities of high-energy excited molecules and ions increase. They reach the cathode and emit the secondary electrons. Positive ions move to the cathode, obtaining energy from the strong electric field and losing most of the energy due to ion-neutral charge-exchange collisions. High-energy excited molecules and ions are produced at the same time. The high-energy excited molecules collide elastically with the other low-energy neutrals. Almost without losing their energy in the cathode sheath, the high-energy excited molecules and the ions bombard the cathode surface and liberate their kinetic energy as heat, which increases the cathode temperature [15,16], and release their potential energy, which produces the secondary electrons emitted at the cathode surface. It is concluded that the increase of high-energy excited molecules is responsible for the increase of gamma.

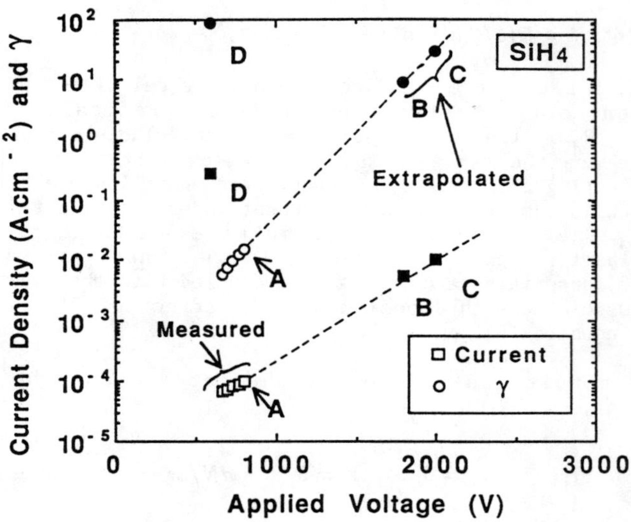

FIGURE 13 Current-voltage characteristics and values of gamma. The data points in A group are obtained based on the experiments of silane plasma.

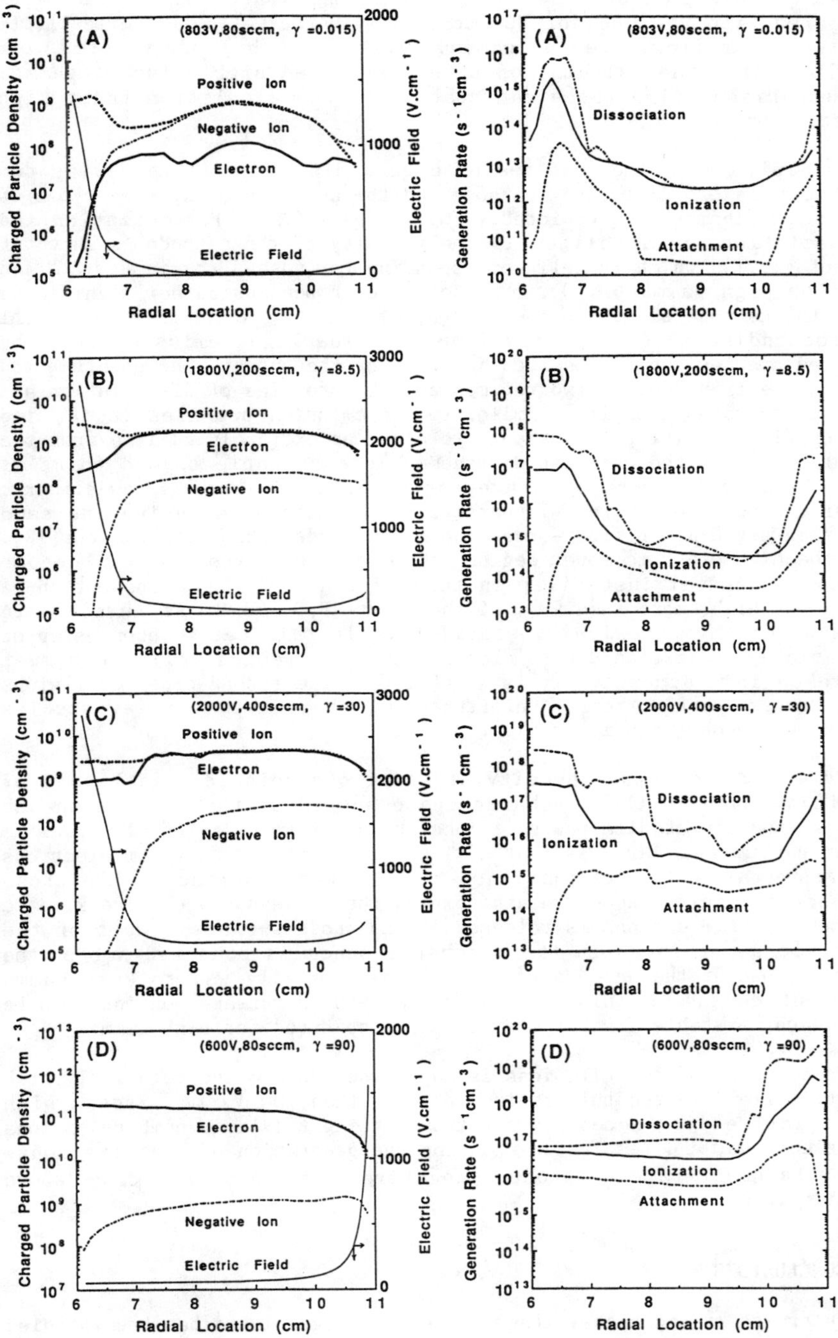

FIGURE 14 One-dimensional spatial profiles of charged particle density, electric field [left column], and generation rates of dissociation, ionization and attachment [right column].

To study the effect of flow rates, gamma calculated for 150 sccm is compared with the one for 80 sccm. They agree well. There is little influence of flow rates on gamma. Therefore, high flow rates are employed for the calculation of extrapolated high value of gamma, because higher flow rates can stabilize the calculation under high electric power.

In general, one of the features of a glow discharge is in having two boundary layers, i.e., the cathode and the anode sheaths. Two types of discharge plasmas are predicted. For low gamma($\gamma < 1$), ionization and dissociation are localized in the vicinity of the cathode as shown in (A) of Fig.14, which is well corresponding to (a) and (b) in Figs.7 and 8. For high gamma($\gamma \geqq 90$), the anode sheath becomes rather significant than the cathode sheath as seen in (D) of Fig.14, which is again well corresponding to (d) in Figs.7 and 8. The light emission from the excited species can be observed in front of the anode. The points B and C show the transient of two regimes and the profiles of dissociation and ionization rates look like similar to the emission profiles in (e) and (f) of Fig.12. The point C seems to cross an extrapolated line from the point D. Since the calculation becomes more unstable beyond point C, the higher values of gamma are not calculated. For the low electric power region, the entire V-I characteristics calculated using the same equation set exhibit first-order and second-order phase transitions[17]. For the high electric power region, this is under investigation because of the numerical instabilty in the calculation. SiH_4 consumption is extremely high, more than 99%, at the points B,C and D. It is the main reason for the numerical instability. In this case, second-step or accumulated ionization due to electron-excited species collisions must be taken into account. It is an effect of the second-order collisions with respect to electrons. Unfortunately, there are very few studies about the second-step ionization.

At point D, the number density of electrons coincides with that of positive ions in the discharge space and then, the cathode sheath shrinks and almost disappears. Since the drift velocity of electrons is about 90 times as large as that of ions, the number density of electrons becomes equal to that of the ions at the cathode surface. Electrons accelerated in the anode sheath bombard the anode surface. The kinetic energy of the electrons is released at the anode surface. Most of the electric power input in the discharge space is consumed not at the cathode, but at the anode surface. It is the only way to keep gamma constant during the discharge. If the cathode surface continues to be heated up, a stable discharge will not be observed for high gamma.

The generation of negative ions is suppressed due to the relatively high electric field in the bulk region, of more than 100 V/cm. Such a high electric field is necessary for transporting the large numbers of ions and electrons, and results in reducing the generation of negative ions. The attachment rate to produce negative ions drastically drops above 100 V/cm.

5. CONCLUSIONS

The ultra-short and high-voltage pulse discharge plasma has been studied focusing on its structure as a promising technology for the active control of radical reactions in the plasma CVD processing, and the following conclusions have been obtained.

(1) Stable and uniform plasmas have been successfully established by the ultra-short and high-voltage pulse discharge plasma at very high applied voltages up to several kilovolts without any plasma non-uniformity on the electrode or substrate surface.

(2) The conditions or criteria to sustain the ultra-short pulse plasma have been clarified. The lag time of current rise from that of voltage increases with the decrease of applied voltage and if it reaches the value of pulse width the initiation of plasma becomes impossible.

(3) Several peculiar features of the ultra-short and high-voltage pulse plasma have been clarified including voltage-current characteristics at very high power regions, lag times of current rise and effects of pressure and pulse width, and radical emission intensity profiles between the anode and the cathode. The instantaneous emission intensity of radicals in the pulse plasma becomes three orders higher than that of the usual DC glow discharge plasma.

(4) In the high power pulse plasma, the most luminescent zone appears near the anode, while it remains in the well-known cathode glow region in the usual DC plasma. This drastic change of plasma structure has been well predicted by the thoretical analysis and the quite new structure of high power pulse plasma could be due to the increase of effective gamma effect on the cathode surface by high energy ions and excited neutrals.

(5) The ultra-short and high voltage pulse plasma has a high possibility in controlling the CVD process by combining it with other power sources of DC, RF and pulse of different hight and width. This system can be also used as a research tool to discriminate the contributions of radicals and ions in the CVD process. Various applications of the pulse plasma in CVD processings is now under investigation as a next step.

Acknowledgement

The authors would like to express their thanks to Prof. K. Hijikata for his valuable comments to the experiments in this study, and also to Mr. Takekoshi and Mr. Tobe for their great help in doing experiments and preparing this paper. This study is partially supported by the Grant-in-Aid for Scientific Research (B) [No.01460117] of The Ministry of Education, Science and Culture.

REFERENCES

1. Tachibana, K. and Matsui, Y., On Homogeneous and Heterogeneous Reactions in Plasma Processing, Shinku (in Japanese), Vol.31, No.3, pp.179-187, 1988.

2. Kline, L.E. and Partlow, W.D., Electron and Chemical Kinetics in Low Pressure RF Discharge: Deposition of Amorphous Silicon-Carbide Thin Films, ISPC-8th, pp.192-196, 1987.

3. Antonie, A.M. and Drevillon, B., Influence of Ion Bombardment on the Nucleation and Growth of Plasma Deposited Amorphous Silicon, Mat. Res. Soc. Symp. Proc., Vol.75, 1987.

4. Gallagher, A., Neutral Radical Deposition from Silane Discharges, J. Appl. Phys., Vol.63, No.7, pp.2406-2413, 1988.

5. Kushner, M.J., A Model for the Discharge Kinetics and Plasma Chemistry during Plasma Enhanced Chemical Vapor Deposition of Amorphous Silicon, J. Appl. Phys., Vol.63, No.8, pp.2532-2551, 1988.

6. Watanabe, Y., Shiratani, M., Kubo, Y., Ogawa, I. and Ogi, S., Modulation on RF Discharge Chemical Vapor Deposition, Appl. Phys. Lett., Vol.53, No.14, pp.1263-1265, 1988.

7. Yoshida, T., Ichikawa, Y. and Sakai, H., a-Si:H Formation by Pulse Discharge CVD, Report of CVD Project Research (in Japanese), Chem. Eng. Soc. Jap., pp.140-142, 1990.

8. Richards, A.D., Thompson ,B.E. and Swain, H.H., Continuum Modelling of Argon Radio Frequency Glow Discharges, Appl. Phys. Lett., Vol.50, No.9, pp.482-494, 1987.

9. Bouef, J.P., Numerical Model of RF Glow Discharges, Phys. Rev., A36, pp.2782-2792, 1987.

10. Graves, D.B., Fluid Model Simulation of a 13.56-MHz RF Discharge: Time and Space Dependence of Rates of Electron Impact Excitation, J. Appl. Phys., Vol.62, pp.88-94, 1987.

11. Okazaki, K., Makabe, T. and Yamaguchi, Y., Modelling of RF Glow Discharge Plasma, Appl. Phys. Lett., Vol.54, pp.1742-1744, 1989.

12. Boeuf, J.P., A Two-Dimensional Model of DC Glow Discharges, J. Appl. Phys., Vol.63, pp.1342-1349, 1988.

13. Yamaguchi, Y., Sumiyama, A., Hattori, R., Morokuma, Y. and Makabe, T., A model of Amorphous Silicon Deposition in DC Glow Discharge in Silane, J. Phys. D, Vol.22, pp.505-511, 1989.

14. Shimozuma, M. and Tagashira, H., Measurement of the Ionization and Attachment Coefficients in Monosilane and Disilane, J. Phys. D: Appl. Phys., Vol.19, L179-182, 1986.

15. Yamaguchi, Y. and Makabe, T., An Analysis of Substrate Heating in DC Glow Discharge in Silane, ISPC-9th, pp.1353-1359, 1989.

16. Yamaguchi, Y., Konno, S., Muramatu, K. and Makabe, T., An Analysis of Substrate Heating during a-Si:H Deposition using Concentric Cylindrical DC Glow Discharge, Kagaku Kogaku Ronbunshu, Vol.16, No.3, pp.605-613, 1990.

17. Yamaguchi, Y. and Makabe, T., Phase Transition of DC Glow Discharge in SiH_4, submitted to Phys. Rev. Lett., 1990.

Heat Flow and Thermal Contraction During Plasma Spray Deposition

T. W. CLYNE and S. C. GILL
Department of Materials Science and Metallurgy
Pembroke Street
Cambridge CB2 3QZ, UK

1. INTRODUCTION

1.1 The Plasma Spray Process

The family of thermal spraying processes is of increasing industrial importance. Plasma spraying is frequently employed when a coating of high integrity and controlled structure is required, as the plasma torch provides a high energy density, leading to very high core temperatures and rapid acceleration of the powder particles [1-5]. When spraying is carried out under reduced pressure in a vacuum chamber (VPS), there are further benefits in terms of chemical and process control [6-10]. This process can be used to produce both conventional coatings and to build up novel materials, such as Ti-based composites [11,12].

A schematic section through a VPS torch is shown in Fig.1. A DC arc is established between a protruding cathode and an annular anode embedded in a converging-diverging nozzle within the body of the torch. During reduced pressure operation, the plasma plume can extend far ahead of the nozzle. A swirling motion is usually imparted to the incoming plasma gases to ensure that the anode arc root rotates, reducing erosion of the anode. Powder particles are injected near the front of the torch, so as to pass through the core of the plasma before being rapidly accelerated by the expanding gas.

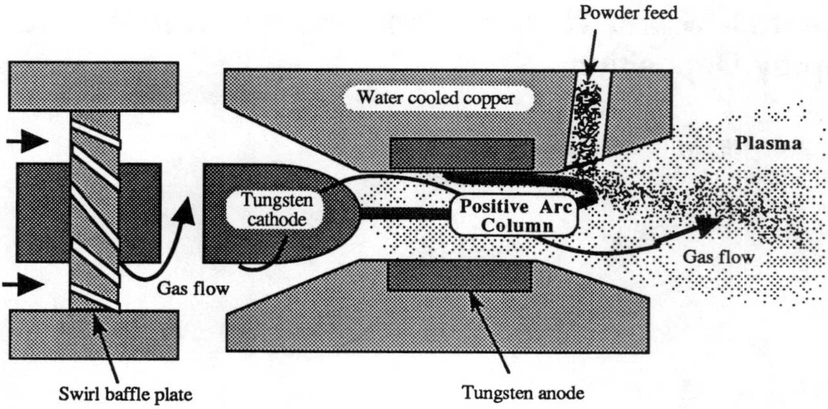

FIGURE 1 A schematic section through a vacuum plasma spray torch

Heat flow during the process can be divided into two distinct regimes, the first involving heat transfer between the particles and their environment along the flight path and the second concerning particle-deposit-substrate heat exchange after impact. The first of these is complex in terms of hydrodynamic features of the gas flow and boundary layers around the particle, but can at least be taken as having the same characteristics throughout the deposition process. Post-impact heat flow, on the other hand, will change in nature as the deposit thickens and is of great importance because it dictates the residual stresses in substrate and coating caused by differential thermal contraction.

1.2 Environment - Particle Heat Transfer

Particle velocities depend strongly on their mass and on the gas flow characteristics in the plasma plume. In practice, peak velocities are in the range 200-500 m s^{-1} [4,6,7,13], higher speeds being attained with lower chamber pressure. In-flight heat transfer between the particle and the plasma gases has been examined theoretically and experimentally by a number of workers [14-20]. Recent work [21] has shown that there are fluctuations in the hydrodynamics of the plume which can influence particle flight histories. An important practical point to note about spraying under reduced pressure is that this leads to shorter particle residence times and to lower energy densities in the plasma core, both effects tending to reduce particle heating. It can be advantageous [17,22] when particle melting is difficult to use higher chamber pressures in order to combat this.

An indication of the type of thermal history experienced by a particle during (atmospheric) plasma spraying can be obtained from the predicted data in Fig.2, following Fauchais [23]. Although histories are shown for Newtonian and non-Newtonian cases, conditions will in practice commonly be Newtonian, as the thermal conductance of most particles will be $\geq 10^6$ W m^{-2} K^{-1} and interfacial heat transfer coefficients in flight would rarely be as large as this value [24]. It is clear that the melting and droplet superheat characteristics are very sensitive to the environment during the initial portion of the trajectory.

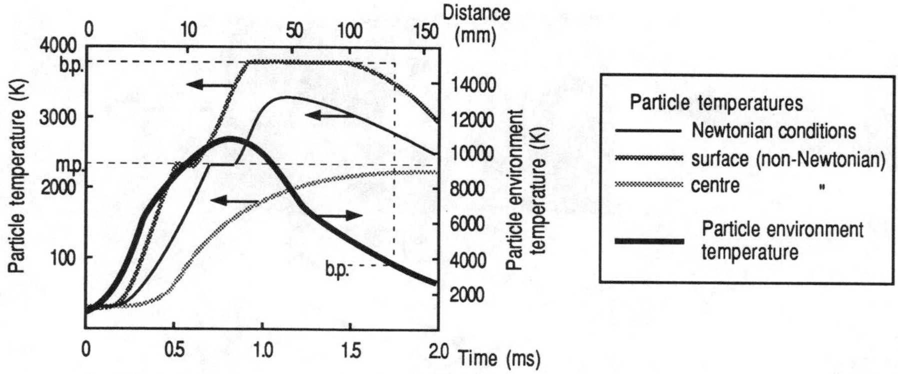

FIGURE 2 Typical predicted in-flight thermal histories for a 100 μm diameter alumina particle during plasma spraying. The temperature of the gaseous environment and of the particle itself are shown as a function of time from the point of injection. Particle temperatures are shown for Newtonian and non-Newtonian conditions (centre and surface histories shown). (After Fauchais [23].)

1.3 Particle - Deposit - Substrate Heat & Mass Transfer

There have been a number of studies covering the impact, spreading and solidification of incoming droplets [25-29]. A schematic depiction of the phenomena involved during deposition of a single splat is shown in Fig.3, together with an indication of typical values for the main parameters characterizing the event. The large value of the Weber number (ratio of inertial to surface tension forces) indicates that there is a danger of the droplet disintegrating on impact; in practice, the contribution of viscous work during flow, heat loss during spreading, surface roughness effects and other phenomena can result in the splats retaining their integrity and this is usually regarded as desirable. The thickness of most splats tend to be around 2-4 μm, across a wide range of conditions and materials

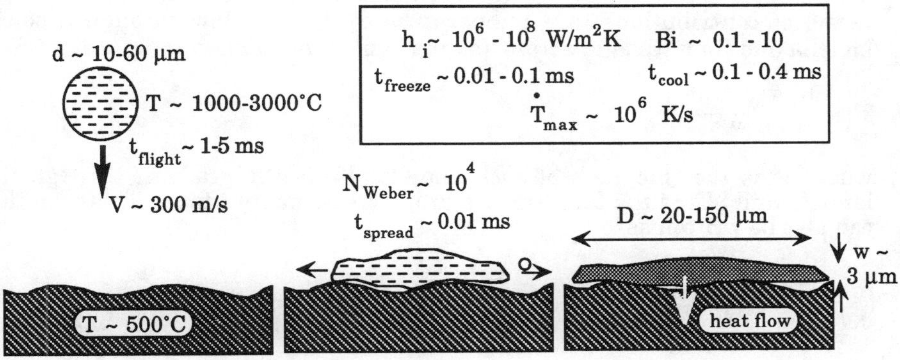

FIGURE 3 A schematic illustration of the impact, spreading and initial quenching of a droplet striking a substrate during plasma spraying, together with an indication of the values expected for the main parameters characterizing these processes.

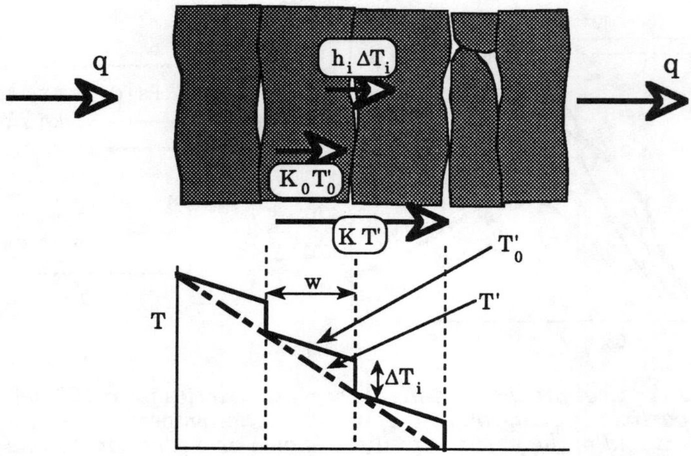

FIGURE 4 A schematic depiction of the thermal profile within a sprayed coating during through-thickness heat flow, with the structure idealized to a set of lamellae of equal thickness, separated by interfaces with a constant heat transfer coefficient, h_i.

The splat-deposit interfacial heat transfer coefficient during quenching, h_i, determines the cooling conditions. This parameter will tend to vary significantly with surface roughness, cleanliness, droplet impact conditions etc. A value of 10^6 W m^{-2} K^{-1} has been assumed [30], in which case cooling will tend to be Newtonian for metal splats but not for those of non-metals. With h_i values of this order, it is readily shown that splat cooling rates will be around 10^6 K s^{-1}. This represents very rapid quenching, which will give rise to a variety of metastable structures, such as γ-alumina [31].

Estimates of h_i can be made from the through-thickness thermal conductivity of the coating, using the simple model represented in Fig.4. During the flow of heat through the coating, the average thermal gradient T' can be written in terms of contributions to a temperature drop from flow through a sound lamella and from transfer across an inter-lamellar boundary

$$T' = \frac{\left(w\,T'_0 + \Delta T_i\right)}{w} \qquad (1)$$

where w is the thickness of each lamella, T'_0 is the gradient through the lamella and ΔT_i is the temperature drop across the interface. This equation can also be written as

$$\left(\frac{q}{K}\right) = \frac{\left(w\left(\frac{q}{K_0}\right) + \left(\frac{q}{h_i}\right)\right)}{w} \qquad (2)$$

where q is the heat flux and K, K_0 are respectively the measured conductivity of the coating and the expected conductivity of sound material. On this basis, h_i is easily estimated from these two conductivities and the splat thickness

$$h_i = \frac{K_0\,K}{w\,(K_0 - K)} \qquad (3)$$

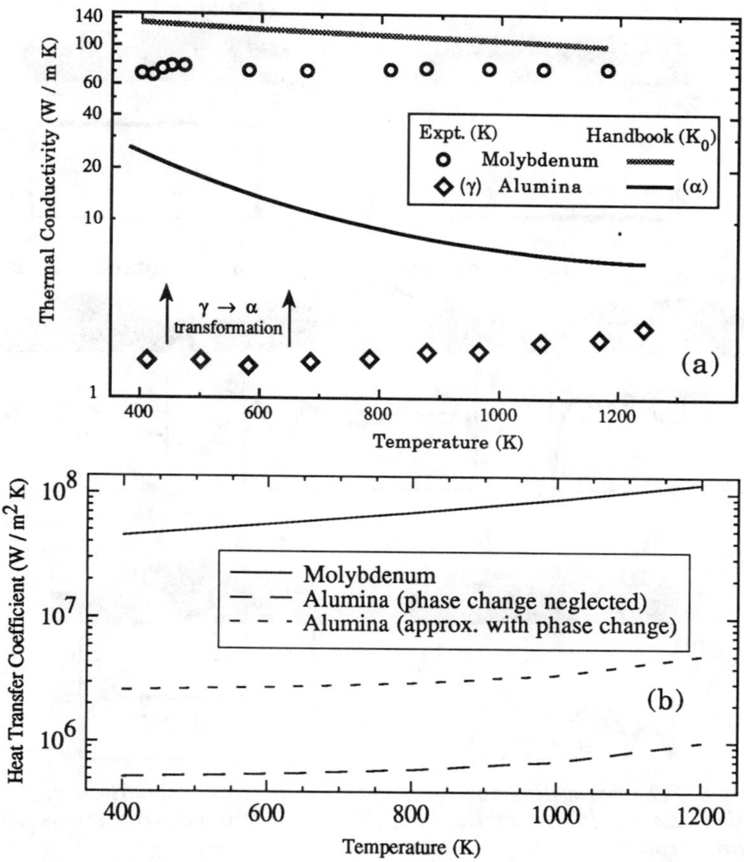

FIGURE 5 (a) Measured [32] through-thickness thermal conductivity data for Mo and Al_2O_3 VPS coatings and (b) derived values of the interfacial heat transfer coefficient, obtained using 3 μm for the splat thickness.

Examples of such an estimate are shown in Fig.5, which refers to Mo and Al_2O_3 coatings produced by VPS and measured [32] using the laser flash method [33]. There is a difficulty with the alumina data, in that the deposits normally freeze predominantly as the metastable γ phase. This phase is thought [34] to have an appreciably lower conductivity than the stable α phase. Unfortunately, experimental data are only available for as-sprayed material, which contains cracks and porosity. Heat treatment causes the phase transformation to α phase and a sintering effect to proceed simultaneously, making it difficult either to obtain an accurate value of the conductivity of dense γ or to retain the same crack and pore pattern with an α structure. The h_i values shown for Al_2O_3 are therefore rather crude estimates. In general, however, h_i appears to fall in the range 10^6 - 10^8 W m^{-2} K^{-1} for VPS coatings. It follows that the corresponding Biot numbers ($Bi = h_i\ w / K$, where K is the thermal conductivity of the splat material) will generally be of the order of unity and cooling conditions are not necessarily Newtonian.

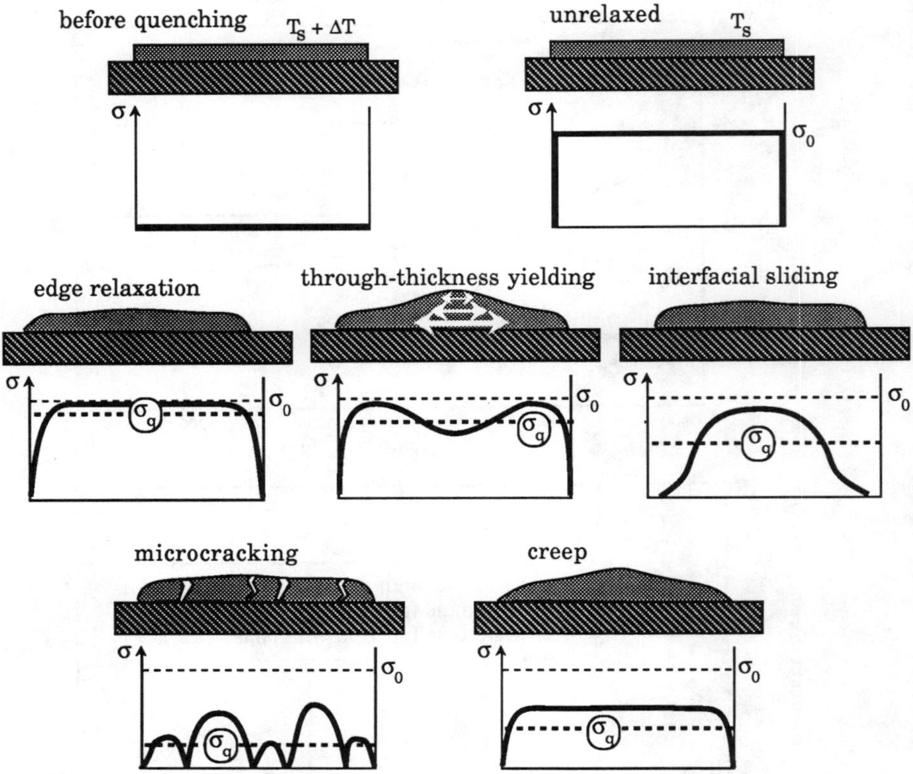

FIGURE 6 Schematic illustration of the various stress relaxation mechanisms [35] which can act to reduce the magnitude of the quenching stress below its maximum theoretical value

1.4 Thermal Contraction Effects

Stresses arise from unequal contraction strains in different parts of the system. These strains can be divided into (a) thermal contraction of an individual splat as it is quenched, (b) differential thermal contraction within the substrate or the deposit as a result of thermal gradients, (c) differential thermal contraction between the substrate and the deposit as a result of different thermal expansion coefficients and (d) strains from any solid state phase transformations. Of these, (a) and (c) are usually the most significant; (b) can only lead to retained residual stress at room temperature as a result of non-uniform stress relaxation processes, while (d) is often entirely absent. Prediction of the final stress state is relatively straightforward when the system is taken as elastic (no stress relaxation), particularly if through-thickness stresses are neglected. In practice, stress relaxation always occurs, especially during quenching of individual splats. A schematic illustration is given in Fig.6 of the various stress relaxation mechanisms which may be operative [35]. The main difficulties in modelling of stress build-up arise in quantification of this and from the need to use appropriate thermophysical properties for the deposit [36].

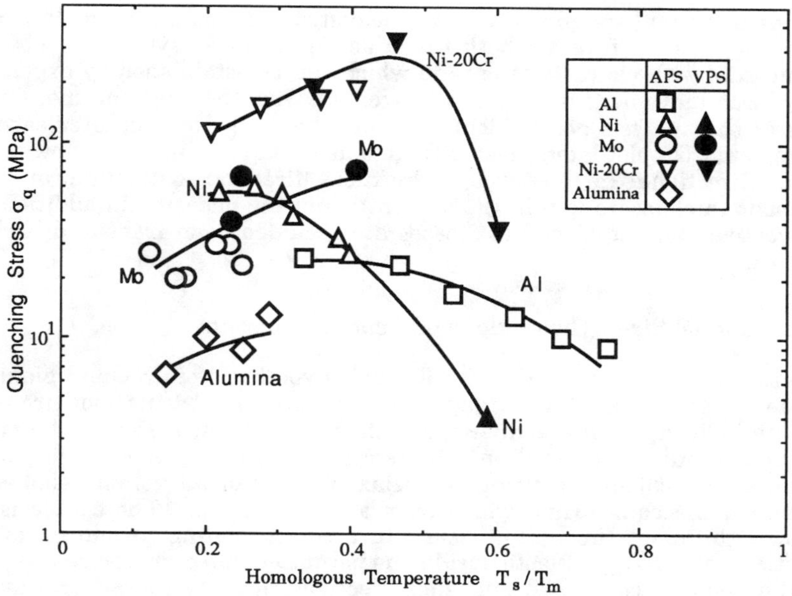

FIGURE 7 Experimental values [35] of the quenching stress for different materials, as a function of the temperature of the substrate, expressed as a fraction of the melting point of the material.

Several techniques have been used to measure residual stresses. These include X-ray diffraction [37-39], although only stresses in a thin surface layer can be measured and the technique is sensitive to surface roughness; most experimental data for thermally sprayed coatings are rather inconclusive, which is probably related to the presence of sharp through-thickness stress gradients. Most destructive techniques, such as hole drilling [40] and slit closure [41], are also incapable of detecting such gradients.

Probably the most sensitive and versatile technique is that based on measuring the curvature induced in the substrate-deposit couple as a result of the internal stresses [42]. The method is particularly powerful if used in conjunction with a stress analysis model which can be used to predict curvature histories during spraying [43]. Both contacting [44] and non-contacting [45] methods have been described for monitoring such curvature histories. While experimental measurements must be made with thin substrates in order to give measurable curvatures, a stress analysis model (once validated in this way for established boundary conditions) can be used to predict the residual stresses expected with thick substrates. Curvature measurements can also be used to isolate and evaluate the splat quenching stress for given materials and substrate temperatures [35]. Some results of such measurements are presented in Fig.7.

2. A COUPLED THERMO-MECHANICAL FINITE DIFFERENCE MODEL

2.1. Heat Transfer during Deposition

A fully implicit finite difference model, utilizing the tri-diagonal matrix algorithm, has been set up to describe heat flow in a substrate/deposit system during a thermal spraying process. The distributions of gas-borne heat and of incoming droplet flux are both taken as Gaussian axisymmetric about the torch axis, with characteristic radii which can be established by experiment. Any torch movement history can be treated, with the heat and mass arrival flux at the analysis point calculated from the torch position after each time increment. Droplet latent heat is treated as a source term. Heat flow is taken as unidirectional in the through-thickness direction, with uniform volume element thicknesses within the substrate and the deposit. In addition, both convective and radiative heat losses are included from front and back free surfaces

2.2. Residual Stress Generation and Bending Moments

Stresses are evaluated using the Kirchoff hypothesis of an equal biaxial in-plane stress state and no through-thickness stress. Calculations are carried out periodically, using a finer mesh than that for heat flow. The current thermal profile is mapped onto the stress mesh using linear interpolation. Stresses are calculated using the relaxed width of all volume elements to establish the equilibrium width from a force balance. The couple is then allowed to adopt the uniform curvature corresponding to minimisation of elastic strain energy. Plastic yielding is permitted, affecting the relaxed width of the element concerned and also effectively moving the neutral axis. A newly-arrived splat is treated as an extra deposit element for one time step, before being incorporated into the outer deposit element. This allows a specified quenching stress effect to be introduced. The thermal and curvature histories, and the stress distribution at any time, can be calculated. Further details of the model are given elsewhere [43,46].

3. THERMO-MECHANICAL BEHAVIOUR

3.1. Experimental Procedure

Vacuum Plasma Spraying was carried out using a Plasma Technik PT800 M1000 unit with a F4 Plasma gun. Substrates were in the form of strips about 1 mm in thickness. A two-pass raster of the gun was used, repeated ten times, in order to build up a uniform deposit about 0.2-0.4 mm in thickness. Details of the conditions are given elsewhere [43]. Temperatures were monitored via thermocouples spot-welded to the rear of the substrate. The curvature histories were studied by video recording a magnified image of the substrate seen from the side through a viewport, with backlighting through a diffuser screen. The video tape was synchronised with a timing signal from a microcomputer, and the curvature history data were obtained subsequently using a simple frame-by-frame image analysis procedure.

3.2. Theory - Experiment Comparisons

A typical thermal history for the rear of a substrate can be seen in Fig.8, which shows both experimental and predicted curves for a particular run. In this case the conditions were designed to maintain the temperature at around 600°C during spraying. It can be seen that good agreement can be obtained between theory and experiment. Most of the thermal boundary conditions used in the model were established from independent experiments.

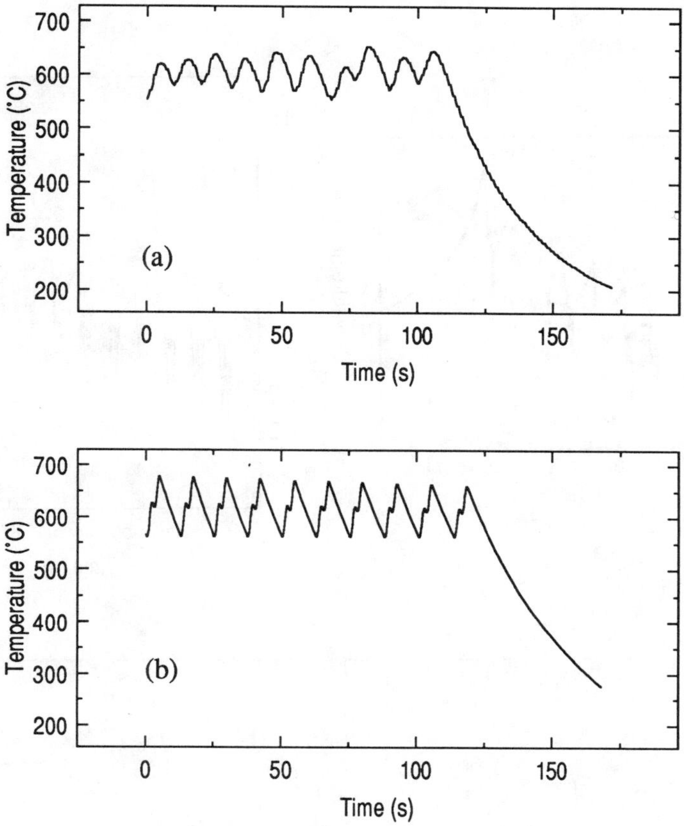

FIGURE 8 Thermal histories for the rear face of a thin Ti-6Al-4V substrate during spraying of alumina, showing (a) a typical experimental trace and (b) the corresponding theoretical prediction, obtained using appropriate thermal boundary conditions.

Curvature histories recorded during spraying can vary considerably in nature for different types of deposit and substrate. Two examples are shown in Fig.9, together with predictions from the model. For the case of alumina, the quenching stress is very low (see Fig.7), primarily as a result of microcracking; a value of 20 MPa was used in the model. There are changes in curvature as the torch passes near the analysis point, which are largely the result of transient thermal gradients; a higher magnification view of a portion of the plot, with notes of the processes occurring within different regimes, is shown in Fig.10. A progressive increase in curvature (towards the torch) is observed, which is largely the cumulative effect of the quenching stress. Post-deposition cooling reverses this trend, as the substrate (a Ti alloy) contracts somewhat more than the deposit. All the changes are relatively small. This is not the case for the spraying of the FeCrAlY, which appears to be exceptionally resistant to all stress relaxation phenomena and has a very high quenching stress. (A value of about 1 GPa was used for the predicted curve; note that the theoretical maximum, given by α E (T_m-T_s), is about 2-3 GPa.) A very sharp curvature is quickly generated. This is little changed during post-deposition cooling, as the expansivities of substrate and deposit are again rather similar.

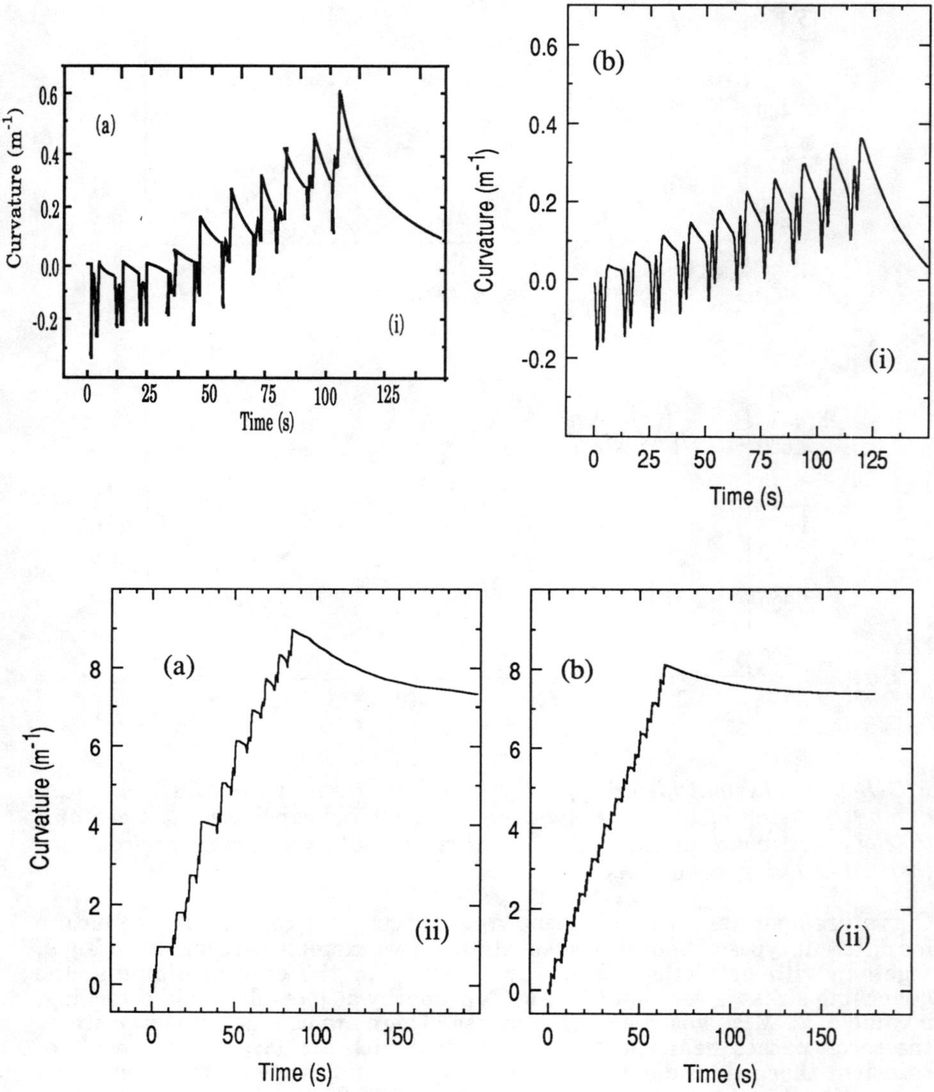

FIGURE 9 Curvature history comparisons showing (a) experimental and (b) predicted variations during and after spraying for deposition of (i) alumina onto Ti-6Al-4V and (ii) an FeCrAlY alloy onto stainless steel.

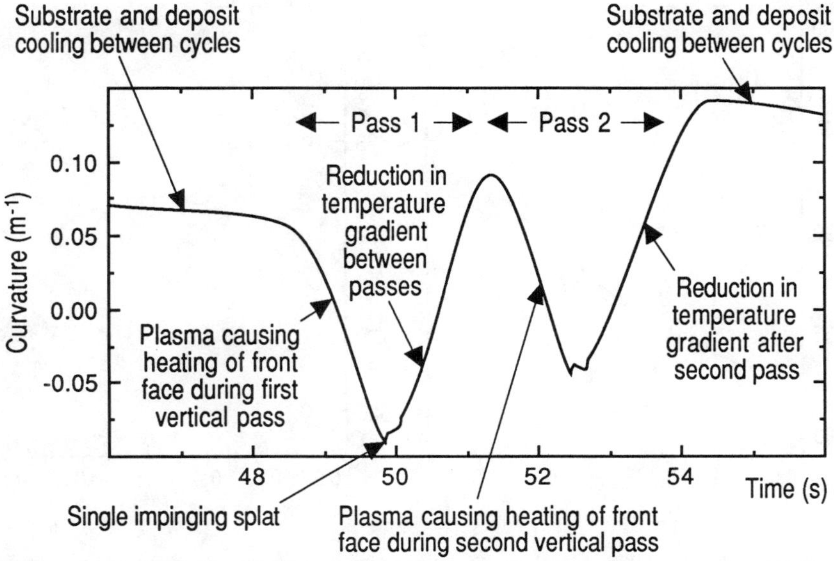

FIGURE 10 Magnified view of a curvature history during passage of the plasma torch, showing the processes responsible for the shape of the curve.

3.3. Predicted System Characteristics

Various features of the process may be explored using the model. For example, Fig.11 shows through-thickness thermal gradients for the alumina on Ti case, while the torch is heating the surface and during subsequent cooling. Note that temperature differences are relatively small, even for these low conductivity materials. This helps to explain why stresses from thermal gradients are usually unimportant. The main interest, however, is in the residual stresses. These are shown in Fig.12 for the alumina and FeCrAlY coatings after cooling to room temperature. The presence of through-thickness gradients of stress is immediately apparent. Note also that the magnitudes are very different for the two cases. For the alumina, microcracking has reduced both the quenching stress and the stiffness of the deposit, with the consequence that stresses will be relatively low, even using deposits having a larger expansivity mismatch than is the case for Ti-6Al-4V. That no yielding has occurred can be inferred from the linear plots. At the other extreme, very large stresses are generated with the FeCrAlY. (These would be even higher if bending were inhibited, such as with a thicker substrate.) This has given rise to some yielding of the steel substrate. These stresses are known to cause problems, and a common consequence is spallation of the coating.

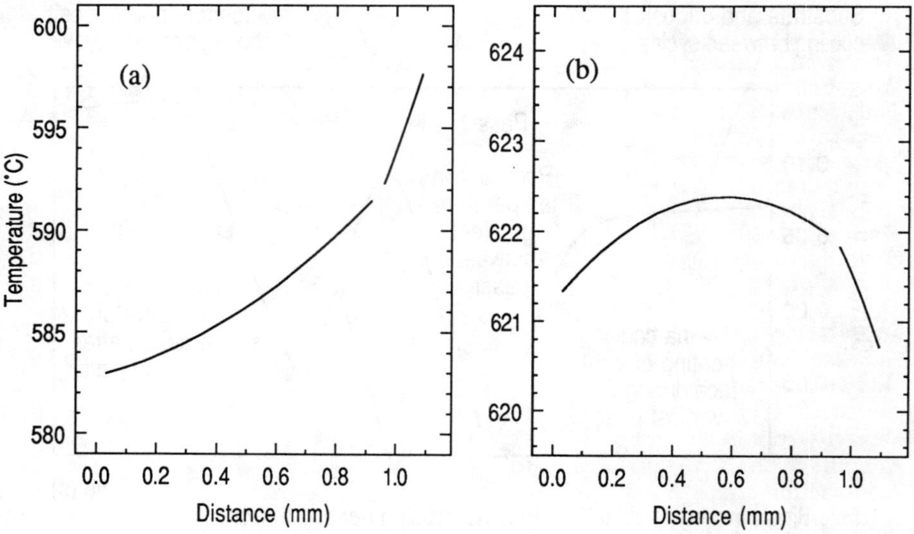

FIGURE 11 Predicted thermal gradients across the substrate-deposit couple for spraying of alumina onto Ti-6Al-4V, (a) while the torch is passing over the point concerned and (b) soon after spraying has been completed

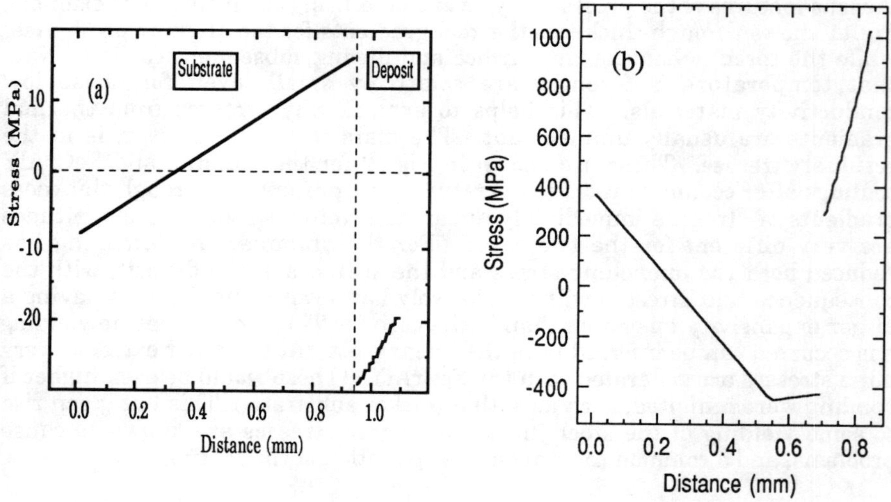

FIGURE 12 Predicted distributions of residual stress in substrate and deposit after cooling to room temperature for (a) alumina and (b) FeCrAlY alloy sprayed onto steel substrates.

4. SUMMARY & CONCLUSIONS

Heat flow and thermal contraction during plasma spraying has been briefly reviewed. This treatment has been primarily concerned with the heat flux at the surface of the growing deposit and its transport through the deposit and substrate. It has been noted that while heat transfer to particles usually occurs under Newtonian conditions in flight, very good thermal contact is normally established between splat and underlying deposit, so that quenching conditions may well be non-Newtonian. A method for estimating the splat-deposit heat transfer coefficient has been outlined. Values appear to be in the range 10^6-10^8 W m^{-2} K^{-1}. Splat cooling rates are of the order of 10^6 K s^{-1}, leading in some cases to metastable phase formation; for example, alumina forms largely as the γ phase.

A numerical model has described for the prediction of thermal contraction stresses during spray deposition. This has been validated by comparison between predicted and observed changes in the curvature adopted by the substrate-deposit couple as deposition proceeds. The residual stresses arise largely from two sources - the quenching stess associated with rapid cooling of a splat to the temperature of the underlying deposit and differential thermal contraction stresses caused by cooling of substrate and deposit when they have different expansivities. Data have been presented for spraying of alumina onto a Ti alloy and a FeCrAlY alloy onto stainless steel. These two cases exhibit very different behaviour, although in both cases the differential thermal contraction strains are relatively small. The alumina undergoes extensive microcracking, which results in reduced stresses, while the FeCrAlY is resistant to all forms of stress relaxation and large quenching stresses are developed during deposition. With recent improvements in understanding of the origins of these stresses, there should be scope for their control by changes in deposition conditions.

ACKNOWLEDGEMENTS

The authors are grateful to Dr.S.Kuroda for helpful discussions and to Mr.K.A.Roberts for his expertise in running the VPS unit at Cambridge.

REFERENCES

[1] J.H.Zaat , "A Quarter of a Century of Plasma Spraying", *Ann Rev Mat Sci* , vol.**13** (1983) 9-42.

[2] E. Lugscheider, "The Family of Plasma Spray Processes - Present Status and Future Prospects", *1st Plasma Technik Symposium*, Lucerne, May 1988, H. Eshnauer et al (eds) 23-48, Plasma Technik, Wohlen, Switzerland (1988).

[3] G.Goueshet, "A review on measurements of particle velocities and diameters by laser techniques, with emphasis on thermal plasmas", *Plas Chem Plas Proc* , vol.**5** (1985) 91-117.

[4] P.Fauchais, A.Vardelle, M.Vardelle, J.F.Coudert & J.Lesinski, "Correlation of the Physical Properties of Sprayed Ceramic coatings to the Temperature and Velocity of the Particles travelling in atmospheric plasma jets: measurements, modelling and comparisons", *Thin Solid Films* , vol.**121** (1984) 303-316.

[5] I.A.Fisher, "Variables influencing the characteristics of Plasma-sprayed coatings", *Int. Met. Rev.* vol.**17** (1972) 117-129.

[6] M.F.Smith & R.C.Dykhuizen, "Effect of Chamber Pressure on Particle Velocities in Low Pressure Plasma Spray Deposition", *Surf. & Coat. Technol.*, vol.**34** (1988) 25-31.

[7] H D Steffens, M.Mack & R.Lauterbach, "Measurement of Particle Velocities for an Analytical Model of Low Pressure Plasma jets", *1st Plasma-Technik Symposium*, Lucerne, May 1988, H.Eshnauer et al (eds.), 67-76, Plasma-Technik, Wohlen, Switzerland (1988)

[8] A.R. Stetson & C.A.Hauck, "Plasma Spraying Techniques for Toxic and Oxidisable Materials" *J. of Metals* , vol.**13** (1961) 479-482.

[9] H.Gruner, A.R.Nicholl & D.Prince, "Princess Quality Control during Vacuum Plasma Spraying with Oxygen-sensitive Materials", in *UK Corrosion 1984*, Inst of Corrosion Sci & Technol (1984).

[10] E.Pfender, "Fundamental Studies association with the plasma spray process", *Surf. & Coat.Technol.*, vol.**34** (1988) 1-14.

[11] P.A.Dearnley, K.A.Roberts & T.W.Clyne, "Some observations on the Microstructure of Boron Carbide Reinforced Titanium Composites Produced by Spray Co-deposition", in *12th International Plansee Seminar '89*, paper C31, p.523-538 H.Bildstein & H.M.Ortner (eds.), (1989).

[12] R.R.Kieschke & T.W.Clyne, "Fibre Barrier Coatings in Plasma-Sprayed Titanium Composites", *Proc. Sixth World Conf. on Titanium*, p.1789-1794, P.Lacombe, R.Tricot & G.Beranger, Sociétè Française de Metallurgie, Cannes, June, 1988.

[13] H.D.Steffens, M.Mack, B.Eckhardt & R.Lauterbach, "Mathematical Simulation of Particle Velocities in a Low Pressure Plasma Jet and its Experimental Verification", *Surf. & Coat. Techn.*, vol.**41** (1990) 117-126

[14] N.El-Kaddah, J.McKelliger & J.Szekely, "Heat Transfer and Fluid Flow in Plasma Spraying" *Metall. Trans* ., vol.**15B** (1984) 59-70.

[15] A.H.Dilawari & J.Szekely, "Fluid Flow and Heat Transfer in Plasma Reactors - I Calculation of Velocities, Temperature Profiles and Mixing", *Int.J.Heat & Mass Transfer*, vol.**30** (1987) 2357-2372

[16] A.H.Dilawari & J.Szekely, J.F.Coudert & P.Fauchais, "Fluid Flow and Heat Transfer in Plasma Reactors - II A Critical Comparison of Experimentally Measured and Theoretically Predicted Temperature Profiles in Plasma Jets in the Absence and Presence of Side Stream Injection", *Int.J.Heat & Mass Transfer*, vol.**32** (1989) 35-46

[17] D.Y.Wei, D.Apelian & B.Farouk, "Particle Melting in High Temperature Supersonic Low Pressure Plasmas", *Metall.Trans.*, vol.**20B** (1989) 251-262

[18] Y.P.Chyou & E.Pfender, "Behaviour of Particulates in Thermal Plasma Flows", *Plasma Chem. Plasma Proc.*, vol.**9** (1989) 45-71

[19] R.Spores & E.Pfender, "Flow Structure of a Turbulent Thermal Plasma Jet", *Surface & Coatings Technol.* vol.**37** (1989) 251-270.

[20] D. Apelian, M.Paliwal, R.W.Smith & W.F.Schilling, "Melting and Solidification in Plasma Spray Deposition - a Phenomenological Review", *Int Met Rev.* vol.**28** (1983) 271-394.

[21] R.R.Kieschke, K.A.Roberts & T.W.Clyne, "Instabilities in the Vacuum Plasma Spray Process", to be published in *Surface & Coatings Technol.* (1990)

[22] R.C.Dykhuizen & M.F.Smith, "Investigations into the Plasma Spray Process", Proc. Nat. Ther. Spray Conf., Oct 1988, Cincinatti, p.61-66

[23] P.Fauchais, M.Vardelle, A.Vardelle & J.F.Coudert, "Plasma Spraying of Ceramic Particles in Argon-Hydrogen DC Plasma Jets: Modelling and Measurements of Particles in flight Correlation with Thermophysical Properties of Sprayed Layers", *Metall.Trans.*, vol.**20B** (1989) 263-276

[24] T.W.Clyne, R.A.Ricks & P.J.Goodhew, "The Production of Rapidly Solidified Aluminium Powder by Ultrasonic Gas Atomization - I Heat and Fluid Flow", *Int.J.Rapid Solidif.*, vol.**1** (1984-85) 59-80

[25] C.A.Williams & H.Jones, "The Effect of Melt Superheat and Impact Velocity on Splat Thickness", *Mat.Sci. & Eng.*, vol.**19** (1975) 293-297.

[26] J.Madejski, "Solidification of Droplets on a Cold Substrate", *Int.J.Heat & Mass Transf.*, vol.**19** (1976) 1009-1013.

[27] Y.A.Kharmalov, M.S.Hassan & R.N.Anderson, "Effect of Substrate Surface Conditions and Impact velocity of Sprayed Particles on Coatings produced by Plasma Spraying", *Thin Solid Films* , vol.**63** (1979) 111-118.

[28] T.W.Clyne, "Numerical Treatment of Rapid Solidification", *Metall.Trans.*, vol.**15B** (1984) 369-381.

[29] R.McPherson, "On the Formation of Thermally Sprayed Alumina Coatings", *J.Mat.Sci.*, vol.**15** (1980) 3141-3149.

[30] E.Garrity, D.Wei & D.Apelian, "Modeling of the Spreading Kinetics during Droplet Solidification Processing", in "*Modelling of Casting & Welding Processes IV*", Eng. Found. (1988)

[31] R.McPherson, "Formation of Metastable Phases in Flame and Plasma-Sprayed Alumina", *J.Mat.Sci.*, vol.**8** (1973) 851-858.

[32] T.W.Clyne & S.Kuroda, "Use of the Eshelby Method to Interpret Thermophysical Properties of Porous Material produced by Spray Deposition", to be submitted to *Acta Met.* (1990)

[33] W.J.Parker, R.J.Jenkins, C.P.Butler & G.L.Abbott, "Flash Method of Determining Thermal diffusivity, Heat Capacity and Thermal Conductivity", *J.Appl.Phys.*, vol.**32** (1961) 1679-1684

[34] H.C.Fielder, "The Effect of Structure on the Thermal Conductivity of Plasma Sprayed Alumina", in "Plasma Processing and Synthesis of Materials", J.Szekely & D.Apelian (eds.), *Mat.Res.Soc.Symp.Proc.*, vol.**30**, p.173-181, (1984)

[35] S.Kuroda & T.W.Clyne, "The Quenching Stress in Thermally Sprayed Coatings", submitted to *Thin Solid Films* (1990)

[36] S.C.Gill & T.W.Clyne, "Property Data Evaluation for the Modelling of Residual Stress Development during Vacuum Plasma Spray Deposition",to be published in *"1st European Conference on Advanced Materials and Processes (Euromat'89)"*, Aachen, Nov.1989, H.E.Exner (ed.), Deutsche Ges. Metallk. (1990)

[37] H.Zhuang & C.Gu, "A Study of Residual Stress of ZrO_2 + MgO Plasma Sprayed Coating", *1st Plasma-Technik Symp.*, H.Eschnauer et al (eds.), 1988, vol.2, p.277-284.

[38] D.S.Rickerby, G.Eckold, K.T.Scott & I.M.Buckley-Golder, "The Interrelationship between Internal Stress, Processing Parameters & Microstructure of Physically Vapour Deposited and Thermally Sprayed Coatings", *Thin Solid Films.* vol.**154** (1987) 125-141

[39] U.Selvadurai & W.Reimers, "Characterisation of Phase Composition and Residual Stress State in Plasma Sprayed Ceramic Coatings", in *7th CIMTEC World Ceramics Conference,* paper S1.2A-LO5, R.Vincenzini (ed.), (1990).

[40] G.S.Shajer, "Application of Finite Element Calculations to Residual Stress Measurement", *J.Eng.Mat.Tech.,* vol.**103** (1981) 157-163

[41] D.S.Rickerby, K.T.Scott, G.Eckold & D.Lloyd-Thomas, "Analysis of the Residual Stresses in Plasma Sprayed Coatings", *1st Plasma-Technik Symp.* H.Eschnauer et al (eds.), 1988, vol.2, pp.267-276.

[42] M.K.Hobbs & H.Reiter, "Residual Stresses in ZrO_2-8%Y_2O_3 Plasma Sprayed Thermal Barrier Coatings", *Surf. & Coatings Techn.* vol.**34** (1988) 33-42

[43] S.C.Gill & T.W.Clyne, "Stress Distributions and Material Response in Thermal Spraying of Metallic and Ceramic Deposits", *Metall. Trans.* vol.**21B** (1990) 377-385

[44] S.Kuroda, T.Fukushima and S.Kitihara, "Simultaneous Measurement of Coating Thickness and Deposition Stress during Thermal Spraying", *Thin Solid Films*, vol.**164** (1988) 157-163.

[45] S.C.Gill & T.W.Clyne, "Residual Stress Modelling and Characterisation of Thermally Sprayed Ceramics", in *7th CIMTEC World Ceramics Conference,* paper S1.2A-LO7, R.Vincenzini (ed.), (1990).

[46] S.C.Gill, "Residual Stresses in Plasma Sprayed Coatings", PhD thesis, University of Cambridge, (1990)

Summary of General Discussion of Session on Plasma Spraying

K. Hijikata, Tokyo Institute of Technology
M. M. Chen, University of Illinois

In the session, two different subject were included: Plasma Spraying (2 papers) and Plasma CVD (1 paper). In both cases, the temperature nonequiliblium between the electron and the gas and the impurities(species and concentration) have great influences on the characteristics of the generated films. There are many thermofluidproblems, as listing below.

Plasma Spraying:

The merit of this method is the ease in coating thick films. Its disadvantages are the weakness of the film (high porosity and low adhesion between the formed layers) and poor impurity control.

Thermofluid Problems Inside the Plasma Gun:

1. Development of cooling technlogy of the electrodes to get a stable plasma.
2. New steady feeding method of very small amount of particles.
3. Non-equilibrium in the discharge process of the plasma.

Thermofluid Problems in the Plasma Jet:

1. Radiation heat transfer between the plasma and particles.
2. Number density distribution of the particles in the jet.
3. Multi-phase dynamics between particles and plasma.

Problems on the Coating Plate:

1. Hydrodynamics of droplet impact with the substrate.
2. Hydrodynamic characterization of the jet.
3. Stress analysis of the coated film.
4. Porosity control and impurity control of the film.
5. Estimation of the thermodynamic physical properties.

Plasma CVD:

A thin homogeneous thin film can be obtained by this method, but many pheneomena are poorly understood. Both thermofluid problems and the eletrical discharge physics must be investigated.

1. Electric potential distribution between the electrodes considering the motion of electrons and ions.
2. Practical CVD applications of ultra-short pulse driven plasma.
3. Effect of ions and radicals on the film characteristics.

LASER AND ELECTRON BEAM PROCESSING

A Study on Metal Melting Process by Laser Heating

WEN-JEI YANG and NENGLI ZHANG
Department of Mechanical Engineering
and Applied Mechanics
University of Michigan
Ann Arbor, Michigan 48109, USA

B.-X. WANG and JIDONG YANG
Department of Thermal Engineering
Tsinghua University
Beijing 100084, PRC

INTRODUCTION

Lasers have increasingly become reliable processing tools for the fabrication industry with various applications. They are especially useful in areas where conventional processing tools can not work well, such as material processing in vaccum chambers, distortionless fine processing and other processing under special circumstances. Laser heating has potentially wide applications in material processing in outer space. Hence, a comprehensive knowledge of the entire physical process of laser heating is needed in extending its applications.

During the metal heating process, the most important physical phenomenon is laser beam absorption at the metallic surface which has been the subject of many investigations [for example, 1 - 3]. It was found that the absorptivity of a CO_2 laser beam by solid or liquid metallic surface is very small, for example 10 ~15% at most in steels. However, if a cavity is formed by focusing a high power laser on a metallic surface, a more complex situation will occur. It is generally accepted that the plasma induced by the laser plays a very important role in the laser-material interaction, but the actual role of the plasma is still poorly understood and has been debated for more than a decade. Some researchers reported considerable loss of laser energy through the plasma but others reported an enhancement in the coupling of the laser energy onto the workpiece by the plasma. Herziger et al. [3 - 5] considered that in the low intensity limit of laser processing, the laser-induced plasma increases the coupling of the laser energy into the workpiece through the generation of laser-supported combustion waves, whereas in the high intensity limit of laser processing, the laser-induced plasma shields the workpiece through the generation of laser-supported detonation waves. A comprehensive review of the literature pertinent to the interaction mechanism of laser processing with special emphasis on the optimum processing results is available in [6]. It is evident that any laser materials processing model should cover the entire process ranging from room temperature to the melting-down stage.

Almost all previous studies have directed efforts on plasma effects on the materials already in the molten state, neglecting the early heating stage and the intermediate melting state. This paper presents both experimental and numerical investigations on the whole process of the laser-material interaction. The temperature-time history at a strategic point discloses the existence of distinct stages.

EXPERIMENTAL APPARATUS

A schematic diagram of the experimental setup is depicted in Fig.1. A carbon dioxide laser of 200 watts, with wavelength 10.6 μm in continuous mode, was used as a power source. The laser beam was reflected by a gold-plat full-reflected mirror, then passing through a gallium arsenide lens to form an irradiation spot of 0.1 - 1 mm diameter on the surface of a test piece. A graphite output switch was used to control the duration of laser irradiation on the test piece. The optical system, consisted of a gold-plating mirror, a Ga-As lens and a shielding gas chamber. It was designed to move vertically up and down over 60 mm. Nitrogen filled the shielding gas to protect the optical system from a damage by the metal vapor of the test piece. Lead disks were used as the test pieces, because of lower melting point, lower thermal conductivity and lower thermal diffusivity. That means, less laser power is needed to melt the test pieces and its slower temperature response permits thermocouples to monitor the temperature change of the test piece with little time lag. Three or five thermocouples were placed on the upper surface of each lead disk in the radial direction at the distances of 2 mm, 6 mm, 11 mm (for disk I) or 2 mm, 6 mm, 11mm, 17 mm and 24 mm (for disk III), respectively from the origin where the laser beam irradiates. Two thermocouples were positioned on the lower surfaces, one at the origin and the other at 11 mm from the origin. All temperatures were monitored using a data logger.

Since the whole process of metal melting by the laser beam was completed within a short time, a computer data acquisition system was provided to record the data. The system consisted of analogue signal amplifiers, a 16-channel multipass switch which was capable of sampling every 0.04 ms, a sample keeper, an A/D converter of 16-channel inputs with maximum conversion speed of 25 KHz, and an IBM/AT-286 computer system with printer and plotter. After passing through amplifiers, the temperature signals were fed into the multipass switch and the sample keeper. These analog signals were converted into digital data using an A/D converter and then stored in the computer. The digital data can be displayed as the temperature-time history on a monitor graphically or numerically. The results can be listed by a printer or graphically illustrated by a plotter. A series of electrical shielding measures were taken to protect the data acquisition system from interference by external noises.

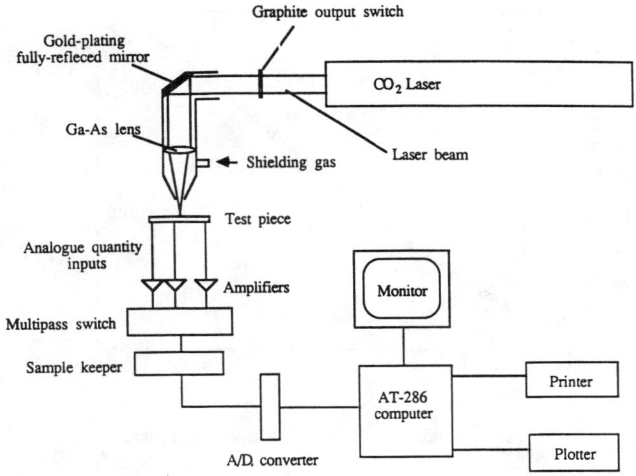

FIGURE 1. A schematic of experimental setup

EXPERIMENTAL PROCEDURE

Prior to test runs, both static and dynamic tests were performed on the computer data acquisition system. The static test was carried out using thermostat. The data acquired by the computer system agreed very well with the readings from a standard thermometer in the thermostat. In the dynamic test, an autobalance measurer was used as an instrument for comparison. A thermocouple placed on the surface of a test piece was connected to both the computer data acquisition system and the autobalance measurer to monitor the warm-up process of the test piece. Figure 2 shows a typical result, indicating a complete agreement of the two measurements.

After the completion of checking thermocouple calibrations, zero adjustment of the amplifiers and the A/D converter, the following procedure was exercised:

1. Turn on the amplifiers for more than a half hour for stablizing;
2. Turn on the CO_2 laser for about 10 minutes to achieve a stable power output;
3. Turn on the graphite output switch to measure the power of the laser beam at the focal plane of the optical focus system;
4. Turn off the output switch and place the test piece on a proper position;
5. Measure the diameter of the laser beam spot;

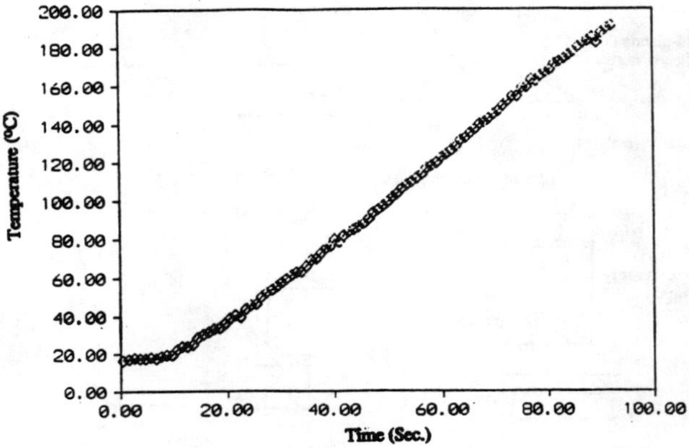

(a) Result by computer data aquisition system

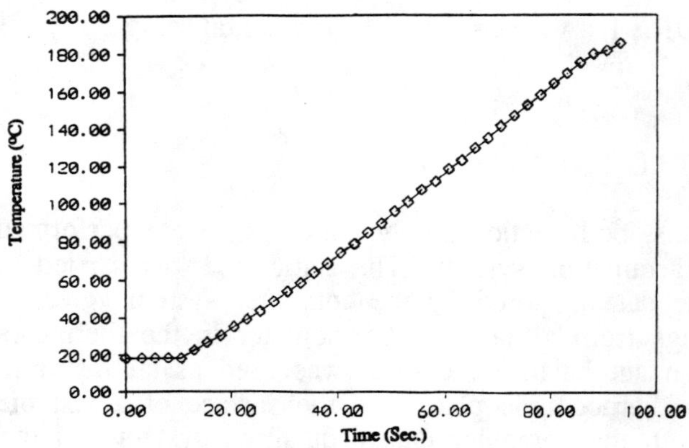

(b) Result by autobalance measurer

FIGURE 2. Test of computer data acquisition system

6. Turn on the computer system to operate a data-logging program;
7. Open the valve of shielding gas for a desired flow rate;
8. Turn on the output switch again and start recording the laser melting process.

EXPERIMENTAL RESULTS AND DISCUSSION

A number of lead disks with different physical dimensions and various surface conditions were used as the test pieces. Figures 3 and 4 show three typical temperature recordings at strategic points. Figure 3 depicts the recordings on disk I of 30 mm diameter and 7 mm thick, with the condition of the irradiation surface at $\nabla 7$, having 1 mm diameter of laser irradiation spot and being irradiated by 100 W of laser power. The recordings in Fig. 3 - a and - b were taken at 11 mm from the disk center on the upper surface and at the disk center on the lower surface, respectively. Figure 4 is the temperature recording at a location 2 mm from the disk center on the upper surface of disk III of 80 mm diameter and 5 mm thick, with the condition of the irradiated surface at $\nabla 4$, having 1 mm diameter of laser irradiation spot and being irradiated by 140 W of laser power. It is observed in the figures that the metal temperature responses can be classified into three stages: solid-heating, melt-heating and vapor plasma. In the first stage immediately following the irradiation of laser beam, the metal spots in the solid phase are heated up at a lower rate. The response curves are slightly concaved downward

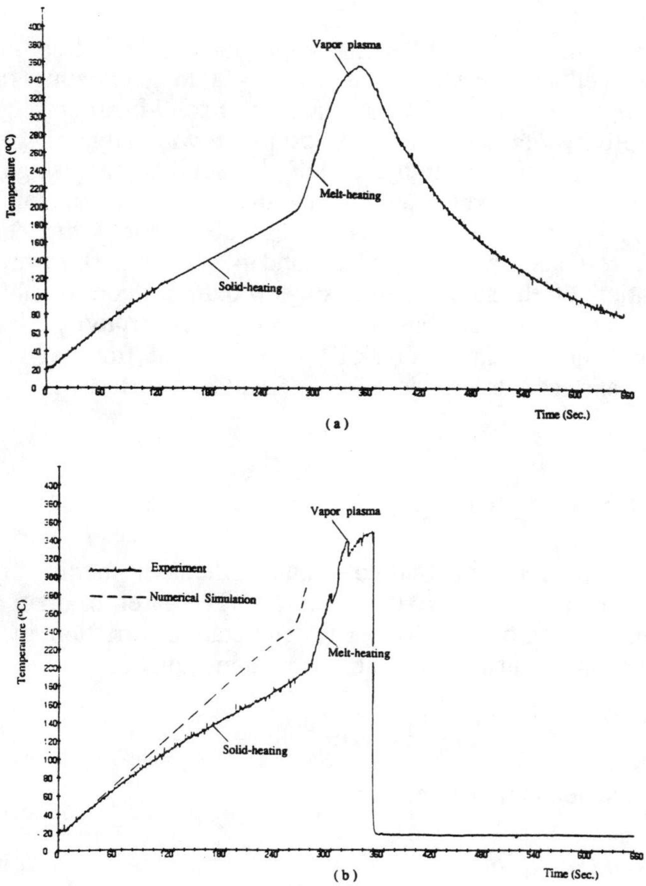

FIGURE 3. Typical temperature-time history of disk I

followed by a linear variation with time. As soon as the metal was melted by the laser beam, the local metal temperatures in the test piece immediately shoot up with a steep gradint. Take Fig.3 for disk I as an example. The temperature response curves of both locations exhibit a marked deflection at about 284 sec. after the initiation of laser irradiation on the disk center. This steep rise in the temperatures suggests that the energy absorbed by the test piece has jumped to a new level since the absorptivity of liquid metal is much larger than that at solid state. The rising speed of these temperatures is retarded with the formation of metal vapor plasma in the cavity. This is because the vapor plasma interrupts the incoming laser beam and the continuous generation of the vapor plasma contributes to a lower absorption of laser radiation energy. After the test piece is punctured by the plasma beam, the temperatures fall rapidly as shown in Fig. 3 - a and 4. In the case of Fig.3 - b, the thermocouple at the origin on the lower surface is burnt, resulting in a sudden termination of temperature reading.

It is obvious that during the solid-heating stage, most of the laser radiation energy is not absorbed by the test metal, causing a longer heating time. A test piece with rough surface would have a shorter solid-heating stage due to a higher absorptivity of the surface. A test piece with large mass would have a lower metal temperature at the end of the solid-heating stage. An increase in the irradiation power would lessen the time durations of both the melt-heating and vapor plasma stages. These observations are derived by comparing the response curves in Figs.3 and 4. If the test surface is oxidized, the duration of the solid-heating stage would be considerably shortened because such a surface has a much higher absorptivity. For example, the solid-heating stages of disk III was reduced from 62 sec. (in Fig. 4 for a clean surface) to 20 sec. (for an oxidized surface, not shown).

NUMERICAL SIMULATION

A simple mathematical model is used to simulate the laser melting process. The lead disk has a radius R with thickness L. The laser beam of radius R' irradiates at the center of the disk. The temperature variation in the test metal is governed by unsteady heat conduction equation:

$$C^*(T)\frac{\partial T}{\partial \tau} = \frac{1}{r}\frac{\partial}{\partial r}\left[r\,k^*(T)\frac{\partial T}{\partial r}\right] + \frac{\partial}{\partial z}\left[k^*(T)\frac{\partial T}{\partial z}\right] \qquad (1)$$

where T denotes the temperature; τ, time; and

$$C^*(T) = \begin{cases} \rho_s C_s(T) & T < T_m \\ \rho_l C_l(T) & T > T_m + \Delta T \end{cases} \qquad (2)$$

Within the phase change range, it yields

$$\int_{Tm}^{Tm+\Delta T} C^*(T)\, dT = \rho h_{sl} + \int_{Tm}^{Tm+\Delta T} \rho C_l(T)\, dT \qquad (3)$$

k* is defined as

$$k^*(T) = \begin{cases} k_s(T) & T < T_m \\ k_l(T) & T > T_m + \Delta T \end{cases} \qquad (4)$$

Within the phase change range, a linear variation of thermal conductivity with temperature is assumed, one obtains

$$k^*(T) = k_s + \frac{k_l - k_s}{\Delta T}(T - T_m) \qquad (5)$$

C is heat capacity; ρ, density; k, conductivity; h_{sl}, heat of phase change; and T_m, melting point. The subscripts s and l denote the solid and liquid states, respectively. Within the phase change range, the heat of phase change is taken as a considerable sensible heat capacity. This is the basic idea of so-called sensible heat capacity method [7, 8]. The appropriate initial and boundary conditions are

$$T = T_0 \text{ (room temperature)}, \qquad \tau = 0 \qquad (6)$$

$$k\left(\frac{\partial T}{\partial z}\right) = \varepsilon P / \pi R'^2, \qquad z = 0,\ r \leq R' \qquad (7)$$

$$\frac{\partial T}{\partial z} = 0, \qquad z = 0, \quad R' \angle r \leq R \qquad (8)$$

and

$$\left(\frac{\partial T}{\partial r}\right)_{r=0} = 0, \qquad \left(\frac{\partial T}{\partial r}\right)_{r=R} = 0, \qquad \left(\frac{\partial T}{\partial z}\right)_{z=1} = 0 \qquad (9)$$

Numerical results were obtained using a TTL (Three Time Level) scheme of finite difference method. The temperature-time history at the center of the lower surface of disk I is superimposed in Fig.3-b for comparison with the experimental result. Theory agrees qualitatively well with experiments. The discrepancy results mainly from the assumption that all boundaries are insulated except the spot where the laser irradiation is applied. No formation of metal vapor plasma is taken into consideration in the model.

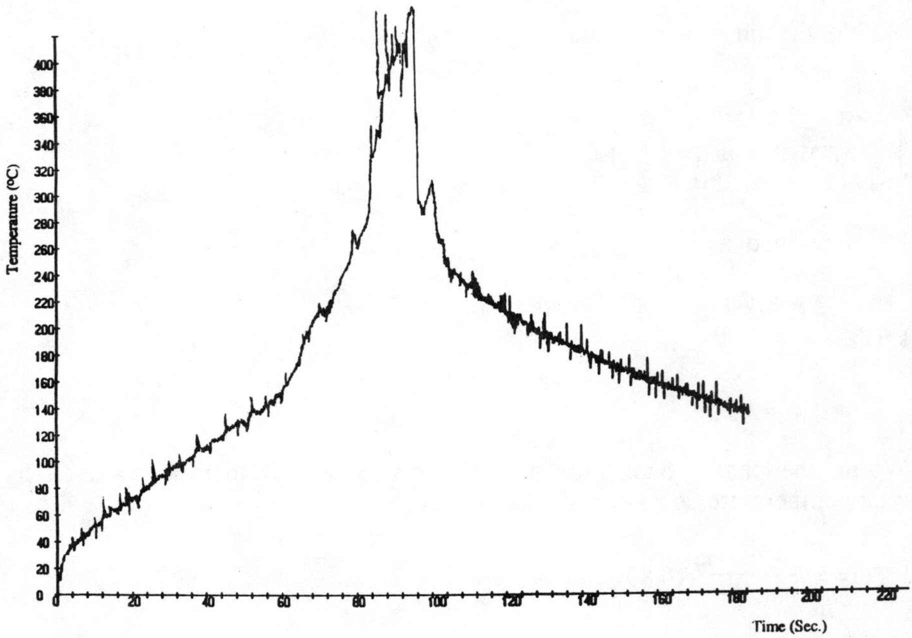

FIGURE 4. Typical temperature-time history of disk III

Neverthless, the transition from the solid-heating stage to the melt-heating stage is clearly shown in the theoretical result. A typical temperature profile is depicted in Fig.5. A sharp spike in the temperature profile at the spot of laser irradiation is observed. Figure 6 is a plot of temperature contours. It reveals a steep temperature change near the spot of laser irradiation as well as the shape of the melt front. Here, the contour of the isotherm at 337 ºC is taken as the solid-liquid phase interface. Some isothermal lines are not perpendicular to the adiabatic and symmetrical surfaces, simply because the grids near these walls are not fine enough.

CONCLUSIONS

The study has disclosed that :

(1) Three distinct stages exist in the entire melting process. This is due to a substantial change of the absorptivity in the course of the process which is induced by the phase change and the formation of metal vapor plasma.

(2) The solid-heating stage shares a substantial fraction of laser heating time. A reduction in the time for metal melting process can be achieved by an oxidization or roughening of the metal surface to increase laser energy absorption by the metal surface. The duration of laser melting process can thereby be reduced.

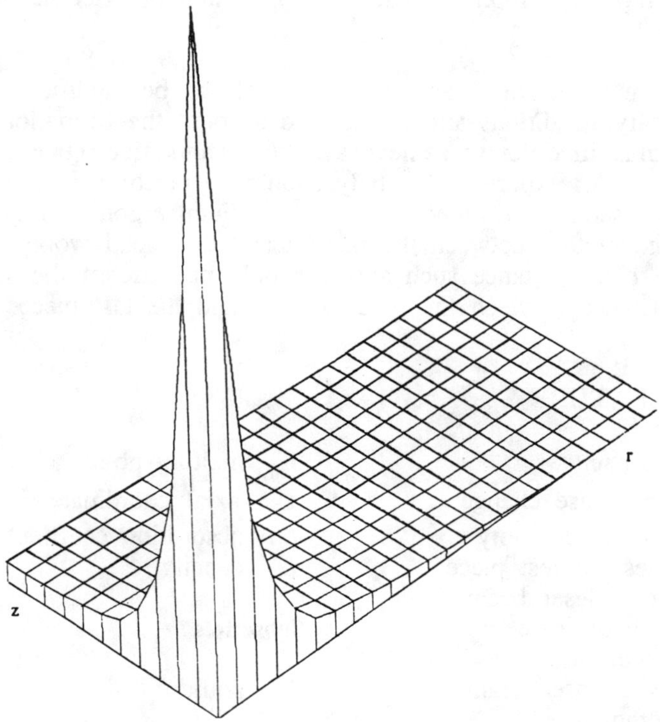

FIGURE 5. Typical temperature profile predicted by theoretical model

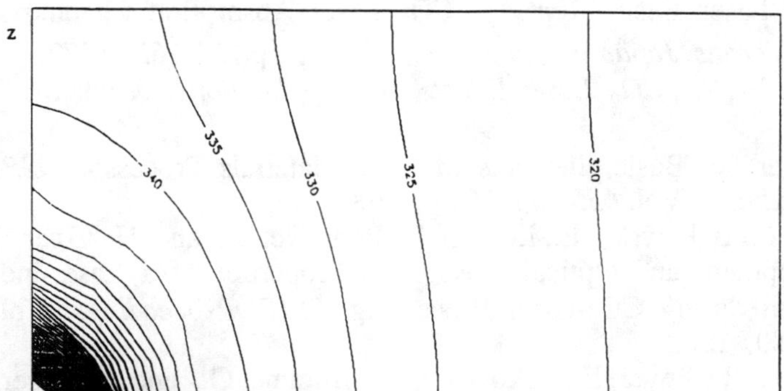

FIGURE 6. Typical constant temperature contours predicted by theoretical model

(3) An evidence of formation of metal vapor plasma is revealed which is important implication to the application of laser heating in metal melting.

In the future study, the mathematical model can be modified by changing the bouindary conditions and taking into account the formation of metal vapor plasma, in order to achieve a better quantitative agreement with experiments. An experimental study should be performed to determine the effects of gaseous environment (such as helium, argon, nitrogen and others) on the interaction between the plasma and the metal workpiece. Other effects are of importance such as the supply pressure of the shielding gas and the distance between the focus system and the test piece.

NOMENCLATURE

C	heat capacity
h_{sl}	heat of phase change
k	thermal conductivity
L	thickness of test piece
P	power of laser beam
r	radial coordinate
R	radius of disk
R'	radius of laser beam spot
T	temperature
T_m	melting point
z	axial coordinate
ε	absorption coeffient
ρ	dencity

Subscripts

l	liquid
s	solid

REFERENCES

1. Atrata, Y., and Miyamoto, I., Some Fundamental Properties of High Power Laser Beam (Rept. 1)-- CO_2 Laser Absorption Charactristics of Metal, *Trans. Japan Welding Soc.,* Vol. 3, pp.152-162, 1972.
2. Duley, W. W., *CO_2 Laser-Effects and Applications,* Academic Press, 1976.
3. Herziger, G., Basic Elements of laser Materials Processing, *SPIE Proceedings,* Vol. 455, pp. 66-74, 1983.
4. Beyer, E., Bakowsky, L., Loosen, P., Poprawe, R., and Herziger, G., Development and Optical Absorption Properties of a Laser-induced Plasman during CO_2-Laser Processing, *SPIE Proceedings,* Vol. 455, pp.75-80, 1983.
5. Poprawe, R., Beyer, E., Bakowsky, L., Brumme, G., and Herziger, G., Limitation of Laser Processing Intensity by Laser-induced Breakdown, *Optoelectronics in Engineering,* Vol. 6, pp. 361-366, 1983.
6. Herziger, G., Kreutz, E. W., and Wissenbach, K., Fundamentals of Laser Processing of Materials, *SPIE Proceedings,* Vol. 668, pp. 2-10, 1986.
7. Bonacina, C., and Comini, G., On the Solution of the Nonlinear Heat

Conduction Equation by Numerical Methods, *Int. J. Heat Mass Transfer* Vol. 16, pp. 581-589, 1973.

8. Basu, B., and Date, A. W., Numerical Study of Steady State and Transient L:aser Melting Problem, *Heat Transfer on the Second International Symposium,* Beijing, China, Preprinted Vol. II, pp. 708 -715, 1988.

Improved Electron Beam Weld Design and Control with Beam Current Profile Measurements

WARREN H. GIEDT
Mechanical, Aeronautical, and Materials Engineering
University of California
Davis, California 95616, USA

ABSTRACT

The determination of machine settings for making an electron beam weld still involves trial and error tests. Also, even after settings are selected, serious variations in penetration may occur. Results are presented to demonstrate that improved weld consistency and quality can be obtained with measurement of the beam size and intensity distribution.

INTRODUCTION

The application of an electron beam for welding occurred during the decade beginning in 1950. The initial development of this new technique was in response to a need in the emerging nuclear energy industry for a method of joining refractory metals [1]. A very intense heat source could be produced in a vacuum which proved to be superior to the use of a shielding gas. Equipment design benefitted from efforts during the previous twenty years to develop equipment for producing a beam of electrons in air, a particular application being the electron microscope [2]. The beam originated in a vacuum chamber and passed through cascaded pressure chambers before exiting to the atmosphere. With the addition of a focusing lens the possibility of melting and vaporization was recognized (see, e.g., Reference 2). The need for greater power soon led to the idea of enclosing the workpiece in a vacuum chamber. Equipment for doing this was introduced before 1960 [1]. Currently most EB welding is done in a vacuum chamber in which the pressure is around 10^{-4} torr or lower.

The key component in an EB welding system is the electron gun. Low voltage welders use a Pierce type gun, which has a diode configuration and produces a rapidly converging beam. Voltages range from 20 to 60 kV. High voltage machines incorporate a Steigerwald telefocus type gun. This has a triode configuration and operation up to 150 kV is common. The major difference between these two systems is that the higher voltage and lower current beams can be focused to smaller minimum areas and higher power densities (e.g., greater than 10^6 W/cm^2).

Electron Beam Welding Characteristics

Electron beams for welding are focused to diameters of 1 mm and smaller. When such a beam is directed on the interface between parts to be joined, a vapor cavity is formed. As the joint is moved under the beam, the effect is a quasi-steady state condition in which the cavity moves through the material at the welding speed. Material in front is melted and flows in a thin layer around the vapor supported cavity to the rear where it solidifies. High speed photographs of EB welding cavity openings and of the cross-sections of cavities as revealed by x-ray beams show that oscillations occur in the molten metal flow, and that under some conditions the cavity may be partly filled and then reopened in a periodic manner [3]. A quasi-steady state model of the EB welding process appears, however, to be a satisfactory average representation of the energy transfer and molten metal flow. It has been the basis for successful prediction of penetration. All analyses for predicting the temperature field and partial penetration have utilized either the two-dimensional moving-line-source solution or the solution for a cylindrical or elliptical cavity moving through an infinite plate. The results of these different approaches are reviewed in Reference 4. A comparison of the predicted trends is shown in Figure 1. The vertical coordinate represents the power P required per unit of penetration d. This quotient is non-dimensionalized by dividing by the product of the average thermal conductivity k_a and the melting temperature θ_m of the material above ambient. This parameter, $P/(dk_a\theta_m)$, is a function of a welding Peclet number vw/α_a. Here v is the weld travel speed, w the maximum width of the fusion zone, and α_a the average thermal diffusivity of the material. Constant thermal properties are assumed in the analyses, and values of k_a and α_a at about one half the melting temperature are commonly used.

The curves in Figure 1 are based on the maximum width of the fusion zone being uniform with depth. However, EB weld fusion zones normal to the welding direction tend to be triangular in shape. To account for this variation the liquid-solid interface during welding is approximated as conical with a maximum width w at the surface. Since weld depth to width ratios are large, heat transfer in the vertical direction is negligible. The temperature distribution and penetration can therefore be determined from the relations for the curves in Figure 1 by integrating from the base of a fusion zone (where w = 0) to the surface where the width = w. The results are plotted in Figure 2. Note that the values of $P/(dk_a\theta_m)$ are lower than in Figure 1 for the same values of vw/α_a. Data for aluminum and steel, for which fusion zone profiles were available, have also been plotted in Figure 2. Average thermal property data used are tabulated in [4].

Although the data points in Figure 2 follow the predicted trend, they do not agree well with any of the individual curves. The possibility of correlating the data with a power curve was therefore investigated. This led to the following relation:

$$P/(dk_a\theta_m) = 3.33\ (vw/\alpha_a)^{0.625} \qquad (1)$$

which is compared with the data in Figure 3. Deviation of the data from this curve was of the order of $\pm$ 40% for values of $(vw/\alpha_a) < 1.0$, improving to around $\pm$ 20% for $(vw/\alpha_a) > 10$. Variations of this magnitude are characteristic of EB welding. Possible reasons for this are (1) differences between actual and indicated machine settings, (2) variations due to operator adjustments, and (3) specific characteristics of different machines. Estimates of the effects of probable errors in machine settings (i.e., voltage, current, travel speed, workpiece distance from

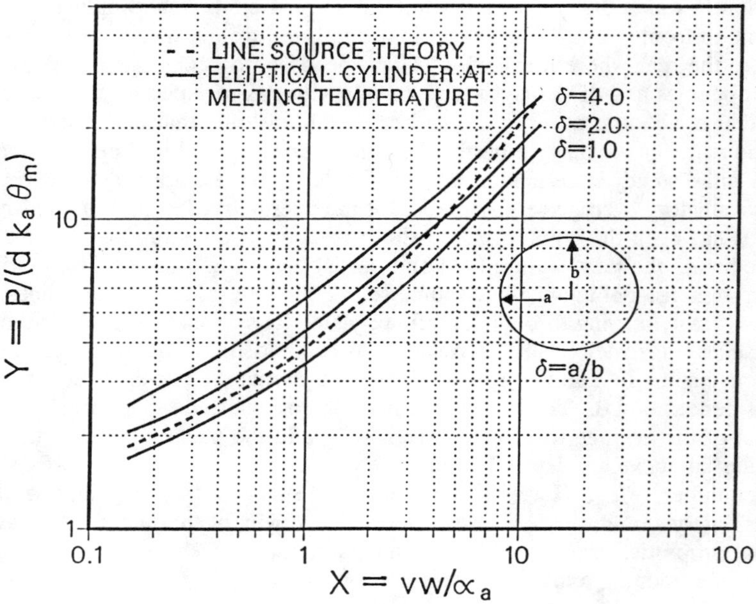

Fig. 1 Power to depth ratio variation with welding velocity based on moving-line source and elliptical cavity solutions (Ref. 4)

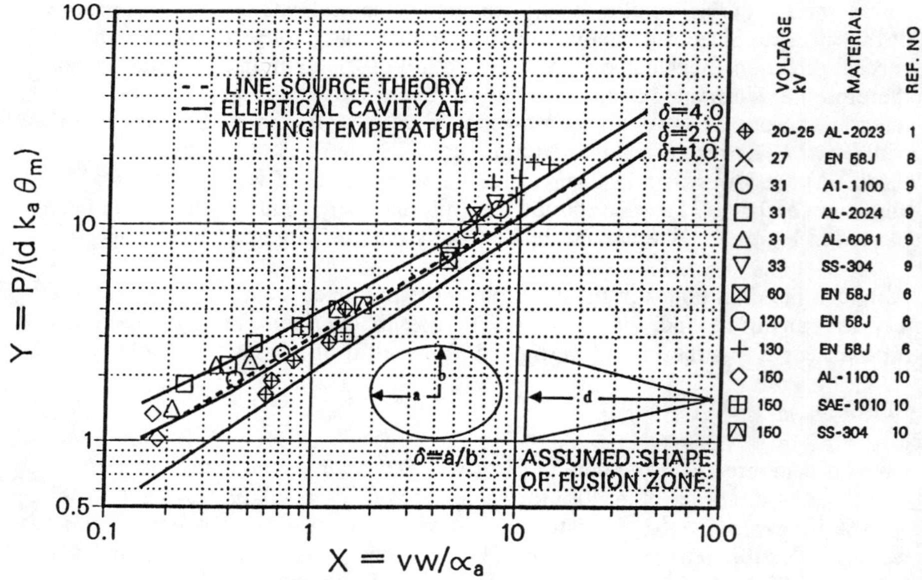

Fig. 2 Comparison of measured with predicted EB welding penetration (Ref. 4)

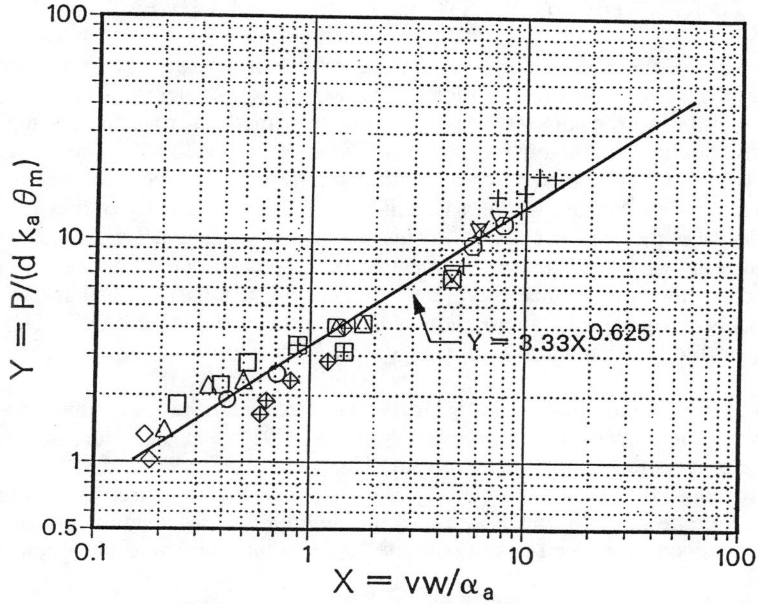

Fig. 3 Correlation of EB welding partial penetration data (Ref. 4)

the focus coil, and focus coil current) showed focus coil current to be the most important [4].

The data points plotted in Figure 3 are for the beam focused to produce the maximum penetration. This is usually achieved with the beam focused at or slightly below the workpiece surface. This condition is determined by adjusting the focus coil current until the apparent size of the beam impinging on a tungsten plate is minimized. This is a qualitative procedure, and is quite likely partly responsible for some of the data scatter in Figure 3. In addition this technique does not provide a quantitative measure of the beam size. This will depend on the accelerating voltage, the current, the distance from the focus coil, and possibly the orientation and condition of the filament in the electron gun. Hence, along with beam focus location, it is important to know the effective size, or more specifically the power density distribution in the beam. Although this is well recognized, accurate determination of the current or energy distribution is difficult and just how an electron beam diameter should be defined has not been established. It was for this reason that the maximum width of the fusion zone was used in correlating the data in Figure 3. (Also this is the dimension that occurs in the line source and elliptical cylinder solutions.) This means that in applying equation (1) to predict penetration d given the beam power P, or to predict the power required to produce a specified penetration d, involves estimating the width of the fusion zone at the surface. This can usually be done by considering the desired depth to width ratio. However, it should be apparent that for predictions of depth it would be desirable to be able to use a characteristic beam dimension. The selection of an appropriate dimension and its measurement will be described in the following sections.

BEAM CURRENT DENSITY DISTRIBUTION MEASUREMENT

Measurement of the current density distribution in an EB essentially provides a measure of the energy flux distribution. This is because the electrons have all been accelerated to the same velocity. A number of methods for investigating the current distribution in a highly focused beam have been developed. The most familiar is to move the beam across a small diameter wire. Moving the beam across a narrow slit located above a Faraday cup provides a similar output [5]. Such a device is pictured in Figure 4. This system can withstand exposure to higher current beams and does not appear to be subject to perturbations introduced by the wire surface curvature. The output of a narrow slit Faraday cup at any location represents a summation of the current through the beam. The resulting profile (such as illustrated in Figure 5) is therefore one-dimensional.

The two-dimensional nature of a low current electron beam has been investigated by recording the current through a small aperture (0.01 cm in diameter) moved through a grid pattern under the beam [6]. This procedure is limited to low currents because of heating and melting of the sides of the aperture. A system based on the alternate approach, i.e., moving the beam across a small aperture, has recently been developed at the Institute for Welding Fabrication in Aachen,

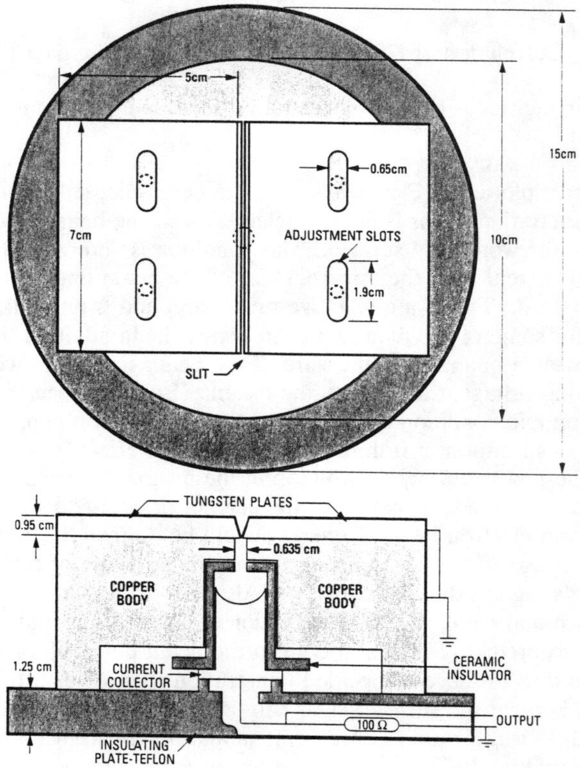

Fig. 4 Narrow slit Faraday cup (Ref. 5)

Germany [7]. Operation at power levels up to 100 kW is possible by using beam sweep speeds on the order of 500 m/s. Water cooling of the sensor surface is, however, required. Computerized presentation of measurements is illustrated in Figure 6. Note that as the voltage is decreased and the current increased to maintain constant power, the beam peak current density decreases and the beam becomes broader. The fusion zone widens and the penetration decreases by about 20%.

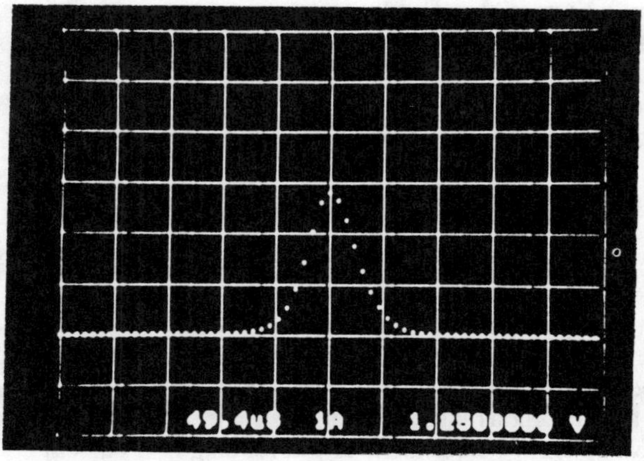

Fig. 5 Oscillograph record of EB swept across a narrow slit Faraday Cup

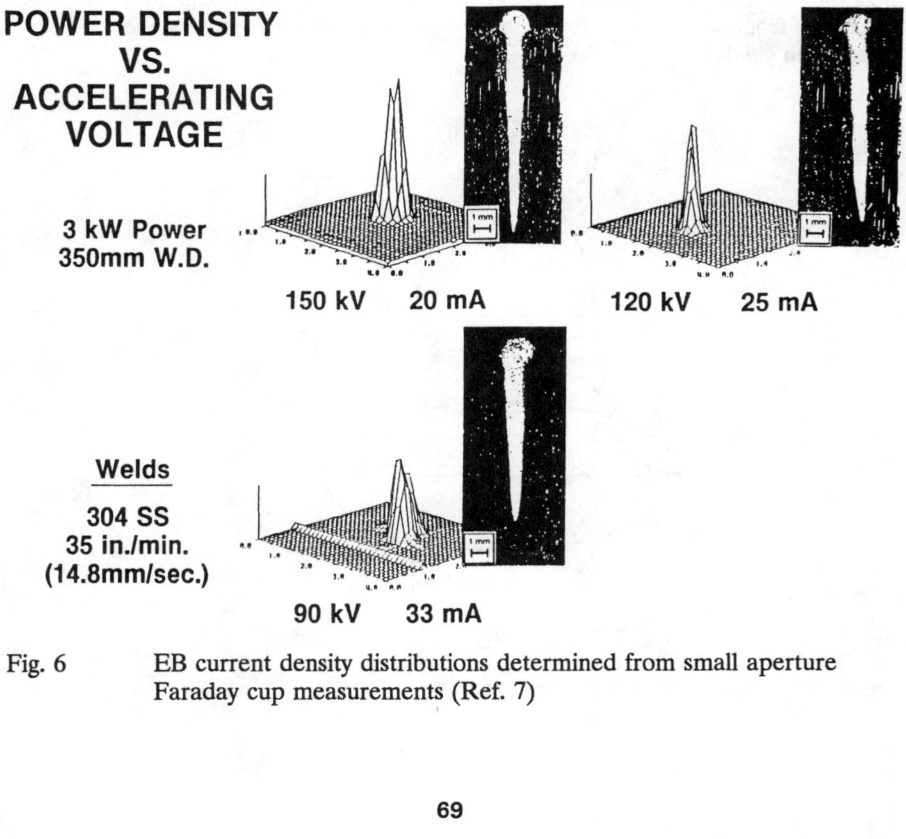

Fig. 6 EB current density distributions determined from small aperture Faraday cup measurements (Ref. 7)

Analytical Representation of Beam Current Density

In order to utilize a beam current distribution measurement it is necessary to be able to predict the welding result it will produce. For example, can the penetration be predicted once the beam profile is determined? An analytical representation of the current or power density profile is needed to accomplish this. For practical purposes the measurement and the functional representation should be as simple as possible. For these reasons a beam profile from a narrow slit would be desirable.

Since the beam current density distribution is anticipated to be similar to a Gaussian probability function, it should be approximately described by

$$I(r) = \frac{I}{2\pi\sigma^2} \exp(-r^2/2\sigma^2) \tag{2}$$

where I(r) denotes the current per unit area at a radius r (see Figure 7), I the total beam current, and σ the standard deviation. Note that the measurements recorded by sweeping the beam across a small aperture yield a profile of the current through a width Δx as illustrated in Figure 7. Integration of the current during a single sweep will be equivalent to the reading of a current collector under a slot of width Δx at the same location (i.e., the same x coordinate). To determine the magnitude of the measurement (assuming a Gaussian distribution) it is necessary to integrate equation (2) in one coordinate direction (e.g., in the y-direction) from $-\infty$ to $+\infty$. Introducing $r^2 = x^2 + y^2$ and carrying out the integration yields

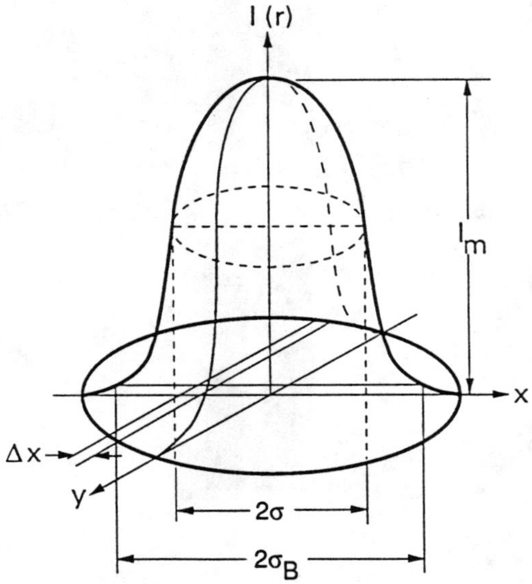

Fig. 7 Gaussian distribution function

$$I_s(x) = I(x)\Delta s = \frac{I}{\sqrt{2\pi}\,\sigma} \exp(-x^2/2\sigma^2)\,\Delta s \quad (3)$$

or

$$I(x) = \frac{I}{\sqrt{2\pi}\,\sigma} \exp(-x^2/2\sigma^2) \quad (4)$$

for which typical units would be mA/mm.

The following procedure is suggested for determining the best value of σ to represent the data. The experimental points are first normalized to obtain a series of $i(x_i) = I_s(x_i)/I\Delta s$ values. Then the first and second moments $M(x)$ and $M(x^2)$ are calculated from

$$M(x) = \sum_{i=1}^{n} x_i\, i(x_i)\, \Delta x \quad (5)$$

$$M(x^2) = \sum_{i=1}^{n} x_i^2\, i(x_i)\, \Delta x \quad (6)$$

where $\Delta x = v\Delta t$ is the distance the beam moves between data points and n is the number of data points. The standard deviation σ is then given by

$$\sigma^2 = [M(x^2) - M(x)^2] \quad (7)$$

Although the current distribution can be represented with the standard deviation σ, the use of a value which is more representative of the effective beam diameter has been proposed [5]. This is the radius σ_B which defines the area within which 95% of the current or energy flows. σ_B is a characterstic dimension commonly used in describing welding heat sources, and is related to σ by

$$\sigma_B^2 = 6\sigma^2 \quad (8)$$

Substituting this in equation (4) gives

$$I(x) = \frac{\sqrt{3/\pi}\, I}{\sigma_B} \exp\{-3(x/\sigma_B)^2\} \quad (9)$$

The one-dimensional slot current distribution predicted from equation (9) is compared in Figure 8 with measurements made in a representative high voltage welder beam. The Faraday cup slit opening was located 0.625 cm above the beam focus location (equivalent to having the beam focused slightly below the workpiece surface). The value of σ_B was evaluated from the data as described above. The agreement is encouragingly good. A similar comparison is shown in Figure 9 for the same beam current but with the Faraday cup slot located 2.5 cm above the beam focus location. At this location the measured distribution is not symmetrical, and the peak of the predicted distribution is about 10% higher than the experimental. Additional measurements indicated that the beam was elliptical in cross section and the major axis is at an angle to the travel direction. Analysis of the weld produced by such an incident distribution would be complex. Hence practical considerations dictate the use of the average Gaussian representation to obtain an acceptable estimate of the action of the beam.

The variation of beam size (in terms of σ_B determined as described) with focus coil current is illustrated in Figure 10. Note that there is little difference between surface focus (518 mA) and at 0.625 cm below (522 mA). The effect of the same range of focus coil current (i.e., focus location) on penetration in 304L stainless steel is shown in Figure 11. From the error band defining a $\pm$ 5% variation in penetration it can be seen that to satisfy such a requirement the beam focus coil current must be controlled to an accuracy of $\pm$ 1%.

CORRELATING PENETRATION WITH DISTRIBUTION PARAMETER

Comparison of measured beam widths with transverse fusion zone sections suggested that a diameter of $2\sigma_B$ was an appropriate width parameter for correlating EB penetration [5]. Data for a series of tests with 304L SS are plotted in Figure 12 with the Peclet parameter defined as (vw/α_a) and as $(2v\sigma_B/\alpha_a)$. The data points represented by circles along the curve defined by equation (2) are from measurements of penetration and fusion zone width at the surface for test runs with the beam focused within 0.625 cm of the workpiece surface. Deviations of the data from the curve are within $\pm$ 20%. As anticipated deviations of results for beam focus at distances greater than 0.625 cm from the surface were greater than $\pm$ 20% and were not included. The points for these focus locations tend to fall above the curve indicating that a higher power per unit of penetration is required.

The points enclosed in squares represent the data in terms of $X_\sigma = 2v\sigma_B/\alpha_a$. A least squares fit of these points yielded

$$Y = P/(dk_a\theta_m) = 6.42\ (2v\sigma_B/\alpha_a)^{0.362} \tag{10}$$

The scatter about this line is similar to that about the line for equation (2). The implication of this is that, for a reasonable range of focus locations near or on a workpiece surface, penetration can be correlated by a measured beam distribution parameter. This hypothesis is supported by the analytical study of Wei, et al., [8]. They assumed a Gaussian distribution for the beam power density, and as can be seen the curve in Figure 12 representing their results essentially coincides with the curve of equation (10) for values of $2v\sigma_B/\alpha_a$ up to 1.0, and rises to about 25% above this curve as $2v\sigma_B/\alpha_a$ approaches 2.0. Extension of predictions to higher values of $2v\sigma_B/\alpha_a$ would be desirable.

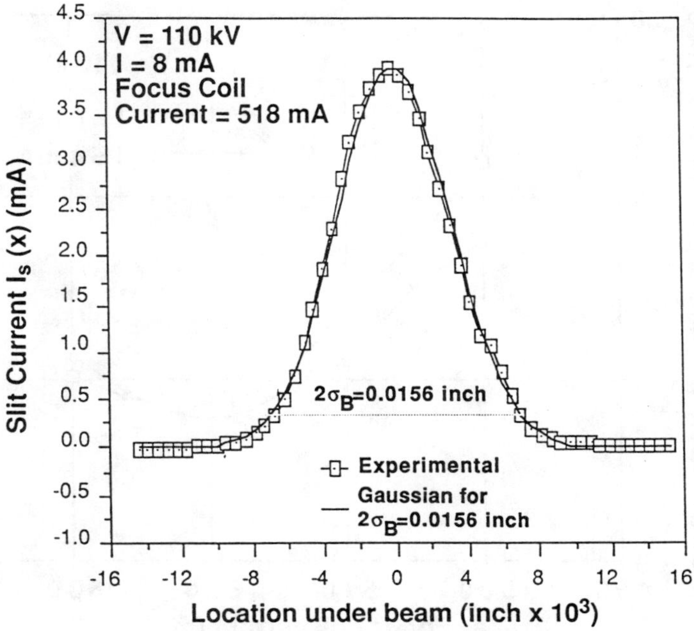

Fig. 8 One-dimensional integrated beam current distribution in travel direction for focus 0.625 cm below workpiece surface (Ref. 5)

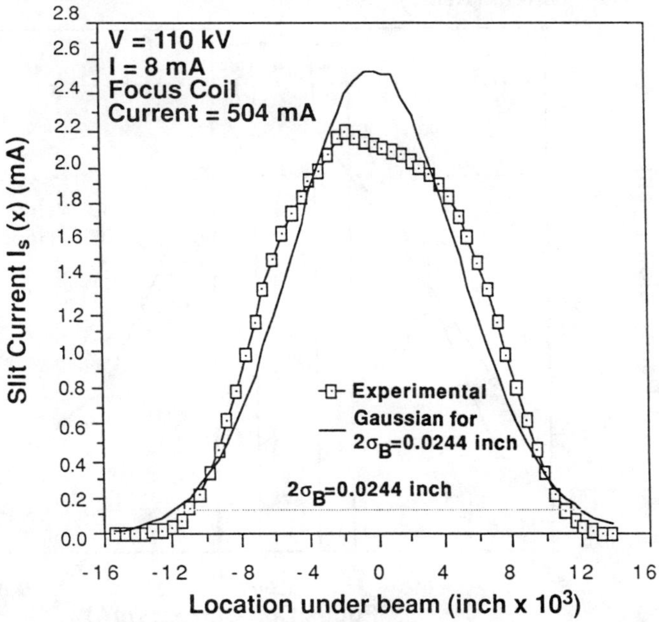

Fig. 9 One-dimensional integrated beam current distribution in travel direction for focus 2.5 cm below workpiece surface (Ref. 5)

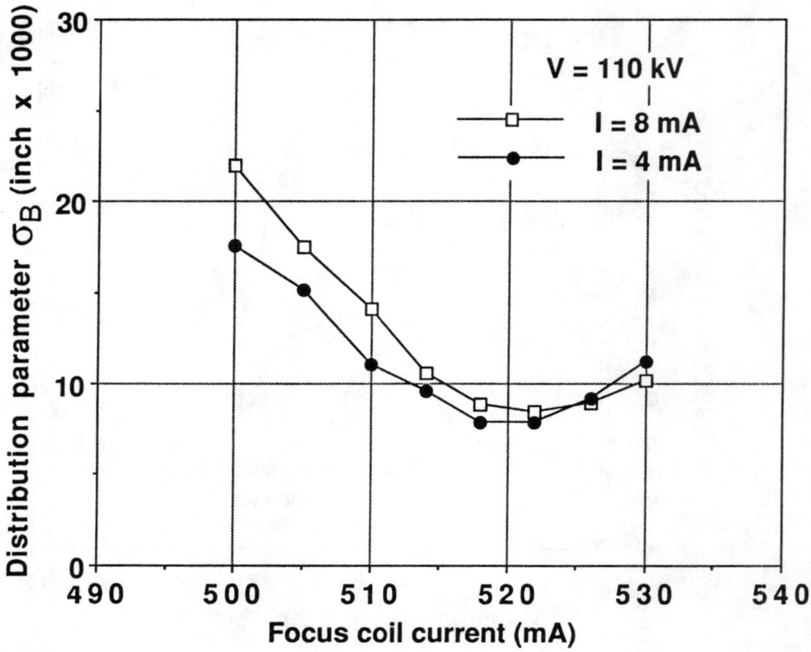

Fig. 10 Beam current distribution parameter σ_B as a function of focus coil current (Ref. 5)

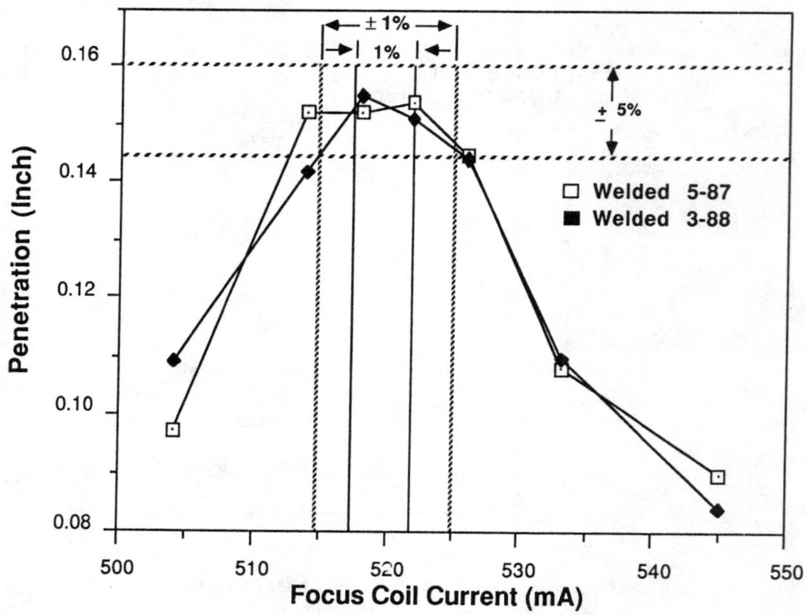

Fig. 11 Variation of penetration with focus coil current (V = 110 kV, I = 8 mA) (Ref. 5)

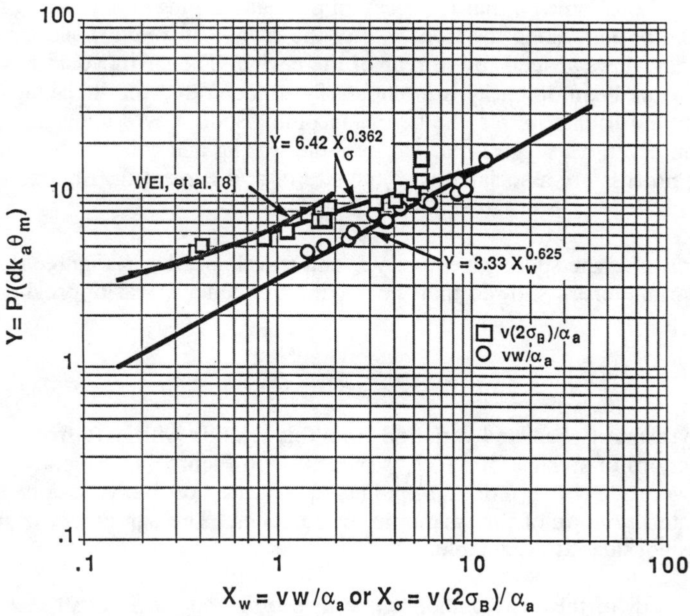

Fig. 12 Correlation of penetration with beam current Gaussian distribution parameter (Ref. 5)

BEAM INTENSITY INFLUENCE ON WELD QUALITY

In addition to providing an operator with quantitative information needed to control penetration, an electron beam current density profile can be of value in avoiding operating conditions which could yield serious welding defects. Figure 13 shows a longitudinal section of a weld in 304L SS made with a voltage of 110 kV and current of 8.5 mA at a travel speed of 1.27 cm/s - typical conditions for high voltage machines. The beam was focused at the workpiece surface and had a diameter ($2\sigma_B$) of less than 0.5 mm. The power density at the center of the beam exceeded 10^6 W/cm^2. Severe root spiking and root porosity, with a definite periodic nature, are clearly evident.

An interpretation of the defects exhibited in Figure 13 has recently been proposed [9]. This is based on periodic movement of the beam down the front face of the EB vapor cavity as suggested by x-ray pinhole camera records combined with beam current density distribution measurements. The severity of spiking is found to depend on the beam Gaussian distribution parameter σ_B and the ratio of material enthalpy at the melting temperature to the vapor enthalpy.

Spiking Frequency Related to Beam Radius

Fundamental information about the periodic nature of the EB welding process has been revealed by recording the x-rays generated by the beam interaction with the material in a cavity with a pinhole camera [10,11]. Results for welding with high intensity beams indicate that the location of beam impingement moves from the top to the bottom along the front wall of the cavity. Incoming solid is rapidly melted and flows around to the rear of the cavity. The cavity volume is increased and the peak beam intensity is momentarily concentrated on the cavity base. The result is the "drilling" of a spike before incoming solid moves under the beam. These observations suggest that the periodic spiking and solidification of the EB welding process are associated with the size of and current distribution in the beam.

Comparison of the spiking frequency f_s and travel speed v in Figure 13 with the beam parameter σ_B showed that the spiking frequency f_s was approximately given by

$$f_s = v/(0.75\ \sigma_B) \tag{11}$$

A comparison of predicted variation from this relation with data from investigations of spiking [10,11,12] is presented in Figure 14. Regions for the data are shown, rather than individual points, to indicate possible variations in σ_B and in the interpretation of the spiking in fusion zones. The agreement with Equation (11) was considered acceptable.

Further study of the x-ray pinhole camera records and longitudinal sections of fusion zones in conjunction with beam radii (i.e., σ_B) indicated that spikes were formed during approximately one third of a cycle. If this time is denoted by Δt_s,

$$\Delta t_s = (1/3)\ (1/f_s) = (1/4)\ \sigma_B/v \tag{12}$$

Fig. 13 Longitudinal section of EB weld in 304L SS. V = 110 kV, I = 8.5 mA, v = 1.27 cm/s, magnification 15X. (Courtesy Joining Laboratory, Sandia National Laboratories, Livermore, CA)

This led to the proposed sequence of events illustrated in Figure 15. A cycle begins with the center of the beam creating a "shoulder" near the workpiece surface. Intense heating causes the shoulder to move downward and melted material flows around to the cavity rear. In another 1/3 of a cycle the shoulder reaches the root region. The beam is so intense that it has melted solid faster than the average rate of travel v. Hence when the beam reaches the root the cavity volume has been increased and the center of the beam impinges directly on the base of the cavity. While the beam is melting the front surface of the cavity liquid from the layer around the rear starts the flow into the empty spike created at the end of the previous cycle. Note, however, in Figure 13, that liquid was unable to fill much of the spike volume. Also no significant solidified layer around the periphery of the spikes appears to be present. These observations suggest that the action of the beam was to melt and vaporize the material in the spike volume.

Ratio of Spiking Penetration to Average Penetration

To estimate the spike penetrating velocity v_s it was assumed that the incident energy flux is approximately that at the center of the beam. For a beam characterized by σ_B the energy flux distribution is

$$P(r) = 3P/(\pi \sigma_B^2) \exp(-3r^2/\sigma_B^2) \tag{13}$$

and at $r = 0$, $P(0) = 3P/(\pi\sigma_B^2)$. If material exposed to this heat rate is melted and evaporated, the rate or velocity of penetration would be

$$v_s = (3P/\pi \sigma_B^2)/\rho h_v \tag{14}$$

where h_v is the enthalpy increase as the solid of density ρ is heated from ambient through melting to boiling and evaporation. Finally the spiking depth would then be given by

$$d_s = v_s \, \Delta t_s = (3P/\pi \sigma_B^2)(\sigma_B/4v)/\rho h_v \tag{15}$$

Note that relations for both d_s and d (Equation (10)) as functions of σ_B are now available.

Solving Eq. (10) for d, the spiking penetration ratio becomes

$$P_s = d_s/d = 1.97\, k_a \theta_m (v\sigma_B/\alpha_a)^{0.362}/v\sigma_B \rho h_v \tag{16}$$

Multiplying by c_p/c_p and substituting h_m for $c_p\, \theta_m$, Eq. (16) becomes

$$P_s = d_s/d = 1.97(h_m/h_v)(v\sigma_B/\alpha_a)^{-0.638} \tag{17}$$

This result shows that the spiking depth is proportional to the ratio of the sensible enthalpy to melting h_m to the vapor enthalpy h_v and inversely proportional to 0.638 power of the Peclet number defined as $Pe = v\sigma_B/\alpha_a$.

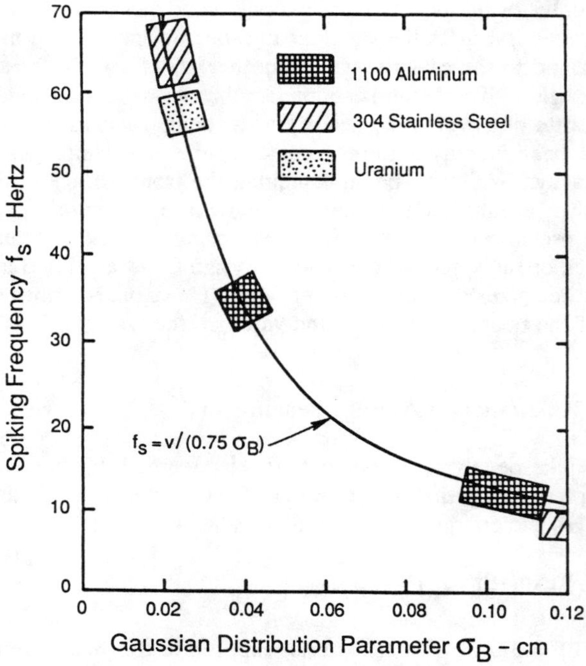

Fig. 14 Comparison of measured with predicted spiking frequency (Ref. 9)

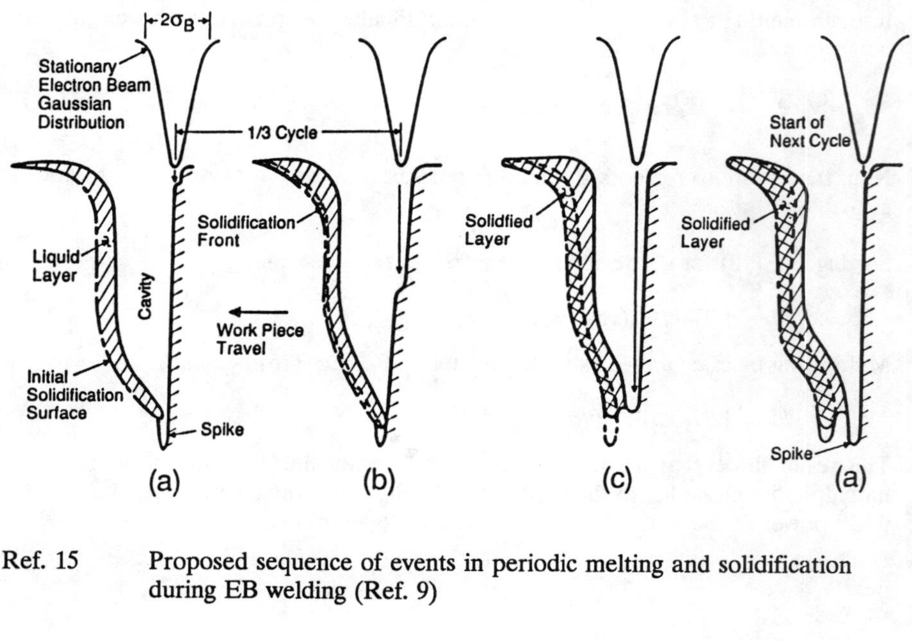

Ref. 15 Proposed sequence of events in periodic melting and solidification during EB welding (Ref. 9)

Available data for spiking in stainless steel, uranium and aluminum 1100 are compared with curves representing Eq. (17) in Figure 16. The predicted dependence of d_s/d on h_m/h_v and $(v\sigma_B/\alpha_a)$ is clearly supported. Again, regions for the data are shown rather than individual points because it is necessary to estimate an average spiking depth and values for σ_B have been estimated in most cases from beam spot size observations. Results tend to be in limited ranges of Pe^{-1} because they are from high or low voltage machines.

CONCLUSIONS

Electron beam power density distribution and size at a workpiece surface have a major influence on the weld produced. Attention is currently being given to developing techniques for measuring actual welding beam size and current density distribution. Limited results indicate that a one-dimensional type of measurement such as obtained by moving the beam across a small aperture or narrow slit Faraday cup can be used to characterize the beam. The use of the radius within which 95% of the current (energy) is included is proposed for correlation of penetration. It is also shown that this Gaussian distribution parameter can be used to estimate the severity of spiking to be anticipated with a specified set of welding conditions.

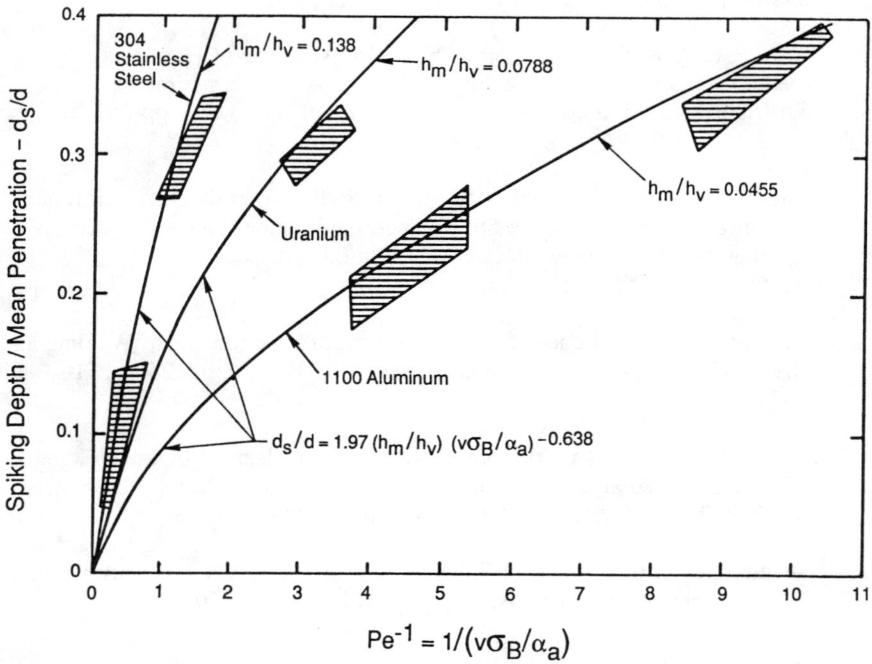

Fig. 16 Comparison of measured with predicted spiking depth ratio for aluminum, uranium, and stainless steel (Ref. 9)

REFERENCES

1. Melika, A. H., Ed., Electron-beam Welding: Principles and Practice, McGraw-Hill, pp. 1-16, 1971.

2. Schumacher, B. W., Optik, vol. 10, pp. 116-131, 1953.

3. Tong, H., and Giedt, W. H., A Dynamic Interpretation of Electron Beam Welding, Welding Jour., vol. 49, no. 6, pp. 259-s-266-s, 1970.

4. Giedt, W. H., and Tallerico, L. N., Prediction of Electron Beam Depth of Penetration, Welding Jour., vol. 67, no. 12, pp. 299-s-305-s, 1988.

5. Hicken, G. K., Giedt, W. H., and Bentley, A. E., Correlation of Weld Penetration with Electron Beam Current Distribution, Welding Jour., forthcoming.

6. Hartwig, C. M. and Hopwood, J. E., Technique for Sampling Focused Electron Beam for Welding Application, Sandia National Labs. Report SAND85-8710, Livermore, CA, 1985.

7. LaFlamme, G. R., and Powers, D. E., Electron Beam Power Density Distribution Measuring Device, Welding Jour., forthcoming.

8. Wei, P. S., Wu, T. H., and Chow, Y. T., Investigation of High-Intensity Beam Characteristics on Welding Cavity Shape and Temperature Distribution, ASME Jour. of Heat Transfer, vol. 112, no. 1, pp 163-169, 1990.

9. Giedt, W. H., A Periodic Melting Model of High Intensity Electron Beam Welding, Proceedings of the 5th Engineering Foundation Conference on Casting, Welding and Advanced Solidification Processes, Davos, Switzerland, 1990, forthcoming.

10. Mara, G. L., et al., Penetration Mechanisms of Electron Beam Welding and the Spiking Phenomenon, Welding Jour., vol. 53, no. 6, pp. 246-s-251-s, 1974.

11. Arata, Y., Abe, N., and Yamamoto, S., Tandem Electron Beam Welding (Report III) - Analysis of Front Wall of Beam Hole by X-ray Observation Method, Trans. JWRI, vol. 9, no. 1, pp. 1-10, 1980.

12. Armstrong, R. E., Control of Spiking in Partial Penetration Electron Beam Welds, Welding Jour., vol. 49, no. 8, pp. 382-s-388-s, 1970.

Transport Phenomena in Laser Materials Processing

J. MAZUMDER and A. KAR
Laser-Aided Materials Processing Laboratory
Department of Mechanical and Industrial Engineering
University of Illinois at Urbana-Champaign
1206 West Green Street
Urbana, Illinois 61801, USA

ABSTRACT

Various transport phenomena occur simultaneously during laser materials processing depending on the laser power density and interaction time. This paper examines the role of energy, momentum and mass transfer for different laser processing techniques. The heat conduction equation for three-dimensional transient case is solved for uniformly moving finite slabs in the context of laser chemical vapor deposition. A two-dimensional energy conservation model based on the Stefan condition at the solid-liquid and liquid-vapor interfaces is presented for drilling or crater formation for deep penetration welding. Various models for studying the dynamics of fluids in laser melted weld pool are also presented.

A three-dimensional transient nonlinear mass diffusion model is developed for the prediction of titanium film deposit profile for laser chemical vapor deposition. Also, the distribution of externally added solute atoms in laser melted pool is examined for laser surface alloying processes. Finally, a one-dimensional model is presented to predict the composition of the extended solid solution formed during laser cladding and to obtain the non-equilibrium phase diagram.

1. INTRODUCTION

Several transport phenomena occur simultaneously in laser processing depending on the incident intensity of the laser beam. For example, in laser surface hardening where power density level is on the order of 10^3 to 10^4 W/cm^2, heat transfer plays the most important role with mass transfer determining the limits of dwell time required for phase transformation. In surface melting (or melt quenching) and welding where power density level is on the order of 10^5 to 10^7 W/cm^2, momentum transfer or convection play a significant role. Convection also dominates the process of laser surface alloying and cladding (power density is on the order of 10^5 to 10^6 W/cm^2) along with mass transport which determines the nonequilibrium microstructure and composition of the solidified materials. Vaporization and plasma formation also play an important role in determining the surface contour, energy partitioning, and deep penetration welding. However, vaporization and gas dynamical effect become predominant in laser processing when power density increases to 10^7 W/cm^2 such as in drilling. Figure 1 summarizes the associated transport phenomena in various laser processing.

It is evident that as the energy density during materials processing increases, the interaction mechanism changes adding complexity to the associated transport phenomena. Some of the important processes and their associated transport phenomena will be discussed in this paper whereas aspects of thermocapillary flow will also be presented in this workshop by Professor M. M. Chen.

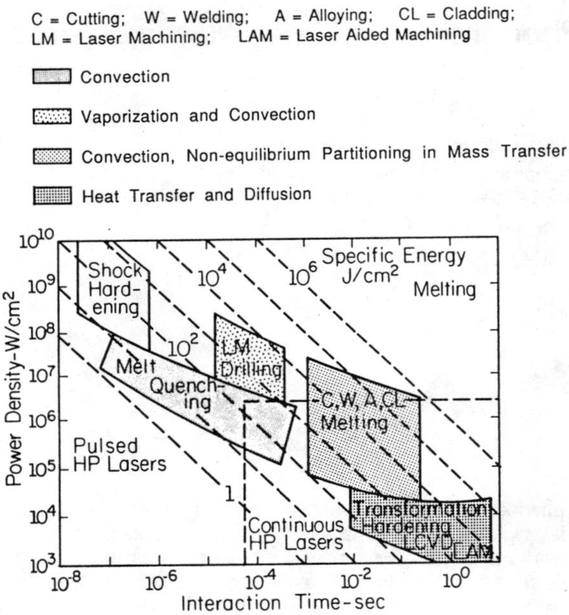

FIGURE 1. Operational regimes and associated transport phenomena for various laser processing techniques.

2. TRANSPORT PHENOMENA THEORY FOR LASER PROCESSING

2.1. Heat Conduction

Heat conduction plays an important role for all laser surface treatments, but for surface hardening this is the single most important phenomenon. Several theoretical investigations have been published with the objective of establishing the relationship between process parameters and temperature and hardness distribution. Also, thermal analysis can be carried out to determine the range of various process parameters for laser chemical vapor deposition.

Analytical solutions. Some of the important analytical models [1-4] published to date are summarized in our earlier paper [5] by the authors. Many of the analytical solutions available are based on one of the many cases solved by Carslaw and Jaeger [1] and modified by Gregson [2] and Sandven [3] to suit the particular cases for laser heat treatment. Gregson reported a one-dimensional transient case for flat semi-infinite body whereas Sandven modeled cylindrical bodies. Perhaps the most realistic analytical model so far is reported by Anthony and Cline [4]. Their model considered a three-dimensional semi-infinite plate and Gaussian heat source. However, often laser processing is carried out with finite bodies where semi-infinite assumption is not valid.

Three-dimensional transient model for finite slabs. In the context of laser chemical vapor deposition, we analyzed the three-dimensional transient heat conduction equation with temperature-dependent thermophysical properties by considering both convective and radiative losses of energy at the boundaries [6]. The geometric configuration for this problem is given in Fig. 2 where the laser beam moves in the x-direction at a constant velocity U.

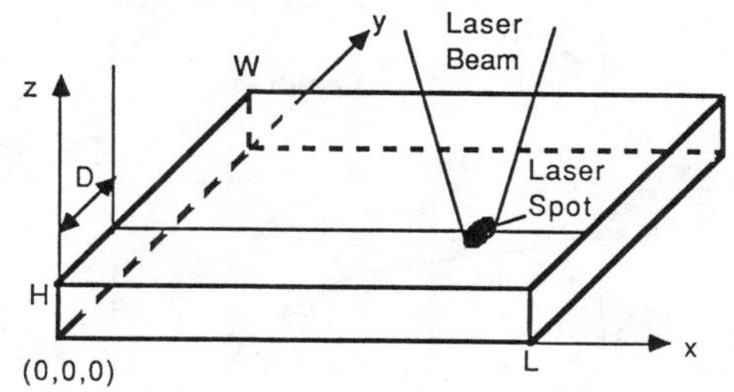

FIGURE 2. Geometric configuration of the substrate and the relative position of the laser spot.

The solution of this thermal problem, which has been discussed in detail in [6], allows us to study the effect of various process parameters on chemically reactive zones and to determine the operating conditions for laser chemical vapor deposition. The chemically reactive zone is defined as the zone where the temperature is more than of equal to that at which the chemical reaction that generates the film forming material takes place. It can be seen from Fig. 3 that the chemically reactive zone becomes narrower as the laser scanning speed increases. In this figure, the chemically reactive zones are bounded by the curves A, B, C, and D for the scanning speeds 0.35, 0.25, 0.167, and 0.125 cm/s, respectively. There will be a critical scanning speed at which the two boundary curves of the chemically reactive zone will collapse into one giving rise to the narrowest possible film deposition region. With any other scanning speed higher than the critical speed, the width of the chemically reactive zone cannot be reduced any further. So the film deposition process has to be operated at a scanning speed lower than the critical speed. The critical scanning speed is defined as the one at which the nondimensional peak temperature, T_p^* is unity, where the peak temperature refers to the temperature at the center of the laser beam on the top surface of the substrate. The line, $T_p^* = 1$ is referred to as the line of the narrowest chemical reaction zone in Fig. 4. This figure is plotted on logarithmic scales which show the linear

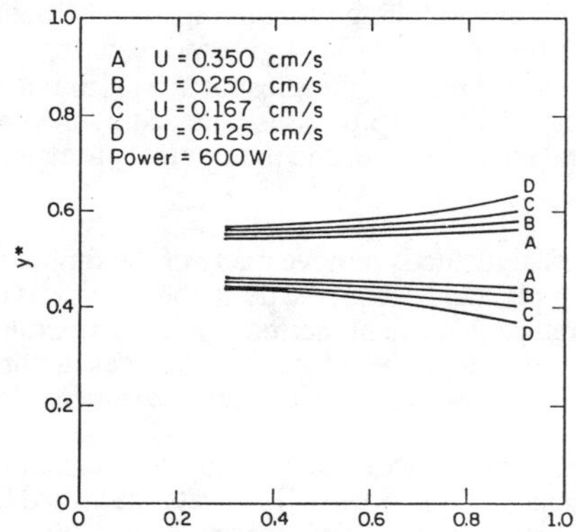

FIGURE 3. Variation of the width of the heated zone at the surface of the substrate along the direction of motion of the laser beam for various scanning speeds of the laser beam relative to the substrate for laser power 600 W.

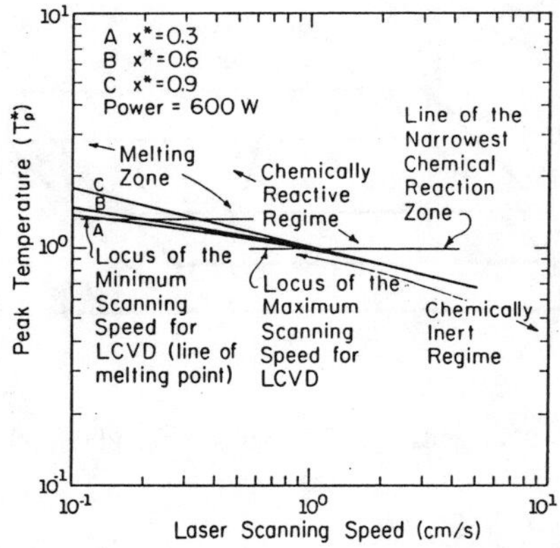

FIGURE 4. Variation of the peak temperature with laser scanning speed relative to the substrate at various locations on the top surface of the substrate for laser power 600 W.

variation of the peak temperature (T_p^*) with the laser scanning speed (U) at various locations, $y^* = y/W = 0.5$, and x^* ($= x/L$), on the top surface of the substrate. The points of intersections of the line of the narrowest chemical reaction zone with the curves of Fig. 4 give the critical scanning speed of the laser beam. The region which is to the right of the critical speed is referred to as the chemically inert regime because the operating conditions of this region do not raise the surface temperature of the substrate to the film forming chemical reaction temperature. The region to the left of the critical speed is referred to as the chemically reactive regime where the operating conditions are such that films of finite width can be deposited.

However, the surface temperature of the substrate can reach its melting temperature at a low scanning speed for a given operating condition. Since melting the substrate is not desirable in LCVD processes, the scanning speed has to be higher than the upper limit of the speed at which melting occurs. The line, $T_p^* = T_m/T_d$, where T_m is the melting temperature of the substrate, is referred to as the line of melting point in Fig. 4. The points of intersections of the line of melting point with the curves of Fig. 4 give the scanning speed above which the process has to be operated to avoid melting the substrate. So, from the thermal considerations the operating regime for an LCVD process is bounded by the line of melting point, the line of narrowest chemical reaction zone, and their points of intersections with the curves of Fig. 4.

Numerical solutions. Numerical methods remove many of the limitations that apply to the analytical methods. For example, the heat source does not have to be concentrated in a point, line or plane. Temperature-dependent thermophysical properties and real boundary conditions may be included. In spite of the inherent advantages in numerical methods, only a few numerical models for heat flow in laser processing have been developed so far.

The numerical solution to the three-dimensional heat-transfer model for laser materials processing developed by Mazumder and Steen [7] and later modified by Chande and Mazumder [8] allows for temperature-dependent thermophysical properties, spatial distribution of the heat source (Gaussian, uniform, or any other known distribution), radiative heat losses, convective heat losses, and latent heat of transformation.

The following assumptions were made for this numerical solution [7,9]:

1. The laser beam is stationary relative to the earth and is incident at right angles at the center of the substrate width. The substrate moves in the positive x direction (along the length) with a uniform velocity u and is infinite in length but has finite width and thickness. A control volume approach is utilized where the heat balance on an axisymmetric control volume as shown in Fig. 5.
2. Quasi-steady state is assumed.
3. The power distribution in the beam is Gaussian.
4. The substrate thermal conductivity, specific heat (C_p), thermal diffusivity, and the surface reflectivity are temperature dependent. The substrate density is assumed to be independent of temperature. Specific heat variation accounts for heats of fusion and vaporization (L) by converting them into equivalent step increases in values of specific heat.
5. Surface reflectivity is considered to be zero when the temperature exceeds the boiling point.
6. There is a radiation loss at the upper and lower surface of the workpiece.
7. Convective losses occur at the upper surface due to shielding gas flow.
8. When the temperature at any location exceeds the boiling point, it is considered to have evaporated but remains in the conducting network at a fictitiously high temperature to simulate the effect of high convection and radiation transfer effects from the fast moving plasma in the keyhole.
9. Radiation penetrating the substrate is absorbed according to the Beer-Lambert law.

Courtney and Steen [10] used this model to predict the depths of the heat-treated zone and thermal cycle time as shown in Figs. 6 and 7. Knowledge of thermal cycle time is essential to calculate the carbon diffusion distance. This model has also been successfully applied for welding [11] and melt quenching experiments. However, the Beer-Lambert coefficient has to be experimentally determined to account for convection which has been neglected.

2.2. Convection

For processes such as melt quenching, alloying, and some welding, convection in the melt pool dominates the process. It is also the single most important factor influencing the geometry of the pool including pool shape, undercut, and ripples, and can result in defects such as variable penetration, porosity, and lack of fusion. Convection is also primarily responsible for mixing, and, therefore affects the composition of the weld pool. The heat transfer and thus the cooling rate is greatly enhanced in the presence of convection [11]. This in turn will affect the microstructure. The homogeneity of solute redistribution during laser surface alloying (LSA) as reported by Chande and Mazumder [12] can only be explained by the presence of convection current.

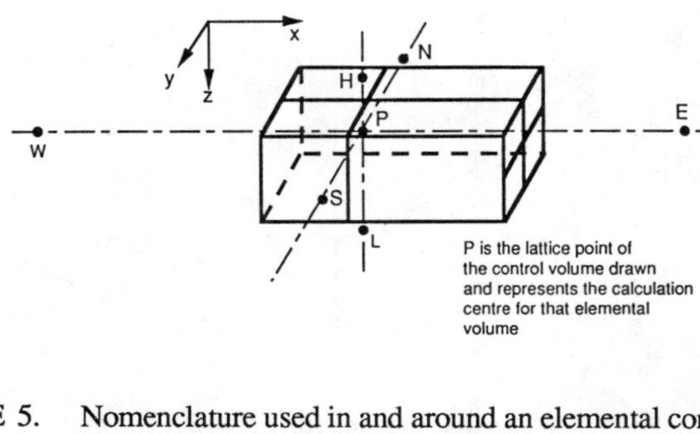

FIGURE 5. Nomenclature used in and around an elemental control volume.

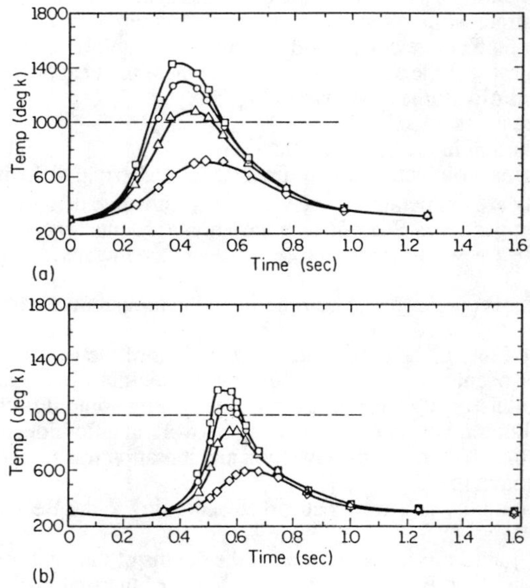

FIGURE 6. Theoretically Predicted Thermal Cycle during Laser Heating of En8 Steel (power = 2 kW, beam radius = 3.00 mm, and reflectivity = 0.4); (a) speed = 22.5 mm/s and (b) speed = 42.5 mm/s (after Ref. [9]).

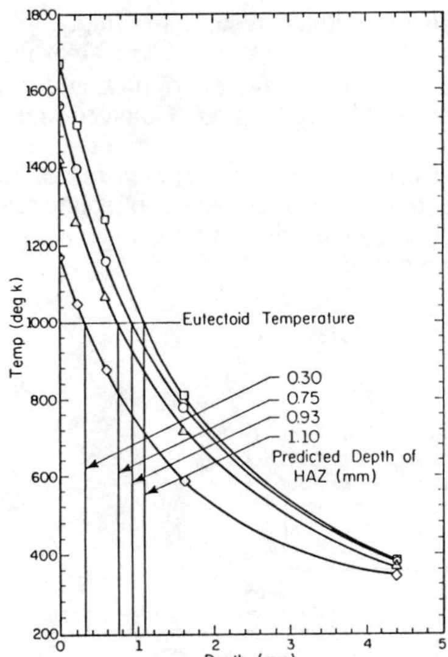

FIGURE 7. Theoretically Predicted Thermal Profiles (power = 2 kW, beam radius = 3.0 mm, reflectivity = 0.4) (after Ref. [9]). Speed, mm/s: 8.0; 15.0; 22.5; 42.5.

Surface tension driven flow has been identified to be responsible for the convection within the molten pool. While most of the work to date has been convection, a quantitative understanding of the effect of convection on pool shape and mixing is of importance. Anthony and Cline [13] did an analysis on the surface tension driven flow within the molten pool. It is essentially a one-dimensional problem and the flow field thus obtained is not coupled to the energy equation. Therefore, no additional information is obtained as far as the heat transfer process is concerned. Oreper, et al. [14] developed a two-dimensional convection model. Buoyancy, electromagnetic, and surface tension forces were considered. Numerical solutions were obtained based on a specified pool profile. It was found that the surface tension gradient is the dominant factor in many cases. However, the solid-liquid interface is not known a priori. It is a part of the problem to be solved. In fact, the solid-liquid interface, i.e. the pool shape, is a piece of information that is of great interest. A two-dimensional transient model for laser surface melted pool was first developed by Chan, Mazumder, and Chen [15,16]. It was shown that surface tension is responsible for the fluid flow and convection. The cooling rate at the edge of the pool is found to be higher than that at the bottom of the pool below the centerline which agrees with the experimental findings that the microstructure is finer at the edge of the pool than at the bottom of the pool. It is also predicted that the recirculating velocity is of one or two orders of magnitude higher than that of the scanning velocity.

Two-dimensional self-consistent transient model for surface tension driven pool. A two-dimensional transient self-consistent (i.e. solid-liquid interface position is a part of the solution) model for laser melted pool was first developed by Chan, Mazumder, and Chen [16]. The geometric configuration of the process is given in Fig. 8. The basic assumptions of this model are as follows:

1. A laser beam having a defined power distribution strikes the surface of an opaque material of infinite width, thickness, and length.
2. Only part of the energy is absorbed.
3. Absorbed energy induces surface tension driven flow due to the high temperature gradient.
4. The liquid metal is considered to be Newtonian so that the Navier-Stokes equation is applicable.
5. All properties of the liquid metal and solid metal are constant, independent of temperature (except the surface tension). This allows simplifications of the model; however, variable properties can be treated with slight modifications. The dependence of surface tension on temperature, the driving force of the flow, is assumed to be linear.

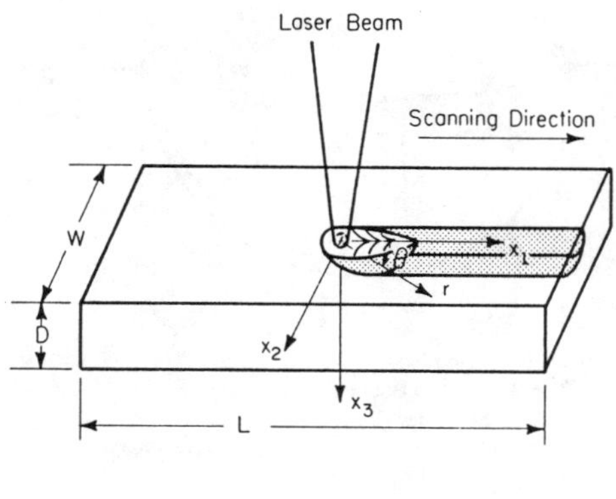

FIGURE 8. Schematic diagram of the process.

6. The latent heat of fusion is neglected since the energy liberated is small compared to total enthalpy change associated with temperature differences.
7. Thermal conductivity is assumed to be the same for both liquid and solid phases for the sake of simplicity of the model.
8. The surface of the melt pool is assumed to be flat to simplify the surface boundary condition and, hence, surface rippling is neglected.

A computer program SOLA [17] is employed. The basic method of the algorithm is presented in Hirt's report [17].

This model provides the details of the flow field and heat transfer on a plane perpendicular to the scanning direction. Flow behavior predicted by this model will be valid for Regions 1, 2, and 5 in Fig. 9. However, the front-to-back motion and its effect on heat transfer also play important roles in the physical process, especially for the fact that the recirculating velocity is so much higher than the scanning velocity. Such a motion cannot be obtained from two-dimensional models.

Three-dimensional self-consistent quasi-steady state models. Recently developed three-dimensional models [18-21] have provided better understanding of the flow behavior. It is expected that better accuracy of the magnitude of flow will be obtained and results should be valid for Regions 1, 2, 4, and 5 of Fig. 9.

Kou and Wang [21] developed a three-dimensional model for convection in laser melted pool using both surface tension and buoyancy driven flow.

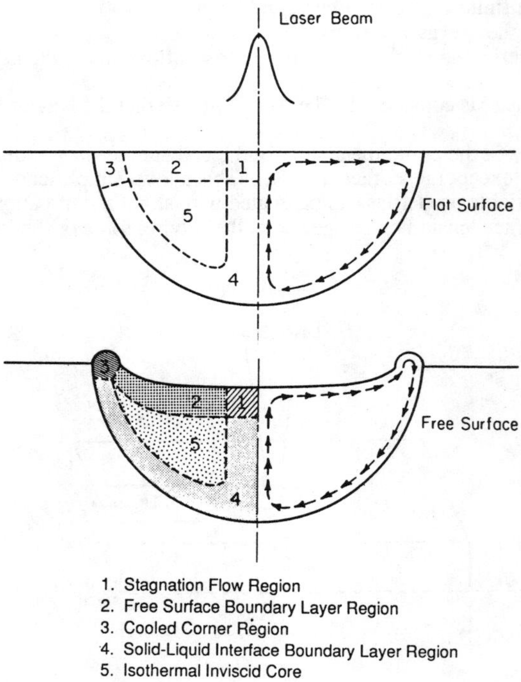

FIGURE 9. General features and various regions of the flow.

The Semi-Implicit Method for Pressure Linked Equation (SIMPLE) algorithm [22] was used to calculate the velocity field. A conjugate heat transfer method was used at the solid-liquid interface for calculation. This model predicted a velocity profile and magnitude (on the order of m/s) similar to that of the transient two-dimensional model developed by Chan, et al. [16]. The conjugate heat transfer method uses discontinuous viscosity variation at the solid-liquid interface. Therefore, flow behavior at the S/L interface (Region 4, Fig. 8) will not be realistic unless an extremely fine grid is used. However, this model also essentially reconfirmed that surface tension induced by temperature gradient significantly affected the flow behavior and pool shape.

The group at the University of Illinois [18-20] is taking a step-by-step approach to construct three-dimensional models for laser melted pools. Initially a stationary spot source axisymmetric case was considered [18]. Subsequently a perturbation solution based on low scanning velocity was sought for the basic axisymmetric case [19,20]. The advantage of seeking a perturbation solution, as it turns out, is that the three-dimensional flow is modeled by two sets of two-dimensional equations which provide considerable computing advantages. However, the perturbation solution is valid only for low scanning velocities. Another model which is a full three-dimensional numerical solution of Navier-Stokes equations, was developed using a point-by-point partially vectorized iteration scheme. This model was then modified to accommodate a free surface [19]. These models were used to study the effect of convection on pool geometry, cooling rate, and solute redistribution.

The assumptions for materials properties for the quasi-steady state three-dimensional models are similar to those of a two-dimensional model. These models also assume a gray body radiative heat loss and flat surface as the starting point. It is important to point out that the location of the solid-liquid interface is a part of the problem to be solved. This type of problem is generally known as the Stefan problem in the literature [23]. An iterative scheme is used to solve for the interface.

The three-dimensional model is the numerical scheme that solves the above set of governing equations. However, the perturbation model requires further mathematical consideration. The basic solution is the axisymmetric case corresponding to a stationary laser beam irradiating on a stationary workpiece. The perturbation is obtained by perturbing the basic solution with a low scanning speed (low as compared to the recirculating flow due to surface tension gradient). As it turns out, the angular dependence of the perturbation is separable. It is more appropriate to use cylindrical coordinates since the basic solution is axisymmetric and the perturbation solution can be reduced to an axisymmetric form. Details of the mathematical derivation for perturbation model are available [19,20,24].

Three important dimensionless groups; Marangoni number, Prandtl number, and Reynolds number, appear in the model. The Marangoni number, Ma, is the ratio of the rate of convection associated with the recirculating velocity and the rate of conduction. Thus, the higher the Marangoni number, the higher the convection. The Prandtl number, Pr, a fluid property, is the ratio of kinematic viscosity and heat diffusivity. It is the ratio of momentum diffusion and energy diffusion. The Reynolds number, Re, is the ratio of the inertia term and the viscous term. For high Reynolds number, there would be a boundary layer phenomenon [25]. In the case of numerical solution, this phenomenon is automatically taken care of by using a fine enough grid. It should be noted that the Marangoni number is a product of the Reynolds and the Prandtl numbers.

Convective heat transfer depends on the recirculating flow field. The higher the recirculating flow field, the higher the convective heat transfer. However, a direct proportional relationship cannot be assumed because of the complexity of the process due to the interaction of temperature field and velocity field. The dimensionless melting temperature defines the solid-liquid interface. For the same amount of heat input, the lower dimensionless melting temperature would imply a larger molten region. The radiation factor governs the heat loss from the surface due to radiation.

Axisymmetric model: numerical analysis and results. A standard Alternating Direction Iterative (ADI) [26] method is employed to solve the governing equation for the axisymmetric case. The effect of convection and Prandtl number on melt pool shape is studied using this model. The very existence of the flow field changes the mechanism of heat transfer to convective. It tends to bring the higher temperature fluid on the surface right underneath the beam sideways. This, in turn, melts down the solid at the edge of the molten pool, creating a wider pool. Because of the conservation of energy, the size of the molten pool must remain roughly unchanged, with or without convection. As a result, the molten pool becomes shallower. The effect of an increase in the Marangoni number on the pool shape is studied. The Marangoni number is a measure of the convection. An increase in the Marangoni number implies an increase in the amount of heat being brought sideways. Consequently, the molten pool becomes wider and wider. The ranges in aspect ratios (width to depth ratio) for different Marangoni numbers are quite large, up to a 150 percent increase as compared to the conduction case.

The effect of changing the Prandtl number is also studied. The Prandtl number is the ratio of the momentum diffusion to heat diffusion. To study its effect, the momentum diffusion must be kept constant and only the heat diffusion is allowed to vary. This corresponds to keeping the Reynolds number (Re = Ma/Pr) constant and changing Pr and Ma. This, in some sense, implies that the flow field is being maintained while changing the thermal diffusion. Caution must be exercised because of the complexity of the process due to the coupling of the energy and momentum equations. An increase in the Prandtl number with the Reynolds number being kept constant will increase the Marangoni number and hence the convective heat transfer. The aspect ratio will, therefore, increase with the Prandtl number.

Numerical analysis of perturbation model. Numerical solutions are sought for the two sets of two-dimensional equations of the perturbation model. Finite difference equations, central difference for the diffusion terms, and upwind difference for the convective terms governing the velocities and temperature are derived. The staggered grid is used. The resultant nonlinear algebraic equations are solved by the standard Alternating Direction Iteration (ADI) method [26].

The computational procedure is as follows. For a certain temperature field where both solid and liquid regions exist, the interface is first determined. The velocities within the molten pool are iterated for a prescribed number of iterations. Using this updated velocity field, the energy equations are iterated next. Thus, the temperature field for the next global iteration is obtained. The initial guesses are the steady-state conduction temperature and zero velocity for the basic solution and zero temperature and velocity for the perturbation solution. The convergent solutions are obtained when the average residual of each equation is smaller than a specified small number, typically 10^{-4}.

Numerical analysis of three-dimensional model. This model is a fully three-dimensional finite difference scheme which iteratively evaluates the energy, momentum, and continuity (modified pressure corrector) equations for transient and steady-state cases. For programming simplicity, a rectangular coordinate system was defined which uses an in-line grid and primitive variable formulations. Because the material is assumed to be scanning in the positive x_1 direction only, symmetry can be assumed about the slab midpoint ($x_2 = 0.0$), thus reducing the calculation domain. Due to the large storage demands of a three-dimensional program, the concept of extending the domain boundaries to sufficiently large distances to simulate the infinite domain conditions is prohibitively expensive computationally. To overcome this difficulty, a finite domain is chosen with given heat transfer coefficients at the boundary. The heat transfer coefficients are chosen with the help of analytical as well as numerical solutions of the conduction problem with moving boundaries for a large domain exterior of the primary computing domain. The results represent an approximate simulation of the infinite domain boundary condition.

To permit computations on a vectorized computer, a mixed Jacobi/Gauss-Seidel iteration scheme was adopted to solve the finite difference equations. The computation proceeds column-by-column according to the Gauss-Seidel scheme. Within each column the Jacobi scheme is used to facilitate vectorization. Pressure correction is based on a point-wise evaluation of the continuity equation, employing a modified artificial compressibility scheme.

The perturbation model has an advantage over the full three-dimensional model because the former has two-dimensional governing equations. This allows finer mesh and shorter computation time for the perturbation model. However, this model is limited to low scanning speeds whereas the three-dimensional model is capable of handling high scanning speeds.

Perturbation model results. This model is used to study surface temperature, velocity field, solute redistribution, temperature of the molten pool, and cooling rate.

a. Surface Temperature The temperature attains its maximum at the center underneath the beam and decreases radially outward. It is interesting to note that the temperature gradient, the driving force, has three distinct regimes. Its magnitude increases from zero at the center in the radial direction roughly to the edge of the laser beam, it then decreases, and finally increases again as the flow approaches the edge of the molten pool. The fluid flow driven by the surface tension gradient becomes important in the second stage where it tends to smooth out the temperature. Finally, the flow approaches the edge of the molten pool and turns around. This phenomenon is very similar to the heat transfer near the stagnation point and, therefore, gives rise to high temperature gradient.

b. Velocity Field The variation of surface temperature due to surface heating induces a surface tension gradient thus pulling the molten materials radially outward. The general pattern of the flow field is that the molten materials are going radially outward on the surface. As the flow approaches the edge of the molten pool, it goes down and turns around. It then moves to the center and comes up to complete the recirculation pattern. In addition to this recirculation, there is also the motion of the workpiece. Therefore, material enters the molten pool from the front portion of the pool and goes through the recirculation pattern, ultimately resolidifying on the trailing edge.

c. Solute Redistribution To help the reader gain a qualitative understanding of the mechanism of solute redistribution, a particle trajectory is plotted in Fig. 10 for radiation factor, RF = 0. This trajectory is presented in three views--front, side, and top. The initial position is $(x_1^*, x_2^*, x_3^*) = (-2.97, 0.423, 0.01)$, a point very close to the surface and before the molten pool. The recirculating pattern of the particle can be clearly observed. This recirculating pattern implies that a particle will travel on a rather long path before it freezes into the resolidifying surface. Consequently, the molten materials can be well mixed. The composition within the melt pool is, therefore, uniform.

d. Temperature Field within the Molten Pool The isotherms in the vicinity of the solid-liquid interface are compressed near the surface of the pool and stretched apart near the bottom due to the existence of the flow field. This distortion is due to recirculation of the molten material. The isotherms and the solid-liquid interface are asymmetrical due to the motion of the workpiece.

e. Cooling Rate The cooling rate right after resolidification is related to the microstructure. The higher the cooling rate, the finer the microstructure. The cooling rate (i.e. the temperature gradient) decreases from maximum at the centerline to zero at the edge of the molten pool and also decreases from maximum on the surface to minimum at the bottom of the molten pool. Such a variation in cooling causes a variation of the resolidified microstructure. Further investigation of the solidification mechanism is required to fully predict the resulting microstructure.

Three-dimensional model results. This model is also used to study surface temperature, velocity field, and cooling rate for higher scanning speed.

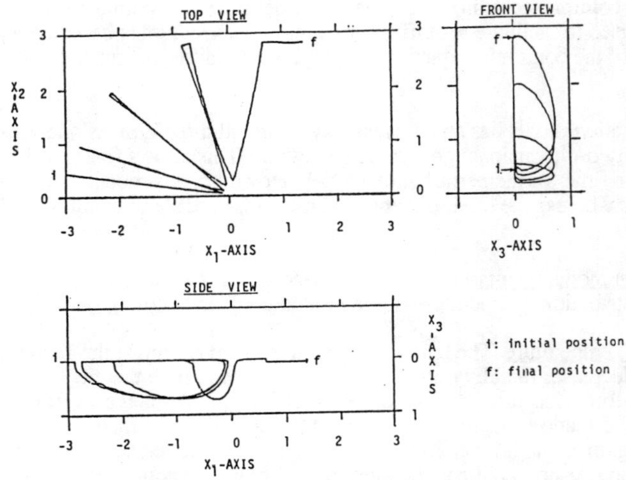

FIGURE 10. Trajectory of a particle as it enters, recirculates, and ultimately freezes. The trajectory is plotted in front, side, and top view in (a), (b), and (c), respectively. Initial position is $(X_1^*, X_3^*, X_3^*) = (-2.97, 0.423, 0.01)$ Ma = 1,500, Pr = 0.15, T_m^* 0.25 and RF = 0.0 (perturbation model).

a. <u>Surface Temperature</u> Isotherms are used to study the temperature distribution at the weld pool surface. The isotherms are compressed near the solid-liquid interface due to the rapid impingement of fluid (convecting heat) against the solid substrate. Circular isotherms near the beam center suggests that the temperature distribution in the center of the pool is only mildly affected by the scanning speed, thus validating the perturbation technique employed previously. However, severe temperature gradients exist beneath the beam and at the solid-liquid interface.

b. <u>Velocity field</u> Variation in surface tension drives the fluid radially outward along the surface from the beam center until it reaches the solid-liquid interface. At this point the flow is turned downward and recirculates along the pool bottom until it reaches the (beam) center where it is forced upwards towards the surface. The large radial surface velocities are a direct consequence of the dramatic surface temperature gradients. The thermocapillary velocities due to surface tension effects are approximately two orders of magnitude larger than the substrate scanning velocity.

c. <u>Cooling Rates</u> The results of three-dimensional model are essentially similar to those of perturbation models although the present results show less fore-aft symmetry due to the higher scanning speed.

<u>Asymptotic analysis</u>. Efforts at the University of Illinois also include asymptotic analysis of thermocapillary flow under the laser beam (corresponds to Region 1 of Fig. 9) [27,28]. This provides algebraic relationship for critical velocity, temperature, and shear stress. Accuracy seems to be around 10 percent of the numerical solution [28]. They may be used for quick estimation in many practical cases.

<u>Effect of flow on surface deformation</u>. Most of the work discussed so far assumes flat surface. The effect of surface rippling induced by surface-tension gradient during laser surface melting and alloying was first studied by Anthony and Cline [13]. They essentially solved the Navier-Stokes equation for one-dimensional steady-state condition with surface tension and buoyancy forces.

Chan, et al. [19], modified the three-dimensional numerical model to study the forced surface deformation. The method shows the following trends. Thermocapillarity drives the surface fluid radially outward at high velocities. These high velocities displace more mass from the central surface region than can be replaced by the recirculating flow, thus causing a depression. The displaced mass builds up at the solid-liquid interface causing the surface to bulge upward where it is then forced downward into the molten pool. For this calculation, the contact angle is unknown. The contact edge height, however, was defined equal to the initial surface height.

<u>Model for flow behavior with vaporization</u>. Processes such as deep penetration welding, cutting, and drilling have vaporization affecting the flow dynamics. One of the major problems of treating the vapor-liquid interface is the discontinuity at the interface and this makes application of continuum mechanics rather difficult. Theoretical understanding of such problems is at its infancy.

Recently Chan and Mazumder [29] developed a one-dimensional steady-state model where discontinuity of a low molecular mean free path at the solid-liquid interface is incorporated using the "Knudsen Layer" jump condition and the Mott-Smith type solution. In this model, the vaporization process creates a recoil pressure which pushes the vapor away from the target and expels the liquid. The materials are, therefore, removed in both the vapor and liquid phases. The materials removal rates are incorporated in the moving boundary immobilization transformation. The vapor phase is assumed to be optically thin so that its absorption of the high energy beam is negligible. However, this is a simplifying assumption at this initial stage. Rockstroh and Mazumder [30] have observed from their spectroscopic work that the beam does interact with the plasma. Due to the simplification, closed form analytical solutions are obtained for this analysis. The effect of heat source power on removal rates, vaporization rate, liquid expulsion rate, surface temperature, and Mach number are examined. Results are obtained for three different materials--aluminum, superalloy, and titanium.

This analysis will be valid for drilling and onset of "keyhole" transformation for deep penetration welding. A typical liquid layer thickness predicted by this analysis is shown in Fig. 11. It seems that for the power and interaction time (1 ms pulse width) the range of liquid layer thickness is on the order of 10 to 30 μm depending on the material. Figure 12

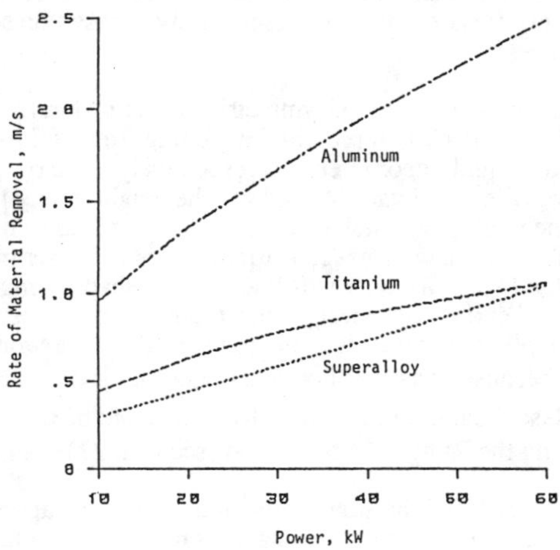

FIGURE 11. Total material removal rate vs. beam power for three different materials: aluminum, superalloy, and titanium.

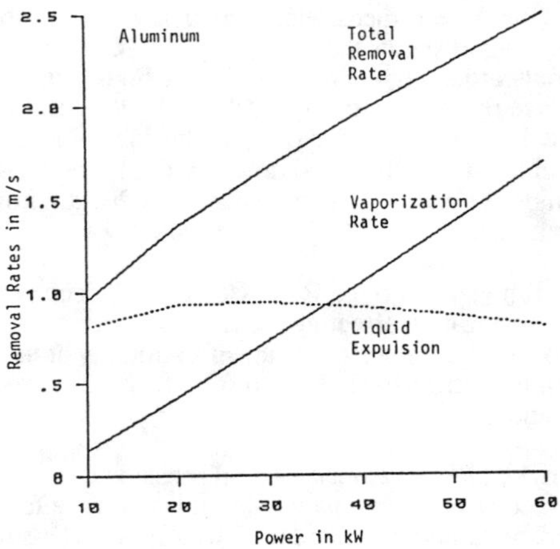

FIGURE 12. Vaporization, liquid expulsion, and total removal rates vs. beam power for aluminum.

shows the materials removal rate for aluminum. The fraction of materials removed by vaporization and liquid expulsion is compared. It can be observed from Fig. 12 that at lower power there is more liquid expulsion than vaporization. At high power the converse is true. At very high power, vaporization is expected to be the dominant form of material removal.

However, this work is extended to a two-dimensional analysis to examine the laser drilling process.

Two-dimensional transient model for laser drilling. The laser drilling model of this study utilizes the conservation of energy, that is the Stefan condition at the solid-liquid and the liquid-vapor interfaces. This condition is written by taking the effects of curvature of the interfaces and the liquid metal flow into account. In order to examine the evolution of the crater, the Stefan condition is applied to various points on the interfaces to determine the displacements of these points and then interpolation between these points is used to obtain the shape of the crater. Besides this, the present study is based on several other assumptions (see Ref. 31).

Figure 13 shows the geometry of an axisymmetrically formed cavity considered in this study. The curves DFH and CEG represent the melting front (solid-liquid interface) and the vaporization front (liquid-vapor interface) respectively. The origin of the coordinate system (r,z) where r is the radial variable and z is the height of any point of the work-piece, is at the bottom of the work-piece such that the work-piece is axisymmetric around the z-axis. The top surface of the work-piece is irradiated with a laser beam propagating vertically downward. Due to this, a cavity is formed whose shape is modeled as shown in Fig. 13. As mentioned earlier, the energy conservation equation is written for the points D, C, E, and G. The point H, which is at the top surface of the work-piece and at $r = R_o$, remains unchanged because of assumption that crater radius does not exceed R_o, where R_o is the radius of the laser beam. In this study, R_o is taken to be the $1/e^2$-point. The problem is solved by using the Runge-Kutta method (see Ref. 31).

Figure 14 shows the profiles of the solid-liquid and the liquid-vapor interfaces in a laser-induced crater. In these figures, the horizontal axis represents the radial distance of the two interfaces from the line of symmetry of the crater and the vertical axis indicates the height of the two interfaces from the bottom of the work-piece. For a given laser power, beam radius, and pulse-on time; the vaporization fronts, dashed lines A, B and C, D of Fig. 14

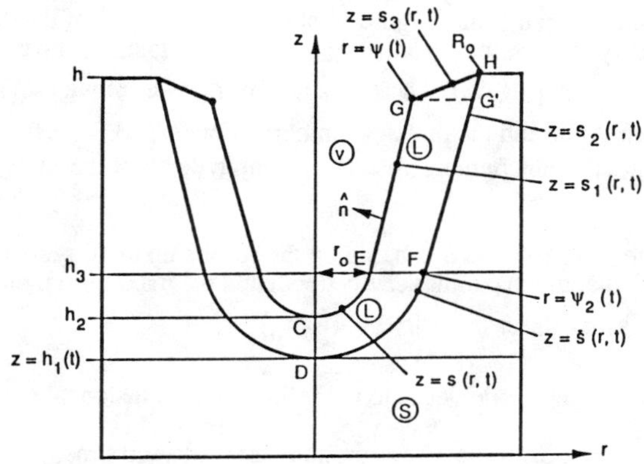

FIGURE 13. Geometric model of the crater formed during laser irradiation. (v) = vapor region, (L) = liquid region, and (s) = solid region.

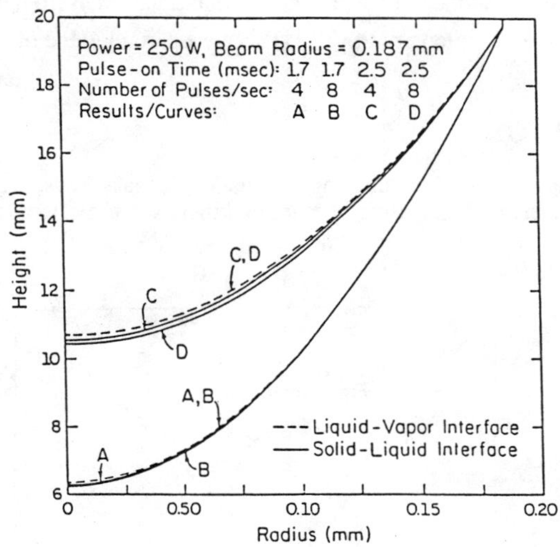

FIGURE 14. Variation of the cavity profile with the radial distance of the solid-liquid and the liquid-vapor interfaces for P = 250 W, R_O = 0.187 mm, and t = 10 sec.

are found to be very close to each other (dashed lines B and D lie very slightly above the dashed lines A and C, respectively) when the number of pulses per second is increased from 4 to 8. Increasing the number of pulses per second decreases the average energy per pulse which lowers the laser intensity and leads to melting dominated process. This causes less vaporization and more melting which means that the dashed lines (vaporization fronts) B and D should be above the dashed lines A and C, respectively, to indicate less vaporization and the solid lines (melting fronts) B and D should be below the solid lines A and C, respectively, to indicate more melting.

An interesting result is obtained by studying the effects of various laser parameters on the depth of cavity and the melt front. Figure 15 shows the variation of the depth of the free surface on the cavity and the melt front at $r = 0$ from the mouth of the cavity with the "gross" laser intensity $Pt_d/(R_0^2\, t_p)$, where P is the laser power in watt, t_d and t_p are respectively the irradiation time and the pulse-on time in seconds, and R_0 is the beam radius in centimeter. Essentially, this figure gives the maximum depth of the cavity and the melt front.

It can be seen from Fig. 15, which is based on the results up to 10 seconds after the irradiation is initiated, that the maximum cavity depth and the maximum melt front depth vary linearly with the "gross" laser intensity $Pt_d/(R_0^2\, t_p)$.

The results of this drilling model leads to the following conclusions:

1. For a given laser power, beam radius, and laser pulse-on time, the depth of the vaporization front decreases and the depth of the melting front increases as the number of pulses per second is increased.
2. For a given laser power, beam radius, and the number of pulses per second, the depth of the vaporization front decreases and the depth of the melting front increases as the laser pulse-on time is increased.
3. Based on the numerical results up to 10 seconds after the irradiation is initiated, it is found that the maximum cavity depth and the maximum melt front depth vary linearly with the "gross" laser intensity $Pt_d/R_0^2\, t_p)$ above a certain value of this intensity.

3. MASS TRANSPORT

For processes such as laser surface alloying and cladding, mass transport is a very important phenomenon. Basic understanding of reaction kinetics is important for laser chemical

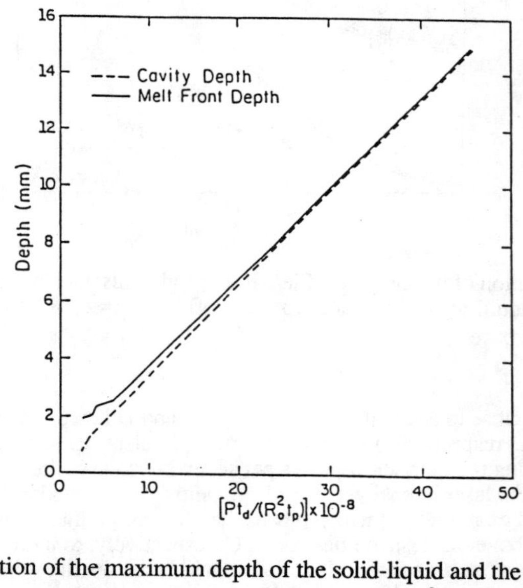

FIGURE 15. Variation of the maximum depth of the solid-liquid and the liquid-vapor interfaces with the "gross" laser intensity $(Pt_d/(R_0^2\, t_p)$.

vapor deposition (LCVD) in order to predict the film thickness and profile. A mathematical model for mass transport during laser surface alloying would help clarify several aspects of the problem. A set of processing conditions could be tested for uniformity of mixing and for the resultant average compositions in liquid state. The mass flux necessary to obtain a desired average composition in the liquid state could be computed. Powder loss during alloying would then be estimable, knowing the actual mass flux during laser processing. The model could be used to simulate the effect of varying the method of supplying alloying elements. Having predicted an average liquid pool composition and measured the actual solid state composition, the effective solute partitioning coefficient C^*_s at the solid liquid interface could be calculated. This effective C^*_s could then be compared to the value determined from the equilibrium phase diagram to check if conditions of local equilibrium existed ahead of the solid liquid interface. This would be a useful check as local equilibrium is assumed in predicting possible compositions during rapid solidification from equilibrium phase diagrams.

3.1. Two-Dimensional Transient Model for Mass Transport in Laser Surface Alloying

Chande and Mazumder [32] used the momentum transfer model described in the last section and coupled that with diffusion equation to estimate the mass transfer during laser surface alloying since convection dominates the process.

For mass transfer calculations, the following assumptions are made:

1. The effect of alloying on solute mass diffusivity was neglected because accurate high temperature data were unavailable; mass transport by diffusion was negligible compared to that by convection ($u_o d/D > 100$) and it simplified the formulation.
2. Mass flux was constant and uniform across the width of the pool at the surface. This rate and distribution could be altered to allow for any other experimental condition such as wire feed or nonuniform powder addition.
3. There was no transfer of alloying elements across the solid liquid interface. This was a very good assumption that was verified experimentally using Electron Probe Micro Analysis (EPMA) technique.

Melting of powder particles introduced into the melt was found to be practically instantaneous [32]. Thus mass flux could be considered as being added in the liquid state.

The mass transfer problem was numerically solved by using the alternate direction implicit method. The results of the mass transfer calculations indicate the importance of fluid flow in determining solute distribution during lase surface alloying. The small effect of changing solute diffusivity on the solute distribution, the uniform mixing even when mass flux was present over only a part of the pool surface and the nature of the three-dimensional plots show that the pattern of fluid flow controls the resultant solute distribution. The effect of laser-substrate interaction time on solute distribution during surface alloying is shown in Fig. 16.

3.2. One-Dimensional Transient Model for Laser Cladding

Laser cladding is a process used for altering the surface properties of substrate materials. Due to high cooling rate in laser cladding, extended solid solution is formed whenever a mixture of several elements is used as cladding powder. Usually, the extended solid solutions have higher melting point, better oxidation and corrosion resistant properties than the alloys with equilibrium compositions. To study the extension of solid solubility during laser cladding, a diffusion model is developed by Kar and Mazumder [33-35] under the following assumptions:

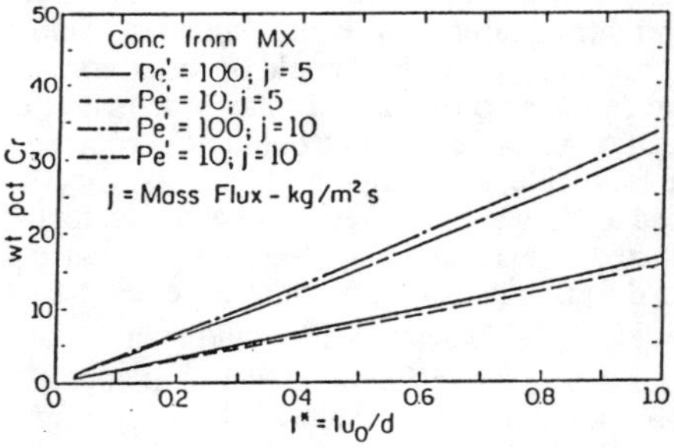

FIGURE 16. Summary of mass transfer calculations showing growth of average composition with increasing interaction time for uniform mass flux of 5 and 10 kg/m^2-s and Pe' values of 100 and 1000.

1. The thermal conductivity and the thermal diffusivity for a mixture is the sum of the volume-averaged value of the respective transport properties of each element of the mixture,
2. The mass diffusivity of each element in the liquid phase is the average value of self-diffusivity over the room temperature and the initial pool mean temperature with modified activation energy for the mixture,
3. The cladding pool and the substrate are in thermally perfect contact,
4. There is no mass diffusion in the solid phase,
5. The solute segregated at the solid-liquid interface moves into the liquid phase by diffusion only. This is because a boundary layer is formed near the interface where diffusion of solute atoms is dominant,
6. The cladding melt forms a uniform solution of composition equal to that of the cladding powder mixture before its solidification begins,
7. Only 50% of the laser energy is absorbed by the cladding material. Studies [36] show that the amount of laser energy absorbed by different materials is 37-60%.

The procedure to solve this cladding problem can be found in Ref. [34] where the model has been discussed in detail and applied to the Ni-Hf system. The importance of studying the Ni-Al system [35] compared to that of Ref. [34] can be understood by analyzing the equilibrium phase diagrams of the Ni-Al and the Ni-Hf systems. The equilibrium phase diagrams of various systems can be divided into several regions based on the sign of the slope of the phase diagrams. Each of these regions can be divided further into smaller sections based on the assumption of constant equilibrium partition coefficient (k_e) which is true over a small range of the concentration around the point of interest on the equilibrium phase diagram. In Ref. [34], we have presented the effects of that regime of the equilibrium phase diagram that has negative slope. The study of Ref. [35] examines the effects of that region of the equilibrium phase diagram that has positive slope. For rapid cooling rate, an expression for the nonequilibrium partition coefficient has been derived by Aziz [38], which has been used in this model.

The results of the above model have been compared with experimental data. Laser cladding was performed on nickel substrates with a mixture of Ni-Hf powder of nominal composition by weight, 74% Ni and 26% Hf in one case and 74% Ni and 26% Al in the other case. STEM analysis of these samples shows that the concentration of Hf in the α phase and Al in the martensitic solid solution regions is in excess of that predicted by the equilibrium phase diagram. The present model predicts the composition of the extended solid solution quite well. These results have been presented in Table 1. A small fraction of the Ni-Hf alloy was found experimentally to contain 9.3 percent (by wt.) Hf in the Ni matrix. Also, the above model was used to obtain the nonequilibrium phase diagrams for Ni-Hf and Ni-Al alloy systems as shown in Figs. 17 and 18. The characteristic parameters for

TABLE 1 Comparison of Theoretical Results with Experimental Results of Extended Solid Solution

Nominal Composition of the cladding, wt.%	74% Ni, 26% Hf	74% Ni, 26% Al
Laser Power, (Kw)	5	5
Laser Beam Diameter, mm	3	3
Workpiece Speed, in/min	50	50
Initial Pool Mean Temperature ($\bar{T}_2$),°K	1862	2420
Composition in the solid solution, wt.%		
Theoretical Results	7.15 Hf	30.76 Al
Experimental Results	6.5 Hf	29 Al

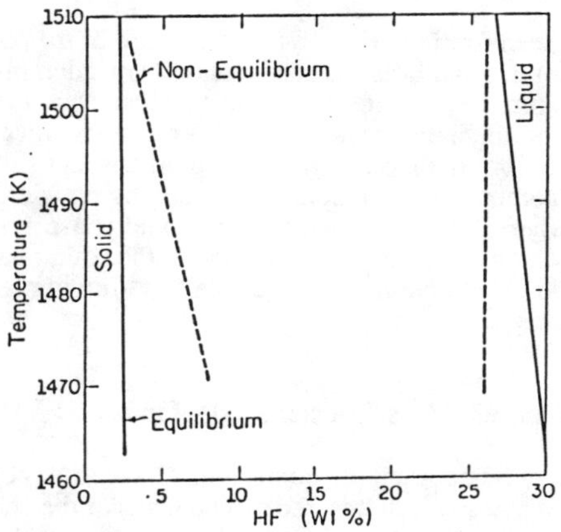

FIGURE 17. Comparison of the nonequilibrium phase diagram of the extended solid solution of nickel-hafnium with its equilibrium phase diagram.

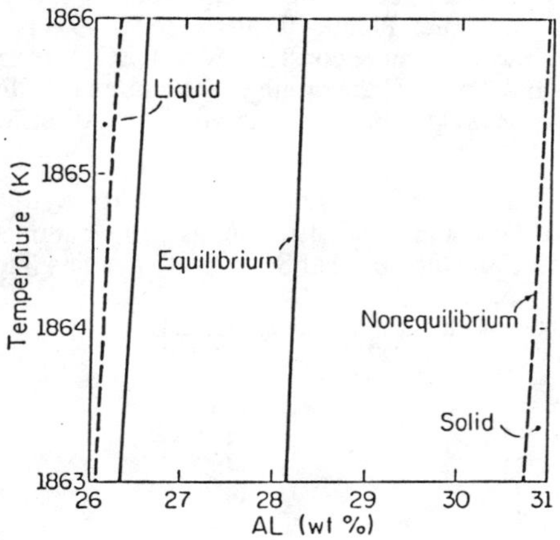

FIGURE 18. Comparison of the nonequilibrium phase diagram of the extended solid solution of nickel-aluminum with its equilibrium phase diagram.

these figures have been presented in Table 2. Deviation of nonequilibrium phase diagram from the equilibrium one shows the extension of solid solubility that can be obtained during laser cladding. The nonequilibrium phase diagram has been plotted in the neighborhood of the melting point of the cladding powder. It can be seen from Fig. 17 that the width of the solid-liquid region between the equilibrium solidus and liquidus has reduced considerably. The nonequilibrium solidus line of this figure shows the extension of Hf concentration in Ni that can be obtained due to the rapid cooling in laser cladding. This phenomenon can be understood from the fact that the equilibrium phase diagram of Ni-Hf has a negative slope at the point which corresponds to the nominal composition of the cladding powder. Due to this, Hf is rejected from the Ni matrix when the solution of Ni-Hf solidifies and this results in increasing the concentration of Hf in the liquid phase. But since the solid phase retains more Hf than its equilibrium composition, the liquid phase will have less Hf than the equilibrium value and hence the extended solid solution (that is, the nonequilibrium) phase diagram will shrink.

On the contrary, it can be seen from the Fig. 18 that the width of the solid-liquid region between the equilibrium solidus and liquidus has increased considerably. The nonequilibrium solidus line of this figure shows the extension of Al concentration in Ni that can be obtained due to the rapid cooling in laser cladding. This phenomenon can also be understood from the fact that equilibrium phase diagram of Ni-Al has a positive slope at the point which corresponds to the nominal composition of the cladding powder. Due to this, Al is retained in the Ni matrix whereas Ni is rejected into the liquid phase as solidification proceeds. This causes an extension of Al concentration in the solid phase and lowers its weight fraction in the liquid phase and thus enlarges the solid-liquid region between the solidus and the liquidus lines.

3.3. Three-dimensional Transient Mass Transfer Model for LCVD

Heat and mass transfer are two important phenomena that have to be considered for modeling the LCVD processes. Besides this, the mechanisms of the chemical reactions occurring during LCVD and the associated chemical kinetics should also be known along with other parameters such as the sticking probability, diffusion coefficient, and absorptivity.

The concentration of various species such as $TiBr_4$, $TiBr_2$, Ti, and Br_2 inside the LCVD chamber and the thickness of the deposited Ti film are determined by solving five three-dimensional transient, and nonlinear mass diffusion equations by using the finite-medium Fourier integral transform technique. Figure 19 shows the LCVD chamber, the substrate, and the Cartesian coordinate system under consideration. OLO'W represents the top surface of the substrate, and the point O the origin of the chosen coordinate system. In Fig. 19, the LCVD chamber is divided into two regions. The top surface of the substrate

TABLE 2 Solidus Composition of Hf, Solidus/Liquidus Temperature at the Substrate-Cladding Interface and the Speed of Solidification at this Interface

Solute Element	Composition of Hf wt%)	Solidus/Liquidus Temperature (K)	Interface Speed (cm/sec)
	3.19	1505	3.26
	4.83	1499	2.88
Hafnium	6.23	1489	2.74
	7.15	1480	2.68
	9.58	1453	2.55
	30.83	1864	6.1
Aluminum	30.74	1863	5.6
	30.56	1862	4.5

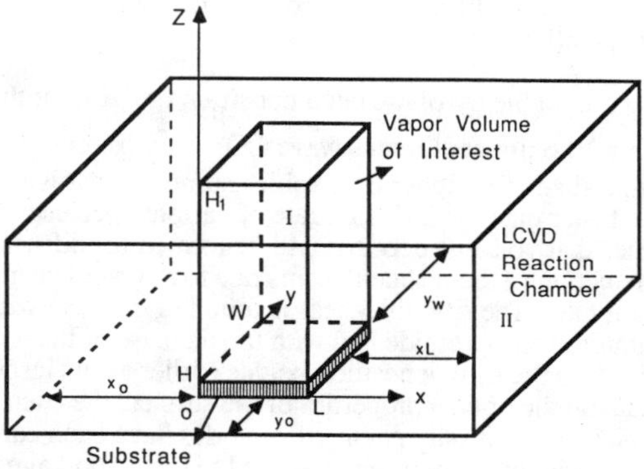

FIGURE 19. Idealized LCVD chamber and the mass transfer region of interest.

forms the base of region I and this region extends up to the top inner surface of the LCVD chamber. Region II constitutes the rest of the chamber; thus, the length, width, and height of the region I are L, W, and H, respectively. The governing equation for mass transfer of the i-th species in the region I, and the solution of this problem can be found in Ref. 39.

The effects of the activation energy, E, are examined in Fig. 20. It can be seen from this figure that the film thickness decreases as the activation energy increases. For a given value of the total pressure and the partial pressure of $TiBr_4$, the Damkohler number which is defined as the ratio of the chemical reaction rate to the diffusion rate, decreases as the activation energy increases. We can, therefore, say that the film thickness decreases as the Damkohler number decreases. The thickness profiles are found to have depressions at $x = 0.5$ cm, that is, at the center of the deposits. This kind of volcanic morphology of the deposits has been reported in many studies. here again, we find that the depression decreases as the Damkohler number decreases. For large Damkohler numbers, the reaction rate is much faster than the diffusion rate and hence there will be depletion of $TiBr_4$ near

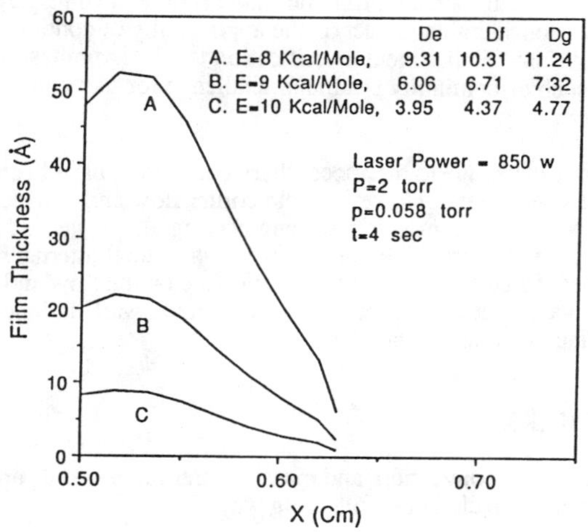

FIGURE 20. Spatial variations of the thickness profile and the morphology of titanium deposits with activation energies.

the center of the deposits as the diffusion limited chemical reaction progresses which will cause volcano-like deposition.

The film is produced due to the pyrolytic decomposition of $TiBr_4$ at the surface of SS 304 which is heated with a 10.6 μm continuous wave CO_2 laser beam. The deposited films are found to have volcanic shape for higher Damkohler numbers, that is, when the chemical reaction rate is much faster than the diffusion rate of various species inside the LCVD chamber. This implies that volcanic deposits will be formed for diffusion-limited chemical reactions in which depletion of the reactant occurs due to the rapid chemical reaction rate and the slow diffusion rate. The film thickness is found to increase with the increase of the partial pressure of titanium tetrabromide and with the increase of the total pressure inside the LCVD chamber. Also the film deposition begins earlier as the laser power increases. Due to the lack of data on the optical properties of the surface, the chemical reaction mechanisms and the chemical kinetics only parametric sutides have been carried out in this work. With proper data, the model predictions are expected to be in good agreement with experimental results.

4. FUTURE CHALLENGES AND NEEDS

Paucity of temperature dependent thermophysical data for materials is a serious problem for development of mechanistic and predictive models for transport phenomena associated with materials processing. This is rather a generic problem for all process modeling. The European community is making an effort based on a joint agreement made at Versailles (France) in the 1980's. Although some data have been produced from the program, an accelerated international effort is necessary to review and verify such data and measure new data where necessary.

Data for surface absorption and beam attenuation due to plasma beam interaction for optically thin and optically thick plasma are of particular interest for laser processing. The absorption coefficient is often an adjustable pararameter for laser process modeling whereas this is probably the most important input parameter for transport models. Plasma effects include both absorption and refraction leading to the uncertainties of input parameters. Another important area needing attention for high intensity laser processing is the discontinuity at the liquid-vapor interface. Prokhorov, et al. [40], presented a hypothesis that during high intensity laser beam and materials interaction, a thin layer at the liquid vapor interface essentially changes their metallic properties and becomes a "transparent wave" of dielectric liquid. This has considerable impact on the applicability of continuum equations under such physical conditions. Experimental verification for the hypothesis and theoretical approach to address such discontinuity is a major challenge for modeling high intensity beam-materials interaction.

Free surface and weldpool shape due to thermocapillary convection have been active areas of modeling for the past seven years. However, cold corner flow and convection due to a concentration gradient, which is common in most engineering alloys, needs further attention. For processes involving vaporization phenomena such as total internal reflection in the keyhole, effect of shear force at the liquid-vapor interface on the flow and materials removal, balance between vaporization and recondensation on the wall, and the discontinuity at the liquid-vapor interface require urgent attention.

5. CONCLUDING REMARKS

Comments on heat conduction, convection, and mass transfer processes as applied to laser processing of materials can be made in the following way:

a. Heat Conduction: Although the transfer of laser energy, that is heat conduction within the substrate materials can be analyzed by elegant mathematical techniques, more work still remains to be done to understand the coupling of laser energy with the substrate materials.

b. Convection: The present three-dimensional model brings additional insight to the convection and fluid flow within the melt pool. The model gives details of three-dimensional velocity and temperature fields. Consequently, a more realistic prediction of pool shape is obtained. The detailed three-dimensional velocity field gives a quantitative explanation of the mechanism of the mixing process within the molten pool. A more realistic prediction of the cooling rate is now available for laser melted pool.

The perturbation model yields better spatial resolution because two dimensionality renders it practical to employ a finer mesh. On the other hand, the three-dimensional model is capable of handling larger scanning speeds than the perturbation model. It is found that in the presence of thermocapillary convection, the physics of the process changes from conduction-dominated to convection-dominated. This changes the pool geometry dramatically, resulting in up to a 150 percent increase in the aspect ratio (width/depth) as compared to the pure conduction case. The temperature gradient on the resolidifying surface, which is directly proportional to the cooling rate, is found to be non-uniform. This, in turn, gives rise to the non-uniform resolidified microstructure. The mechanisms of solute redistribution are discussed qualitatively. It is found that a particle recirculates many, many times in the molten pool before it resolidifies. Consequently, the solute can be well mixed.

Although understanding of flow dynamics of laser melted pure metal pool has improved significantly, a vast amount of work is still needed to understand the effect of convection on free surface and alloying element addition. Also, combined vaporization and liquid convection needs further attention to understand keyhole formation, cutting, and drilling.

c. Mass Transfer: Although laser surface treatments are complicated processes, proper understanding of mechanisms can be obtained by mathematical modeling. The problem can be first approached in simplified form with proper assumptions. Each step will be useful for a range in spite of its limitation. With gradual removal of simplifying assumption, any complicated process such as this can be studied.

ACKNOWLEDGMENT

This paper includes many years of a group (consisting of Professor J. Mazumder, Professor M. M. Chen, Dr. T. Chande, Dr. C. L. Chan, Dr. A. Kar, Mr. R. Zehr, Mr. D. Voelkel and Mr. M. Azer) effort at the University of Illinois sponsored by U.S. Office of Naval Research (Grant N00014-84-K0315), U.S. Air Force Office of the Scientific Research (Grant AFOSR: 85-0333), and the U.S. National Science Foundation (Grant NSF MSM 84-12118).

REFERENCES

1. Carslaw, H. S., and Jaeger, J. C., Conduction of Heat in Solids, 2nd. ed. Oxford Univ. Press, 1959.

2. Gregson, V., Laser Heat Treatment, Laser Materials Processing, M. Bass (ed.), North Holland, 1983.

3. Sandven, O. A., Heat Flow in Cylindrical Bodies during Laser Transformation Hardening, Proc. SPIE, vol. 198, pp. 138-143, 1979.

4. Cline, H. E., and Anthony, T. R., Heat Treatment and Melting Material with a Scanning Laser or Electron Beam, J. Appl. Phys. vol. 48, pp. 3895-3900, 1977.

5. Mazumder, J., and A. Kar, Mathematical Modeling of Transport Phenomena in Laser Materials Processing, Proc. of Indo-US Workshop on Transport Phenomena in Materials Processing, R. Shah (ed.), to be published by Hemisphere Publishing Co., 1988.

6. Kar, A., and Mazumder, J., Three-Dimensional Transient Thermal Analysis for Laser Chemical Vapor Deposition on Uniformly Moving Finite Slabs, J. Appl. Phys., vol. 65, pp. 2923-2934, 1989.

7. Mazumder, J., and Steen, W. M., Heat Transfer Model for CW Laser Materials Processing, J. Appl. Phys., vol. 51, pp. 941-947, 1980.

8. Chande, T., and Mazumder, J., Lasers in Metallurgy, K. Mukherjee and J. Mazumder (eds.), The Metallurgical Society of AIME, Warrendale, PA, pp. 165-178, 1981.

9. Mazumder, J., Ph.D., Laser Welding of Titanium 6Al-4V-Ti, London University, 1978.

10. Courtney, C. G. H., and Steen, W. M., Surface Heat Treatment of En8 Steel using a 2 KW Continuous-Wave CO_2 Laser, Met. Technol., vol. 6, pp. 456-462, 1979.

11. Chande, T., and Mazumder, J., Estimating Effects of Processing Conditions and Variable Properties Upon Pool Shape, Cooling Rates, and Absorption Coefficient in Laser Welding, J. Appl. Phys., pp. 1981, 1984.

12. Chande, T., and Mazumder, J., Mass Transport in Laser Surface Alloying: Iron-Nickel System, Appl. Phys. Lett., vol. 41, pp. 42-43.

13. Anthony T. R., and Cline, H. F., Surface Rippling Induced by Surface Tension Gradients during Laser Surface Melting and Alloying, J. Appl. Phys., vol. 48, pp. 3888-3894, 1977.

14. Oreper, G. M. Eagar, T. W., and Szekely, J., Convection in Arc Weld Pools, Welding Journal, vol. 63, pp. 307-312, 1983.

15. Chan, C. L., Mazumder, J., and Chen, M. M., Applications of Lasers in Materials Processing, E. A. Metzbower (ed.), ASM, p. 150, 1983.

16. Chan, C. L., Mazumder, J., and Chen, M. M., A Two-Dimensional Transient Model for Convection in Laser Melted Pools, Metall. Trans., vol. 15A, pp. 2175-2184, 1984.

17. Hirt, C. W., Nichols, B. D., and Romero, N. C., A Numerical Algorithm for Transient Fluid Flows, UC-34 and UC-79d, April 1975.

18. Chan, C. L., Mazumder, J., and Chen, M. M., A Three-Dimensional Axisymmetric Model for Convection in Laser Melted Pool, Mat. Sci. Engineering, vol. 3, pp. 306-311, 1987.

19. Chan, C. L., Zehr, R., Mazumder, J., and Chen, M. M., Three-Dimensional Model for Laser Weld Pool, Proc., 3rd Engineering Foundation Conference on Modeling and Control of Casting and Welding Processes, eds. B. Kuo and R. Mehrabian, TMS-AIME, Warrendale, PA, pp. 229-246, 1986.

20. Chan, C. L., Mazumder, J., and Chen, M. M., Effect of Surface Tension Gradient Driven Convection in a Laser Melt Pool, J. Appl. Phys., vol. 64, no. 11, pp. 6166-6174, 1988.

21. Kou, S., and Wang, J. H., Three-Dimensional Convection in Laser Melted Pools, Metall. Trans., vol. 17A, pp. 2265-2270, 1986.

22. Patankar, S. V., and Spalding, D. B., A Calculation Procedure for Heat, Mass, and Momentum Transfer, Int. J. Heat and Mass Transfer, vol. 15, pp. 1787-1806, 1972.

23. Viskanta, R., Solar Heat Storage: Latent Heat Materials, G. A. Lane (ed.), CRC Press, New York, pp. 153-222, 1984.

24. Chan, C. L., Ph.D. Thesis, Thermocapillary Convection during Laser Surface Heating, University of Illinois at Urbana-Champaign, 1986.

25. Schlichting, H., Boundary Layer Theory, McGraw Hill, New York, 1960.

26. Carnahan, B., Luther, H. A., and Wilkes, J. O., Applied Numerical Methods, John Wiley & Sons, New York, 1969.

27. Chan, C. L., Chen, M. M., and Mazumder, J., Asymptotic Solution for Thermocapillary Flow at High and Low Prandtl Numbers due to Concentrated Surface Heating, ASME J. of Heat Transfer, vol. 110, pp. 140-146, 1988.

28. Chan, C. L., Chen, M. M., and Mazumder, J., ASME/AIChE National Heat Transfer Conference, ASME, New York, 85-HT-21, 7 p., 1985.

29. Chan, C. L., and Mazumder, J., One-Dimensional Steady-State Model for Damage by Vaporization and Liquid Expulsion during Laser Materials Interaction, J. Appl. Phys., vol. 62, pp. 4579-4586, 1987.

30. Rockstroh, T. J., and Mazumder, J., Spectroscopic Studies of Plasma during Laser Materials Interaction, J. Appl. Phys., vol. 61, pp. 917-923, 1987.

31. Kar, A., and Mazumder, J., Two-Dimensional Model for Material Damage due to Melting and Vaporization during Laser Irradiation, J. Appl. Phys., accepted for publication.

32. Chande, T., and Mazumder, J., Two-Dimensional, Transient Model for Mass Transport in Laser Surface Alloying, J. Appl. Phys., vol. 57, pp. 2226-2232, 1985.

33. Kar, A., and Mazumder, J., One-Dimensional Diffusion Model for Extended Solid Solution in Laser Cladding, J. Appl. Phys., vol. 61, pp. 2645-2655, 1987.

34. Kar, A., and Mazumder, J., One-Dimensional Finite-Medium Diffusion Model for Extended Solid Solution in Laser Cadding of Hf on Nickel, Acta Metall., vol. 36, pp. 701-712, 1988.

35. Kar, A., and Mazumder, J., Extended Solid Solution and Nonequilibrium Phase Diagram for Ni-Al Alloy Formed during Laser Cladding, Metall. Trans. A., vol. 20A, pp. 363-371, 1989.

36. Li, L. J., and Mazumder, J., A Study of the Mechanism of Laser Cladding Processes, Laser Processing of Materials, K. Mukherjee and J. Mazumder (eds.), The Metallurgical Society of AIME, Los Angeles, CA, pp. 35-50, 1984.

37. Boettinger, W. J., and Perepezko, J. H., Fundamentals of Rapid Solidification, Rapidly Solidified Crystalline Alloys, S. K. Das, B. H. Kear, and C. M. Adam (eds.), The Metallurgical Society of AIME, Warrendale, PA, pp. 21-58, 1985.

38. Aziz, M. J., Model for Solute Redistribution during Rapid Solidification, J. Appl. Phys., vol. 53, pp. 1158-1168, 1982.

39. Kar, A., Azer, M. N., and Mazumder, J., Three-Dimensional Transient Mass Transfer Model for LCVD, J. Appl. Phys., accepted for publication.

40. Prokhorov, A., M., Batanov, V. A., Bunkin, F. V., and Fedorov, V. B., "Metal Evaporation under Powerful Optical Radiation," IEEE J. of Quantum Electronics, vol. QE-9, no. 5, pp. 503-510, 1973.

Formation of Metal and Non-Metal Clusters Through Laser and Electron Beam Vaporization

SUSUMU KOTAKE
Faculty of Engineering
The University of Tokyo
7-3-1 Hongo, Tokyo 113, Japan

ABSTRACT

Because of peculiar features of atomic or molecular clusters for the condensation process, the size distribution can be expected to be one of the factors of controlling the process. In order to control the condensation process of vapor atoms of metal and non-metal elements, the size distributions of their clusters were experimentally investigated with relation to the film condensation of these elements. Solid materials of metal and non-metal elements were vaporized from the solid state by the irradiation of laser and electron beams. The elements having small values of the cohesive energy such as the 2b group form clusters of widely distributed sizes, whereas the elements of large cohesive energy such as the 5a group are hardly evaporated and clustered into large sizes. In cases of non-metal elements, the size distributions of clusters are not always related with the cohesive energy. Concerning the condensation process of the vapor atoms, qualitatively, the metal elements associated with clusters of widely distributed sizes are to have high deposition rates of forming thin films. For non-metal elements, the deposition rates are not always accounted for by the clustering features such as the size distribution. With more quantitative knowledge of cluster formation, it will be possible to control the condensation process of metal and non-metal atoms by changing the size distribution of their clusters which is a function of the species of the carrier gas, its density and/or temperature.

INTRODUCTION

Clusters consisting of molecules or atoms of tens to hundreds have attracted increased attention due to their physical and chemical features in such diverse areas as phase-changes and chemical reactions. The interest stems from many sources. Small clusters have been found very useful in the study of intermolecular forces, physical and chemical dynamics of molecular collisions and energy transfers. Large clusters have been studied for understanding the fundamental process of nucleation and condensation. These clusters are expected to have many possibilities in engineering applications such as manufacturing fine particles and films, controlling mass and heat transfer in nucleation and condensation, and finding new materials of special properties in physical and chemical behavior.

Metal or non-metal clusters can be obtained from their vapors prepared simply by heating the solid material above the boiling point. These hot oven methods are limited, however, to relatively low boiling materials and to a rather modest extent of cooling of the internal degrees of freedom. Other methods for vaporization of materials can be accomplished by the irradiation of laser or electron beams. Employing these methods of vaporization sources, metal and non-metal clusters have been studied especially in the state of supersonic beams followed by expansion for many of the elements[1-6]. However, these supersonic beams of metal and non-metal clusters are to pass through many complicated clustering processes followed by vaporization such as cooling by the carrier gas, expansion to the vacuum, and skimming through the skimmer. This leads to the results influenced by all these factors of the vaporization mechanisms, the kinetics of clustering, the dynamic stability of clusters formed, and the method of detecting the clusters. Any conclusions from these experiments, therefore, should include careful consideration of these factors.

In this study, metal and non-metal vapors are vaporized from their solid material by laser and electron irradiation in a rather free vacuum space of effusion, and the clusters formed are introduced directly into the mass analyzing system to measure the size distribution. The materials of metal and non-metal elements are selected systematically according to the periodic table. Using these results of the size distribution of clusters, the deposition rates of their vapors onto the condensate are measured and compared with the clustering features to study the fundamental behavior of controlling the condensation process.

VAPORIZATION TO CLUSTERS

Vaporization of solid materials by irradiation of laser or electron beams takes the following processes(FIGURE 1);

(1) Solid atoms absorb the beam energy irradiated to excite their vibrational modes of atomic motion if the beam energy is in the range lower than that of infrared light.

(2) If the beam energy is higher than the visible light, it is first absorbed in the form of electronical excitation or ionization of solid atoms and then transferred to the vibrational modes between the solid atoms.

(3) Vibrational motions between solid atoms with high enough energy lead to overcome the interatomic binding forces, and to make an atom or atoms desorb from the solid material.

(4) The desorption of an atom or atoms depends largely on the solid configuration; crystal configurations are associated with single atom desorption, whereas structural or partial crystal configurations vaporize atoms in groups.

(5) Atoms desorbed from the solid surface collide with each other and exchange their translational energies to form clusters of larger sizes.

(6) Atoms evaporated as a group grow up to larger clusters or decompose into smaller clusters.

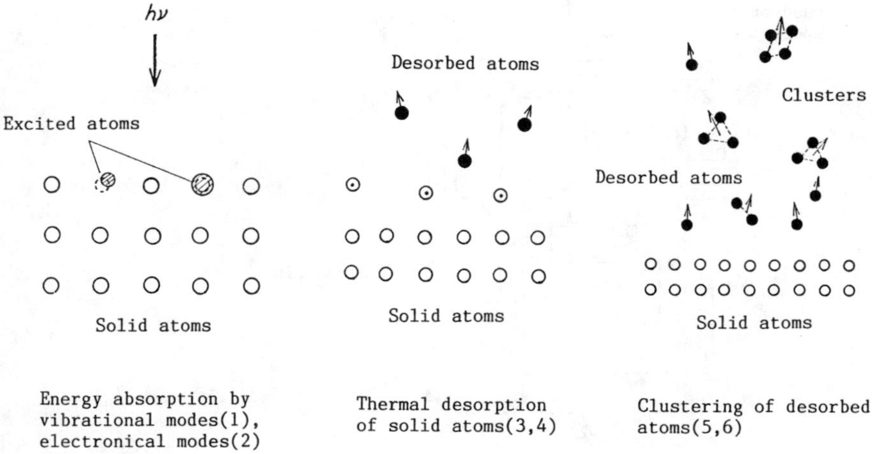

FIGURE 1. Thermal desorption of solid atoms by irradiation of laser and electron beams.

These processes of energy absorption of solid atoms for desorption or evaporation from the solid state are thus controlled by the factors of the vibrational energy modes of solid atoms, the interatomic binding forces, the ionization potential, the statical atomic configuration, and the dynamical state of atomic motions. The vibrational excitation energy of solid atoms affects the absorption process of the energy as well as the dissipation process through thermal conduction. Larger values of the vibrational excitation energy means larger amount of irradiation energy to be required for atomic desorption and faster dissipation of the absorbed energy, that is, less evaporation of atoms from the solid materials with less clustering of evaporated atoms. The interatomic binding forces can be related with the cohesive energy of atom which is the difference between the energy of atoms in the isolated state and that in the solid state.

Atoms desorbed from the solid surface are to form their clusters of which the size distribution is determined by the processes of their collision frequency and energy exchange. The dynamic processes of clustering are influenced by the dynamic state of the atoms, the kind of colliding atoms or molecules, and their dynamic state of motion. But if the latter conditions of the surrounding are kept in the same ones, the size distribution of clusters of desorbed or vaporized atoms are determined by the desorption process of atoms from the solid phase.

In order to investigate how many and of what sizes clusters are formed through the evaporation of solid atoms by laser irradiation, solid materials are irradiated in a vacuum chamber as shown in FIGURE 2. The chamber evacuated by a diffusion pump up to 10^{-7} torr contains the source devices of metal and non-metal vapors, the quadrupole mass spectrometer. Solid material of metal and non-metal elements are vaporized by a pulse beam of Nd:YAG laser(100MJ,100ms), and the size distribution of clusters formed are measured by the mass spectrometer.

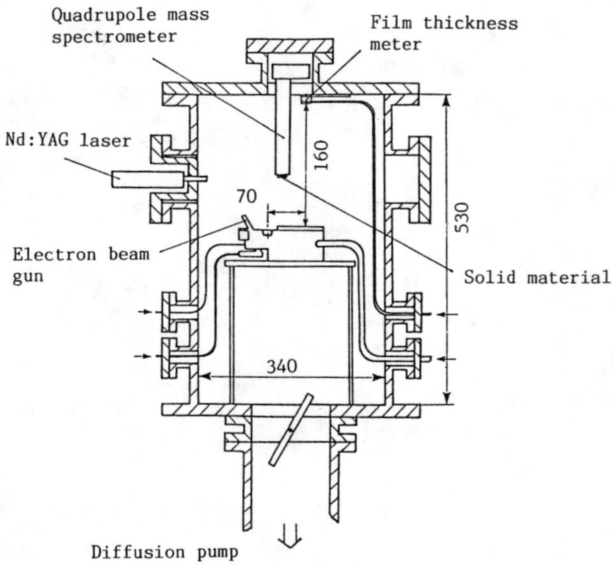

FIGURE 2. Experimental apparatus.

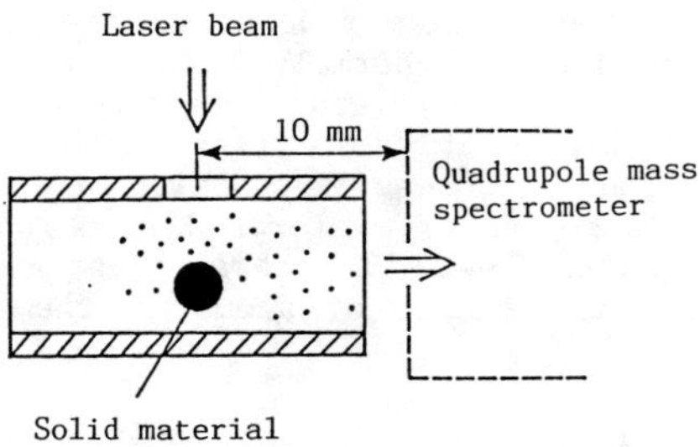

FIGURE 3. Laser vaporization of solid material.

The material is suspended on the inner wall of a small pipe of 4.5mm diameter and irradiated by the laser through a small hole as shown in FIGURE 3. Vaporized atoms diffuse effusively into their surrounding space, forming their clusters. The fundamental frequency of the laser, 1.06μm, is employed for the irradiation for the purpose of thermal desorption of atoms. The material elements studied are selected according to the periodic table as shown in TABLE 1 of which physical properties are shown in TABLE 2.

TABLE 1. Metal and non-metal elements tested.

	Metal									Nonmetal	
	4a	5a	6a	7a	8	8	8	1b	2b	3b	4b
2											C
3										Al	Si
4	Ti	V	Cr	Mn	Fe	Co	Ni	Cu	Zn		
5	Zr	Nb	Mo				Pd	Ag	Cd	In	Sn
6	Hf	Ta	W				Pt	Au			Pb

TABLE 2. Physical properties of metal and non-metal elements tested.

Element	Cohesive energy	Melting Heat	Temp.	Vaporization Heat	Temp.	Heat conductivity	Ionization energy
Ti	4.85	4.14	1155	18.6	1933	21.9	6.83
V	5.31	17.2	1770	393	3470	31.3	6.74
Cr	4.10	14.6	2173	349	2915	90.3	6.76
Mn	2.92	14.6	1517	22.5	2360	7.8	7.43
Fe	4.28	15.1	1808	354	3008	80.3	7.90
Co	4.39	16.2	1768	373	3170	99.2	7.86
Ni	4.44	17.6	1728	381	3073	90.5	7.63
Cu	3.49	13.3	1356	305	2848	398	7.72
Zn	1.35	6.6	693	115	1180	121	9.39
Zr	6.25	5.0	2125	141	4777	22.7	6.95
Nb	7.57	26.8	2710	695	5170	53.7	6.77
Mo	6.82	27.6	2892	590	4919	138	7.18
Pd	3.89	16.7	1823			75.5	8.33
Ag	2.95	11.3	1234			427	7.57
Cd	1.16	6.1	594	100	1040	96.8	8.99
Hf	6.44					23.0	7.
Ta	8.10	31.4	3279	753	5673	57.5	7.88
W	8.90	35.4	3663	799	5803	178	7.98
Pt	5.84	21.7	2042	447	4583	71.4	8.96
Au	3.81	12.7	1336	310	2933	315	9.22
Al	3.39	10.7	933	291	2767	237	5.98
In	2.52	3.3	430	226	2320	81.7	5.78
C	7.37						11.26
Si	4.63						8.15
Sn	3.14	7.1	505	290	2960		7.34
Pb	2.03	4.8	600	180	2024	35.2	7.41
	eV/atom	kJ/mol	K	kJ/mol	K	kW/m.K	eV

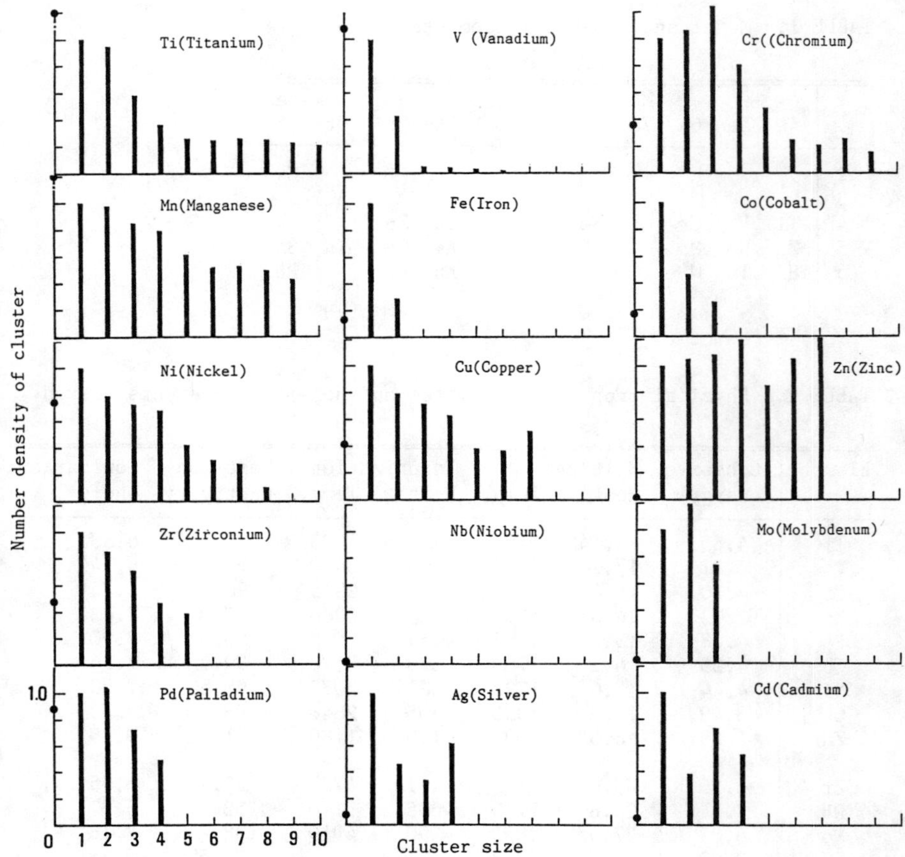

(a) Metal elements.

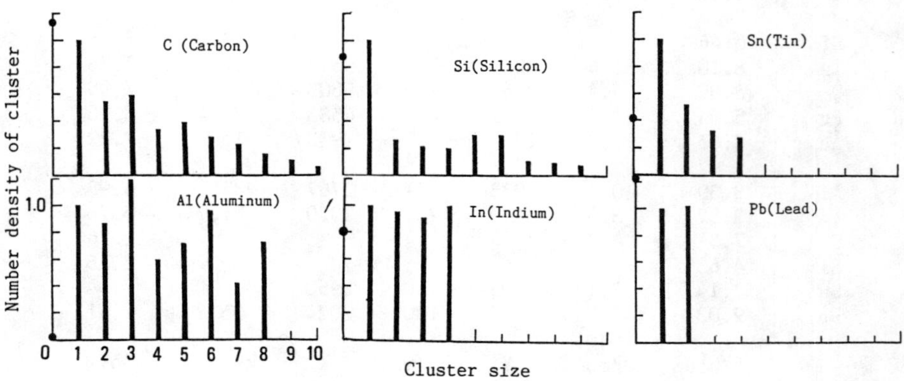

(b) Non-metal element.

FIGURE 4. Size distribution of clusters: The number density of cluster is measured relatively to that of the monomer and the relative density of the monomer is shown on an arbitrary scale by the solid circle.

In FIGURE 4, the size distributions of metal and non-metal clusters formed are shown in order of the periodic table. The x-axis is the size of cluster, that is, the number of atoms consisting the cluster. The y-axis is the ratio of the number density of the cluster to that of the monomer. The number density of the monomer is indicated by a solid circle on the y-axis with an arbitrary scale.

The metal materials of 4a, 6a, 7a, 8, 1b and 2b elements are to be considerably clusterized. Especially 6a(Cr, Mo) and 2b(Zn, Cd) elements have clusters of which the number density is larger than the monomer, taking the maximum at the size larger than the monomer. For elements of 5a(V, Nb) and 8(Fe, Co), however, they are hardly clusterized, although their monomers are not sufficiently observed except for Vanadium. Of the same group of the 8-th element, Ni and Pd have appreciable clusters with monomers of high number density. These overall features of clustering can not been accounted for simply by the physical properties of Table 2, but one of the most relevant factors is the cohesive energy. The cohesive energy takes the maximum values at 5a(V, Nb) and 8(Fe, Co, Ni) elements, and the minimum values at 2b(Zn, Cd). Large values of the cohesive energy for 5a(V, Nb) and 8(Fe, Co) can be related with the fact that the monomer and clusters of these elements are formed too small in number to be observed. Small cohesive energies of 2b(Zn, Cd) also account for the observation of large size clusters of these elements.

In the case of nonmetal elements, appreciable clustering is observed for all materials tested. Carbon and silicon have large number densities of the monomer and clusters of large size. Aluminum has a small amount of the monomer but widely distributed clusters. In the case of nonmetal elements, the cohesive energy takes the maximum at the 4b(C, Si, Ge, Sn, Pb) and the energy decreases in increasing order of the periodic row as C, Si, Ge, Sn, Pb. Although not so strictly confirmed, the clusters formed are widely distributed in this order related to the cohesive energy, but the monomer density is decreased in this order, except for Pb(lead). For non-metal elements, thus, the size distributions of clusters are not always well related to the relative amount of the cohesive energy.

CONDENSATION FROM CLUSTERS

In the condensation process of vapor atoms or molecules, their clusters play an important role due to their surface reactivity resulted from the internal degrees of freedom of motion. The condensation process of monomers and clusters onto the condensate can be described as follows(FIGURE 5).

(1) Vapor atoms approaching the condensate surface exchange their translational energy with those of the condensate atoms through the potential interaction, and

(2) Only atoms of which translational energy is decreased sufficiently can be trapped onto the condensate surface.

(3) With the condensate atoms, clusters can exchange their dynamical energies of not only the translational energy but also the vibrational and rotational energies, and

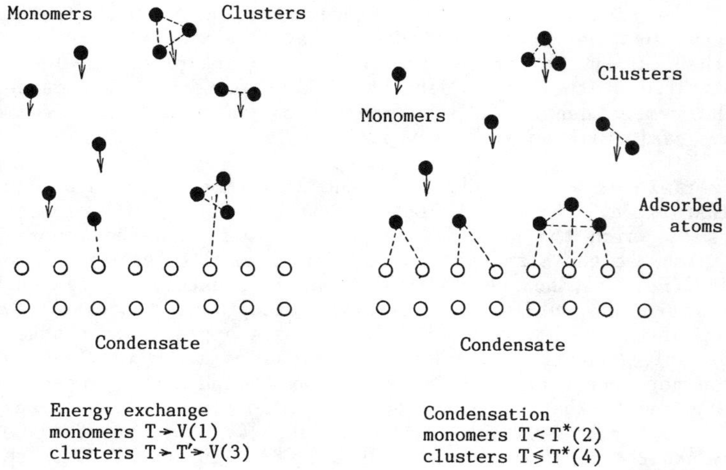

FIGURE 5. Condensation of monomers and clusters onto condensates.

(4) Even if the translational energy is not sufficiently decreased, atoms of the cluster have a possibility to stick to the condensate atoms by self-adjusting the distribution or partition of the internal energies.

These mean that elements which form easily their clusters of large sizes are to stick or to be deposited onto the condensate surfaces more easily. Vapor atoms in clusters of widely distributed sizes are more favorable for condensation or sticking to the condensate than those of the monomer. From the results of the size distribution of clusters, thus, the condensation features such as the condensation rate may be predicted qualitatively. Elements having clusters of widely distributed sizes will have higher values of the condensation rates, or deposition rate. According to the above results of clustering of evaporated materials, the 2b elements should have the largest rate of deposition onto solid surfaces to form thin films, and 5a(Nb, V) and 8(Fe, Co) elements have the smaller rates of condensation. Carbon, silicon and lead should have larger values of the deposition rate in this order.

In order to study the condensation rates or the deposition rates of vapor atoms, solid materials are irradiated by an electron beam. The laser vaporization makes too small density of vapor atoms to form a thin film of which thickness can be measured. The electron beam has higher energies of which absorption mechanism of the solid atoms is different from that of the laser beam, but the long-time continuous irradiation of the electron beam leads to a state of thermal heating of solid atoms of which vibrational motions are excited sufficiently for thermal desorption or evaporation of atoms. Thus, the size distributions of clusters formed in both methods can be considered to have the almost same features.

The solid material is placed on the anode which electrons of 4 to 10 kV strike to vaporize the consisting atoms. The evaporated atoms diffuse

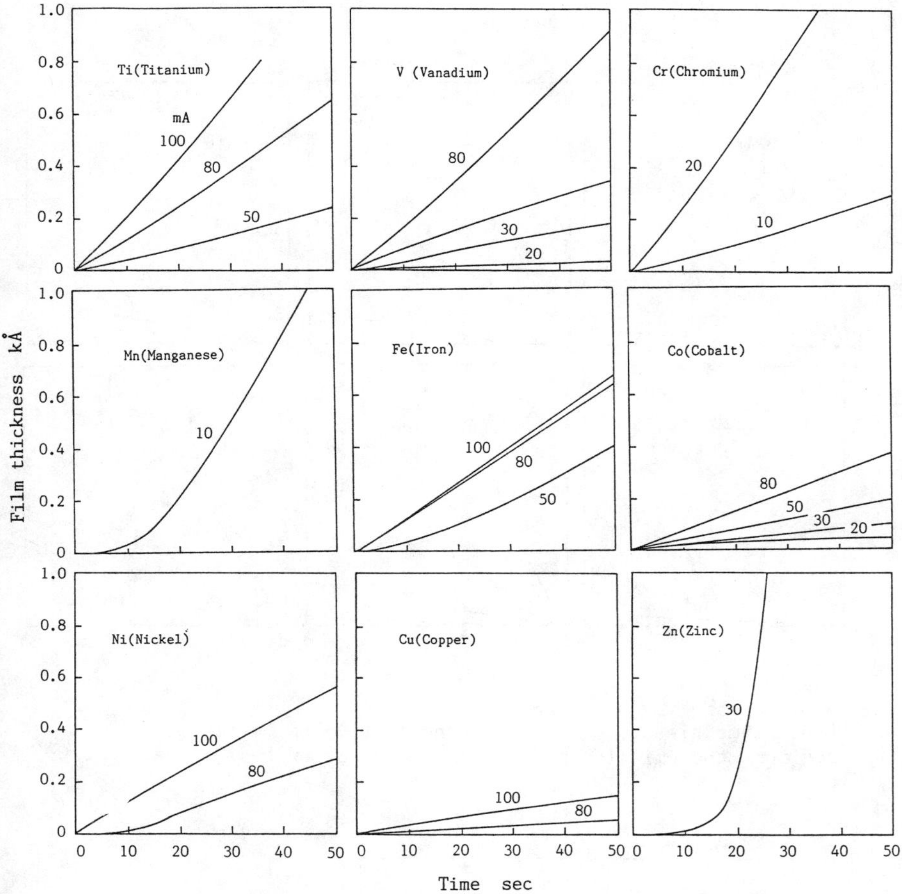

FIGURE 6. Time history of film thickness; (a) Metal elements. (Figures mean the emission current of the electron beam in mA.)

effusively and are deposited onto the surface of the thickness meter to form a thin film. The thickness of the deposited film is measured with the method of measuring the resonance frequency of the transverse elastic waves of 5MHz due to the change in the film thickness under the condition of controlled temperature.

The deposition rate can be calculated by measuring the thickness of the film formed as a function of time, being defined as the time derivative of the film thickness. FIGURE 6 shows the time history of the film thickness for the materials of metal and nonmetal elements. The growth rate of the film thickness depends upon the emission current, that is, the number density of electron which strikes the solid material. As the emission current increases, the film thickness increases in more than linear relation with it. Except for the initial period of deposition, the deposition rates are almost constant. Comparing the results of con-

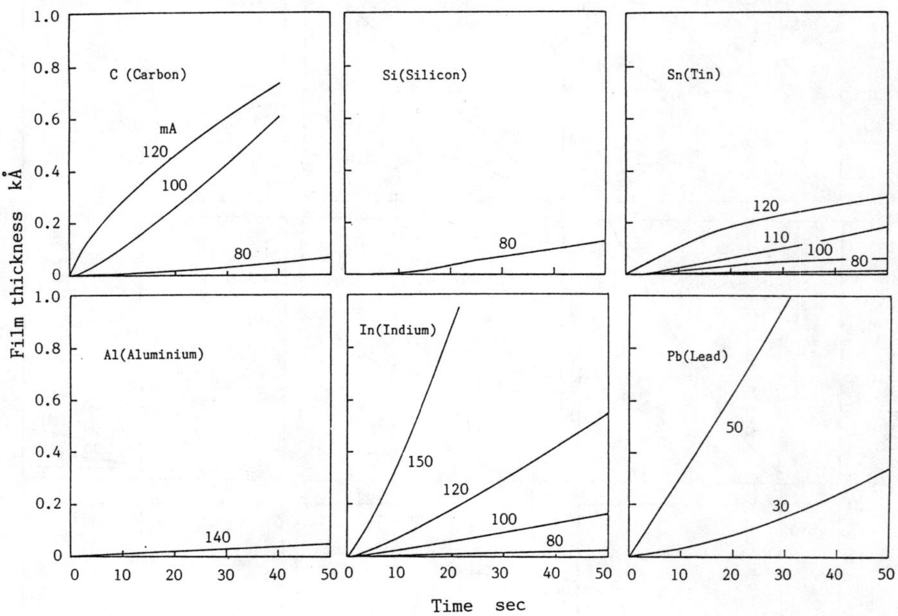

FIGURE 6. Time history of film thickness; (b) Non-metal elements. (Figures mean the emission current of the electron beam in mA.)

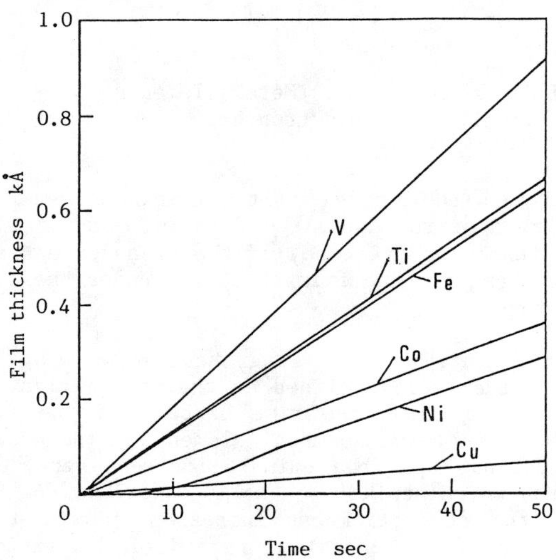

FIGURE 7. Comparison of film growth for the fourth row elements.

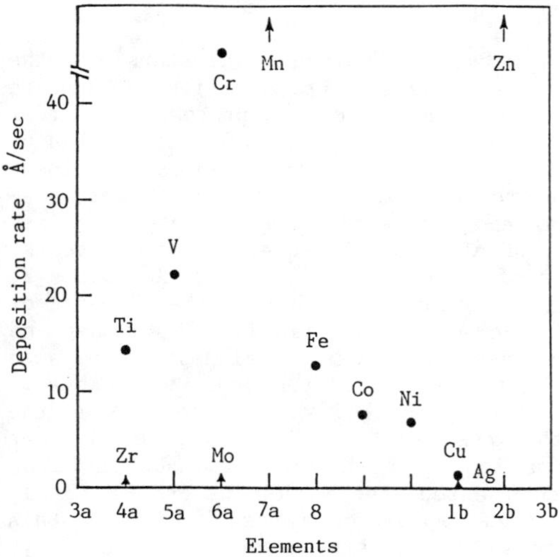

FIGURE 8. Comparison of deposition rates of the fourth row elements.

stant values of the emission current, it is seen from these figures, as expected, that the 2b(Zn) has a large value of the deposition rate and the 5a(V) and 8(Fe, Co, Ni) take small values. But, Mn and Cr have larger values than that of 2b(Zn), and the 1b(Cu, Ag) elements take much smaller rates than those of 5a(V) and 8(Fe, Co). These cannot be expected from the clustering features shown in FIGURE 4. In FIGURES 7 and 8, the film growth and the deposition rate, respectively, are shown for the fourth row elements(V, Ti, Cr, Mn, Fe, Co, Ni, Cu, and Zn).

In the case of nonmetal elements, the general features of the deposition rate are not so comprehensible. The rates are always smaller than those of metal elements except for lead. Carbon and silicon take almost the same rate of deposition larger than that of tin. Although aluminum has the same features of clustering as Zn, it has a very small deposition rate, smaller than Cu and Ag. The elements of small deposition rates such as Cu, Ag, and Al have higher values of the thermal conductivity which may affect the readjusting process of the internal energy of clusters as well as the process of energy exchange between vapor and condensate atoms. Even if they have the same clustering features, metal and non-metal atoms may have different behavior in the deposition process due to the difference of energy exchange.

Although the film formation of these vaporized atoms is related to the processes of vaporization and clustering, it would not be the same for atoms vaporized by laser and electron beam irradiations. In the range of thermal excitation of the solid atoms both give similar features of vaporization and clustering. But, the latter has a much higher energy than the laser beam to vaporize the atoms from the material of solid state, being ionized in some parts. This may also lead to some differences in the features of metal and nonmetal elements concerning the clustering and film deposition of their vapors vaporized by the laser and electron beams.

CONCLUSION

Because of peculiar features of atomic or molecular clusters for the condensation process, the characteristics of size distribution can be expected to be one of the controlling factors of the process. In order to control the condensation process of vapor atoms of metal and non-metal elements, the size distributions of their clusters were experimentally investigated with relation to the film condensation of these elements. Solid materials of metal and non-metal elements were vaporized from the solid state by the irradiation of laser and electron beams. The elements having small values of the cohesive energy such as the 2b group form clusters of widely distributed sizes, whereas the elements of large cohesive energy such as 5a and 8 are hardly evaporated and clustered into large sizes. In cases of non-metal elements, the size distributions of clusters are not always related with the cohesive energy. The deposition rate of the evaporated vapors, defined as the time derivative of the film thickness with respect to time, can be largely related with the formation of both monomer and clusters. Qualitatively, the metal elements associated with clusters of widely distributed sizes are to have high deposition rates of forming thin films. But, for non-metal elements, the deposition rates are not always accounted for by the clustering features.

To obtain more quantitative prediction and control of the condensation process of metal and non-metal atomic vapors, it is necessary first to relate the size distribution of clusters to the condensation rate more quantitatively, and then to find a method of controlling the size distribution of clusters by changing the species of the carrier gas, its density and/or temperature which affect substantially the clustering process.

REFERENCES

1. Dietz,T.G., Duncan,D., Powers,D.E., and Smalley, R.E., "Laser production of supersonic cluster beams," J. Chem. Phys. 74(11), 6511-6512(1981).

2. Powers,D.E. and Smalley,R.E., "Supersonic copper clusters," J. Chem. Phys. 78(6), 2866-2881(1983).

3. Hopkins,J.B., Langridges-Smith,P.R.R., Morse,M.D., and Smalley,R.E., "Supersonic metal cluster beams of refractory metals," J. Chem. Phys. 78(4), 1627-1637(1983).

4. Rohlfing,E.A., Cox,D.M., and Kalder,A., "Production and characterization of supersonic carbon cluster beams," J. Chem. Phys. 81(7), 3322-3330(1984).

5. O'Brien,S.C., Zang,Y.L.Q. and Smalley,R.E.,"Supersonic cluster beams of 3-5 semiconductors: GaxAsy," J. Chem. Phys. 84(7), 4074-4079 (1986).

6. Rohlfing,E.A.,"Optical emission studies of atomic, molecular and particulate carbon produced from a laser vaporization cluster sources," J. Chem. Phys. 89(10), 6103-6112(1988).

Summary of General Discussion of Session on Laser and Electron Beam Processing

Warren H. Giedt, University of California, Davis
Susumu Kotake, University of Tokyo

Papers presented in this Session reviewed experimental and analytical studies of laser and electron beam heating, melting and evaporation, and of laser assisted vapor deposition. Results presented showed that substantial progress has been achieved, and indicated where further effort is needed. General comments and conclusions included:

1) Although the effect of the energy transferred to a material from a laser or electron beam can be analyzed by numerical methods, much more fundamental study must be carried out to understand the mechanism of the energy transfer from a laser beam to the material. This will require resolving the interaction between surface absorption, melting and evaporation, and then beam attenuation and reradiation by the plasma formed. Although additional thermophysical property data would be very helpful, it was pointed out that the application of scaling laws could provide adequate guidance for carrying out studies.

2) Experimental evidence indicates that very intense laser and electron beams produce non-steady or even non-equilibrium conditions. Examples are the discontinuity at the liquid-vapor interface during laser heating, in the formation of clusters during vaporization and in laser chemical vapor deposition. Experimental techniques to identify the characteristics of these conditions must be developed, and then models developed to enable analyses to be conducted.

CRYSTAL GROWTH (1)

Numerical Study of the Czochralski Growth of Oxide Single Crystal

NOBUYUKI IMAISHI
Institute of Advanced Material Study, Kyushu University
6-1 Kasuga-Koen, Kasuga 816, Japan

TAKAO TSUKADA and MITSUNORI HOZAWA
Chemical Research Institute of Non-Aqueous Solutions
Tohoku University, 2-1-1 Katahira, Aoba-ku, Sendai, Japan

YASUNORI OKANO
Institute for Materials Research, Tohoku University
2-1-1 Katahira, Aoba-ku, Sendai, Japan

AKIRA HIRATA
Department of Chemical Engineering, Waseda University
3-4-1 Okubo, Shinjuku-ku, Tokyo, Japan

ABSTRACT

The Czochralski process of oxide single crystal growth is modeled and numerically analysed from a thermal engineer's point of view. By taking the over–all heat balance and using the finite element method, the shape of the grown crystal is calculated under given processing conditions. The inversion of the melt-crystal interface is successfully predicted by the model, if the thermophysical properties are known. Thermal stress distribution in the grown crystal is also calculated. These results can help the detailed crystallographic discussions on defect formation and crystal quality.

INTRODUCTION

Single crystal is one of the most important raw materials in modern acoust–opt–electronics industries of "the High–Tech. Age", since single crystals alone can provide low scattering and high mobility path for electrons, photons and phonons with the desired long range order of atomic configurations. Most of the single crystals are now prepared by melt growth techniques, unless it can not be grown from melt, e.g. quartz crystals are grown only by the hydrothermal synthesis.

Among the various melt growth techniques, the Czochralski (CZ) technique is the most widely utilized for semiconductor and oxide crystals. Large and high quality single crystal of silicon is now one of the most important material, and the silicon crystal engineering has become highly sophisticated, and enjoys its large business market. On the other hand, the business market of oxide single crystals is yet much smaller. The largest market among them may be the SAW filter devices which is found in VTR, TV and mobile telephone sets. The most common SAW materials are thin wafers of single crystal of $LiNbO_3$ or $LiTaO_3$. For efficient manufacturing, large diameter (3 or 4 inches in diameter) wafers are required. Although, in comparison with the single crystals used for optical devices, such as SHG,THG generator and solid state laser, the required quality of the SAW crystal is much lower, pulling of large oxide crystal is still challenging, since oxide is less thermally conductive in solid and melt phases, more viscous in melt phase and much more fragile in crystal phase than silicon. CZ technique of oxides has many serious problems to be solved. But because of its smaller market (?), and the crystallographical difficulties involved, modern engineering analysis of oxide CZ process has achieved less remarkable developement than those in semiconductor CZ processes.

From thermal engineer's view point, oxide CZ process involves more interesting problems. In cases of semiconductors, the Prandtle number is very small (in the order of 10^{-2}), then the convections exert less effect on temperature distribution in melt and crystal. In these cases, the conduction dominated model[1,2], in which the conduction is assumed the only heat transfer mechanism in the melt, is conveniently used to predict the general behaviour of CZ and LEC pullers to the first approximation. Experiences[3,4,5] have shown that the quality of oxide crystals grown by the CZ technique reflects the flow field in the melt. The critical (crystal rotation) Reynolds number, at which the well-known inversion of the melt-crystal interface occurrs, is one of the principal and argent informations for oxide crystal pullers. There are few empirical equations[6,7,8] which correlate the critical Reynolds number with Grashof number, crystal diameter and other opperating parameters. But so far there has been no correlaion which enables us to predict the diameter of the grown crystal, these equations can not predict the criteria of the inversion. Numerical simulation is expected to contribute greatly to this subject by clarifying the details of transport phenomena in the melt and complex, nonlinear interactions between them.

The Prandtle number of oxide melt ranges in 0.1 to 10 or larger. Thus the fluid motion and heat transfer are tightly couppled and the conduction dominated model is no longer justified. The couppled heat transfer and fluid dynamic equations conjugated with phase change, solid phase heat conduction and radiative heat exchange with ambient surfaces must be solved simultaneously in order to know the characteristics of the oxide CZ process,the behaviour of the growing crystal in it and the effects of the factors which determine quality of the grown crystal.

Such attempts appeared at the end of the 80's, and the first simulation result was reported by Tsukada et al[10](1987) on CZ growth of sapphire and silicon crystals. The second was reported by Chan et al.[11](1988) on silicon. Later, Sackinger and Brown[12] (1989) reported their analysis on silicon and Gadolinium Gallium Garnet ($Gd_3Ga_5O_{12}$). Kopetsch[13,14] analysed the melt-crystal interface shape for silicon (1988) and sapphire(1990). Chan et al. and Kopetsch used the finite difference scheme and the crystal diameter was assumed constant, on the other hand Tsukada et al. and Sackinger et al. used the finite element method in order to include the crystal diameter into one of the variables in the analysis.

In this paper, the crystal diameter, melt-crystal interface shape, shape of the melt surface, temperature, velocity and thermal stress fields in a model CZ system for sapphire single crystal are discussed.

MATHEMATICAL MODEL

The essencial features of the mathematical modeling and the numerical solution procedure of the analysis are summerized here. The CZ furnace is modeled as shown in Fig. 1. The following assumptions are introduced.

1. The crucible and the ambient wall temperatures are constant.
2. The system is rotationally symmetric.
3. The crystal is pulled up at a constant and very slow velocity and the system is in a pseudo steady state.
4. The melt flow is laminar.
5. The heat loss from each surface is due only to radiation.
6. The crystal-melt interface is at the melting temperature.
7. The contact angles of the melt against the crystal and the crucible are constant.

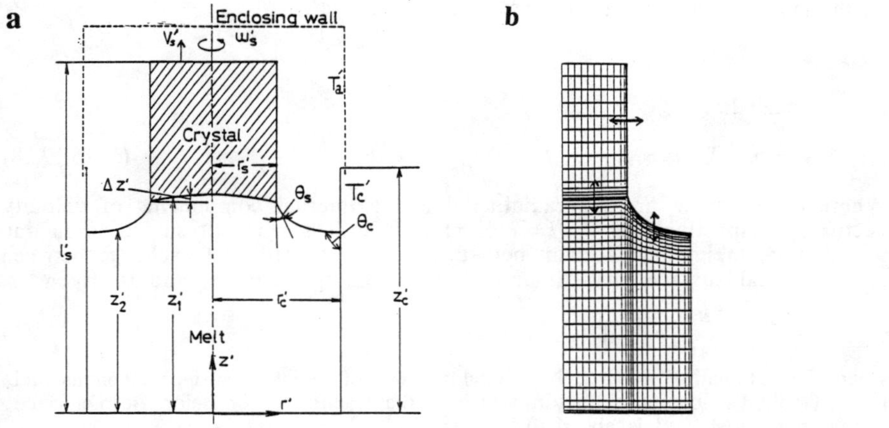

FIGURE 1. Model CZ furnace. a) System coordinate and configuration. b) Discretization of the calculation domain for the finite element method.

The thermophysical properties are constant, except for the density and the surface tension, which are dependent on temperature to allow for the effects of buoyancy and thermocapillary forces.

Under these assumptions, the dimensionless form of the Navier–Stokes equation with the Boussinesq approximation, continuity equation, and the energy equations are described by Eqs.(1) – (4).

$$\boldsymbol{v}_l \cdot \nabla \boldsymbol{v}_l = -\nabla p_l - \nabla \cdot \boldsymbol{\tau}_l + Gr(T_l - 1)\boldsymbol{e}_z \quad \text{where} \quad \boldsymbol{\tau}_l = -(\nabla \boldsymbol{v}_l + \nabla \boldsymbol{v}_l^T) \tag{1}$$

$$\nabla \cdot \boldsymbol{v}_l = 0 \tag{2}$$

$$Pr\ \boldsymbol{v}_l \cdot \nabla T_l = \nabla^2 T_l \tag{3}$$

$$Pe\ \boldsymbol{e}_z \cdot \nabla T_s = \kappa \nabla^2 T_s \tag{4}$$

Where the suffixes l and s stand melt and single crystal, respectively. The boundary conditions are expressed as follows.
At both the side and the bottom wall of the crucible:

$$u_l = v_l = 0,\ w_l = 0,\ T_l = T_c. \tag{5–1,2,3,4}$$

At the melt–gas interface:

$$\boldsymbol{n} \cdot \boldsymbol{v}_l = 0,\ \boldsymbol{\tau}_l\boldsymbol{:nt} = Ma \nabla T_l \cdot \boldsymbol{t},\ \boldsymbol{\tau}_l\boldsymbol{:n}\ \boldsymbol{e}_\theta = 0,\ -\boldsymbol{n}.\nabla T_l = q_{rad.l} \tag{5–5,6,7,8}$$

At the melt–crystal interface:

$$u_l = v_l = 0, \tag{5–9,10}$$

$$w_l = r\ Re_s, \tag{5–11}$$

$$\boldsymbol{n} \cdot \nabla T_l - \kappa\ \boldsymbol{n} \cdot \nabla T_s = Pe\ St\ \boldsymbol{n} \cdot \boldsymbol{e}_z \tag{5–12}$$

$$T_l = T_s = 1 \tag{5–13,14}$$

At the side and the top of the crystal:

$$-\kappa \boldsymbol{n} \cdot \nabla T_s = q_{rad,s} \tag{5-15}$$

At the center line:

$$\boldsymbol{n} \cdot \nabla v_l = \boldsymbol{n} \cdot \nabla T_l = \boldsymbol{n} \cdot \nabla T_s = 0. \tag{5-16,17,18}$$

Where u_l, v_l and w_l are the radial,axial and azimuthal components of velocity vector $\boldsymbol{v}_l$, respectively. $q_{rad,k}$ ($k= l$ or s) is heat loss per unit surface area due to radiation, taking into account both direct and reflected heat exchange between an individual surface element and all surrounding surfaces, and is given as,

$$q_{rad,k} = R_k \left(T_k^4 - \left(\sum_{j=1}^{N} \varepsilon_j A_j T_j^4 G_{jk}\right) / A_k\right) \varepsilon_k \tag{6}$$

where R_k is radiation number, N is total number of radiation surface elements and G_{jk} is Gebhart's absorption factor, which is the fraction of emission from surface A_j that reaches A_k and is absorbed[15].

The melt-gas interface shape z_2 and the crystal radius are obtained by solving the Young-Laplace equation Eq. (7) under the following conditions.

$$2H = Bo\, z_2 + \lambda \tag{7}$$

The boundary conditions are:

$$\theta = \theta_s \quad \text{at } r = r_s \;;\; \theta = \theta_c \quad \text{at } r = 1 \tag{8-1,2}$$

The melt volume is given by Eq. (9).

$$V_m = \int_0^{r_1} z_1\, r\, dr + \int_{r_1}^{1} z_2\, r\, dr \tag{9}$$

The melt-crystal interface shape (z_1) is determined such that Eq. (5-13) is satisfied, i.e. the interface is at the melting point temperature.

The set of equations listed above are solved by use of the Galerkin finite element method. The melt and crystal are discretized by isoparametric elements consisting of a nine-noded quadrilateral as shown in Fig.2, where the total number of elements and nodal points are 719 and 2999, respectively.

In each element, velocity vector **v** and temperature T are approximated with the biquadratic trial function ϕ_i,and the pressure is by the bilinear trial function ψ_k.

$$\mathbf{v}(r,z) = \sum \phi_i \mathbf{v}_i \tag{10}$$
$$T(r,z) = \sum \phi_i T_i \tag{11}$$
$$p(r,z) = \sum \psi_k p_k \tag{12}$$

The shape of the melt-gas interface is expressed by the Hermite cubic trial function.

$$z_2(r) = \sum \chi_j z_{2j} \tag{13}$$

By substituting Eqs. (10) - (13) into Eqs. (1) - (4) and (7), the Galerkin procedure gives the set of non-linear algebraic equations. The equations are solved by use of the Newton-Raphson scheme. The calculations for the temperature and velocity fields, crystal radius and interfacial shapes are repeated until the cut off error is less than a given value.

Once the crystal shape and the temperature distribution in it are determined, axisymmetric thermal stress field is easily calculated under the following assumptions.

1. The material is linearly elastic and isotropic.
2. The body forces applied to the material are zero.
3. The crystal is free from surface traction.

The constituitive equation for linear elastic material in case of axisymmetry is expressed as follows. Hereafter, all variables have their original unit.

$$\sigma = \begin{Bmatrix} \sigma_r \\ \sigma_z \\ \sigma_\theta \\ \tau_{rz} \end{Bmatrix} = \frac{E}{(1+\nu_p)(1-2\nu_p)} \times \begin{bmatrix} 1-\nu_p & \nu_p & \nu_p & 0 \\ \nu_p & 1-\nu_p & \nu_p & 0 \\ \nu_p & \nu_p & 1-\nu_p & 0 \\ 0 & 0 & 0 & \frac{1-2\nu_p}{2} \end{bmatrix} \begin{pmatrix} \varepsilon_r - \varepsilon_{r0} \\ \varepsilon_z - \varepsilon_{z0} \\ \varepsilon_\theta - \varepsilon_{\theta 0} \\ \gamma_{rz} - \gamma_{rz0} \end{pmatrix}$$

$$= \boldsymbol{D} (\boldsymbol{\varepsilon} - \boldsymbol{\varepsilon}_0) \tag{14}$$

where E is Young's modulus and ν_p is Poisson's ratio. ε_0 is the initial strain which corresponds to free expansion or contraction by temperature variations.

$$\boldsymbol{\varepsilon}_0 = \beta_s (T_s' - T'_{s,ref})(1,1,1,0)^T \tag{15}$$

$T_s'(= T_m \cdot T_s)$ is the local temperature.
The strain-displacement equation is given by

$$\boldsymbol{\varepsilon} = \begin{bmatrix} \partial/\partial r, & 0 \\ 0, & \partial/\partial z \\ 1/r, & 0 \\ \partial/\partial z, & \partial/\partial r \end{bmatrix} \begin{bmatrix} d_r \\ d_z \end{bmatrix} = \boldsymbol{Ad} \tag{16}$$

The finite element method is applied to the thermal stress analysis, in which four-node quadrilateral elements are used. In each element, the displacement vector is approximated by

$$\boldsymbol{d} = \boldsymbol{\phi a} \tag{17}$$

where $\boldsymbol{a}$ is the nodal displacement vector. Thus, the strain is described by nodal displacements as

$$\boldsymbol{\varepsilon} = \boldsymbol{A \phi a} = \boldsymbol{Ba} \tag{18}$$

where $\boldsymbol{B}$ is the matrix derived from the derivatives of shape functions. Under the assumptions of (2) and (3), the virtual work principle gives the following equations.

$$\int \delta \boldsymbol{\varepsilon}^T \boldsymbol{\sigma} \, dV = 0 \tag{19}$$

Substituting Eqs. (14) and (18) into (19), the set of algebraic equations is given by the next equation.

$$\int \boldsymbol{B}^T \boldsymbol{DB} dV \boldsymbol{a} = \int \boldsymbol{B}^T \boldsymbol{D} \boldsymbol{\varepsilon}_0 dV \tag{20}$$

TABLE 1. Thermophysical properties[16,17] and processing parameters used for simulation.

1. Thermophysical properties			
Melting point		2316.	K
Heat of solidification		1046.	J/g
Thermal conductivity			
melt	k_l	0.1	W/cm/K
crystal	k_s	0.1	W/cm/K
Density			
melt	ρ_l	3.05	g/cm^3
crystal	ρ_s	4.0	g/cm^3
Kinematic viscosity		0.00892	cm^2/s
Heat capacity			
melt	Cp_l	1.26	J/g/K
crystal	Cp_s	1.26	J/g/K
Emissivity	ε		
melt		0.33	
crystal		0.33	
crucible temperature		0.3	
ambient wall		0.3	
Thermal expansion coefficient			
melt	β_l (volumetric)	3.0×10^{-5}	K^{-1}
crystal	β_s (linear)	$5.927E-6+4.284E-9\times Ts'-6.621E-13\times Ts'^2$	K^{-1}
Surface tension	γ	700.	dyn/cm
$(\partial\gamma/\partial T')$		0.06	dyn/cm/K
Contact angle			
against crystal	θ_s	17.	degree
against crucible	θ_c	90.	degree
Prandtl number	Pr	0.34	
Poisson's ratio	ν_p	0.25	
Young's modulus	E	$-60.94(Ts'-273.15)+3.62E5$	MPa
2. Processing parameters			
crucible rudius	r'_c	1.0	cm
crucible height	z'_c	1.9	cm
crystal height	z'_s	2.7	cm
Melt volume	V'_m	4.7	cm^3
Ambient wall area	A'_a	10.0	cm^2
Ambient wall temperature	T'_a	1737.	K
Crucible temperature	T'_c	2473.	K
Crystal rotation rate	n^c	0 ~ 68.	rpm
Pull rate	v'_s	0.1	mm/min

Eq. (20) is solved for nodal displacements under the boundary condition that the radial component of displacements is zero at the z axis. Finally, the thermal stress field can be obtained by Eq. (21).

$$\sigma = D\,(Ba - \varepsilon_0) \tag{21}$$

In this work, the final expression of the thermal stress distribution is shown in the form of the Tresca stress (the maximum shear stress) defined as

$$\tau_{max} = (\sigma_1 - \sigma_3)/2 \tag{22}$$

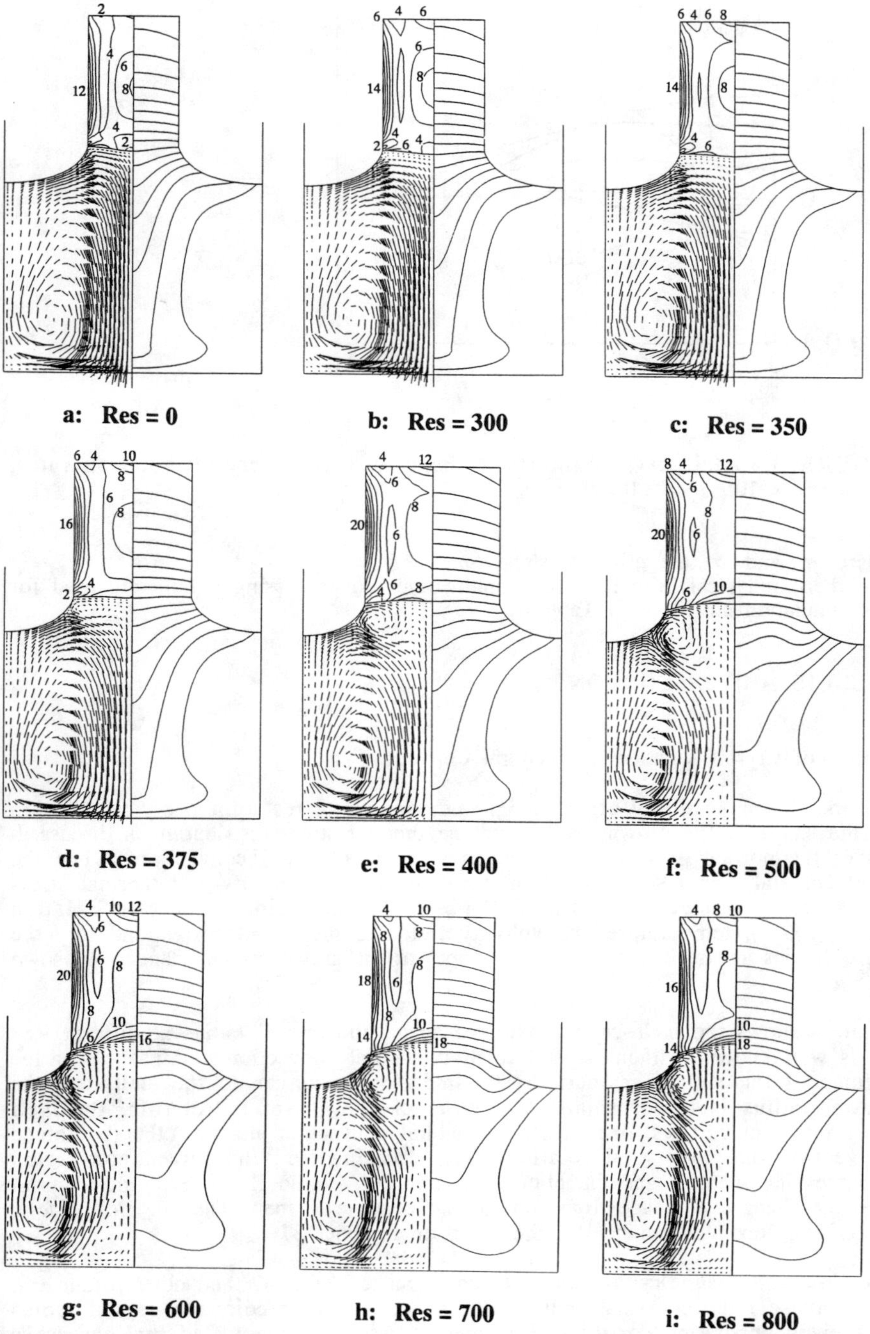

FIGURE 2. Effect of crystal rotation on velocity, temperature, and thermal stress fields. Thermocapillary effect is ignored, $Gr = 8.56\times 10^5$, $T_c = 1.068$. $\Delta T = 0.01$, $\Delta \tau_{max} = 2$ MPa

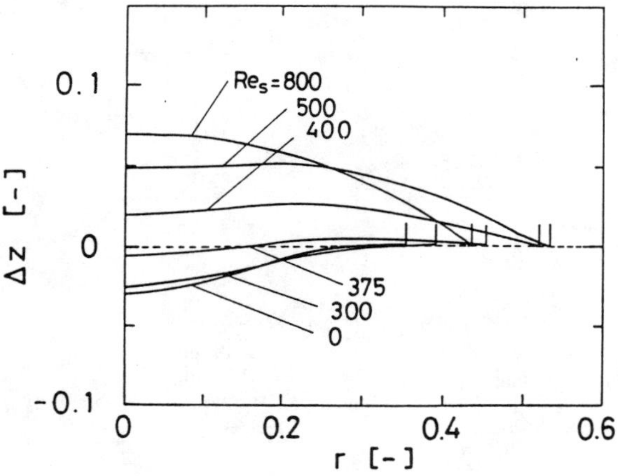

FIGURE 3. Melt–crystal interface shape at various crystal rotation rates. $Gr = 8.56 \times 10^5$, $T_c = 1.068$.

where σ_1 and σ_3 are principal stress ($\sigma_1 > \sigma_2 > \sigma_3$)
The thermophysical properties of sapphire and the processing parameters used for this simulation are listed in Table 1.

RESULTS AND DISCUSSION

Effects of Crystal Rotation and Natural Convection

At first, thermocapillary effect is ignored, and the effect of the crystal rotation on the shape of the grown crystal and the thermal stress distribution is discussed. Under the normal gravity field, buoyancy triggers natural convection. When the single crystal is not set in rotation, the temperature, velocity and thermal stress fields are calculated as shown in Fig.2–**a**. As increasing the crystal rotation speed (Re_s), temperature and velocity fields in melt, and correspondingly the shape (radius and melt–crystal interfce shape) of the grown crystal changes as shown in Fig.2.

Figure.3 shows the melt–crystal interface shape and crystal radius at differnt Re_s. As is well known, without crystal rotation, natural convection dominates over the volume of the melt. The melt–crystal interface is convex to the melt, and the crystal radius is rather small. When the crystal is set in rotation, the melt splits into two vortices, one caused by natural convection and the other by forced convection due to crystal rotation. As increasing Re_s, the forced convection increases its intensity and the circulation area. At around Re_s = 375, melt–crystal interface changes its shape from convex into concave to melt. Then, in this system, the critical Reynolds number for the inversion is about 375.

There are some empirical equations which correlates Re_s to Gr and other parameters, all of these require crystal radius. But no one could predict the crystal radius at a given opperating condition. The present result was compared to Kobayashi's equation[9], which is rewritten in our notation,

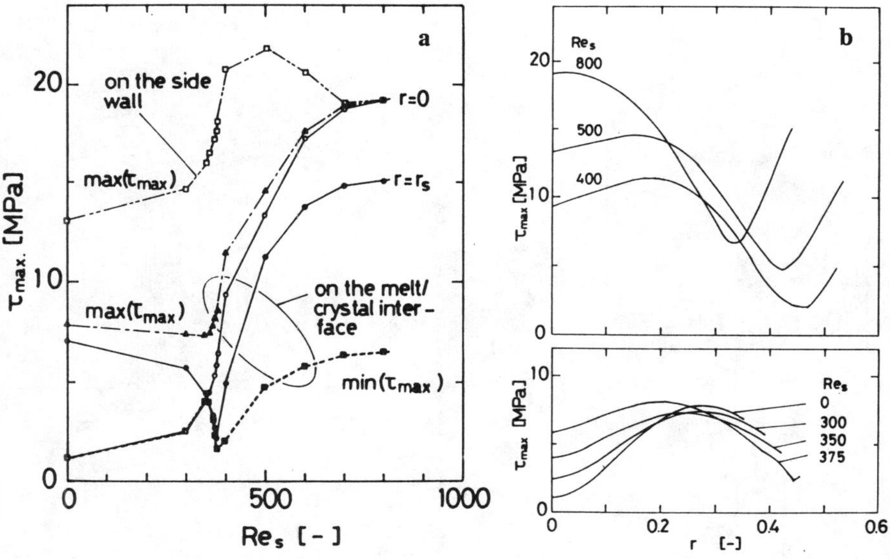

a Effect of Re_s on the thermal stress level in the crystal.

b Distributions on the growing interface.

FIGURE 4. Thermal stress in growing crystals.

$$Re_{s,c} = 2.51 Gr^{0.4} r_s^{-0.8} (T_c - 1)^{0.4} \qquad (23)$$

Using the calculated values of r_s, T_c and Gr, Eq.(23) gives $Re_{s,c}$ = 380. This reveals that the simulation results are reasonable and reproduces the experimental results which Kobayashi's equation based on.

Fig.4-**a** shows the effect of Re_s on the maximum and minimum values of thermal stress in the growing crystal. Thermal stress in the crystal increases as Re_s exceeds $Re_{s,c}$, corresponding to the change of crystal radius and the deformation of the interface. As seen in Fig.2, maximum thermal stress appears on the crystal wall slightly above the brim of the crucible, where the radial heat flux shows its maximum. The thermal stress distribution along the melt-crystal interface shows a W-shape as shown in Fig.4–**b**. The maximum value of the thermal stress corresponds to the breakage, and the stress on the interface might be responsible to the crystal quality,through yet unknown way. Thermal stress in the crystal rapidly increases as Re_s exceeds $Re_{s,c}$,corresponding to the change of crystal radius.

Thermocapillary Effect

In addition to the natural convection (buoyancy effect), the thermocapillary flow (Marangoni effect) should be also taken into account. In the present simulation, $(\partial \gamma / \partial T_1')$ is assumed 1.05×10^{-3} dyn/cm/K, which is about 1/57 of the reported experimental value. The crystal shape, temperature, velocity and thermal stress distributions are shown in Fig. 5, the parameters, other than Ma, are the same as those in Fig. 2.

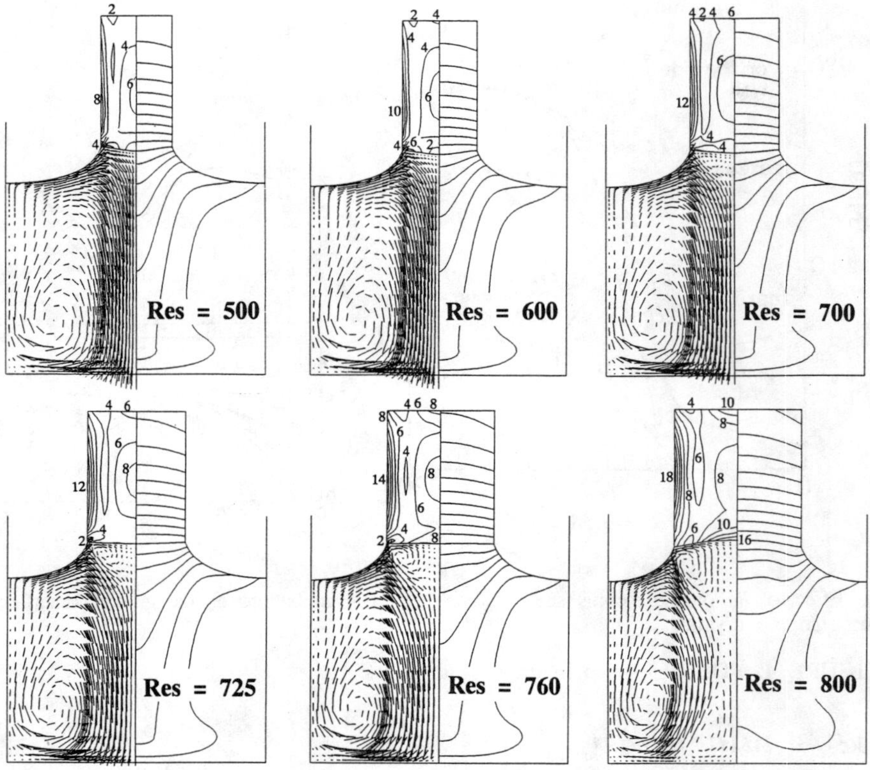

FIGURE 5. Effect of crystal rotation on velocity, temperature and thermal stress fields when the thermocapillary effect is taken into account.
$Gr = 8.56\times10^5, T_c = 1.068$ and $Ma = 1.0\times10^4$. $\Delta T = 0.01$, $\Delta\tau_{max} = 2$ MPa

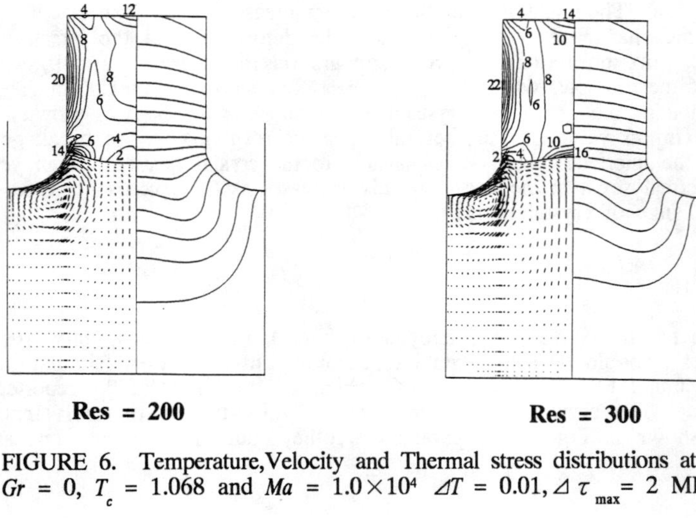

FIGURE 6. Temperature,Velocity and Thermal stress distributions at zero gravity limit.
$Gr = 0$, $T_c = 1.068$ and $Ma = 1.0\times10^4$ $\Delta T = 0.01$, $\Delta\tau_{max} = 2$ MPa

Thermocapillary effect strengthens the natural convection, and the critical Reynolds number shifts up to about 760, and at this Reynolds number, the forced convection vortex is clearly seen under the crystal although the melt–crystal interface is almost flat.

The roll of the thermocapillary flow in CZ melt convection has been argued for long years[18,19,20], however, there still remains some doubt on its enrollment. It is necessary to set up an experiment on the critical Reynolds number, with accurate temperature measurements, and the result should be examined with the corresponding simulation to elucidate the thermocapillary effect in the melt. Our research group have tried some experiments on sapphire CZ pulling, but to date, uncertainties in measurements do not allow qualitative discussions.

Effect of Gravity Level.

It is not plausible to launch a CZ puller to space, however, we made some additional calculations under reduced gravity levels. Since the details of the results were presented elsewhere[19], then here we show the results at the zero-gravity limit in Fig.6.

Even at zero gravity limit, the thermocapillary effect induces melt convection. The convection induced by thermocapillary effect circulates in the same direction as that of the natural convection in Fig. 2, there again appears the critical Reynolds number at which the interface shape becomes flat.

CONCLUSION

A simple mathematical model of CZ furnace was proposed and the model calculations were done for sapphire single crystal pulling. The model can express the crystal radius and melt–ctystal interface shape, temperature, velocity, thermal stress fields under a given operation condition. Since oxide melt has large Prandtl number, temperature field and crystal shapes are strongly dependent on the melt convection.

The model predicts thermocapillary effect takes important rolls in melt convection in space and on the earth.

The model predicts the critical Reynolds number at which the interface inversion takes place.

The thermal stress field in the crystal is affected strongly by the shape of the melt–crystal interface and crystal radius. The maximum shear stress on the melt–crystal interface increases at high crystal rotation rate.

NOMENCLATURE

$\boldsymbol{A}$	=matrix defined by eq.(16)	[m^{-1}]
A_k'	=area of surface element	[m^2]
A_k	$=A_k'/2\pi r_c'^2$	[–]
$\boldsymbol{a}$	=nodal displacement vector	[m]
$\boldsymbol{B}$	=strain–nodal displacement matrix	[m^{-1}]
Bo	=Bond number $(=\rho_l r_c'^2 g/\gamma)$	[–]
C_p	=heat capacity	[$J\ kg^{-1}\ K^{-1}$]
$\boldsymbol{D}$	=elastic stress–strain matrix	[Pa]
$\boldsymbol{d}$	=displacement vector	[m]

E	=Young's modulus	[Pa]
$\mathbf{e}$	=unit vector	[−]
G_{jk}	=Gebhart's absorption factor	[−]
Gr	=Grashof number(=$r_c'^3 \beta_l T_m' g/\nu_l^2$)	[−]
g	=graviational acceleration	[m s^{-2}]
H'	=mean curvature	[m^{-1}]
H	=$r_c' H'$	[−]
ΔH_s	=latent heat of solidification	[J kg^{-1}]
k	=thermal conductivity	[W m^{-1} K^{-1}]
l_s'	=crystal length	[m]
l_s	=l_s'/r_c'	[−]
Ma	=Marangoni number(=$(-\mathrm{d}\gamma/\mathrm{d}T')r_c' T_m'/\rho_l \nu_l^2$)	[−]
$\mathbf{n}$	=outward-pointing normal at the boundary	[−]
n'	=rotation rate	[s^{-1}]
Pe	=Peclet number(=$\rho_s C_{ps} V_s' r_c'/k_s$)	[−]
Pr	=Prandtl number(=$\mu C_{pl}/k_l$)	[−]
p'	=pressure	[Pa]
p_G'	=pressure in gas phase	[Pa]
p_0'	=pressure at reference position	[Pa]
p	=$(p'-p_0'-\rho_l g z')r_c'^2/\rho_l \nu_l^2$	[−]
q_{rad}'	=heat flux due to radiation	[W m^{-2}]
q_{rad}	=$q_{rad}' r_c'/k' T_m'^{-1}$	[−]
R_k	=radiation number(=$\varepsilon_k \sigma r_c' T_m'^3/k_k$)	[−]
Re	=Reynolds number(=$r_c'^2 \omega'/\nu_l$)	[−]
r'	=radial distance in cylindrical coordinates	[m]
r_c'	=crucible radius	[m]
r_s'	=crystal radius	[m]
r, r_s	=$r'/r_c', r_s'/r_c'$	[−]
St	=Stefan number(=$\Delta H_s/C_{ps} T_m'$)	[−]
T'	=temperature	[K]
T_c'	=crucible temperature	[K]
T_m'	=melting-point temperature	[K]
T_a'	=enclosing-wall temperature	[K]
T, T_c, T_a	=$T'/T_m', T_c'/T_m', T_a'/T_m'$	[−]
ΔT	=interval of isotherms	[−]
$\mathbf{t}$	=tangential unit vector	[−]
u'	=radial component of velocity vectors	[m s^{-1}]
V_m'	=melt volume	[m^3]
V_m	=$V_m'/2\pi r_c'^3$	[−]
V_s'	=crystal pull rate	[m s^{-1}]
$\mathbf{v}'$	=velocity vectors	[m s^{-1}]
v'	=axial component of velocity vectors	[m s^{-1}]
w'	=azimuthal component of velocity vectors	[m s^{-1}]
$u, \mathbf{v}, v, w$	=$u'r_c'/\nu_l, \mathbf{v}'r_c'/\nu_l, v'r_c'/\nu_l, w'r_c'/\nu_l$	[−]
z'	=axial distance in cylindrical coordinates	[m]
z_c'	=crucible hight	[m]
z, z_c	=$z'/r_c', z_c'/r_c'$	[−]
z_1'	=distance between crucible bottom and melt/crystal interface	[m]
z_2'	=distance between crucible bottom and melt/gas interface	[m]
z_1, z_2	=$z_1'/r_c', z_2'/r_c'$	[−]
$\Delta z'$	=$z_1' - z_1'^{(r'=r_s')}$	[m]
Δz	=$\Delta z'/r_c'$	[−]
β_l	=volumetric thermal expansion coefficient	[K^{-1}]
β_s	=linear thermal expansion coefficient	[K^{-1}]

γ	=surface tension	[N m^{-1}]
ε	=emissivity	[–]
$\boldsymbol{\varepsilon}$	=strain	[–]
$\boldsymbol{\varepsilon}_0$	=initial strain	[–]
θ^0_s, θ_c	=contact angle	[deg]
κ	$=k_s/k_l$	[–]
λ	$=(p_0-p_G')r_c'/\gamma$	[–]
μ	=viscosity	[Pa s]
ν	=kinematic viscosity	[m^2 s^{-1}]
ν_p	=Poisson's ratio	[–]
ρ	=density	[kg m^{-3}]
σ	=Stefan–Boltzman constant	[W m^{-2} K^{-4}]
$\boldsymbol{\sigma}'$	=stress	[Pa]
$\boldsymbol{\tau}'$	=stress tensor	[Pa]
$\boldsymbol{\tau}$	$=\boldsymbol{\tau}'r_c'^2/\rho_1\nu_1^2$	[–]
τ'_{ms}	=maximum shear stress (Tresca stress)	[Pa]
ω	=angular velocity($=2\pi$ n')	[rad s^{-1}]
$\phi, \psi, \chi, \boldsymbol{\phi}$	=trial function	[–]

SUBSCRIPTS

l	=melt
s	=crystal
r	=radial
z	=z directional
θ	=azimuthal

LITERATURE CITED

1. Derby, J. J. and Brown, R. A., *Thermal-capillary analysis of Czochralski and liquid encapsulated Czochralski crystal growth (I .Simulation)*,J.Crystal Growth, vol. 74, pp. 605–624, 1986

2. Tsukada, T., Imaishi, N., Hozawa, M., and Fujinawa, K., *Effenct of a radiation shield on silicon CZ growth*, J. Chem. Eng. Japan, vol. 20, no. 2, pp. 146–151, 1987

3. Cockayne, B., and Gates, M. P., *Growth striations in vertically pulled oxide and fluoride single crystals*, J. Mater. Sci., vol. 2, pp. 118–123, 1967

4. Cockayne, B., Chesswas, M., and Gasson, D. B., *Facetting and optical perfection in Czochralski grown garnets and ruby*, J. Mater. Sci., vol. 4, pp. 450–454, 1969

5. Miller, D. C., Valentino, A. J., and Shick, L. K., *The effect of melt flow phenomena on the perfection of Czochralski grown gadolinium gallium garnet*, J. Crystal Growth, vol. 44, pp. 121–134, 1978

6. Carruthers, J. R., *Flow transitions and interface shapes in the Czochralski growth of oxide crystals*, J. Crystal Growth, vol. 36, pp. 212–214, 1976

7. Brice, J. C., Bruton, T. M., Hill, O.F., and Whiffin, P. A. C., *The Czochralski growth of $Bi_{12}SiO_{20}$ crystals*, J. Crystal Growth, vol. 24/25, pp. 429–431, 1975

8. Kobayashi, N., *Effect of fluid flow on the formation of gas bubbles in oxide crystals grown by the Czochralski method,* J. Crystal Growth, vol.54, pp. 414–416,1981

9. Nikolov, V., Iliev, K., and Peshev, P., *Relationship between the hydrody-*

namics in the melt and the shape of the crystal melt interface during Czochralski growth of oxide single crystals.
II.Determination of the critical rotation rate from physical simulation data on growth in the presence of simultaneous free and forced convections;comparison with experimental data on crystal growth, J. Crystal Growth, vol.89, pp. 324–330,1988

10. Tsukada, T., Imaishi, N., and Hozawa, M.,*Theoretical study of the flow and temperature fields in CZ single crystal growth,* J. Chem. Eng. Japan, vol. 21, no. 2, pp. 184–191, 1988

11. Chan, Y. T., Giebeling, H. J., and Grubin, H. L., *Numerical simulations of Czochralski silicon growth,* J. Appl. Phys., vol. 64, no. 3, pp.1425–1439, 1988

12. Sackinger, P. A., Brown, R. A., and Derby, J. J.,*A finite element method for analysis of fluid flow, heat transfer and free interfaces in Czochralski crystal growth,* Int. J. Numer. Method in Fluids, vol. 9, pp. 453–492, 1989

13. Kopetsch, H., *A numerical method for the time-dependent Stefan problem in Czochralski crystal growth,* J.. Crystal Growth, vol. 88, pp. 71– 86, 1988

14. Kopetsch, H., *Numerical simulation of the interface inversion in Czochralski growth of oxide crystals,* J. Crystal Growth, vol. 102, pp. 505–528, 1990

15. Siegel, R., and Howell, J. R., Thermal Radiation Heat Transfer, p. 247, McGraw-Hill, New York, 1972

16. Langlois, W. E., *Digital simulation of Czochralski bulk flow in microgravity,* J. Crystal Growth, vol. 48, pp. 25–28, 1980

17. Kalejs, J. P., Ettouney, H. M., and Brown, R. A., *Comparison of growth characteristics of saphire and silicon ribbon produced by EFG,* J. Crystal Growth, vol. 65, pp. 316–323, 1983

18. Chang, E. C., and Wilcox, W. R., *Analysis of surface tension driven flow in a floating zone melting.*,Int.J. Heat Mass Transfer, vol. 19, pp.355–366, 1976

19. Balasubramaniam, R., and Ostrach, S., *Transport phenomena near the interface of a Czochralski-grown crystal,* J. Crystal Growth, vol. 88, pp. 263–281, 1988

20. Okano, Y., Tachibana, M., Hatano, A., and Hirata, A., *Control of crystal-melt interface shape during Czochralski growth of oxide single crystals*, J. Chem. Eng. Japan, vol. 22, no. 4, pp.389–394, 1989

21. Okano, Y., Fukuda, T., Hirata,A., Takano, N., Tsukada, T., Hozawa, M., and Imaishi, N., *Numerical study on Czochralski growth of oxide single crystals,* forthcoming in J. Crystal Growth

Perspectives on Integrated Modeling of Transport Processes in Semiconductor Crystal Growth

ROBERT A. BROWN
Department of Chemical Engineering and
Materials Processing Center
Massachusetts Institute of Technology
Cambridge, Massachusetts 02139, USA

1. Introduction

Today a myriad of semiconductor and metal alloy crystals are produced by solidification from the melt by variations of the age old idea of directional colidification. There is little need to motivate the desire to understand the interactions of heat and mass transfer during solidification with the crystallographic, morphological and compositional details of the crystals. It is well established that the mechanical and electrical properties of bulk metals and semiconductors produced this way are tightly coupled to the processing conditions. Indeed, this coupling has launched the enormous interest in understanding the physics and chemistry of material processing in a wide range of solidification systems. However, there have been two barriers to the development of a complete understanding of this coupling in solidification processing. First, the hostile thermal and chemical environments present in these systems destroys traditional sensors for temperature and composition that are placed in the active zones of crystal growth furnaces. Moreover, the sensitivity of properly shielded sensors is reduced to the point that the meaning of the measurements are unclear. This experimental opaqueness has reduced researchers to the position of inferring the mechanisms for crystal formation from post mortem analysis or characterization of the final material. It is in this arena that modelling has played such a profound role in understanding solidification systems.

Modelling provides the connections and mechanistic interpretations that linkthe final material properties to processing variables, such as the magnitude of the temperature gradient and the furnace design. The introduction of modelling to understand bulk crystal growth is relatively new, but is leading the way in the understanding of materials processing. The ideas for analyzing heat and solute transport in directional solidification started in the early 1950's in Bruce Chalmer's laboratory at the University of Toronto. As the utility spread of the new simple, but incisive theoretical results and the elegant experiments that verified them, solidification moved from being a totally empirical technology, to an engineering science with a distinct theoretical basis for description of many complex experimental observations. The advancement of the field was fantastic. Simple theories emerged to explain solute segregation caused by convection, the onset of morphological structure formation, and the coarsening of dendritic structures, to mention just a few of the developments. These ideas lead to new concepts, such as the effective segregation coefficient, the diffusion layer thickness, constitutional supercooling, morphological stability theory, and cellular and secondary dendritic spacing. The state of these theories and the experimental basis for their interpretation was elegantly captured in Fleming's text in 1970 (1) and in the more recent review by Kurz (2).

Although these simple theories are the cornerstone for the development of solidification science, most stop short of being quantitative predictions for linking processing conditions to the final material properties. This shortcoming is due to the many simplifying assumptions that must be made so that closed-form solutions can be found and because of the tremendous complexity in traditional crystal growth systems. Large-scale numerical simulation has the potential to remove many of these limitations by incorporating a *variety* of the physics and chemistry important to making quantitative predictions.

Even with the most powerful supercomputers and the most efficient algorithms, present numerical simulations cannot capture *all* the processes important in an industrial scale solidification system. The length and time scales represented in these systems are too great to be captured by a single model; however, models that include a wide range of physics are feasible and are being applied to analysis of several aspects of crystal growth processes.

This wide range of length and time scales is demonstrated here by considering the Czochralski process for the growth of large diameter silicon crystals that become the substrate material for modern microelectronic devices. A schematic diagram of the length and time scales important in the Czochralski (CZ) process are shown in Figure 1. Scales range in time from microseconds to thousands of seconds and in space from microns to meters. The physics and chemistry needed to model processes on these different length scales is reviewed in the following sections.

2. Transport Processes

The batchwise nature of the CZ process and the physical size of the system dictate the modelling of macroscopic heat and mass transfer on a nearly meter length scale and on a time scale of 10^3-10^4 seconds. We refer to this scale of modelling as the *Integrated System*, named after the simulations in references (3-6). Analyses that focus on this largest length scale include those aimed at understanding heat transport through the CZ system and the resulting effects on the thermoelastic stresses in the crystal and on the mode of convection in the melt.

The shortest length and time scales in the system are set differently in the melt and crystal. In the melt, convection driven by buoyancy and surface tension forces and by crystal and crucible rotation leads to fine ($O(10^{-4}$ m)) viscous and diffusive boundary layers for momentum and species transport (7). Moreover, convection in these systems is typically three-dimensional and chaotic in time, with time scales for fluctuation on the order of seconds. The contrast between this *Hydrodynamic Model* of transport in the melt and the *Integrated System Model* shows the already tremendous difficulty with incorporating the 10^4 variation in length and time scales into any simulation. The only simulations to attempt this analysis either simplify the heat transfer system and the solidification model (8), assume that the heat transfer and convection are axisymmetric (4,5), or do both (9).

Detailed simulations of heat and species transport can give quantitative predictions of the temperature field everywhere in the system, the shape of the melt/crystal interface, and the distribution of dopants and impurities in the grown crystal. We have demonstrated the ability of calculations to quantitatively predict properties of the final crystal in analyses of convection and dopant segregation in small-scale directional solidification of a dilute alloy (10), in the simulation of the transient processes in seeding and growth of gallium arsenide by the liquid encapsulated Czochralski method (11) and in the prediction of the temperature field throughout a large-scale CZ system for growth of silicon (5). Sample results from several of these studies are described below.

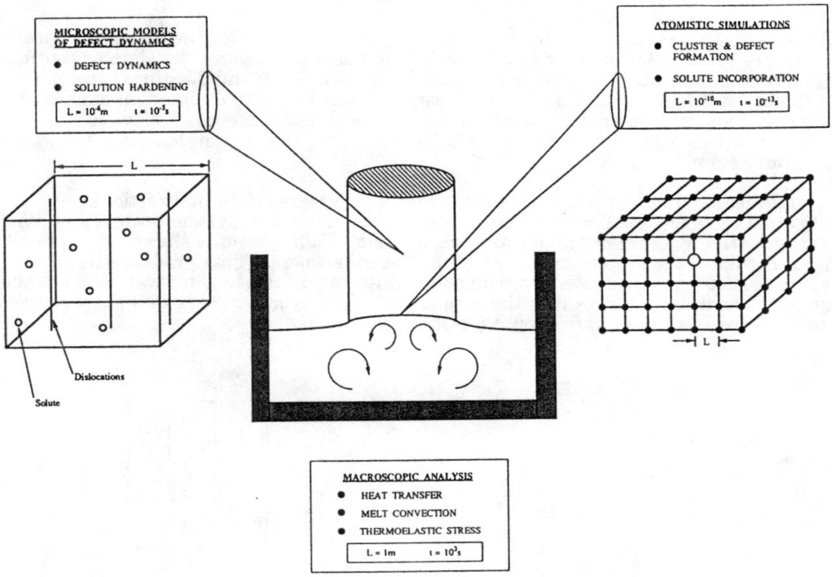

Figure 1. Schematic description of the connections between microscopic and macroscopic modelling of transport processes and defect formation in crystal growth by the Czochralski method. The length and time scales for analysis of each process are shown.

Each of these calculations are performed using variations of a common numerical methodology, the finite-element/Newton method, that has proved to be very powerful for solving the free- and moving-boundary problems that arise in modelling solidification systems and in other materials processing problems (13-16). The methods are based on rigorous Galerkin finite element discretizations of each field variable, e.g. temperature, velocity and species composition, and explicit calculation of the shapes of unknown interface shapes, e.g. the melt/crystal, melt/gas and crystal shapes in the CZ system, as boundaries in the finite element mesh. These discretizations lead to accurate approximations of each variable with the flexibility of applying local mesh refinement to improve the resolution of boundary and internal layers. The finite element formulations lead to large sets of nonlinear algebraic equations for steady-state models and systems of differential-algebraic equations for time-dependent descriptions. The robustness of the numerical method comes from the use of Newton's method for solution of the nonlinear algebraic equations and of fully implicit integration techniques for ordinary differential systems.

2.1 Analysis of Small-Scale Directional Solidification

Small-scale directional solidification experiments have been used for the last three decades as laboratory systems for growing new materials and as a test-bed for the simple theories for convection and solute segregation advanced by Burton, Prim and Slichter (17). Nevertheless, a direct comparison has not been attempted between numerical simulation of such a system, experimental data and correlation with the concepts of effective segregation coefficient and diffusion layer thickness. This comparison has been the goal of some of our recent research.

Our simulations of heat transfer, convection and solute segregation in directional solidification have focussed on the small-scale vertical Bridgman system designed by Wang (18) for growth of gallium-doped germanium crystals. This system is shown schematically in Fig. 2; details of the system are in ref. (10). The experimental data produced from this system is particulary well suited for comparison with calculations. Care was taken in the design of the furnace to insure that the temperature field everywhere is well characterized and so that steady-state solidification conditions are obtained.

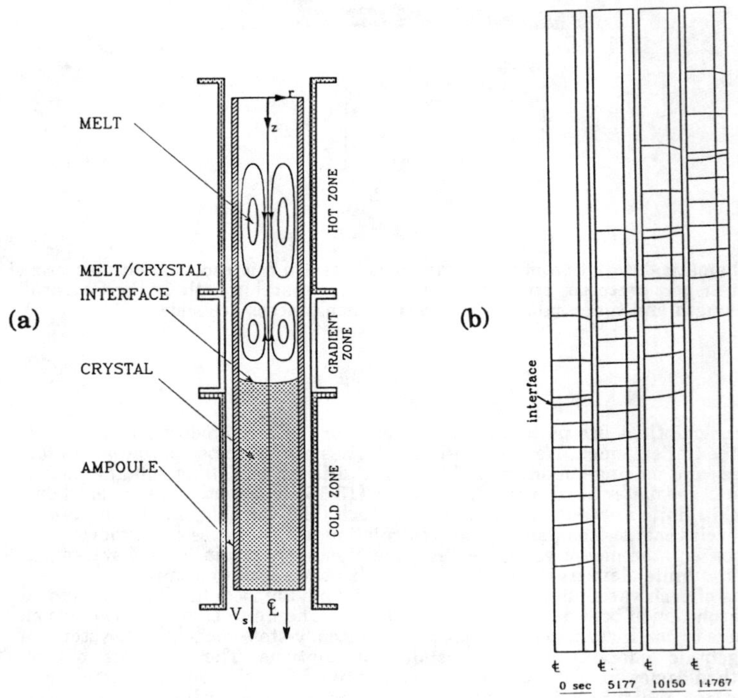

Figure 2. (a) Schematic of vertical Bridgman system used by Wang for growth of gallium-doped germanium (GaGe). (b) Evolution of temperature field with solidification in the melt for GaGe growth; ampoule translation rate is 4 μm/s.

The calculations described here are for a fully-transient description of the evolution of the system as a function of the time of solidification. Other results for analysis of a quasi-steady-state model of the vertical Bridgman system are described in the references given in (7,10). For the small-scale system, the temperature field in the melt is almost independent of convective transport and is shown in Fig. 2 as a function of time during solidification. The evolution of the melt/crystal interface is marked by the heavy curve. The shape of the interface is self-similar throughout most of the crystal growth process; this is a result of the careful system design, which leads to an almost constant temperature field near the melt/crystal interface. As demonstrated in (12), the microscopic growth rate and the interface deflection are independent of time for a significant portion of the growth process. The axial temperature gradient is always positive in the system and hence stabilizing for buoyancy-driven convection. However, significant radial temperature gradients exist and drive convection.

The flow patterns that evolve during the solidification process are shown in Fig. 3 for four different times; the pull rate of the ampoule is 4 μm/s. The reasons for the appearance of the two axially stacked toriodal flow cells is easily understood from the radial temperature gradients. Near the melt/crystal interface, the differences in the thermal conductivities of the melt, crystal and ampoule cause the melt to be colder at the wall than at the axis of the ampoule. The density differences caused by this radial gradient drives a convection cell that moves melt down at the wall and up along the axis of the ampoule. The radial temperature gradients are reversed at the junction between the adiabatic and hot zones of the vertical Bridgman system; higher temperatures are present at the wall and melt moves upward along the ampoule. This change in the radial gradient with increasing axial position causes the presence of the two flow cells *and* a region of melt between them where convection is weak. This flow structure has a dramatic effect on dopant segregation in the melt and incorporation in the crystal.

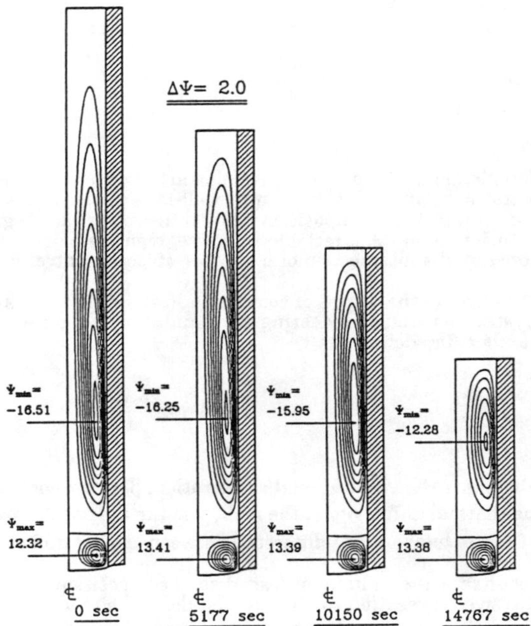

Figure 3. Evolution of melt convection with solidification in the melt for GaGe growth in vertical Bridgman crystal growth system; ampoule translation rate is 4 μm/s.

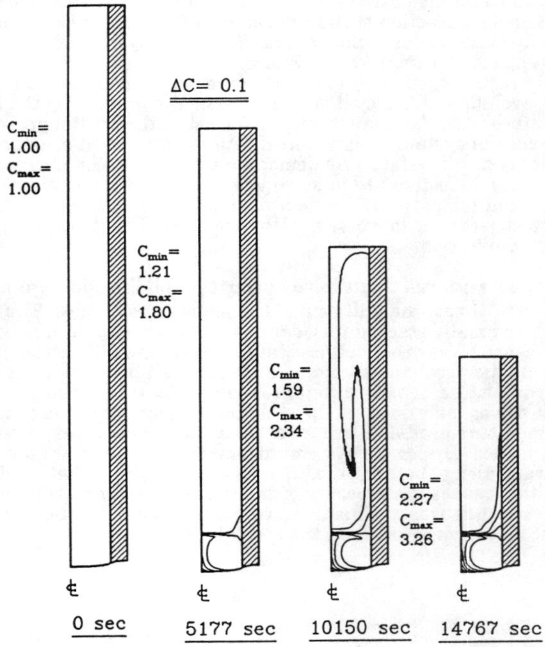

Figure 4. Sample gallium concentration fields for growth of GaGe in MIT furnace at several dimensional times; ampoule translation rate is 4 μm/s.

Sample gallium concentration fields are shown in Fig. 4 that correspond to the convection pattern shown in Fig. 3. Almost well-mixed melt forms in the two convection cells, with the average bulk composition being different in each region, separated by an internal diffusion layer. Appreciable radial segregation exists along the melt/crystal interface, as noted by the intersection of a number of isoconcentration curves with the interface.

The effect of the degree of convection in the melt on axial and radial segregation in the crystal was studied by caring out simulations for different values of the thermal Rayleigh number Ra_T defined as

$$Ra_T \equiv \frac{g\beta_t \Delta T R_c^3}{\alpha_m \nu} \qquad [1]$$

where g is the gravitational acceleration, β_t is the coefficient of thermal expansion, α_m is the thermal diffusivity of the melt, ν is the kinematic viscosity, R_c is the crystal radius and ΔT is the temperature difference between the hot and cold zones of the furnace.

Results for the radially-averaged solid composition, the radial segregation, and the effective segregation coefficient k_{eff} (defined as the ratio of the radially-averaged interface composition in the solid to the bulk composition of the melt times the equilibrium segregation coefficient k) are shown in Fig. 5 as a function of the length of the crystal solidification and Ra_T. The thermal Rayleigh number Ra_T ranges from the value ($2x10^5$) appropriate for the MIT experimental system to zero, where dopant transport is diffusion controlled.

Several results are obvious. First, there is appreciable radial segregation is this system. The radial segregation shown in Fig. 5b is largest for intermediate levels of convection, as predicted by the quasi-steady-state analysis (7). The transition in the axial segregation between diffusion-controlled growth and a well-mixed bulk is best demonstrated by examining the results for the solid composition c_s. At low convection levels the simulations agree well with the analysis of Smith et al. (19) for the evolution of the composition in a one-dimensional, diffusion-controlled system. As Ra_T is increased, the axial composition profile evolves toward the results of Scheil (20) for a well mixed bulk melt with no resistance to solute transport at the interface, i.e. a diffusion layer of zero thickness.

The values of the effective segregation coefficient computed as a function of the fraction solidified are displayed in Fig. 5c and demonstrate the inappropriateness of this concept when the convection level is low, approaching diffusion-controlled crystal growth. In this limit, the computed value of k_{eff} is not constant during the run, indicating that the composition in the melt is not nearly well-mixed. Only for the higher convection levels is k_{eff} nearly constant and is the stagnant film model appropriate.

A direct comparison between the numerical simulations and the stagnant film model of Burton et al. (17) is shown in Fig. 6. The calculations are fit extremely well by this model with $k_{eff} = 0.106$, compared to $k = 0.087$. This comparison between the stagnant film model and the direct numerical simulation of convection and dopant segregation accentuates the global nature of the concept of the effective segregation coefficient. The expression for k_{eff} in the stagnant film model is derived for a melt with a uniformly thick diffusion layer adjacent to the melt/crystal interface and a well-mixed bulk. The model correlates the composition in the solid c_s even though the solute field in the melt does not correspond to the simple picture used to derive the stagnant film model. For example, the stagnant film analysis misses the two-cell structure of the flow and ignores the radial variation of the composition along the interface.

The numerical simulations were compared directly with the experimental results of Wang (18) for the growth of gallium-doped germanium crystals. The simulations were connected to the furnace designed by Wang by comparing calculations and experiments for the heat transfer in a solid graphite ampoule in the furnace. No adjustments were made in the simulations in an attempt to fit the experimental data.

The comparison shown in Fig. 7 is based on an experiment in which the ampoule pull rate was changed from 2 to 4 μm/s after 0.3 of the ampoule length was solidified. The calculations reproduced well the local variation in the microscopic solidification rate, as measured by Wang. The temporal response for the deflection of the melt/crystal interface is correctly predicted, but the absolute value of the deflection is over-estimated by approximately 5 percent.

The comparisons between the predicted and measured values of the gallium composition in the crystal are shown in Fig. 7 in terms of c_s, Δc and k_{eff}. The predicted and computed values of the solid composition agree to within 5 percent everywhere in the crystal. The accuracy is similar for the prediction of the effective segregation coefficient. The error in the prediction of the radial segregation is larger, ranging between a factor of two at low fractions of solidification to less than 30 percent toward the end of the growth run. This difference may be caused by the same sources that lead to overestimation of the radial deflection of the melt/crystal interface.

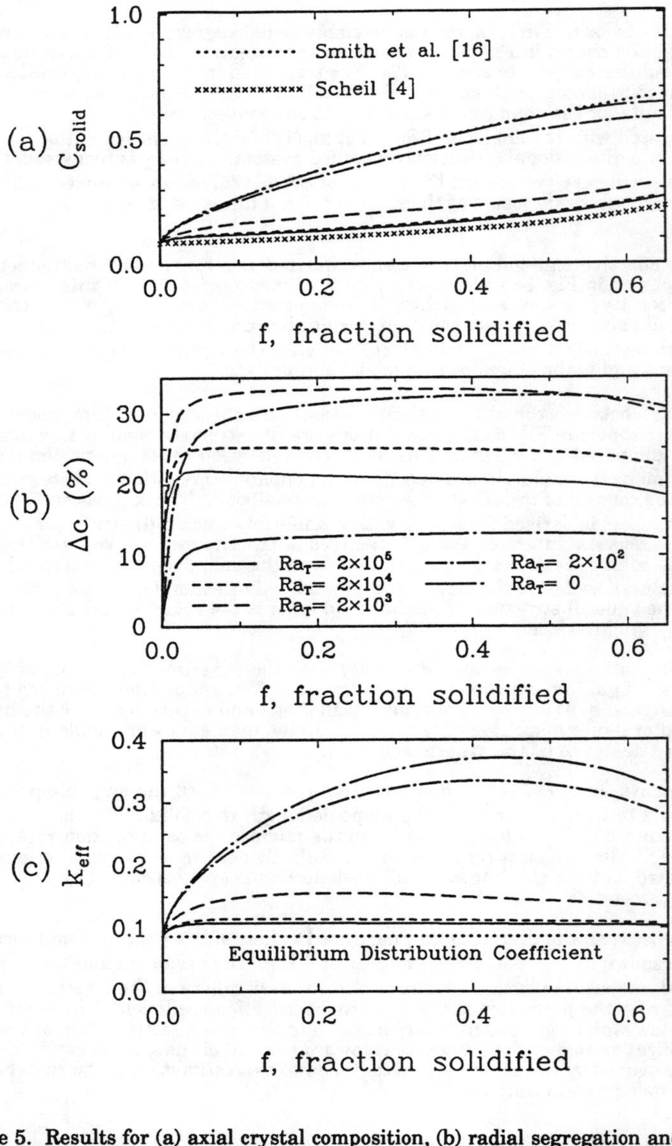

Figure 5. Results for (a) axial crystal composition, (b) radial segregation and (c) effective segregation coefficient as a function of fraction solidified f and thermal Rayleigh number Ra_T. The predictions of Smith et al. [19] for diffusion-controlled growth and Scheil [20] for a well-mixed melt are shown for comparison.

When taken together, the comparison between the calculations and experiments for the growth of GaGe in the small-scale vertical Bridgman system give the first indication of the power of direct numerical simulation to predict the details of the performance of well-controlled crystal growth systems. Many of the issues that follow from this comparison revolve around the feasibility of extending this type of modelling to the analysis of larger scale, more complex crystal growth systems.

2.2 Integrated System Model of Czochralski Crystal Growth of Silicon

We have developed very detailed Integrated System and Hydrodynamic models of Czochralski (CZ) and Liquid Encapsulated Czochralski (LEC) growth in order to extend the concept of predictive modelling to commercially important systems.

The Integrated Hydrodynamic Model (IHM) of Kinney et al. (5) includes conduction between all the components of a classical CZ furnace, diffuse gray radiation throughout the system, steady axlsymmetric convection in melt and the couplings of this physics with the three deforming boundaries in the system, the crystal shape, the melt/crystal interface and the melt/ambient meniscus. The IHM is described in detail in (5) along with the finite-element/Newton solution algorithm.

The temperature field computed using the IHM for a Siltec CZ system using a 7.25 cm diameter crucible is shown in Fig. 8. This solution was computed using only geometrical input from the system and the estimated thermophysical properties of the materials in the system and of molten and solid silicon. The calculation is based on the growth of a crystal of a specified diameter, where the power to the resistance heater that surrounds the crucible is computed as part of the solution.

The temperature field shows the relatively small temperature difference that exists across the melt and the relatively large radial temperature gradient at the melt/crystal interface. The interface has the gull-winged shape that is typical of the influence of hot melt impacting the outer edge of the crystal due to convection. The temperature field computed throughout the Siltec system has been compared to measurements made by George Kim at IBM for the identical system. The agreement is more than satisfactory, considering the large uncertainties in the thermophysical properties of many of the materials used in the system. This comparison is described in (5).

An important part of the results of the IHM is the understanding of the effect of modifications of the thermal environment of the system on convection in the melt and on the temperature field and thermoelastic stresses in the crystal. The utility of the calculations toward these goals is demonstrated here by calculation of the flow field in Fig. 8, which depicts the large toroidal roll cell created by the radial temperature gradients in the system. Interestingly, the small inversion of the temperature gradients near the upper corner of the crucible causes a separate flow cell there with the opposite circulation. This flow may affect the transport of oxygen from the crucible to the crystal in the case of silicon growth.

3. Crystal Defect Formation

Understanding the links between processing conditions and the crystallographic defects and microscale morphology of the material is one the outstanding problems in solidification theory. Here the thermal history of the crystal, the stoichiometry of the crystal, the local composition of dopants and impurities, and the type and number of defects produced at the solidification interface all interact to produce distributions of interacting point defects, microdefects, impurity clusters and dislocations in the crystal.

Figure 6. Predicted axial segregation of gallium, compared to prediction of diffusion layer model with $k_{eff} = 0.1067$; $k = 0.087$ for gallium in germanium.

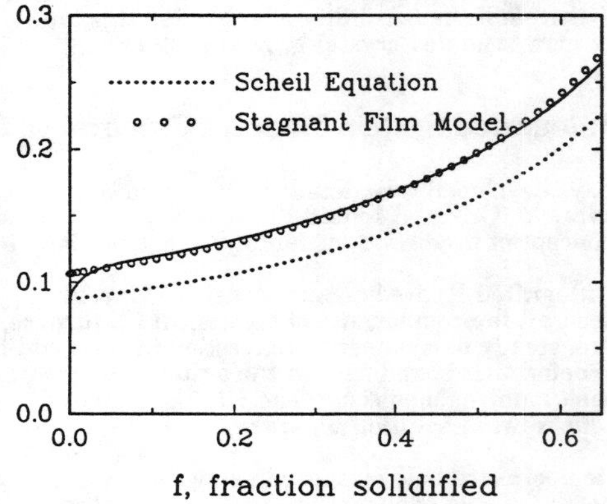

Figure 7. Comparison between computed and experimentally determined values of (a) solid concentration, (b) radial concentration variation and (c) effective segregation coefficient k_{eff} for the system of Wang [13] and the case of a step change in ampoule translation rate from 2 to 4 μm/s.

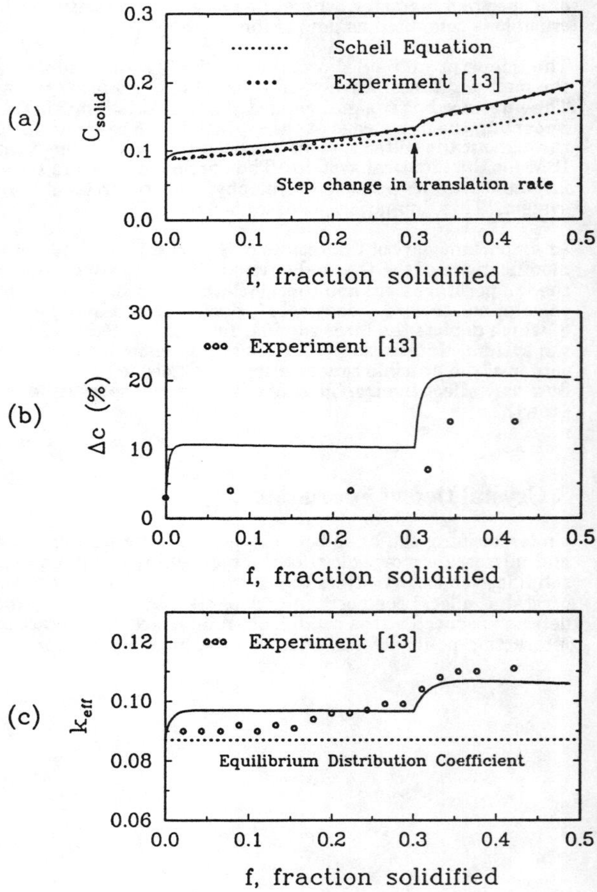

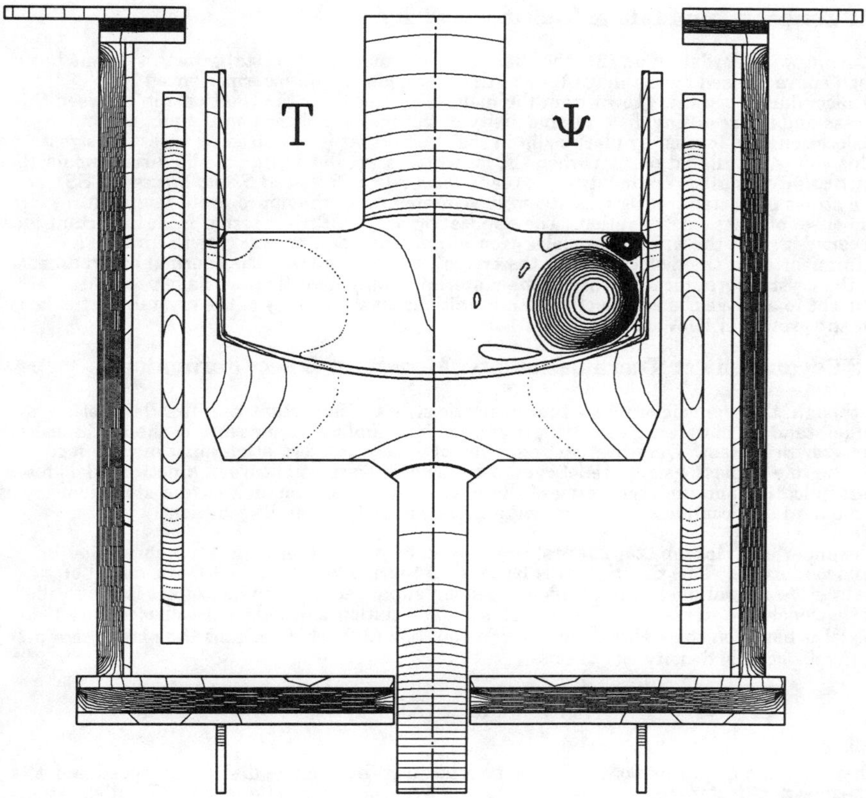

Figure 8. Computed temperature field for Integrated System Model of Siltec Czochralski system for silicon crystal growth. The temperature field throughout the system and the streamlines of the axisymmetric flow field are shown.

The difficulties in understanding the mechanisms for defect formation and the role of processing conditions revolve around the tremendously disparate ranges of length and time scales. These range from the 10^{-10} m interatomic spacing of a typical crystal lattice with a characteristic time for an interatomic jump of 10^{-13} s to the macroscales of the crystal. Quantitative prediction of defect formation and concurrent optimization of a crystal growth system requires of the processes for defect creation, and motion and interaction and atomistic, understanding statistical mechanical models that relate these mechanisms to continuum models for the dynamics of defect densities. We are at the very beginning of building such an understanding.

3.1 Concepts from Linear Thermoelasticity

Starting with Jordan et al. (21) the thermal environment for crystal growth systems has been characterized by the magnitude of the thermoelastic stress experienced by semiconductor crystals grown from the melt. The model for the relationship between this stress and the resulting dislocation density in the crystal is based on the picture that dislocations are formed by plastic slip in the crystal that is produced by excessive stress. This slip is postulated to start when the local stress, resolved into the slip directions for the particular crystallographic lattice, exceeds the *Critical Resolved Shear Stress* (CRSS), or the stress predicted from stress-strain measurements at the appropriate temperature for the onset of plastic deformation. The application of the CRSS criterion is an important idea because it gives the first methodology for optimization of a crystal growth furnace for minimization of the dislocations in the crystal, thereby relating the thermal environment to the crystallographic structure of the material. Many investigators have used this concept to analyze the effects of growth conditions on the quality of the crystal, particularly for the growth of III-V materials (6, 21-23).

3.2 Beyond Linear Thermoelasticity: Models for Defect Formation

Although, the analysis of linear thermoelastic stress is important as a first level of understanding in defect generation, it gives only a qualitative measure of the harshness of the growth system. Quantitative prediction of dislocation formation must include models for the role of plastic stress relief caused by dislocation multiplication, kinetic models for multiplication and understanding of the roles of vacancies, impurities and stoichiometry of compound semiconductors on the dynamics of the dislocation distribution.

Alexander and Haasen (24, 25) first proposed such a model for dislocation formation in semiconductors. This description is based on a thermally-activated, kinetic model of dislocation migration and multiplication along slip directions, and accounts for the relief of thermoelastic stress caused by the plastic deformation induced by the dislocations. The model is based on the well-known Orowan equation (26), which relates the plastic strain ε^p to the dislocation density $N(\mathbf{x},t)$ as

$$\frac{D\varepsilon^p}{Dt} = \phi b u N(\mathbf{x},t) \qquad [2]$$

where b is the magnitude of the Burgers vector, u is the average dislocation speed and ϕ is a geometrical factor that relates shear to compression and tensile strain. The dislocation speed is modelled as a thermally-activated process and is proportional to the effective stress τ_{eff} experienced by the dislocations, which is taken as the applied stress minus the back stress created by the dislocation (25):

$$u = B_o \exp\left(-\frac{Q}{kT}\right)\tau_{eff}^m \qquad [3]$$

where Q is the activation energy related to the Peierls barrier for dislocation glide, k is the Boltzmann constant, B_o is an empirically measured constant and m is the strain hardening exponent.

The population of dislocations is modelled by a modification of the multiplication law originally proposed by Johnston and Gilman (27), which relates the rate of increase of $N(\mathbf{x},t)$ to the density as

$$\frac{DN}{Dt} = KuN(\mathbf{x},t)\tau_{eff} \qquad [4]$$

where K is a kinetic constant determined from deformation experiments. The model is described in detail by Maroudas and Brown (28) using the same notation introduced here. Alexander and Haasen (25) demonstrated the appropriateness of the model for fitting deformation experiments for silicon and germanium. Sumino and coworkers (29) extended the results for silicon and have used the ideas of plastic flow in analysis of dislocation formation in gallium arsenide. Recently, Völkl and Müller (30) have measured the constants needed for InP and have shown that the predictions of the model fit dislocations measurements for dislocations created during LEC growth of InP.

The structure of the Haasen model has very important implications for the growth of low dislocation semiconductor crystals. We have shown (28) that proper scaling of the Haasen model gives rise to an important new dimensionless group for understanding the physics of dislocation multiplication in semiconductors, the Dislocation Damköhler Number D:

$$D = \frac{KB_oR_c}{V_p} \exp\left(-\frac{Q}{kT}\right)\sigma_o^{m+1} \qquad [5]$$

The parameter D scales the time scale for generating dislocations with the time scale for motion of the crystal through the region with applied stress σ_o and temperature T. For $D \ll 1$, dislocations are formed slowly relative to the motion of the crystal. For $D \gg 1$, dislocations are formed rapidly and the dislocation density causes rapid plastic relief, which eventually halts the formation of more dislocations. The analysis in (28) shows that thin boundary layers of thickness D^{-1} form if $D \gg 1$. In these layers the dislocation density varies rapidly from some initial value to the value necessary for a zero effective stress. It is demonstrated in (28) that CZ and LEC crystal growth conditions for silicon and GaAs and InP, respectively, all correspond to high Damköhlker number conditions. In these systems, near the melt/crystal interface the initial density corresponds to the number of dislocations grown into the crystal during solidification.

The importance of these predictions is demonstrated by calculations in (6) for the thermoelastic stresses in two different configurations of the heat transfer system in a CZ furnace for growth of 6 inch diameter silicon crystals. These results are shown in Fig. 9 and include calculations for a conventional CZ puller and one with a radiative heat shield that extends across the opening above the crucible.

The temperature field and the von Mises stress in the crystal are shown for both cases. There are two significant differences. First, the maximum stress in the crystal is lowered significantly by including the shield. Without the shield, this stress occurs at the edge of the crystal near the solidification interface. The value of the stress 1.42×10^{-4}G (where G is the shear modulus) is greater than the CRSS for silicon at high temperatures. If dislocations nucleate anywhere in the crystal rapid multiplication is expected. Solution of the Haasen model predicts a density of nearly 1×10^4 dislocations/cm^2 for this geometry! Including the heat shield lowers the maximum stress in the crystal by almost a factor of four from the value without the shield. Just as significantly, the location of the maximum stress shifts to be near the heat shield where the crystal is colder and more able to withstand the stress. The lower stress coupled with its gradual increase from the interface to the shield yields predictions of dislocation densities from the Haasen model that are 50 times lower than the prediction without the shield. Heat transfer configurations that lower the stress and dislocation density are described in (28).

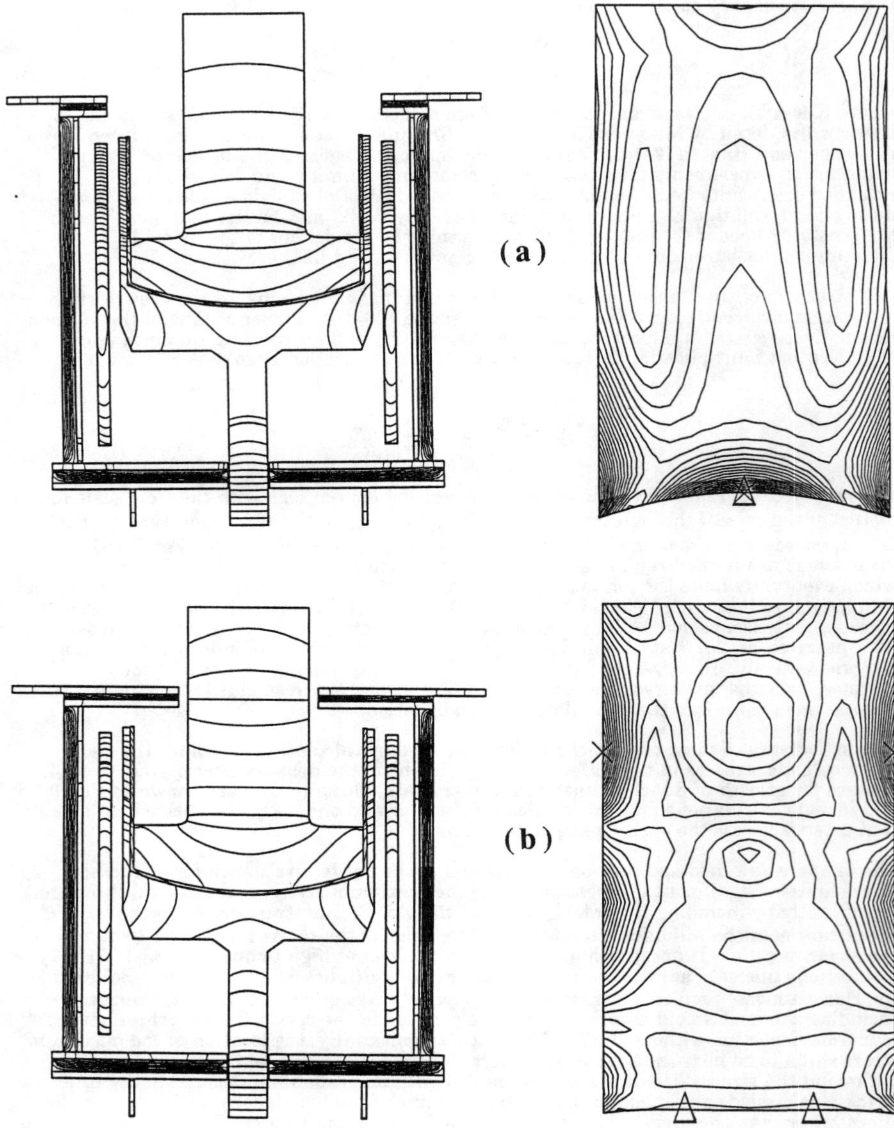

Figure 9. Comparison of thermoelastic stress fields predicted for analysis of silicon growth in CZ systems (a) without and (b) with thermal shields. Including the thermal shield decreases the thermoelastic stress by nearly an order-of-magnitude and decreases the maximum stress by the same amount. The maximum stress also moves from the melt/crystal interface without the shield to the point of closest approach of the shield to the crystal. The Alexander-Haasen model predicts that the dislocation density is lower by over an order-of-magnitude [6].

An interesting paradox that arises from these calculations is that typical silicon crystals grown in furnaces like the one shown in Fig. 9a are dislocation free, even though the dislocation density predicted from the Haasen model is still $1x10^3$! We believe that the explanation for this difference lies in the absence of dislocations at the melt/crystal interface, so that *no* dislocations are available for multiplication. The need for a zero dislocation density at the interface is well known in silicon technology and is the basis for the Dash procedure for necking a crystal at the start of growth to remove dislocations in the seed (31). In fact, dislocation densities of 10^3 - 10^4 km^{-3} in silicon were common before Dash's invention and are still observed in germanium.

4. The Next Steps

Although substantial progress has been made, there are still a myriad of unresolved and significant problems. I believe that most of the challenges involve pushing the limitations on the length scales in the modelling to expand the interconnections of transport processes with final material properties. Two types of problems deserve specific note.

The first involves detailed understanding of the fluid dynamical transitions that occur in the melt and relating the critical conditions for these changes with operating conditions. Because of the low Prandtl number of typical semiconductor melts and the moderate system sizes, the convection in the melt is only weakly turbulent with complex boundary and internal layer structure that changes in time. Direct simulation of these flows is presently at the leading edge of numerical simulations and has been attempted only for very idealized models of these flow geometries (32). Although knowledge of flow transitions may prove crucial for the ultimate design and optimization of semiconductor crystal growth, other limitations may prevent extensive calculations from being justified. Two problems are obvious. First, the sensitivity of the flow transitions and the microscopic structure of the flow to the details of the heat transfer environment may make the calculations meaningless, except for the best characterized and controlled systems. Secondly, the database of thermophysical properties of the material presently is too inaccurate to support this level of analysis; the characterization of the important semiconductor materials is crucial to the quantitative advancement of our understanding of material processing.

The challenge in modelling of convection in large-scale semiconductor melts is the development and application of appropriate turbulence models which capture the macroscopic flow structure and dopant and heat transfer caused by the flow *without* resolution of the fine details of the temporal and spatial structure of the flow. This problem is as challenging as any that faces the turbulence modelling community.

Many of the challenges in the modern theory of crystal growth will focus on linking material processing conditions with the microscopic structure of defects in the solid. Advancing this theory will involve the combination of atomistic analysis of the formation and interaction of defects and impurities within a crystal, the development of continuum theories for defect motion and kinetics, and the relating of these theories to the macroscopic processing conditions in crystal growth systems.

Czochralski-grown silicon is a wonderful example of the need for modelling of defect dynamics. Although almost free of dislocations, the typical crystal is a heterogenous mixture of point and microdefects caused by the high processing temperatures and by impurities, mainly oxygen. Oxygen is present in supersaturation and so precipitates form clusters and microdefects within the lattice and at dislocations. It is well known that the mechanical and electrical properties of CZ silicon are strongly influenced by the annealing cycle for the crystal. For example, oxygen preferentially diffuses to dislocations (32) and inhibits dislocation motion (33, 34). Models for the transport and reactions of oxygen in the crystal are needed to quantitatively predict the effect of processing cycles on the microstructure of silicon.

Acknowledgements

We are grateful to the Microgravity Sciences and Applications Program of the United States National Aeronautics and Space Administration, the National Science Foundation, the Defense Advanced Research Projects Agency, and the IBM Corporation for support of the research described in this paper. The author is indebted to D.E. Bornside, D.H Kim, T.K. Kinney and D. Maroudas for their contributions to the research.

References

1. M.C. Flemings, *Solidification Processing*, McGraw-Hill, New York (1974).
2. W. Kurz and D.J. Fisher, *Fundamentals of Solidification*, Trans-Tech Publications, Switzerland (1985).
3. J.J. Derby, L.J. Atherton, P.M. Gresho, J. Crystal Growth, 97, 792 (1989).
4. R.A. Brown, P.A. Sackinger, T.A. Kinney, D. Bornside, J. Crystal Growth 97, 99 (1989).
5. T.A. Kinney, D.E. Bornside, R.A. Brown, and G. Kim, J. Crystal Growth, submitted (1990).
6. D.E. Bornside, T.A. Kinney and R.A. Brown, J. Crystal Growth, in press (1990).
7. R.A. Brown, AIChE Journal 34, 881 (1988).
8. M. Mihelcic, K. Winegrath, K. and Chr. Pirron, J. Crystal Growth 69, 473 (1985).
9. K.M. Kim and W.E. Langlois, J. Electrochem. Soc. 136, 1137 (1989).
10. R.A. Brown and D.H. Kim J. Crystal Growth, in press (1990).
11. P.D. Thomas, R.A. Brown, J.J. Derby, L.J. Atherton, and M.J. Wargo, J. Crystal Growth 96, 135 (1989).
12. H.M. Ettouney and R.A. Brown, J. Computat. Phys. 49, 118 (1983).
13. J.J. Derby, L.J. Atherton, P.D. Thomas, and R.A. Brown, J. Sci. Computing 2, 297 (1987).
14. P.A. Sackinger, J.J. Derby, and R.A. Brown, Inter. J. Numer. Meths, Fluids 9, 453 (1989).
15. D.E. Bornside, T.A. Kinney, and R.A. Brown, R.A. Inter. J. Meth. Engng. **xx**, xxx (1990).
16. D.H. Kim , Ph.D. Thesis, Massachusetts Institute of Technology, 1990.
17. J.A. Burton, R.C. Prim, and W.P. Slichter, J. Chem. Phys. 21, 1987 (1953).
18. C.D. Wang, Ph.D. Thesis, Massachusetts Institute of Technology, 1984.
19. V.G. Smith, W.A. Tiller, and J.W. Rutter, Can. J. Phys. 33, 723 (1955).
20. E. Scheil, Z. Metallk. 34, 70 (1942).

21. A.S. Jordan, A.R. Caruso, and A.R. Von Nieda, Bell Systems Tech. J. 59, 593 (1980).

22. C. Schvezov, I.V. Samarasekera and F.Weinberg, J. Crystal Growth 84, 219 (1987).

23. S. Motakeff and A.F. Witt, J. Crystal Growth 80, 37 (1987).

24. P. Haasen, in: *Dislocation Dynamics*, Eds. A.R. Rosenfield, G.T. Hahn, A.L. Bement and I Jaffee, Battlell Inst. Materials Colloq., p. 701 (1967).

25. H. Alexander and H. Haasen, So. Stat. Phys. 22, 27 (1968).

26. E. Orowan, Proc. Phys. Soc. Lond. 52, 8 (1940).

27. W.G. Johnston and J.J. Gilman, J. Appl. Phys. 30, 159 (1959).

28. D. Maroudas and R.A. Brown J. Crystal Growth in press (1990).

29. I. Yonenaga and K. Sumino, Phys. Stat. Solidi (a) 50, 685 (1978).

30. J.Völkl and G. Müller, J. Crystal Growth 97, 136 (1989).

31. W.C. Dash, J. Appl. Phys. 30, 459 (1950).

32. B. Roux, in *Numerical Simulation of Oscillatory Convection in Low-Pr Fluids*, Vieweg, Braunschweig, 1990.

33. D. Maroudas and R.A. Brown, Phys. Rev. B, submitted (1990).

34. D. Maroudas and R.A. Brown, J. App. Phys. submitted (1990).

35. K. Sumino, in *Defects and Properties of Semiconductors: Defect Engineering*, J. chikawa, K. Sumino and K. Wada, KTK Scientific Publishers, Tokyo, 1987.

Magnetic Suppression of Natural Convection of Liquid Metal

HIROYUKI OZOE
Institute of Advanced Material Study, Kyushu University
Kasuga-Koen 6-1, Kasuga 816, Japan

KAZUTO OKADA
Graduate School of Interdisciplinary Study, Kyushu University
Kasuga-Koen 6-1, Kasuga 816, Japan

INTRODUCTION

The convection of liquid metal has been one of the major interesting research subjects due to its direct application to various physical phenomena as well as to recent steel industry and semiconducting industry. Fluid flow and heat transfer of liquid metal are controlled by various ways including an external magnetic field. The magnetic control for an electroconducting fluid such as liquid metal is an attractive and unique method for corrosive and high temperature materials due to its indirect non-contact characteristics. In this paper, we mainly concern the application of magnetic field to the Czochralski crystallization method employed in semiconducting industry because of its current importance.

1. LITERATURE SURVEY ON MAGNETIC CZOCHRALSKI CONVECTION

Chandrasekhar [1] summarized the works on hydrodynamic and hydromagnetic stability of fluid in 1961 and his book appears to have surveyed the most of the previous works. The effects of rotation and magnetic field on the convection of electroconducting fluid were discussed quite in detail. It provided the theoretical aspect of the magnetic convection. Followings are those related to magnetic Czochralski convection.

Kakutani [2] extended the Karman's problem to the hydromagnetic case and introduced a new parameter for the rotational field under a magnetic force. Utech and Flemings [3] employed an external magnetic field for the solidification process of tellurium-doped indium antimonide in a horizontal boat in 1966 and reported the elimination of solute banding in the crystals. The application of magnetic field of 1735 Gs suppressed temperature fluctuations of 2°C and the fluctuation period of about 3 seconds disappeared. Witt et al. [4] reported the Czochralski crystal growth of InSb in a transverse magnetic field of 4000 Gs and insisted the advantage of the use of the magnetic field in crystal growth. Hoshi et al. [5] employed the magnetic field in the industrial process of Czochralski crystallization of silicon. They succeeded to produce silicon crystal rods with the control of not only the crystal diameter but also many aspects of the characteristics of crystal including oxygen concentration. The development process was described by Hoshi et al. [6].

Fukuda & Terashima [7] reported the application of a magnetic force for the liquid encapsulated Czochralski crystallization for Gallium Arsenide which enabled the control of the defect and accelerated the pulling rate. Langlois & Lee [8] reported the numerical computation of the bulk convection with an axial magnetic field for an actual size of a Czochralski crystallization of 1000 Gs. They concluded the joule heating is negligible. Oreper & Szekely [9] reported the numerical computation of natural convection in a square enclosure with lateral (horizontal) magnetic field. Their 11 x 11 mesh points computation was for molten silicon in a 100 mm wide cavity with temperature difference of 10 K.

Langlois [10] reviewed earlier works on the magnetic convection applied for Czochralski growth. Hoshikawa et al. [11] obtained homogeneous dopant distribution of silicon crystal grown by 1 KOe vertical magnetic field-applied Czochralski method. They observed the most homogeneous radial oxygen distribution for the uniform diffusion boundary layer. Hurle and Series[12] discussed the dependence of the effective distribution coefficient on growth and crystal rotation rates under an axial magnetic field in Czochralski growth and suggested that the magnetic field would be to make the crystal stoichiometry more sensitive to melt stoichiometry. Organ [13] carried out axi-symmetric three-dimensional computation for a magnetic Czochralski growth system and pointed out that the effect of the magnetic field is to enhance the azimuthal motion in the lower part of the crucible and the crystal rotation effect penetrates to a much greater depth under an axial magnetic field. Mihelcic and Wingerath[14] carried out the axi-symmetric numerical computation for an axial magnetic field up to 3000 Gs in a large scale crucible of 22.1 cm radius and obtained a non-fluctuating velocity and temperature field. Hjellming and Walker[15-20] published series papers with the division of convection and stagnant regimes under a strong axial magnetic field of 1kGs or over. For modest magnetic field (500 Gs), they modeled the thin boundary layer around a crucible which enabled them to expand the asymptotic approximation. They also presumed conduction dominance. They studied various limiting cases such as isothermal motion and motion due to buoyancy and thermocapillarity and then mass transfer. Kim and Langlois [21] tried to explain the more pronounced radial distribution of dopant under an axial magnetic field than that in nonmagnetic growth which was observed by Hoshikawa et al.[11] experimentally. Under a strong magnetic field of 2kG the diffusion boundary layer was found to be responsible for the radial distribution of Boron in a silicon crystal.

Kobayashi [22] derived model equations for solute distribution in magnetic Czochralski grown crystals. He analyzed both an axial and a transverse magnetic field and concluded that an axial magnetic field gives an axial uniformity of the solute distribution and the transverse magnetic field suppresses the thermal convection. Kobayashi [23] subsequently carried out series expansion analysis to interpret the different effects of the transverse (low content of oxygen) and axial (high content) magnetic field on the content of oxygen in a silicon crystal.

Langlois [24] et al. reported the effect of finite electrical conductivity of the crystal and found that in the immediate neighborhood of the growth interface the meridional melt circulation depends on the conductivity of the crystal. Riley [25] discussed the boundary layer approximation with using Kakutani's [2] interaction parameter for a rotational magnetic Cz growth and concluded that the boundary layer model may only be expected to provide a reasonable quantitative description of the flow. Ozoe and Maruo [26] solved both the magnetic field and the natural convection equations for liquid metal under a vertical magnetic field in a square enclosure heated and cooled from opposing vertical side walls and showed the transient characteristics. Brown [27] reported an extensive review on the crystal growth including magnetic effect from the point of transport processes. Cartwright et al. [28] made an order-of-magnitude and numerical analyses and concluded that the radial segregation of flow would become more difficult in the presence of a strong axial magnetic field and horizontal magnetic fields would cause spatially non-uniform boundary layer thickness. Hicks et al. [29], Kerr and Wheeler [30], Series [31] and Riley [32] focussed the analyses on the species transport and distribution in a magnetic Czochralski growth.

Munakata and Tanasawa [33] carried out both numerical and analytical approach on the magnetic Czochralski convection and found fluctuating phenomena is essential in the boundary condition of Czochralski mixed convection. Brown et al. [34] again reviewed the Czochralski growth including thermal-capillary model and the finite-element/Newton method to simulate the entire system. Ozoe and Okada [35] reported the effect of the direction of the external magnetic field on the three-dimensional natural convection in a cubical enclosure and found that the magnetic field parallel to the heated wall was least effective. Hirata & Hoshikawa [36] reported an ingenious new cusp magnetic field in silicon growth to homogenize oxygen concentration along the growth axes.

Ozoe and Toh [37] carried out numerical calculation for a Czochralski bulk flow of liquid metals under a vertical external magnetic field and found enhancement in the rate of heat transfer on a crystal interface at weak magnetic field of Ha = 100. Langlois [38] reviewed the fluid dynamics in crystal growth and pointed out four interesting fields such as three-dimensional and turbulence, magnetic damping, multi-compornent nutrient and heat and mass transfer in complete systems.

These are chronological (with some exception) review on the works reported on the magnetic Czochralski convection. Followings are introduction of our recent works related to the magnetic suppression of liquid metal convection.

2. TWO-DIMENSIONAL NUMERICAL ANALYSES OF LIQUID METAL IN AN ENCLOSURE UNDER A MAGNETIC FIELD

2.1 Vertical External Magnetic Field [26]

The system considered is shown in Fig. 1. The computational grid points are drawn. The liquid metal of Pr=0.054 was heated from a left side wall and cooled from a right wall and otherwise insulated. The external magnetic field was impressed vertically & uniformly. The magnetic vector potential equation was solved to satisfy the induction equation and a Maxwell equation. The magnetic Prandtl number was arbitrarily assumed to be unity.

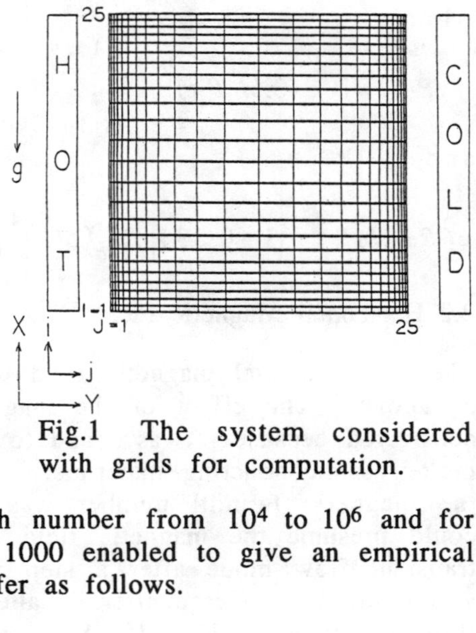

Fig.1 The system considered with grids for computation.

The computation for the Rayleigh number from 10^4 to 10^6 and for the Hartmann number from 0 to 1000 enabled to give an empirical equation for the rate of heat transfer as follows.

$$\frac{Nu-1}{Nu(Ha=0)-1} = 1 - \frac{1}{\left[1+\left(\frac{2.2\,Ra^{1/3}}{Ha}\right)^{9.012}\right]^{1/5.62}} \tag{1}$$

Table 1 Converged results

Ra	Ha	Pr_m	Ψ_c	Nu
10^6	10	10^1	9.2970	5.848
			9.0863	
10^6	10	10^2	9.2463	5.863
			9.2456	5.862
10^6	10	10^3	9.2452	5.863
			9.2445	5.862
10^6	10	10^4	9.2604	5.866
			9.2596	5.866

$$Nu(Ha=0) = \overline{Nu_l}$$

$$\overline{Nu_l^3} = \left\{ \left[1 + \left(\frac{Ra_l}{4472} \right) \frac{l}{b} \right]^{-8} + \left[\frac{Ra_l (l/b) f(Pr)}{57 \phi(Ra_b, b/l)} \right]^{-2} \right\}^{-3/8} + \frac{Ra_l f(Pr)}{8000} \quad (2)$$

where

$$f(Pr) = \left[1 + (0.5/Pr)^{9/16} \right]^{-16/9} \quad (3)$$

and

$$\phi(Ra_b, b/l) = \left\{ 1 + 0.231 \left[Ra_b (b/l)^4 \right]^{-1/4} \right\}^2 . \quad (4)$$

2.2 Horizontal Magnetic Field [39]

Then, a horizontal magnetic field was considered otherwise the same as above. The effect of the magnetic Prandtl number was at first considered because it was found to be in the order of 10^4 to 10^5 for molten semiconducting materials. As seen in Table 1, the effect of the magnetic Prandtl number was found to be negligible and we could presume the magnetic field is uniform in liquid metal. The transient flow mode after a step increase of the magnetic strength was found to converge after an alternative change in the direction of flow as shown in Fig. 2. Sometimes even the secondary flow was found to occur as seen at τ = 680, 840 and 980.

2.3. Transient Natural Convection in a Tall Enclosure under a Horizontal Magnetic Field [40].

The physical properties of molten semiconducting materials are scarce in open literature. The thermal conductivity of molten liquid metal in a hot furnace is especially difficult to be measured. We

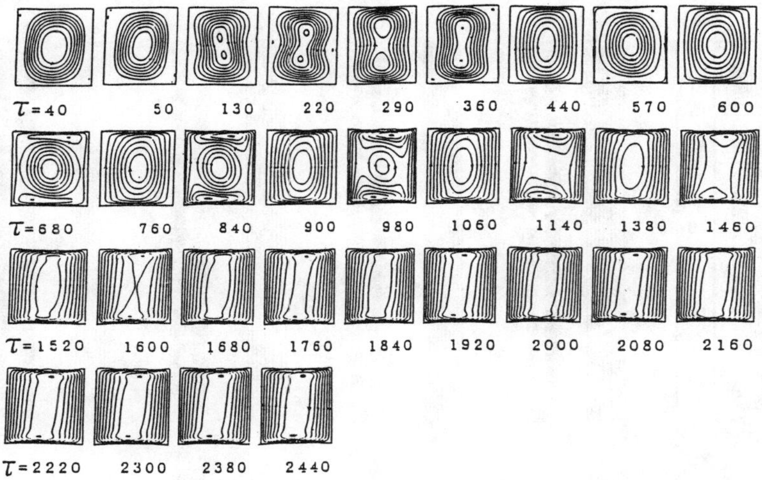

Fig.2 Transient profiles of stream function after a step change in the Hartmann number from 100 to 500. $Ra=10^6$, $Pr=0.054$ and $Pr_m=1$.

proposed to employ an external magnetic field in measuring the thermal conductivity of electro-conducting material transiently or in a steady state wise even heated from side. This idea was practically employed to measure the thermal conductivity of mercury by Nakamura et al.[41] afterwards.

Figure 3 shows the system considered. At a steady state the isotherms and stream functions are shown in Fig.4. The Nusselt number and the central value of stream function differ very much depending on the magnitude in the Hartmann number as shown in Table 2.

The effect of the Prandtl number was also studied on the transient characteristics for convergence. When there is no magnetic field, the Nusselt numbers decreases in the order of Prandtl number as seen in Fig. 5. However, at $Ra=1000$ and $Ha=10^{3/2}$, the order of Nusselt number was reversed dependent on the Prandtl number although the absolute value of the Nusselt number approached to unity as seen in Fig. 6.

3. THREE-DIMENSIONAL NUMERICAL ANALYSES OF NATURAL CONVECTION OF LIQUID METAL UNDER A MAGNETIC FIELD EITHER IN X-, Y- OR Z-DIRECTION.[42]

The system considered is shown in Fig. 7. The numerical computations for three directions of a magnetic field were carried

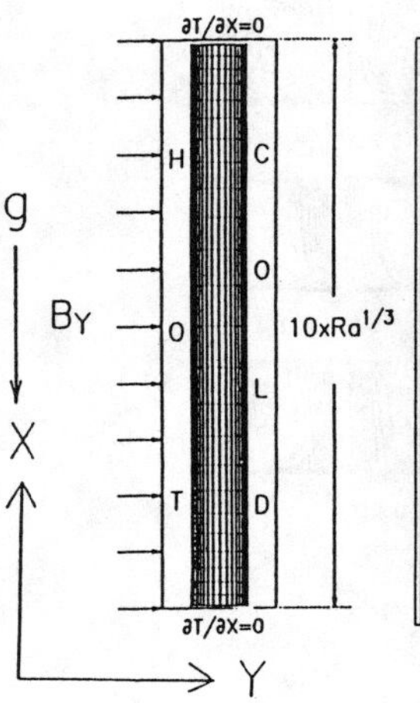

Fig.3 The tall system considered with grids.

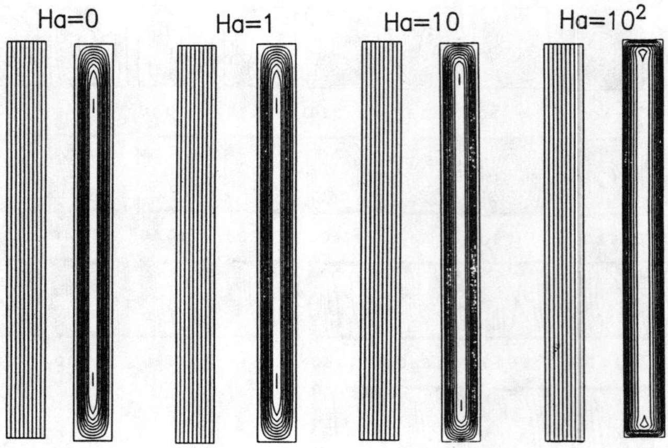

Fig.4 Steady state profile of the isotherms and stream function at Ra=10 and Pr=0.054

Table.2 Computed results for Ra=100 & Pr=0.054

Ha	0	$10^{1/2}$	10	10^2	10^3
$Nu\|_{Y=0}-1$	4.02×10^{-4}	2.86×10^{-4}	4.80×10^{-5}	1.165×10^{-8}	
Ψ_c	2.62×10^{-1}	2.14×10^{-1}	7.88×10^{-2}	1.215×10^{-3}	1.09×10^{-4}
$Max(Nu_{loc}\|_{Y=0})-1$	7.02×10^{-2}	5.97×10^{-2}	2.51×10^{-2}	4.27×10^{-4}	

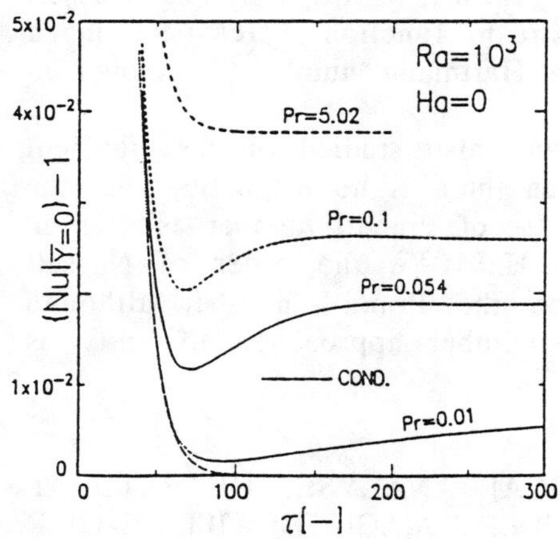

Fig.5 Transient responses of the average Nusselt number minus unity at various Prandtl numbers at Ha=0 and $Ra=10^3$.

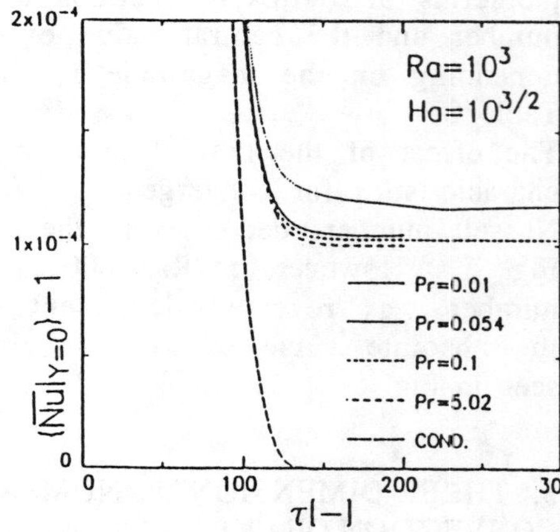

Fig.6 Transient responses of the average Nusselt number minus unity at various Prandtl numbers at $Ha=10^{3/2}$ and $Ra=10^3$.

out. At Ra=10^7, Ha=100 and Pr=0.054, the approximately converged average Nusselt numbers were 9.65, 10.4 and 10.0 under a magnetic field in the x-, y- or z-direction in contrast to 10.5 at Ha=0. This suggests that the y-directional magnetic field is least effective even if it is perpendicular to both the horizontal and vertical flow. Computed velocity vectors at Ra=10^6 and Pr=0.054 and (a) at Ha=0 and (b) at Ha=100 are compared in Fig. 8. The maximum vector decreased in (b).

4. EXPERIMENTAL MEASUREMENTS OF THE HEAT TRANSFER RATE OF NATURAL CONVECTION IN A CUBIC ENCLOSURE UNDER AN EXTERNAL MAGNETIC FIELD EITHER IN X-, Y- OR Z-DIRECTIONS.[43]

4.1. Experimental Apparatus

An experimental schematic is shown in Fig. 9. An experimental apparatus no.3 was placed between an electro magnet no.2 and both of them were kept in a temperature controlled compact room. The experimental apparatus was made of a plexiglass cubic enclosure whose vertical side wall was a 1 mm thick copper plate. The copper plate was heated electrically from its rear side. An opposing side wall was isothermally cooled. Temperatures were measured

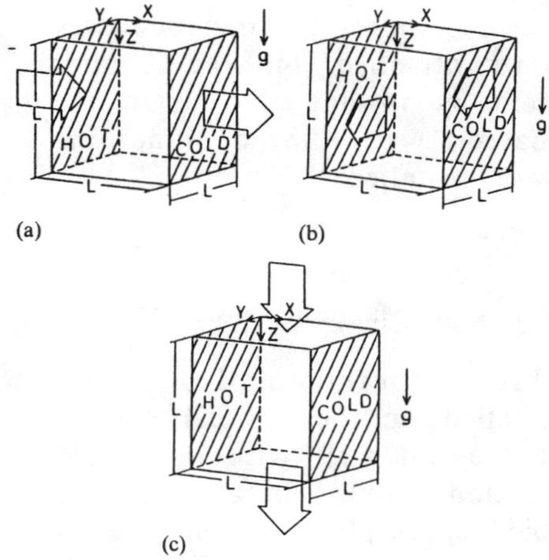

Fig.7 Schematics of the system: (a)magnetic field in the x-direction only, (b)magnetic field in the y-direction only, (c)magnetic field in the z-direction only.

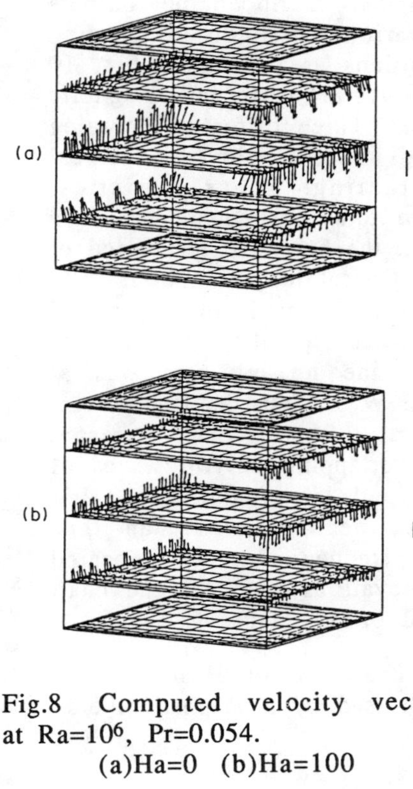

Fig.8 Computed velocity vectors at Ra=10^6, Pr=0.054.
(a)Ha=0 (b)Ha=100

with copper-constantan thermocouples. Thermocouples were attached on the copper plate by instant glue. On the heated wall, three junctions were installed.

4.2 Temperature Fluctuation without a Magnetic Field [44]

Three thermocouples were installed on the vertical surface of the hot wall with their cold junctions dipped in a hole of a cold copper plate at a mid height on the opposing vertical wall. Three thermocouples were located at a mid-height (z=o), 10mm upper (z=10mm) and lower levels (z=−10mm) on a 30mm high vertical hot wall. Transient thermocouple outputs at various strengths of heating conditions are shown in Fig. 10 without an external magnetic field. Three different recording speeds were employed to reveal fluctuating characteristics. When total heat input was 2.99 W on the hot wall, the period of the fluctuation is about 120 seconds. The fluctuating period decreased to about 7 seconds when the heating rate was at 14.93 W. When the heating rate was at 7.96 W, there appeared an overlapping of two waves of high and low frequencies. For higher heating rates than this, high frequency waves appeared to prevail as seen for the chart speed of 20 cm/hr.

4.3 The Rate of Heat Transfer Suppressed by a Magnetic Field

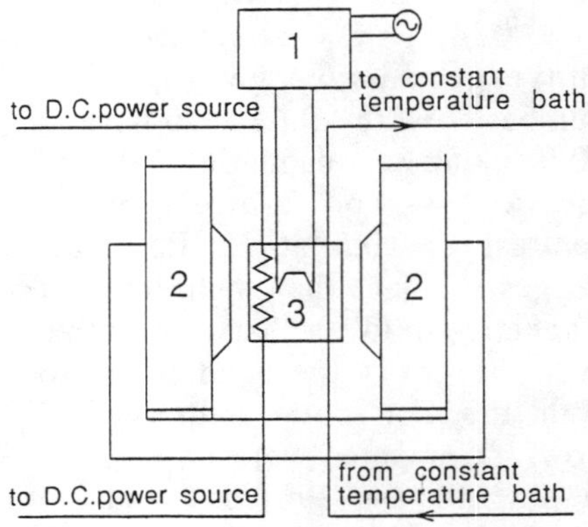

Fig.9 Schematics of the experimental set-up.

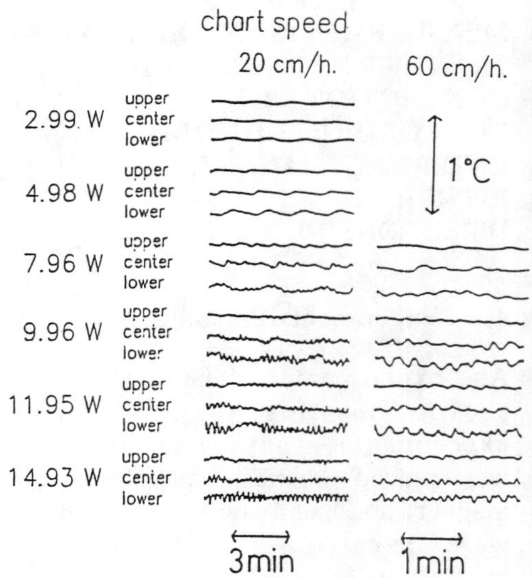

Fig.10 Temperature fluctuations on the heated wall recorded with three sets of thermocouples under various rates of heating.
Upper is at 10mm from the middle height, center is at the middle height and lower is at 10mm below the middle height.

Figure 11 shows steady state temperature differences $\Delta \theta$ [°C] at three locations of thermocouple (center and ± 10mm from a center) as a function of the strength of a magnetic field B_0 [Gs] when the total heat input was 2.99 W with a x-directional magnetic field. The temperature at 10 mm higher than the center on the hot wall was higher than those at middle or 10 mm lower level when the magnetic field was zero or weak. However, the difference became almost negligible when the magnetic field became 3,000 to 4,000 [Gs].

Similar graphs were obtained for higher heat inputs. Figure 12 shows a summary of these graphs under different heat inputs. The temperature difference is that between a central height of the hot and the cold walls for these plots. Hatched lines were obtained by separate experiments carried out with the convection enclosure heated from above and cooled from below. The data points nicely converged to the conduction lines as the strength of a magnetic field attained to 4,000 [Gs]. When the magnetic field is in the y-direction (horizontal and parallel to the heated wall), the temperature differences between a hot wall and a cold wall are shown in Fig. 13 for various heat inputs. Even at 4,000 [Gs], the temperature differences between the hot and

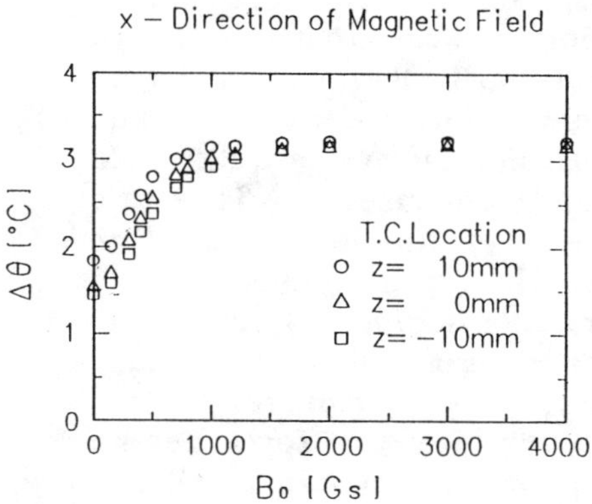

Fig.11 Measured temperature differences between a hot wall and a cold wall at a central height for a total heat input 2.99W with a magnetic field in the x-direction.

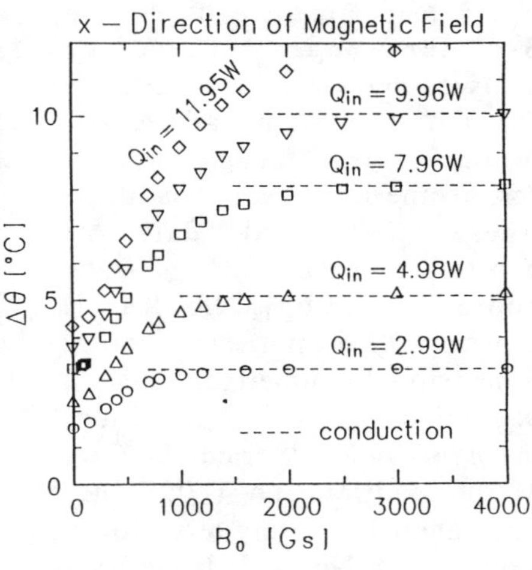

Fig.12 Steady state temperature differences between hot and cold walls under a x-directional magnetic field. Hatched lines are those obtained by other experiment heated from above.

cold walls are increasing, which indicates that convection becomes weak (but not to be zero) due to the magnetic force. These results are translated into dimensionless values Ra*, Nu and Ha numbers. They are plotted as Nu versus Ra* with the Hartmann number as a parameter in Fig. 14 for a case of a x-directional magnetic field. These experimental results can be employed for design purposes if they are correlated. Following Ozoe & Maruo [26], the average heat transfer rate Nu in a shape (Nu−1)/[Nu(Ha=0)−1] are plotted versus $Log_{10}(Ha/Ra^{*1/3})$ in Fig. 15. Symbols No.1 to No.3 are for the x-, y- or z-directional magnetic fields measured in ref. 43. Symbols No. 4 to 6 are those obtained by 3-dimensional numerical analyses by Ozoe & Okada [42] for the x-, y- or z-directional magnetic field though they are for isothermal wall conditions. These theoretical data are plotted versus the upper coordinate $Log_{10}(Ha/Ra^{1/3})$. Symbol No.7 indicates the 2-dimensional numerical analysis equivalent to the x-directional magnetic field reported by Ozoe, Maruo and Matsuo [39]. These data agreeds closely each other. Dotted line No.11 indicates heat transfer rates correlated from a two-dimensional numerical analyses by Ozoe & Maruo [26] for a z-directional magnetic field. It is to be noted that the z-directional magnetic field is almost equivalent to that in the x-directional one in suppressing

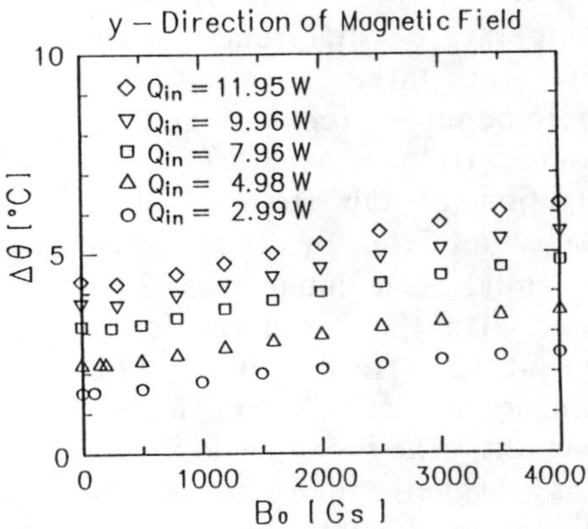

Fig.13 Measured temperature differences at various heat inputs under a y-directional magnetic field.

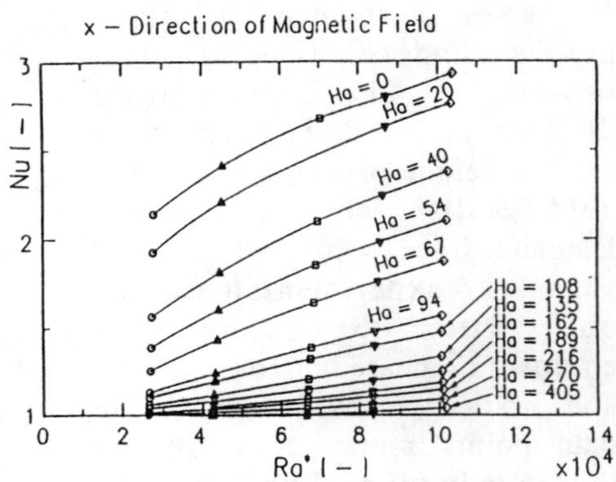

Fig.14 The effect of the Rayleigh and the Hartmann number on the average Nusselt number under a x-directional magnetic field.

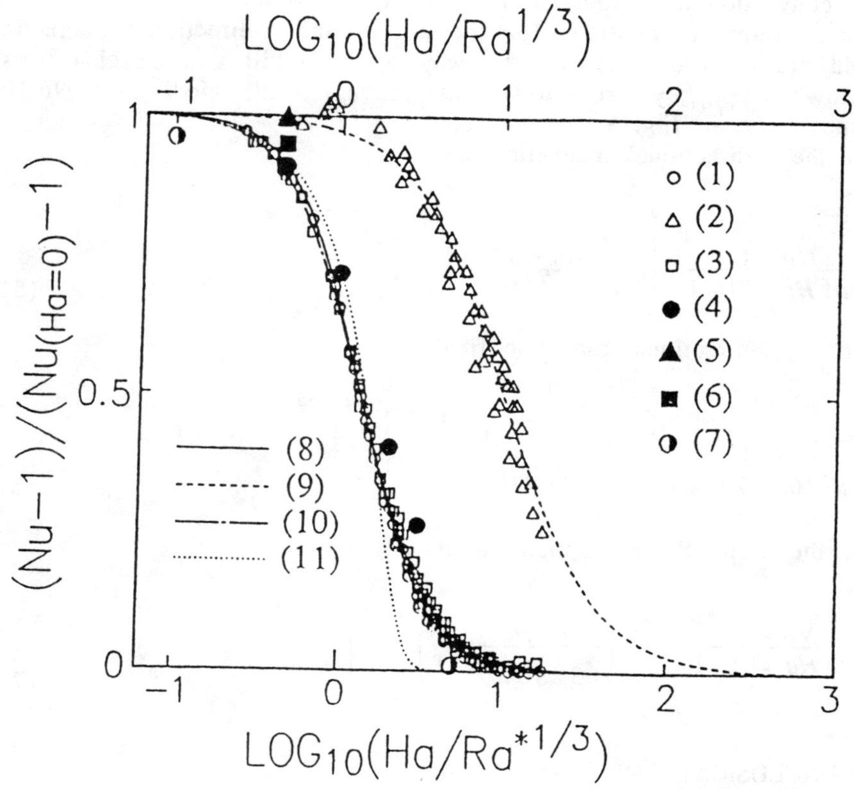

Fig.15 Universal plot for the rate of heat transfer under both the effect of the Rayleigh and the Hartmann Number.

(1) the x-directional magnetic field (Okada & Ozoe [43])
(2) the y-directional magnetic field (Okada & Ozoe [43])
(3) the z-directional magnetic field (Okada & Ozoe [43])
(4) the x-directional magnetic field (3/D analysis by Ozoe & Okada [42])
(5) the y-directional magnetic field (3/D analysis by Ozoe & Okada [42])
(6) the z-directional magnetic field (3/D analysis by Ozoe & Okada [42])
(7) the x-directional magnetic field (2/D analysis by Ozoe & Matsuo [40])
(8) the x-directional magnetic field (Okada & Ozoe [43], Eq.(5))
(9) the y-directional magnetic field (Okada & Ozoe [43], Eq.(6))
(10) the z-directional magnetic field (Okada & Ozoe [43], Eq.(7))
(11) the x-directional magnetic field (2/D analyses by Ozoe & Maruo [26])

the convection in contrast to that in the y-direction.
The correlation equations for the x-, y- and z-directional magnetic field are derived following the way by Churchill and Usagi [45] as follows and they are plotted as curves No.8, No.9 and No.10 respectively, in Fig.15.
For the x-directional magnetic field,

$$\frac{Nu-1}{Nu(Ha=0)-1}=\left[1+\left(\frac{Ha/Ra^{*1/3}}{1.3109}\right)^{2.19}\right]^{-1/1.029} \tag{5}$$

For the y-directional magnetic field,

$$\frac{Nu-1}{Nu(Ha=0)-1}=\left[1+\left(\frac{Ha/Ra^{*1/3}}{10.107}\right)^{1.758}\right]^{-1/1.165} \tag{6}$$

For the z-directional magnetic field,

$$\frac{Nu-1}{Nu(Ha=0)-1}=\left[1+\left(\frac{Ha}{1.3280\,Ra^{*1/3}}\right)^{1.9758}\right]^{-1} \tag{7}$$

5. CONCLUSIONS

Recent literatures on the magnetic Czochralski method were surveyed at the beginning and then our recent works on the related subjects were introduced for a rectangular enclosure. Two- and three-dimensional numerical analyses were primarily carried out for liquid metal especially on the effect of the direction of magnetic field and compared with experimental results.
The rates of heat transfer were measured and correlated as a function of Ra* and Ha number for the direction of a magnetic field either in perpendicular to the vertical hot wall (x-direction) or in parallel and horizontal to the hot wall (y-direction) or in vertical (z-direction).
The x- or z-directional magnetic fields were almost 10 times effective in suppressing the rate of heat transfer than that in the y-direction. The rates of heat transfer were correlated as a function of Ra* and Ha.

NOMENCLATURE

B_O	magnetic field	[Weber·m^{-2}]

b	height of an enclosure in Eq.(2)	[m]
B_Y	dimensionless magnetic field in y-direction	[–]
f	function	[–]
g	acceleration due to gravity	[m/s^2]
Ha	Hartmann number=$(\sigma_o/\mu)^{1/2}B_o l$	[–]
k	thermal conductivity of fluid	[J/(msk)]
l	distance between hot and cold wall	[m]
Nu	Nusselt number = $ql/[k(T_h-T_l)]$	[–]
Pr	Prandtl number = ν/α	[–]
Pr_m	magnetic Prandtl number = ν_m/α	[–]
q	heat flux	[J/(m^{2}s)]
Q_{in}	total heat supply	[J/s]
Ra	Rayleigh number = $g\beta(T_h-T_l)l^3/(\alpha\nu)$	[–]
Ra^*	Rayleigh number = $g\beta l^4 q/(\alpha\nu k)$	[–]
t	time	[s]
t_o	reference time = $x_o{}^2/\alpha$	[s]
T_h	hot wall temperature	[K]
T_l	cold wall temperature	[K]
U	u/u_o	[–]
u	velocity component in x-direction	[m/s]
u_o	α/x_o	[m/s]
V	v/u_o	[–]
v	velocity component in y-direction	[m/s]
W	w/u_o	[–]
w	velocity component in z-direction	[m/s]
X	x/x_o	[–]
x	coordinate	[m]
x_o	$=[g\beta(T_h-T_l)/(\alpha\nu)]^{-1/3}$	[m]
Y	y/x_o	[–]
y	coordinate	[m]
Z	z/x_o	[–]
z	coordinate	[m]

Greek letters

α	thermal diffusivity	[m^2/s]
β	volumetric coefficient of expansion	[1/K]
$\Delta\theta$	temperature difference between hot and cold walls	[K]
μ	viscosity	[kg/(m·s)]
μ_m	magnetic permeability	[henry/m]
ν	kinematic viscosity	[m^2/s]
ν_m	$=1/(\sigma_0\mu_m)$	[ohm·m^2/henry]

σ_0	electric conductivity	[1/(ohm·m)]
ϕ	function, see Eq.(4)	[–]
τ	$=t/t_0$	[–]
Ψ_c	central value of stream function	[–]

REFERENCES

1. Chandrasekhar, S., Hydrodynamic and hydromagnetic stability, *Oxford Univ. Press*, 1961.

2. Kakutani, T., Hydromagnetic flow due to a rotating disk, *J. Phys. Soc. Japan*, vol.17, No.9, pp.1496-1506, 1961.

3. Utech, H. P. and Flemings, M. C., Elimination of solute banding in indium antimonide crystals by growth in a magnetic field, *J. Appl. Phys.*, vol.37, No.5, pp.2021-2024,1966.

4. Witt, A. F., Herman, C. J. and Gatos, H. C., Czochralski-type crystal growth in transverse magnetic field, *J. Mat. Sci.*, vol.5, pp.822-824, 1970.

5. Hoshi, K., Suzuki, T., Okubo, Y. and Isawa, N., *The Electrochemical Society Extended Abstracts*, *St. Louis*, pp.811, 1980.

6. Hoshi, K., Isawa, N. and Suzuki, T., Growth of silicon monocrystals in a magnetic field, *Applied Physics(Japanese)*, vol.53,.No.1, pp.38-41, 1984.

7. Fukuda, T. and Terashima, K., Growth of GaAs single crystal by magnetic field applied LEC Technique, *Applied Physics (Japanese)*, vol.53, No.1, pp.42-47, 1984.

8. Langlois, W. E. and Lee, Ki-Jun, Czochralski crystal growth in an axial magnetic field: Effects of joule heating, *J. Crystal Growth*, vol.62, pp.481-486, 1983.

9. Oreper, G. M. and Szekely, J., The effect of an externally imposed magnetic field on buoyancy driven flow in a rectangular cavity, *J. Crystal Growth*, vol.64, pp.505-515, 1983.

10. Langlois, W. E., Computer simulation of Czochralski melt convection in a magnetic field, *J. Cryst. Growth*, vol.70, pp.73-77, 1984.

11. Hoshikawa, K., Kohda, H., and Hirata, H., Homogeneous dopant distribution of silicon crystal grown by vertical magnetic field -

applied Czochralski method, *Japanese J. Appl. Phys.*, vol.23, No.1, pp.L37-L39, 1984.

12. Hurle, D. T. J. and Series, R. W., Effective distribution coefficient in magnetic Czochralski growth, *J. Cryst. Growth*, vol.73, pp.1-9, 1985.

13. Organ, A. E., Flow patterns in a magnetic Czochralski crystal growth system, *J. Cryst. Growth*, vol.73, pp.571-582, 1985.

14. Mihelcic, M. and Wingerath, K., Numerical simulation of the Czochralski bulk flow in an axial magnetic field: Effects on the flow and temperature oscillations in the melt, *J. Cryst. Growth*, vol.71, pp.163-168, 1985.

15. Hjelming, L. N. and Walker, J. S., Modelling of melt motion in a Czochralski crystal puller with an axial magnetic field, ibid, vol.79, pp.96-98, 1986.

16. ibid, Melt motion in a Czochralski crystal puller with an axial magnetic field: isothermal motion, *J. Fluid Mech.*, vol.164, pp.237-273, 1986.

17. ibid, Melt motion in a Czochralski crystal puller with an axial magnetic field: motion due to buoyancy and thermocapillarity, ibid, vol.182, pp.335-368, 1987.

18. ibid, Mass transport in a Czochralski puller with a strong magnetic field, *J. Crystal Growth*, vol.85, pp.25-31, 1987.

19. ibid, Melt motion in a Czochralski crystal puller with an axial magnetic field: Uncertainty in the thermal constants, ibid, vol.87, pp.18-32, 1988.

20. ibid, Mass transport in a Czochralski crystal puller with an axial magnetic field: Melt motion due to crystal growth and buoyancy, ibid, vol.92, pp.371-389, 1988.

21. Kim, K. M. and Langlois, W. E., Computer simulation of Boron transport in magnetic Czochralski growth of silicon, *J. Electrochem. Soc.*, vol.133, No.12, pp.2586-2590, 1986.

22. Kobayashi, S., Effects of an external magnetic field on solute distribution in Czochralski grown crystals - A theoretical analysis, *J. Crystal Growth*, vol.75, pp.301-308, 1986.

23. ibid, Numerical analysis of oxygen transport in magnetic Czochralski growth of silicon, ibid, vol.85, pp.69-74, 1987.

24. Langlois, W. E., Effects of the finite electrical conductivity of the crystal on hydromagnetic Czochralski flow, ibid, vol.83, pp.51-61 1987.

25. Riley, N., Boundary-layer models of magnetic Czochralski crystal growth, ibid, vol.85, pp.417-421, 1987.

26. Ozoe, H. and Maruo, E., Magnetic and gravitational natural convection of melted silicon - Two-dimensional numerical computations for the rate of heat transfer, *JSME International J.*, vol.30, No.263, pp.774-784, 1987.

27. Brown, R. A., Theory of transport processes in single crystal growth from the melt, *AIChE Jl.*, vol.34, No.6, pp.881-911. 1988.

28. Cartwright, R., Ilegbusi, O. J. and Szekely, J., A comparison of order-of-magnitude and numerical analyses of flow phenomena in Czochralski and magnetic Czochralski systems, *J. Cryst. Growth*, vol.94, pp.321-333, 1989.

29. Hicks, T. W., Organ, A. E. and Riley N., Oxygen transport in magnetic Czochralski growth of silicon with a non-uniform magnetic field, ibid, vol.94, pp.213-228, 1989.

30. Kerr, O. S. and Wheeler A. A., The effect of a magnetic field on the flux of a contaminant dissolving into the crucible wall boundary layer in Czochralski crystal growth, ibid, vol.96, pp.915-932, 1989.

31. Series, R. W., Czochralski growth of silicon under an axial magnetic field, ibid, vol.97, pp.85-91, 1989.

32. Riley, N., Species transport in magnetic field Czochralski growth, ibid, vol.97, pp.76-84, 1989.

33. Munakata, T. and Tanasawa, I., A study of the onset of oscillatory flow in a Czochralski growth melt and its suppression by a magnetic field, *Nihon Kikai Gakkai Ronbunshu(Transaction of JSME)*, series B, vol.55, No.517, pp.2610-2617, 1989.

34. Brown, R. A., Kinney, T. A., Sackinger, P. A. and Bornside, D. E., Toward an integrated analysis of Czochralski growth, *J. Crystal Growth*, vol.97, pp.99-115, 1989.

35. Ozoe H. and Okada, K., The effect of the direction of the external magnetic field on the three-dimensional convection in a cubical enclosure, *Int. J. Heat Mass Transfer*, vol.32, No.10, pp.1939-1954, 1989.

36. Hirata, H. and Hoshikawa, K., Homogeneous increase in oxygen concentration in Czochralski silicon crystals by a cusp magnetic field, *J. Cryst. Growth*, vol.98, pp.777-781, 1989.

37. Ozoe, H. and Toh, K., Numerical computation for a Czochralski bulk flow of liquid metals under a vertical external magnetic field, *Proc. 9th Int. Heat Transfer Conf., Jerusalem,* vol.6, pp.311-316, 1990.

38. Langlois, W. E., Fluid dynamics in crystal growth, *Proc. of 22nd ICHMT International Symposium on Manufacturing and materials processing, Dubrovnik, Keynote Lecture,* Aug., 1990.

39. Ozoe, H. & Maruo, E. and Matuo, H., Numerical analyses of transient natural convection of liquid metals after a step change in an external lateral magnetic field, *Kagaku Kogaku Ronbunshu,* vol.16, No.5, pp.990-997, 1990.

40. Ozoe, H. and Matsuo, H., Transient numerical computations for natural convection of liquid metal in a tall enclosure heated from side and with a lateral external magnetic field, *Num. Methods in Thermal Problems, Swansea*, vol.6, part 1, pp.440-450, 1989.

41. Nakamura, S., Hibiya, T., Yamamoto, F. and Yokota, T., Thermal conductivity measurement of mercury under a magnetic field, *10th Japan Symp. on thermophysical Properties,* pp.91-94, 1989.

42. Ozoe. H., and Okada, K., The effect of the direction of the external magnetic field on the three-dimensional natural convection in a cubical enclosure, *Int. J. Heat Mass Transfer*, vol.32, No.10, pp.1939-1954, 1989.

43. Okada, K., and Ozoe. H., in review.

44. Okada, K., and Ozoe. H., in review.

45. Churchill, S. W. & Usagi, R., A general expression for the correlation of rates of transfer and other phenomena, *AIChE Journal*, vol.18, No.6, pp.1121-1128, 1972.

Heat Transfer During Liquid Encapsulated Czochralski Growth of GaAs and Associated Defect Formation in the Crystal

KEVIN W. KELLY, KEITH KOAI, and SHAHRYAR MOTAKEF
Department of Mechanical Engineering
Massachusetts Institute of Technology
Cambridge, Massachusetts 02139, USA

1. ABSTRACT

Modelling of heat transfer and the associated mechanisms of dislocation multiplication during LEC growth of GaAs are reviewed. Results obtained from the numerical simulation of LEC process, factors influencing the accuracy of modelling results, and impact of modelling on crystal growth activities are discussed.

2. INTRODUCTION

Modelling of heat transfer during liquid encapsulated Czochralski (LEC) of GaAs, Fig. 1, is motivated by the need to understand the genesis of, and control generation of temperature and temperature-gradient related defects in the grown material. The primary crystalline defect in melt-grown GaAs crystals is high dislocation densities of the order of 10^4 - 10^5 cm^{-2}. These values of dislocation densities are much higher than those present in, for example, Si or Ge crystals solidified by the Czochralski technique. This difference may be attributed to several factors: (a) nucleation of dislocation sources by coalescence of point defects is more readily possible in compound semiconductors where the melt stoichiometry has a significant influence on the point defect density in the grown solid, (b) thermal gradients (and thus thermal stresses) are much higher in GaAs than in similar-size Si and Ge crystals grown from the melt, partly because of the lower thermal conductivity of GaAs and partly due to liquid encapsulation of the melt by a semi-transparent medium, and (c) the resistance of GaAs to plastic deformation is much lower than its elemental counterparts Si and Ge. Thermal modelling of LEC growth of GaAs aims at identifying the magnitude of thermal stresses prevailing in the crystal during growth, and establishing the effectiveness of various approaches to reducing these stresses to levels considered to be sufficiently low.

In this document we begin with a background summary outlining: (a) influence of dislocations on the performance of various electronic devices, (b) the relationship between thermal field and dislocation multiplication during growth of semiconductors, and (c) various models of LEC growth process. We then provide a more detailed description of our numerical model of the LEC process, its predictions regarding the stress distribution in LEC-grown crystals, and the thermal field requirements for lowering of stresses in the crystal. The influence of uncertainties in the thermophysical property data-base, thermal boundary conditions, and convection in the melt on the accuracy of modelling predictions are presented.

3. BACKGROUND

3.1 The Influence of Dislocations on Device Performance

Gallium arsenide (GaAs) is a III-V compound semiconductor with a high electron mobility (five times larger than that of Si at room temperature) and a direct band gap. These two properties make it the material of choice in the manufacture of electro-optical devices (such as solid state lasers and detectors), and high speed computing structures.

In addition to controlling the plastic deformation of semiconductors, dislocations may also be expected to influence the electronic properties of the matrix: wherever dislocations are present the lattice is strained and atomic bonds are broken. For example, when a plane of arsenic atoms is removed from the lattice (most easily visualized by arsenic vacancy collapse) an edge dislocation is formed. The remaining gallium atoms surrounding the dislocation strongly attract local free electrons and alter the free carrier density in their vicinity. An analogous example could be made for the case when a plane of gallium atoms is removed from the matrix. The numerous broken bonds generated at each dislocation might serve as midgap energy levels and change the electrical characteristics of the material. Also, the disruption in the lattice caused by the presence of dislocations might be expected to reduce the material's carrier mobility. Therefore an uneven dislocation density distribution across the wafer surface might cause variations in device performance.

A number of investigations have focused on the effect of dislocations on the performance of field effect transistors (FET). Miyasawa et al. [1] measured the threshold voltage of an array of FETs, etched the array in KOH etch to determine the location of dislocations on the wafer surface, and determined that the threshold voltage was a strong function of the FET channel proximity to dislocations. Matsumura et al. [2] showed that the resistivity of a GaAs wafer could decrease by as much as an order of magnitude in undoped semi-insulating GaAs near both dislocation-clustered networks and slip lines. Nanishi [3] examined FET performance distribution in low chromium doped LEC GaAs and showed that, for a given voltage, the source to drain current value of a particular FET corresponded to the local dislocation density of the wafer. Winston et al. [4] also examined the effect of dislocations on GaAs FETs and proposed that the threshold voltage varied with the dislocation density distribution. This study, however, noted that the threshold voltage of a FET was *not* a function of its distance to the nearest neighboring dislocation as was proposed by similar studies. Furthermore, O'Neill et al. [5] compared FET characteristics on wafers with very different dislocation densities and determined that dislocation density did not effect device performance.

In contrast to the inconclusive results of the above studies on majority carrier devices, investigations on minority-carrier devices have unanimously concluded that the performance characteristics of lasers and light emitting diodes (LEDs) are adversely effected by the presence of dislocations. For example, Roedel et al [6] have shown that the efficiency of GaAs LEDs was inversely proportional to the density of dislocations within the material. This result suggested that dislocations act as indirect recombination centers. Furthermore, it established that dislocations propagate from the substrate into the epitaxial material deposited on the substrate surface. Thus suggesting that epitaxial growth of GaAs layers on dislocated substrates does not totally eliminate the influence of substrate property on device performance. Another investigation, Petroff and Hartman [7], noted that the location of black lines appearing in GaAs lasers, as well as degradation of their performance, are related to the presence of dislocations in the laser structure.

Dislocations also exert a deleterious influence on wafer-scale yield of GaAs devices. For example, ref. [8] reports that the low yield of 16 kbit static random access memory (SRAM) structures using In-alloyed substrates with the relatively low dislocation density of 300 cm^{-2} can be increased ten-fold from the current rate of 1% if the dislocation density is reduced to lower than 10 cm^{-2}.

3.2 Models of LEC Growth of GaAs

The fact that thermal stresses cause dislocation multiplication in semiconductor crystals was first established by Penning [9]. In this study, he slowly cooled germanium crystals and noted that the dislocation density within the crystal varied radially from a maximum at the edge, through a minimum about halfway across the crystal with another local maximum in the center. Penning also noted that the dislocation patterns reflected the crystallographic orientation of the growth direction. He proposed that thermal stresses induced by radial temperature gradients in the crystal boule were relieved plastically once they exceeded some critical value on the slip planes of the crystal. Thus plastic deformation was considered to be a source of localized dislocation generation and multiplication. Jordan, Caruso and vonNeida [10] used this framework, in their pioneering work, to relate the elastic thermal stresses calculated to be present in the GaAs crystal during growth to its dislocation density. Whereas the role of thermal stresses on dislocation multiplication is qualitatively well established, the mathematical model of this process is not well formulated. A short description of two complementary models is given in the following.

The, so-called, thermoelastic model of Jordan et al. [10] is based on Schmid's rule which states that glide along the crystallographic slip planes of the matrix occurs whenever the shear stress on any of the slip systems exceeds the critical resolved shear stress (CRSS) of the material. In the thermoelastic approach the elastic stresses, obtained from the calculated temperature distribution in the crystal, are projected onto the 12 slip systems of the material ($\{111\} < 1\bar{1}1 >$ for growth along the <100> direction). Whenever these stresses exceed the CRSS of the matrix, plastic deformation and dislocation multiplication is expected. Mathematically this can be written as:

$$\sigma_e^i = | \sigma_{rs}^i - CRSS | \quad | \sigma_{rs}^i | > CRSS \quad i = 1, 12 \quad (1)$$
$$\sigma_e^i = 0 \quad | \sigma_{rs}^i | < CRSS \quad i = 1, 12 \quad (2)$$

In the above σ_{rs}^i is the resolved shear stress on the ith slip system, and σ_e^i is the excess stress at the ith slip system. The total local driving force for plastic deformation of the crystal is taken to be proportional to the sum of excess stresses in all glide operations:

$$\overline{\sigma_e} = \sum_{i=1}^{12} \sigma_e^i \quad (3)$$

Jordan and colleagues extended the above formulation and proposed a relationship between the dislocation density and $\overline{\sigma_e}$.

An alternate formulation relating thermal stresses to dislocation multiplication can be based on the micro-yield mechanism of dislocation multiplication in diamond crystals as proposed by Alexander and Haasen (AH) for elemental semiconductors [11]. In this approach dislocation multiplication is a continuous function of stresses in the crystal and there are no 'critical' threshold stress levels. The AH creep model relates the creep strain rate to the velocity of mobile dislocations; the velocity of dislocations are experimentally observed to be related to the shear stress τ on the slip plane. The creep strain rate can be written as:

$$\frac{d\epsilon_c}{dt} = \phi N b B_o e^{\frac{-Q}{kT}} \cdot (\tau - A\sqrt{N})^m \quad (4)$$

In the above ϕ is a geometrical factor relating shear strain to normal strain, B_o can be considered to be the dislocation mobility, b the length of the burgers vector, N the number of mobile dislocations, Q the activation energy, T the temperature, A the self-stress coefficient related to presence of dislocations in the matrix, and m an exponent of the order 1. Dislocation multiplication rate for these materials has been found to be proportional to the density and velocity of mobile dislocations and the 'effective' stress $(\tau - A\sqrt{N})$ to some exponent p:

$$\frac{dN}{dt} = K B_o N e^{\frac{-Q}{kT}} \cdot (\tau - A\sqrt{N})^{m+p} \quad (5)$$

In the above K is the dislocation multiplication factor. The self-stress term $A\sqrt{N}$ in eqs. (4) and (5) can be easily shown to be negligible for typical single-crystal semiconductors and processing parameters, and can be neglected. The exponent p has been found to be equal to 1 for Si, Ge, and InP. The primary obstacle in applying the AH model to melt growth of GaAs was related to absence of numerical values for the material properties appearing in eqs. (4) and (5). Motakef has recently obtained these values, with the exception of dislocation multiplication factor K, for GaAs [12].

The two dislocation-stress models described above have been shown to be mutually consistent by Motakef [13]. However, as the dislocation density is related to the time-integral of stresses in the crystal, the excess stresses calculated on the basis of the Schmid's rule should be considered to be diagnostic indicators of plastic deformation of the crystal, and cannot be related to the magnitude of dislocation density in the matrix. In this context it is important to note that Volkl and Muller [14] have experimentally obtained the material constants of the AH model for InP, and have shown a remarkably good agreement between the dislocation density predicted by the AH model and that observed in a grown sample. Maroudas and Brown [15] have obtained asymptotic solutions to the AH model and predicted significant dislocation multiplication at the solid/liquid interface. Motakef has performed a similar analysis and obtained close-form solutions for dislocation density of the material both at the growth interface and subsequent stages. Motakef [13] disagrees with Maroudas and Brown on the applicability of the AH model to the material undergoing phase change and suggests that the massive dislocation multiplication suggested in ref. [15] to occur at the growth interface cannot be supported by the available experimental data.

Regardless of which model of dislocation multiplication is used for relating thermal stresses to dislocation multiplication in the crystal, the predictive reliability of the process model in contingent on the accurate calculation of temperature field, and thus thermal strains, in the growing crystal. The first quantitative thermal model of LEC growth of GaAs is due to Jordan, Caruso and vonNeida [10]. Using a flat-ended cylinder as the geometric model of the crystal, they modelled heat transfer from the crystal to consist of convective loss to the encapsulant and the ambient gas. Each of these two heat loss mechanisms were characterized by a heat transfer coefficient and sink temperature. The obtained analytical temperature distribution in the crystal was used to calculate stresses in the crystal from the plane-strain formulation of linear thermoelasticity equations. The thermal model of ref. [10] has been shown to be deficient in several respects. First, Motakef [16] and later Hicks [17] showed that convective heat loss to the encapsulant is much smaller than the simple one-dimensional correlation used in ref. [10] suggests. Furthermore, Motakef and Witt [18] showed that the infra-red transmittance of the encapsulant plays a significant role in generation of temperature gradients in the crystal. Subsequent measurement of infra-red spectral transmittance of the encapsulant showed that the encapsulant is semi-transparent to infra-red radiation. These considerations, in addition to others, have prompted development of more sophisticated mathematical models of the growth system.

Models of thermal transport during LEC growth of GaAs can be separated into three categories. The first category considers the thermal interaction of the crystal with its environment in terms of average heat transfer coefficients and single point sink/source temperatures or temperature distributions. The second category recognizes the complex thermal coupling in the growth process, and models the entire growth system including the crystal, melt encapsulant, crucible support assembly, and heater; in this category of models convective heat transfer in the melt and encapsulant are ignored. The third category of models include convection in the melt. The first category is essentially based on the work of Jordan and colleagues [10]. Duseaux and Jacob [19], Duseaux [20], Schvezov et al. [21-23], and Meduoye et al. [24] have calculated temperature and stress fields in the crystal using the heat transfer coefficients of Jordan et al., and experimentally measured temperature profiles of the growth furnace. The accuracy of these works is limited, as explained in the above, by the inaccuracy of values of heat transfer coefficients assumed to be present in the growth system. In addition, this approach suffers from its innate shortcoming in dealing with the solid-liquid interface shape, in the sense that the growth morphology cannot be calculated self-consistently, and thus calculated stress close to the growth front cannot reflect the state of the crystal. The second class of models captures all elements of the growth process which are considered to be influence the temperature field in the crystal. System-scale modelling of LEC and CZ processes have been reported in refs. [25-30]. These models aim at self-consistent calculation of temperature field in the growth furnace. They all ignore convective heat transfer in the melt, and thus the solid-liquid interface shape can be expected be inaccurate. We will later show that this inaccuracy introduces a second order effect on the stress distribution in the crystal. The third class of problems extends the capabilities of the previous category through inclusion of convection in the melt [31,32]. Natural convection in melt sizes of interest to crystal growth is highly turbulent, and three dimensional. Therefore, the extra effort needed to correctly model convection in melt-growth process is both large and computationally intensive. It is not clear at this point whether such an expenditure of effort will provide a commensurate-scale understanding of LEC and CZ growth processes.

4. MODELLING OF LEC GROWTH OF GaAs

4.1 The Model

In this section a brief description of the numerical model of LEC growth of GaAs is presented. A more detailed description is given in ref. [33]. The LEC model is intended to capture all thermal interactions present in the growth system, and provide an accurate tool for control of thermal stresses during growth of GaAs. The model of the LEC system is shown in Fig. 2. It consists of all elements of the growth system including the crystal, melt, encapsulant, crucible support system, water-cooled crucible shaft, and graphite heater. The transport of energy within the melt, crystal and all other solid elements is considered to be by conduction and, thus, convective heat transfer in the melt is not considered. Radiative interaction between the crystal, melt, semi-transparent encapsulant, crucible support assembly, the base plate, water-cooled shaft, and the heater are modelled as radiation between diffuse gray surfaces, and incorporated in the model as boundary conditions. The crystal-melt interface morphology is calculated consistently through use of Hurle's capillarity equations [34]. The elastic stress field is calculated from the obtained temperature distribution in the crystal by solving the axi-symmetric form of the linear thermoelasticity equations. In order to use the Schmid's law, the stresses are resolved onto the slip system of the crystal which has 12 possible active directions. The excess of the resolved shear stresses over the CRSS of the matrix, eqs. (1)-(3), is taken as the measure of plastic flow and dislocation multiplication in the crystal. In order to quantify the extent of plastic deformation forces in relation to the resistance of the matrix to slip, $\overline{\sigma_e}$ is non-dimensionalized by the product of number of active slip systems and the local (temperature dependent) value of CRSS:

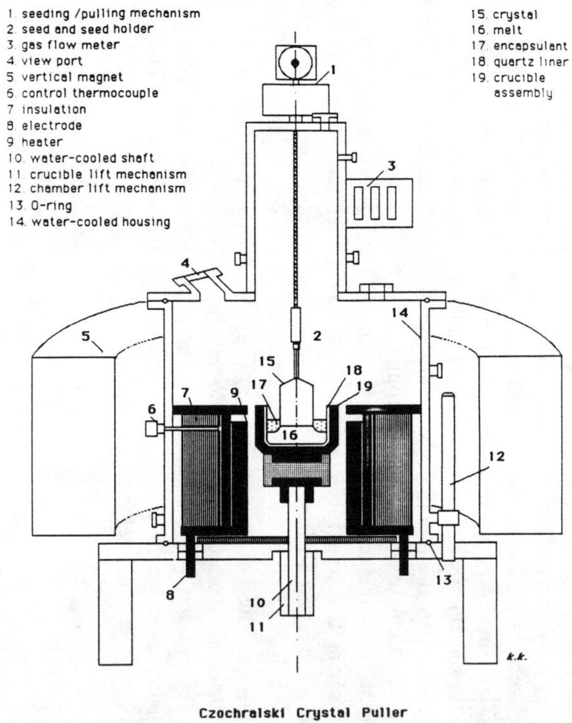

Fig. 1 Schematic of LEC growth process.

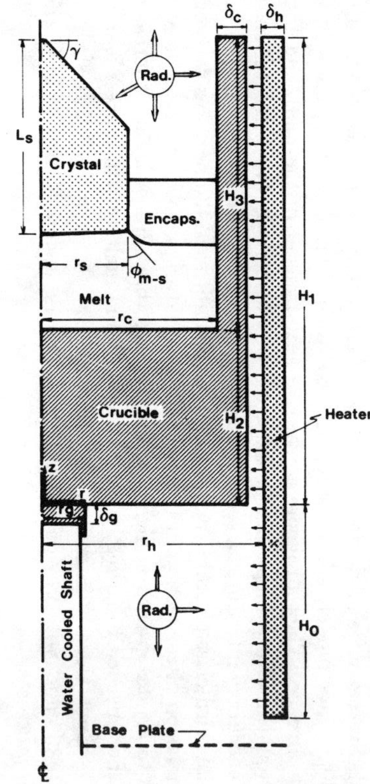

Fig. 2 Computational domain of the LEC model.

$$\sigma_e(r,z) = \frac{\overline{\sigma_e}(r,z)}{n \times CRSS[T(r,z)]} \tag{6}$$

4.2 Some Modelling Results

A more detailed presentation of LEC modelling results is given in refs. [33,35]. In this section we briefly enumerate some of the salient findings. In Figs. 3(a)-(c) the calculated temperature and excess stress fields at three crystal growth stages are shown. The temperature field is shown in non-dimensional form ($\theta = T/T_f$ where T_f is the melting point of the matrix), and the regions of non-zero excess stress are dotted with the iso-excess stress line of 1 shown as a dotted line. The isotherms are plotted at $\Delta\theta = 0.02$ (30K) intervals. These calculations correspond to growth of 7.5 cm crystals from a 5 Kg charge, and assume adiabatic conditions at the bottom face of the crucible support assembly. During the early stages of growth (Fig. 3(a)) large excess stresses are observed close to the growth interface and the cone region protruding out of the encapsulant. The former is associated with the non-planarity of the growth interface, and the latter with the step-wise change in the mode of heat transfer from the crystal as it emerges out of the encapsulant. Similar results are observed at the intermediate stages of growth, Fig. 3(b). At the final stages of growth, Fig. 3(c), the crystal is observed to experience stresses in its mid-section due to *overheating* by the crucible. This finding indicates that excessive heat input into as well as excessive heat extraction from the crystal can result in plastic deformation of the crystal.

In summary, the modelling results indicate that in addition to heat transfer at the crystal periphery, the morphology of the growth interface and heat transfer at the cone region, especially as it leaves the encapsulant, control the extent of thermal stresses in the crystal. The results also indicate the inadequacy of conventional LEC systems for growth of crystals with stresses below the CRSS of the matrix.

4.3 Required Degree of Thermal Control

The required thermal conditions for control of stresses in LEC-grown GaAs crystals at below the CRSS of the solid are identified in refs. [36-38]. The maximum allowable heat transfer rate at the crystal periphery, q_s^{max}, can be identified through solution of an inverse thermo-elastic problem where the maximum stress in the crystal is set at equal to CRSS of the matrix. The following relationship is, then, obtained:

$$q_s^{max} = \frac{1}{c}\frac{(1-\nu)k}{\alpha E r_c} CRSS \tag{7}$$

In the above c is a numerical constant which relates the stresses to the temperature field in the material and reflects the slip system of the material ($c = 1./8.8$), r_c crystal radius, α thermal expansion coefficient, ν Poisson's ratio, E Young's modulus, and k thermal conductivity of the crystal. Results indicate that with increasing crystal length (lower matrix temperatures) the magnitude of allowable heat loss from the periphery of the crystal increases. Nevertheless, the magnitude of this heat loss is not appreciable and the ratio of allowable radial heat loss to the total heat loss (sum of heat loss from the cone and periphery of the boule) is nearly zero. The maximum allowable radial temperature difference between the center and periphery of the crystal, ΔT_r, can be related to q_s^{max} as:

$$q_s^{max} = k\frac{\Delta T_r}{r_c} \tag{8}$$

Introducing eq. (8) into eq. (9) yields the following for ΔT_r :

$$\Delta T_r = \frac{1}{c}\frac{(1-\nu)}{\alpha E}CRSS\,, \tag{9}$$

Equation (9) is of the order of 3-5 K for GaAs at elevated temperatures. This clearly indicates that nearly adiabatic conditions must be established at the crystal periphery during growth.

Other issues related to the influence of the cone-angle and solid-liquid interface morphology on the stresses in the crystal are given in refs. [36,37].

5. FACTORS INFLUENCING ACCURACY OF MODEL CALCULATIONS

Three factors influence the accuracy of model calculations. These are (i) thermophysical properties of materials present in the growth system, (ii) thermal boundary conditions, and (iii) the assumption of low Peclet number in the melt. In this section we investigate the influence of these three factors.

5.1 Thermophysical Properties The uncertainty in the values of thermophysical properties used in the model calculations, in addition to their possible evolution during a growth cycle (e.g. changes in the surface emissivity of the crystal associated with evaporation of As) may adversely influence the accuracy of modelling predictions. We have investigated this issue by varying the property values in a reasonable range and observing the influence on the calculated stresses. To this end a standard case is defined with thermophysical property values reported in [35]. Figure 4 summarizes the results of this sensitivity analysis. The maximum stresses are shown for each case corresponding to a change in the value of the property by the indicated factor. It can be readily observed that the most important factor influencing the modelling results is the emissivity of the unencapsulated portion of the crystal. A study involving parameters other than the ones reported here are given in ref. [39].

Another property which dramatically influences the modelling results is the infra-red transparency of the liquid encapsulant. The importance of this parameter was first identified in ref. [18] which motivated its experimental evaluation. In Fig. 5 the

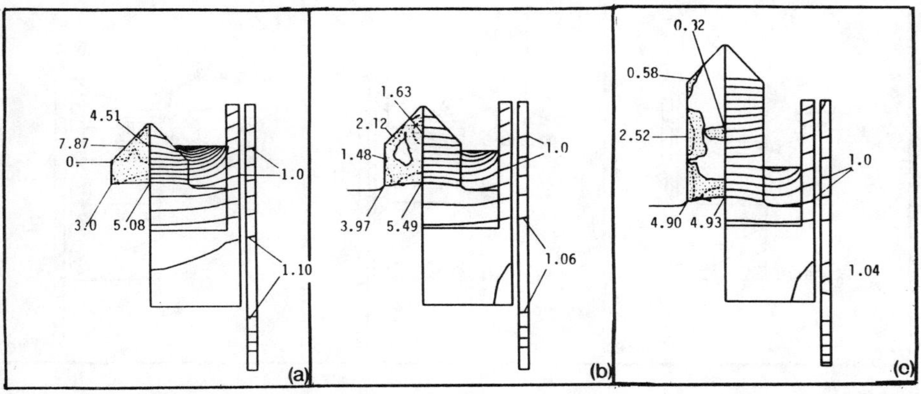

Fig. 3 Temperature and stress distribution in the crystal at three stages of growth.

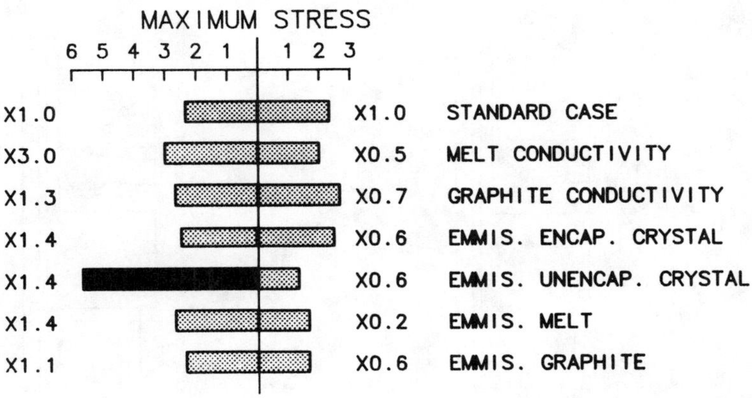

Fig. 4 Influence of uncertainty in various properties on maximum value of calculated stresses.

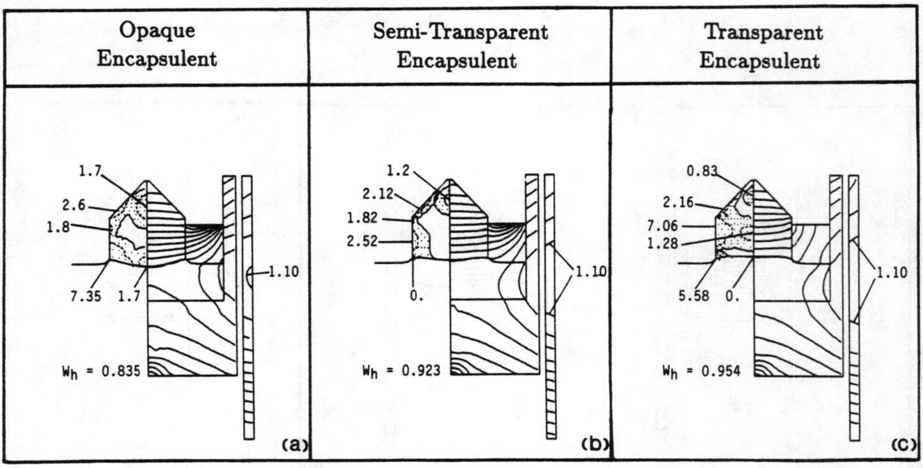

Fig. 5 Influence of infra-red transmittance of the encapsulant on temperature and stress field in the crystal.

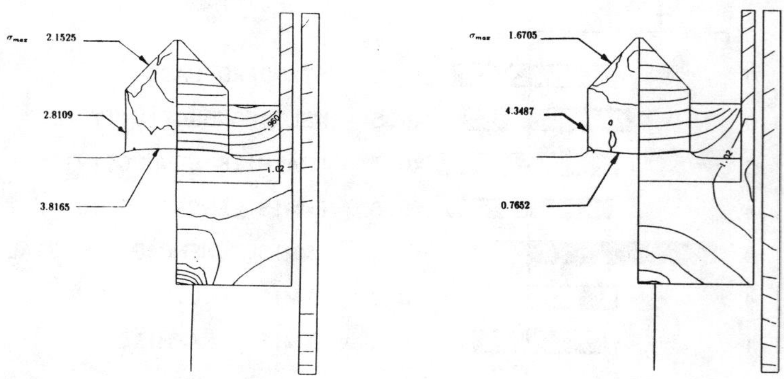

"Best Guess" Old
Boundary Conditions

New Measured
Boundary Conditions

Fig. 6 The calculated temperature and stress field associated with two different thermal boundary conditions at the bottom of the growth interface.

stress and temperature distributions in a GaAs crystal are shown for the two limiting cases of thermally-transparent and thermally-opaque encapsulants, and for the semi-transparent boric oxide. Results clearly indicate the influence of this parameter.

5.2 Thermal Boundary Conditions

The thermal boundary conditions in LEC growth system are difficult to quantify. Although in system-scale models, such as the present one, many of the ambiguities associated with the thermal characteristics of the growth environment are eliminated through inclusion of all major furnace elements in the model (e.g. crucible support assembly), inaccuracies associated with thermal boundary conditions at the lower part of the crucible support assembly (the enclosure consisting of the water-cooled support shaft, lower part of the heater and the bottom of the graphite susceptor) can significantly influence the results. In one set of experiments we measured the temperature of the graphite felt covering the lower part of the water-cooled furnace housing (this is the fourth side of the enclosure described above). We had originally estimated this to be close to the temperature of the water-cooled housing, and thus, have negligible radiative emission. Experiments, however, showed this to be a wrong assumption. The temperature was measured, at growth conditions, to be in the 900-1100 K range. Figure 6 shows the difference in the calculated temperature and stress distributions for the two cases of 'best guess' and measured boundary conditions. The difference in the stress states, as well as the morphology of the growth interface are significant.

5.3 Convective Effects

The last factor that may influence the accuracy of modelling results is convective heat transfer in the melt. This effect clearly influences the morphology of the solid-liquid interface, overall temperature distribution in the system, and the stresses in the crystal close to the growth interface. As convection in the melt is highly turbulent and three-dimensional, the possible influence of convection in the melt on the stresses in the crystal cannot be accurately investigated through numerical simulation. To address this issue, we resorted to comparison of the experimentally observed solid-liquid interface (assumed to be identical to the radial dopant striation) and results of a numerical simulation performed at conditions *identical* to the growth experiment. The calculated solid-liquid interface shape and stresses in that area are shown in Fig. 7(a). The measured solid-liquid interface shape of the grown crystal was used in the companion numerical simulation. In this simulation the heat flux associated with release of solidification energy at the growth interface was artificially modulated to yield a solid-liquid interface shape similar to the one experimentally obtained. The local stress distribution was, then, calculated. These results are shown in Fig. 7(b). The interface morphology of the grown solid is doubly-curved reflecting the influence of forced and natural convection on heat transfer at the growth interface, whereas the calculated one, Fig. 7(a), is concave. The discrepancy on the overall non-planarity of the interface is, however, very small. The stress distribution calculated based

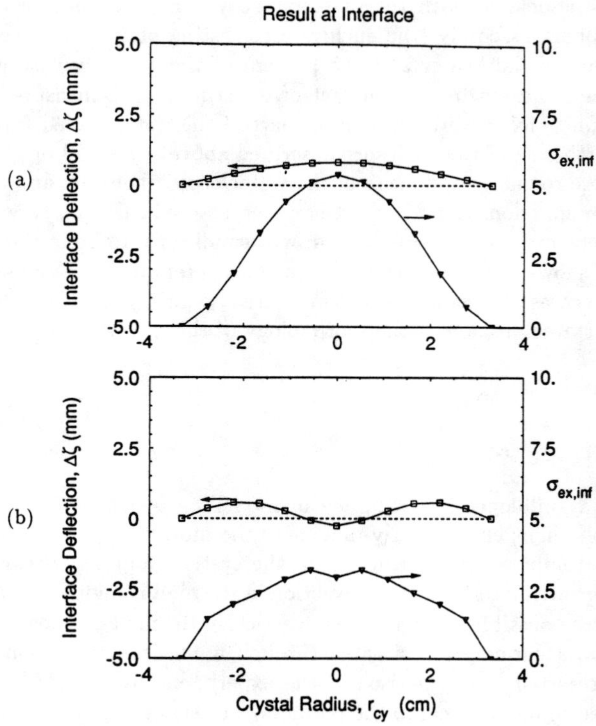

Fig. 7 The morphology of the growth interface and the associated stresses: (a) direct model prediction, (b) stresses calculated using the experimental shape of interface.

on the observed interface morphology is also not significantly different from the one obtained directly from the model. This evidence, therefore, suggests that although the assumption of conduction-dominated heat transfer is not rigorously correct, it is justifiably accurate for stress calculations.

5. MODEL BASED CONTROL OF LEC PROCESS

The ultimate vindication of the modelling effort is its ability to provide for growth of better crystals. We have implemented a control structure built around the numerical model reported here. A very brief description of this effort is given here, and a complete discussion given in refs. [39,40]. In this approach the growth process is numerically simulated, on-line and nearly at real time, and conditions necessary to minimize thermal stresses during growth are indicated by the model. These conditions are implemented in a modified LEC system instantaneously. Using this approach we have succeeded in growth of 2-inch diameter GaAs crystals with dislocation densities as low as 2000 cm^{-2}. This is more than one order of magnitude reduction in the dislocation density of typical LEC-grown GaAs crystals. It clearly indicates the potential impact of numerical modelling on crystal growth processes.

6. ACKNOWLEDGEMENT

The authors wish to express their appreciation for the support provided by Defense Advanced Research Project Agency (contract No. F49620-87-C-0106)to this project.

7. REFERENCES

1. Miyasawa, S., et al., Appl. Phys. Lett. 43 (1983) 853.
2. Matsumura, T. et al., J. Appl. Phys. 57 (1985) 118.
3. Nanishi, Y., et al., Jap. J. Appl. Phys., 22 (1983) L54.
4. Winston, H.V., et al., Appl. Phys. Lett. 45 (1984) 447.
5. O'Neill, T., et al., Personal Communication.
6. Roedel, R.J., et al., J. Electrochem. Soc: Solid-State Sci and Tech, 126 (79) 637.
7. Petroff, P., and L. Hartmann, Appl. Phys. Lett, 23 (1973) 469.
8. Jordan, A.S., and J.M. Parsey, MRS Bulletin, (Oct. 1988) 36
9. Penning, P., Philips Res. Repts., 13 (1958) 79.
10. Jordan, A.S., R. Caruso, and A.R. vonNeida, Bell Syst. Tech. J., 59, (April 1980) 593

11. Alexander, H., and P. Haasen, Solid State Phys., 22 (1968) 27

12. Motakef, S., J. Cryst. Growth, (1990), in press

13. Motakef, S., J. Cryst. Growth, (1990), in review

14. Volkl, J., and G. Muller, J. Cryst. Growth, 97 (1989) 136

15. Maroudas, D., and R.A. Brown, J. Cryst. Growth, (1990), in review.

16. Motakef, S., Int. J. Heat and Mass. Trans., 30, (1987) 1487

17. Hicks, T.W., J. Cryst. Growth, 84 (1987) 598.

18. Motakef, S. and A.F. Witt, J. Cryst. Growth, 80 (1987) 37

19. Duseaux, M., and G. Jacob, Appl. Phys. Letters 40 (1982) 790

20. Duseaux, M., J. Cryst. Growth, 61 (1983) 576

21. Schvezov, C., I.V. Samarasekera, and F. Weinberg, J. Cryst. Growth 84 (1987) 212

22. Schvezov, C., I.V. Samarasekera, and F. Weinberg, J. Cryst. Growth 84 (1987) 219

23. Schvezov, C., I.V. Samarasekera, and F. Weinberg, J. Cryst. Growth 85 (1987) 142

24. Meduoye, G.O., D. J. Bacon, and K.E. Evans, J. Cryst. Growth, 88 (1988) 397

25. Derby, J.J., and R.A. Brown, J. Cryst. Growth, 74 (1986) 605.

26. Kobayashi, N., in:Heat Transfer in Czochralski Crystal Growth System, Ed. W.R. Wilcox (Dekker, New York 1981)

27. Srivastava, R.K., P.A. Ramachandra, and M.P. Dudukovic, J. Cryst. Growth, 73 (1985) 487.

28. Tsukada, T. et al, J. Chem. Eng. of Japan, 20 (1987) 146

29. Nicodeme, P., F. Dupret, and M.J. Crochet, 5th Conf. on Semi-Insulating III-V Materials, Malmo, Sweden, IOP Publishing, (1988) 465

30. Motakef, S., J. Cryst. Growth, 88 (1988) 341.

31. Sackinger, P.A., R.A. Brown, and J.J. Derby, Int. J. Num. Methods, 9 (1989) 453.

32. Nicodeme, P., et al, 5th Conf. on Semi-Insulating III-V Materials, Malmo, Sweden, IOP Publishing, (1988)471

33. Cazin-Bourguignon, A.M., and S. Motakef, J. Cryst. Growth, 96 (1989) 390

34. Hurle, D.T.J., J. Cryst. Growth, 63 (1983) 13

35. Motakef, S., J. Cryst. Growth, 88 (1988) 341.

36. Motakef, S., J. Cryst. Growth, 98 (1989) 711.

37. Motakef, S., J. Appl. Mech., 54 (1987) 813.

38. Motakef, S., and H. Ozoe, Japan. J. Soc. Mech. Eng., (1990) in press.

39. Koai, K., Ph.D. Dissertation, MIT, (1990)

40. Kelly,K.W., Ph.D. Dissertation, MIT, (1990)

Rapporteur's Report of Session on Crystal Growth (1)

Wataru Nakayama, Tokyo Institute of Technology

The session "Crystal growth (1)" had four presentations. The common feature in all these presentations is the mobilization of enormous computing resources. It was amply demonstrated that computers could produce detailed predictions of the system level heat transfer, the crystal–melt interface shape, dislocation generations in the crystal, and the effect of the magnetic field application. My observation of the current state of the art in the industry tells that the dramatic advances in the computer performance and numerical simulation techniques have not yet produced significant impacts on the crystal grower's practice in gap between the numerical simulation performed by academia and the practice of growing crystals in the industry.

1. It seems to me that some important aspects of the scaling analysis are still missing in most of the simulation work. The one among them is the consideration on the time scale. The growth of a single crystal rod takes more than 12 hours. The thermalhistory that the crystal experiences during this long hour converges to the quality of the product (grown crystal). For example, the change of the interface shape from the one convex toward the melt to the concave shape requires a period of time at least as longas that for a crystal to grow by a height comparable to the crystal radius. What happens in and around the crystal during this transition period is likely to produce a significant influence on the crystal quality. Also important is the effect of annealing that the crystal undergoes during the growth and after the growth. Understandably, due to an enormous requirement of the computing time, most simulations are made on the assumption of a quasi–steady state. In a quasi–steady state analysis everything determined by the thermal boundary condition, not by the initial condition. The real growth process involves a wide spectrum of time scales, so that it is necessary to examine in what time frame the quasi–steady state analysis is meaningful. Professor Simon Ostrach, co–chairman of the session, emphasized the importance of the scaling analysis. Much of the controversy over the assumptions made in the analysis can be settled by the scaling analysis needs be extended over not only the space and temperature fields but also the time frame.

2. As the computer becomes more powerful, it churns out increasingly detailed features of thermal field. For practitioners such detailed information are hard to digest. For example, realistic pictures of melt convection containing details of turbulent motion are useless to the practitioner, unless they are presented in terms of statistical terms such as turbulence intensity. More important is the correlation between the environment parameters of the system level and the turbulence intensity in the melt. Knowing such correlations the crystal grower can improve the structural design and operation of the apparatus. On the part of academia it may be hard to work on problems where all system parameters are specified according to a real design on the apparatus. Such problem settings often cause the loss of generic nature in the analytical result. At the present time there seems little concensus between academia and practitioners on the issue of what details in the modeling of the growth apparatus are needed to produce useful information from numerical simulation.

3. On the part of the industry there is a need to educate engineers for retrieving useful insight from numerical simulations reported by the academia. If we expect from the fundamental research some quantitative directives to improve the product yield or the crystal quality, we wouid almost certainly get disappointed. The present state of the art of numerical simulation, although impressing the laymen as an exact science,is far short of providing the practitioner with all quantitative answers he or she demands.It should be recognized, however, that a quantitative physical insight could be greatly beneficial for the improvement of the technology, if it is ingeneouslly applied to the industrial process.

Summary of General Discussion of Session on Crystal Growth (1)

S. Ostrach, Case Western Reserve University

The four papers presented in the Crystal Growth 1 session all made use of numerical methods to obtain solutions. Although considerable emphasis was given to the numerical solutions, insufficient attention was given to formulation of the problems which would enable direct relation to be made between the numerical models and the real physical situations.

It is clear that the complexity of the phenomena involved in crystal growth will demand that numerical solutions will be required. However, numerical results are not useful unless the models contain the essential physics and chemistry and have the appropriate resolution to describe the true phenomena in sufficient detail. Proper scaling analysis can provide the means of achieving all these objectives, but, unfortunately, despite its power and utility such analysis is either ignored or improperly used in most numerical studies. Inappropriate scaling cannot only lead to numerical problems but can also misrepresent the physics (see Ref. 1).

Considerable mention was made in the papers of the disparate time and length scales in crystal growth. Whereas global scale analysis, based on physical dimensions, can indicate the qualitative features of the phenomena, multiple-scale analysis is required to deal with disparate scales and with flow sub-regions (such as cells and layers) which are imbedded in the overall region of interest with extents different from the physical dimensions. This is also discussed in Ref. 1.

Proper scaling analysis requires that appropriate reference quantities (velocities, lengths, and times) be used so that the variables are all of unit order of magnitude. Only if this is done, will the dimensionless parameters correctly represent the ratio of forces, transfer modes, etc. Then those parameters when evaluated for the prototype conditions, will indicate the dominant aspects of the problem, the proper mathematical simplifications, how best to correlate the data, and the true relation to the real situation. It is distressing to see arbitrary and inappropriate reference quantities used indiscriminately. Every choice of reference quantities implies specific physical conditions. They are not arbitrary but have definite implications. In view of all of this it is just as disturbing to find numerical results presented with no delineation of the range of parameters for which the solutions are determined.

In summary, the enormous efforts being expended on numerical solutions would be more meaningful and useful if they were preceded by careful scaling analysis. The additional time spent in such a priori analysis is warranted to ensure that the solutions are related to reality.

Reference

1. Ostrach, S.: Natural Convection in Enclosures, Jour. Heat Transfer, 50th Anniversary Issue, Vol. 110, No. 4-B, 1988.

CRYSTAL GROWTH (2)

Energy Economy in Melt Convection Control by Thermal and Magnetic Methods in Czochralski Single-Crystal Growth Apparatus

WATARU NAKAYAMA
Department of Mechanical Engineering for Production
Tokyo Institute of Technology
2-12-1 Oh-Okayama, Meguro-ku, Tokyo, Japan

KAZUMASA FUJIOKA
Mechanical Engineering Research Laboratory
Hitachi, Ltd.
502 Kandatsu, Tsuchiura-shi, Ibaraki-ken, Japan

ABSTRACT

Alteration of convection pattern in the melt is attemped by projecting a beam of CO_2 laser on the melt free surface near the growing crystal. Local heating produced the expected effects in the laboratory experiment; however, the results of the test conducted with the real apparatus were inconclusive. The initial idea was to develop a more energy-thrift method than the magnetic field Czochralski method (MCZ). There is a prospect of requiring very powerful lasers that consume as much electricity as MCZ.

INTRODUCTION

The progress of the microelectronics technology depends in large part on the art of growing high-quality single crystals of silicon. Silicon is dominating the microelectronics technology and will continue to do so, because of its useful characteristics that are required to support high-density integration of microelectronic circuits. Today,a silicon chip of 1 cm^2 area can store 4 million bits of information in the finely meshed network of transistors and conductors. This 4-megabit DRAM(Dynamic Random Access Memory)chip has a circuit density that is highest among commercially available chips as of 1990. There are already announcements from some manufacturers that 16-megabit chips will go into commercial market by 1992. The key to the success of such high-dencity circuit integration is the improvement of the product yield,that is becoming increasingly difficullt with miniaturization of circuits.

Among the factors that affect the product yield the quality of silicon crystals is one of the most important. The quality of crystal is measured in terms of several parameters;the density of crystal defects,the presence of dislocations,the non-uniformity of dopants,and the density of impurities. All these quality parameters are significantly influenced by the thermal history that the crystal experiences during growth in the apparatus(Figure 1). In recent art of silicon crystal growth the problem of dislocation has been virtually solved. The rest of the quality parameters are concerned with the mechanics of microscopic scales,and they are closely coupled. Defects in crystal structure are often produced by the agglomeration of dopant and impurity atoms. In silicon crystals most of impurity atoms are oxygen atoms that are dissolved from the silica crucible wall and find their way into the growing crystal through the crystal-melt interface. In recent years much attention has been directed to the role of

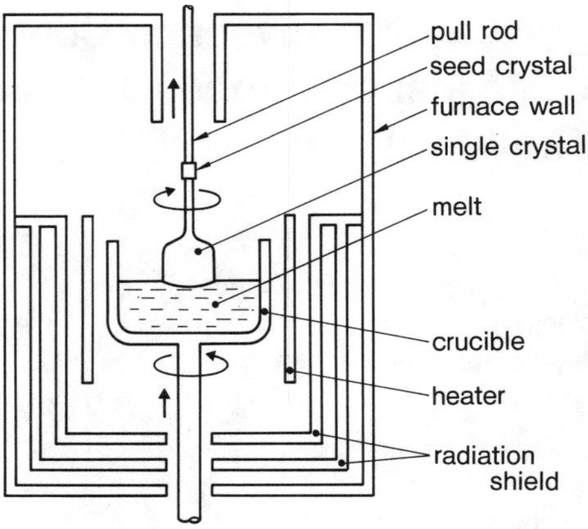

FIGURE 1 Czochralski apparatus for growth of single crystal

oxygens in determining the crystal quality. A proper amount of oxygen atoms in the crystal are said to give certain strength to the crystal structure;however an excessive concentration of oxygen does harm to the crystal quality. The miniaturization of electronic circuits demands the reduction of crystal defects,that eventually demands lowering the oxygen atom concentration.

The control of oxygen concentration in the silicon crystal is synonymous with the control of melt convection,since oxygen atoms are carried by the melt convection from the crucible wall to the crystal-melt interface. One of the most significant developments in the history of crystal growth is the introduction of melt convection control by the magnetic field. This,so called MCZ,the abbreviation of magnetic Czochralski growth method,was first proven effective in suppression of striations in the grown crystal, the indicator of thermal fluctuations in the melt,by Witt in 1970[1]. In 1980 Hoshi et al.[2] of Sony Corporation built the first MCZ apparatus of an industrial size. During the 1980s there were attempts to improve the MCZ by changing the orientation of magnetic field[3,4,5],employing superconducting magnets[6],and pursuing the optimum magnet shape[7]. However, all the demonstrations of apparent benefits to be gained by MCZ have not prompted a widespread use of MCZ in the industry. The reasons are high costs of equipment and operation. The increase of operation cost is caused by large electricity consumption by magnets.

In MCZ the bulk of melt is put in a magnetic field,wheras the most critical part of the melt is the boundary layer next to the crystal-melt interface. In an attempt to control the melt convection near the crystal the authors studied the method of laser beam projection[8]. The scheme is based on the expectation that the intense energy input to the free surface of the melt near the crystal may eventually alter and hopefully suppress the oscillation of convection near the crystal-melt interface. Also expected is a possible reduction of required energy to alter the convection pattern on the ground that the zone of the melt subject to control is small. This paper reports the result of our preliminary study and rather discouraging prospects for improving the energy economy compared to that in MCZ.

EFFECT OF TEMPERATURE OSCILLATION IN THE MELT ON THE CRYSTAL DEFECT DENSITY

Kuroda et al.[9] provide the data of how sensitive the density of defects in grown crystal is to the temperature oscillation in the melt. A part of their experimental results are reproduced in Figure 2. The vertical axis shows the density of microdefects in cm^{-3}. The horizontal axis shows the amplitude of temperature oscillations recorded by the thermocouples attached to the pull rod. The temperature measurements demand an elaborate arrangement that produces the data on the horizontal axis of Fig.2. The thermocouples are tied to the pull rod; they are initially in the melt, pulled up with the rod, and embedded in the growing crystal. The temperature data near the crystal-melt interface are obtained, as the thermocouple beads are about to be absorbed in the crystal. The experiment is expensive, so that it was discontinued and has never been resumed elsewhere.

Figure 2 shows two curves for different sets of operative parameters. The details of experimental condition may not be necessary for the present discussion. The point to be noted is the sensitivity of the defect density to the temperature oscillation. A reduction of oscillation amplitude from 1.5 °C to 0.5 °C produces a decrease of defect density by almost two decades.

At the present state of knowledge it is difficult to relate the oxygen concentration to the crystal defect density in definite and quantitative terms. However, the existence of certain correlations among the oxygen concentration, its spatial variations in paticular, the amplitude of temperature oscillation, and the crystal defect density is widely recognized by crystal growers. The temperature amplitude in the melt is the most commonly used index to gauge the effectiveness of control measures such as magnetic field application or laser beam projection in the dynamic process of crystal growth.

METHODS OF MELT CONVECTION CONTROL

Magnetic Field Czochralski Growth: Figure 3 shows the schematic of a MCZ apparatus equipped with superconducting magnets developed by Takasu et al.[6], where computed magnetic field lines are also shown. The diameter of the crucible is in a range 400-450 mm. The magnetic flux at the center of

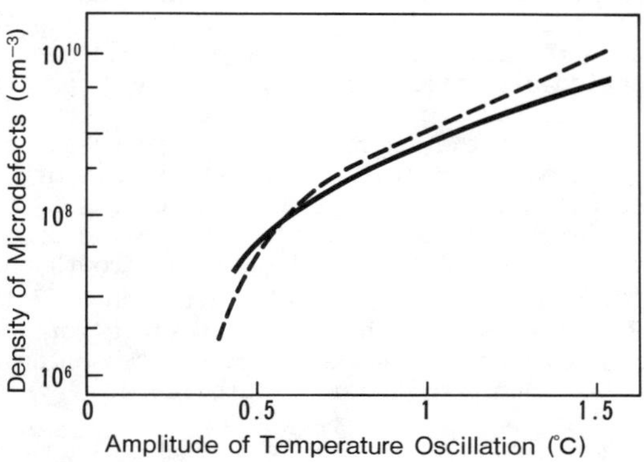

FIGURE 2 Effect of temperature oscillation in the melt on the crystal defect density

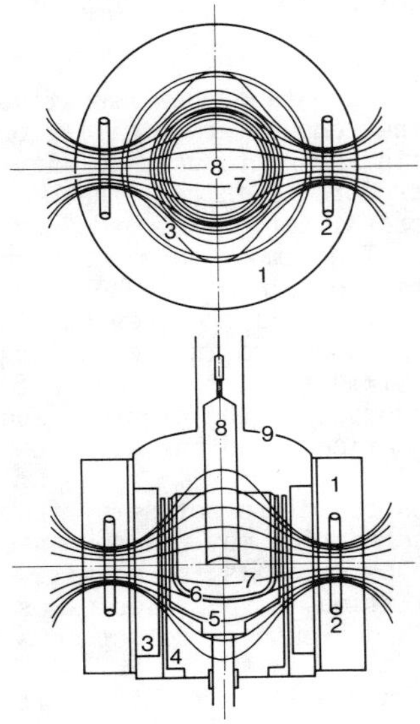

FIGURE 3 Magnetic Czochralski growth apparatus using superconducting magnets:lines shows magnetic field;1 cryostat;2 coil;3 thermal insulation;4 heater;5 graphite crucible;6 silica crucible;7 melt; 8 grown crystal (Takasu et al.1990)

the crucible is computed as 0.3 Tesla. In the magnet coils the flux intensity becomes as high as 1.5 Tesla. A large gap between the magnet poles, where the container wall,the thermal insulation,the heater,the crucibles, the air gap,and the space filled with argon gas at a low pressure affect magnetic field lines,and poses a large resistance to the magnetic flux. This resistance requires a high level of power energizing magnets.

Figure 4 shows an example of the data showing the effect of magnetic field on the amplitude of temperature oscillation in the melt. The curve is drawn referring to the temperature oscillograms reported in [10]. In this case the test apparatus is small,designed for the growth of crystals of 76 mm in diameter(3 inch crystals). Due to the small size of the apparatus the applied magnetic field at the magnet coils can be as small as 0.1 Tesla to have a noticeable effect on the temperature oscillation. The potential of MCZ in suppressing the temperature oscillation is obvious. However,as the diameter of the growth apparatus increases,the required flux intensity at the magnet rapidly increases. A rule of thumb estimate indicates that the required flux at the magnet increases in proportion to the square of the diameter of the apparatus. Since the diameter of the crystal is almost proportional to the size of the growth apparatus,the growth of,say,an 8-inch (200 mm) crystal requires about 7 times the field intensity for a 3-inch crystal to have the same level of benefits by MCZ. Actual requirements are sure to be much higher,because the melt in a large crucible tends to have a higher level of turbulence.

Laser Beam Projection: The idea to use extra energy source to control the growth of crystal has been popular among researchers. For instance,Ekhult and Carlberg[11] used infrared lamps to control the diameter of crystals

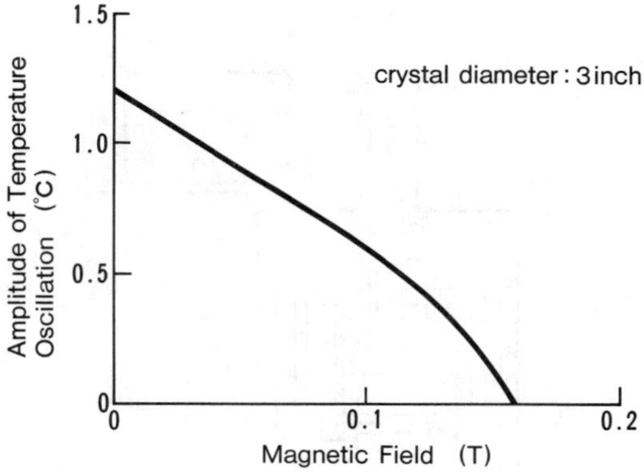

FIGURE 4 Amplitude of temperature oscillation vs.magnetic field strength

during growth. In spite of apparent effects of local heating around the crystal on the stability of crystal growth, the methods have not found industry-wide uses. The reason is the concern over possible contamination of crystal materials by impurity atoms evaporated from extraneous components such as lamps and heaters that have to serve in high temperature environments.

The same concern may arise if one attempts to use lamps or heaters for the control of melt convection. The concern can be resolved by the use of energy beams introduced from remote sources and the elimination of extra components from the hot zone of the growth apparatus. There are two types of energy beams that can be used for the present purpose; electron beam and laser beam. The electron beam, however, is more difficult to be deflected and reflected than the laser beam, so that its use requires considerable modifications of the existing growth apparatus. The laser beam can be introduced into the existing apparatus with small modifications; installation of a window on the apparatus wall and a deflecting mirror in the upper cold zone, as shown in Figure 5. Among a few candidates of lasers, carbon-dioxide lasers are superior in easy management and service life.

The potential of the laser beam to affect flows of the melt has not been well established. The idea is based on the conjecture that heating of the melt's free surface around the crystal may affect the momentum of convection approaching the crystal. Heating should be strong enough to produce substantial changes in the density and surface tension forces on the spot of beam projection. In the apparatus shown in Fig.5 the beam is fixed on the stationary frame, while the melt is rotating. When viewed in the rotating coordinate system, the beam spot travels around the crystal on a concentric circle. A proper combination of the rotating speed and the beam intensity would produce a heated ring on the free surface of the melt. Where the time for the beam spot to make a full circle is much shorter than the time of thermal diffusion around the spot, the heat input to the ring approaches axisymmetry. Meanwhile, in most practical situations a global convection of the melt sweeps the vertical crucible wall upward, then radially inward near the free surface. The radial flow near the free surface is oxygen-rich at the periphery, and near the crystal the oxygen concentration depends on the rate of evaporation on the free surface and the diffusion of oxygen from the bulk of melt toward the free surface. Since the diffusion coefficient of oxygen in the melt is very small, the

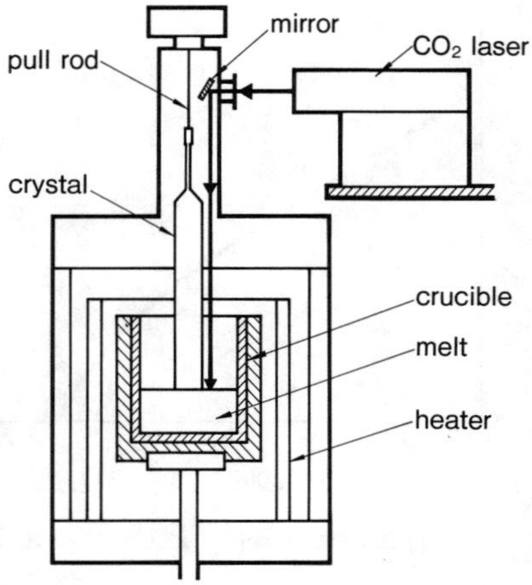

FIGURE 5 Laser beam projection on the melt

diffusion boundary layer below the free surface is extremely thin,of the order of 0.1 mm[12]. Hence,an alteration of the flow,the thermocapillary convection in particular,could significantly disturb local balance of oxygen concentration. The imposition of a hot ring around the crystal is expected to produce a certain possible effect concerning the oxygen concentration. Local heating is likely to weaken the thermocapillary force around the hot ring,thereby decelerating the radial inward flow or causing recirculating flows. This shielding effect may produce an oxygen-lean ring around the crystal,because the approaching convection toward the crystal is given a more time for evaporation of oxygen atoms from the free surface.

The next question to be addressed is the relationship between the flow beneath the crystal-melt interface and that around the crystal periphery. For this we need to know the global pattern of melt convection,that is the subject of many investigators,for,example,Carruthers et al.[13]. In the case of silicon crystal growth,the numerical analysis of the melt flow conducted by the authors[14] shows that the effect of crystal rotation is contained in a cylindrical zone beneath the crystal and extending to the bottom of the crucible. Hence,it is likely to find two recirculating cells in the vertical cross section of the melt,one near the central axis, and another in the bulk zone below the free surface. The oxygen concentration in the central zone,the most important factor affecting the crystal quality,is determined by the balance between the dissolution rate from the section of the crucible's bottom wall,the rate of incorporation into the crystal,and the rate of exchange with the outer zone. The last mechanism could have a significant effect on the oxygen concentration in the central zone. It is likely that the mass transfer process in the outer zone has a predominant influence on the average oxygen concentration for the whole melt in the crucible,because almost 99 percent of dissolved oxygen evaporates from the melt's free surface[15]. If the local heating around the crystal has a certain effect on the evaporation rate on the free surface, then,the average concentration could be lowered although indirectly.

The above conjecture assumes an axisymmetry of melt convection. In actual growth apparatuses there are sources that cause non-axisymmetric modes in the crystal. In fact, the initial objective of the present study was to annihilate non-axisymmetric modes that take ordered structures in the rotating system. Typical ordered structures are wavy meandering streams on the free surface. The melt of oxide materials facilitates the observation of such patterns, and the analogy to baroclinic waves was pointed out in [16]. The onset of non-axisymmetric modes can be explained by the instability of thermocapillary flow in the rotating system. It is difficult to confirm that ordered non-axisymmetric structures are present on the free surface of silicon melt. However, there is a high level of possibility of finding instability of thermocapillary convection in the case of silicon also. Ostrach[17] pointed out the importance of thermocapillary effects on the melt flow of silicon. The thickness of the shear layer for thermocapillary flows is as thin as 0.5 mm. The momentum of thermocapillary flow may not be large enough to move floating objects on the melt surface, so that visual confirmation of its presence on the actual melt surface is very difficult. However, as discussed in the previous paragraph, the mode of thermocapillary flow is important for the balance of oxygen concentration in the melt.

The authors conducted a preliminary experiment to confirm the effectiveness of laser beam projection in altering the convection mode. The same apparatus as that used for baroclinic waves[18] was used, where silicone oil served as the working fluid. A CO_2 laser of 10W capacity was used as the beam source. It was found that, where the combination of the rotational speed of the system and the temperature difference between the warmer outer wall(corresponding to the crucible wall)and the cooler inner wall(the crystal)is in a certain parametric domain, a non-axisymmetric mode appeared. It was also confirmed that in certain cases, the beam projection was effective in restoring the axisymmetry of convection, and with it the suppression of temperature oscillatioins that were measured near the inner wall before the beam projection.

Encouraged by the result of this model experiment, the authors worked with the crystal growth department of Kofu Works, Hitachi, Ltd. installing 500W CO_2 laser on the existing growth apparatus. The experiment with this real working apparatus proved somewhat discouraging. The growth rate of crystal was affected by laser beam projection as it should be. However, the effect of beam projection on oxygen concentration in grown crystals was not discernible. A reason can be sought to the presense of other factors affecting the oxygen concentration. Also, the indecisive results may be attributed to insufficient power of the laser used in the experiment.

The experience described in the preceding paragraph prompted the authors back to the laboratory, this time to conduct numerical simulation of the process. The computer program has not yet been completed to the extent that produces a prediction of oxygen concentration. As for the thermofluid analysis, however, the program is most comprehensive in incorporating thermal factors in the industrial growth apparatus. As outlined in the authors' previous reports[19,20], the boundary conditions for the melt flow and heat transfer are given on the basis of thermal analysis of the industrial apparatus. Figure 6 shows samples of the numerical analysis. Fig.6 (a) explains the symbols; the radius of the crystal R, the point of beam projection on the melt free surface P, the point on the crucible bottom on the vertical extension of the beam P', the distance between the projection point and the crystal's periphery Δr, and the vertical coordinate z. The crucible diameter and the melt depth are set at 40 cm and 12 cm, respectively. The heat flux is specified on the melt surface as -6.50 W/cm^2, 10.0 W/cm^2 on the crucible side wall, and -1.74 W/cm^2 on the crucible bottom.

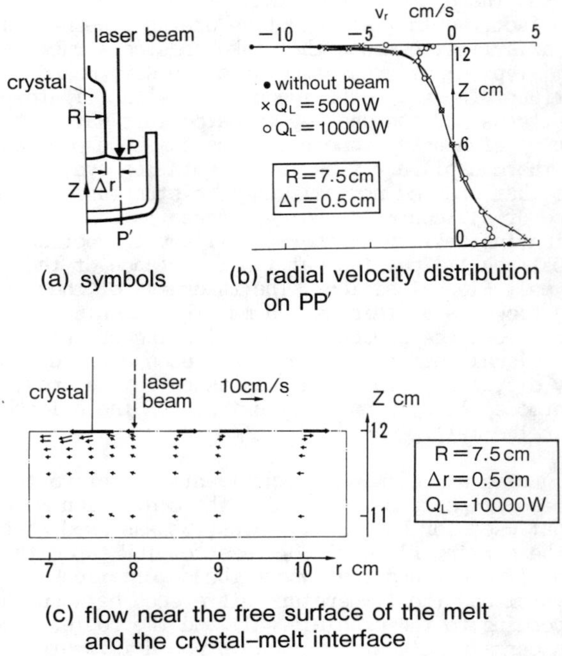

FIGURE 6 Effect of laser beam projection: the result of numerical analysis

These values are the results of the system-level thermal analysis[20]. The results shown in Figs.6(b) and (c) were obtained for the following cases of operative parameters; the rotation rate of the crucible 0.105 rad/s, and that of the crystal 0.838 rad/s. The melt zone is discretized into 40×40 meshes, with finer cells near the free surface, the crystal-melt interface, and the crucible wall. The coefficient of absorption of projected energy by silicon melt is set at 0.3. The Q_L in Figs.6(b) and (c) is the beam energy at the source, so that 30 percent of the specified value is assumed to be absorbed by silicon melt. The transport properties of silicon melt are; the density 2500 kg/m^3 , the dynamic viscosity 5.3×10^{-7} m^2/s, the thermal conductivity 29 W/m K, the specific heat 0.975 kJ/kg K, the Prandtl number 0.045, the coefficient of volumetric expansion 1.41×10^{-5} 1/K, and the temperature coefficient of surface tension force -0.43×10^{-3} N/m K. The details of the analysis will be reported elsewhere.

Figure 6(b) shows the radial velocity distribution on PP' for the case of R=7.5 cm and Δr=0.5 cm. The hot ring produced by the beam projection has a width of 1 cm, its center being at P. Within this band the heat input is assumed uniform. The radial velocity is appreciably affected by the beam projection near the free surface. Near the crucible bottom the effect of the beam projection is apparent only when the power(Q_L) is 10000W. The 10000W power is also found to produce a reverse flow near the free surface. The reverse flow near the free surface and the decelerated flow on the crucible wall are expected to work favorably for the reduction of oxygen concentration in the melt. From this view point a 500W laser, used in the experiment in the plant, is deemed insufficiently low. Fig.6(c) shows an expanded view of melt convection near the free surface and the crystal-melt interface.

The conclusion reached by the analysis is a tentative one,because the mechanism leading to the process of oxygen diffusion in the melt is not yet fully understood. Besides,there are certain uncertainties regarding the thermophysical properties of molten silicon,particulary the absorption coefficient. In spite of these uncertainties we see the need to use more powerful lasers to study the feasibility of our idea in the real apparatus. The requirement for more powerful lasers diminishes an initially expected benefit of melt convection control by energy beams.

POWER CONSUMPTION BY CZ APPARATUSES

Figure 7 shows a general guide for the electric power required by conventional Czochralski apparatuses. The horizontal axis shows the crucible diameter which has been increased over the years to grow larger diameter crystals. The vertical axis shows the capacity of heaters. The actual electric power varies during the growth period,following a programmed pattern and keeping the temperature in the apparatus in designed ranges.

Takasu et al.[8] reported the design data for the MCZ apparatuses equipped with iron-core magnets and superconducting magnets. The growth of 150-200 mm diameter crystals requires a 400 mm crucible. The capacity of DC power source to energize the iron-core magnets for the apparatus of the above size is estimated as 300 kVA. This required power capacity is comparable to or exceeds the heater capacity in the conventional CZ apparatus. The superconducting magnets need far less energizing power,0.5 kVA for the above apparatus size ;however,we must take into account the power consumed by the helium liquefier. Also the costs of helium and equipment are large factors in the economy. Referring to the estimate by Takasu et al.[8] we see the cost of helium supply dominating the operation cost,although the consumption of liquid helium has been reduced rapidly by improvements on the equipment. Also pointed out is a large share occupied by the large DC-current source in the equipment cost. With all the economy factors combined,the equipment cost is almost comparable in conventional and superconducting magnet technologies. The superconducting magnet technology will

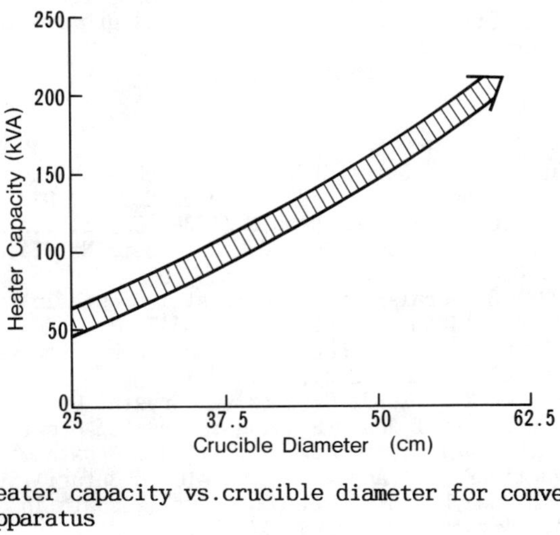

FIGURE 7 Heater capacity vs.crucible diameter for conventional growth apparatus

increase its advantage in the future with the advancement in the designs of cryostat and helium liquefier.

As for the laser method described in the previous section,we still have large uncertainties about its effectiveness. At the present time we can present only a guesswork. If a laser power of 10 kW is sufficient for our purpose,assuming the efficiency of CO_2 laser at 5 %,we need a power source capacity of 200 kVA. This is almost comparable to the power level required by the conventional magnet technology.

CONCLUSIONS

This paper describes our attempt to develop an alternative to the method of melt convection control by magnetic field application. The idea is based on the effect of laser beam projection on the melt's free surface on thermocapillary convection which plays supposedly important roles in the determination of oxygen concentration and temperature oscillations near the cyrstal. The laboratory experiment conducted using a fundamental model has produced results supporting our idea. However,the experiment conducted with the real growth apparatus and a 500W CO_2 laser turned out inconclusive results. This was attributed to a too low power level of the laser. The numerical analysis of melt convection revealed that a 10 kW of beam power is required to cause a reverse flow near the point of beam projection and eventually decelerate the melt convection.

The next stage of investigation will be further development of numerical simulation programs that include the analysis of oxygen concentration in the melt. After confirmation of the feasibility of the present concept and the required laser power the test at the plant will be resumed.

The investigation conducted up to the present time identifies a set of research and development problems for the melt convection control technologies. For MCZ using conventional magnets more improvements in the design of the system are needed to reduce the required electricity. Also needed is the improvement of the system using superconducting magnets to reduce the consumption of liquid helium and the equipment cost. Attempts to use thermal methods such as local heating of the melt described in this paper should be continued to offer more options to crystal growers in their endeavor to improve crystal quality.

REFERENCES

1. Witt,A.F.,Herman,C.J.,and Gatos,H.C.(1970)'Czochralski-Type Crystal Growth in Transverse Magnetic Fields',J.Mater.Sci.,37,822-824.
2. Hoshi,K.,Suzuki,T.,Okubo,Y.,and Isawa,N.(1980)'CZ Silicon Crystal Grown in Transverse Magnetic Fields',ESC Ext.Abstr.No.324,80-1,ECS, Princeton,pp.811-813.
3. Hoshikawa,K.(1982)'Czochralski Silicon Crystal Growth in the Vertical Magnetic Field',Japan.J.Appl.Phys.,21,L545-L547.
4. Kim,K.M.(1982)'Suppression of Thermal Convection by Transverse Magnetic Field',J.Electrochem.Soc.,129,427-429.
5. Langlois,W.E.and Lee,K.J.(1983)'Czochralski Crystal Growth in an Axial Magnetic Field: Effects of Joule Heating',J.Crystal Growth,62,481-486.
6. Takasu,S.,Ohwa,M.,Suzuki,O.,and Higuchi,T.(1990)'Growth of Silicon Crystals in Magnetic Field',Applied Physics(Ohyo Buturi),59,1044-1050.
7. Hirata,H.and Hoshikawa,K.(1989)'Silicon Crystal Growth in a Cusp Magnetic Field',J.Crystal Growth,96,747-755.

8. Nakayama,W.,Fujioka,K.,and Ohyanagi,H.(1990)'Transition of Convection Mode Near the Free Surface of Rotating Liquid Produced by Laser Beam Projection',Heat Transfer 1990,Proc.9th International Heat Transfer Conf.Jerusalem,Vol.5,pp.359-364.
9. Kuroda,E.,Kozuka,H.,and Takano,Y.(1984)'The Effect of Temperature Oscillations at the Growth Interface on Crystal Perfection',J.Crystal Growth,68,613-623.
10. Suzuki,T.,Isawa,N.,Okubo,Y.,and Hoshi,K.(1981)'CZ Silicon Crystals Grown in a Transverse Magnetic Field',Semiconductor Silicon 1981, H.R.Huff,R.J.Kriegler,and Y.Takeishi,ed.,Electrochem.Soc.,Pergamon, pp.90-100.
11. Ekhult,U.and Carlberg,T.(1986)'Czochralski Growth of Tin Crystals Under Constant Pull Rate and IR Diameter Control',J.Crystal Growth,76, 317-322.
12. Hirata,H.and Hoshikawa,K.(1988)'Oxygen Incorporation and Melt Convection in CZ Silicon Crystal Growth',J.Japan Crystal Growth Soc.,15,61-70.
13. Carruthers,J.R.and Nassau,K.(1968)'Nonmixing Cells due to Crucible Rotation during Czochralski Crystal Growth',J.Appl.Phys.,39,5205-5214.
14. Nakayama,W.and Fujioka,K.,to be reported elsewhere.
15. Hoshikawa,K.,Hirata,H.,Nakanishi,H.,and Ikuta,K.(1981)'Control of Oxygen Concentration in CZ Silicon Growth',Semiconductor Silicon 1981, H.R.Huff,R.J.Kriegler,and Y.Takeishi,ed.,Electrochem.Soc.,Pergamon, pp.101-112.
16. Whiffin,P.A.C.,Burton,T.M.,and Brice,J.C.(1976)'Simulated Rotational Instabilities in Molten Bismuth Silicon Oxide',J.Crystal Growth,32, 205-210.
17. Ostrach,S.(1982)'Fluid Mechanics in Crystal Growth',1982 Freeman Scholar Lecture,1982 ASME Winter Annual Meeting.
18. Koschmieder,E.L.(1972)'Convection in a Rotating Laterally-Heated Annulus',J.Fluid Mechanics,51,637-656.
19. Nakayama,W.and Masaki,H.(1986)'Influence of the Thermal Environment on the Growth of Single Crystals in a Czochralski Apparatus',Heat Transfer 1986,Vol.4,Hemisphere Publishing Corp.,pp.1755-1760.
20. Nakayama,W.(1986)'Heat Transfer Engineering for the Growth of Large Single Crystals by the Czochralski Process',Heat Transfer in High Technology and Power Engineering,W.J.Yang and Y.Mori,ed.,Hemisphere Publishing Corp.,pp.94-105.

Numerical Simulation of Natural Convection in Crystal Growth in Space and on the Earth

HIROYUKI UCHIDA, JUN-ICHI OCHIAI, KEIICHI KUWAHARA, SAWAKO YOKOHAMA, and SHINTARO ENYA
Ishikawajima-Harima Heavy Industries Co., Ltd.
1 Shin-nakahara-cho, Isogo-ku, Yokohama 235, Japan

INTRODUCTION

In the crystal growth from melt, natural convection of the fluid plays an important role. To control the fluid flow is essential to obtain high quality materials. In this chapter, we will review the research on the numerical simulation of the natural convection, particularly focussed on the surface tension driven flow, i.e. Marangoni convection. In the following chapters, we will discuss on our recent study on the Marangoni convection at low Prandtl number.

Pioneering works on the Marangoni convection have been done by Ostrach and co-workers [1]-[4]. Ostrach pointed out the importance of the Marangoni convection in low-gravity environments. They have been presenting numerous results mainly related to floating-zone method of crystal growth, particularly emphasizing the importance of oscillatory convection.

In 1984, Maekawa et al. [5] and Morioka et al. [6] presented papers on both theoretical and experimental studies of natural convection driven by buoyancy and surface tension. They derived analytical solutions of the convection in an infinite horizontal fluid layer, and also obtained numerical solutions for rectangular cell with a free surface. Those solutions showed good agreements with their experimental data.

Uchida et al. [7] obtained numerical solutions of Marangoni convection in Bridgman crystal growth. They showed possible flow patterns in the material processing in space. They [8] also analyzed Marangoni convection in a square cell using two different numerical method and compared the performances. The flow solution was also shown by an animation movie, which helped understand the result.

Recently, Kuwahara et al. [9] carried out an experiment under artificial low gravity condition attained by using the free flight of an airplane. They obtained valuable experimental data under low gravity condition. They also showed an numerical result for unsteady Marangoni convection. They showed a good agreement with the experimental data.

Maekawa and Tanasawa [10] presented their pioneer paper on the flow control by the external magnetic field. They analytically solved the natural convection problem in an infinite horizontal layer driven by both buoyancy and surface tension.

Hirata et al. [11] conducted a fundamental study of complex transport phenomena accompanied with heat, mass and momentum transport through phase interface. Boundary layer approximation was adopted to solve two-dimesnsional problem.

In the following chapter, we will discuss on the Marangoni convection at low Prandtl number. A motivation of this investigation is that the Prandtl numbers of the semiconductor materials are generally small, for example, about 0.01 for Si and about 0.1 for GaAs. To discuss on the natural convection of the melt of these materials, it must be significant to obtain the solutions in the limit of Pr=0. Deviation of the solution at some Prandtl number will be evaluated by referring the limit solution. As the result of dimensional analysis, it will be shown that modified Marangoni number $Ma^* = Ma/Pr$ is the only parameter of the present problem for Pr=0, while the Prandtl number itself is no longer an independent parameter.

From the experimental point of view, this theory will suggest the possibility of simulation experiment for low Prandtl number fluid by using alternative fluid of another Prandtl number. Because low Pr fluid is generally opaque, it is difficult to observe the fluid flow directly. Therefore, the usage of the alternative fluid will be helpful for the visualization experiment.

Governing Equations

Governing equations for steady incompressible flows are given by following Navier-Stokes equations.

$$\nabla \cdot v = 0 \tag{1}$$

$$(v \cdot \nabla) v = -\rho^{-1} \nabla p + \nu \Delta v \tag{2}$$

$$(v \cdot \nabla) T = \Delta T \tag{3}$$

Equations (1), (2) and (3) are continuity equation, momentum transport equations and energy transport equation, respectively. No driving forces such as buoyancy force are considered in these equations. The only driving force considered in this analysis is the surface tension, which appear in th boundary condition as mentioned below.

The length, velocity, pressure and temperature are scaled as

$$x^* = x / H, \quad y^* = y / H, \quad v^* = v / (\nu / H)$$
$$p^* = p / (\nu / H)^2, \quad T^* = (T - Tc) / (Th - Tc) \tag{4}$$

in which the variables with asterisk are nondimensional. By using these equations, the governing equations (1) ~ (3) are rewritten as

$$\nabla \cdot v = 0 \tag{5}$$

$$(v \cdot \nabla) v = -\nabla p + \Delta v \tag{6}$$

$$(v \cdot \nabla) T = Pr^{-1} \Delta T \tag{7}$$

In the equations (5) ~ (7), the asterisks which denote the nondimensional variables were abbriviated for simplicity. In the limit of Pr=0, the convective term in the energy equation can be neglected, and Laplace's equation is obtained as

$$\triangle T = 0 \quad (8)$$

Therefore, all the nondimensional parameters are eliminated from the governing equations for Pr=0.

Boundary Conditions

Surface tension driven convection in a square cell with a free surface is analyzed in the present study. The shape of the free surface is assumed to be flat, therefore asymptotic solutions for microgravity condition will be obtained. The left and the right boundary are hot and cold wall, where the temperature are set to unity and zero, respectively. The bottom boundary is adiabatic wall, where the temperature gradient normal to the wall is set to zero. The top boundary is free surface, where the adiabatic boundary condition is given.

On the wall boundaries, the velocity components are specified to zero. On the free surface, nondimensional form of the boundary condition for the velocity is given as

$$\partial u / \partial y = - Ma/Pr \ \partial T / \partial x \quad (9)$$

since the surface tension balances with the shear stress. In eq.(9), suffix x and y denote partial derivatives, and Marangoni number Ma is defined as

$$Ma = \sigma_T H (T_h - T_c) / \mu \ \alpha \quad (10)$$

Marangoni number and Prandtl number are essential parameters in this study, but in the limit of Pr=0, the term Ma/Pr in eq.(9) becomes the only parameter. We define modified Marangoni number Ma^* as

$$Ma^* = Ma / Pr \quad (11)$$

Numerical Method

Finite element method was employed in the present analysis. Biquadratic rectangular elements were used to discretize the governing equations. Newton-Raphson method was adopted to solve the nonlinear equations.

Frontal solution algorism was used to reduce the computer memory and computer time to solve large scale linear system. The computer code used in the present analysis was developed based on the code developed by Yamaguchi et al.[12].

Results and Discussion

Figure 1 shows the results for Pr=0. The distributions of the temperature, the velocity and the pressure at four modified Marangoni numbers Ma^*, defined by eq.(11), are shown in each row of this figure.

It is obvious that these temperature distributions are conduction dominated. Actually, the analytical solution for Pr=0 is easy to obtain for the present boundary conditions. It has only one dimensional linear distribution.

At $Ma^*=1$, the flow field is almost symmetric, but the symmetry is broken as Ma^* increases. The center of the vortex moves toward the cold wall.

The symmetry of the pressure field is also broken as Ma^* increases, and an obvious peak of the pressure apears at the corner of the cold wall and the free surface.

Figure 2 shows the results for constant $Ma^*=10^3$. The solutions at four Prandtl numbers Pr are shown in each row of this figure. It should be noted that the original Marangoni number Ma is different for each case, where the ratio Ma/Pr is kept constant.

The deformation of the isothermal lines is obvious as Pr increases. This is the result of the convective transport of energy.

The distributions of the velocity and the pressure at Prandtl number less than 0.1 are very similar. This justifies the result of the dimensional analysis for low Prandtl number fluid in the preceding section, that is, the solutions should be similar for equal modified Marangoni numbers. These results suggest a possibility of simulation experiment for low Prandtl number fluid by using alternative fluid whose Prandtl number is different from that one.

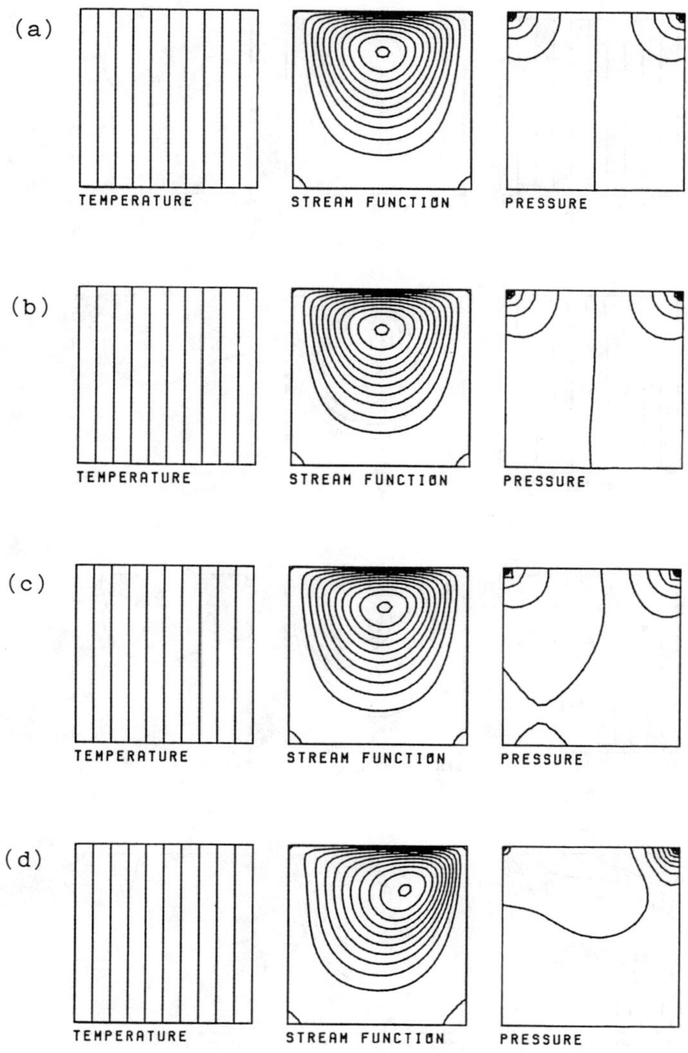

FIGURE 1. The distributions of the temperature, the velocity and the pressure at Pr=0 for Ma/Pr= (a)1, (b)10, (c)100 and (d)1000.

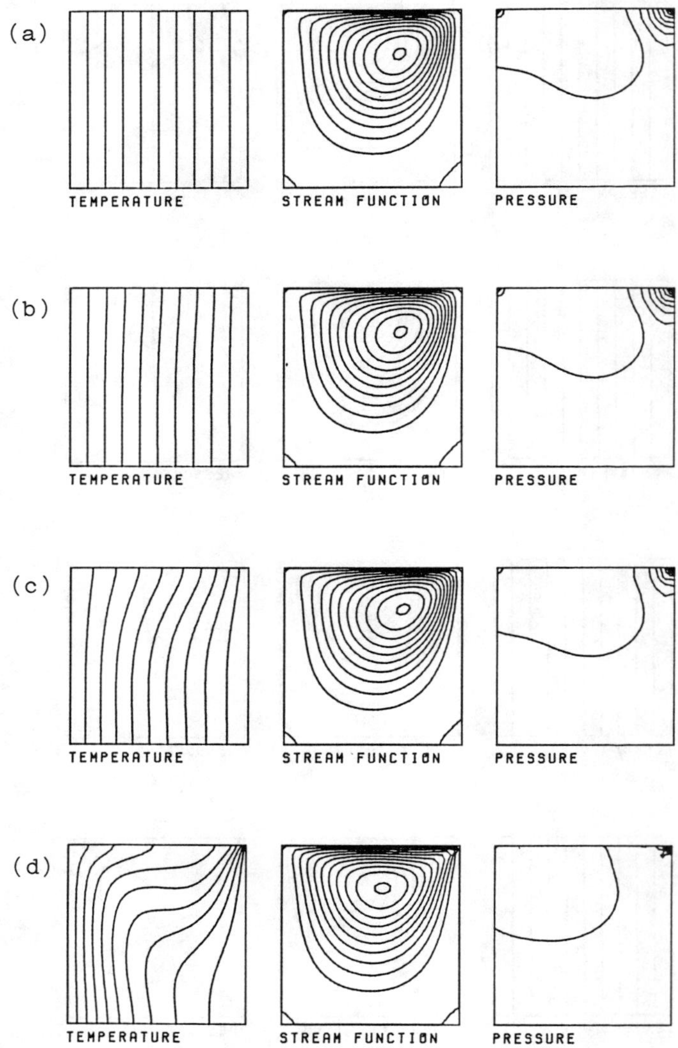

FIGURE 2. The distributions of the temperature, the velocity and the pressure at Ma/Pr=1000 for Pr= (a)0, (b)0.01, (c)0.1 and (d)1.

In the pressure distribution at Pr=1, some wiggles can be seen at the corner of the cold wall and the free surface. It must be a result of the very steep change in the solution at the corner. This problem may be solved by condensing the finite element mesh toward the corner, though an even mesh is used in this case. From physical point of view, such a peak of the pressure will make a deformation of the free surface. An analysis considering the deformation of the free surface may be necessary.

In order to investigate the deviation of the solutions from Pr=0 solutions, a series of computation was carried out for some sets of parameters Ma/Pr (1, 10, 100 and 1000) and Pr (0, 0.01, 0.1, 1 and 10).

Maximum velocity $u_{s,max}$ on the free surface is plotted in Figure 3 as a function of modified Marangoni number Ma/Pr. The curve for Pr=0 is almost linear within the range of Ma/Pr in the present computations. Other curves slightly deviate from the curve for Pr=0 as Ma/Pr or Pr increases, but the amount of the deviation is very little for Prandtl number less than unity.

Figure 4 shows the maximum velocity divided by Ma/Pr as a function of Ma/Pr. This normalized maximum velocity approaches to a constant 0.14 as Ma/Pr decreases, that is,

$$u_{s,max} = 0.14\ Ma/Pr \qquad \text{for}\ \ Ma/Pr \rightarrow 0 \qquad (12)$$

This figure also shows that the solutions for Prandtl number less than 0.1 is very close to the solutions for Pr=0.

Figure 5 shows the maximum velocity divided by the maximum velocity $u_{s,o}$ for Pr=0 as a function of modified Marangoni number Ma/Pr. As Ma/Pr increases, each curve deviates from unity (curve for Pr=0), and it seems that the curve returns to unity after taking a minimum value. However, the number of the data is not enough to conclude at present. Further computations are being done.

Figure 6 shows the position $x_{s,max}$ where the maximum velocity appear on the free surface as a function of Ma/Pr. As Ma/Pr or Pr decreases, $x_{s,max}$ approaches to a constant 0.5, which correspond with the symmetry of the flow field. As Ma/Pr increases, $x_{s,max}$ increases and approaches to unity, that is, the center of the vortex approaches to the corner of the cold wall and the free surface.

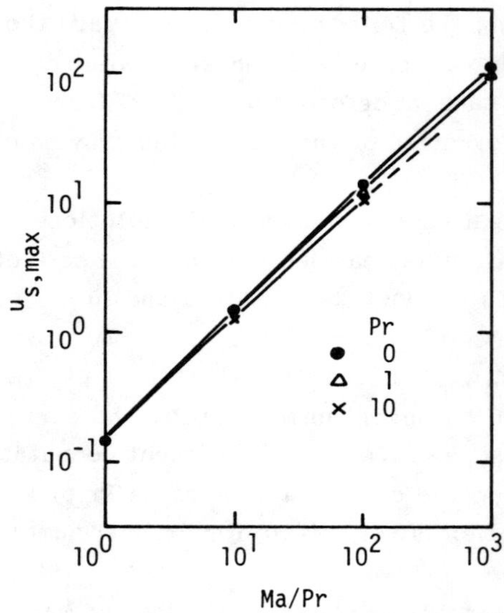

FIGURE 3. Maximum velocity $u_{s,max}$ on the free surface as a function of modified Marangoni number Ma/Pr.

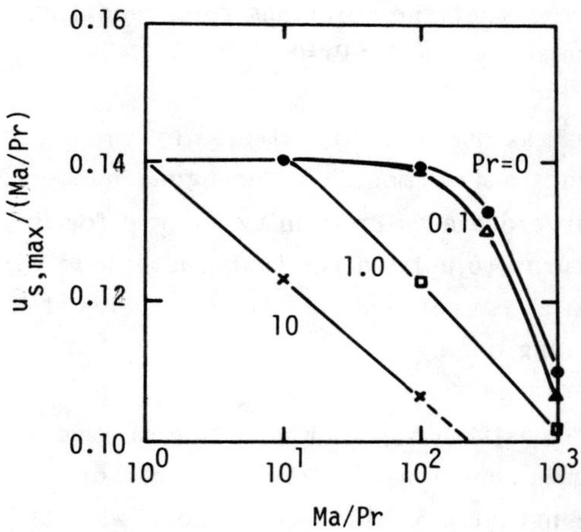

FIGURE 4. Maximum velocity divided by Ma/Pr as a function of modified Marangoni number Ma/Pr

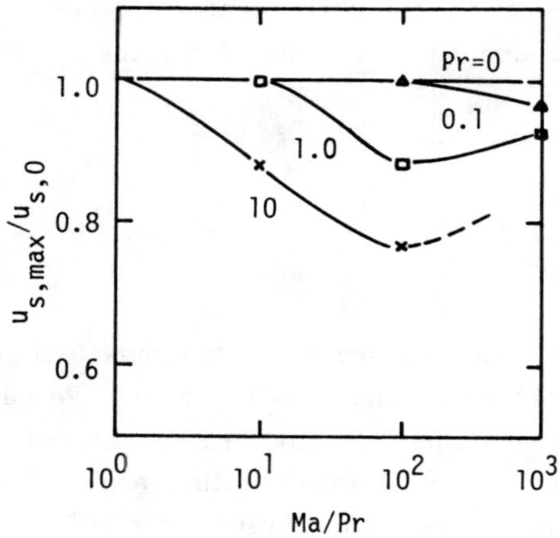

FIGURE 5. Maximum velocity divided by the velocity $u_{s,0}$ for Pr=0 as a function of modified Marangoni number Ma/Pr.

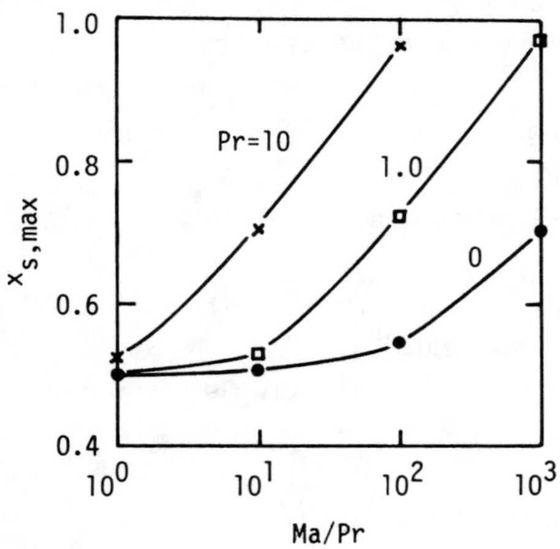

FIGURE 6. Position where the maximum velocity appear on the free surface as a function of modified Marangoni number Ma/Pr.

In the present analysis, the computational domain was devided into 20x20 rectangular elememts. Basically the even mesh was employed though, grading meshes were used to suppress the wiggles in the pressure distributions. Solutions were converged within 10 times Newton iterations, typically 5 or 6 times, where the convergence criterion for residuals was given as 10^{-6}. Computation time was about 2 minutes for an iteration on a computer FACOM VP-50.

CONCLUSION

A review was made about the researches on the numerical simulation of the natural convection related to the crystal growth. Recent study on the Marangoni convection analysis for low Prandtl number fluid was also presented. The solutions for Pr=0 were obtained and used as the reference solutions to discuss the results at low Prandtl number.

NOMENCLATURE

H	height of the cell
Ma	Marangoni number
Ma^*	modified Marangoni number, Ma/Pr
p	pressure
Pr	Prandtl number
u, v	fluid velocity
$u_{s,max}$	maximum velocity on the free surface
$u_{s,o}$	$u_{s,max}$ for Pr=0
T	temperature
x, y	Cartesian coordinates
$x_{s,max}$	the position where the velocity is maximum on the free surface

Greek Letters

α	thermal diffusivity
ν	kinematic viscosity
ρ	density
σ_T	temperature coefficient of surface tension, $d\sigma/dT$

Subscripts

c	cold wall
h	hot wall

REFERENCES

1. Ostrach, S., Low-Gravity Fluid Flows, Ann. Rev. Fluid Mech., vol. 14, pp. 313-345, 1982.

2. Fu, B. -I. and Ostrach, S., Numerical Solutions of Floating-Zone Thermocapillary Flows, Proceedings of the 4th European Symposium on Materials Sciences under Microgravity, pp. 239-245, 1983.

3. Kamotani, Y., Ostrach, S. and Vargas, M., Oscillatory Thermocapillary Convection in a Simulated Floating Configuration, J. Crystal Growth, vol. 66, pp. 83-90, 1984.

4. Lai, C. L., Ostrach, S. and Kamotani, Y., The Role of Free-Surface Deformation in Unsteady Thermocapillary Flow, Heat Transfer in High Technology and Power Engineering, Hemisphere, pp. 15-38, 1987.

5. Maekawa, T., Tanasawa, I., Ochiai, J., Kuwahara, K., Morioka, M. and Enya, S., Two-Dimensional Marangoni and Buoyancy Convection Related to Crystal Growth Techniques in Space, Adv. Space Res., vol. 4, no. 5, pp. 63-66, 1984.

6. Morioka, M., Ochiai, J., Kuwahara, K., Enya, S., Maekawa, T., Tanasawa, I. and Nakahata Y., Two-Dimensional Natural Convection Driven by Surface Tension and Buoyancy Generated in Liquid Layers Contained in a Horizontal Rectangular Cell, Proceedings of the 14th International Symposium on Space Technology and Science, Tokyo, pp. 1591-1597, 1984.

7. Uchida, H., Kuwahara, K. and Enya, S., Numerical Analysis of Marangoni Convection in Bridgman Method, Proceedings of the Symposium on Mechanics for Space Flight, 1986.

8. Uchida, H., Kuwahara, K. and Enya, S., Numerical Simulation on the Flow Field of Material Processing in Space, JSME Special Session for 948th Lecture Meeting, pp. 25-28, 1987.

9. Kuwahara, K., Enya, S., Akiyoshi, R., Ochiai, J. and Uchida, H., Fundamental Study on Marangoni Convection Related to Bridgman Crystal Growth, Proceedings of the 17th International Symposium on Space Technology and Science, 1990.

10. Maekawa, T. and Tanasawa, I., Natural Convection Driven by Buoyancy and Surface Tension Forces under External Magnetic Field, Applied Microgravity Tech., vol. 1, no. 2, pp. 77-84, 1988.

11. Hirata, A., Okano, Y. and Nishizawa, S., Numerical Study of Transport Phenomena through Fluid-Liquid Interfaces, Bulletin of the Centre for Informatics, Waseda University, vol. 8, autumn, 1988.

12. Yamaguchi, Y., Chang, C. J. and Brown, R. A., Multiple Buoyancy-Driven Flows in a Vertical Cylinder Heated from Below, Phil. Trans. R. Soc. Lond., A312, pp. 519-552, 1984.

On Conjugate Heat Transfer in Crystal Growth

NOAM LIOR and YOSHIHISA FUJII
Department of Mechanical Engineering and Applied Mechanics
University of Pennsylvania
Philadelphia, Pennsylvania 19104-6315, USA

INTRODUCTION

Natural convection has a very important role in crystal growth, and it is strongly influenced by the heat transfer in the walls enclosing the melt. In process modeling, the true influence of the heat transfer in these walls can only be found by solving the conjugate heat transfer problem in the melt, crystal, and enclosure. The liquid/enclosure system is solved here as a conjugate heat transfer problem in the range of parameters relevant to crystal growth, and the influence of the walls on the natural convection in the melt is elucidated.

Crystal growth has attracted significant attention and research over the past two decades, and several books and excellent reviews and analyses of the process have been published (cf. Brice 1976, Rosenberger 1979, Kobayashi 1981, Pimputkar and Ostrach 1981, Brown 1988, Brown et al. 1989, Dupret et. al. 1990).

Two of the most popular crystal growth processes are depicted in Figs. 1a and 1b. Fig. 1a shows a confined system (cf. Flemings, 1974), in this case the Bridgman-Stockbarger method. Alternatively, the furnace heat output can be varied with time around a stationary ampule, which is the gradient-freeze method.

Fig. 1b shows a meniscus-defined crystal growth system, the Czochralski (CZ) process. Here the material is melted in a crucible open at the top, and a seed crystal attached to a puller is dipped into it and then pulled upwards, typically with simultaneous rotation of both puller and crucible. The crystal is continuously grown from the seed-melt interface, with corresponding depletion of the melt.

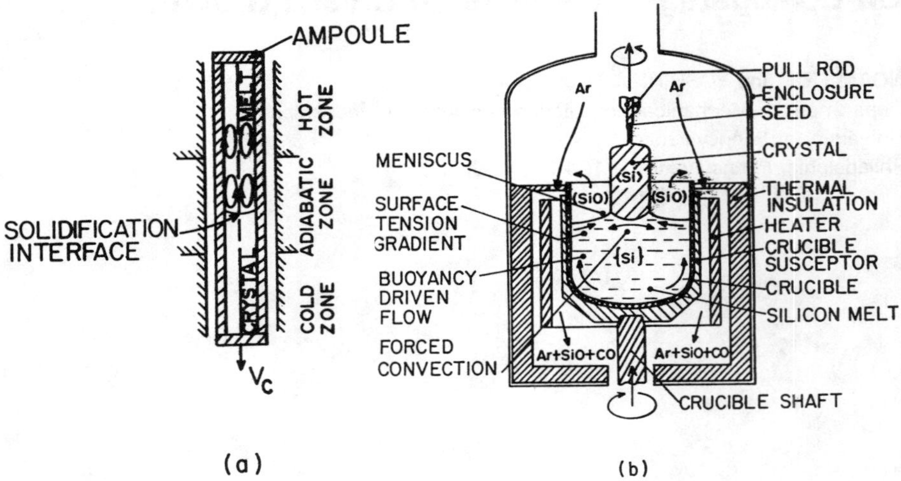

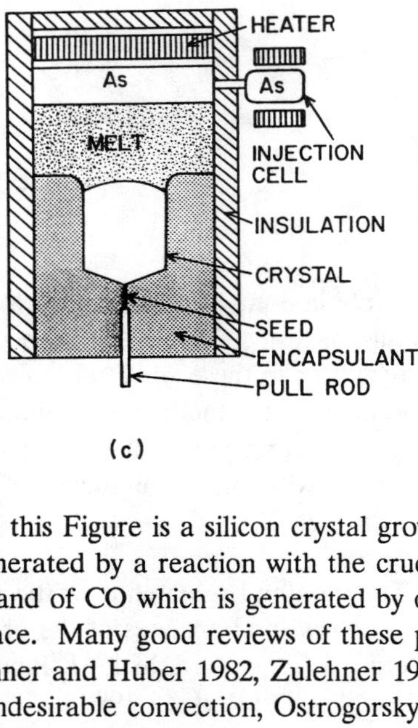

FIGURE 1. Crystal growth processes. (a) Vertical Bridgman-Stockbarger, (b) Czochralski, (c) Inverted encapsulated Czochralski.

Depicted in this Figure is a silicon crystal growth system, with evolution of SiO gas which is generated by a reaction with the crucible, and with vigorous argon purging of the SiO and of CO which is generated by oxidation of hot graphite components of the furnace. Many good reviews of these processes are available (cf. Kobayashi 1981, Zulehner and Huber 1982, Zulehner 1983, Lin and Benson 1987). To minimize undesirable convection, Ostrogorsky (1989) recently proposed an upside-down LEC CZ, in which the crystal is under the melt and drawn downwards, with heating from the top (through the melt) in a system which is thus thermally stably stratified, as shown in Fig. 1c.

All practical melting/solidification processes, crystal growth included, incur temperature and concentration gradients within the melt, and between the melt and the enclosure walls, which under the effect of a force field (most typically gravity) give rise to natural convection, unless it is suppressed intentionally by annulling or counteracting the effects of the driving force field. When convection occurs, it affects the crystal growth process in a number of possible ways:

(1) It influences the **temperature and its gradients** in the melt, perhaps most importantly at the solid-liquid interface where the crystal is formed (cf. Woodruff 1973, Glicksman et al. 1986). The magnitudes of the local temperature gradients are important in determining crystal growth rates, and thus also uniformity, in driving natural convection, in determining the microscale solidification structure, and in the generation of internal stresses and dislocations. A more minor effect is Soret diffusion of dopants and contaminants. The temperature level itself affects the compositions of the solid and liquid.

The shape of the melt-crystal interface is not given, is a function of local temperature and concentration, and varies with time. It thus needs to be determined simultaneously with the solution of the governing equations.

The interaction between the solid microstructure and the temperature and concentration fields at the melt/crystal interface is quite complex. The crystal interface forms dendritic and cellular patterns due to microscale instabilities, arising from the penetration of small solid protuberances into melt zones in which the concentration is lower than that at the freezing interface, while the temperature is not significantly higher. This "constitutional subcooling" (cf. Tiller et al. 1953) causes freezing and further growth of the protuberance. Schaefer and Coriell (1984) observed that interfacial distortions are clearly related to the nature of the convection. The radial outflow shown in Fig. 1b tends to reduce dopant concentration in the center and thus create a crystal with more uniform properties. For example, Zulehener (1983) has shown a 24% improvement in the radial uniformity of resistivity when the crystal rotation speed was increased from 5 RPM to 25 RPM.

At the free surface of the melt, which exists in some crystal growth methods, such as CZ (Fig. 1b), three phases interface at the same location, giving rise to surface tension forces at a part of the melt boundary, which take part in defining the shape of the boundary and the convection inside it. It is noteworthy that the interfacial tension is a function of both temperature and concentration (and hence a function of location along the interface, and time), both of which are also strongly influenced by convection in the melt.

Instabilities at the solidification front, including cellular and dendritic growth phenomena, have been analyzed primarily for the cases in which convection was negligible (cf. Mullins and Sekerka 1963, 1964, Coriell et al. 1981, Ungar et al. 1985, Langer and Hong 1986, Karma 1986, Bennett et al. 1988). Convection complicates the solution and was shown to create additional instability modes (cf. Coriell et al. 1980, Davis et al. 1984).

(2) In the absence of stronger forced convection, it is typically the strongest **driving force for mass transport** of dopants and contaminants, which affects their

concentration distribution in the crystal. Convection of the melt, as seen in Fig. 1a,b, moves contaminants from the crucible wall, along the free surface, and to the growing crystal. Along the path, and particularly while near the free surface, some of the contaminants evaporate, and reduce their concentration by the time they arrive at the solidifying interface. On the other hand, the radial convection towards the crystal creates a pronounced radial gradient of dopants in the crystal, which is undesirable. Furthermore, the thermal conditions in the process keep changing as the crystal grows and the melt is depleted, bringing rise to different convection patterns which typically increase the oxygen concentration and diminish the dopant concentration gradient in the crystal.

The influence of convection on the radial and axial segregation coefficients was summarized by Brown (1988). If the only convection is due to crystal growth and thus diffusion dominated, the radial segregation coefficient is low but the axial segregation coefficient is high due to gradual change in the solute concentration in the unmixed melt. Onset of cellular convection makes radial segregation higher, because of poor interaction with the diffusion layer at the curved crystal interface, but reduces axial desegregation. More intense convection reduces all segregation, but at some critical Rayleigh number brings up transition to unstable flows. Fully turbulent flow minimizes segregation. With the commonly-used rotation of both crucible and puller in CZ, the flow patterns are, naturally, different from those in which natural convection alone occurs, (cf. Carruthers and Nassau 1968). Natural convection was found to sustain an effect on the flow field at all but the highest rotational velocities used.

(3) Convective processes might become unstable, exhibiting **oscillations and transitions** to different flow regimes. They create corresponding oscillations in temperature (cf. Caruthers 1977, Kuroda et al. 1984, Mihelčić et al. 1984) and concentration (cf. Wilson 1980a,b). This is particularly likely when, as in crystal growth, several driving forces act at the same time. Coexisting here may be buoyancy forces due to temperature and concentration gradients, forced motion of the melt boundaries arising from crystal and crucible rotation and axial movement, and at free surfaces Marangoni motions due to surface tension, normal velocities due to evaporation, and velocities due to the flow of purging gas. Instabilities create both radial and axial nonuniformities in the crystal (cf. Hurle et al. 1968, 1974, K.M. Kim et al. 1972, ,1978) and measures need therefore to be taken to prevent them.

It is noteworthy that even systems which are thermally stably stratified (having purely vertical temperature gradients which increase, for fluids with $\beta>0$, in the direction of gravity) will in most cases become unstable. Coriell et al. (1980) reminded us of the role of concentration distributions in the melt which might oppose the stable stratification due to temperature distribution and lead to convection in the fluid. Viskanta and Hale (1978), Kim and Viskanta (1984), and Gopalakrishna et. al. (1987) have shown by numerical analysis how thermally-conducting side walls of the enclosure bring up natural convection in fluids which are initially stably stratified, with the exception of the improbable case in which the thermal conductivity and diffusivity of the fluid and walls are equal. This is particularly relevant in the vertical Bridgman crystal growth process and in the inverted CZ process (Ostrogorsky, 1989) and is elaborated further below as the main new research presented in this paper.

THE EFFECTS OF THE MELT CONTAINER

Natural convection is strongly sensitive to the nature of the boundary conditions, especially when the flow is driven by temperature differences between the enclosure wall and the fluid (cf. El Sherbiny et al. 1982, Gebhart et al. 1988). It is also obvious that the heat transfer in the walls is affected by the nature and strength of the natural convection. For simplicity, past analyses of natural convection assumed a prescribed temperature or heat flux on the wall and analyzed the fluid domain only; this is also true for most of the extensive analyses of crystal growth processes (cf. Langlois 1985), such as BS (cf. Chang and Wilcox 1974, Naumann 1983, Jasinski et al. 1983, Chang and Brown 1983) or CZ (cf. Langlois 1977, 1982, Kobayashi 1978, 1981, Mihelcic et al. 1984, Williams and Reusser 1983, Crochet et al. 1983, Derby et al. 1985, Derby and Brown 1986, 1987, Sabhapathy and Salcudean 1989, Hurle et al. 1990).

Effects of the heat transfer in the container walls or those of the heater have been studied, first in a non-conjugated manner, (cf. Fu and Wilcox 1980, Naumann and Lehoczky 1983, Jasinski et al. 1985) and most recently also with conjugate consideration of the walls (Adornato and Brown 1987a,b, Atherton et al. 1987, Derby et al. 1988, Sackinger et al. 1989, Brown et al. 1989, Crochet et al. 1989, Dupret et al. 1990, all using finite element methods).

Proper modeling of the container effects also allows the development and evaluation of ways to affect the crystal growth process, by controlling temperature gradients, heating, cooling and other container-communicated parameters (cf. Cockayne et al. 1972, Fu and Wilcox 1980, Adornato and Brown 1987a,b).

The container must be selected to withstand the temperatures incurred during crystal growth, and the reactive nature of the melt. Molten silicon, for example, reacts with all crucible materials at least to some extent because of its high binding force to light elements and high dissolving power (cf. Zulehner 1983). It is therefore necessary to select container materials which react slowly with the melt, and which do not give off products which impair the crystal electronically or mechanically. Some of the materials used are dense (such as vitreous) carbon, carbides, nitrides, and oxides (such as vitreous silica), which do not include metals or dopants. Consequently, silicon carbides, nitrides, and almost exclusively oxide, are used. The product of the reaction of the latter material with the silicon melt produces silicon oxide which evaporates readily from the melt and only about 1% of the oxygen gets captured in the crystal. Since reaction of the melt with the container is the major source of contaminants which migrate into the crystal (cf.

Carlberg et al. 1982, Murgai 1985), complete description of the crystallization process should include the equations which describe this reaction.

The stress field arising in crystals is related to the formation of dislocations. It is created primarily due to thermal effects: the stresses increase with the magnitude of the temperature gradients (cf. Iwaki and Kobayashi 1981, 1986, Jordan et al. 1984). The modification of the temperature gradients in the crystal by the presence of the walls (cf. Adornato and Brown 1987), and the effect of the walls on the natural convection in the melt both have an effect on the stress field (cf. Motakef and Witt 1987, Motakef 1987b, 1988, Lambropoulos 1987). Brandle (1977) observed that the nature of the dislocation distribution changes with transition from natural convection dominated to forced convection dominated melt flows. In confined growth systems, such as BS, the container also has an important role in determining the stress field in the confined crystal, and thus its thermoelastic model must be included in the analysis (cf. Motakef 1987a).

The adaption of process controls which maximize crystal uniformity despite the various changes in the system during the pulling, such as its configuration (due to growth of the crystal and consequent depletion of the melt), and potential of onset of instabilities, require good dynamic models which represent all of the relevant phenomena correctly. Solution of the conjugate problem allows consideration of not only the effects of the wall on the convection, but also proper accounting of the thermal capacity of the enclosure.

In summary, a realistic, conjugate inclusion of the melt "container" in the heat and mass transfer analysis is quite important in determining not only the natural convection driven by the temperature difference between the container and the melt, but also the mass transport from the container to the melt, the stress field in the crystal, the mechanical behavior of the container, and overall process dynamics and control, all of which affect the ultimate quality of the crystal.

As reviewed above, the crystallization process is described by a set of multi-dimensional highly nonlinear partial differential equations, and more generally by integro-differential equations when radiation is included. Furthermore, the geometry may be irregular, and at least the melt/crystal interface, of transient shape which is part of the problem solution. The addition of the conjugate problem to the model requires additional programming and computer time and memory, and, most significantly, poses numerical difficulties at the interface between the media (cf. Bontroux et al. 1980). Robust methods for the solution of conjugate problems are still in their infancy, but effective iteration philosophies in some problems have been developed (cf. Jones and Lior 1987).

THE EFFECTS OF WALL CONDUCTION ON NATURAL CONVECTION IN ENCLOSURES

Preface

To gain some fundamental insights into the effects of wall conduction on natural convection in enclosures, we present the results of a numerical analysis of a conjugate natural convection in an enclosure without phase change, with the liquid and enclosure having properties in a range of parameters typical of crystal growth.

Previous work on conjugate heat transfer with natural convection includes the studies by Gdalevich et al. (1979), Viskanta and Lankford (1981), Burch et al. (1985), Jones and Lior (1987), and dealing specifically with enclosures, Viskanta and Hale (1978), Hess and Miller (1982), Kim and Viskanta (1984, they also included radiative exchange between the internal walls), Shyu and Hsieh (1985), Gopalakrishna, Himasekhar and Lior (1987), and Joubert and Le Quéré (1989).

Formulation

Figure 2 shows the configuration of the problems solved, one a 2-d rectangular and the other an axisymmetric cylindrical enclosure, where the top and bottom surfaces are perfectly insulated, and the side walls are conductive. The exterior surface of the side wall is subjected to a generalized convective boundary condition. Initially, the temperatures in the enclosure is stratified stably: the upper part of the enclosure is at higher and the lower part at lower temperature, with a linear distribution between them. The initial temperatures in the side wall is assumed to be same as that in the fluid. These initial conditions may occur in the vertical BS

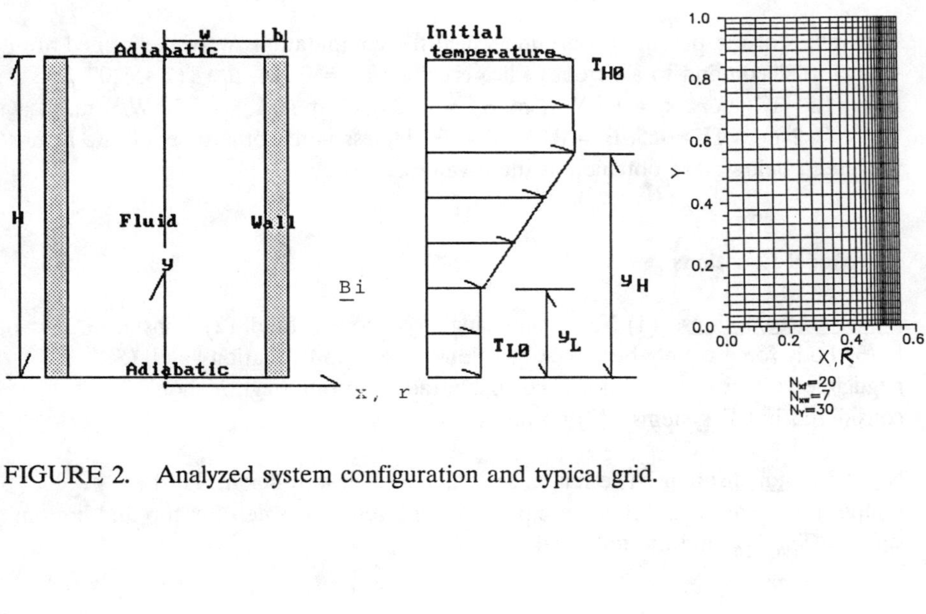

FIGURE 2. Analyzed system configuration and typical grid.

process, in the inverted encapsulated CZ process, or in the melt of the CZ process prior to crystal pulling. We assume that there are initially no fluid motions in the enclosure.

Even though the fluid is initially stable, destratification will occur as time passes, because of conduction and ensuing convection. As Fig. 3 shows, there are two kinds of heat flows though the side walls. One is the heat flow in the normal direction to the wall, producing a temperature gradient between the fluid and the wall in a direction perpendicular to gravity (horizontal), and thus natural convection. This heat flow and ensuing convection can be largely prevented by

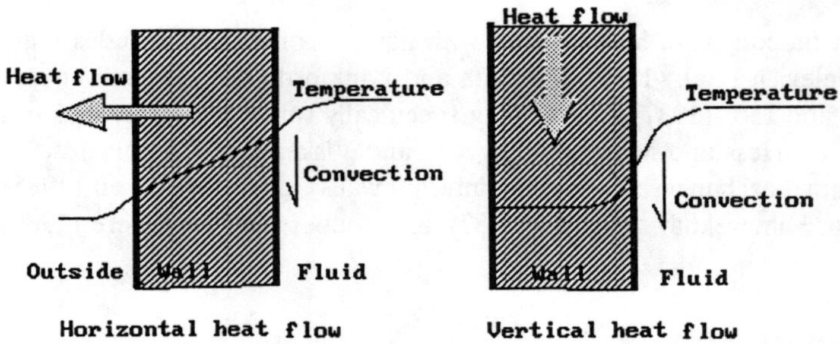

FIGURE 3. Ways in which heat conduction in the walls initiates natural convection in a stably-stratified fluid.

good insulation. However, even when the outside of the wall is perfectly insulated, heat still flows though the walls vertically, and if the thermophysical transport properties of fluid differ from those of the side walls, a temperature difference will establish itself between the fluid and the wall and will consequently produce natural convection.

The values of the parameters around which the computations were performed are typical to silicon and to a vitreous silica crucible: Pr = 0.024, $\beta_f = (1.41)10^{-4}$ 1/K, $\alpha_f = (1.25)10^{-5}$ m^2/s, $k_f = 64$ W/K m, $\alpha_w = (9.33)10^{-7}$ m^2/s, $k_w = 1.59$ W/K m, H = 0.05 to 0.2 m, w/H = 0.5, B = 0.05, Bi = 0. Unless noted otherwise, all the results presented below were obtained at these values.

The Model Equations

General assumptions. (1) Newtonian, single component fluid, (2) Constant μ and k, (3) body force is only buoyancy, (4) Boussinesq approximation valid, (5) negligible viscous dissipation, (6) negligible radiation (although it should be considered in CZ systems), (7) no phase change.

Nondimensionalization. The length scale is the enclosure height, H, the temperature scale is the initial temperature difference between the top and bottom surface, T_{H0}-T_{L0}, and the time scale is H^2/α_f.

The dimensionless variables are therefore

$$X=x/H \qquad U = uH/\alpha_f \qquad \theta = \frac{T - T_{LO}}{T_{HO} - T_{LO}}$$
$$R=r/H \qquad V = vh/\alpha_f \qquad \alpha_R = \alpha_w/\alpha_f \tag{1}$$
$$Y=y/H \qquad \tau = t\alpha_f/H^2 \qquad k_R = k_w/k_f$$

$$Pr=\nu_f/\alpha_f$$
$$Ra = GrPr = \frac{H^3 g\beta (T_{HO} - T_{LO})}{Pr\alpha_f^2} \tag{2}$$
$$Bi = h_e H/k_w$$

<u>The governing equations</u>. Continuity:

$$\nabla\cdot\vec{V} = 0 \tag{3}$$

momentum:

$$\frac{D\vec{V}}{D\tau} = -\nabla\vec{P} + RaPr\theta_f\vec{i} + Pr\nabla^2\vec{V}, \tag{4}$$

energy equation in the fluid:

$$\frac{D\theta_f}{D\tau} = \nabla^2\theta_f, \tag{5}$$

and the energy equation in the side walls:

$$\frac{\partial\theta_w}{\partial\tau} = \alpha_R\nabla^2\theta_w. \tag{6}$$

<u>The governing equations in vorticity-stream function form</u>. The vorticity-stream formulation is used to reduce the number of unknowns. With

$$\zeta\equiv\nabla\times\vec{V}, \tag{7}$$

the vorticity transport equation is

$$\frac{D\zeta}{D\tau} = Pr\nabla^2\zeta + RaPr\theta_f\nabla\times\vec{i}. \tag{8}$$

The stream function is related to the vorticity by

$$\nabla^2\psi = -\zeta, \tag{9}$$

where, in Cartesian and cylindrical coordinates,

$$U = \frac{\partial\psi}{\partial Y} \qquad U = \frac{1}{R}\frac{\partial\psi}{\partial Y}$$
$$V = -\frac{\partial\psi}{\partial X} \qquad V = -\frac{1}{R}\frac{\partial\psi}{\partial R}, \tag{10}$$

respectively. We now have three unknown parameters, ζ, ψ, and θ. The equations of vorticity transport (8), stream function (9), and energy (5 and 6) are the set of basic equations describing this problem.

Initial conditions. The fluid is assumed to be initially stagnant in the entire fluid region, so

$$\zeta|_{\tau=0} = 0$$
$$\psi|_{\tau=0} = 0 \quad at\ 0 \le R \le W\ and\ 0 \le Y \le 1 \tag{11}$$

For temperature

$$\theta = 0 \quad at\ 0 \le R \le (W+B),\ 0 \le Y \le Y_L \quad and\ at\ Y_H \le Y \le 1, \tag{12}$$

$$\theta = \frac{Y - Y_L}{Y_H - Y_L} \quad for\ 0 \le R \le (W+B),\ Y_L \le Y \le Y_H, \tag{13}$$

Boundary conditions. Velocities:
From symmetry:

$$U = 0 \quad at\ R = 0,\ 0 \le Y \le 1. \tag{14}$$

and no-slip on the fluid-wall interfaces:

$$\psi = 0 \quad \begin{array}{l} at\ R = 0,\ 0 \le Y \le 1\ (the\ center\ line) \\ 0 \le R \le W,\ Y = 0\ (the\ bottom\ surface) \\ 0 \le R \le W,\ Y = 1\ (the\ top\ surface) \\ R = 1,\ 0 \le Y \le 1\ (the\ side\ wall) \end{array} . \tag{15}$$

Vorticity:

$$\zeta = 0 \quad at\ R = 0,\ 0 \le Y \le 1, \tag{16}$$

and the boundary vorticities on the no-slip surfaces are calculated from the stream functions by Equation (9).
Temperatures:

$$\frac{\partial\theta}{\partial n} = 0 \quad \begin{array}{l} at\ R = 0,\ 0 \le Y \le 1\ (the\ center\ line) \\ 0 \le R \le W,\ Y = 0\ (the\ bottom\ surface) \\ 0 \le R \le W,\ Y = 1\ (the\ top\ surface) \end{array} . \tag{17}$$

continuity on the inside surface:

$$\theta_f = \theta_w \quad and \quad \frac{\partial\theta_f}{\partial n} = K_R \frac{\partial\theta_w}{\partial n} \qquad at\ R = W, \quad 0 \le Y \le 1. \tag{18}$$

On the outside surface of the side wall we assume a constant heat transfer coefficient, h_e. Then the heat balance is

$$-\frac{\partial\theta_w}{\partial n} = Bi(\theta_w - \theta_a) \qquad at\ R = W + B, \quad 0 \le Y \le 1 \tag{19}$$

where Bi is the Biot number.

Numerical Analysis

Coordinate Transformation. The gradients of temperature and velocities are particularly large at the fluid/solid interfaces, and thus a very fine grid is required in this regions. To conserve computation time and memory, a non-uniform grid is employed in the fluid, gradually becoming finer as the solid boundaries are approached. This is accomplished by a coordinate transformation in which the new variables, ξ and η, are functions of X and Y, related by a logarithmic function which generates a smooth non-uniform grid system (Anderson et al. 1984). A uniform grid is used in the side wall. N_{rw} is the number of grid points in the wall, and N_{rf} is the number of grid points in the fluid, along the radius. A typical grid system used for computations with silicon is shown in Fig. 2.

The finite difference method. The vorticity transport and energy equations (8) and (5), (6) are second order partial differential equations which are parabolic in time and elliptic in space. As amply tested (cf. Ozoe et al. 1983, Chao et al. 1983, Ozoe et al. 1985), the Alternative-Direction-Implicit method (ADI) is a good choice for the solution of this type of equations, since it is unconditionally stable and it has second order accuracy in time and space. The convective terms in these equations are treated by the hybrid scheme (Patankar 1980) in which central differencing is applied in the slow velocity regions where $|Pe| < 2$, and the second order upwind scheme is employed where $|Pe| \ge 2$.

The stream function equation (9) is second-order elliptic partial differential, and the Successive-Line-Overrelaxation method (SLOR) was chosen for its solution because its fast convergence, especially in problems which have sharp gradients in one direction.

This coupled system of the governing equations was decoupled by the "lagged" method: (1) the temperature field at the (n+1) time step is calculated by the ADI method using the velocities at their values of the n-th step, (2) the vorticities at the (n+1) time step are calculated using the temperatures at the same time step; the stream function values of the n-th time step are used to determine the vorticities at the no-slip boundaries, (3) using the vorticities obtained in the previous step 2, the stream function equation for the (n+1) step is solved by the SLOR method, (4) the velocities for that time step are computed by eq. (10), (5) the procedure (1) to (4) is repeated until the final time of interest is reached.

The finite difference control volumes. For vorticities and stream functions, the calculation points are distributed uniformly in the coordinate-transformed fluid region. To avoid abrupt temperature transitions and consequent false heat transfer resistances at the fluid/solid interface, the interfacial control volumes for the energy equation were placed on both sides of the interface, one entirely in the fluid and one entirely in the solid, and the heat flux was evaluated by the harmonic mean heat conductivity as suggested by Patankar (1980). Consequently, the calculation points for temperatures are shifted by half grid size from those for vorticities and stream functions. To restore the spatial match, the velocities used in the energy equation were interpolated between those obtained from the solution of the stream function equation.

Because of the centerline singularity in cylindrical coordinates, the velocity there was computed from the weighted stream function determined by Parmentier et al. (1975) to satisfy the circulation theorem.

Accuracy of the Solution Method

The computational grid which gave good results is described by $N_{xf} = 20$, $N_{rf} = 30$, $N_y = 30$, $N_{rw} = 7$, $\gamma_{x,r} = 1.05$, and $\gamma_y=1.2$, resulting in a minimal grid size of 0.0056. Extrapolation to zero grid size indicated that this grid produces a maximal error of 4%.

Due to the "lagged" solution method used, there is a critical computational time step, $\Delta\tau_c$, above which the solution does not converge (cf. Roache 1982). As Pearson (1971) suggested, $\Delta\tau_c$ depends primarily on grid spacing Δx and Prandtl number Pr, and it is independent from Ra. We have developed from numerical experiments the correlation

$$\Delta\tau_c = \frac{0.8}{Pr}\Delta x^2 \qquad (20)$$

In comparison, the analytically-obtained equation by Bontoux et al.(1980) was found to underestimate the critical time step at Pr $\not\approx$ 4.64. $\Delta\tau_c$ in our case is $4(10)^{-4}$, and the computations were performed with time steps of 10^{-4}, corresponding here to real time steps of 0.08 s.

Results

Flow patterns and temperature distributions. For cylindrical geometry (R,Y) Fig. 4 shows stream function contours and isotherms at τ=0.001, 0.1, and 0.3 (t=0.8, 80, and, 240 s, respectively), Fig. 5 shows the vertical and horizontal velocity component profiles, and Fig. 6 shows the vertical temperature distributions in the

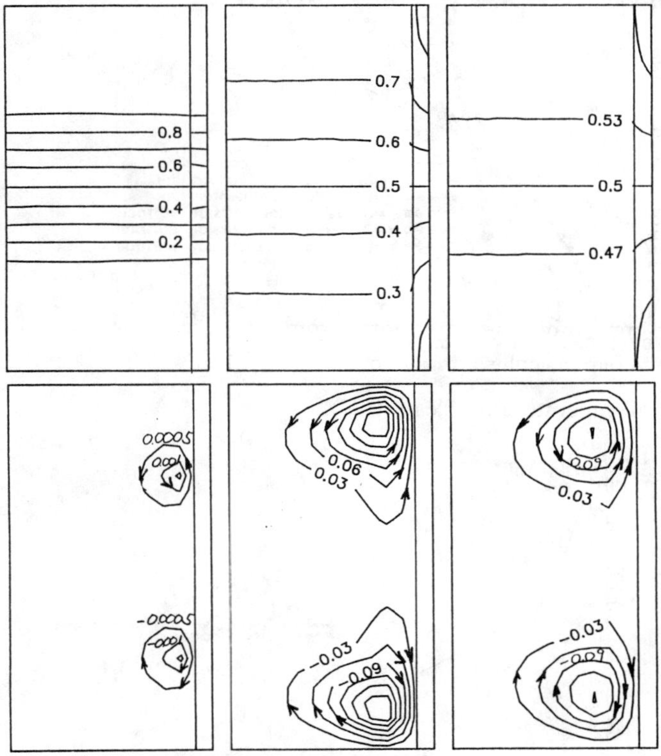

τ = 0.001 τ = 0.1 τ = 0.3

FIGURE 4. Isotherms and streamlines with increasing time. (R,Y), Ra=10^7.

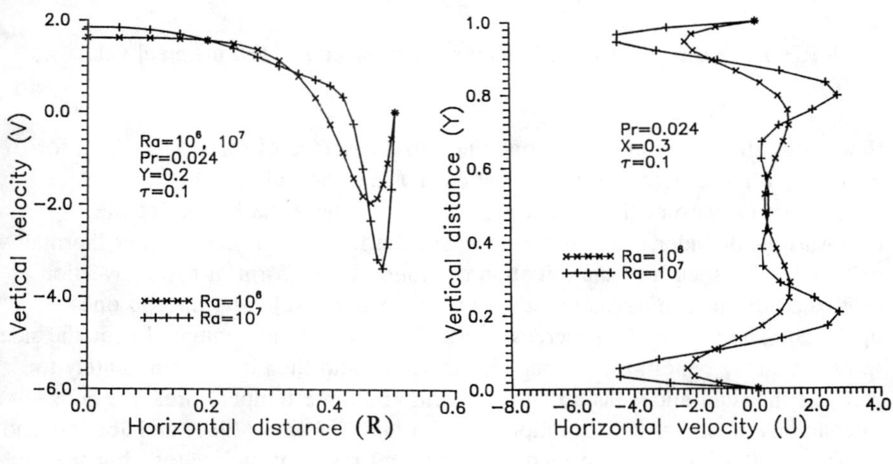

FIGURE 5. Vertical and horizontal velocity component profiles. (R,Y).

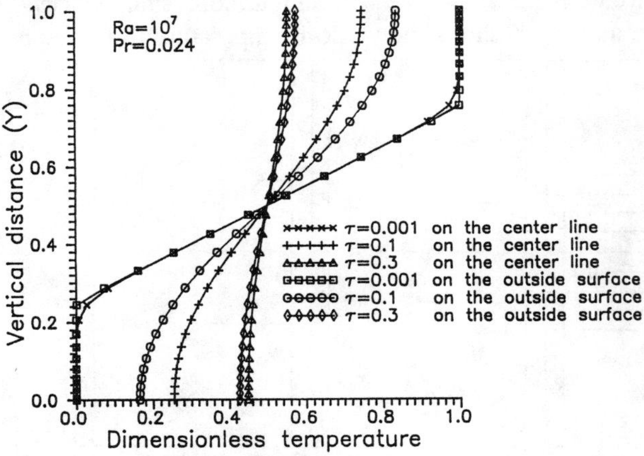

FIGURE 6. Temperature profiles. (R,Y).

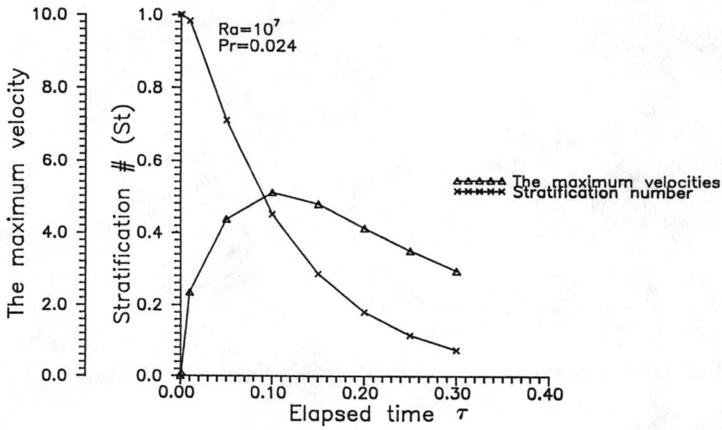

FIGURE 7. History of the stratification number and of the maximal velocity. (X,Y).

fluid along the centerline, and along the outside surface of the side wall, all for Bi = 0. From these Figures one can see clearly that the fluid
soon becomes warmer than the wall adjacent to it near the bottom of the enclosure, and colder near the top, because the fluid has a much higher thermal diffusivity. This leads to the initiation of motion in the form of two convection cells, the bottom one circulation downwards near the wall, and the top one upwards. The velocity first increases with time, and motion, initiated near the side walls, gradually penetrates throughout the fluid, and then decays (ultimately to zero) as the fluid and walls attain a uniform and same temperature. Previous calculations of similar configurations by Kim and Viskanta (1984) and Joubert and Le Quéré (1989) were performed in a diffferent range of parameters, but the results are qualitatively very similar.

The stratification number. As an index of the temperature stratification in the enclosure, Gopalakrishna et al. (1987) have defined a stratification number as the ratio of the mean temperature gradient in the vertical direction in the entire volume of the fluid at a given time, to the initial one:

$$S=\frac{\overline{(\partial\theta/\partial Y)_{\tau}}}{\overline{(\partial\theta/\partial Y)_{0}}}. \tag{21}$$

For the rectangular configuration (X,Y), Fig. 7 shows how the stratification number decays with time due to the conduction and natural convection in the fluid, and the same Figure also shows the change in maximal velocity. In this particular case it was found that the behavior of the stratification number with time is practically indistinguishable from that calculated by assuming pure conduction (no convection) in the fluid. This is reasonable, since the thermal diffusivity of the fluid is high, the velocities are very small, and Ra is relatively small. Similarly,. the thickness of the wall has a small effect (typically <10% for an increase from B = 0 to 0.1) in reducing the rate of stratification decay. The behavior of the fluid in the cylindrical and rectangular geometries was basically similar. The primary difference between them was in that the velocity near the centerline was higher for the cylindrical geometry, as expected.

The effects of Pr. Computations for Pr from 0.01 to 1.0 have shown a small increase in boundary layer thickness with Pr, but almost identical velocity maxima, and negligible effect on S.

The effects of Rayleigh number and of α_R and k_R. It should be noted that, strictly speaking, the Rayleigh number defined in this case for a stably stratified fluid is proportional to the degree of stability rather than, as most commonly applied to unstable systems, to the intensity of motion. Nevertheless, since large Ra imply the presence of stronger driving forces for conduction through the wall and then for the ensuing convection, the ultimate influence on natural convection is of the same nature. For example, Fig. 8, computed for (X,Y) and $\tau = 0.1$, shows that destratification is more rapid as Ra increases, but this influence diminishes as α_R decreases, becoming insignificant at around $\alpha_R = 1$.

With Bi = 0, $\alpha_R = 1$ in Fig. 8 also corresponds to the special case in which the fluid and wall have the same diffusivity and conductivity, and is equivalent to the non-conjugate, purely conductive problem solution. This allows direct evaluation of the significance of solving the problem in a conjugate manner, i.e. the role of heat conduction in the wall on the heat transfer and convection in the enclosure. Clearly, if the conduction in the wall was not taken into account, no convection would have been predicted and the destratification would have taken place by conduction alone. For $\alpha_R>1$ the rate of destratification would have been severely underpredicted. For $\alpha_R<1$ motion in the fluid exists, but it is very slow due to the relatively low valu of α_w and k_w, and destratification is consequently slow. It is interesting to note that when $\alpha_R=0.1$ and 0.01, the rate of destratification is reduced compared to a purely conductive case even though convection occurs. The explanation is that the thermal capacity of the walls is high in these cases of $\alpha_R << 1$, and the wall, which retains its original stabilizing temperature profile transfers heat into the liquid and tends to stabilize it thereby. This tendency is not monotonic with the decrease of α_R: the stratification is weaker for $\alpha_R = 0.01$ than it is for 0.1 (but still stronger than that for $\alpha_R = 1$). This reversal in influence occurs because k_w also decreases with α_R, and the low rate of heat transfer between the wall and the fluid reduces the wall's stabilization effect. Further reduction of α_R would cause the destratification rate to converge to essentially the same one occurring when $\alpha_R = 1$ (pure conduction in fluid).

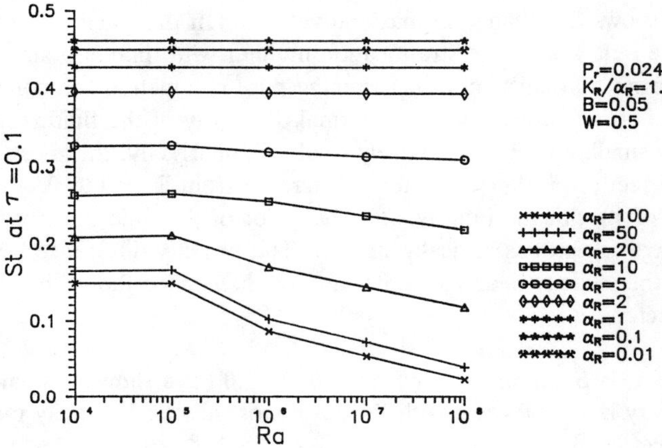

FIGURE 8. The Stratification Number at $\tau = 0.1$ as a function of the Rayleigh Number for different α_R. (X,Y).

Figure 9 shows that increasing the conductivity (or k_R) of a high diffusivity wall ($\alpha_R = 100$), also increases the rate of destratification. For a low diffusivity wall ($\alpha_R = 0.01$), large k_R implies high heat capacity of the wall, which improves its ability to stabilize the fluid.

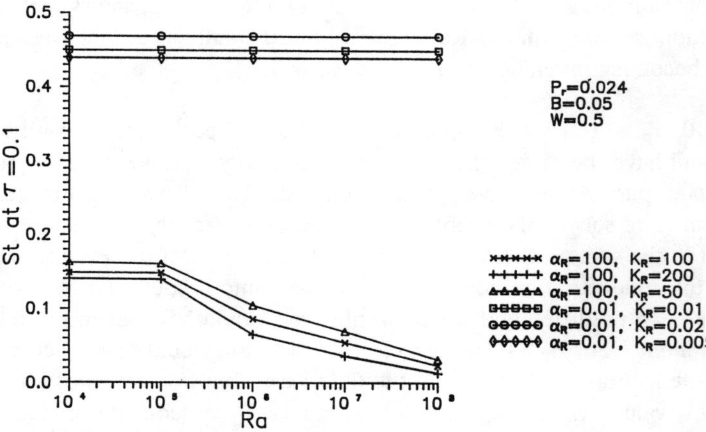

FIGURE 9. Effect of k_R on the Stratification Number at $\tau = 0.1$, as a function of the Rayleigh Number. (X,Y).

The effects of Biot number. Figure 10 shows the streamlines for Bi = 1000, a very high exterior heat loss condition which makes the exterior surface in effect isothermal at the ambient temperature $T_a = 0$. Compared to Fig. 4 the circulation is much stronger because of the stronger radial temperature gradient in the lower part of the enclosure, which depend to lesser extent on the axial conduction in the wall. The earlier two convection cells merge with time into one, descending near the wall. As expected, the destratification increases with Bi, the effect increasing with time. For example, S is about 25% lower for Bi = 1000 as compared with that for Bi = 0, at $\tau = 0.3$.

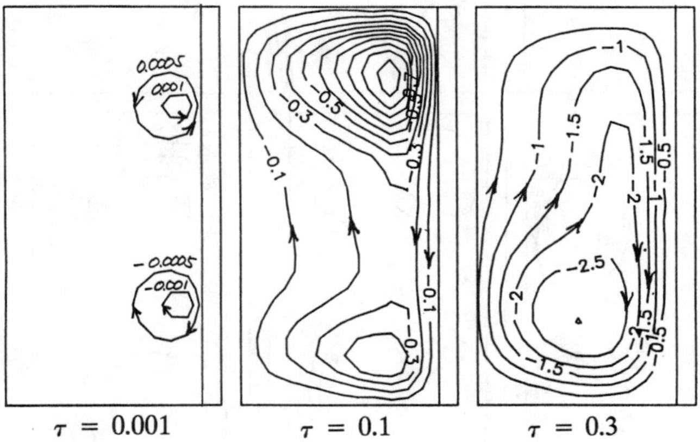

FIGURE 10. Streamlines with increasing time, Bi = 1000, (X,Y)

Unstably stratified fluid. When the fluid is initially unstably stratified thermally (such as is the case during CZ growth), the solution of the conjugate problem shows the ensuing convection directly, without need to introduce artificial disturbances to trigger the flow. The streamlines and isotherms are shown in Fig. 11 for Bi = 0. As expected, Fig. 12 shows that destratification occurs in this case much more rapidly than in the formerly-discussed cases in which the fluid was initially stably stratified. It is noteworthy, however, that the flow patterns and rate of destratification depend on the location and magnitude of the numerical disturbance introduced, even for disturbances which are less than 1% of the overall temperature difference. Clearly, divergence of α_R from 1 accelerates destratification.

CONCLUSIONS AND RECOMMENDATIONS FOR FUTURE RESEARCH

The Conjugate Problem

- An effective finite difference program was developed to solve 2-d cartesian and axisymmetric cylindrical conjugate laminar natural convection problems up to transition to turbulence ($Ra < 10^9$) and for $Pr \leq 1$.

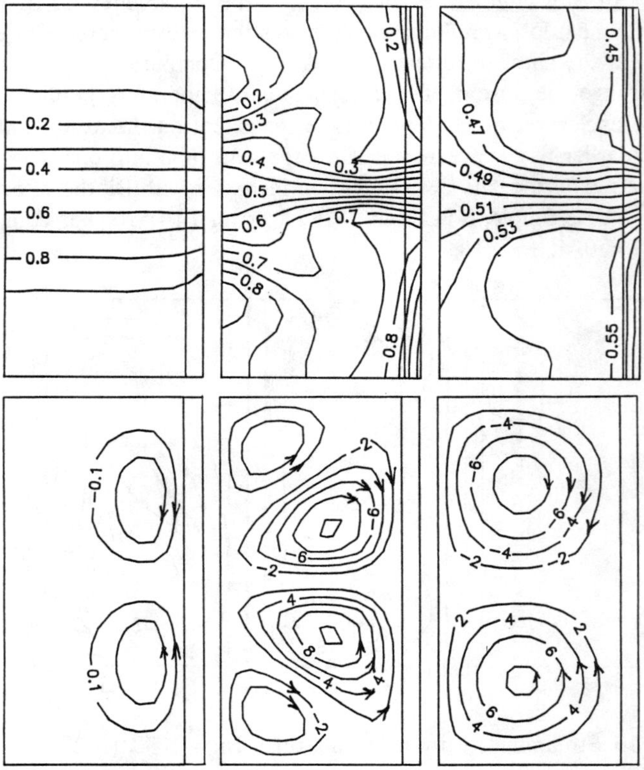

τ = 0.001   τ = 0.1   τ = 0.3

FIGURE 11. Isotherms and streamlines for an initially unstably stratified fluid, $Ra=10^7$, (X,Y).

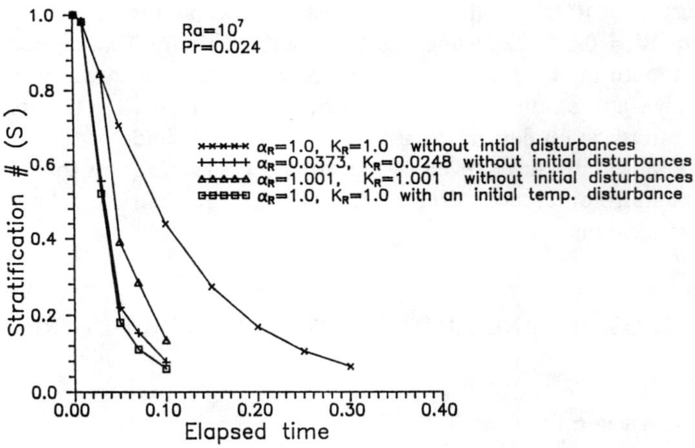

FIGURE 12. History of the stratification number for initially unstably stratified fluids, (X,Y).

- Solution of the conjugate problem is necessary for revealing and evaluating the natural convection that arises in stably stratified fluids, and for starting the solution in fluids which are initially unstably stratified.

- Although the velocities for molten silicon in a vitreous silica crucible are typically of the order of only 1 mm/s, the circulation arising in stably stratified fluids is capable of moving contaminants from the wall to the solidifying crystal surface and of creating radial segregation of solute there because it is oriented downwards along the walls in the lower part of the enclosure and upwards along the walls in its upper part, and then inward to the center at both bottom and top surfaces. In initially unstably stratified fluids the velocities are at least an order of magnitude higher.

- The velocities become larger as α_R, and to somewhat lesser extent k_R, increase, significantly so for $\alpha_R >> 1$.

- It is recommended that mixed convection (with translational and rotational velocity components) be studied with conjugate heat transfer, and that the relative effects of natural and forced convection be determined.

- Extension of the computer program to turbulent natural convection (it is ongoing by the authors), to three-dimensional domains, and then more specifically to crystal growing processes is recommended.

General

- Practically all of the heat and mass transfer modes as well as surface tension, and system components, play in crystal growth a role which can not be neglected or significantly simplified in modeling and analysis. The current trend to model the system as a set of conjugate problems is correct and should be continued.

- The solution of such models is difficult to attain, and more robust and time-effective numerical techniques are needed.

- System modeling requires good knowledge of the melt/crystal interfacial transport and transient morphology, particularly on the microscale. This too is a conjugate problem.

- To assure process stability and investigate the onset and effects of oscillations on crystal quality, the effects of melt container walls must be considered as part of a conjugate problem.

- Reactions of the container walls with the melt should be included in the models to improves the description of mass transfer. This indicates the need for a conjugate heat/mass transfer solution of the wall/melt.

- Mass transfer at free surfaces of the melt, due to such phenomena as oxide evaporation and gas (such as CO and purge gas) absorption, should also be included in describing mass transfer.

- The study of the link between the processing conditions and the formation of crystalline defects must continue; of particular importance is the microscale establishment of the relationship between the stresses and the formation and movement of dislocations.

- The difficulties in measuring temperatures, velocities, concentration and other relevant parameters in the melt and crystal pose severe difficulties in modeling and in validation of model results. New sensors and measurement techniques (for an innovative method see Kakimoto et al. 1989) are needed to produce relevant and sufficient experimental data.

- There is a great need for correct values of thermophysical (including radiative), thermodynamic, and mechanical properties of the materials in both phases.

NOMENCLATURE (for symbols not defined in the text)

k Heat conductivity
n Normal direction
P Pressure
t Time
T Temperature
T_a Ambient temperature
u Velocity in x or r direction
v Velocity in y direction
α Thermal diffusivity
β Volumetric coefficient of expansion
ν Kinematic viscosity
ρ Density

Subscripts

f Fluid
w Wall
0 Initial values

REFERENCES

Adornato, P. M. and Brown, R. A., The Effect of Ampoule on Convection and Segregation During Vertical Bridgman Growth of Dilute and Nondilute Binary Alloys, J. Crystal Growth, 80, 155, 1987a.

Adornato, P. M. and Brown, R. A., Petrov-Galerkin Methods for Natural Convection in Directional Solidification of Binary Alloys,"Int. J. Num. Meth. Fluids, 7, 5, 1987b.

Anderson, D. A., Tannehill, J. C. and Pletcher, E. H., Computational Fluid Mechanics and Heat Transfer, McGraw-Hill, NY, 1984.

Atherton, L. J., Derby, J. J., and Brown, R. A., Radiative Heat Exchange in Czochralski Crystal Growth, J. Crystal Growth, vol. 84, pp. 57-78, 1987.

Bennett, M. J., Brown, R. A., and Ungar, L. H., Nonlinear Interactions of Interface Structures of Differing Wavelength in Directional Solidification, Pro. 1986 Symp. Physics of Structure Formation, Springer, Berlin, 1988)

Bontroux, P., Gilly, B., and Roux, B., Analysis of the Effect of Boundary Conditions on Numerical Stability of Solutions of Navier-Stokes Equations, J. of Comp. Phys. 36, 65, 1980.

Brandle, C. D., Simulation of Fluid Flow in G_{d3} G_{a5} O_{12} Melts, J. Crystal Growth, vol. 42, 400-404, 1977.

Brice, J. C., The Growth of Crystals from Liquids; North-Holland, New York 1973.

Brown, R. A., Convection and Solidification in Melt Crystal Growth, Advanced Crystal Growth, P. Dryburgh, ed., Prentice Hall, New York, 41, 1987a.

Brown, R. A., Theory of Transport Processes in Single Crystal Growth from the Melt, A.I.Ch..E.J. no. 6, 34, 8881, 1988.

Brown, R. A., Kinney, T.A., Sackinger, P. A., and Bornside, D. E., Toward an Integrated Analysis of Czochralski Growth, J. Crystal Growth, vol. 97, pp. 99-115, 1989.

Burch, T., Rhodes, T., and Acharya, A., Laminar Natural Convection between Finitely Conducting Vertical Plates, Int. J. Heat Mass Transfer, vol. 28, 1173, 1986.

Carlberg, T., King, T. B., and Witt, A. F., Dynamic Oxygen Equilibrium in Silicon Melts During Crystal Growth by the Czochralski Technique, J. Electrochem. Soc., 129, 189, 1982.

Carruthers, J. R., an Nassau, K., J. Appl. Phys., vol. 39, 205, 1968.

Carruthers, J. R., Thermal Convection Instabilities Relevant to Crystal Growth from Liquids, Preparation and Properties of Solid State Materials, 3, W. R. Wilcox, R. A. Lefever, eds., Dekker, 1977.

Chang, C. E. and Wilcox, W. R., Control of Interface Shape in Vertical Bridgman-Stockbarger Technique, J. Crystal Growth, vol. 21, 135-140, 1974.

Chang, C. J. and Brown, R. A., Radial Segregation Induced by Natural Convection and Melt Solid Interface Shape in Vertical Bridgman Growth, J. Crystal Growth, vol. 63, pp. 343-364, 1983.

Chao, P. K., Ozoe, Lior, N., and Churchill, S. W., The Effect of Partial Baffles on Natural Convection in an Inclined Rectangular Enclosure. Part II: Numerical Solution, Chem. Engr. Fund., no. 2, 2, 38, 1983.

Cockayne, B., Plant, J. G. and Vere, A. W., Crucible-Base Cooling: An Aid to Czochralski Crystal Growth, J. Crystal Growth, 15, 167, 1972.

Coriell, S. R., Cordes, M. R., Boettinger, W. J., and Sekerka, R. F., Convective and Interfacial Stabilities During Unidirectional Solidification, J. Crystal Growth, vol. 49, 13, 1980.

Coriell, S. R., Boisvert, R. F., Rehm, R. G., and Sekerka, R. F., Lateral Solute Segregation During Unidirectional Solidification and Interface Instabilities During Unidirectional Solidification of a Binary Alloy, J. Crystal Growth, 54, vol. 167, 1981.

Crochet, M.J., Dupret, F., Ryckman, Y., Geyling, F.T., and Monberg, E.M., Numerical Simulation of Crystal Growth in a Vertical Bridgman Furnace, J. Crystal Growth, vol. 97, pp. 173-185, 1989.

Crochet, M. J., Wouters, P. J., Geyling, F. T., and Jordan, A. S., Finite-Element Simulation of Czochralski Bulk Flow, J. Crystal Growth, vol. 65, pp. 153-165, 1983.

Davis, S. H., Muller, U. and Dietsche, C., Pattern Selection in Single-Component Systems Coupling Benard Convection and Solidification, J. F. M., vol. 147, 1, 1984.

Derby, J. J., Brown, R. A., Geyling, F. T., Jordan, A. S., and Nikolakopoulou, G. A., Finite Element Analysis of a Thermal-Capillary Model for Liquid-Encapsulated Czochralski Growth, J. Electrochem. Soc. vol. 132, pp. 470-482, 1985.

Derby, J. J. and Brown, R. A., Thermal-Capillary Analysis of Czochralski and Liquid-Encapsulated Czochralski Crystal Growth. I: Steady State Simulation, J. Crystal Growth, vol. 74, 605, 1986.

Derby. J. J. and Brown, R. A., The Dynamics of Czochralski Crystal Growth, J. Crystal Growth, vol. 83, 137, 1987.

Derby. J. J., Atherton, L. J., Thomas, P. D., and Brown, R. M., Finite-Element Methods for Analysis of the Dynamics and Control of Czochralski Crystal Growth, J. Sci. Computing, 2, 297, 1988.

Dupret, F., Nicodème, P., Ryckmans, Y., Wouters, P., and Crochet, M. J., Global Modeling of Heat Transfer in Crystal Growth Furnaces, Int. J. Heat Mass Transfer, vol. 33, no. 9, pp. 1849-1872, 1990.

El Sherbiny, S.M., Hollands, K.G.T. and Raithby, G.D., Effect of Thermal Boundary Conditions on Natural Convection in Vertical and Inclined air Layers, J. Heat Transfer, vol. 104, pp.515-520, 1982.

Flemings, M. C., Solidification Processing, McGraw-Hill, New York, 1974.

Fu, T.-W. and Wilcox, W.R., Influence of Insulation on Stability of Interface Shape and Position in the Vertical Bridgman-Stockbarger Technique, J. Crystal Growth, vol. 48, pp. 416-424, 1980.

Gdalevich, L. B., Nogotov, E. F., and Fertman, V. E., Effect of Side Walls on Heat Transfer through a Vertical Air Layer in Laminar Natural Convection, Int. J. Heat Mass Transfer, vol. 22, 1601, 1979.

Gebhart, B., Jaluria, Y., Mahajan, R. L. and Sammakia, B., Buoyancy-Induced Flows and Transport, Hemisphere, New York, 1988.

Glicksman, M. E., Coriell, S. R., and McFadden, G. B., Interaction of Flows with the Crystal-Melt Interface, Ann. Rev. Fluid Mech., vol. 18, pp. 307-335, 1986.

Gopalakrishna, S., Himasekhar, K., and Lior, N., A Numerical Study of Transient Conjugate Natural Convection in an Enclosed Stratified Medium, A.I.Ch.E. Symp. Series, vol. 83, 177, 1987.

Hess, C.F. and Miller, C.W., Decay of Thermal Stratification in a Water Body for Thermal Storage, Solar Energy, vol. 28, pp. 137-144, 1982.

Hurle, D.T.J., Jakeman, E., and Pike, E. R., Striated Solute Distributions Produced by Temperature Oscillations During Crystal Growth from the Melt, J. Crystal Growth, vol. 34, 633, 1968.

Hurle, D.T.J., Jakeman E., and Johnson, C. P., Convection Temperature Oscillations in Molten Gallium, J. F. M., vol. 64, 565, 1974.

Hurle, D.T.J., Joyce, G. C., Ghassempoory, M., Crowley, A. B., and Stern, E. J., The Dynamics of Czochralski Growth, J. Crystal Growth, vol. 100, pp. 11-25, 1990.

Iwaki, T. and Kobayashi, N., Thermal and Residual Stresses of Czochralski-Grown Semiconductor Material, Int. J. Solid Struct., vol. 25, no. 3, pp. 307-314, 1986.

Jasinski, T. J., Rohsenow, W. M., and Witt, A. F., Heat Transfer Analysis of the Bridgman-Stockbarger Configuration for Crystal Growth. I: Analytical Treatment of the Axial Temperature Profile, J. Crystal Growth, vol. 61, pp. 339-354, 1983.

Jasinski, T.J., Witt, A.F. and Rohsenow, W.M., Heat Transfer Analysis of the Bridgman-Stockbarger Configuration for Crystal Growth, II. Analytical Treatment of radial temperature Variations, J. Crystal Growth, vol. 67, pp. 173-184, 1984.

Jasinski, T. J. and Witt, A. F.,, On Control of the Interface Shape During Growth in Vertical Bridgman Configuration, J. Crystal Growth, 71, 295, 1985.

Jones, G. F. and Lior, N., Conjugate Heat Transfer and Flow Distribution in an Assembly of Manifolded Finned Tubes, Maldistribution of Flow and Its Effect on Heat Exchanger Performance, Kitto, J. B., Jr., and Robertson, Eds., ASME HTD-75, 127, 1987.

Jordan, A. S., Von Neida A. R., and Caruso, A. R., The Theory and Practice of Dislocation Reduction in GaAs and InP, J. Crystal Growth, vol. 70, 555-573, 1984.

Joubert, P. and Le Quéré, P., Numerical Study of Thermal Coupling between Conductive Walls and a Boussinesq Stratified Fluid, Num. Heat Transfer, Part A, vol. 16, pp.489-506, 1989.

Kakimoto, K., Eguchi, M., Watanabe, H., and Hibiya, T., Natural and Forced Convection of Molten Silicon during Czochralski Single Crystal Growth, J. Crystal Growth, vol. 94, pp. 412-420, 1989.

Karma, A., Solidification Cells at Low Velocity: The Moving Symmetric Model, Phys. Rev. A., 34, 4353, 1986.

Kim, D.M. and Viskanta, R., Study of the Effects of Wall Conductance on Natural Convection in Differently Oriented Square Cavities, J.F.M., vol. 144, pp. 153-176, 1984.

Kim, K. M., Witt, A. F., and Gatus, H., Crystal Growth from the Melt Under Destabilizing Thermal Gradients, J. Electronic Soc., vol. 119, 1218, 1972.

Kim, K. M., Witt, A. F., Lichtensteiger, M., and Gatos, H. C., Quantitative Analysis of the Effects of Destabilizing Vertical Gradients on Crystal Growth and Segregation: Ga-doped Ge, J. Electrochem. Soc., vol. 125, 475, 1978.

Kobayashi, N., Heat Transfer in Czochralski Crystal Growth, Preparation and Properties of Solid State Materials, 6, W. R. Wilcox, ed., Dekker, New York, 1981.

Kobayashi, N., Hydrodynamics in Czochralski Growth-Computer Analysis and Experiments, J. Crystal Growth, vol. 52, pp. 425-434, 1981.

Kobayashi, N., Computational Simulation of the Melt Flow during Czochralski Growth, J. Crystal Growth, vol. 43, pp. 357-363, 1978.

Kuehn, T. H., Kwon, S. S. and Tolpadi, A. K., Similarity Solution for Conjugate Natural Convection Heat Transfer from a Long Vertical Plate Fin, Int. J. Heat Mass Transfer, 26,6 1718, 1983.

Kuroda, K, Kozuka H., and Takano, W., The Effect of Temperature Oscillations at the Growth Interface on Crystal Perfection, J. Crystal Growth, 68, 613, 1984.

Lambropoulos, J. C., Stresses Near the Solid-Liquid Interface During the Growth of a Czochralski Crystal, J. Crystal Growth, vol. 80, 245, 1987.

Langer, J. S. and Hong, D. C., Solvability Conditions for Dendritic Growth in a Boundary-Layer Model with Capillary Anisotropy, Phys. Rev. A., vol. 34, 1462, 1986.

Langlois, W. E., A Parameter Sensitivity Study for Czochralski Bulk Flow of Silicon, J. Crystal Growth, vol. 56, 15, 1982.

Langlois, W. E., Digital Simulation of Czochralski Bulk Flow in a Parameter Range Appropriate for Liquid Semiconductors, J. Crystal Growth, vol. 42, pp. 386-399, 1977.

Langlois, W. E.,Buoyancy-Driven Flows in Crystal Growth Melts," Ann. Rev. Fluid Mech., vol. 17, 191, 1985.

Lin, W. and Benson, K. E., The Science and Engineering of Large-Diameter Czochralski Silicon Crystal Growth, Ann. Rev. Mater. Sci., vol. 17, 293, 1987.

Mihelčić, M., Winegrath, K., and Pirron, Chr., Three-dimensional Simulations of the Czochralski Bulk Flow, J. Crystal Growth, vol. 69, 473, 1984.

Motakef, S., Effect of Natural Convection and Thermal Transparency of Liquid Encapsulant on Thermal Stresses during LEC Growth of GaAs, Int. J. Heat Mass Transfer, vol. 30, pp. 1487-1495, 1987a.

Motakef, S., Analysis of the Thermoelastic Stresses in GaAs Crystals Grown by the Bridgman-Stockbarger and the Liquid Encapsulated Czochralski Methods, ASME Paper 87-WA/HT-2, ASME, New York, 1987b.

Motakef, S. and Witt, A.F., Thermoelastic Analysis of GaAs in LEC Growth Configuration. I. Effect of Liquid Encapsulation on Thermal Stresses, J. Crystal Growth, vol. 80, pp. 37-50, 1987.

Motakef, S., Thermoelastic Analysis of GaAs in LEC Growth Configuration. II. Thermal Evolution of the Stress Field, J. Crystal Growth, vol. 88, pp. 341-352, 1988.

Mullins, W. W. and Sekerka, R. F., Morphological Stability of a Particle Growing by Diffusion of Heat Flow, J. Appl. Phys., vol. 34, 323, 1963.

Mullins, W. W. an Sekerka, R. F., Stability of a Planar Interface During Solidification of a Dilute Binary Alloy, J. Appl. Phys., vol. 35, 444, 1964.

Murgai, A., Oxygen Incorporation in Czochralski Silicon, Crystal Growth of Electronic Materials, E. Kaldis, ed., Elsevier, New York, 211, 1985.

Naumann, R. J., An Analytic Approach to Thermal Modeling of Bridgman-type Crystal Growth. I: One-dimensional Analysis, J. Crystal Growth, vol. 58, 589, 1983.

Naumann, R. J. and Lehoczky, S. L., Effect of Variable Thermal Conductivity on Isotherms in Bridgman Growth, J. Crystal Growth, vol. 61, pp. 707-710, 1983.

Ostrogorsky, A. G., Numerical Simulation of Crystal Pulling Downward, from the Lower Interface of the Floating Melt, J. Crystal Growth, vol. 97, pp. 18-22, 1989.

Ozoe, H., Chao, P.K.B., Churchill, S. W., and Lior, N., Laminar Natural Convection in an Inclined Rectangular Box with the Lower Surface Half Hearing and Half Insulated, ASME J. Heat Transfer. 105, 425, 1983.

Ozoe, H., Mouri, M. A., Ohmuro, M., Churchill, S. W., and Lior, N., Numerical Calculations of Laminar and Turbulent Natural Convection in Water in Rectangular Channels Heated and Cooled Isothermally on the Opposing Vertical Walls, Int. J. Heat Mass Transfer, Vol. 28, 1, 125, 1985.

Patankar, S. V., Numerical Heat Transfer and Fluid Flow, Hemisphere Pub. Corp., Washington, 1980.

Parmentier, E.M. and Torrance, K.E, Kinematically Consistent Velocity Fields for Hydrodynamic Calculations in Curvilinear Coordinates", J. Comp. Phys., vol. 19, pp.404-417, 1975.

Pearson, C.E., A Numerical Method for Incompressible Viscous Flow Problems in a Spherical Geometry", J. F. M., vol.28, pp.323-336, 1967.

Pimputkar, S. M. and Ostrach, S., Convective Effects in Crystal Growth from Melt," J. Crystal Growth, 55, 614, 1981.

Roache, P. J., Computational Fluid Dynamics, Hermosa, Albuquerque, 1982.

Rosenberger, F., Fundamentals of Crystal Growth. Springer, New York, 1979.

Sabhapathy, P. and Salcudean, M. E., Numerical Analysis of Heat Transfer in LEC Growth of GaAs, J. Crystal Growth, vol. 97, pp. 125-135, 1989.

Sackinger, P. A., Derby, J. J. and Brown, R., Axisymmetric Convection in a Complete Thermal-Capillary Model for Czochralski Crystal Growth, Int. J. Num. Meths. Fluids, vol. 9, 453, 1989.

Schaefer, R. J. and Coriell, S. R., Convection-Induced Distortion of a Solid-Liquid Interface, Metallurgical Transactions A, vol. 15A, pp. 2109-2115, 1984.

Shyu, R. J. and Hsieh, C. K., Unsteady Natural Convection in Enclosure with Stratified Medium, ASME Paper, 85-HT-33, 1985.

Tiller, W. A., Jackson, K. A., Rutter, J. W., and Chalmers, B., The Redistribution of Solute Atoms During the Solidification of Metals, Acta Met., vol. 1, 428, 1953.

Tsukada, T., Imaishi, N., and Hozawa, M., Theoretical Study of the Flow and Temperature Fields in CZ Single Crystal Growth, J. Chem. Engrg. Japan, vol. 21, no.2, pp. 184-191, 1988.

Ungar, L. H., Bennett, M.J. and Brown, R.A., Cellular Interface Morphologies in Directional Solidification. 3: The Effect of coupled Heat Transfer, Phys. Rev. B, vol. 31, 5923, 1985.

Viskanta, R. and Hale, N.W., Jr., Free Convection Circulation in a Thermal energy Storage System Driven by Heat Loss from the Wall, Proc. 14th Southeastern Seminar on Thermal Sciences, pp. 29-43, 1978.

Viskanta, R. and Lankford, D. W., Coupling of Heat Transfer between Two Natural Convection Systems Separated by a Vertical Wall, Int. J. Heat Mass Transfer, 24, 1171, 1981.

Williams, G. and Reusser, W. E., Heat Transfer in Silicon Czochralski Crystal Growth," J. Crystal Growth, vol. 64, 448, 1983.

Wilson, L. O., The Effect of Fluctuating Growth Rates on Segregation in Crystals Grown from the Melt. I: No Backmelting, J. Crystal Growth, vol.48, 435, 1980a.

Wilson, L. O., The Effect of Fluctuating Growth Rates on Segregation in Crystals Grown from the Melt. II: Backmelting," J. Crystal Growth, vol. 48, 451, 1980b.

Woodruff, D. P., The Solid-Liquid Interface; Cambridge Press, Cambridge, 1973.

Zulehner, W. and Huber, D., Czochralski-Grown Silicon, Crystals Growth, Properties and Applications, 8, J. Grabmaier, ed., Springer, New York, I, 1982.

Zulehner, W., Czochralski Growth of Silicon, J. Crystal Growth, vol. 65, pp. 189-213, 1983.

Summary of General Discussion of Session on Crystal Growth (2)

Nobuyuki Imaishi, Kyushu University

The session "Crystal Growth (2)" had three presentations. Two of them concerned with the thermocapillary flow in low Prandtl number melt, and the third one treated the conjugated heat transfer in crystal growth processes. There still is an argument whether the thermocapillary flow exists in industrial silicon CZ furnaces or not. In most of the industrial Si CZ furnaces, the radial motion of the melt surface, which must be caused by the thermocapillary effect and even by the natural convection, is not observable. On the other hand, strong surface convections have been reported in FZ experiments. These discrepancies might correspond to the amount of the disolved solutes (CZ melt containes about 10 ppm of oxygen). Some of these solutes possibly may become a kind of surface contaminant which makes the liquid surface rigid. In many cases, both the thermocapillary and the solutal-capillary effects must be taken into account. Substantial data base of the thermophysical properties of melts is required.

Thermal engineer's attension on the crystal growth tends to concentrate on the melt convections. Radiative heat transfer also takes very important rolls in such high temperature operations. The radiative heat transfer rate from the crystal surface to the ambient wall often determines the pulling rate. (In the silicon CZ industry, some types of radiation shields installed into the furnace become popular in order to increase the production rate.) And the internal radiative heat transfer in transparent and semi-transparent materials affects the steady and transient behaviour of growing crystals. It is nessessary to develope efficient numerical methods applicable to systems with complex geometry and external/internal radiative heat exchange.

Crystal engineering is now in its infancy, and thermal engineers can greatly contribute to develop qualitative and quantitative understandings of macro and micro characteristics of the growing and grown crystals through the hydrodynamic, thermal, and stress-strain analyses.

SOLIDIFICATION

Transport Phenomena in Melting and Solidification in Manufacturing Process

G. J. HWANG
Department of Power Mechanical Engineering
National Tsing Hua University
Hsinchu, Taiwan 30043, ROC

ABSTRACT

Transport phenomena in an internal flow associated with melting and solidification are frequently encountered in a manufacturing process. Studies on liquid solidification in laminar, transitional and turbulent flows inside tubes or channels, solidification of an advancing pipe flow, solidification shut, and continuous casting are reviewed. General discussions on thermal physical properties of molten metals, heat conduction in solid shell, energy balance at liquid-solid interface, conservation of mass, momentum and thermal energy in liquid phase are made. This paper summarizes both the theoretical and experimental results on solidification-free zone, liquid-solid interface profile, heat transfer and pressure drop. Possible future studies are also suggested.

LITERATURE REVIEW

Casting is a major manufacturing process because of its cost effectiveness and wide range of applications [1]. The process used for making a casting depends on the quantity to be produced, the metal to be cast, and the intricacy of the part. The methods of casting can be classified as casting in metallic and sand molds, centrifugal casting, precision or investment casting, and continuous casting. The mechanics of casting process generally involves strong thermal/flow coupling, gas-liquid free surface, and/or liquid-solid moving interface with melting and solidification.

In Figure 1(a), a schematic diagram of die-casting machine is shown. Molten metal is driven by the plunger and flows through the pipe to the die cavity. Transport phenomena of momentum and thermal energy in a pipe and solidification in the die cavity occur. Figure 1(b) shows the casting of lead grid for a secondary battery. The molten lead pours into an iron mold with a temperature below the lead melting temperature. The mold temperature increases when it contacts with the molten lead. The flow stops after the lead temperature drops to the freezing temperature and the latent heat is removed. Then the solidification occurs, and the Liquid-solid interface moves backwards to the upstream. Figure 1(c) depicts the casting of aluminum ingots by a continuous process. The molten aluminum liquid is fed from a furnace. The thermal energy of the molten aluminum is removed by a direct chill process. Melting and solidification phenomena in internal flows are frequently encountered in a casting process. This paper reviews literature concerning the liquid solidification in laminar, transitional and turbulent flows in tubes or channels, solidification of an advancing flow, solidification shut and continuous casting.

Liquid Solidification in Laminar Flows

It is known for a quite long period of time that liquid metals can serve as excellent heat transfer media for their high thermal conductivities and high evaporation temperatures

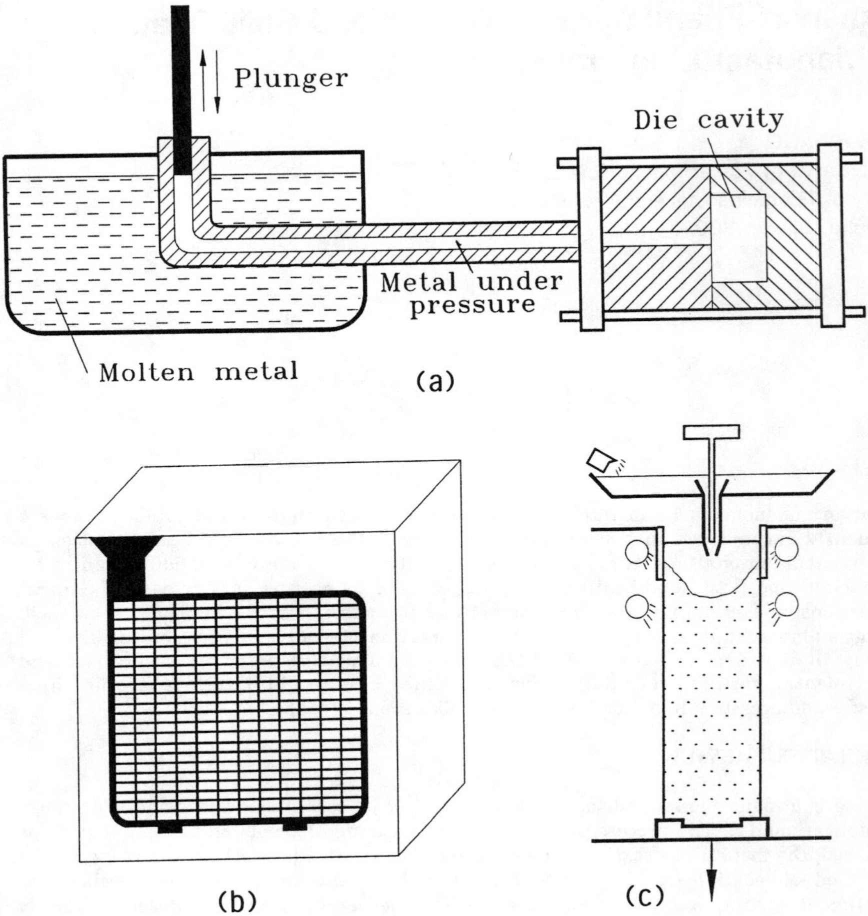

FIGURE 1. Melting and solidification in casting

and applies extensively to nuclear power generation [2]. But until two decades ago, Zerkle and Sunderland [3] analyzed the effect of liquid solidification in a tube upon laminar flow heat transfer and pressure drop. They utilized a parabolic axial velocity to obtain heat transfer coefficients and also carried out an experiment by using water as a working fluid in a test section of 3.81 cm diameter. A discrepancy due to natural convection effect was found. With an posed heat flux to the liquid-solid interface and a finite ambient heat transfer, the solidification of fluids flowing along a plane wall or through a pipe was calculated by Stephan [4]. In contrast to the parabolic velocity profile used in [3]. Özisik and Mulligan [5] employed a rectangular velocity profile in the analysis of an transient solidification problem in a tube with the Dirichlet thermal boundary condition. The slug flow velocity may be used to simulate a low Prandtl number fluid flow of liquid metals. They predicted greater values of heat transfer coefficient, and the results coincidently agreed with the experimental results of Zerkle and Sunderland [3]. Depew and Zenter [6] studied experimentally the laminar flow heat transfer and pressure drop in a test section of 2.00 cm diameter with freezing at wall.

As shown in Figure 2, external convective cooling used in the solidification of a molten metal flow was first proposed by Lock et al. [7] for a circular tube. This analysis yielded a solidification-free zone and provided a description of the solid-liquid interface and its effect on pressure drop. Bilenas and Jiji [8] re-examined the problem in [3] by a variational technique and the solution was expressed in terms of the short-time, asymptotic and steady-state components. Lock and Nyren [9] extended the work in [7] in a long circular tube cooled by external nonuniform convection, therefore, a peripherally nonuniform solid shell was obtained. Hwang and Yih [10] corrected the length of the solidification-free zone in [7] by recalculating numerically the eigenvalues and eigenfunctions. Ou and Cheng [11] extended further the solidification-free zone in a convectively cooled horizontal square channel with the effect of free convection. It was suggested that free convection effects should be included in the analysis of liquid solidification in a convectively-cooled pipe or channel when the Rayleigh number is greater than say 10^4.

Considering a gradual development of axial velocity from a uniform distribution, Hwang and Sheu [12] presented a theoretical and experimental investigation of liquid solidification in the combined hydrodynamic and thermal entrance of a circular tube with a uniform wall temperature. The theoretical solution assumed that axial variation in liquid-solid interface is gradual and considered that effects of radial velocity component are significant. Experimental work using water as a working medium was carried out to verify the theoretical result. a 0.95 cm ID test tube was used to reduced the free convection effect. Mulligan and Jones [13] also run the same experiment to obtain heat transfer and pressure drop data by using a 1.45 cm diameter test section. Using a 95 cm long tube before cooling, Liu and Hwang [14] carried out an experimental investigation on the effect of liquid solidification upon heat transfer in the thermal entrance region of a circular tube. The results are similar to that of [12].

Cheng and Wong [15] made a theoretical analysis of liquid solidification in the thermal entrance region of a parallel-plate channel with a uniform external convection. The analysis was carried out by using the confluent hyper-geometric function in the solution for the laminar flow and steady state condition. Lee and Hwang [16] examined the solidification of a low Peclet number fluid flow in the thermal entrance region of a round tube. The velocity was assumed to be laminar and fully developed throughout the pipe. The axial conduction was considered and the fluid temperature was taken to be uniform at $Z=-\infty$. Without the axial conduction effect, Sadeghipour, Özisik and Mulligan [17] studied the

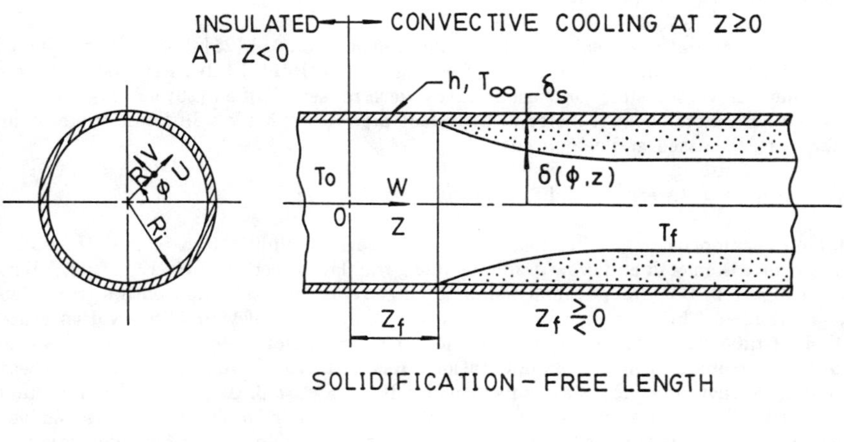

FIGURE 2. Configuration and coordinate system for internal solidification

transient solidification of liquid metals in the thermal entry region of a circular tube by using a slug flow velocity profile. Epstein and Cheung [18] made an extensive review on the complex freezing-melting interfaces in fluid flow. Literature on external forced flow concerning simple phase change, transition flow over a meltable surface, wavy interfaces in turbulent flow, impingement melting and simultaneous melting and freezing, internal forced flow concerning crust profiles and pressure drops in steady-state pipe flow, wavy interfaces in internal solidification, freezing of an advancing pipe flow, and simultaneous melting and freezing, and natural convection relating simple phase change, water-ice system, embedded heat source, melting in a density gradient and frozen-crust stability in liquid layers. Among these topics, the subjects in internal forced flow are of interest in the present study.

Hwang and Chen [19] reviewed literature on liquid solidification in laminar, transitional and turbulent flows in ducts. The theoretical and experimental results on solidification-free zone, liquid-solid interface profile, heat transfer rate and pressure drop in liquid phase were summarized. Recently, combined effects of axial conduction and solidification on heat transfer and pressure drop in low Peclet number pipe flows were investigated by Lee and Hwang [20], and an order of magnitude analysis on liquid solidification in pipe flows was presented by Hwang and Chen [21] to give an insight on the heat transfer mechanism in internal solidification.

Liquid Solidification in Turbulent Flows

The problem of internal solidification under turbulent flow conditions is even more involved than that of corresponding laminar flow case because of the complicated transport mechanism. Arora and Howell [22] investigated the freezing of supercooled liquid in forced turbulent flow inside circular tubes. Shibani and Özisik [23, 24] studied the steady-state freezing of liquids of low Prandtl number in turbulent flow between parallel plates and in a circular tube. Thomason et al. [25] presented a simple analysis of the steady-state heat transfer and pressure drop in turbulent flow in a tube with a thin steady-state frozen deposit on the inside tube wall. Experimental heat transfer and pressure drop data were also presented for comparison. The transient freezing of liquids in internal turbulent flow is important in liquid metal technology and the other engineering applications. Cho and Özisik [26] investigated the transient freezing of liquids in turbulent flow inside tubes. The radius of the solid-liquid interface and the local wall heat flux were determined as a function of time and position along the tube for several different values of the Prandtl number and the freezing parameter.

Thomason and Mulligan [27] extended their experimental work in [25] to the observation of the flow instability during turbulent flow freezing in a horizontal tube. Periodic pressure-drop oscillations versus time for various cases were presented. Lee [28] summarized the results of liquid metal heat transfer in turbulent pipe flows. Models in liquid metal heat transfer were discussed and compared.

Wavy Interfaces in Internal Solidification

A common assumption made in most studies of internal solidification is that the solid-liquid interface is smooth and grow gradually along the flow direction [7,10, 12, 16. 17, 20] as shown in Figure 2. But this phenomenon is no longer exist in a tube flow with a transition Reynolds number. The first experimental study by Gilpin [29] of visual observation of ice growth in a tube flow was carried out at laminar or transition Reynolds number. Due to occurrence of transition and relaminarization process, a wavy solid-liquid interface was observed. The wavy interface is shown schematically in Figure 3. Gilpin [30] further found that the ice-band structure reported previously [29] for Reynolds number near transition was observed to be of steady state ice structure formed in a pipe for flows with Reynolds numbers over the entire range of the investigation (up to $Re_D=1.4x10^4$). A flow passage with a dramatic cyclic variation in cross-section along the length of the pipe was observed. The

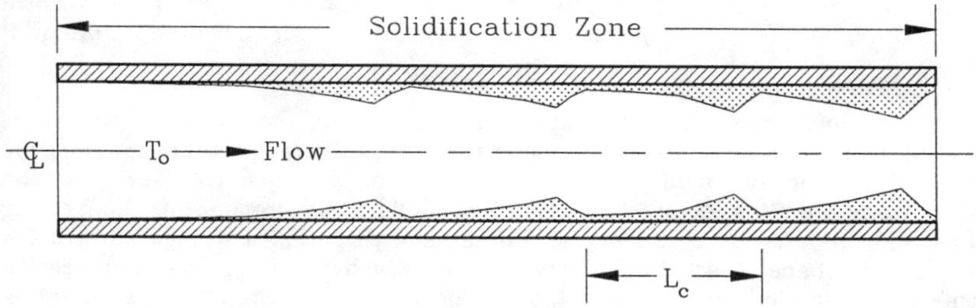

FIGURE 3. Liquid-solid interface band structure

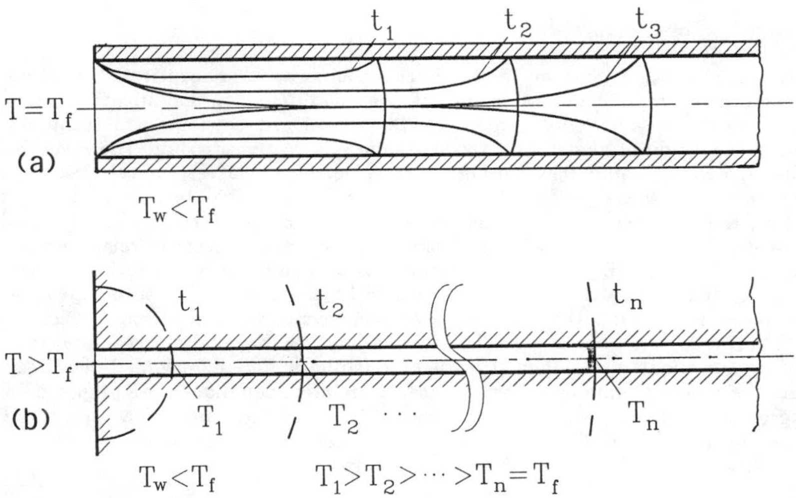

FIGURE 4. Thermal boundary conditions for advancing pipe flow

influence of the ice-band structure on pressure drop and on pipe freeze-off were also investigated.

Solidification of an Advancing Pipe Flow

As shown in Figure 4, there are two types of transient solidification of an advancing pipe flow depending on the molten metal superheat ratio and thermal boundary conditions. In Figure 4(a), The solidification of molten metal without superheat may occur on the tube wall cooled with a high value of heat transfer coefficient. The first study of solidification of a penetrating tube flow was performed by Cheung and Baker [31] for a liquid-metal fast breeder reactor. They conducted a series of tests in which various liquids were allowed to flow under gravity into long cooper tubes cooled by liquid nitrogen or by a bath of dry ice and acetone. Different constant driving heads in the liquid reservoir were used and the penetration distances were measured. An empirical correlation of the penetration length in terms of the measured average flow velocity was found. Epstein, Yim and Cheung [32] studied theoretically and experimentally the freezing-controlled penetration of a saturated liquid into a cold tube. They obtained an approximate closed form solution for the flow penetration length before solidification is complete. Epstein and Hauser [33] verified this approximate crust profile method [32] theoretically by using a numerical approach in which the governing integro-differential equation of liquid motion is rigorous solved on a digital computer.

In Figure 4(b), a molten metal with a temperature higher than the melting temperature enters the passage of a mold with a temperature slightly lower than the melting temperature. Occurrence of solidification may not begin at the inlet. It is seen that the solidification starts after the temperature of leading liquid metal drops below the freezing temperature and the latent heat is removed. Kolling and Grigull [34] studied numerically the instantaneous shape of a solidified layer in an advancing tube flow. Penetration experiments were also conducted by the authors by pouring molten Sn, Pb, Zn, or Al into a thick-walled, horizontal casting tube. A good agreement was observed by assuming an concentric air gap between the crust and the tube wall in their theoretical prediction. Unfortunately, It was not possible to confirm the development of such an air gap with available measurement techniques.

Solidification Shut

Although solidification shut or freeze-off condition was also mentioned in literature reviewed previously, the determination of the conditions for solidification shut was discussed here. DesRuisseaux and Zerkle [35] developed a theoretical technique for predicting the condition under which a hydraulic system freezes shut. The investigation considered laminar liquid flow through a tube between two reservoirs. Freezing was assumed to occur within a portion of the tube having a uniform wall temperature lower than the liquid freezing temperature. Analytical results indicated the minimum allowable pressure drop which must be maintained across the system to prevent it from freezing shut. Epstein and Cheung [36] predicted the conditions of pipe free-shut in turbulent flow. By ignoring the thermal entrance region, known to be short for highly turbulent flow and Prandtl number above 0.5, they obtained a simple, asymptotic closed-form solution for the problem in [24, 25] and used it to predict the conditions for the onset of freeze-shut in a pipe containing turbulent water flow. Hirata and Ishihara [37] determined the freeze-off conditions of a pipe containing a flow of water. correlation equations were proposed for the spacing of the ice-bands, the heat transfer coefficient in the contraction region and the friction factor in the pipe. The conditions for pipe freeze-off were obtained by introducing a modified Reynolds number based on a total pressure.

Continuous Casting

Continuous casting is a method of cost effective and quality in producing bars and billets. The thermal fluid flow of the feed molten metal and the induced buoyancy flow may affect the profile of liquid-solid interface and also the quality of product. Kroeger [38] made a heat transfer analysis of solidification of pure metals in continuous casting processes. He pointed out that the simplified assumption without considering the conduction can introduce a serious error in the solution of pool depth. Sfeir and Clumpner [39] applied a heat balance integral method (HBIM) to solve for the continuous casting of cylindrical ingots. This paper developed a HBIM which includes conduction in the direction of casting as well as the effect of superheat in the feed metal. The pool depth prediction by the present analysis can be as much as 40 percent shorter. Yao [40] investigated the natural convection effect in the continuous casting of a horizontal cylinder. Due to natural convection, the solidification rate along the bottom of a horizontal cylinder is shown to be faster than that along the top.

THERMAL PHYSICAL PROPERTIES OF LIQUID METALS AND ALLOYS

It is well known that physical properties i.e. density, viscosity, surface tension, thermal conductivity and specific heat are closely related to the transport phenomena of liquid metals and alloys in melting and solidification. The properties are in general functions of pressure and temperature. The pressure dependence of physical properties is negligible in practice. Therefore, the variations of physical properties with temperature is discussed in this section and simple correction formulae is suggested for the analyses of transport phenomena in melting and solidification of liquid metals and alloys.

Density

In examining the nature and behavior of liquid metals and alloys, density is an indispensable basic thermodynamic property. Density data provide essential information for topics including mass balance, slag and metal reactions and thermal natural convection. Of the several empirical expressions for thermal density variations of liquids, the most familiar is the law of Cailletet and Mathias [41]. The law states that the average density of a liquid and its vapor decreases linearly as the temperature rises to the critical point. This is

$$(\rho_l + \rho_g)/2 = a - bT \tag{1}$$

In most of manufacturing processes, temperatures of the liquid metals and alloys are much less than their critical values, thus the vapor density is much less than the liquid density. One may simply use a linear relationship for the liquid density and the temperature with a fairly high degree of accuracy. One may check it easily with liquid metal data in a heat transfer text book [42]. It is found that the values of b/a for most of liquid metals are of order of 10^{-4} in comparison with 10^{-2}-10^{-3} for air and 10^{-3}-10^{-4} for water.

The majority of metals exhibit a few percents density decrease during melting. Density increase during melting occurs in the semi-metals gallium, silicon, germanium, plus cerium and plutonium. The density decrease is unusually small for antimony, lanthanum, praseodymium, neodymium and neptunium.

Surface Tension

In melting and refining of metals, the surface tension or interfacial tension is a dominating factor for phenomena such as gas absorption, nucleation of gas bubbles, nucleation growth of non-metallic inclusion, and slag/metal reactions. Unfortunately, due to the difficulty of precise experimental determinations of surface tensions and interfacial tensions, there are no satisfactory theoretical and experimental investigations of the nature and behavior of surfaces and interfaces between phases. However, the dependence of surface tension on the absolute temperature can be expressed by Eotvos' law [41]

$$\gamma = \kappa_v (T_c - T) / V^{2/3} \tag{2}$$

where κ_v is approximately equal to $6.4\text{x}10^{-8}$ ($JK^{-1}mol^{-2/3}$) for all liquid metals, T_c is the critical temperature, and V is the atomic volume (m^3/mole). After differentiating equation (2) with respect to temperature, one will find that the surface tension decreases with the increase in temperature.

Viscosity

In momentum transport processes of liquid metal and alloys, viscosity plays an important role as a key to solve quantitatively problems in fluid behavior as well as those related to the kinetics of reactions in metallurgical processes. Although experimental measurements of liquid metal viscosities have been made over hundred years, accurate and reliable data are still not for some of liquid metals. If one examines the temperature dependence of the viscosity for temperature higher than the melting point, an Arrhenius type relationship [41] can be used.

$$\mu = A \exp(H_\mu/RT) \tag{3}$$

where A and H_μ are constants, the numerical values depending on the liquid metal used. This relationship shows a good approximation for liquid metals over a wide temperature range.

Thermal Conductivity

At present , few experimental data on the thermal conductivity of liquid metals and alloys are available. In addition, the accuracy of measurements is relatively poor, and large discrepancies exist among different set of data. The thermal conductivity of a metal normally decreases on melting because of the disordered arrangement of ions in the liquid state. The temperature dependence of thermal conductivities is quite complex, and there is no simple correlation curve available. For instance, thermal conductivities of Hg, Pb, Al, Sn-Pb and Bi-Pb increase with the increase in temperature, but thermal conductivities of Na, Zn and K decrease with the increase in temperature. In addition to these two behaviors, the thermal conductivity of K-Na alloy increase and then decrease with the increase in temperature. It is suggested that the correlation of temperature dependence of thermal conductivities should be made case by case for different liquid metals or alloys.

Physical Properties and Buoyancy Effect

The relative importance of natural and forced convection is closely related to the physical properties. From a scale analysis of momentum equation, the ratio of natural and forced convection is indicated by $Gr^{1/2}/Re=(g\beta\Delta TL)^{1/2}/U$. For fixed values of ΔT, L and U, one is able to estimate the ratio of relative importance of natural and forced convection for air, water and a liquid metal flow. The ratio is approximately 5:2:1 if liquid lead is selected for comparison. It is suggested that if natural convection is considered in a forced water flow, the natural convection should be considered for a forced liquid metal flow as well.

REVIEW OF THEORETICAL ANALYSES

Consider a laminar or turbulent convection in the hydrodynamic entrance region or fully-developed region of a duct. The duct cross-section can be circular or rectangular with a finite or infinite aspect ratio. When the duct wall is cooled convectively by an external cold stream with a temperature below the solidification temperature of the liquid in the duct. Solidification may occur at a certain distance along the duct. If the external convection is strong enough, the distance may close to or in the upstream of the inlet. The thickness of the solid shell grows on the duct wall along the axial direction. The growth of the solid shell may be gradual or rapid enough to shut off the flow depending on the relative magnitude of the axial flow and the external heat transfer rate. Review of past literature on theoretical analyses is made as follows:

Solid Phase

As noted in references [7, 10, 11, 15-17], the solid shell grows at $Z \geq Z_f$ where Z_f may be positive, negative or zero is the solidification-free length. The heat conduction equation in the solid shell in cylindrical coordinate is

$$\rho C\partial T/\partial t = k[\partial^2T/\partial R^2+(1/R)\partial T/\partial R+(1/R^2)\partial^2T/\partial\phi^2+\partial^2T/\partial Z^2] \tag{4}$$

All papers reviewed here considered an axi-symmetric condition i.e. $\partial^2T/\partial\phi^2=0$, even in the analysis of solidification in a circular tube with a nonuniform external cooling [9]. Furthermore the analysis without the axial conduction term was found in [4, 8, 9]. In addition to the axi-symmetric condition and the neglection of axial conduction term, a quasi-steady condition was also posed in [3, 5, 7, 10, 12, 15-17, 24, 26, 31] by considering a small Stefan number. There is only one paper [20] reviewed without the neglection of axial conduction term.

Liquid Phase

Liquid phase molten metals exist before the beginning of solidification or coexist with the solid phase. As shown in Figure 5, different flow regimes may appear in the region with a different characteristic velocity and length. In this figure, typical values of physical properties for liquid metals and temperature difference in casting process are chosen for demonstrating the different flow regimes. It is seen that turbulent flow, laminar flow and conduction regimes are separated by the inclined lines $Re \sim 10^3$ and 10 for forced convection and by the horizontal lines $Gr \sim 10^9$ and 10^2 for natural convection. Between the inclined lines $Re=0.1\ Gr^{1/2}$ and $Re=10\ Gr^{1/2}$, regimes for mixed convection exist. All the regimes appear in this figure may be encountered in engineering practice. In this section, the governing equations in laminar and turbulent forced convection will be discussed.

Laminar flow analysis. To facilitate the theoretical analysis in the liquid phase, assumptions of incompressible, constant physical properties and negligible viscous

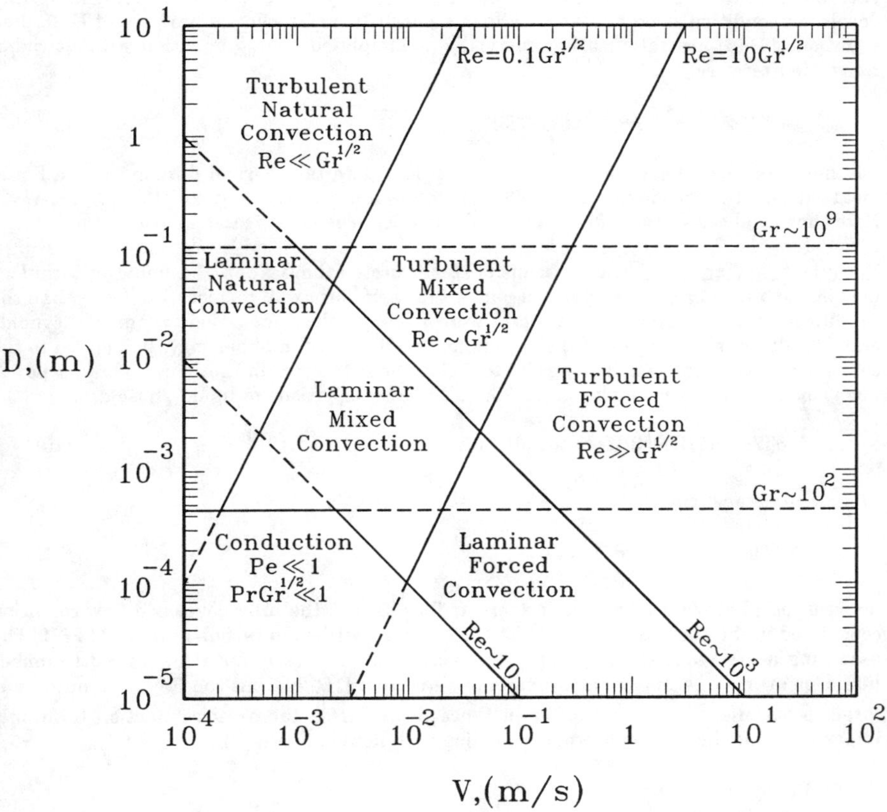

FIGURE 5. Flow regimes based on $Pr=0.01$, $\Delta T=100K$, $\nu=10^{-7} m^2/s$ and $g\beta/\nu^2=10^{10} m^{-3} K^{-1}$

dissipation are made. Assumption of axi-symmetric condition was employed in most of paper reviewed except in [11, 40] the buoyancy effect on solidification was studied. By considering a laminar forced convection of a molten metal in a convectively cooled tube, the axi-symmetric continuity, momentum and energy equations without the axial viscous and conduction terms are:

$$\partial U/\partial R+U/R+\partial W/\partial Z=0 \tag{5}$$

$$U\partial U/\partial R+W\partial U/\partial Z=-(1/\rho)\partial P/\partial R+\nu[\partial^2U/\partial R^2+(1/R)\partial U/\partial R-U/R^2] \tag{6}$$

$$U\partial W/\partial R+W\partial W/\partial Z=-(1/\rho)\partial P/\partial Z+\nu[\partial^2W/\partial R^2+(1/R)\partial W/\partial R] \tag{7}$$

$$U\partial T/\partial R+W\partial T/\partial Z=\alpha[\partial^2T/\partial R^2+(1/R)\partial T/\partial R] \tag{8}$$

The solution of the above equations in the hydrodynamical and thermal entrance region can be obtained without too much difficulty by using a numerical technique, but the boundary involving the liquid-solid interface coupled with the heat conduction in the solid phase is quite tricky. Therefore a transformation of $R=\delta r$ and $Z=(W_{mo}\delta_o^2/\nu)z$ was carried out in literature [3,12]. By assuming a small growth rate of solid shell along the flow direction [2, 5, 7, 8, 12, 16, 20, 24, 25, 26, 31, 35], the difficulty in solving equations (4) to (7) in a variable liquid core region can be overcome readily. If the hydrodynamically fully developed condition is posed, We will have a parabolic axial velocity profile and U=0, then equations (4)-(6) are satisfied automatically. A simplified energy equation without radial convection term results.

$$W\partial T/\partial Z=\alpha[\partial^2T/\partial R^2+(1/R)\partial T/\partial R] \tag{9}$$

Furthermore, the kinematic viscosity is much less than the thermal diffusivity in a liquid metal. It may be a good reason to use the slug flow approximation in equation (8), i.e. W=C. Note that axial conduction effect on the solidification was investigated in [16].

<u>Turbulent flow analysis</u>. The problems of steady state solidification of a liquid in turbulent flow inside a tube or between parallel plates with a uniform wall temperature lower than the solidification temperature of the liquid were usually solved for a wide range of Reynolds and Prandtl numbers. Some of the results for small Prandtl number may be applied to the case for molten metals. In solving the turbulent heat transfer in the liquid phase, a power law velocity profile was employed [23-26, 28]. Energy equations in liquid phase are:

$$W\partial T/\partial Z=(1/R)\partial[R(\alpha+\varepsilon_H)\partial T/\partial R]/\partial R \tag{10}$$

in a circular tube, and

$$W\partial T/\partial Z=\partial[(\alpha+\varepsilon_H)\partial T/\partial Y]/\partial Y \tag{11}$$

between parallel plates. As done in laminar flow [3, 12], the fully developed flow condition considered in the above equations (9) and (10) is also justified in turbulent flow [23-26, 28] by assuming a low varying solid shell. The axial conduction term for a low Prandtl number liquid metal may be important in laminar convection [16, 20], but the Reynolds number is large in turbulent flow regime, for instance $Re=10^4$-10^6, the axial conduction term only important near the inlet in a problem dealing with liquid metals [28].

Solid-Liquid Interface

Dendrite morphology are found in casting and ingot-making processes of alloys. The dendrite structure becomes finer as heat is extracted at greater rate, but becomes coarser as

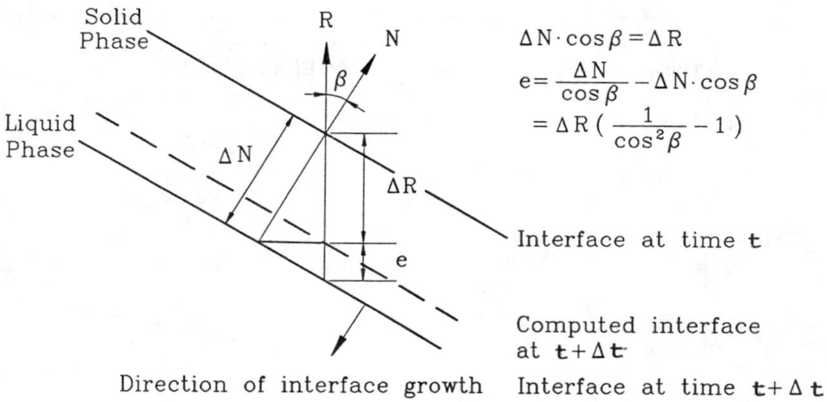

FIGURE 6 Moving liquid-solid interface

the solidification time increases. As described in [43, 44], the sizes of structure are about 10-100 μm. The solid-liquid interface with the dendrite morphology may be in some sense considered as a smooth surface or a surface with small roughness.

As shown in Figure 6, the moving of liquid-solid interface is described by an energy balance equation on the interface

$$\rho_s L dN/dt = k_s(\partial T_s/\partial N)_i - k_l(\partial T_l/\partial N)_i \qquad (12)$$

where L is the latent heat of the material and N is a coordinate normal to the interface with an angle β with respect to the radial coordinate R. Note that the coordinate N is not necessary on the R-Z plane as shown in Figure 2. All the papers [3, 5, 7-12, 15, 16,20, 35] used an approximate form of interface equation by using the coordinate R instead of N in eq. (11). The error depending on the value of β is shown in Figure 6. If there is no superheat [9], the temperature gradient in liquid phase reduces to zero. If the interface is stationary [3, 15, 16, 20], one can drop the time dependent term in eq. (11). In a transient analysis, a characteristic time can be obtained from eq. (11) to evaluate the relative magnitude of the unsteady term in solid phase heat conduction equation by using a parameter called Stefan number which is a ratio of the sensible heat to the latent heat of the solid.

SUMMARY OF RESULTS AND DISCUSSION

Length of Solidification-Free Zone

In an engineering problem shown in Figure 2, it is interesting to see where the solidification starts. Due to a finite Biot number employed in the analyses [7, 10, 11, 15-17], the wall temperature drops gradually from the inlet temperature at Z=0 to the solidification temperature at $Z=Z_f$, where Z_f is the length of solidification-free zone. The solid shell grows at $Z \geq Z_f$ for a large Peclet number fluid flow. Figure 7 presents the length of the solidification-free zone versus the superheat ratio ε with the Biot number as a parameter for a circular tube [7, 10],a parallel plate channel [15] and a square channel with the effect of natural convection [11]. Lock et al. [7] made a first attempt to analyze the liquid solidification in a convectively cooled pipe. Hwang and Yih [10] recompute the value by using a power series solution and a numerical technique. Cheng and Wong [15] also studied the liquid solidification in a convectively cooled parallel-plate channel. Ou and

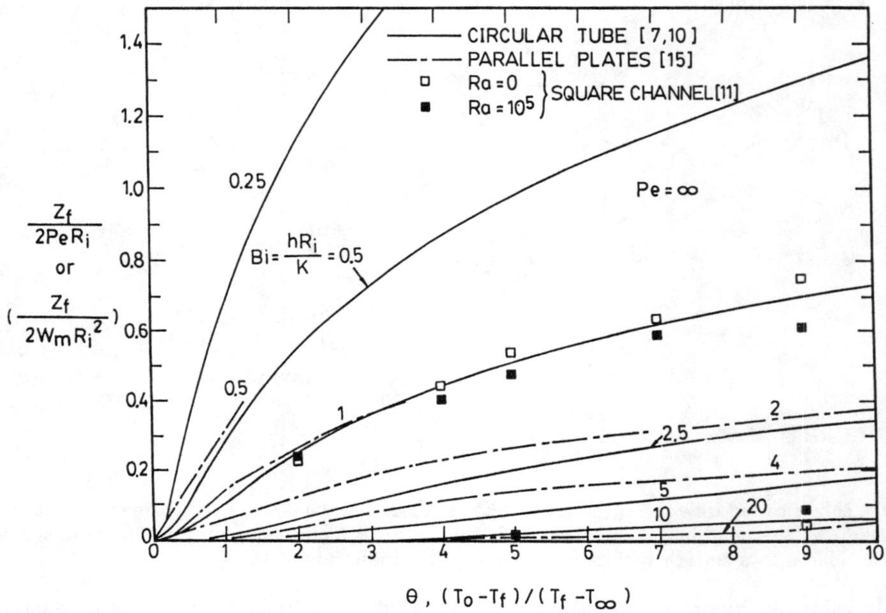

FIGURE 7. Length of solidification-free zone versus superheat ratio

Cheng [11] made an analysis on the buoyancy effect on the solidification-free zone in a convectively cooled horizontal square channel. If one uses R_i as a hydraulic radius (=2 area/perimeter) for circular tube, parallel plate and square channel in this figure, the curves for these ducts with different cross section lie close with each other for Bi=0.5, 1.0 and 10. For the other Biot number the trend of these curves is reasonable. In [11], the data with buoyancy effect $Ra=10^5$ present shorter solidification-free length than that with Ra=0. Although it is not shown here the trend is opposite for $Z_f/(2PeR_i)<0.2$ [11].

Due to axial conduction effect the solidification of liquid can start upstream of Z=0 for a small Pe and a large Bi. The studies of the solidification-free zone for liquid metals or low Peclet number fluid flow are shown in Figure 8. Lee and Hwang [16] kept the axial conduction term and examined the length of the solidification-free zone with the Peclet number as a parameter. Sadeghipour et al. [17] investigated the solidification of liquid metals without the axial conduction term. The comparison of these results are made in Figure 8. It is seen that negative values of solidification-free length are found for a small Peclet number and a Large Biot number say $Pe \lesssim 1.5$ and $Bi \gtrsim 2$. The curves Bi=0.1 and 0.5 of [17] show a reasonable trend in comparison with the data of [16]. It is also noted that the results of [16] with the axial conduction effect are in excellent agreement with that of [10] with a large Pe.

Liquid-Solid Interface

The steady state liquid-solid interface profiles with laminar and turbulent forced convection are summarized. Figure 9 shows the steady state liquid-solid interface profile with laminar forced convection reported in [3, 5, 7, 8, 12]. The result of Zerkle and Sunderland [3] was from the solution with a parabolic axial velocity profile. But the result of Özisik and Mulligan [5] with a slug flow showed a larger value in the radius of liquid-solid

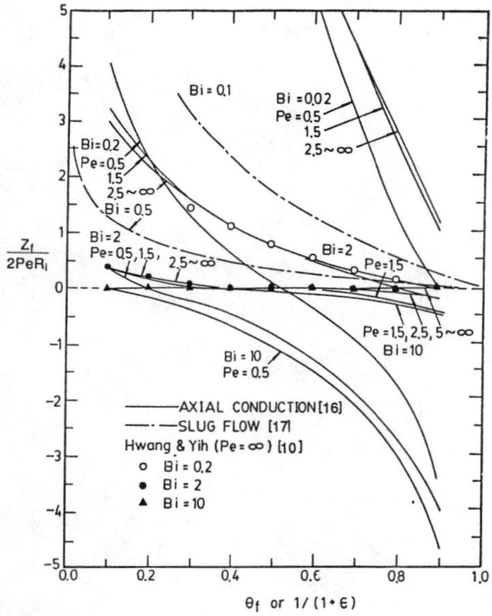

FIGURE 8. $Z_f/(2PeR_i)$ versus θ_f for various Pe and Bi

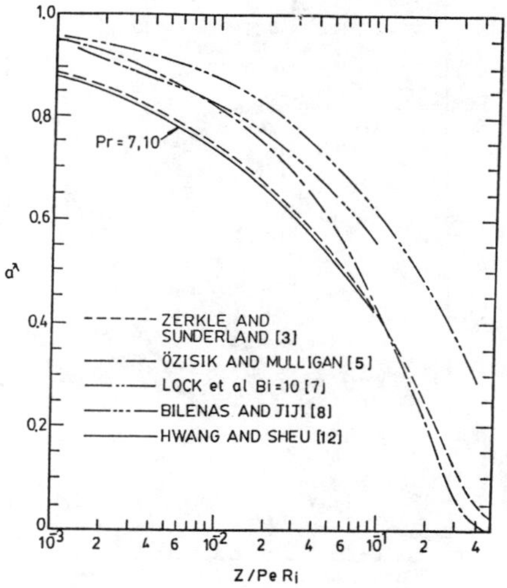

FIGURE 9. Liquid-solid interface

interface. The variational solution using a parabolic velocity by Bilenas and Jiji [8] showed a discrepancy of 0.1 in the value of a^{λ} from the solution of Zerkle and Sunderland [3]. Hwang and Sheu [12] presented the solution in the combined hydrodynamic and thermal entrance region with a developing axial velocity from a uniform velocity profile.The solution coincided well with that obtained with a parabolic velocity [3]. This indicated that the rapid growth of the momentum boundary layer for a liquid of a large Prandtl number, say $Pr \gtrsim 7$, in the hydrodynamic entrance region. But this is not true for liquid metals. Note that in [3, 5, 8, 12] thermal boundary condition of constant wall temperature was used. This boundary condition gives the steady liquid-solid interface as

$$(\delta/R_i)^{\lambda}=f(Z/PeR_i) \tag{13}$$

The solution of [7] for ε=1 and Bi=10 is also shown. This case presents a thinner solid shell along the Z direction. Although it is not shown in this figure, the curve for ε=0.1 and Bi=10 of [7] lies close to the curve of [3, 12] due to a small superheat and a strong convective cooling.

Heat Transfer Rates and Pressure Drop

Due to different flow conditions, there are three types of heat transfer characteristics:

1. Laminar forced convection with slug, parabolic or developing axial velocity as reported in [3, 5, 7, 8, 10, 12, 14-16, 20].
2. Laminar forced convection with natural convection effect as presented in [6, 11, 35, 40]
3. Transition to turbulent convection as shown in [23-30, 37].

Figure 10 shows variations in the dimensionless laminar heat transfer rate with the dimensionless axial position for uniform wall temperature. One can see that the rate of heat

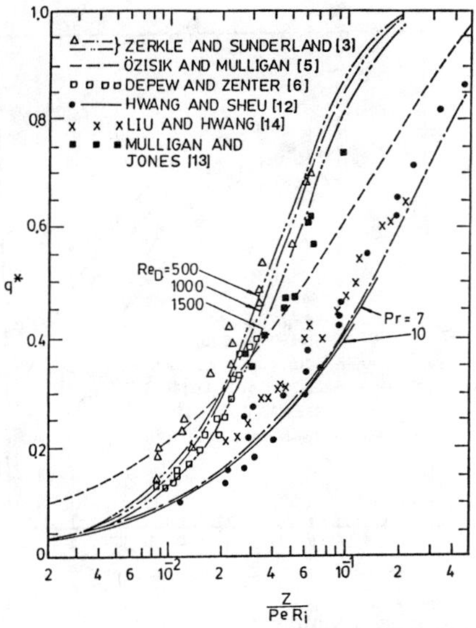

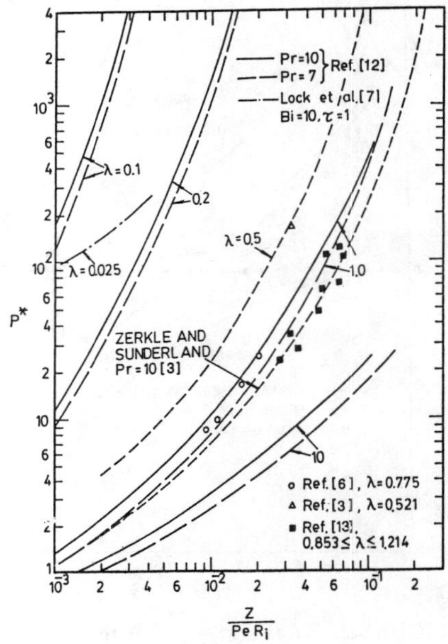

FIGURE 10. Heat transfer rate in laminar flow

FIGURE 11. Pressure drops in laminar flow

transfer from the liquid is independent of tube wall temperature and the thickness of the solid phase shell. The theoretical result of [3] is in an excellent agreement with that obtained with a developing axial velocity for Pr=7 and 10. This indicates that the result using the parabolic velocity profile represents a limiting case of developing flow [12] for large Prandtl number fluids, and the slug flow result [5] can be realized in the combined hydrodynamic and thermal entrance region for fluids of small Prandtl number. The experimental data [3, 6, 12, 13, 14] are also plotted for comparison. Due to free convection effects, the data [3, 6, 13] lie considerably above the theoretical curve. It is noted that the Grashof number is of order of 10^6 in [12] and is of order of 10^5 in [5] but it is only of order of 10^4 in [12, 14].

The dimensionless pressure drop with various Prandtl numbers and dimensionless parameter λ are shown in Figure 11. From the results of [12], one sees that the dimensionless pressure drop at all position Z/PeR_i will increase as the Prandtl number increases and the parameter λ decreases. The result of Zerkle and Sunderland [3] for the case of Pr=10 and λ=1 in the hydrodynamically fully developed region is also plotted in this figure. Due to the effect of the hydrodynamic entrance length, the value of the pressure drop from [12] is higher than that obtained by Zerkle and Sunderland [3]. The curve of [7] shows that the dimensionless pressure drop is less than that of [12] due to a finite Biot number. The experimental data [3, 6, 13] are also plotted for comparison.

RECOMMENDATION OF FUTURE STUDIES

1. The profile of liquid-solid interface controls the direction of metal solidification and also the direction of crystal growth which is important to the material properties. The growth of solid shell in general is not gradual and its surface is not axi-symmetric and stationary

also. Therefore a technique for dealing with the problem with moving interface should be developed. The enthalpy formulation [45] may be one of possibilities to tackle this problem.

2. Most of papers using water as a working medium to predict the flow and heat transfer characteristics with the effect of liquid solidification. The properties of water are quite different from molten metals. Although it is difficult to handle high temperature problem and not cost effective, experimental work using molten metals should be carried out to verify the theoretical predictions.

3. More studies related to the effect of dendrite morphology on the flow and heat transfer characteristics should be taken. The interface with dendrite structure may have a definite effect on the turbulent flow structure near the liquid-solid interface.

NOMENCLATURE

a,b	constants, in eq. (1)
A	constant, in eq. (3)
Bi	Biot number, hR_i/k
C	heat capacity
D	hydraulic diameter, 4 area/perimeter
e	error
Gr	Grashof number, $g\beta\Delta T D^3/\nu^2$
h	convection heat transfer coefficient
H_μ	constant, in eq. (3)
k	thermal conductivity or thermal conductivity of liquid
L	latent heat in eq. (12)
L_c	wavy interface band width
N	normal direction coordinate
P	pressure
Pe	Peclet number, $PrRe=W_mR_i/\alpha$
Pr	Prandtl, ν/α
P^*	dimensionless pressure drop, $(P_o-P)/(\rho W^2_{mo}/2)$
q*	dimensionless heat transfer rate, $(q/\dot{m}C_l)(T_o-T_f)$
R	gas constant
Re	Reynolds number, W_mR_i/ν
R_i	Hydraulic radius, $D_h/2$=2 area/perimeter
R,ϕ,Z	cylindrical coordinate system
t	time
T	temperature
U,V,W	velocity components, in Figure 2
V	atomic volume (m^3/mole), in eq. (2)
Y,Z	rectangular coordinates
α	thermal diffusivity
β	angle, in Figure 6
γ	surface tension, in eq. (2)
δ	radius of liquid-solid interface
Δ	difference
ε	superheat ratio, $(T_o-T_f)/(T_f-T_w)$ or $(T_o-T_f)/(T_f-T_\infty)$
ε_H	turbulent thermal diffusivity
θ	dimensionless temperature, $(T-T_f)/(T_o-T_f)$
λ	product of conductivity and superheat ratio, $(k_l/k_s)(T_o-T_f)/(T_f-T_w)$
κ_v	constant, in eq. (2)

μ	dynamic viscosity, in eq. (3)
ν	kinematic viscosity
ρ	density

Subscripts

o	inlet
f	freezing or solidification
g	gas
i	interface
l	liquid
m	mean value
s	solid
w	wall
∞	ambient

REFERENCES

1. Amstead, B. H., Ostwald, P. F. and Begeman, M. L., *Manufacturing Processes,* 8th ed,. pp. 104-130, John Wiley & Sons, New York, 1987.

2. Kutateladze, S. S., Borishanskii, V. M., Novikov, I. I. and Fedynskii, O. S., Liquid-Metal Heat Transfer Media, Supplement No. 2, *Soviet Journal of Atomic Energy,* Consultants Bureau, New York, 1959.

3. Zerkle, R. D. and Sunderland, J. E., The Effect of Liquid Solidification in a Tube Upon Laminar-Flow Heat Transfer and Pressure Drop, *J. Heat Transfer,* vol. 90, pp. 183-190, 1968.

4. Stephan, L., Influence of Heat Transfer on Melting and Solidification in Forced Flow, *Int. J. Heat Mass Transfer,* vol. 12, pp. 199-214, 1969.

5. Özisik, M. N. and Mulligan, J. C., Transient Freezing of Liquids in Forced Flow Inside Circular Tubes, *J. Heat Transfer,* vol. 91, pp. 385-390, 1969.

6. Depew, C. A. and Zenter, R.C., Laminar Flow Heat Transfer and Pressure Drop with Freezing at the Wall, *Int. J. Heat Mass Transfer,* vol. 12, pp. 1710-1714, 1969.

7. Lock, G. S. H., Freeborn, R. D. J. and Nyren, R. H., An Analysis of Ice Formation in a Convectively-Cooled Pipe, *Proc. Int. Heat Transfer Conf., 4th.,* Paris-Versailles, (Cu 2.9), pp. 1-9, 1970.

8. Bilenas, J. A. and Jiji, L. M., Variational Solution of Axisymmetric Fluid Flow in Tubes with Surface Solidification, *J. Franklin Institute,* vol. 289, no. 4 pp. 265-279, 1970.

9. Lock, G. S. H. and Nyren, R. H., Analysis of Fully-Developed Ice Formation in a Convectively-Cooled Circular Tube, *Int. J. Heat Mass Transfer,* vol. 14, pp. 825-834, 1971.

10. Hwang, G. J. and Yih I., Correction on the Length of Ice-Free Zone in a Convectively-Cooled Pipe, *Int. J. Heat Mass Transfer,* vol. 16, pp. 681-683, 1973.

11. Ou, J. W. and Cheng K. C., Buoyancy Effects on Heat Transfer and Liquid Solidification-Free Zone in a Convectively-Cooled Horizontal Square Channel, *Appl. Sci. Res.* vol. 30, pp. 355-366, 1975.

12. Hwang, G. J. and Sheu, J. P., Liquid Solidification in Combined Hydrodynamic and Thermal Entrance Region of a Circular Tube, *Canadian J. Chemical Engineering,* vol. 54, pp. 66-71, 1976.

13. Mulligan, J. C. and Jones, D. D., Experiments on Heat Transfer and Pressure Drop in a Horizontal Tube with Internal Solidification, *Int. J. Heat Mass Transfer,* vol. 19, pp. 213-219, 1976.

14. Liu, H. L. and Hwang, G. J., An Experiment on Liquid Solidification in Thermal Entrance Region of a Circular Tube, *Letters Heat Mass Transfer,* vol. 4, pp. 437-444, 1977.

15. Cheng, K. C. and Wong, S. L., Liquid Solidification in a Convectively-Cooled Parallel-Plate *Channel, Canadian J. Chemical Engineering,* vol. 55, pp. 149-155, 1977.

16. Lee, S. L. and Hwang, G. J., Solidification of a Low Peclet Number Fluid Flow in a Round Pipe with the Boundary Condition of the Third Kind, *Canadian J. Chemical Engineering,* vol. 58, pp. 177-184, 1980.

17. Sadeghipour, M. S., Özisik, M.N. and Mulligan, J. C., Transient Solidification of Liquid Metals in Thermal Entry Region of a Circular Tube, *Nuclear Science and Engineering,* vol. 79, pp. 9-18, 1981.

18. Epstein, M. and Cheung F. B., Complex Freezing-Melting Interfaces in Fluid Flow, *Ann. Rev. Fluid Mech.* vol. 15, pp. 293-319, 1983.

19. Hwang, G. J. and Chen, R. S., Liquid Solidification in Internal Flows, Invited Review Paper, *Proc. Natl. Sci. Counc. ROC(A),* vol. 11, no. 1, pp. 15-27, 1987.

20. Lee, S. L. and Hwang, G. J., Liquid Solidification in Low Peclet Number Pipe Flows, *Canadian J. Chemical Engineering,* vol. 67, pp. 569-578, 1989.

21. Hwang, G. J. and Chen, R. S., Order of Magnitude Analysis on Liquid Solidification in Pipe Flows, *Freezing and Melting Heat Transfer in Engineering,* edited by Cheng and Seki, Hemisphere, New York, 1990.

22. Arora, A. P. S. and Howell, J. R., An Investigation of the Freezing of Supercooled Liquid in Forced Turbulent Flow Inside Circular Tubes, *Int. J. Heat Mass Transfer,* vol. 16, pp. 2077-2085, 1973.

23. Shibani, A. A. and Özisik, M. M., A Solution of Freezing of liquid of Low Prandtl Number in Turbulent Flow Between Parallel Plates, *J. Heat Transfer,* vol. 99, pp. 20-24, 1977.

24. Shibani, A. A. and Özisik, M. M., Freezing of Liquids in Turbulent Flow Inside Tubes, *Canadian J. Chemical Engineering,* vol. 55, pp. 672-677, 1977.

25. Thomason, S. B., Mulligan, J. C. and Everhart, J., The Effect of Internal Solidification on Turbulent Flow Heat Transfer and Pressure Drop in a Horizontal Tube, *J. Heat Transfer,* vol. 100, pp. 387-393, 1978.

26. Cho, C. and Özisik, M. N., Transient Freezing of Liquids in Turbulent Flow inside Tubes, *J. Heat Transfer,* vol. 101, pp. 465-468, 1979.

27. Thomason, S. B. and Mulligan, J. C., Experimental Observations of Flow Instability During Turbulent Flow Freezing in a Horizontal Tube, *J. Heat Transfer,* vol. 102, pp. 782-784, 1980.

28. Lee, S. L., Liquid Metal Heat Transfer in Turbulent Pipe Flows,*Heat Transfer Operations,* edited by Cheremisinoff, N. P., vol. 1, chap. 12, pp. 405-427, 1986.

29. Gilpin, R. R., The Morphology of Ice Structure in a Pipe at or near Transition Reynolds numbers, *AIChE, Symp. Ser.,* vol. 189, pp. 89-94, 1979.

30. Gilpin, R. R., Ice Formation in a Pipe Containing Flows in the Transition and Turbulent Regimes, *J. Heat Transfer,* vol. 103, pp. 363-367, 1981.

31 Cheung, F. B. and Baker, L. Jr., Transient Freezing of Liquids in Tube Flow, *Nuclear Science and Engineering,* vol. 60, pp. 1-9, 1976.

32 Epstein, M., Yim, A. and Cheung, F. B., Freezing-Controlled Penetration of a Saturated Liquid into a Cold Tube, *J. Heat Transfer,* vol. 99, pp. 233-238, 1977.

33. Epstein, M. and Hauser, G. M., Freezing of an Advancing Tube Flow, *J. Heat Transfer,* vol. 99, pp. 687-689, 1977.

34. Kolling, M. and Grigull, U., Unsteady Heat Transfer with Solidification Applied to the Fluidity of Pure Metals, *Waerme Stoffubertrag,* vol. 14, pp. 231-244. 1980.

35. DesRuisseaux, N. and Zerkle, R. D., Freezing of Hydraulic Systems, *Canadian J. Chemical Engineering,* vol. 47, pp. 233-237, 1969.

36. Epstein, M. and Cheung, F. B., On the Prediction of Pipe Freeze-Shut in Turbulent Flow, *J. Heat Transfer,* vol 104, pp. 381-384, 1982.

37. Hirata, T. and Ishihara, M., Freeze-off Conditions of a Pipe Containing a Flow of Water, *Int. J. Heat Mass Transfer,* vol.38, no. 2, pp.331-337, 1985.

38. Kroeger, P. G., A Heat Transfer Analysis of Solidification of Pure Metals in Continuous Casting Processes, in Heat Transfer 1970, *Proc., 4th Int. Heat Transfer Conference,* vol. 1, Cu. 2.7, Paris, 1970.

39. Sfeir, A. A. and Clumpner, J. A., Continuous Casting of Cylindrical Ingots, *J. Heat Transfer,* vol. 99, pp. 29-34, 1977.

40. Yao, L. S., Natural Convection Effects in the Continuous Casting of a Horizontal Cylinder, *Int. J. Heat Mass Transfer,* vol. 27, no. 5, pp. 697-704, 1984.

41. Iida, T. and Guthrie, R. I. L., *The Physical Properties of Liquid Metals,* Clarendon Press, Oxford, 1988.

42. White, F. M., *Heat Transfer,* pp. 547-548, Addison-Wesley, London, 1984.

43. Flemings, M. C., *Solidification Processing,* McGraw Hill, New York, Chap. 5, 1974.

44. Kurz, W. and Fisher, D. J., *Fundamentals of Solidification,* Trans Tech Publications, Switzerland, 1986.

45. Lee, S. L. and Tzong, R. Y., An Enthalpy Formulation for Phase Change Problem with Large Thermal Diffusivity Jump Across the Interface, *Int. J. Heat Mass Transfer,* in press, 1990.

Solidification of Mixtures with Supercooling

YUJIRO HAYASHI
Department of Mechanical Systems Engineering
Kanazawa University
2-40-20, Kodatsuno, Kanazawa-shi, Japan

KANJI KUNIMINE
Department of Mechanical Engineering
Akashi College of Technology
679-3, Nishioka, Uozumi-cho, Akashi-shi, Japan

ABSTRACT

The solidification of supercooled mixtures was studied for the system of the conventional solidification processing. In this system, solidification doesn't occur massively at the nucleation stage, and the second stage of solidification continues under a local equilibrium conditions. On the basis of the physical model presented by simulating experiments using aqueous solution, the relaxation process of thermodynamically unstable fields were clarified theoretically as the thermal and solutal diffusion problem with phase change in mushy zone. Freezing of solutions in other systems was also studied experimentally; supercooled water, constitutionally supercooled solution, and cellular system.

INTRODUCTION

In material processing, increasing attention has been focused on "rapid solidification processing (RSP)". Melt spinning and self- quenching manufacture rapidly solidified alloy in the forms of thin films and wires. Gas atomization and centrifugal atomization produce powders from molten droplets. Explosive atomization by using pulsation boiling is in the development stage [1,2]. In RSP, whose cooling rates are order of magnitude faster than 10^2 K/sec, supercooling state appears, and the solidification at the nucleation stage under a non-equilibrium condition occurs resulting into fine-scale microstructure features. Supercooling may also result into the formation of new meta-stable phases of crystalline or of amorphous [3,4]. Producing these novel- and micro-structure relevant to high properties, system dimension must be extremely small to attain uniform cooling state.

In conventional solidification processing like casting of alloys, supercooling state also appears. In this case, the cooling rate is not so high and system dimension is also not very small relative to those for RSP. So, temperature is un-uniformly distributed in molten material and if supercooling state exists, then it is located in close region from the supercooled surface to certain internal position being in equilibrium. Namely solidification doesn't occur massively, at the nucleation stage in the system, the second solidification stage must continue under a local equilibrium condition [5]. Comparisons of the rapid and conventional solidification processing are shown in Table 1.

The motivation of this work is a basic understanding of the essential phenomena; mechanism and kinetics of nucleation and growth, and its resulting micro-structure.

Practically, micro-structure produced in the conventional solidification processing, may be important in relating with surface properties of solidified material. Furthermore, we expect that this work pushes the foundation of transport theory of solidification process with the unstable mushy zone. Solidification of mixtures is always accompanied by a mushy zone, and this region is generally treated as a conduction problem with heat generation over a range of temperatures. However, pursuing high-lebeled heat transfer control in material processing, more precise treatment must be essential.

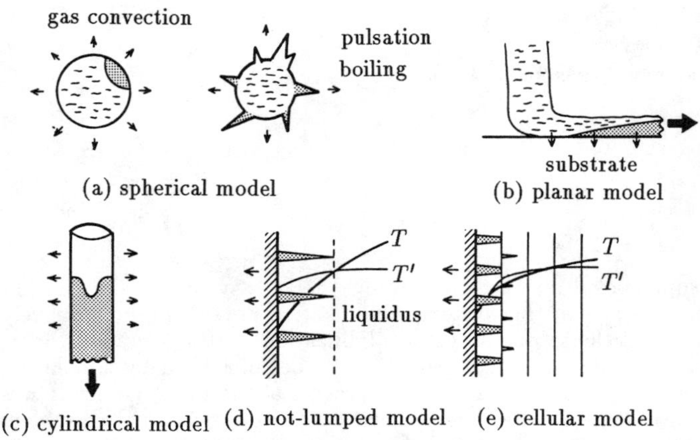

Figure 1 Schematic of physical systems of material processing

Table 1 Comparison between RSP and CSP

	RAPID SOLIDIFICATION PROCESSING		CONVENTIONAL SOLIDIFICATION PROCESSING
COOLING RATE (HEAT TRANSFER COEFF., h)	$10^2 \sim 10^8$ K/sec	HIGH	LOW
DEGREE OF SUPERCOOLING : ΔT		LARGE	SMALL
DIMENSION ; l	$10 \sim 20\ \mu$ m	SMALL	LARGE
UNIFORMITY ; $Bi = hl/\lambda$	< 0.01	UNIFORM	NON-UNIFORM
EXTRACTION OF LATENT HEAT	$\Delta T' = L_H/C_p < \Delta T$	ADIABATIC	SURFACE COOLING

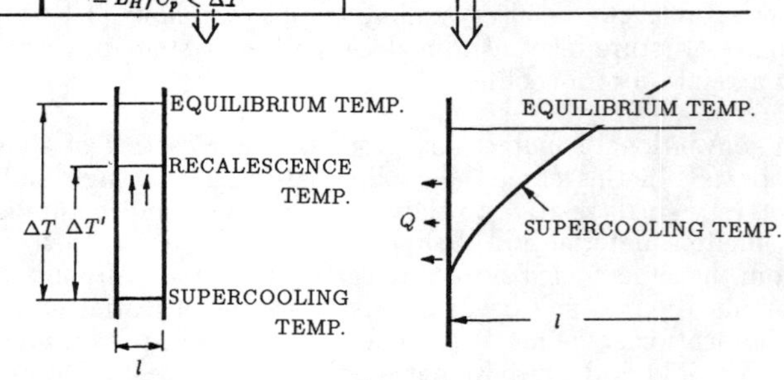

In this paper, the relaxation process of thermodynamically unstable fields were clarified theoretically as the thermal and solutal diffusion problem with phase change, on the basis of the physical model presented by the simulating experiments using aqueous solutions. Freezing of solutions in other systems were also studied experimentally; supercooled water, constitutionally supercooled solution, and cellular system.

FREEZING OF SUPERCOOLED SOLUTION AND ITS RELAXATION PROCESS

Experimental Results [5]

The freezing of an supercooled solution may be discussed on the basis of experimental results obtained for NaCl solutions of 2 wt% concentration. Figure 2 shows the example results of changes in advancing freezing fronts and temperatures. Points a', b', c' and d' in Fig. 2-(a) are the temperatures corresponding to the advancing front of points a, b, c and d in Fig. 2-(b). According to these results and observation, the freezing process may be explained as follows.

When the solution was cooled uni-directionally in the test cell, the temperature near the cooling surface fell below the liquidus temperature, -1.15°C, and the supercooled zone appeared. At some supercooled temperature near the surface, about -10°C, many ice crystals suddenly were generated on the cooling surface and spread over the supercooled zone momentarily. Due to the rapid growth of ice crystals from point a to point b, the temperature increased abruptly from points a' to b' due to the liberation of latent

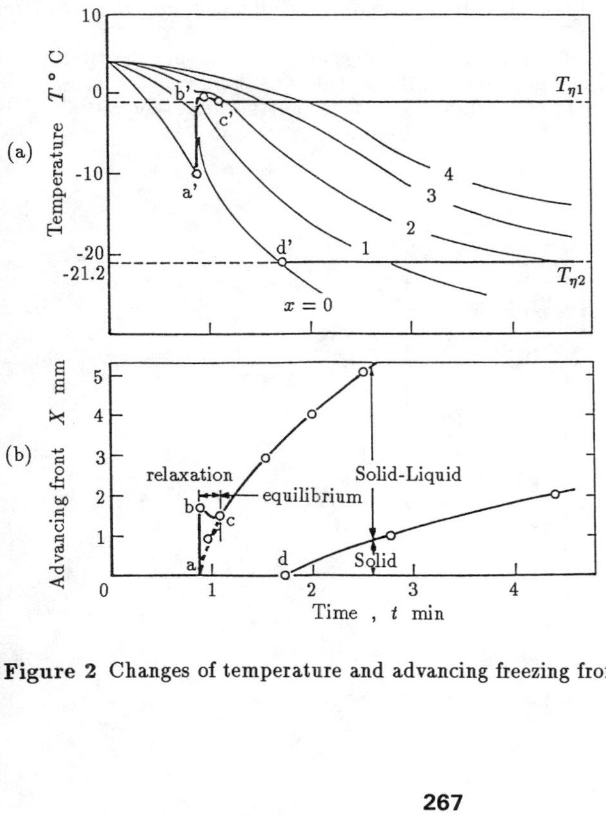

Figure 2 Changes of temperature and advancing freezing front

heat-of-fusion, and the initial supercooling seemed to relax to some extent. The tips of the ice crystals had grown fully in the supercooled zone, melted and retired their position from points b to c due to the recalescence temperature above 0°C. At the same time, the crystals grew gradually in width from the root to the top, and relaxed the residual supercooled state. After exceeding point c, the freezing front advanced at the liquidus temperature, and freezing supposedly continued in the solid-liquid and liquid regions under a thermodynamically stable state. Beyond point d, the temperature of the cooling surface descended below the eutectic temperature, -21.2°C, creating the solid, solid-liquid and liquid regions.

Theoretical Analysis

According to the previous experiments, the following matters were clarified. Freezing of supercooled solution being unstable thermodynamically, is succeeded under the following two stages, and shifts to stable freezing with fully mixed mushy zone; (1) heterogeneous nucleation at the cooling surface, and the uni-directional rapid crystallization. (2) fattening of ice crystals and fully mixturization of concentrated solution among those crystals. Here, the second stage is discussed theoretically, and its relaxation process may be clarified as the thermal and solutal diffusion problems with freezing. To avoid unnecessary complications for present purpose, morphologically simplified physical model of mushy zone formed at the first stage is presented shown in Fig. 3.

The model can be characterized by the following notes;
(1) heterogeneous nucleation on the cooling surface occurs at some supercooling temperature.
(2) ice-crystallization following after nucleation occurs momentally in the supercooled region.
(3) ice crystals stand normally on the surface at constant intervals, and those counter is assumed to be straight line.
(4) latent heat of freezing is liberated adiabatically, and temperature distribution in x-direction is given by a recalescence temperature at its position.
(5) local equilibrium exists at the solid/liquid interface. Furthermore, the initial solute concentration is maintained at the centerline between ice crystals.

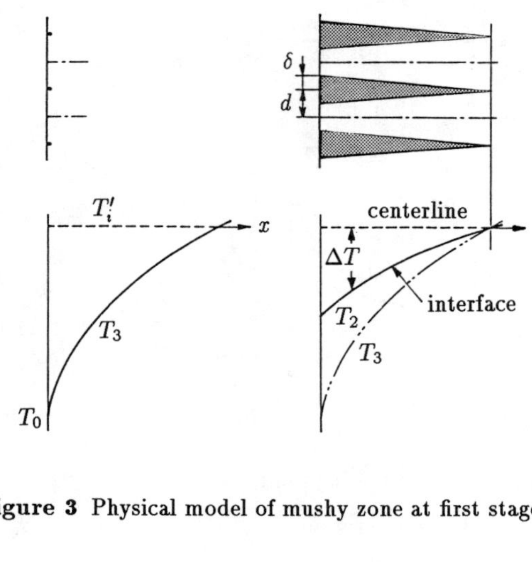

Figure 3 Physical model of mushy zone at first stage

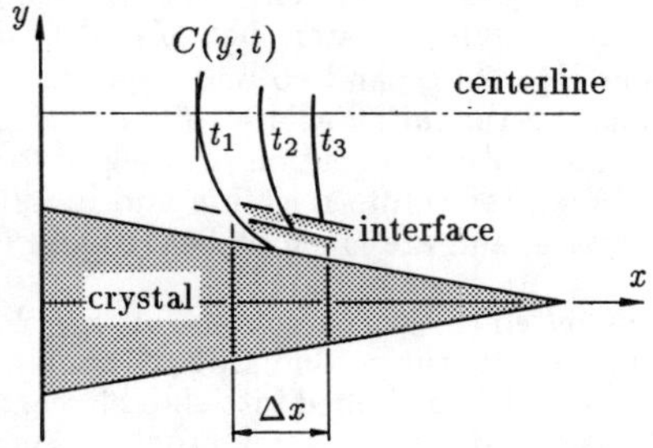

Figure 4 Physical system and coordinate

Generally, cristal morphology at the first stage is not so simple like a needle-shaped one shown in Fig. 3, but is very complicated such as dendritic and feather-like crystals. As for the supercooling temperature, Ichimiya [6] studied on the critical value at which heterogeneous nucleation occured, but we didn't refer to it in this paper. Figure 4 illustrates the physical model of the system. The additional assumptions implied in the model are as follows:

(6) freezing of two-dimensional mushy zone proceeds in y-direction with the liberation of latent heat and solute.

(7) the bulk velocity of the liquid due to the difference in densities of the two phases is not considered.

(8) heat flows in only x-direction under the condition: constant heat flux at the cooling surface, and constant equilibrium temperature corresponding to initial solute concentration at the front of mushy zone.

(9) diffusion of solute occurs normally in the gap of ice crystals, since the local equilibrium condition at the interface prevails along x-direction.

The basic energy- and mass-balance equations and these initial and boundary conditions are:

Energy:

$$(C_p\rho)_2\frac{\partial T}{\partial t} = \frac{\partial}{\partial x}(\lambda_2\frac{\partial T}{\partial x}) + \rho_3 L_H\frac{\partial f(x)}{\partial t} \quad (1)$$

$$t = 0\ ;\quad T = T_i(x) + \frac{\rho_3 L_H}{(\rho C_p)_2}f(x) \quad (2)$$

$$x = 0\ ;\quad -\lambda_2(\frac{\partial T}{\partial x}) = q_0 = const. \quad (3)$$

$$x = \eta\ ;\quad T = T'(C_i) = const. \quad (4)$$

Mass:

$$\int_{\delta(x)}^{d} C_L dy = C_i d \quad (5)$$

$$y = \delta(x)\ ;\quad -D_3\frac{dC_2}{dy} = C_2\frac{d\delta}{dt} \quad (6)$$

$$C_L = C'\{T(x)\} \quad (7)$$

$$y = d\ ;\quad \frac{dC_L}{dy} = 0 \quad (8)$$

where, subscripts 2 and 3 denote the mushy and liquid phase. The other nomenclatures are; specific heat C_p, density ρ, thermal conductivity λ, latent heat L_H, diffusion coefficient D, and solid fraction in mushy zone $f(=\delta/d)$, and so on. Equation (5) represented in integral form, is obtained by equating the initial solute to that which is in the concentrated solution. Equation (5) is mass balance equation in integral form. Condition of local equilibrium and relation between the solute rejection and its diffusion come into existence at the solid-liquid interface, and are given by eqs.(6) and (7).

Using recalescence temperature T_i represented by eq.(2) and simplified morphology δ, d, η, obtained experimentally at the end of first stage, the second stage of solidification will be numerically calculated for mushy zone which is divided into slab-element in width Δx. For the first element, temperature $T(\Delta x)$ and concentration profile $C(\Delta x, y)$ were calculated for assumed solidification rate $\Delta\delta$ in time-increment δt. Here, concentration profile of a quadratic form was introduced and determined by eqs.(6), (8) and (5). Obtained $\Delta\delta$ satisfying the local equilibrium condition represented by eq.(7), calculation may carried out in succession for next element until the coexisting fields are completely relaxed.

Results and Discussion

Predictios for crystal growth, solute diffusion in y-direction, and temperature distribution as a function of longitudinal position with time are shown in Fig. 5. Enlargening of ice crystals and solute segregation occur more rapidly at the place being close to the cooling surface. The comparison between predicted and measured values show good agreement in the temperature distribution. Relaxation characterized by the fully mixturization of solute in the gap of ice crystals, was considered that the difference of concentration distribution become less than 1.0 percent, and its prediction was shown by dotted line. From these simulating results, freezing of supercooled solution and its relaxation process are recognized precisely, and it was clarified that the model presented here was reasonable.

Finally, on the basis of prediction, the timewise evolution of the freezing-front morphology is shown schematically along with the actual- and liquidus-temperature distributions in Fig. 6. Firstly, the supercooling state represented by the hatched zone between the thin and solid lines, collapsed due to the generation of the ice crystals on the cooling surface in the opposite direction of the heat flow. The generation of the ice crystals rejected not only the latent heat-of-fusion, but also the solute, and the supercooled fields relaxed to some extent. Then the actual and liquidus temperatures were brought closer to each other, as shown in Fig. 6-(ii). Next, these crystals gradually grew width-wise and captured the concentrated solution among their spaces to reach a equilibrium state in the solid-liquid region (Fig. 6-(iii)). Namely, the dotted line between points a and c in Fig. 2-(b) represent the relaxation process. The freezing process after point c is the same as described above.

OTHER SYSTEMS WITH SUPERCOOLING

The freezing of supercooled solution was clarified experimentally and theoretically. Particularly, the second stage of freezing characterized by the relaxation of thermodynamically unstable field could be explained in relating with the fattening of ice crystals and mixturization of concentrated solution. In this chapter, freezing of solutions in other systems may be studied experimentally; supercooled water, constitutionally supercooled solution, and cellular system. Theoretical discussion may be omitted here, since the fundamental mechanisms and kinetics are considered to be same with those in the system of supercooled solution.

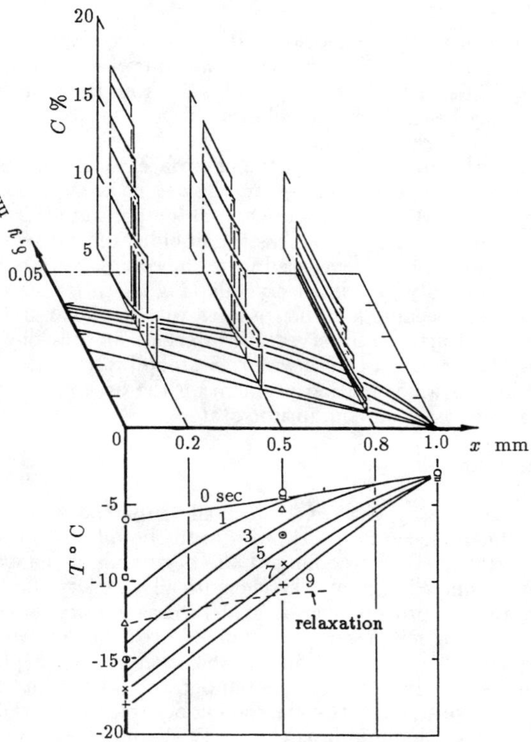

Figure 5 Relaxation of the temperature and concentration fields with crystal growth

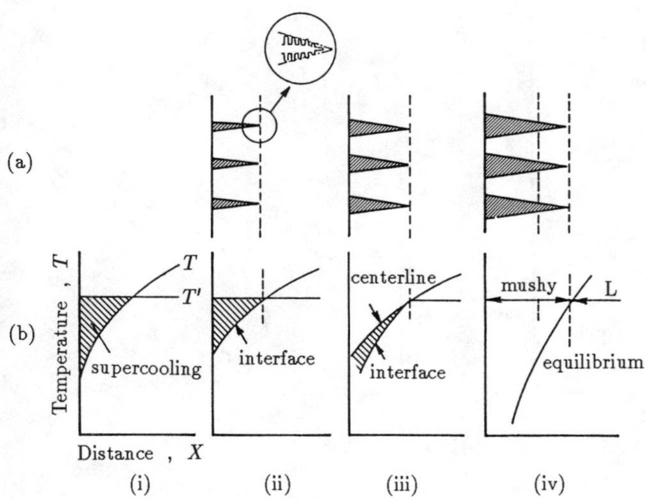

Figure 6 Schematic of freezing process of supercooled solution

Supercooled Water

In order to exemplify the freezing of the supercooled melt without receiving constitutional effect, a pure water was used as a test material. The timewise evolution of the freezing-front morphology are schematically shown in Fig. 7. From the figure, the freezing process of a pure water may be explained as follows.

The supercooled state appeared at the vicinity of the cooling surface, and collapsed in the same manner as for the case of the solution. However, the degree of supercooling remained the same, as shown in Fig. 7-(ii), since only the actual temperature increased due to the liberation of the latent heat-of-fusion, whereas the liquidus temperature remained at constant. Next, a dense solid phase developed from the cooling surface, and finally the solid layer covered the initially generated crystals (Fig. 7-(iii) and -(iv)). After this point, the freezing process became a stable process with a planar freezing front maintained at 0°C. Tiller and Rutter[7] observed the interface morphology and crystal texture produced during the freezing of pure water. Saitoh[8] also studied the freezing of water with supercooling as a conduction problem on the assumption that the solid-liquid phase region was always in an equilibrium state.

Constitutionally Supercooled Solution

Here, the relaxation process of the freezing of a constitutionally supercooled solution will be discussed. To this end, a small piece of ice was put into the liquid just ahead of the cooling surface, and the occurrence of supercooling at the beginning of the experiment was restrained artificially. Using Fig. 8, which schematically depicts the basis of the experimental results, the freezing process may be explained as follows. Freezing first occurred around the nucleus of the ice piece, and extended over the cold surface to form a thin ice layer segregating the solute. Namely, the layer having a planar surface advanced one-dimensionally, keeping the liquidus temperature at the interface, Fig. 8-(i). After a short period, the constitutional supercooling occurred as a gradient of supercooling from the interface into the liquid, as shown by the hatched region in Fig. 8-(ii). This made the planar interface unstable and induced the dendritic crystal growth with needle-shaped structures perpendicular to the interface. This was due to the fact that the growth rate would be increased in any localized region that advanced ahead of the general interface. These crystals grew gradually both in height and in

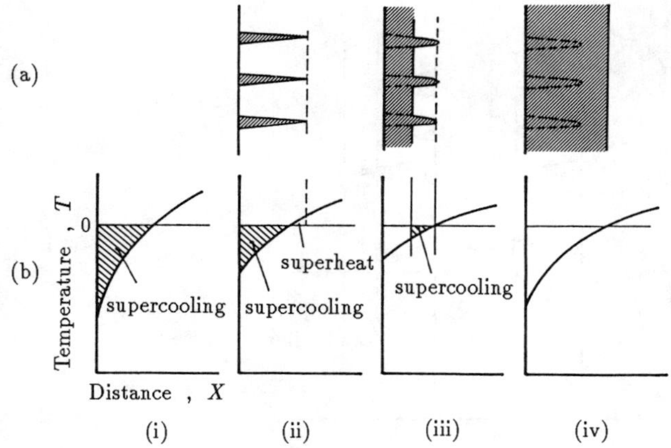

Figure 7 Schematic of freezing process of supercooled water

width, capturing the concentrated solution between them to relax the thermodynamically unstable state, Figs. 8-(ii) and (iii). After this period, freezing advanced in a stable state, however, the interface between the solid and the solid-liquid regions remained stationary at the same location. Finally, this interface began to move quasi-steadily in all three regions, Fig. 8-(iv), as the freezing continued.

Constitutional supercooling induces instabilities on the planar interface to form more complicated yet more stable dendritic structures. However, constitutional supercooling is a necessary condition for morphology instability. Mullins and Sekerka[9,10] calculated the variation of sinusoidal perturbation introduced into the shape of the interface, and obtained a sufficient condition, taking account the temperature and concentration distributions in the vicinity of the interface. O'Callagham et al. [11] discussed the instability of the planar interface during the freezing of binary aqueous solutions by using the Mullins-Sekerka stability criterion. Their results indicate that the planar interface rapidly becomes unstable except at very low freezing rates or solute concentrations. Terwilliger and Dizio [12] analyzed solute rejection from a planar solid-liquid interface, and obtained the criterion for the onset of constitutional supercooling. Levin[13,14] presented a generalized analysis of the unidirectional freezing assuming that the solid-liquid interface remains planar in spite of the appearance of the constitutional supercooling. Furthermore, O'Callagham et al. [15,16] analyzed energy and mass transport near the dendrite-tip and -basal plane regions, and predicted the dendrites shapes, lengths and spacing, together with the temperature and concentration fields. Kurz and Fisher [17] studied dendrite growth using a simplified solution for the wavelength of instability, and obtained a relationship between growth conditions and primary arm spacing. Based on their calculations, Roosz et al. [18] developed a method of solute distribution in the liquid between dendrite arms in order to produce alloys with uniform properties by controlling the process.

Supercooled Solution in A Cellular System

For freezing of biological substances, generally, extracellular ice formation occurs initially with an associated concentration difference between the extra- and intra-cell. And slow cooling decreases the probability of intracellular freezing with permitting water to leave the cell. Fast cooling induces supercooling state and the freezing at the nucleation stage brings numerous fine ice particles to prevent the injury of living cells [19,20]. Here, freezing of supercooled solution in the cellular system becomes one of subjects studying here.

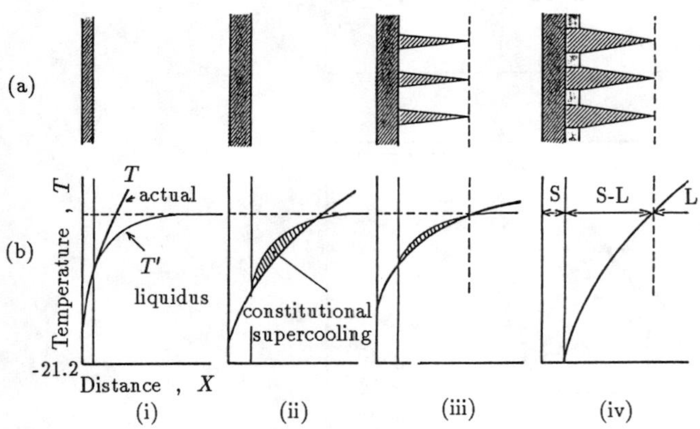

Figure 8 Schematic of freezing process of solution with constitutionally supercooling

Figure 9 shows the representative results of changes in the advancing freezing front and temperatures at the fixed positions. Timewise evolution of the interface morphology and the relaxation of the supercooling state are schematically shown in Fig. 10. From these figures, the freezing of an supercooled solution in a cellular system will be explained as follows.

Firstly, the supercooling zone appeared at the vicinity of the cooling surface, and this unstable state was broken down due to the generation of needle-shaped ice crystals. These crystals grew in width and relaxed the supercooling state to reach an equilibrium state. These relaxation processes were basically same as those for the freezing of an supercooled solution without cells. However, these processes occurred individually in each cell, since the growth of ice crystals in the supercooling zone was physically restrained by the cell wall. Some portion of supercooling zone remained in the next cell after completing relaxation process in the former cell. In actuality, this characteristic behavior does not always appear, as is the case in this experiment, since the necessary condition for supercooling is not same in each cell. However, it is reasonable to consider here that the phenomena described as the relaxation processes in a cell occurred one after another in the following cells.

Relaxation Process on Phase Diagram

Finally, these relaxation processes are explained by introducing an equilibrium phase diagram. Changes of state in Figs. 11-(a) and (b) correspond to the relaxation processes of the supercooled solution and that in the cellular system, respectively. In both figures, the initial supercooling state is represented by a segment of vertical lines ranging from the maximum supercooling temperature to the liquidus temperature. This implies that microstructural freezing may result from supercooling at the nucleation stage at the vicinity of the cooling surface. In case of (a), the relaxation progresses from the states of (i) to (iii), and then freezing advances in the equilibrium state along

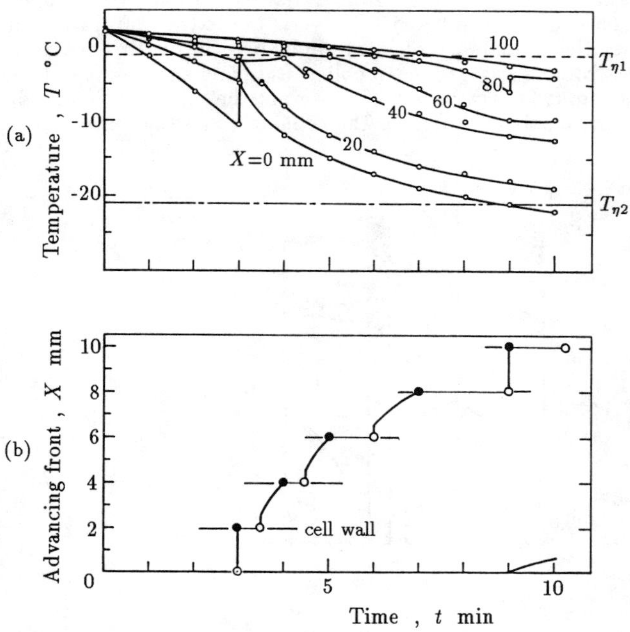

Figure 9 Changes of temperature and advancing freezing front

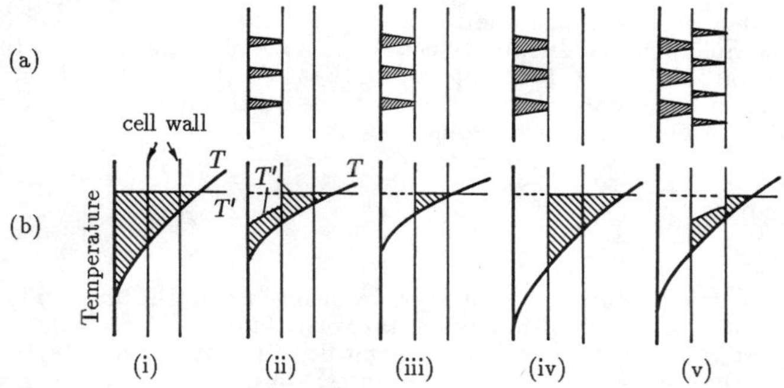

Figure 10 Schematic of freezing process of supercooled solution in a cellular system

the liquidus line, (iv). In the case of (b), the relaxation progresses only partially in each individual cell. Firstly, the supercooled portion of the first cell moves to the liquidus line. During freezing it advances equilibriumly and the temperature decreases along the liquidus line in the first cell, and then the temperature descends and the false-equilibrium state collapses in the following second cell.

Figure 11-(c) shows the relaxation of supercooled pure water. Figure 11-(d) shows the relaxation process of the constitutional supercooling. In this case, the dendritic crystals breaking down the thermodynamical unstable region, generate an equilibrium condition on the planar ice layer. This suggests that the conditions of local equilibrium are supposed to prevail at the solid-liquid interface during the relaxation process.

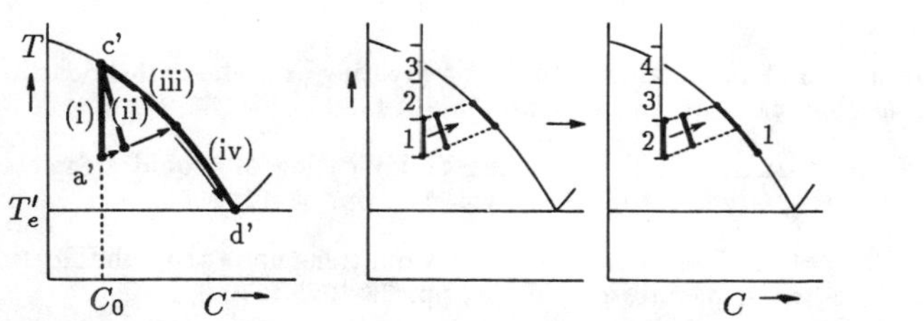

Figure 11 Change of state on phase diagram during solidification of supercooled mixtures

Figure 11-(e) shows instances of the rapid solidification of an alloy. Hypercooling may come about in a small temperature range due to the rapid cooling of the melt with infinitesimal dimensions. Solidification due to nucleation can take place massively with the recalescence temperature. In case 1, the temperature rises into the liquid field, and some dendritic structures may be produced. In case 2, on the other hand, the massive solidification yields grains of uniform composition[3].

CONCLUDING REMARKS

The freezing of an supercooled solution was studied by using a water-sodium chloride solution in a system with a non-uniform temperature distribution. The relaxation processes of thermodynamically unstable fields were clarified in connection with the interface morphology in three cases of supercooling states. In order to advance the rapid solidification processing, for the relevant science and engineering applications, a more basic understanding of the relationship among supercooling states, resulting microstructures, and properties, is essential.

REFERENCES

1. R. Akiyama, S. Nishio, and I. Tanasawa, An Attempt at Production of Rapidly Solidified Particles by Thermal Interaction in Molten Metal/Water System, *Proc. of 25th Heat Transf. Sympo., Japan*, pp. 322-324, 1988.

2. M. Shoji, et al., Quenching of a Copper Shpere in Freon 113 and Water, *Proc. of 25th Heat Transf. Sympo., Japan*, pp. 319-321, 1988.

3. M. Cohe, B. H. Kear, and R. Mehrabian,*Rapid Solidification Processing II*, Claitor's pub. div., 1980.

4. T. W. Clyne, Numerical Treatment of Rapid Solidification, *Met. Trans.*, vol. 15B, pp. 369- ,1984.

5. Y. Hayashi, K. Kunimine and T. Nagamoto, Study on Freezing of Undercooled Solution, *Int. Sympo. on Cold Regions Heat Transfer*, pp. 85-90, 1989.

6. K. Ichimiya and Y. Mochizuki, The Supercooling Solidification of Liquid Droplets - Aqueous Solution of NaCl-, *Trans. of the JAR*, vol. 1-1, pp. 37-42, 1984.

7. W. A. Tiller and J. W. Rutter, The Effect of Growth Condtions upon the Solidification of a Binary Alloy, *Canadian J. of Physics*, vol. 34, pp. 96-107, 1956.

8. A. Saitoh, et al., Heat Conduction with Supercooled Solidification, *Bulletin of the JSME*, vol. 25, No. 202, pp. 591-598, 1982.

9. W. W. Mullins and R. F. Sekerka, Stability of a Planar Interface during Solidification of a Dilute Binary Alloy, *J. Appl. Phys.*, vol. 35, pp. 126-131, 1963.

10. W. W. Mullins and R. F. Sekerka,*J. Appl. Phys.*, vol. 35, pp. 444- , 1964.

11. M. G. O'Callagham, et al., Instability of the Planar Freeze Front during Solidification of an Aqueous Solution, *ASME J.of Heat Transf.*, vol. 102, pp. 673-677, 1980.

12. J. P. Terwillinger and S. F. Dizio, Salt Rejection Phenomena in the Freezing of Saline Solutions, *Chem. Engng. Sci.*, vol. 25, pp. 1331-1349, 1970.

13. R. L. Levin, Generalized Analytical Solution for the Freezing of a Super-cooled Aqueous Solution in a Finite Domain, *Int. J. Heat Mass Transfer*, vol. 23, pp. 951-959, 1980.

14. R. L. Levin, The Freezing of Finite Domain Aqueous Solution: Solute Redistribution, *Int. J. Heat Mass Transfer*, vol. 24, pp. 1443-1455, 1981.

15. M. G. O'Callagham, et al., An Analysis of the Heat and Solute Transport during Solidification of an Aqueous Binary Solution-I. Basal Plane Region, *Int. J. Heat Mass Transfer*, vol. 23, pp. 553-561, 1982.

16. M. G. O'Callagham, et al., An Analysis of the Heat and Solute Transport during Solidification of an Aqueous Binary Solution-II. Dendrite Tip Region, *Int. J. Heat Mass Transfer*, vol. 25, pp. 563-573, 1982.

17. W. Kurz and D. J. Fisher, Dendrite Growth at the Limit of Stability: Tip Radius and Spacing, *Acta Metall.*, vol. 29, pp.11-20, 1981.

18. A. Roosz, et al., Solute Redistribution during Solidification and Homogenization of Binary Solid Solution, *Acta Metall.*, vol. 32, pp. 1745-1754, 1984.

19. Y. Hayashi and T. Komori, Investigation of Freezing of Salt Solutions in Cell, *ASME J. of Heat Transfer*, vol. 101, pp. 459-165, 1979.

20. Y. Hayashi, M. Sekiya and K. Kunimine, Heat Conduction Model of Freezing of Food, *Proc. of 26th Heat Transf. Sympo., Japan*, pp. 223-225, 1989.

Mushy Zone Modeling

MARTIN E. GLICKSMAN[1], RICHARD N. SMITH[2], STEVEN P. MARSH[3], and ROBERT KUKLINSKI[4]
Rensselaer Polytechnic Institute
Troy, New York 12180-3590, USA

ABSTRACT

A key element of mushy zone modeling is the description of the microscopic evolution of the lengthscales within the mushy zone and the influence of macroscopic transport processes. This paper describes some recent progress in developing a mean–field statistical theory of phase coarsening in adiabatic mushy zones. The main theoretical predictions are temporal scaling laws that indicate that average lengthscale increases as time$^{1/3}$, a self–similar distribution of mushy zone lengthscales based on spherical solid particle shapes, and kinetic rate constants which provide the dependences of the coarsening process on material parameters and the volume fraction of the solid phase. High precision thermal decay experiments are described which verify aspects of the theory in pure material mushy zones held under adiabatic conditions. The microscopic coarsening theory is then integrated within a macroscopic heat transfer model of one–dimensional alloy solidification, using the Double Integral Method. The method demonstrates an ability to predict the influence of macroscopic heat transfer on the evolution of primary and secondary dendrite arm spacings in Al–Cu alloys. Finally, some suggestions are made for future experimental and theoretical studies required in developing comprehensive solidification processing models.

INTRODUCTION

Phase coarsening, or Ostwald ripening, is a microscale process occurring in all finely–divided two–phase mixtures. The driving force for coarsening is the reduction of the free energy stored in the interfaces separating the phases. The net result of phase coarsening is an increase in the average microstructural lengthscale and a concomitant decrease in the amount of interfacial area per unit volume of microstructure. Coarsening can occur with or without net phase transformation, because both species and enthalpy are conserved among the participating phases. Thus, the coarsening process induces microscopic transport fields that communicate solutal and thermal fluxes over active regions of the interface. In the case of solid–liquid systems, which will be specifically treated in this paper, the microscale transport fields (1–100 micrometers) are virtually independent of the long–range macro–transport fields (0.01–1 meters) responsible for net solidification. The challenge facing current heat transfer modeling of solidification is to bridge the gap in the lengthscales between the microscopic and the macroscopic.

---

[1]John Tod Horton Professor of Engineering

[2]Associate Professor of Mechanical Engineering

[3]Naval Research Laboaratory, Washington, DC 20375–5000 USA

[4]Naval Underwater Systems Center, Newport, RI 02841 USA

Coarsening effects may be observed easily in casting and welding processes, alloy heat treatment, liquid–phase sintering, powder atomization, as well as in many other material processes which involve more than one phase. Inasmuch as coarsening determines the distribution of microstructural lengthscales and their local chemical composition, it has a profound effect on the physical and mechanical properties of the resultant material. A quantitative understanding and prediction of the kinetics of phase coarsening is essential for designing and optimizing most processes for which prediction and control of the final structure is important.

Recently, Marsh (1989) has developed a statistical mean–field theory to describe diffusion–limited coarsening in two–phase materials. Although space does not permit a review of the many theoretical details in this paper, a brief summary of the salient results follows.

Specifically, Marsh applied scale invariance, via the continuum continuity equation, to the particle volume associated with each size class of dispersed phase. Particle volumes and their statistically associated transport field–volume were both specified through a global constraint specifying the (constant) volume fraction of the microstructure over lengthscales that are large compared to those within the microstructure. In addition, the critical particle volume, which increases linearly with time, was for the first time defined self–consistently, using appropriate spatial and ensemble averages taken over the distribution of the transport fields in the continuous matrix phase. Marsh's theory predicts the scaled rate constants for the growth of the critical volume which, in turn, predicts the growth of the metric lengthscale for any volume fraction of the particulate phase. Reintroduction of the materials constants (chemical diffusivity, interfacial energy, etc.) for the alloy system, which define the appropriate capillary lengthscale and transport timescale, yields dimensional rate constants for coarsening that depend on temperature and alloy composition.

The results of this statistical theory have been carefully checked against lengthscale measurements in the literature, and show good agreement regarding the variation of these constants with respect to the volume fraction parameter, V_V, over a wide range this parameter. For example, Figure 1 shows the comparison of Marsh's theory with liquid phase sintering experiments performed on several binary metallic alloys studied over a wide range of volume fractions. The statistical theory provides an accurate prediction of the coarsening rate constant for volume fractions as high as about 0.8, and then tends increasingly to overpredict the rate. This result is expected, since Marsh's theory assumes spherical particles, whereas the actual particle shapes encountered in liquid phase sintering deviate from spheres at high volume fractions due to shape accommodation.

In addition to the comparison with diffusion–controlled coarsening during liquid phase sintering, some new experimental methods have been developed in our laboratory to monitor, in situ, the coarsening kinetics of pure solid–liquid mixtures. We have measured the mean temperature, $\langle T \rangle$, to a precision of 0.1 mK, of pure "mushy zones" comprised of cubic, elemental, α–P_4 and pure ethylene carbonate (ETC) undergoing phase coarsening. The average supercooling, $T_m - \langle T \rangle$, represents the temperature offset from the thermodynamic equilibrium melting point, T_m, caused by the excess free energy of the solid–liquid interfaces. The capillary influence—also called the Gibbs–Thomson effect—decays as the system coarsens, wherein the lengthscales increase and the interfacial area decreases, thus permitting the continuous monitoring of the kinetics of the process. The thermal measurements in α–P_4 and ETC are representative of mushy zone coarsening behavior in systems with complex, nonspherical solid–liquid morphologies. As will be described later in this paper, the thermal measurements also suggest that the theory described above applies to diffusion–controlled dendritic mushy zones in binary alloys.

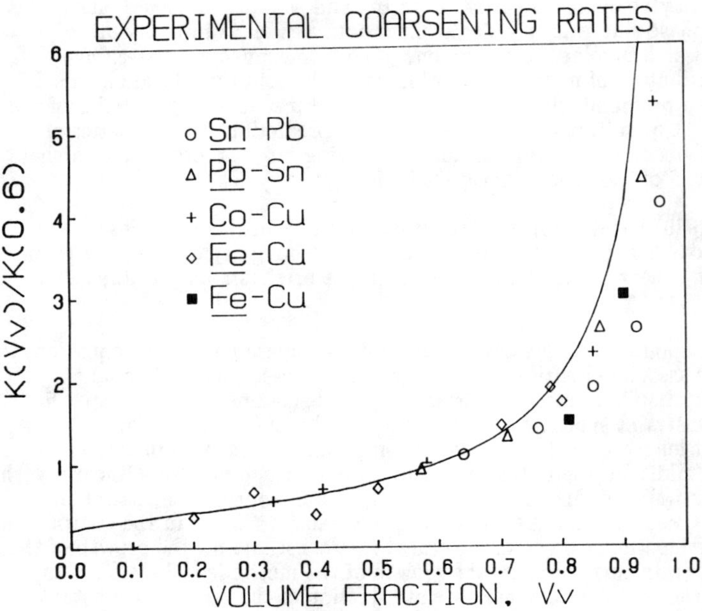

FIGURE 1. Comparison of Marsh (1989) coarsening theory with available data.

THERMAL RELAXATION EXPERIMENTS—PURE MUSHY ZONES

Most coarsening experiments require extensive quantitative analysis of the aged microstructures that must be performed to obtain kinetic data. Such experiments often employ numerous samples which are aged simultaneously and interrupted at various times for quantitative microstructural analysis. An experimental technique has been developed by Marsh and Glicksman (1986) which allows one to measure the extent of coarsening without interruption of the process. This technique involves the direct measurement of the average temperature of a solid–liquid mixture which is held adiabatically near the melting point. Thermodynamic considerations indicate that the average temperature in a pure two–phase system is depressed below the bulk melting point, T_m, by an amount which is related to the average interfacial curvature in the microstructure (mushy zone). This undercooling decays (and the average temperature rises) as the two–phase structure coarsens and the curvatures and total interfacial area decrease. This technique, simply stated, involves measuring the change in the average Gibbs–Thomson effect directly as the mushy zone coarsens.

Experiment

Material Selection. Thermal coarsening experiments were performed on two pure materials, white phosphorus (α–P_4) and ethylene carbonate (ETC), to quantify the effect of solid–phase volume fraction on the coarsening rate. Phosphorus solidifies dendritically from the supercooled melt in the α–manganese cubic form, whereas ETC forms a needle–like structure. These materials have melting points convenient for precision thermal experiments (44.1 C for phosphorus and 36.3 C for ETC), and bulk liquid samples of these substances can be routinely supercooled well below these temperatures. This allows the formation of mushy zones through the adiabatic partial solidification of the supercooled melt. The volume fraction of solid, V_V, which is formed during this solidification can be calculated from a thermal energy balance as

$$V_v = \frac{\Delta T_0}{Q/c} \quad , \qquad (1)$$

where ΔT_0 is the initial supercooling below the bulk melting point. Q/c is the ratio of the latent heat to the liquid heat capacity, and has units of temperature. The Q/c ratio is about 108 K for ETC, and is 25.4 K for α–P_4, which represents, respectively, the supercoolings needed to form 100% solid at T_m through adiabatic solidification. An initial supercooling of 12.7 K would thus form a microstructure containing 50% solid in α–P_4 and 12% solid in ETC. The latent heat given off by formation of this solid is equal to the amount needed to warm the remaining liquid to T_m, at which point further solidification does not occur. Thus one can control the volume fraction of the mushy zone formed by choosing an appropriate initial undercooling ΔT_0.

A highly pure material is desired for these thermal experiments to minimize solutal effects on the melting point and on the thermal diffusion mechanism. The materials used in these experiments were zone–refined using a two–stage method described elsewhere (Rubinstein, et al., 1988). The samples were transferred under vacuum to borosilicate glass ampules containing a platinum resistance thermometer were then hermetically sealed.

Equipment and Procedure. The temperature sensor employed is a wire–wound 50–ohm platinum resistance temperature detector (RTD). The resistance of the RTD is measured using a Tinsley automatic AC digital bridge. This equipment was chosen because it measures resistance to 10^{-5} ohms and has a negligible zero–drift. The resulting temperature resolution of this system is about 0.1 mK, which is needed to resolve the small capillary temperature effects in the mushy zone.

Experimental runs are performed by first placing the ampule into a hot water bath until the enclosed material is completely melted. The ampule is then transferred to a controlled–temperature bath which is maintained at the desired initial temperature, T_0, and allowed to cool until the temperature stabilizes to within 0.01 K. The ampule is removed from the supercooling bath and a short burst of cold aerosol is immediately applied to the bottom tip. This induces nucleation of the solid phase and leads to the rapid partial solidification of the material. The ampule is immediately transferred to a second controlled–temperature bath maintained about 2 mK below the melting point. The mean temperature of the coarsening mixture is then monitored and recorded for about eight hours. At the end of each experimental run the sample is removed from the bath and remelted. Further details of the equipment and procedure are presented elsewhere (Marsh, 1989).

Results

Data Analysis. The experimental coarsening rates are evaluated directly from the time$^{-1/3}$–temperature data. Although the mean undercooling, $\langle \Delta T \rangle$, is proportional to a reciprocal lengthscale, it is not advantageous to analyze the data in the form $\langle \Delta T \rangle^{-3}$ vs. time, which corresponds to the growth of the cube of an average lengthscale, which is the classical mean–field scaling law first introduced by Lifshitz and Slyozov (1961). Specifically, the $\langle \Delta T \rangle$ values are very small (on the order of a few mK or less) and the value of T_m is not known with enough precision to obtain reliable values of $\langle \Delta T \rangle$, where $\langle \Delta T \rangle = T_m - \langle T \rangle$, and $\langle T \rangle$ is the measured average temperature of the mushy zone. Raising small values of $\langle \Delta T \rangle$ to the power -3 would magnify the error inherent in calculating $\langle \Delta T \rangle$. However, the rate of change of $\langle \Delta T \rangle$ is equal to minus the rate of change of the temperature $\langle T \rangle$ because T_m is constant and need not be determined explicitly.

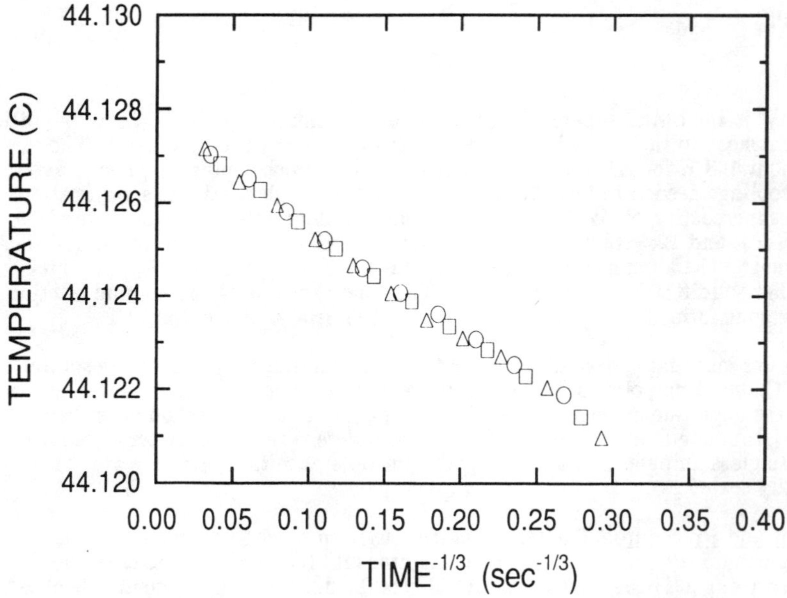

FIGURE 2. Thermal data for three successive coarsening runs in α–P_4 (V_v = 0.27).

The thermal decay rate, K_t, is defined by

$$K_t = \frac{d\,(<\Delta T>)}{d(t^{-\frac{1}{3}})} = \frac{-d<T>}{d(t^{-\frac{1}{3}})}, \tag{2}$$

where K_t has units of mK–sec$^{1/3}$. This definition implies the kinetic equation

$$<\Delta T> = K_t \cdot t^{-\frac{1}{3}} . \tag{3}$$

This relationship represents the asymptotic scaling behavior of the mushy zone. The advantage of using this form of the decay rate is that all data points have a similar random error associated with them, because the directly measured temperature is the quantity for which the rate of change is being analyzed. The decay rate K_t is obtained from the raw time–temperature measurements by regressing the late–time linear range of the <T> vs. $t^{-1/3}$ data for each run.

Reproducibility. Typical coarsening data for α–P_4 are shown in Figure 2 for three successive runs. The initial undercooling for all three runs is 6.28 K which corresponds to a volume fraction of about 0.27. The horizontal axis represents $t^{-1/3}$, so that time increases non–linearly from right–to–left. The vertical axis is the measured temperature. Only every twelfth data point is shown in each run for clarity. At late times it is evident that the temperature rise becomes linear with $t^{-1/3}$ and the slope of this line is the thermal decay rate K_t. The slopes remain similar for the different runs. Regressions performed on the linear regions of these three data sets indicate that K_t varies from 21.2 to 21.7 mK–sec$^{1/3}$. Multiple runs (3 or 4) were performed at

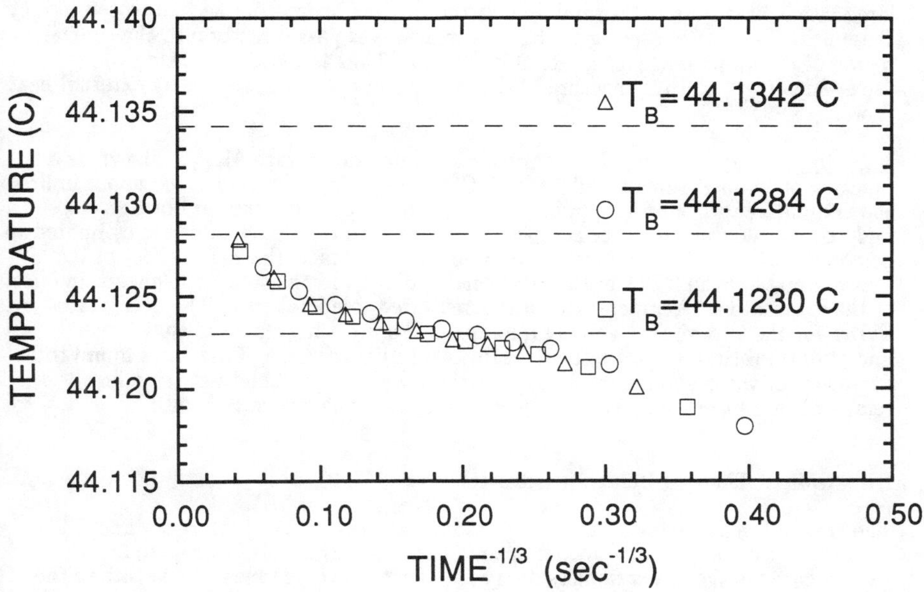

FIGURE 3. α–P_4 coarsening data at three different bath temperatures (V_v=0.11).

each volume fraction considered, and similar reproducibility of the results was observed. The data from all runs at a given volume fraction were first averaged together, and the decay rate was then determined from a regression of this averaged data. This procedure tends to smooth out any minor fluctuations in the individual data sets.

<u>Effect of Bath Temperature on Coarsening Rates</u>. The thermal relaxation measurements involve resolving temperature changes which are on the order of 0.0001 K. The effect of the surrounding bath temperature on the temperature of the coarsening mixture must be carefully considered. In most experimental runs, the bath temperature was maintained about 2–3 mK below T_m. This value of T_m is simply the ordinate intercept of the extrapolated regression of the temperature–time$^{-1/3}$ data, and corresponds to the undercooling at infinite time. In all cases the coarsening leads to an eventual rise above this temperature, indicating that the measured temperature changes are not simply a warming effect from heat leaking into the ampule from the external bath.

A more critical test was performed for the experimental runs with α–P_4 at an undercooling of 2.53 K (V_v=0.11). Three sets of three runs each were performed at different bath temperatures of 44.123 C, 44.128 C, and 44.134 C. For consistency, the bath temperatures were set using the RTD in the ampule. These bath temperatures and the corresponding data sets are shown in Figure 3. These results clearly indicate that the external bath temperature has little effect on the measured coarsening rate. The decay rates for these three data sets vary from 16.6 to 17.6 mK–sec$^{1/3}$, which is only a 6% variation in K_t. This insensitivity to the external temperature is a result of the active nature of the solid–liquid mixture. A temperature difference of only a few mK excites very small heat fluxes to or from the ampule, which tend to be absorbed near the outer wall of the ampule through local freezing or melting. The relatively large latent heat limits to a small extent the melting or freezing needed to counteract these heat flows, and insuring that no thermal waves pass directly into the central region of the ampule where the RTD is located. Thus the system is largely "self–shielding", and the measured thermal behavior is not governed by external heat flows.

Volume–Fraction Effects. The thermal coarsening decay rate, K_t, is shown as a function of V_v in Figures 4 and 5 for ETC and α–P_4, respectively. The upper limits on the volume fractions in these figures correspond to the maximum uniform undercoolings obtained in the samples. The errors in the decay rates are estimated to be about ±5%, which arise primarily from such factors as the slight effect of the thermal mass of the RTD on the local surroundings and the small but uncertain effect of the external bath temperature on the measured temperatures. The decay rates shown for the two highest volume fractions in Figure 4 have a somewhat larger uncertainty, estimated to be about ±10%. The linear range of data was found to be narrower for these volume fractions, and there was a correspondingly smaller temperature range from which to determine the thermal relaxation rate.

Discussion of Thermal Relaxation Results

The thermal relaxation rate was found to increase with V_v in both α–P_4 and ETC. Two theoretical curves are plotted against the phosphorus data in Figure 5. The dashed curve is the dimensionless decay rate for the average mean curvature in the microstructure, K_H. This rate is defined through the relation

$$K_H = \frac{d(\langle H \rangle)}{d(t^{-\frac{1}{3}})} , \tag{4}$$

where <H> is the average mean curvature of the interface. Physical reasoning suggests that the average temperature in the thermally coarsening system, <T>, corresponds to the temperature of the interface having the average curvature. Thus, the volume–fraction dependence of K_t should be similar to that for the dashed curve. This is not observed experimentally, as K_t clearly increases with V_v in both materials. (See again Figures 4 and 5.)

The curve which better fits the data is that representing the theoretical surface–area decay rate, K_S, shown as the solid curve in Figure 5. This rate is defined through the relation

$$K_S = \frac{d(S_V)}{d(t^{-\frac{1}{3}})} , \tag{5}$$

where S_V is the total interfacial area per unit volume of the mushy zone. K_S exhibits the same V_v–dependence as the experimental rate K_t, except for a slight deviation at low volume fractions. The only adjustable parameter in this fit is the vertical scale for the K_S curve, which was chosen to fit the data best. The surface energy γ calculated from this scale factor is close to the value obtained from grain–boundary groove measurements (Jones, 1973). A more precise comparison is not possible because the K_S curve in Figure 5 is based on the coarsening behavior of a collection of spheres. A statistical treatment of more general geometries (Marsh, Glicksman, and Zwillinger, 1988) indicates that K_S has a similar V_v dependence in more complex morphologies typical of mushy zones.

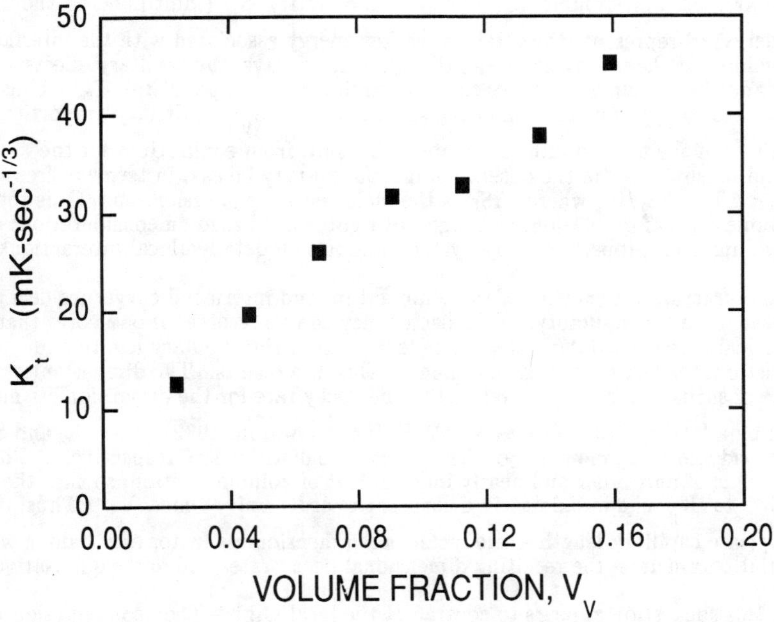

FIGURE 4. Volume fraction dependence of thermal coarsening rates in ETC.

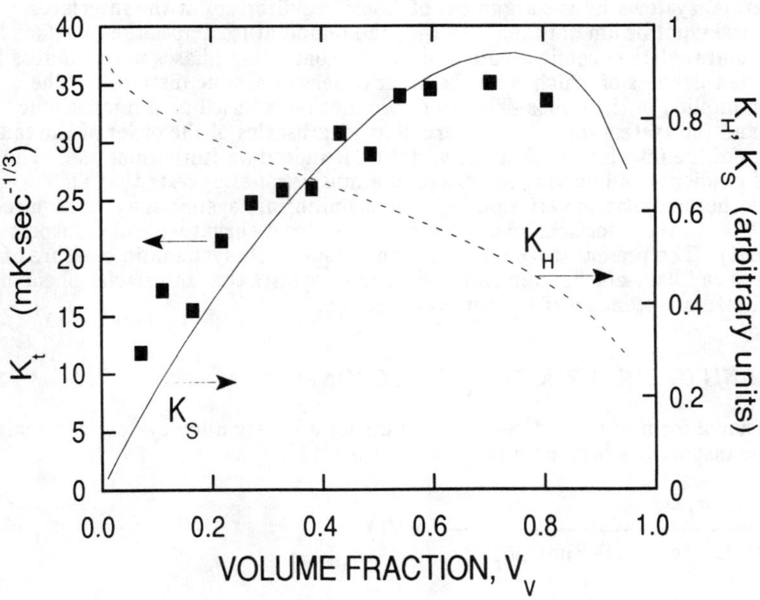

FIGURE 5. Comparison of experimental coarsening rates in α–P_4 with theory.

The volume–fraction dependence of the measured thermal decay rates indicates that the mean undercooling of the solid–liquid mixture, $<\Delta T>$, is proportional to the interfacial area per unit volume, S_V . The global quantity S_V (multiplied by the surface energy, γ) represents the total excess free energy associated with the interface per unit volume of the mushy zone. As this quantity decays, the capillary effects diminish, and the system approaches the bulk equilibrium temperature, T_m . Thus the volumetric free energy contribution of the surface area, $\gamma \cdot S_V$, is directly proportional to the shift from the bulk equilibrium state. This shift from equilibrium for the solid–liquid mushy zone (in the absence of interfaces) may be cast in terms of free energy as $<\Delta T> \cdot \Delta S_f/\Omega$, where ΔS_f is the molar entropy of fusion, and Ω is the molar volume. This formulation, although not rigorous, is based on considerations of specific (volumetric) properties of the system, and not on detailed local interactions.

The volume–fraction dependence of the temperature and interfacial curvature decay rates appear to be contradictory. This discrepancy can be resolved if one notes that the theoretical curvature decay rates are independent of the capillary length and characteristic transport time used to dimensionalize the results. The dimensionless decay rate of surface area, K_S , is related to the decay rate for the dimensionless mean curvature by a factor of $(3/2)(S_2/S_3)V_v$ (DeHoff and Iswaran, 1982), where S_2 and S_3 are the second and third moments of the particle size distribution, respectively. The ratio (S_2/S_3) is of unit order and nearly independent of volume fraction, so that the ratio of K_S to K_H is approximately a constant of order unity times V_v . Thus, if the appropriate capillary length–scale includes a dimensionless factor of V_v along with the material parameters, the resulting dimensional decay rates will agree qualitatively.

Although this suggestion appears to contradict the local Gibbs–Thomson equation by modifying the expression for the capillary length, it should be noted that the Gibbs–Thomson relationship is based on equilibrium considerations and the variation used to derive this relation is an isothermal one. It is often justified for non–isothermal systems by the argument of "local" equilibrium at the interfaces. However, local equilibrium only requires that the temperature across the interface be continuous and that the chemical potentials of the contacting phases match across the interface, the thickness of which is on the order of tens of atomic distances. The equilibrium implied in the Gibbs–Thomson formulation is based on a microscopic variation, and the system must be isothermal at lengthscales of the order of the radius of curvature of the interface so that the variation is indeed an isothermal one. The presence of gradients and moving interfaces in a mushy zone suggests that an appropriate thermodynamic variation in the nonequilibrium system may be required to relate $<\Delta T>$ to the interfacial characteristics (i.e., local curvature and transport field gradient). The present work represents the first direct, systematic measure of the magnitude of capillary effects in mushy zones, and suggests that interfacial phenomena in nonequilibrium systems bear further investigation.

COARSENING IN BINARY ALLOY MUSHY ZONES

The dimensional form of the coarsening constant for a binary alloy system coarsening by species transport has been described by Voorhees (1990) as

$$K = \frac{8T\alpha_L D\gamma}{9Q\{\ DQ/c \ - \ (1-k)mC_L\alpha_L \ \}} \cdot f(V_v) \ . \tag{6}$$

In Eq. (6), T , D , γ are the temperature, diffusion coefficient, and solid–liquid interface energy within the mushy zone, respectively; α_L and C_L are the thermal

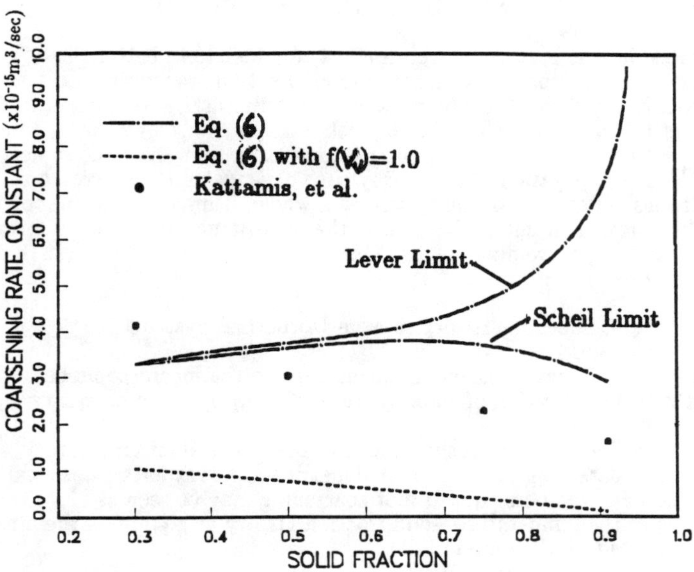

FIGURE 6. Comparison of experimental Al–4.5%Cu dendrite coarsening rates with predictions.

diffusivity and solute concentration in the liquid phase; and k and m are the distribution coefficient and liquidus slope from the phase diagram of the alloy. The final term, $f(V_V)$, is the theoretical result of Marsh (1989), based on the statistical self–similarity for a pure material mushy zone interacting with its mean field.

The coarsening constant itself (for spherically shaped particles) is a strong function of the temperature and solid fraction for an alloy, since the solid fraction as well as material properties depend on the mushy zone temperature. Figure 6 shows a comparison of coarsening constants calculated from data of Kattamis, et al. (1967) with Eq. (6) for Al–4.5%Cu primary dendrites, initially with a primary dendrite spacing of 190 μm, at four different solid fractions (and hence temperatures). The solid fraction effect, $f(V_V)$, is taken directly from Marsh (1989). The "no volume fraction effect" case sets $f(V_V) = 1.0$, and is included to help gauge the influence of the solid fraction effect on the coarsening rate constant.

Solid phase diffusion of solute has a further important effect on the predicted rate constant, through the dependence of C_L on temperature, as is seen by comparing the Scheil and lever–law results. For Al–4.5%Cu, the lever–law leads to a much higher solid fraction than might be expected at a given mushy zone temperature, and results in a considerable disparity with the data. The more realistic Scheil limit yields a prediction which is within a factor of two over the solid fraction range, which corresponds to approximately a 26% variation in the predicted lengthscale. Given the considerable uncertainty in the temperature dependence of the coarsening rate constant, these limited data are in reasonable agreement with the theoretical prediction.

DEVELOPMENT OF A MICRO/MACRO SOLIDIFICATION MODEL

The research described here is part of an overall effort to understand better the relationship among the macroscopic heat, mass, and momentum transport processes occurring during solidification and the dynamics of microstructural development in the mushy zone. It is useful to apply the statistically self–similar coarsening theory to a case in which the solid volume fraction at a point is increasing with time due to net solidification. Since the propagation of the mushy zone depends on the macroscopic heat transfer conditions which control the system as a whole, it may be possible to gain additional understanding of how coarsening affects the microstructure of a solid formed in a process, such as casting or welding.

Adaptation of Adiabatic Coarsening Theory to Non–Isothermal Systems

To couple the analysis of adiabatic coarsening phenomena to the macroscopic heat transfer associated with solidification of an alloy, three assumptions are necessary:

- At each "point" within the mushy zone, the local solid–liquid mixture experiences a coarsening process which depends on the instantaneous local solid fraction and temperature, so that coarsening may be seen as a convolution of the temperature–solid fraction history at a point as the mushy zone propagates into the liquid phase.

- The physical structure of the mushy zone (typically involving lengthscales from 1 – 100 μm) does not affect the macroscopic conduction heat transfer (typically involving lengthscales from 1 – 100 cm) associated with the solidification process. This permits the thermal field to be determined independently of the coarsening process.

- The initial lengthscale, R(0), for which the self–similar coarsening process is described by $R^3 = R^3(0) + K \cdot t$, is located sufficiently close to the liquidus isotherm that it may be considered to exist at the liquidus.

For a casting process, the final lengthscale which would result at a point in the solid may be interpreted from the third assumption above in the following useful form:

$$R^3(t_{final}) = R^3(t_{initial}) + \int_{t_c=t_{initial}}^{t_c=t_{final}} K(T,V_V)\, dt_c \, , \qquad (7)$$

where t_c is the coarsening time, beginning at $t_{initial}$ when the liquidus temperature is reached at a point and ending at t_{final} when the solidus temperature is reached. In other words the coarsening begins immediately when freezing starts and continues until the solidification process is completed.

Macroscopic Heat Transfer

The task remaining is to determinine at each point in a casting the temperature and solid fraction history as a function of time and then to proceed with an evaluation of Eq. (7). Kuklinski and Smith (1989) have provided a one–dimensional analysis for the temperature distribution and growth of the mushy zone using a double integral method (DIM). The model of Clyne and Kurz (1981) was used to relate the mushy zone temperature to the local solid fraction, by considering varying degrees of solid phase diffusion of solute locally within the mushy zone from the lever to the Scheil limits.

The analysis permits closed–form solutions for the solidus and liquidus locations and the temperature field. The results are lengthy and the details are presented elsewhere (Kuklinski, 1990).

For the case of solidification adjacent to a constant temperature boundary, the upper and lower limits appearing in the integral of Eq. (7) may be written as

$$t_{\text{initial}} = \frac{1}{\alpha}\left[\frac{x}{\zeta}\right]^2 , \tag{8a}$$

$$t_{\text{final}} = \frac{1}{\alpha}\left[\frac{x}{\beta}\right]^2 , \tag{8b}$$

where ζ and β are the solidus and liquidus "parabolic freezing constants", α is the solid phase thermal diffusivity, and x is the distance away from the constant temperature wall, $x=0$.

If there is no liquid phase superheating, ζ and β are each functions of the liquid–to–solid thermal diffusivity ratio, dimensionless liquidus and solidus temperatures, the liquidus slope and distribution coefficient from the phase diagram, a solutal Fourier number, μ , used in the Clyne and Kurz model, and the eutectic liquid fraction (for the non–lever–law cases).

Initial Lengthscale Determination

The initial lengthscale, $R(0)$, is itself a function of the macroscopic thermal environment in which it forms. Although the coarsening theory to be applied here is appropriate for an unambiguous microstructural metric, such as the volume–to–surface area ratio, or a statistically consistent inverse curvature, neither data nor theory are available. We may with reasonable justification assume that these values are proportional to a more easily interpreted microstructural scale, such as the primary and secondary dendrite arm spacings, for which data are available.

Young and Kirkwood (1975) measured primary and secondary dendrite arm spacings in directional solidification experiments on Al–4.5wt%Cu and related them to the liquidus velocity and the local temperature gradient. They represented the primary arm spacing in the form $R(0) = e_3\, G^{-e_1}\, \mathcal{V}^{-e_2}$, where e_1 , e_2 , and e_3 depend on the concentration of Cu. Both the temperature gradient G and the velocity $\mathcal{V}$ may be approximated from the macroscopic DIM solution. The velocity is simply the time rate of change of the liquidus position, viz., $\mathcal{V} = \zeta^2 \alpha / 2x$. An approximate temperature gradient which may be applied in the present situation is the liquidus–to–solidus temperature difference divided by the mushy zone thickness:

$$G = \frac{T_L - T_S}{x}\,\frac{\zeta}{\zeta - \beta} . \tag{9}$$

Both G and $\mathcal{V}$ decrease as $1/x$ for the present macroscopic cooling situation, so that the prediction of the initial primary dendrite arm spacing increases with distance from the mold wall.

The initial value of secondary arm spacing was small enough that significant coarsening of secondary arms might have occurred before an accurate measurement could be taken. As a result, an arbitrary constant value of $R(0) = 15\ \mu m$ was used for all cases studied here. This choice was found to have little influence on the final value predicted.

Discussion of Micro/Macro Mushy Zone Modeling

Figure 7 highlights the effects of variation in heat transfer rate on the evolution of initial primary dendrites, final primary dendrites and final secondary dendrites for an Al–4.5%Cu alloy. A low wall temperature (293 K) leads to a high cooling rate, for which the liquidus and solidus isotherms propagate rapidly through the casting, and the local solidification times are brief. As a result, little coarsening occurs. The final primary dendrite lengthscale then is determined almost entirely by its initial value. However, the secondary arms well away from the cooled surface increase in size approximately 7 times under the same conditions. The final solid will contain two distinct lengthscales that resulted from the primary and secondary dendrite arm formation. For a much lower cooling rate (wall temperature of 800 K), for which local solidification times are much longer, coarsening is significant. The secondary arms coarsen rapidly and are nearly as large as the primary arms in the final solid. These results show how different cooling rates may produce dramatically different solid microstructures in the same alloy.

The effects of back diffusion of solute in Al–4.5%Cu under high and low cooling conditions are examined in Figure 8. The secondary arm spacing is more sensitive to the effects of coarsening and is examined here. For a rapidly cooled casting, the assumption about the amount of diffusion in the solid phase has little impact on the size of the final secondary dendrite lengthscale in the solid. Much more coarsening will occur with the higher wall temperature. Diffusion in this case has an important effect.

The Clyne and Kurz model predicts that the solidus temperature is depressed as the solid phase diffusion is decreased from the lever limit to the Scheil limit. This results in a mushy zone bounded by a wider range of temperatures, and hence, more coarsening occurs. The Clyne and Kurz model also predicts that, if the eutectic temperature is reached, decreasing solid phase diffusion will result, not in the further widening of the temperature range of the mushy zone, but in an increased amount of latent heat release at the eutectic point. Decreasing the solid phase diffusion for such a case then lowers the latent heat release in the mushy zone and leads to less coarsening, because the isotherms in the mushy zone will propagate relatively faster. Eutectic heat release is instantaneous and does not contribute to coarsening. The net result then is that if the eutectic temperature is reached, decreasing solid phase diffusion acts to produce less coarsening.

In the case of a low cooling rate, for an Al–4.5wt%Cu alloy, both of these effects may be important. Decreasing the solid phase diffusion from the lever rule to the intermediate case ($\mu = 0.35$) caused the solidus temperature to be depressed and hence more coarsening occurs. The eutectic temperature is reached near ($\mu = 0.35$), and decreasing the solid phase diffusion further produces less coarsening as the Scheil limit is approached ($\mu = 0$).

The effect of the concentration of Cu on the primary dendrite arm spacing in the Al–Cu system is illustrated in Figure 9, for a 293 K wall temperature boundary condition. Decreasing the copper concentration leads to increased coarsening of the secondary arms. The Scheil limit predicts a significant depression of the solidus. For very dilute alloys this leads to a wide mushy zone and a significant amount of coarsening. This behavior of increased coarsening with decreased copper concentration was found in the other test cases as well. Pure materials with minor impurities often act like very dilute binary alloy systems. As a result, coarsening behavior of pure materials with minor impurities could be significant.hese results provide insight into the expected lengthscale distribution in Al–Cu alloys. This distribution will be a strong function of the cooling conditions during the solidification process. In general, the finest grain structure will be found near the mold wall where the total solidification time is relatively short. For most cooling conditions significant coarsening of the

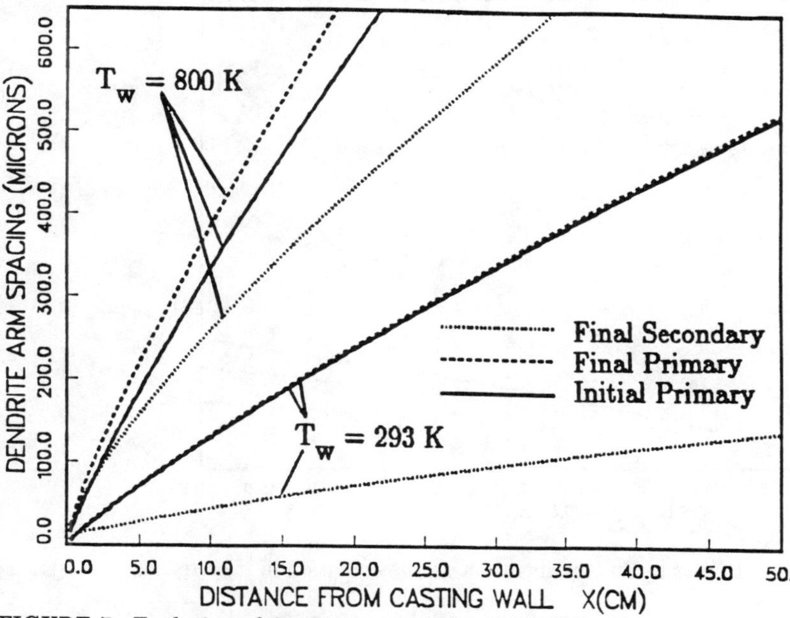

FIGURE 7. Evolution of dendrite spacings in Al–4.5%Cu.

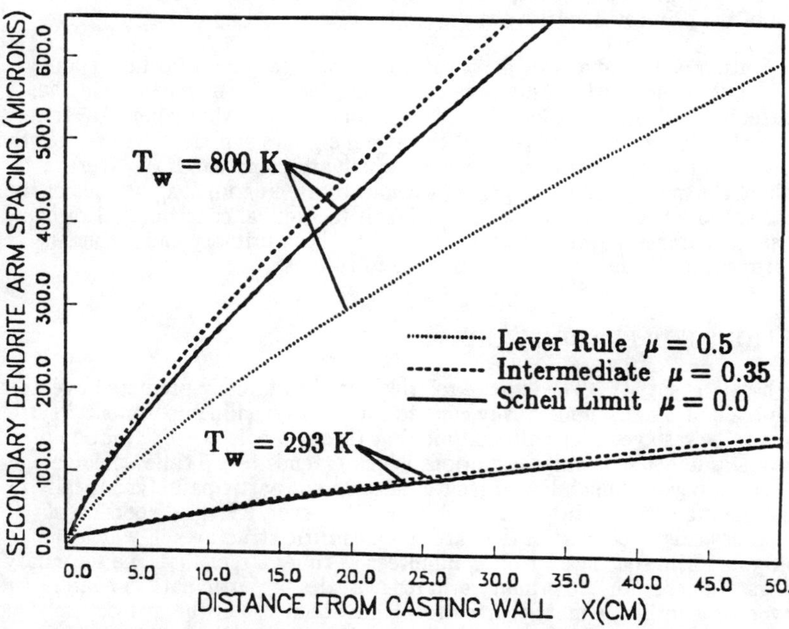

FIGURE 8. Effects of solid phase diffusion on the evolution of secondary dendrite spacing in Al–4.5%Cu.

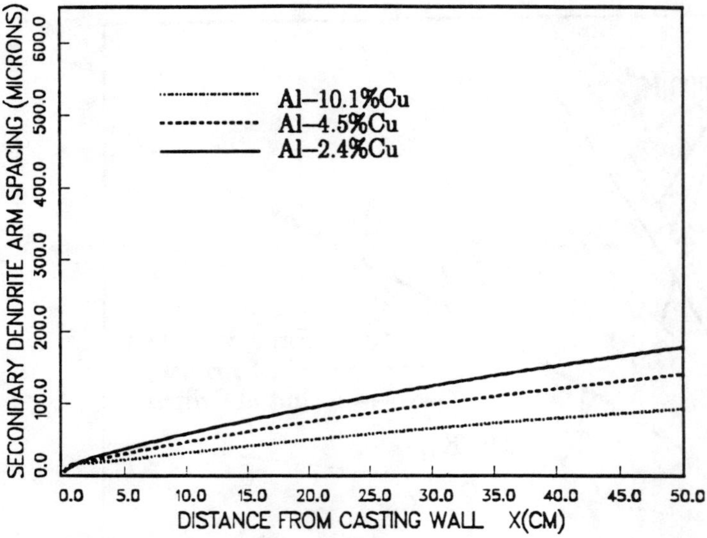

FIGURE 9. Effects of Cu concentration on the evolution of primary dendrite spacing for Al–Cu.

secondary dendrite arms will occur, whereas for similar cooling conditions minimal coarsening of the primary dendrite arms is likely.

A note of caution is needed at this point regarding the treatment of primary and secondary arm spacings with mushy zone coarsening theory. The mean–field theory of phase coarsening indicates that lengthscales which are greater than about 1.5 times the average lengthscale remain inactive until the average grows sufficiently to reduce that ratio below 1.5. Thus, primary arm spacings, which are many times the average lengthscale of the mushy zone, can persist without coarsening for long periods of time. The results shown above seem consistent with this theoretical constraint. However, since the same coarsening rate constant was used for both primary and secondary dendrite arm spacings, this consistency may be fortuitous.

SUGGESTIONS FOR FUTURE WORK

- Lengthscale Disparity. The issue of relating the theoretically predicted average length scale of a coarsening mushy zone to those observed in experimental systems has yet to be resolved. The difficulty is that theory predicts a distribution of lengthscales under selfsimilar conditions which extends to 1.5 times the average lengthscale, beyond which longer lengthscales do not participate dynamically in the coarsening process. In a real mushy zone, however, a large dispersion of lengthscales exists, because of the nature of dendritic structures in which the primary dendrite spacings are often hundreds of times larger than the secondary spacings. In order for the primary arm lengthscales to participate in coarsening, the average lengthscale must coarsen to within about 2/3 of the primary spacing. As a practical matter, this may never happen in castings which are solidified under typical foundry process conditions. Experimental data on lengthscales in mushy zones contain all observable lengthscales rather than the theoretically–required "active" lengthscales. In the future it is hoped that consistent experimental measurements will be checked against theory.

- Microstructological Measurements. Few studies have been reported in the literature to date that compare quantitative microstructural measurements (microstructological data) with coarsening predictions. Such studies would involve preparation of binary alloy mushy zones aged under known temperature–time conditions. These data would describe a microstructural "pathway" that characterizes the evolution of the mushy zone under a given set of thermal conditions.

- Micro/Macro Heat Transport. The relationship between microstructural evolution and the macroscopic heat and mass transfer which occur within the mushy zone remains an open issue. Virtually no definitive experiments have been conducted to check the influence of microscopic behavior on macroscopic transport processes. This represents a central issue for future development of engineering heat transfer codes for solidification processing.

- Verification of Micro/Macro Modeling. With the advent of micro/macro modeling (Rappaz, 1989; Rappaz and Stefanescu, 1988; and Thevoz, et al., 1989), carefully controlled net solidification experiments in well–prescribed geometries become essential. Features of the solidification process including micro– and macro–chemical segregation, phase distributions, enthalpy distributions, mushy zone thickness and dynamics, and grain size must all be compared against the model predictions. Although this might be achieved many years in the future, continued progress toward this goal remains essential for the future of engineering process design.

ACKNOWLEDGEMENT

The research reported in this paper resulted from support from the National Science Foundation under Grant Nos. DMR86–11302, DMR88–07432, and DMR89–21435 for which the authors express their appreciation.

REFERENCES

Clyne, T.W. and Kurz, W., 1981, "Solute Redistribution During Solidification with Rapid Solid State Diffusion," Metallurgical Transactions A, Vol. 12A, pp. 965–971.

DeHoff, R.T. and Iswaran, C.V., 1982, Met. Trans. A, 13A, p. 1389.

Jones, D.R.H., 1973, Phil. Mag., 27, p. 569.

Kattamis, T.Z., Coughlin, J.C. and Flemings, M.C., 1967, "Influence of Coarsening on Dendrite Arm Spacing of Aluminum–Copper Alloys," Transactions of the Metallurgical Society of AIME, Vol. 239, pp. 1504–1511.

Kuklinski, R., 1990, "The Effects of Heat Transfer on the Evolution of Microstructural Length Scales During Solidification of a Binary Alloy," PhD Thesis, Rensselaer Polytechnic Institute, Troy, NY.

Kuklinski, R. and Smith, R.N., 1989, "Use of a Double Intergral Technique to Study Mushy Zone Growth During Alloy Solidification," 1989 Heat Transfer Conference: Heat Transfer in Manufacturing and Materials Processing, Vol. HTD – 113, pp. 55–61.

Lifshitz, I.M. and Slyozov, V.V., 1961, J. Phys. Chem. Solids, 19, p. 35.

Marsh, S.P., 1989, "Kinetics of Diffusion–Limited Microstructural Coarsening," PhD Thesis, Rensselaer Polytechnic Institute, Troy, NY.

Marsh, S.P. and Glicksman, M.E., 1986, Modeling and Control of Casting and Welding Processes III, S. Kou and R. Mehrabian, Eds., The Metallurgical Society, Warrendale, PA, p. 579.

Marsh, S.P., Glicksman, M.E., and Zwillinger, D.I., 1988, Modeling and Control of Casting and Welding Processes IV , A.F. Giamei and G.J. Abbaschian, Eds., The Metallurgical Society, Warrendale, PA, p. 15.

Rappaz, M., 1989, "Micro/Macroscopic Modelling of Solidification," International Materials Review, Vol. 34, pp. 93–123.

Rappaz, M. and Stefanescu, D.M., 1988, "Modeling of Equiaxed Primary and Eutectic Solidification," Solidification Processing of Eutectic Alloys, D.M. Stefanescu, G.J. Abbaschian and R.J. Bayuzick Eds., The Metallurgical Society, pp. 133–151.

Rubinstein, E., Glicksman, M.E., Mangum, B.W., Fang, Q.T., and Singh, N.B., 1988, J. Crystal Growth, 89, p.101.

Thevoz, Ph., Desbiolles, J.L. and Rappaz, M., 1989, "Modeling of Equiaxed Microstructure Formation in Casting," Metallurgical Transactions A, Vol. 20A, pp. 311–321.

Voorhees, P.W., 1990, "Coarsening in Binary Solid–LiquidMixtures," Metallurgical Transactions A, 21A, pp. 27–37.

Young, K.P. and Kirkwood, D.H., 1975, "The Dendrite Arm Spacings of Aluminum–Copper Alloys Solified Under Steady–State Conditions," Metallurigical Transactions A, Vol. 6A, pp. 197–205.

Effects of Convection on the Solidification of Binary Mixtures

F. P. INCROPERA and R. VISKANTA
Heat Transfer Laboratory
School of Mechanical Engineering
Purdue University
W. Lafayette, Indiana 47907, USA

ABSTRACT

The process of solidification in a noneutectic binary system can be strongly influenced by convection occurring in the liquid and solid/liquid (mushy) regions. Physical features which are influenced by convection have been delineated experimentally and include development of an irregular liquidus front, remelting of solid, development of flow channels in the mushy region, and the establishment of characteristic macrosegregation patterns (regions of significantly different composition) in the final solid. In recent years, significant advancements have been made in the development of mathematical models which successfully predict the relationship of convection to the foregoing features. In this paper pertinent literature is reviewed, key results are presented, and future needs and directions are assessed.

INTRODUCTION

Physical Considerations

Solidification occurs in a wide range of manufacturing processes, and in each case successful implementation depends critically on proper control of transport mechanisms in the liquid phase. For example, many modern electro-optical technologies place stringent requirements on the homogeneity of large, single crystals to which impurities (dopants) are added to achieve desired properties. However, when grown from a melt, release of the latent heat of fusion and compositional differences between the crystal and the melt cause the rate of growth and quality of the crystal to be strongly influenced by heat and mass transfer processes. In turn, these processes affect and are affected by convection in the liquid [1]. Just as convection may influence the structure and distribution of inclusions in a crystal, it can also lead to undesirable macroscopic constituent inhomogeneities, known as macrosegregation [2], in alloyed metal castings.

Convection in a melt may originate from external agents, the density difference between phases, surface tension gradients at a free surface, and/or the action of gravity on a density gradient. Externally driven convection may be due to an imposed flow, system rotation, and/or an applied magnetic field. Density differences between the solid and liquid phases typically cause the mixture to contract during solidification, thereby inducing convection in the melt. Surface tension gradients at a free surface may arise from temperature and/or concentration gradients and may induce thermo/diffusocapillary flows. Similarly, in a gravitational field, density gradients due to temperature and/or concentration gradients may yield a buoyancy driven flow.

As determined by the phase diagram for a binary system, solidification occurs over a temperature range which depends on the composition of the melt. Solidification typically occurs dendritically in a two-phase mushy region, which is bounded by liquidus and solidus isotherms and in which solid is formed as a permeable, crystalline-like matrix. Dendritic microstructures may be *columnar,* for which there is preferred growth along dendrite stalks aligned with a dominant direction for heat flow, or *equiaxed,* for which heat flow is multidimensional and growth occurs in several directions. Spacing between dendrite arms ranges from 10 to 100 μm, and the specific morphology of the microstructures depends on the nature and composition of the melt, as well as the solidification geometry and rate.

Another important feature of solidification in a binary system relates to the redistribution of solute which is accumulated or depleted as a consequence of different solubilities in the liquid and solid phases. Solute which is rejected from solidifying dendrites accumulates in the adjoining liquid, establishing local concentration gradients. On a microscopic scale, solute may be redistributed by diffusion, resulting in constituent *microsegregation.* However, on a macroscopic scale, solute may also be redistributed by convection in the mushy region, causing constituent *macrosegregation.* Specifically, large zones may form in which the solute concentration is significantly greater or less than the intended composition.

During solidification of a noneutectic binary liquid, density gradients result from both temperature and composition gradients and may induce augmenting or opposing buoyancy forces. Depending on the orientation relative to gravity, the *solutal* contribution to the liquid density gradient may enhance or retard the *thermal* contribution. If, for example, solidification occurs along a vertical boundary, the downward buoyancy force due to the horizontal temperature gradient may be opposed (or augmented) by the upward (or downward) buoyancy force due to the composition gradient. In contrast, if a binary melt is frozen from below, rejection of a lighter component results in an unstable composition gradient which opposes the stable temperature gradient and may induce convection in the form of thin plumes ascending from the surface.

Through the extent to which they offset or dampen thermo/solutal convection, external forces may be applied to reduce or eliminate macrosegregation. For example, in metals, which are electrically conducting, magnetic fields may be used to dampen convection by providing Lorentz forces which oppose fluid motion transverse to the magnetic field vectors. In addition, for both metals and aqueous solutions, centrifugal forces induced by system rotation may be used to significantly alter fluid motion.

Literature Review: General

The most extensive body of literature concerning the effects of thermo/solutal convection on the solidification of binary mixtures pertains to metallic alloys. Early studies considered upward, downward, or horizontal solidification in metal alloys [1-3] and demonstrated that significant macrosegregation can be caused by thermally and solutally driven motions of interdendritic fluids. Results obtained for axisymmetric systems confirmed the dependence of macrosegregation on interdendritic flow [4-6].

To observe convection phenomena, experiments have also been performed in transparent salt solutions [7-12]. With freezing from below, macrosegregation was found to be strongly influenced by the ascending motion of lighter fluid from the mushy zone. The motion was driven by solutal buoyancy forces, was concentrated in discrete flow channels which appeared to originate at the growth front, and was sustained by the continuous entrainment of descending liquid into adjoining dendritic regions. Similar behavior was observed for side wall freezing, where solute fingers ejected from the advancing solidification front were identified to be the dominant segregation mechanism. Flow visualization revealed the formation of channels in the mushy region, which provided angled paths for the flow of lighter, solute-deficient fluid. It was concluded that A- and V-segregate formation in the final casting is directly attributable to the channeling effect. Side wall freezing has also been shown to yield double-diffusive layering within the melt [12].

Fewer studies have been performed to assess the effects of magnetic fields on convection and macrosegregation. Although early studies found that magnetic stirring could be used to reduce columnar structure in favor of equiaxed zones [13] and that magnetic damping delayed the transition from columnar to equiaxed growth [14], neither study considered the effects of a magnetic field on macrosegregation. Investigations of the effect of a stationary magnetic field on Czochralski crystal growth [15,16] have shown that melt convection may be reduced and a more homogeneous dopant distribution may be obtained through application of the field. Recently, Vives and Perry found that the transition to equiaxed grain structure could be advanced through magnetic stirring [17] and that fluid motion in the melt could be suppressed by a stationary magnetic field [18]. However, effects on interdendritic fluid flow and macrosegregation were not addressed.

Transport phenomena in a solidifying system may also be influenced through imposition of a centrifugal force. Cylindrical systems are of particular interest, where solidification may be induced by chilling from below or from a sidewall, while simultaneously rotating the system. However, apart from studies pertinent to the Czochralski crystal growth process, little has been done to determine the effects of rotation on convection and macrosegregation in binary systems. By using sidewall cooling to cast a Sn-Pb ingot, while simultaneously rotating the mold, Kou et al. [5] did find a significant reduction in macrosegregation, relative to that obtained without rotation. In addition, Sample and Hellawell [10] concluded that channel formation in the mushy region of $NH_4C\ell - H_2O$ and Sn-Pb systems chilled at their base could be suppressed by rotating the ingot about an axis making a small angle with the gravity vector.

Literature Review: Mathematical Models

The development of mathematical models for the solidification of binary mixtures is complicated by the coupling between velocity, temperature and composition fields within the melt and mushy regions, as well as by the coupling which exists between the two regions. Within the melt, for example, velocity, temperature, and concentration fields are mutually coupled through advection terms and buoyancy forces in the corresponding conservation equations. However, through interaction between fluid in the melt and mushy regions, conditions in one region may also be strongly influenced by conditions in the other. For example, since solidification of a binary liquid depends on the local temperature and solute concentration, penetration of bulk liquid into the two-phase mushy region will influence the solidification process. Conversely, solute rejection due to solidification in the mushy region can strongly influence buoyancy driven flows within the melt. Matters are further complicated by the highly irregular nature of the solid/liquid interfaces in the permeable mushy region, where dendritic growth precludes establishing the precise shape and movement of the interfaces.

In a numerical simulation of binary solidification, it is impractical, if not impossible, to implement a macroscopic formulation of the conservation equations which would provide a detailed description of transport phenomena in and around the complex dendritic microstructures existing in the mushy region. Length scales associated with the microstructures ($\sim$10-100 μm) are much smaller than the smallest computational control volumes ($\gtrsim$ 1 mm) which could be used to resolve macroscopic behavior in systems of practical interest. Hence, although it is desirable to predict morphology and field variables at the microscale level, realization is impractical for the foreseeable future. However, macroscopic behavior depends on dendritic morphology and any macroscopic analysis must, in some way, incorporate the effect of microstructures in the mushy region. The customary approach to macroscopic modeling therefore involves development of the conservation equations for an averaging volume which contains both solid and liquid and is much larger than microstructural features. Within this averaging volume, microstructural effects may be accommodated by introducing empirical or semi-empirical results for the interaction between phases.

The first attempts to circumvent difficulties associated with irregular dendritic growth are described in the pioneering papers of Flemings and co-workers [19-21], who computed interdendritic flows by treating the mushy zone as a porous medium for which porosity varied with solid fraction. Solute segregation in the mushy region was predicted, subject to a prescribed flow which was assumed to be due only to solidification contraction. A major extension of the model was made by Mehrabian et al. [22], who included the effect of gravity acting on a fluid of variable density, as well as solidification contraction, in performing hydrodynamic calculations for the mushy region. Darcy's law was used to describe two-dimensional velocity and pressure fields, although temperature gradients and freezing rates were prescribed and the average concentration of the solid and liquid was assumed to be constant. The model was extended to axisymmetric castings by Kou et al. [5], who also considered the effects of mold rotation, and by Ridder et al. [4], who also solved the coupled energy and momentum equations to explicitly evaluate the temperature field in the mushy region. Additional improvements to the model were made by Fujii et al. [23], who solved the coupled momentum, energy and species equations for the mushy zone. However, motion of the bulk liquid was not explicitly considered.

Although their model only allowed for liquid density variations with temperature, Szekely and Jassal [11] appear to have been the first to solve the two-dimensional momentum and energy equations for both the mushy zone and the bulk liquid. Predictions were in good agreement with experiments for $NH_4C\ell$ $-H_2O$ solutions, and an important conclusion was that convection in the bulk liquid had a marked effect on the velocity field in the mushy region. Similarly, in an extension of prior work, Ridder et al. [24] accounted for the effects of convection in the melt, where, again, the flow was driven only by temperature gradients. In both of these studies, separate conservation equations were written for the melt and mushy zones, and each region was discretized with a separate grid. The numerical solution to such a *multidomain* model requires explicit tracking of the moving liquidus interface, for which coordinate transformation or adaptive remeshing techniques must be used [25]. Although both procedures have been used to simulate isothermal phase change in pure substances, they are poorly suited for binary substances, in which solidification is not confined to a single solid/liquid (s/ℓ) interface and the liquidus interface is highly irregular. In such cases, it is preferable to use a *single domain* model for which conservation equations concurrently apply to all regions (solid, mush and liquid) and a fixed numerical grid with a single set of boundary conditions is used to effect a solution.

In recent years, there has been considerable interest in developing single domain, fixed grid models for binary solidification. In one approach [26-29], which may be termed a continuum, mixture, or one phase model, the mushy zone is viewed as a solid-liquid mixture with macroscopic properties. Conservation principles are applied to a macroscopically small control volume to obtain individual phase conservation equations which include source terms to account for interactions between phases. The phase equations are then summed to obtain mixture equations expressed in terms of mean field variables. In the original development of the mixture momentum equation [26], the restriction associated with Newton's third law was not imposed, and, on intuitive grounds, a net phase interaction force was postulated to exist. In a recent reassessment of this issue [30], Newton's law was imposed and the phase interaction force was introduced in a rigorous manner to yield a slightly different form of the momentum equation. Auxilliary relations are also needed to accommodate coupling between the thermal and solutal fields. Although such relations are typically obtained by assuming local thermochemical equilibrium and using the phase diagram for the binary system, the assumption of chemical equilibrium may be relaxed to account for local nonequilibrium associated with the weak diffusion of solute in the solid phase [29,31].

An alternative approach to single domain, fixed grid modelling of binary solidification involves the use volume-averaging procedures conventionalized by Hassanizadeh and Gray [32] to obtain separate macroscopic conservation equations for each phase [33,34]. Continuum relationships are assumed to be valid within each phase at the microscopic level, and the relations are integrated over a control volume which is large relative to characteristic microstructures. The resulting conservation equations are expressed in terms of macroscopic quantities, and phase interactions are described by integrals evaluated over the microscopic interfacial surfaces. Although governing equations based on the continuum and volume-average approaches are mutually consistent [30], the more direct manner in which the volume-average method relates microscopic and macroscopic parameters may render it physically more appealing and better suited for the advancement of micro-macroscopic models [35-37].

Rappaz [35] advocates a micro-macroscopic model in which microscopic features are linked to a macroscopic model of transport phenomena through expressions for the solid volume fraction, while Ni and Beckermann [36] develop a model for which coupling of microscopic phenomena to the macroscopic transport equations is much more extensive. Following the procedures of Drew [38], Ni and Beckermann develop a macroscopic model which can accommodate complete nonequilibrium (mechanical, thermal and chemical) by solving separate phase conservation equations for phase velocities, temperatures and compositions, rather than summing the phase equations to obtain mixture relations. Phase interactions are modelled using constitutive relations for interfacial stresses and fluxes, which provide the specific link between micro-morphological conditions and the macroscopic model. In particular, the constitutive relations are couched in terms of a s/ℓ interfacial area concentration (surface area of microstructures per unit volume), an interfacial drag coefficient, and interfacial heat and mass transfer convection coefficients.

Using single domain models, recent computational studies have revealed the significant effect which coupled convection in the melt and mushy regions can have on binary solidification. For thermal/solutal convection in a rectangular cavity with differentially cooled sidewalls, calculations and supporting experiments have revealed important features such as liquidus interface irregularities, remelting and channeling in the mushy zone, and double-diffusive convection in the melt [33,39-41], as well as the establishment of characteristic macrosegregation patterns [42,43]. Pronounced double-diffusive convection effects have also been determined for a horizontal, annular cavity with solidification induced at the inner surface [44,45]. The effects of thermo/diffuso capillary convection [46,47] and thermal/solutal mixed convection [48,49] have been theoretically and experimentally considered for rectangular cavities and vertical channels, respectively. Although micro-macroscopic models have recently been used to compute binary solidification in one, two and three-dimensional castings, convection effects were neglected. Results of the calculations have been reviewed by Viskanta [37].

EFFECTS OF CONVECTION: SELECTED RESULTS

In this section selected experimental and predicted results obtained by the authors and their students are reviewed. The predictions are based on single domain, fixed grid macroscopic models for which assumptions, governing equations, and numerical procedures are described elsewhere [25,26,33]. The following results highlight some of the rich and robust characteristics of convection in binary solidification, as well as the effects of convection on the nature of the final casting.

Solidification in Cavities with Thermal/Solutal Convection due to Differential Thermal Conditions at Opposing Walls

Solidification induced by chilling one of the vertical walls of a 47.6 mm tall by 38.1 mm wide rectangular cavity has been studied both experimentally and theoretically for an aqueous $NH_4C\ell$ solution [33]. With an initial water mass fraction of $f_o^{H_2O}$

= 0.700 and a temperature of T_o = 34.4 ° C, one of the vertical walls was chilled to T_c = 0 ° C, while the opposite wall was maintained at the initial temperature ($T_h = T_o$). Because T_c exceeded the eutectic temperature of T_e = -15.4 ° C, solid was only formed in a mushy region.

Development of the liquidus front and double-diffusive convection in the melt are revealed by the shadowgraphs of Fig. 1. Initially, some of the solid dendrites which form at the chilled right wall are detached due to shear forces imposed by thermally driven recirculation in the melt and fall to the bottom. After approximately 3 minutes, however, the dendrites become firmly attached and solidification proceeds almost exclusively from the right wall (Fig. 1a). Lighter, water-enriched liquid resulting from solidification ascends through the mushy region to the top of the cavity, as well as along the liquidus interface. Accumulation of compositionally lighter (water-rich) but thermally heavier (colder) fluid in a horizontal layer at the top of the test cell is conducive to establishment of a diffusive interface which separates the layer from underlying fluid which is compositionally heavier and thermally lighter. As the top layer grows, the interface moves downward, and at 15 minutes (Fig. 1b) a second, weaker interface has formed above the first. However, the weaker interface is eventually eroded, and the original interface continues to propagate downward (Fig. 1c), until it also begins to erode at t $\gtrsim$ 100 min. The accumulation of water-rich fluid above a double-diffusive interface depresses the local liquidus temperature, rendering solidification more difficult and causing solid formed earlier in the solidification process to be remelted. Remelting is manifested by less pronounced growth of the liquidus interface in upper regions of the test cell, as well as by irregularities in the overall shape of the liquidus front. Supporting calculations revealed the development of clockwise recirculating flows in the melt above and below a double-diffusive interface, with counter-clockwise recirculation occurring in the mushy region.

Calculations and experiments performed for the solidification of $NH_4C\ell - H_2O$ in a 100 mm tall by 25 mm wide cavity have revealed similar trends with respect to double-diffusive convection, remelting, and liquidus front development [39-41]. An extension of this work [42] also revealed the significant effect which solutally driven,

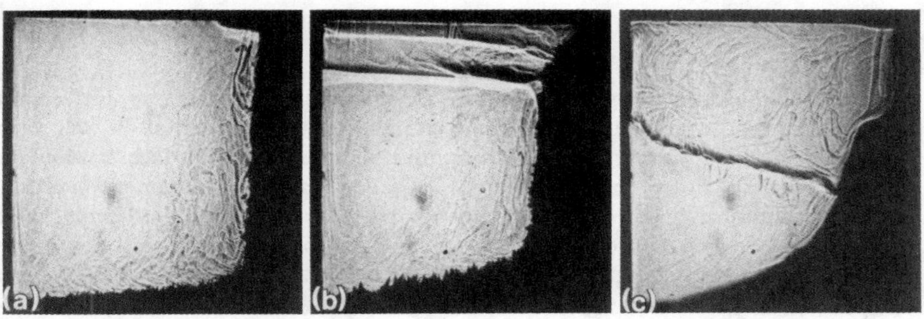

FIGURE 1. Shadowgraphs showing mushy region growth and development of double-diffusive layers for solidification of $NH_4C\ell - H_2O$ in a rectangular cavity: (a) t = 3 min; (b) t = 15 min; (c) t = 100 min.

mushy region flow can have on the pattern of macrosegregation in the final casting. Figure 2 illustrates predictions of the velocity field and macrosegregation profiles when the left sidewall is chilled to $T_c = -30\,^{\circ}C$, while the right wall is adiabatic.

Initial conditions corresponded to $T_o = 38\,^{\circ}C$ and $f_o^{H_2O} = 0.700$. With $T_c < T_e$, both solid (s) and mushy ($s + \ell$) regions are formed and complete solidification occurs at t = 6540s. At t = 180s (Fig. 2a), solutally driven upflows of H_2O-rich interdendritic fluids in the mushy region are seen to penetrate the liquidus interface at the top of the cavity, creating the region of water-enriched fluid revealed by the macrosegregation pattern at 240s (Fig. 2d). At 240s, the velocity field reveals discrete flow channels in the mushy region, which originate from local remelting due to water enrichment and are highlighted by the dashed lines. The channels result in subsequent development of water-rich bands ("A" segregates) in the mushy region along the flow paths (Fig. 2e), and additional channels (Fig. 2c) and segregates (Fig. 2f) form with increasing time. Sustained discharge of interdendritic fluids into the bulk liquid (ℓ) results in downward progression of the water-rich zone, and complete solidification (Fig. 2g) is characterized by a large "cone" segregate extending along the right (adiabatic) wall from the top to the bottom of the cavity, as well as by "A" segregates extending downward and outward from the cone.

Calculations and experiments have also been performed for aqueous Na_2CO_3 in a horizontal annular cavity, with solidification induced at an inner cylindrical surface [44,45]. Double-diffusive convection effects were found to be extensive, and representative features are shown in Fig. 3. Flow conditions were inferred by injecting fluorescent dye along the solidification front and illuminating the dye with a sheet of light from an argon-ion laser. Due to the sustained rejection of cold, water-rich fluid into the melt, double-diffusive interfaces and convection layers develop to the side of the liquidus front. Convection cells between double-diffusive interfaces are well formed and are characterized by a thermally induced, clockwise recirculation. Fluid is warmed as it flows along the bottom interface toward the outer wall and cooled as it returns along the top interface. Hence, convection layers are inclined slightly upward toward the outer wall. Rejection of water-rich fluid is also manifested by the ejection of mushroom-capped plumes from the top of the cylinder.

Effects of Thermocapillary Convection

In most solidification processes, the melt is characterized by a free surface and convection may be influenced by related surface tension gradients, as well as by thermal/solutal buoyancy forces. The combined influence of surface tension and buoyancy driven flows on solidification in a one-gravity environment has been assessed theoretically and experimentally for $NH_4C\ell - H_2O$ in a 20 mm square cavity, with one-side wall chilled ($T_c = -23\,^{\circ}C$) and the opposing wall maintained at the initial melt temperature ($T_h = T_o = 20\,^{\circ}C$) [46,47]. The initial composition of $f_o^{H_2O} = 0.730$ provided hypoeutectic conditions, for which lighter, water-rich fluid was rejected interdendritically.

Results of the calculations are shown in the velocity vector and streamline plots of Fig. 4, where solidus and liquidus fronts have been designated by solid lines on the velocity field. Early in the solidification process, flow is characterized by a large

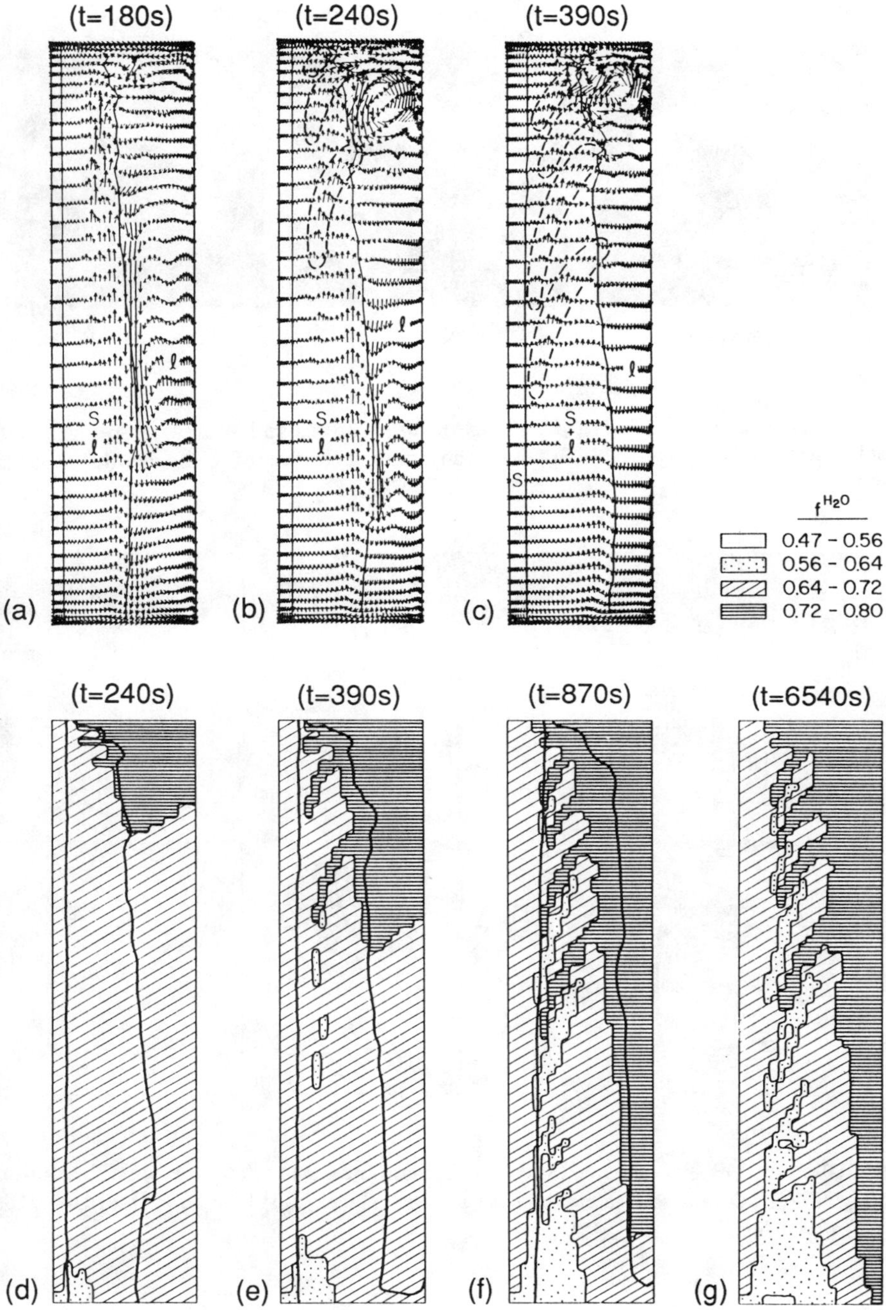

FIGURE 2. Sequential development of velocity vectors and macrosegregation profiles for solidification of $NH_4C\ell - H_2O$ in a rectangular cavity.

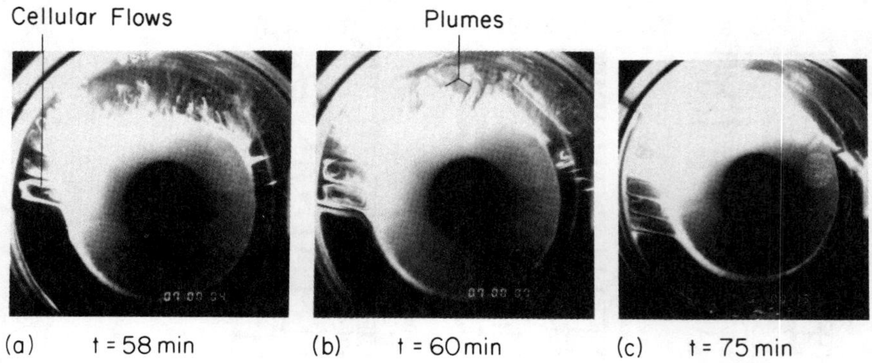

FIGURE 3. Double-diffusive interfaces, cellular flows, and plumes associated with solidification of $Na_2CO_3 - H_2O$ at the inner cylinder of a horizontal annular cavity.

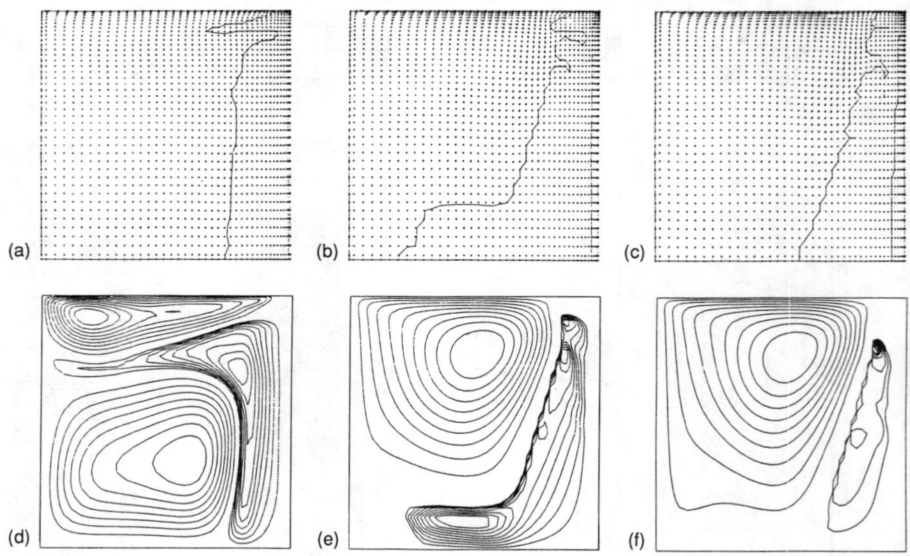

FIGURE 4. Velocity vector (a-c) and streamline (d-f) plots for solidification of $NH_4C\ell - H_2O$ in a square cavity with a free surface: (a,d) t = 30 s; (b,e) t = 210s; (c,f) t = 644s.

clockwise rotating cell in the melt and a counter-clockwise rotating cell in the mushy region. The cell in the melt is influenced by surface tension forces, which significantly enhance flow along the free surface, particularly near the liquidus front. Liquidus front development is thereby inhibited at the top of the mold, as is penetration of the liquidus by water-rich, interdendritic fluid ascending in the mushy region. However, as shown in Fig. 4a,d, the momentum of this fluid is sufficient to cause outflow from the mushy region to occur below the interface. The outflow of interdendritic fluid divides the melt into two clockwise rotating cells. While the lower cell is driven by thermal buoyancy, the upper cell is driven by both surface tension gradients and thermal buoyancy. The extremely irregular liquidus front morphology at the top of the cavity is due to features of the upper cellular flow. Warm, water-deficient fluid advected by this cell reduces solid growth at the free surface. However, with a Lewis number which is much larger than unity, the fluid cools more rapidly than it is enriched with water, as it interacts with the mushy region and is turned 180 deg. The liquidus temperature therefore remains large, providing a local ambient which is conducive to expansion of the liquidus front. Retardation of solidification just below this expansion is due to the reduced liquidus temperature in the underlying solutally driven outflow from the mushy region.

With increasing time, water-rich fluid from the intermediate cell is entrained into the top cell. The attendant liquidus temperature depression increases the strength of thermal buoyancy and surface tension forces in the top cell, thereby increasing the size and strength of the cell (Fig. 4b,e). Enhanced motion in the top cell has displaced the location of pronounced liquidus expansion downward, and due to reductions in the mushy region permeability and the strength of the solutally driven cell, the cell is once again confined to the mushy region. At near steady-state conditions (Fig. 4c,f), flow in the melt is dominated by a single cell and water enrichment of the entire liquid has enhanced remelting, particularly near the bottom of the cavity where much of the protrusion has been eliminated. Recirculation in the melt is driven by thermal buoyancy and surface tension forces, and the incidence of warm fluid from the heated sidewall is responsible for the final liquidus shape. Although flow visualization confirmed many of the foregoing predictions [47], significant differences were associated with the chronological pattern of cellular flow development and the time required to achieve steady-state.

Bottom-Chilled, Directional Solidification

It is generally understood that unidirectional (vertically upward), binary solidification can induce finger-type, double-diffusive convection, which is responsible for a severe form of macrosegregation known as freckling [8,50]. The freckles have been linked to flow channels which develop in the mushy region and from which ascending plumes penetrate the overlying bulk liquid. From an experimental study involving an aqueous $NH_4C\ell$ model alloy [10], it was concluded that freckle nucleation occurs at the growth (liquidus) front and that, concomitant with establishment of an ascending plume, there is downward propagation of the channel toward the chilled surface. Although morphological and hydrodynamic instabilities associated with unidirectional solidification have been analyzed [51-53], it is only recently that a macroscopic model has been used to predict detailed transport phenomena and, in particular, to delineate the origin of freckles [43].

Using the continuum model, calculations have been performed for directional solidification of superheated $NH_4C\ell-H_2O$ in a cylindrical mold ($H \approx 100$ mm, $D \approx 125$ mm) with an initial temperature and salt mass fraction of $T_o = 50\,^\circ C$ and $f_o^{NH_4C\ell} = 0.32$, respectively [43]. With a chill plate temperature ($T_c = -14\,^\circ C$)

slightly larger than the eutectic temperature and a liquidus temperature of T_{liq} = 42 °C, conditions (an extended freezing range and gradual solidification) favored the development of freckles. The calculations revealed that, following a period of diffusion-dominated solidification ($t \lesssim 1$ min), unstable liquidus interface conditions were manifested by nonplanar isotherms and an irregular front morphology (Fig. 5a) which became more distorted with increasing time (Fig. 5b). Figure 5b reveals the nucleation of multiple channels along the liquidus front, where each channel correlates with an ascending plume of cold, $NH_4C\ell$ depleted fluid. Channel nucleation results from a perturbation in the local fluid composition which, in sequence, causes a reduction in the liquidus temperature, remelting, an increase in the permeability, and hence a local reduction in the flow resistance of the mushy region. This behavior is consistent with the perturbation sequence which Sample and Hellawell [10] suggested is necessary for nucleation of a channel.

With increasing time, there is continued growth of the mushy region, downward propagation of the channels, and large scale fluid motion throughout the melt (Fig. 6a,b). Fluid from the melt penetrates the liquidus front and enters the mushy region to sustain the outflow of interdendritic fluid at each channel. Fluid moves downward and towards each channel, seeking a path of least resistance. Velocities in the mushy region are typically 2 to 3 orders of magnitude smaller than those in the mush. The channel locations correspond to spikes in the otherwise nearly uniform $NH_4C\ell$ composition (Fig. 6c) and hence represent sites of local freezing point depression. As colder solute ($NH_4C\ell$) depleted liquid in a channel ascends into warmer, solute-richer regions, it acquires energy far more readily than solute and becomes more resistant to freezing.

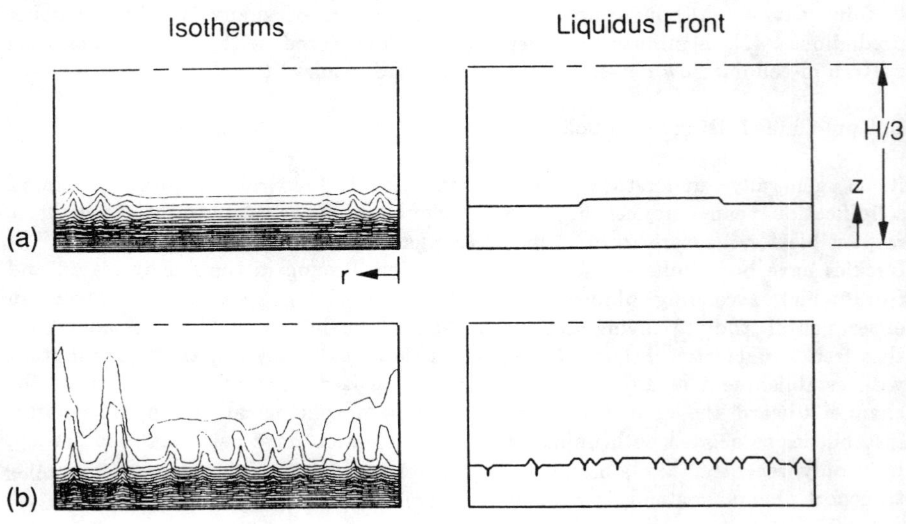

FIGURE 5. Predicted isotherms and liquidus fronts for directional solidification of $NH_4C\ell$ $-H_2O$: (a) = 1.5 min and (b) t = 2.5 min.

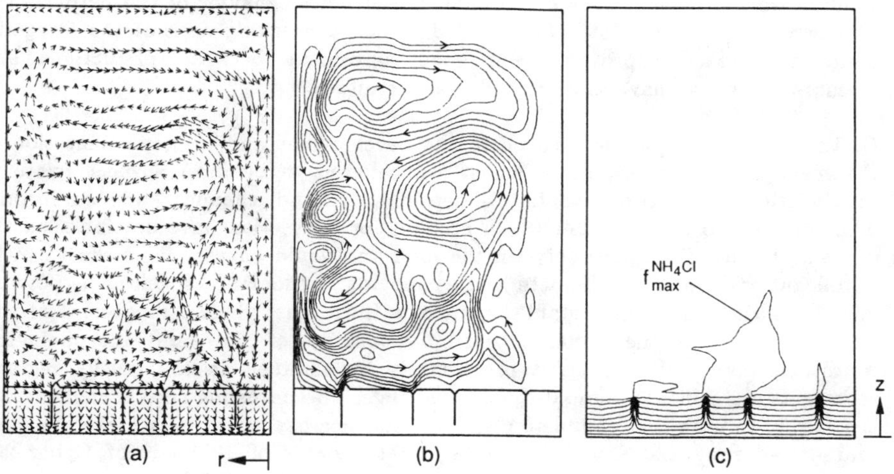

FIGURE 6. Predictions for directional solidification of $NH_4C\ell - H_2O$ at t = 10 min: (a) velocity vectors, (b) streamlines, and (c) liquid isocomposition lines ($f_\ell^{NH_4C\ell}$).

SUMMARY

Macroscopic models which account for convection have recently been extremely successful in predicting important features of solidification in binary systems. Such features, which had heretofore been observed but had eluded prediction, include: (i) double-diffusive convection in the melt; (ii) remelting, with attendant formation of channels in the mushy region and irregularities in the liquidus interface; and (iii) macrosegregation patterns such as A-segregates and freckles. However, despite such successes, difficulties have been experienced in achieving good quantitative agreement between predictions and measurables, such as solidus and liquidus front locations and temperature and composition fields [33,41,45]. Such difficulties may be a consequence of several model assumptions, as well as deficiencies in the requisite material property data base.

Model predictions depend strongly on inputs related to thermophysical properties of the solid and liquid phases (including their temperature and composition dependence), characteristics of the equilibrium phase diagram, and the permeability of the mushy region. Even for common surrogate alloys, such as aqueous solutions of $NH_4C\ell$ and Na_2CO_3, large gaps exist in the thermophysical property data base. Although characterization of mixture phase diagrams appears to be reasonably well established, small errors in key features, such as the slope of the liquidus interface, can strongly influence predicted results [41,49]. Predictions are also extremely sensitive to the permeability, which represents the major link between the macroscopic conservation equations and microscopic morphology. Unfortunately, there is considerable uncertainty in the dependence of permeability on parameters such as primary/secondary dendrite arm spacing, liquid volume fraction and solidification rate, each of which varies spatially and temporally during solidification. Moreover, the directional (anisotropic) behavior of the permeability is poorly understood. Until the knowledge base concerning each of the foregoing issues is greatly improved, even good agreement between predictions and measurements may have to be regarded as fortuitous.

Of the many assumptions invoked in the development and/or implementation of the macroscopic models, those which are believed to be most suspect relate to solidification contraction, spatial dimensionality, and chemical nonequilibrium. Thus far, macroscopic models which account for convection have not included the effect of density differences between the solid and liquid phases. This effect can be significant and would be characterized by moving boundaries, as well as by possible void formation and hence establishment of a third (gas) phase. Treatment of this effect, as well as the actual three-dimensionality of flow and morphological conditions (as manifested most prominently by channels in the mushy region), represents a significant computational challenge. In addressing such a challenge, it is sobering to note that complete grid independence has yet to be achieved in fixed grid, two-dimensional simulations using on the order of 10 hours of Cyber 205 supercomputer time.

The assumption of chemical (compositional) equilibrium at s/ℓ interfaces in the mushy region presumes that, locally, solute is completely mixed within both the solid and liquid phases. In principle, satisfaction of such a requirement necessitates infinite solute diffusivity in the solid, as well as liquid, phase. However, due to the extremely inefficient diffusion of solute in the solid, dendrites are not chemically saturated and an equilibrium freezing assumption may poorly represent local solute segregation between solid and liquid phases. A first attempt at addressing this issue [29] has involved use of the nonequilibrium Scheil equation (in lieu of the equilibrium Lever rule) to relate the solute composition of the mixture to compositions of the two phases. Although predicted macrosegregation patterns varied only slightly with interchange of the equilibrium and nonequilibrium expressions in a macroscopic mixture model, the issue is by no means resolved and nonequilibrium effects may, in fact, be significant.

Further development of the recently proposed micro-macroscopic models and, in particular, of the substantial knowledge base required for implementation of the models offers some hope for obtaining improved quantitative agreement between measured and predicted results. At the very least, such models, should contribute to an improved understanding of the relationship between macroscopic behavior and microstructural conditions and of the degree of thermo/chemical nonequilibrium. However, impediments to expedient and reliable implementation of such models should not be underestimated. For example, as noted by Viskanta [37], implementation of the Ni and Beckermann model [36] would require an extensive knowledge base for the influence of microstructural features on the area concentration and the drag and convection coefficients used to evaluate phase interactions. Acquisition of such information is not a trivial experimental task, particularly in view of the fact that microscopic morphology varies spatially and temporally as solidification occurs, as well as with the initial and boundary conditions of the process. Hence, the area concentration and the interaction coefficients will depend on overall process conditions and, for prescribed conditions, will be a complex function of position and time. Also, treatment of thermo/chemical nonequilibrium effects would require additional energy and species conservation equations, thereby significantly increasing computational requirements.

Although significant advancements have been made in resolving the effects of convection on binary solidification, much remains to be done before such effects can be accurately modelled for technically important materials and casting conditions.

ACKNOWLEDGEMENTS

The authors wish to acknowledge support of their research on binary solidification by the National Science Foundation under Grant CBT-8313573 and by the Department of Energy under Grant DEFG02-87ER13759.

REFERENCES

1. Sharp, R.M. and Hellawell, A., Solute Distribution at Non-Planar Solid-Liquid Growth Fronts, *J. Crystal Growth*, vol. 8, pp. 29-32, 1970.

2. Streat, N. and Weinberg, F., Macrosegregation during Solidification Resulting from Density Differences in the Liquid, *Metall. Trans.*, vol. 5, pp. 2539-2548, 1974.

3. Fisher, K.M. and Hunt, J.D., Observation on the Nature and Extent of Gravitational Interdendritic Fluid Flow Mechanism to Produce Appreciable Macrosegregation, in *Solidification and Casting of Metals*, Proc. Int. Conf., Sheffield, The Metals Society, 1979.

4. Ridder, S.D., Reyes, F.C., Chakravorty, S., Mehrabian, R., Nauman, J.D., Chen, J.H., and Klein, H.J., Steady-State Segregation and Heat Flow in ESR, *Met. Trans. B.*, vol. 9B, pp. 415-425, 1978.

5. Kou, S., Poirer, D.R. and Flemings, M.C., Macrosegregation in Rotated Remelted Ingots, *Met. Trans. B.*, vol. 9B, pp. 711-719, 1978.

6. Ridder, S.D., Kou, S. and Mehrabian, R., Effect of Fluid Flow on Macrosegregation in Axisymmetric Ingots, *Met. Trans. B*, vol. 12B, pp. 435-447, 1981.

7. McDonald, R.J. and Hunt, J.D., Convective Fluid Motion within the Interdendritic Liquid of a Casting, *Trans. TMS-AIME*, vol. 1, pp. 1787-1788, 1970.

8. Copley, S.M., Giamei, A.F., Johnson, S.M. and Hornbecker, M.F., The Origins of Freckles in Unidirectionally Solidified Castings, *Met. Trans.*, vol. 1, pp. 2193-2204, 1970.

9. Asai, S. and Muchi, I., Theoretical Analysis and Model Experiments on the Formation Mechanism of Channel-Type Segregation, *Trans. Iron & Steel Inst. Japan*, vol. 18, pp. 90-98, 1978.

10. Sample, A.K. and Hellawell, A., The Mechanism of Formation and Prevention of Channel Segregation during Alloy Solidification, *Met. Trans. A,* vol. 15A, pp. 2163-2173, 1984.

11. Szekely, J. and Jassal, A.S., An Experimental and Analytical Study of Solidification of a Binary Dendritic System, *Met. Trans. B*, vol. 9B, pp. 389-398, 1978.

12. Thompson, M.E. and Szekely, J., Mathematical and Physical Modelling of Double-Diffusive Convection of Aqueous Solutions Crystallizing at a Vertical Wall, *J. Fluid Mech.*, vol. 187, pp. 409-433, 1988.

13. Cole, G.S. and Bolling, G.F., Augmented Natural Convection and Equiaxed Grain Structure in Castings, *TMS-AIME*, vol. 236, pp. 1366-1368, 1966.

14. Uhlmann, D.R., Seward, T.P., III and Chalmers, B., The Effect of Magnetic Fields on the Structure of Metal Alloy Castings, *TMS-AIME*, vol. 236, pp. 527-531, 1966.

15. Hirata, H. and Inoue, N., Macroscopic Axial Dopant Distribution in Czochralski Silicon Crystals Grown in a Vertical Magnetic Field, *Jap. J. Appl. Phys.*, vol. 24, pp. 1399-1403, 1985.

16. Kim, K.M and Smetana, P., Striations in CZ Silicon Crystals Grown under Various Axial Magnetic Field Strengths, *J. Appl. Phys.*, vol. 58, pp. 2731-2735, 1985.

17. Vives, C. and Perry, C., Effects of Electromagnetic Stirring during the Controlled Solidification of Tin, *Int. J. Heat Mass Transfer*, vol. 29, pp. 21-33, 1986.

18. Vives, C. and Perry C., Effects of Magnetically Damped Convection during the Controlled Solidification of Metals and Alloys, *Int. J. Heat Mass Transfer*, vol. 30, pp. 479-496, 1987.

19. Flemings, M.C. and Nereo, G.E., Macrosegregation, Part I, *Trans. AIME*, vol. 239, pp. 1449-1461, 1967.

20. Flemings, M.C., Mehrabian, R and Nereo, G.E., Macrosegregation, Part II, *Trans. AIME*, vol. 242, pp. 41-49, 1968.

21. Flemings, M.C. and Nereo, G.E., Macrosegregation, Part III, *Trans. AIME*, vol. 242, pp. 50-55, 1968.

22. Mehrabian, R., Keane, M. and Flemings, M.C., Interdendritic Fluid Flow and Macrosegregation: Influence of Gravity, *Met. Trans.*, vol. 1, pp. 1209-1220, 1970.

23. Fujii, T., Poirer, P.R. and Flemings, M.C., Macrosegregation in a Multi-Component Low Alloy Steel, *Met. Trans. B*, vol. 10B, pp. 331-339, 1979.

24. Ridder, S.D., Kou, S., and Mehrabian, R., Effects of Fluid Flow on Macrosegregation in Axisymmetric Ingots, *Met. Trans. B*, vol. 9B, pp. 435-447, 1981.

25. Viskanta, R. and Beckermann, C., Mathematical Modeling of Solidification, *Interdisciplinary Issues in Materials Processing and Manufacturing,* S.K. Samanta, R. Komanduri, R. McMeeking, M.M. Chen and A. Tseng, eds., vol. 2, pp. 501-526, ASME, New York, 1987.

26. Bennon, W.D. and Incropera, F.P., A Continuum Model for Momentum, Heat and Species Transport in Binary Solid-Liquid Phase Change Systems: I. Model Formulation, *Int. J. Heat Mass Transfer*, vol. 30, pp. 2161-2170, 1987.

27. Bennon, W.D. and Incropera, F.P., Numerical Analysis of Binary Solid-Liquid Phase Change using a Continuum Model, *Numerical Heat Transfer*, vol. 13, pp. 277-296, 1988.

28. Voller, V.R. and Prakash, C., A Fixed Grid Numerical Modelling Methodology for Convection-Diffusion Mushy Region Phase-Change Problems, *Int. J. Heat Mass Transfer,* vol. 30, pp. 1709-1719, 1987.

29. Voller, V.R., Brent, A.D., and Prakash, C., The Modelling of Heat, Mass and Solute Transport in Solidificaton Systems, *Int. J. Heat Mass Transfer,* vol. 32, pp. 1719-1731, 1989.

30. Prescott, P.J., Incropera, F.P., and Bennon, W.D., Modeling of Dendritic Solidification Systems: Reassessment of the Continuum Momentum Equation, *Int. J. Heat Mass Transfer,* In Review, 1990.

31. Rappaz, M. and Voller, V., Modeling of Micro-Macrosegregation in Solidification Processes, *Metall. Trans. A,* vol. 21A, pp. 749-753, 1990.

32. Hassanizadeh, M. and Gray, W., General Conservation Equations for Multiphase Systems: 1. Averaging Procedures, *Adv. Water Resour.*, vol. 2, pp. 131-144, 1979.

33. Beckermann, C. and Viskanta, R., Double-diffusive Convection during Dendritic Solidification of a Binary Mixture, *Physico Chem. Hydro.,* vol. 10, pp. 195-213, 1988.

34. Ganesan, S. and Poirer, D.R., Conservation of Mass and Momentum for the Flow of Interdendritic Liquid during Solidification, *Met. Trans. B,* vol. 21B, pp. 173-181, 1990.

35. Rappaz, M., Modelling of Microstructure Formation in Solidification Processes, *Int. Mat. Rev.*, vol. 34, pp. 93-123, 1989.

36. Ni, J. and Beckermann, C., A Two-Phase Model for Mass, Momentum, Heat and Species Transport during Solidification, 5th AIAA/ASME Thermophysics and Heat Transfer Conference, Seattle, Washington, 1990.

37. Viskanta, R., Mathematical Modeling of Transport Phenomena during Solidification of Binary Systems, 22nd Int. Center Heat Mass Trans. Symposium on Manf. and Mat. Processing, Dubrovnik, Yugoslavia, 1990.

38. Drew, D.A., Mathematical Modeling of Two-Phase Flow, *Ann. Rev. Fluid Mech.,* vol. 15, pp. 261-291, 1983.

39. Bennon, W.D. and Incropera, F.P., A Continuum Model for Momentum Heat and Species Transport in Binary Solid Liquid Phase Change Systems: II. Application to Solidification in a Rectangular Cavity, *Int. J. Heat Mass Transfer*, vol. 30, pp. 2171-2187, 1987.

40. Christenson, M.S. and Incropera, F.P., Solidification of an Aqueous Ammonium Chloride Solution in a Rectangular Cavity-I. Experimental Study, *Int. J. Heat Mass Transfer*, vol. 32, pp. 47-68, 1989.

41. Christenson, M.S., Bennon, W.D., and Incropera, F.P., Solidification of an Aqueous Ammonium Chloride Solution in a Rectangular Cavity-II. Comparison of Predicted and Measured Results, *Int. J. Heat Mass Transfer* vol. 32, pp. 69-80, 1989.

42. Bennon, W.D. and Incropera, F.P., The Evolution of Macrosegregation in Statically Cast Binary Ingots, *Metallurgical Transactions B,* vol. 18B, pp. 611-616, 1987.

43. Neilson, D.G. and Incropera, F.P., Unidirectional Solidification of a Binary Alloy and the Effects of Induced Fluid Motion, *Int. J. Heat Mass Transfer,* In Press, 1990.

44. Christenson, M.S. and Incropera, F.P., Experiments on Solidification of an Aqueous Sodium Carbonate Solution in a Horizontal Cylindrical Cavity, *ASME Transactions, J. Heat Transfer*, vol. 111, pp. 998-1005, 1989.

45. Neilson, D.G., Incropera, F.P., and Bennon, W.D., Numerical Simulation of Solidification in a Horizontal Cylindrical Annulus Charged with an Aqueous Salt Solution, *Int. J. Heat Mass Transfer*, vol. 33, pp. 367-380, 1990.

46. Incropera, F.P., Engel, A.H.H., and Bennon, W.D., Numerical Analysis of Binary Solid-Liquid Phase Change with Buoyancy and Surface Tension Driven Convection, *Numerical Heat Transfer, Part A*, vol. 16, pp. 407-427, 1989.

47. Engel, A.H.H. and Incropera, F.P., Solidification of a Binary Mixture in a Square Cavity with a Free Surface, *Warme und Stoffubertragung*, vol. 24, pp. 279-288, 1989.

48. Bennon, W.D. and Incropera, F.P., Numerical Simulation of Binary Solidification in a Vertical Channel with Thermal and Solutal Mixed Convection, *Int. J. Heat Mass Transfer*, vol. 31, pp. 2147-2160, 1988.

49. Bennon, W.D. and Incropera, F.P., An Experimental Investigation of Binary Solidification in a Vertical Channel with Thermal and Solutal Mixed Convection, *ASME Transactions, J. Heat Transfer*, vol. 111, pp. 706-712, 1989.

50. Giamei, A.F. and Kear, B.H., On the Nature of Freckles in Nickel Base Superalloys, *Met. Trans.,* vol. 1, pp. 2185-2192, 1970.

51. McFadden, G.B., Rehm, R.G., Coriell, S.R., Chuck, W., and Morrish, K.A., Thermosolutal Convection during Dendritic Solidification, *Met. Trans. A,* vol. 15A, pp. 2125-2137, 1984.

52. Jenkins, D.R., Nonlinear Analysis of Convective and Morphological Instability during Solidification of a Dilute Binary Alloy, *Phys. Chem. Hydro.,* vol. 6(5/6), pp. 521-537, 1985.

53. Sarazin, J.R., and Hellawell, A., Channel Formation in Pb-Sn, Pb-Sb, and Pb-Sn-Sb Alloy Ingots and Comparison with the System $NH_4C\ell - H_2O$, *Met. Trans. A,* vol. 19A, pp. 1861-1871, 1988.

NEW MATERIALS BY RAPID QUENCHING

Heat Transfer in In-Rotating-Liquid-Spinning Process

ITSUO OHNAKA
Department of Materials Science and Processing
Faculty of Engineering, Osaka University
Yamadaoka 2-1, Suita-shi, Osaka 565, Japan

MITSUYOSHI SHIMAOKA
Department of Mechanical Engineering
Nara National College of Technology
Yata-machi 22, Yamato-Koriyama-shi, Nara 639-11, Japan

INTRODUCTION

New materials has been developed by using the rapid solidification processing, where molten metal is rapidly solidified, resulting in unique solidified structures leading to unique properties. This process includes single roller process, water and gas atomization, etc. The In-Rotating-Liquid-Spinning Process(INROLISP) is one of this process and can produce fibers with unique properties directly from melt without drawing. In this process melt is cooled by liquid cooling medium and the cooling rate is limited by boiling, though the boiling heat transfer has not been well understood.

In this paper the INROLISP and some examples developed by this process are presented and the heat transfer between melt and cooling medium, mainly water, is discussed.

IN-ROTATING-LIQUID-SPINNING PROCESS[1-4]

Figure 1 shows the principle of the INROLISP, where a melt jet is ejected into a liquid coolant layer formed by centrifugal force in a rotating drum and solidifies into a fiber with round cross-section. The process using water as the coolant is called "In-Rotating-Water-Spinning Process (INROWASP)".

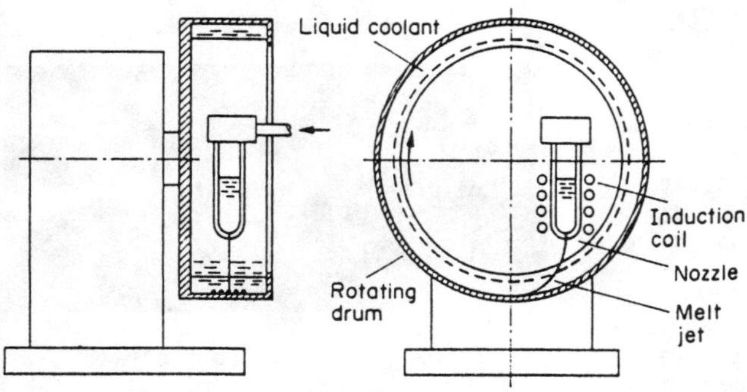

FIGURE 1. In-Rotating-Liquid-Spinning Process

The key parameters of this process are: (1) Melt ejection speed matched with the coolant speed. (2) Nozzle shape, size and material to obtain a stable melt jet. (3) Selection of chemical composition of alloy to form a stable melt jet. Oxide former such as Be and rare earth metals is effective in stabilizing the melt jet. (4) Size and configuration of the drum to realize stable coolant rotation with a minimum surface wave. (5) Kind and temperature of the coolant. Water is most economical but reacts with some metals. (6) Ejection angle and distance between nozzle and coolant surface.

The spinning condition of fibers is usually as follows: Coolant; water, Drum diameter; 600mm, Ejection speed; 5-10m/s, Water layer thickness; 20-30mm, Nozzle material and diameter;quartz, 40-200 μm, Water temperature; 5-25 ℃, Ejection angle; 50-60 degree, Nozzle distance; 2-5 mm.

Some Examples of Newly Developed Metallic Fibers

Amorphous alloy fibers[3-8]. Because the cooling rate of this process is high, it can produce amorphous fibers such as shown in Fig.2 (To produce amorphous fibers the melt should be cooled at higher cooling rates than a critical one determined by its chemical composition). Properties of the fibers are almost equal to those of amorphous ribbons obtained by the single roller process.Some as-spun fibers have higher strength(even 4GPa) than the piano wire which is cold-drown many times. Amorphous fibers such as Co-Si-B fiber have superior corrosion resistance and higher fatigue strength than the piano wire or stainless fiber. Further, fibers such as Co-Si-B and Fe-Si-B have good magnetic properties.

Amorphous fibers have following advantages over the amorphous ribbons: (1) Better cold-drawability, resulting in various diameters and improved properties. (2) Higher squareness of ferromagnetic B-H hysteresis loop. (3) More uniform properties than the fiber formed by splitting a ribbon. (4) More stable and sensitive large Barkhausen jump and Matteucci effects [9].

By making use of these characteristics, amorphous fibers, especially Co-Si-B and Fe-Si-B fibers, are now commercially used for sensors for moving bodies,screens, digitizers, and fiber-reinforced composites such as fishing rods.

Crystalline fibers. Various high strength crystalline metallic fibers can be produced. For example, Fe-20Ni-10Si-1C(mass%) fiber has tensile strength over 1.2 GPa and elongation 15% in as-spun condition. Further, Fe-8Ni-10Cr-7.5Al-3C fiber cold-drawn 90 % has tensile strength of 3.7 GPa and elongation of 0.7%. It is stronger and more corrosion-resistive than the piano wire containing 0.8%C[10].

Further, a Co-base alloy fiber having very fine grains less than 1 μm has very good cor-

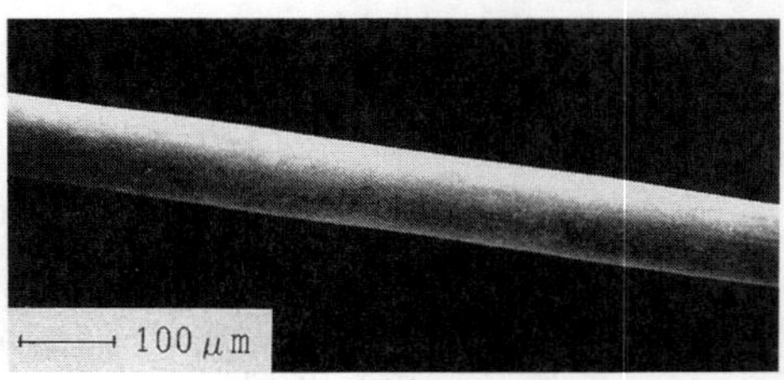

FIGURE 2. Amorphous fiber ($Co_{59}Cr_{15}P_5B_{15}Be_6$)

rosion- and wear-resistances and is now testing for dot printer wire.

These alloys are brittle when they have solidified in conventional processes and hence it is difficult to draw them. These improved mechanical properties are caused by formation of metastable phases and refining of grain and substructure by rapid solidification.

FIGURE 3. Bamboo structure of heusler alloy($Cu_{50}Mn_{25}Al_{25}$)[12]

The INROWASP also can produce fibers with large grains. For example, FIGURE 3 shows large grains formed in a heusler alloy,Cu_2MnAl, which is ferromagnetic in spite of having no ferromagnetic elements. This bamboo structure, where long single crystals are connected in series, may be formed by a higher undercooling and a high crystal growth rate.

Numerical simulation of solidification of melt jet showed that it cools more rapidly and solidifies longitudinally with decreasing jet diameter [11]. Because a higher cooling rate causes a larger undercooling resulting in a higher crystal growth rate, it is predicted that single fibers can be produced if the spinning velocity is equal to or less than the crystal growth rate. Actually single crystal fiber longer than several 100 mm can be produced[13].

Stress-strain curves of the heusler alloy fiber show an ideal elastic-plastic behavior. The heusler fiber with the bamboo structure has large elongation, while conventionally solidified heusler alloy is brittle. It is the same for Fe-6.5mass%Si fiber. The change in ductility may be explained by the easy sliding out of dislocation through the fiber surface.

One of the very interesting characteristics of the single crystal Fe-6.5%Si fiber is a high squareness of ferromagnetic hysteresis loop as the amorphous fiber,while the former has higher saturated magnetic induction. Such squareness may come from the shape effect and homogeneity of the structure.

Besides these fibers it is possible to produce other unique ones such as $Cu_{70}Fe_{30}$ fiber where γ-phase particles less than 1μm in diameter are finely dispersed, and amorphous-copper fiber with layered structure[14].

HEAT TRANSFER IN THE INROWASP

In this process, molten metal jet may be cooled rapidly by boiling heat transfer under forced convection. Although many works on the forced convectional boiling have been reported[15-23], it is not sure whether they can be applied to this case.

Since it is difficult to measure directly the melt jet temperature during cooling in the rotating drum, as a simulation of this process a high temperature filament was dipped into water in order to examine the boiling heat transfer. Further, cooling phenomena of a melt jet were examined in actual spinning condition.

Simulation by Dipping A High Temperature Filament[24]

Experimental apparatus and procedures. Platinum filaments(D=0.1,0.15 and 0.2 mm) heated directly by stabilized direct current in the air were dipped horizontally into subcooled water(ΔT_{sub}=40-90 K) at velocities ranging from 0.25 to 1.0 m/s. The heating of the filament was stopped by a photo sensor and relay unit just before the filament contacted with water. Simultaneously, the electric current of 50 mA was supplied and the voltage drop between the edges of the filament was measured by a transient recorder and was output to a penrecorder. The platinum filament of about 55 mm long was held horizontally by elastic copper electrodes. The temperature averaged over the filament, T_w, were determined by the electrical resistance of the platinum filament and heat flux averaged over the filament was calculated by Eq.(1);

$$q_w = \rho_p C_p D\{T_w(t+\Delta t)-T_w\}/(4\Delta t) \quad (1)$$

where $T_w(t)$ and $T_w(t+\Delta t)$ denote the temperature at a time t and after a time step, Δt, respectively. About 90 points in an obtained cooling curve were used for calculating the heat flux by using B-spline function.

The boiling phenomena around the filament were observed with a high speed camera at the frame rate of 1000 pictures per second.

Boiling phenomena during water quenching. Sketches of boiling observed by the high speed camera are shown in FIGURE 4. Two types of the boiling phenomena were observed;

(1) Type (A): Initially, separation of water was observed and the gas-liquid interface, which is referred to as gas film, connected with the water surface. The gas inside the film may be not only vapor phase. The collapsing site of the gas film propagated from the edges towards the center of the filament. The gas film separated from the filament, followed by covering with a bubble which may be vapor phase((c),in FIGURE 4). Finally,the bubble around the filament split up into several small bubbles and disappeared ((d), in FIGURE 4).

(2) Type(B):The gas film separated earlier at the shallower position from the water surface((e) in FIGURE 4) than the Type(A). Further,the propagating rate of the collapsing site of the vapor film was lower. It seemed that the collapsing sites moved back a little towards the edges of the filament((g), (h)).

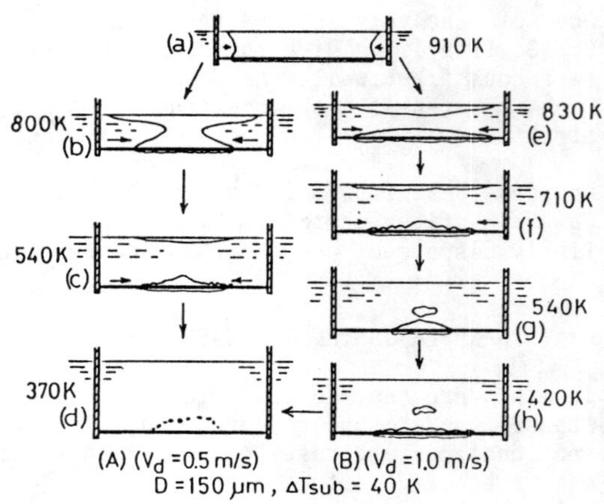

FIGURE 4. Sketches of boiling phenomena around the filament observed by a high speed camera.

The boiling of Type (B) was often observed at higher plunging velocities. At higher plunging velocities, the gas film separated earlier, followed by vapor bubble growth around the filament. A part of the bubble was separated by the action of buoyant force. The phenomenon that the collapsing sites propagate towards the edges of the filament may be explained by the growth of vapor phase.

Propagation of collapsing site of vapor film. The variation in the solid-liquid contact length x_{fc} with time until the gas film separated ((a)-(c) and (a)-(e) in FIGURE 4) was shown in FIGURE 5. The gradient of x_{fc}, i.e., the propagating rate of the collapsing site was about 1 m/s and increased with time. The increase in the propagating rate is probably caused by the filament temperature drop and by the large temperature difference in the middle region of the filament and that at the edges. This difference arises from the different heat transfer on the filament surfaces covered with liquid and vapor film.

The propagating rate increased with increasing water subcooling and the gas film separated earlier from the filament under larger subcoolings. This is because that the generation of vapor decreases with increasing subcooling and that the gas film is pressed and separates earlier due to the pressure difference between the inside and outside of the gas film.

The propagating rate increased with increasing plunging velocity, because a higher plunging velocity results in higher heat transfer coefficient, hence higher cooling.

Although the propagating rate depended little on the diameter of the filament, the gas film separated earlier with decreasing diameter, because the vapor generation decreases with decreasing heat capacity of the filament which depends on diameter.

After the gas film separated((c) or (e),(f),(g) and (h)in FIGURE 4), the propagating rate of the collapsing site was about 0.25 m/s and increased to about 5 or 6 m/s just before the vapor disappeared ((d)).

Relation between heat flux and collapsing of vapor film. Heat fluxes were calculated numerically by using the propagating rates of the collapsing site of the gas films measured by the high speed camera. Namely, the filament was divided into elements of 0.5mm length. The elements were separated into two groups: one is that cooled by the forced convectional heat transfer and another is that cooled by the forced convectional film boiling heat transfer. The number of the element belongs to each group depends on the change of x_{fc} with time. The electrical resistance of each element was calculated by considering temperature dependency of the electrical resistance of the

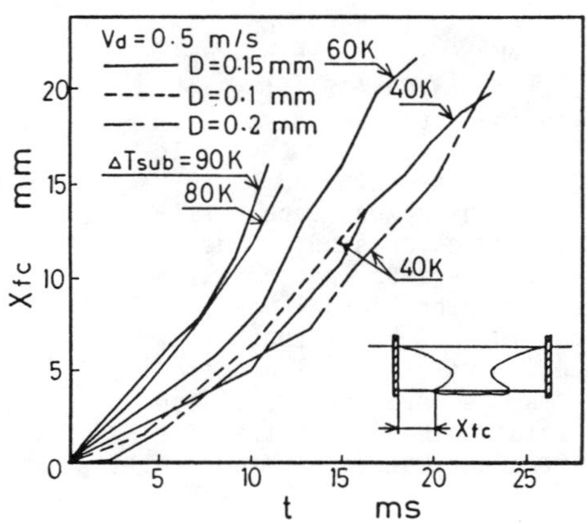

FIGURE 5. Propagation of vapor film collapsing site.

platinum filament. The temperature averaged over the filament is equivalent to the total electrical resistance over the filament.

The heat transfer coefficients in the solid-liquid contact region and in the region covered with gas film have been estimated by using the correlations of Fand[26] and Epstein et al.[20], respectively. These correlations are expressed by Eq.(2) and Eq.(3).

$$h_l = \lambda_l Pr_l^{0.3}(0.35+0.34Re_l^{0.5}+0.15Re_l^{0.58})/D \tag{2}$$

$$h_v = 2.5C\lambda_v Re_l^{0.5}\{1/(24A)+(2/\pi)^2(B/A)^4\}^{0.25}/(\beta D) \tag{3}$$

The film boiling heat transfer coefficient has been corrected by the constant C, because non-condensable gases such as air might be mixed inside the gas film and the bubble. Further, the following assumptions have been made: (1) The filament temperature is uniform radially. (2) The edges of the filament are thermally insulated. (3) The initial filament temperature just before the filament contacts the water varies longitudinally and the temperature distribution is similar to that in the filament heated in the air. Measurement with a two-color optical pyrometer showed that the temperature in the center part of the filament(about 30 mm in length) was uniform and about 1200 K, decreasing to about 800 K near the edges. (4) Radiation heat transfer is neglected.

FIGURE 6 compares calculated heat fluxes with experimental ones. Here the constant C was 0.2 and the points of (a) to (c) correspond to the scenes (a) to (c) in FIGURE 4. Although calculated heat fluxes are higher than the experimental ones in the region of higher filament temperatures, the variation of the heat flux agrees well with experimental ones.

The heat flux for the filament of 200 μm in diameter is higher than that for the filament of 100 μm at surface temperatures below 823 K (the surface temperature of 823 K corresponds to the surface superheat, ΔT_{sat}, of 450K), but the cooling rate is lower because the heat capacity of the filament is four times larger than that of the filament of 100 μm.

It was also found that the change in the heat flux corresponds to the change of the propagating rate of the collapsing site of the gas film.

Equation (3) is based on Eq.(35)* which is an approximate solution of Eq.(32)* for the dimensionless vapor film thickness in the Ref.[20]. In this experimental conditions, the dimensionless vapor film thickness calculated by using Eq.(32)* agrees better with that by using Eq.(33)* within about 10 % of error than that by using Eq.(35)*. Therefore, it may be more practical to calculate the heat transfer coefficient by Eq.(4) based on Eq.(33)*.

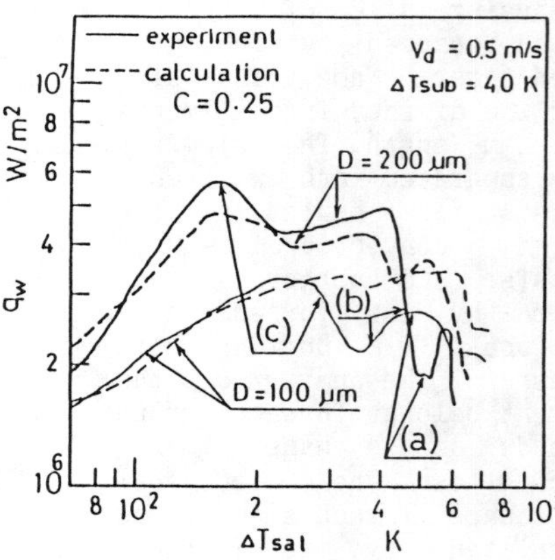

FIGURE 6. Apparent boiling curves

* Number of equation in Ref.[20]

$$h_V = Ce\,\lambda_V Re_l^{0.5}(3/\pi)^{0.5}(B/A)/(\beta D) \qquad (4)$$

where the constant Ce was determined to be 0.5 by comparing calculated and measured heat fluxes. This value is acceptable, because the correlation of Epstein and Hauser was obtained by comparing the theoretical equation with experimental data which are different from the present case where non-condensable gases may be mixed inside the gas films.

Temperature at collapsing site of vapor film. Figure 7 shows the relation between the temperature at the collapsing site of the vapor film and the propagating rate of the collapsing site. It also compares the collapsing site temperature with the minimum heat flux point temperature in film boiling reported. Because it is difficult to measure the temperature at the site, calculated one is shown in this figure. Although considerable scatter is observed, at water subcooling of 40K, mean values agree approximately with the results for the minimum heat flux point temperature obtained by Dhir and Purohit[25](chain lines). On the other hand, Nishio and Sakaguchi[27] pointed out that the minimum heat flux point temperature had dropped with increasing propagating rate and that the real minimum temperature should be determined at the moment when the vapor film collapses at the propagating rate above a few meter per second. This may be one of the reasons why present values are higher than the values reported by Nishio et al.(continuous lines in FIGURE 7).

Further, in the experiment of Nishio and Sakaguchi, a U-shaped filament was used and the edges were kept in the gas phase to prevent the collapsing of the vapor film near the edges of the filament, and the propagating rate was from 10^{-2} to 10 m/s. However, in our experiment, it was from 0.5 to 2 m/s. This may be due to the fact that the vapor film collapsed always at the edges of the filament because of the longitudinal temperature distribution in our experiment.

Although Yilmaz and Westwater[28] reported that the minimum heat flux point temperature had risen with increasing liquid velocity, which is equivalent to the plunging velocity in this paper, at velocities above 2 m/s, it did not depend on the plunging velocity in our experiment.

COOLING OF MOLTEN MELT JET DURING SPINNING[29]

Experimental procedure. A jet of molten $Be_{11}Fe_4Cu$ alloy was ejected into the rotating water layer as shown in FIGURE 1. This alloy was selected because of its good spinnability : good stability of jet and uniform fiber formation. The behavior of the molten metal jet, especially the separation of water around the jet, was observed by a stroboscopic photograph. The separated water flow formed around a stainless wire (SUS304, 0.2mm in diameter) at room temperature which simulates

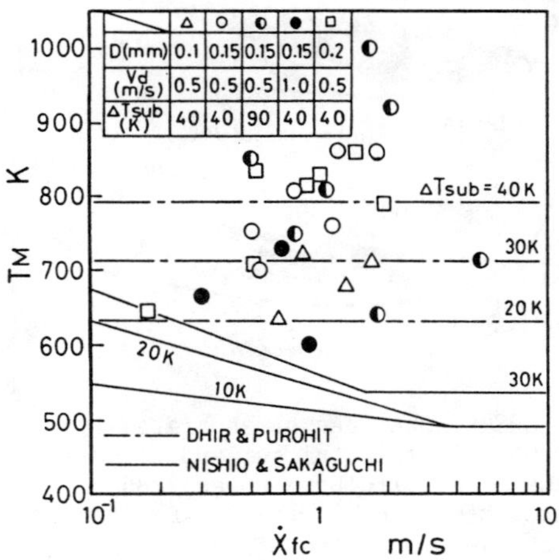

FIGURE 7. Relation between collapsing site temperature and propagating rate.

the molten metal jet was compared with that formed around the melt jet. The stainless steel wire was used, because the rigidity of the Cu-Be-Fe alloy fiber produced by this process was so low that it was difficult to dip it into the rotating water.

To estimate the jet temperature during cooling, brightness of the jet on the photographic paper taken by using a day light color film was measured with an image analyzing system. Although the brightness changes with exposing, developing, fixing and printing conditions, a careful calibration by using 99.9% Cu sphere of 6 mm in diameter showed a linear relationship between the temperature measured by a thermocouple and the brightness in the temperature range from 1080 to 1280K(the range of dark and light gradation is 256 levels from low level 0 (dark) to high level 255(light)).

The experiment was performed under various ejection temperatures, subcooling of water and the nozzle diameter, in the ranges of 1340-1540 K, 60-90 K and 100-190 μm, respectively.

Water flow around molten metal jet. Sketches from the stroboscopic photograph around the molten metal jet and the stainless steel wire are shown in FIGURE 8. In this experimental condition, although Reynolds number for the jet was about 1000, cavitation number($2\Delta p/(\rho_1 V_\infty^2)$; where, Δp is the difference between static and vapor pressures, ρ_1 is the density of water and V_∞ is the mean velocity of water) was 0.013, resulting in the large separation both around the jet and the stainless steel wire. The water surface separated by the jet attached again to the jet or fiber at a more lower stream position. However, the formation of gas-liquid interface in the cases of (a) and (b) differed remarkably.

If a molten metal jet is covered with vapor film due to boiling, the wettability between the jet and water becomes worse. The effect of wettability between filament and water on the separated flow is shown in FIGURE 9. In the cases B and C in FIGURE 9, the stainless filament was cleaned by

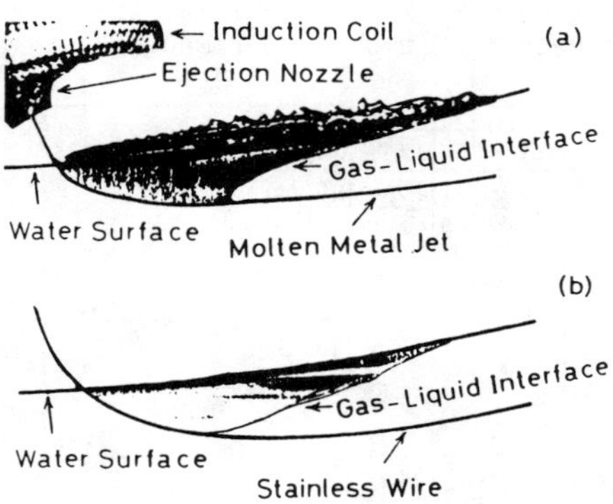

FIGURE 8. Sketches of separated flow around molten metal jet and stainless steel wire(Drum speed:7.85m/s, water temp.:285K). (a) $Cu_{85}Be_{11}Fe_4$ jet. Ejection temperature: 1333K, (b) Stainless steel wire at room temperature.

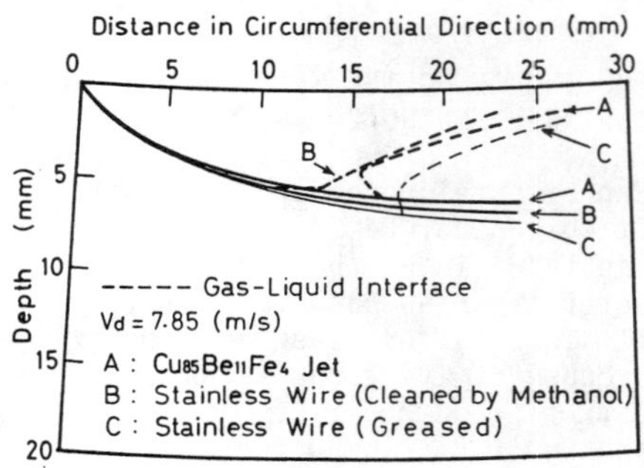

FIGURE 9. Effect of wettability between filament and water on the separated flow.

methanol and coated by silicone grease, respectively. The wettability between the filament and water of the case B is better than that of the case C. The gas-liquid interface shape in the cases B and C differed clearly and that in the latter case was similar to that in case of molten metal jet, A.

From comparison of the length of the water separation along the stainless filament and the molten Cu-Be-Fe alloy jet, it is estimated that the wettability between the alloy jet and water is better than that between the coated stainless filament with silicone grease(on the other hand a measurement of contact angle at room temperature showed that the wettability between the Cu-Be-Fe alloy plate and water was worse than that between clean SUS304 plate and water).

The length of the water separation along the molten metal jet changed also with the jet orbit, especially the depth of jet in the water layer.

To determine the boiling heat transfer region in the jet, enlarged photographs of about 6 magnification of the jet were observed. However,boiling phenomena around the jet were not confirmed except the water separation.

Cooling curve of molten metal jet. The cooling curve of the melt jet was calculated by assuming that the molten metal jet was cooled under forced convectional boiling heat transfer(Eq.(4)) and forced convectional heat transfer(Eq.(2)) in the water separation region and other region, respectively. It was also assumed that a jet divided into cylindrical elements moves along the measured jet orbit with time. The temperature dependency of thermal properties, and thermal radiation from the jet(emissivity was assumed to be 0.1) were considered.

It was also assumed that the solidification proceeded with the crystal growth rate, R, proportional to square of undercooling:

$$R = k(\Delta T)^2 \tag{5}$$

where k is a constant.

Because the position of crystal nucleation and the direction of crystal growth are not known, it was assumed that the nucleation occurs at the lower edge of elements of which temperature becomes below a nucleation temperature and the crystal grows uniformly toward the other edge of the elements. When the ratio of the volume in solid phase to the total volume of the elements reached one, the element was considered to have solidified completely.

Other assumptions of crystal growth rate, $R \propto \Delta T^{1.8}$ and $\propto \Delta T^{2.2}$,resulted in larger discrepancy between the temperature profile of the jet estimated from the brightness of the jet and calculated one. Further, simulation without undercooling was performed by the temperature recovery method[30].

Calculated cooling curves of the molten metal jet in the rotating water are shown in FIGURE 10. The jet temperature estimated from the brightness of the jet(open circles) decreased slowly just after the jet entered the water. It may be due to the low heat transfer coefficient of the boiling heat transfer. After passing through the water separation region, the jet should be cooled more rapidly by the forced convectional heat transfer. However, the rapid temperature change was not observed from the brightness measurement. This may be due to initiation of solidification, namely due to the liberation of latent heat of fusion.

Another case with jet ejection temperature of about 250K above liquidus temperature showed the similar result to the above. It seems that the solidification of the jet starts mainly after passing through the water separation region even if the jet is ejected at temperatures slightly higher than liquidus temperature.

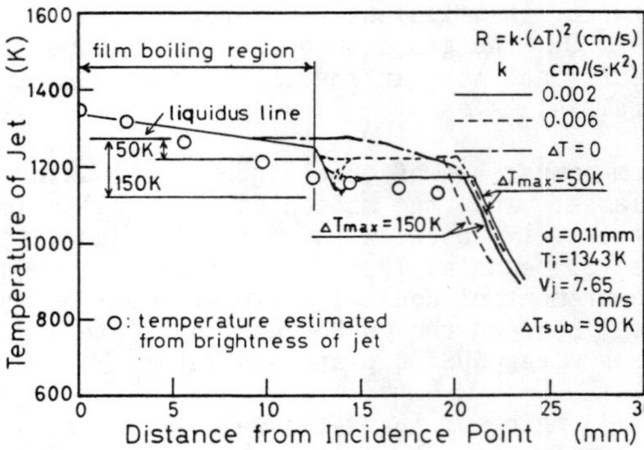

FIGURE 10. Cooling curves of the melt jet in the rotating water layer. Ejection temperature was 1343K.

The temperatures at the end of the water separation for the cases of ejection temperature of 1343 and 1528 K were 1168 and 1336 K, respectively, even or the same subcooling. If this temperature is the minimum heat flux point temperature in film boiling, it is estimated to be about 1193K from the experiment by Dhir and Purohit [25] and about 670K from a more detailed experiment by Nishio et al.[27]. Therefore, these temperatures may be not the minimum heat flux point temperature. Further, it was found that the length of the water separation along the jet depended not on the ejection temperature but mainly on the diameter of the jet,d, and the velocity of jet,V_j. Because the kinematic energy of the jet decreases with decreasing d and V_j, the jet orbit becomes parallel faster to the water surface,resulting in a shorter water separation length. This is supported by the fact that the more the stainless steel wire was tilted in the direction parallel to the rotating water surface, the shorter the length of the water separation along the wire. Therefore, if the wettability between the melt jet and water is same, the length of the water separation along the jet may depends mainly on the jet orbit rather than the jet temperature.

Chain line in FIGURE 10 shows calculated cooling curves assuming no-undercooling. Although undercooling of the jet was not found from the brightness of the jet, undercooling of about 50K may be possible. This is because if it is assumed that the undercooling is 50K and k in Eq.(5) is 0.02 or 0.06($mmK^{-2}s^{-1}$), the calculation showed no recalescence. At least it can be said that the large undercoolings such as 100 K didn't occur in this experiment.

Although the behavior around the jet in water must be observed in detail in order to confirm the phenomena of film boiling, transitional boiling and nucleate boiling,it can be said that the coefficient of heat transfer in the water separation region is determined to be in the order of that estimated from Eq.(4).

Cooling rate during solidification interval. The relation between the cooling rate during solidification interval obtained by numerical calculation and the fiber diameter is shown in FIGURE 11.

Because the specific surface area per unit volume decreases with increasing the diameter of the jet and the heat transfer coefficient doesn't

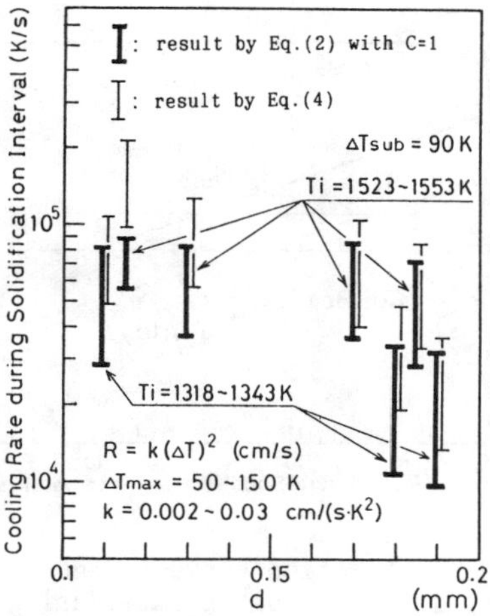

FIGURE 11. Relation between the cooling rate during solidification interval and diameter of fiber.

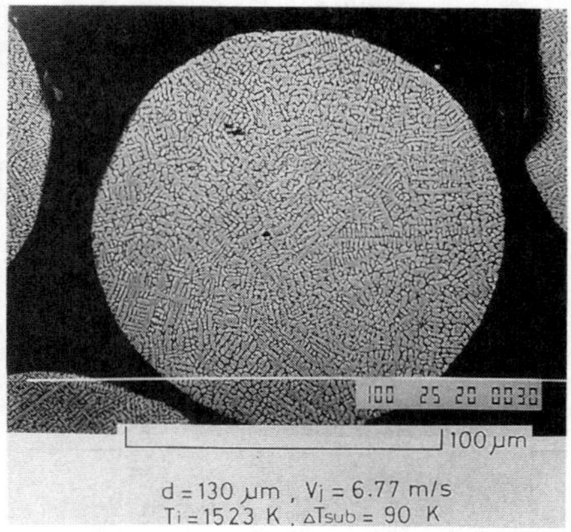

FIGURE 12. SEM view of cross-section of $Cu_{85}Be_{11}Fe_4$ filament

change so much, the cooling rate of the jet decreases with increasing the diameter. In addition,the cooling rate of the jet increases as the ejection temperature rises.This is because that a higher ejection temperature prevents solidification in the water separation region where cooling rate is low and the melt solidifies in the forced convectional heat transfer region with the higher heat transfer coefficient. Further, the cooling rate is higher when the heat transfer coefficient in the water separation region is low. This also shows that solidification may proceed mainly in the forced convectional heat transfer region.

Even if the crystal growth rate increased by 10 times, the cooling rate increased only by 2 or 3 times. Because solidification is determined not only by the crystal growth rate but also by the heat transfer, the cooling rate is not simply proportional to the crystal growth rate.

Dendrite arm spacing. FIGURE 12 shows one of SEM views of the cross-section of the wire. The dendrite arm spacing increased with decreasing subcooling of water, as shown in FIGURE 13.The coefficient of forced convectional heat transfer does not depend on subcooling of water, but that of forced convectional film boiling heat transfer depends on the subcooling remarkably. Namely, it is estimated from Eq.(4) that the value of the coefficient for subcooling of 60K decreases about 30% as compared with that for 90K.Therefore,if the jet has been covered with very thin film of vapor, which is hardly recognizable with naked eyes, and if the solidification of the jet has been started in that region covered with vapor film, the solidification time becomes longer and the dendrite arm spacing becomes wider. This is contradictory to the above calculation. It is suggested that the solidification starts in the separation region or the film boiling occurs even after the separation region.

The dendrite arm spacing was in the range of 0.9 to 1.5 μm, corresponding to the cooling rates from 10^5 to 10^6K/s which was calculated by using Sarin and Grant's data[31] for Cu-Zr alloys and Matyja et al.'s data[32] for Al alloys as shown in FIGURE 14. Thus, although the cooling rate

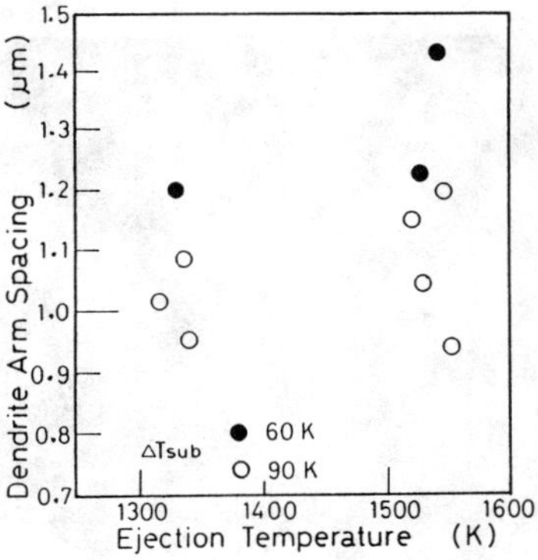

FIGURE 13. Relation between dendrite arm spacing and ejection temperature of the melt.

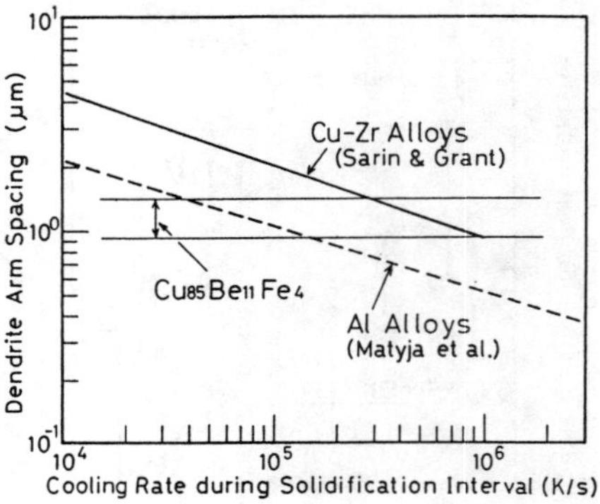

FIGURE 14. Relation between dendrite arm spacing and cooling rate during solidification interval.

depends on the alloy composition, it can be said that the cooling rate of this Cu-Be-Fe alloy fiber of 130 μm in diameter is in the order of 10^5K/s. Further, even if the jet is cooled only under forced convectional heat transfer without the water separation and the liberation of latent heat of fusion, the cooling rate calculated with the heat transfer coefficient estimated by Eq.(4) cannot exceed 10^6K/s.Therefore,the calculated cooling rates during solidification interval as shown in FIGURE 11 may not contain remarkable errors.

CONCLUDING REMARKS

The INROWASP can produce fibers directly from melts and the solidified structure varies from amorphous phase to single crystal,including layered structure. Those unique structures come from not only the rapid cooling but also the feature of containerless solidification.

In this process,the melt ejected into water separates the water, followed by boiling and convectional heat transfers, though the boiling phenomena could not clearly observed by storoboscopic photograph.

The separated water flow is caused not only by boiling but mainly hydrodynamically. This water separation decreases the cooling rate of the melt jet and if the wettability between the filament and water is made better, the length of the water separation along the filament can be made shorter.

The temperature distribution along the melt jet can be calculated by using the modified coefficient of forced convectional film boiling heat transfer by Epstein and Hauser, and that of forced convectional heat transfer by Fand. However, the temperature at the end point of the water separation did not agree with the minimum heat flux point temperature of film boiling and the length of the water separation along the jet didn't depend on the temperature.

$Cu_{85}Be_{11}Fe_4$ fiber of about 130μm in diameter solidified dendritically and the cooling rate during solidification interval was estimated to be about 10^5 K/s.

In order to increase the cooling rate during solidification interval, higher subcooling of water and a rather higher ejection temperature which causes solidification after the water separation region are effective. However, the increase is less than a few hundred percent and it is desired to seek other methods. Replacement of water with other liquid cooling medium such as oil did not result in considerable improvement[33].

From fundamental view point, the more detailed observation of boiling phenomena around the molten metal jet and establishment of the method to measure the jet temperature more accurately are necessary to make clear the cooling process during liquid quenching of micro-sized materials.

Further, it is also desired to make clear the relation among crystal growth rate, undercooling and chemical composition to control the solidification structure.

ACKNOWLEDGMENTS

The authors wish thank Dr. I. Yamauchi, Mr.T. Ohmichi of Osaka University and Mr.T.Mitsui,former graduate student,now at Mitsubishi Cable Ind.Ltd., for their help in experiments. The authors also thank Dr. S. Nishio of Institute of Industrial Science,the University of Tokyo, for many helpful suggestions.

NOMENCLATURE

A	$:C_{pv}\Delta T_{sat}/(Pr_v \Delta H_{lv})$
B	$: C_{pv}\Delta T_{sub}Pr_l^{0.5}\lambda_l/(\lambda_v Pr_v \Delta H_{lv})$
C, Ce	:constants
C_p	:specific heat at constant pressure
d	:diameter of molten metal jet
D	:diameter of the platinum filament
ΔH_{lv}	:latent heat of vaporization of water
h_l	:forced convectional heat transfer coefficient
h_v	:film boiling heat transfer coefficient under forced convection
Pr	:Prandtl number
q_w	:heat flux averaged over the surface of the filament
R	:crystal growth rate
Re	:Reynolds number, $V_d D/\nu_l$
T_f	:water temperature
T_M	:temperature at the collapsing site of the vapor film or minimum heat flux point temperature
T_s	:saturation temperature of water
T_w	:temperature averaged over the filament
ΔT	:undercooling on solidification
ΔT_{sat}	:superheat which is the difference between the surface temperature and the saturation temperature, T_w-T_s
ΔT_{sub}	:subcooling
t	:time
V_d	:plunging velocity of the filament into the water from the air
x_{fc}	:solid-liquid contact length (see Fig.3)
x_{fc}	:change of the collapsing site (x_{fc}) with time
β	$:\{(\rho_v/\rho_l)^{0.5}(\nu_v/\nu_l)\}^{0.5}$
λ	:thermal conductivity

ν :kinematic viscosity
ρ :density
Subscripts
l :liquid phase of water
p :platinum
v :vapor phase of water

REFERENCES

1. Ohnaka,I.,Fukusako,T.,and Ohmichi,T., Production of metallic filaments by In-Rotating-Water Spinning method,J.Japan Inst.Metals,vol.45, no.7, pp.751-758, 1981
2. Ohnaka,I., Japanese Patent,No.1331366
3. Ohnaka,I., Melt spinning into a liquid cooling medium, Int.J. Rapid Solidification,vol.1, no.3, pp.219-236,1985
4. Ohnaka,I.,Wires:Rapid Solidification, Encyclopedia of Materials Science & Engineering,Supplementary Volume 1, ed.by R.W. Cahn, pp.584-587, Pergamon Press,1988
5. Masumoto,T.,Ohnaka,I.,Inoue,A.,and Hagiwara,M., Production of Pd-Cu-Si amorphous wires by melt spinning method using rotating water, Scripta Metal.,vol.15,pp.293-296,1981
6. Ohnaka,I.,Fukusako,T.,Ohmichi,T.,Masumoto,T.,Inoue,A.,and Hagiwara,M., Production of amorphous filament by In-Rotating-Liquid spinning method, Rapidly Quenched Metals, eds.Masumoto T,and Suzuki K, The Japan Institute of Metals,pp.31-34,1981
7. Hagiwara,M.,Inoue,A.,and Masumoto,T., Mechanical properties of Fe-Si-B Amorphous wires produced by In-Rotating-Water Spinning Method, Met. Trans.A,13A,pp.373-382,1982
8. Inoue,A.,Hagiwara,M., and Masumoto,T., Fatigue strength of amorphous alloy wires, Sci.Rep.Res.Inst.Tohoku Univ., vol.34,pp.48-78,1988
9. Mohri,K.,Yamasaki,J.,and Okamura,K., Jitterless Pulse Generator Elements Using amorphous Magnetostrictive Wires, J.Appl.Mag.Japan, vol.8,no.2,pp.233-236,1984
10. Inoue,A., Yano,N.,Tomioka,H.,and Masumoto,T., Mechanical properties of Fe-Ni-S-C and Fe-Ni-Cr-Si-C Wires Prepared by melt spinning, Trans.Iron and Steel Inst.Japan, vol.26,pp.993-1001,1986
11. Ichiryu,T.,Ohnaka,I.,Tateoka,H.and Yamauchi,I., Production and Properties of Rapidly Solidified Fe-Si Alloy Wires by the In-Rotating-Water Spinning Process, J.Japan Inst.Metals,vol.50,no.10,pp.927-934,1986
12. Ono,Y.,Ohnaka,I.,and Yamauchi,I., Production of Cu-Mn-Al Heusler Alloy Wire by the In-Rotating-Water-Spinning Process and Its Properties, J.Japan Inst. Metals,vo.51,no.8,pp.755-761,1987
13. Ichiryu,T.,and Ono,Y., Structure and magnetic properties of melt spun Fe-Si alloy filament, Proc. of the 12th Ann.Conf.on Magnetics in Japan, 30_pE-4, pp.72, 1988
14. Ohnaka,I. and Yamauchi,I.,Production of new materials by In-Rotating-Liquid-Spinning Process, Proc.Japan-U.S.seminar on Solidification Processing of Advanced Materials, pp.107-115,Oiso,Japan,1989.5
15. Bromley,L.A., Leroy,N.R.,and Robbers,J.A., Heat transfer in forced convection film boiling, Ind. Eng. Chem.,vol.45,pp.2639-2646,1953
16. Motte,E.I., and Bromley,L.A., Film boiling of flowing subcooled liquids, Ind. Eng. Chem.,vol.49,pp.1921-1928,1957
17. Kobayashi,K., Film boiling heat transfer around a sphere in forced convection, J. Nucl. Sci. Technol.,vol.2,no.2,pp.62-67,1965
18. Witte,L.C.,Baker,L.Jr.,and Haworth,D.R., Heat transfer from spheres into subcooled liquid sodium during forced convection, Trans. ASME, J. Heat Transf., vol.90,pp.394-398,1968

19. Walford,F.J., Transient heat transfer from a hot nickel sphere moving through water, Int. J. Heat Mass Transf., vol.12, pp.1621-1625,1969
20. Epstein,M. and Hauser,G.M., Subcooled forced-convection film boiling in the forward stagnation region of a sphere or cylinder, Int. J. Heat Mass Transf., vol.23,pp.179-189,1980
21. Ito,T.,Nishikawa,K.,and Shigechi,T., Forced convection film boiling heat transfer from a horizontal cylinder to liquid cross-flowing upward(1st Rept.Saturated Liquid), Trans. JSME, Series B,vol.47, no.416,pp.666-673,1981
22. Shigechi,T.,Ito,T. and Nishikawa,K., Forced convection film boiling heat transfer from a horizontal cylinder to liquid cross-flowing upward(2nd Rept.Subcooled water), Trans.JSME,Series B,vol.48,no.432, pp.1539-1545,1982
23. Cheng,S.C.,Lau P.W.K.,and Poon,K.T., Measurements of true quench temperature of subcooled water under forced convective conditions, Int.J. Heat Mass Transf.,vol.28,pp.235-243,1985.
24. Shimaoka,M.,Ohnaka,I.,Menju,A.,and Ohtomo,K., Boiling heat transfer during water quenching of high temperature platinum filament, Trans. Japan Soci.Mech.Eng.(B), vol.54,no.499,pp.704-708,1988
25. Dhir,V.K. and Purohit,G.P., Subcooled film-boiling heat transfer from spheres, Nucl. Eng. Des.,vol.47,pp.49-66,1978
26. Fand,R.M.,Heat transfer by forced convection from a cylinder to water in crossflow, Int. J. Heat Mass Transf.,vol.8,pp.995-1010,1965
27. Nishio,S. and Sakaguchi,K.,Study on film-boiling heat transfer and minimum heat flux conditions for subcooled boiling(2nd report: Pool boiling of water at subatmospheric pressures from horizontal platinum cylinder), Trans.Japan Soci.Mech.Eng.(B),vol.53,no.490,pp.1781-1787, 1987
28. Yilmaz,S. and Westwater,J.W., Effect of velocity on heat transfer to boiling Freon-113, Trans. ASME,J. Heat Transf.,vol.102,pp.26-31,1980
29. Shimaoka,M. and Ohnaka,I.,Cooling and solidification of molten metal jet in the Rotating-Water-Spinning Process, J.Japan Inst.Metals, vol.53,no.6,pp.621-627,1989
30. Ohnaka,I. and Fukusako,T., Calculation of solidification of casting by a matrix method, Trans. Iron and Steel Inst. Japan, vol.17, pp.410-418,1977
31. Sarin,V.K.and Grant,N.J., Cu-Zr and Cu-Zr-Cr alloys produced from rapidly quenched powders, Metal.Trans., vol.3,pp.875-878,1972
32. Matyja,H.,Giessen,B.C., and Grant,N.J., The effect of cooling rate on the dendrite spacing in splat-cooled aluminium alloys, J.Inst. Metals, vol.96,pp.30-32,1968
33. Ohnaka,I.,Sato,M.,Tanimura,A.,and Yasuhara,K., Water quenching of platinum filament, Prepr.21th Nat.Heat Transf.Symp.Jpn.,pp.484-486, 1984

An Attempt to Produce Particles of Amorphous Materials Utilizing Steam Explosion

RYO AKIYOSHI
Ishikawajima-Harima Heavy Industries
1 Nakaharamachi, Isogo-ku, Yokohama, Japan

SHIGEFUMI NISHIO and ICHIRO TANASAWA
Institute of Industrial Science
University of Tokyo
7-22-1 Roppongi, Minato-ku, Tokyo, Japan

ABSTRACT

We have attempted to produce particles of amorphous phase by ultrarapid quenching of a molten metal making use of thermal interaction, or steam explosion, between drops of the molten metal and water. The principle is based on the facts that the liquid metal is fragmented into numerous small particles as the result of steam explosion, and that a very high cooling rate is attained during the process of solidification. As the results of preliminary experiment using molten tin, we found that particles with smaller sizes were obtained when the atmosphere above the water surface was filled with steam and the water temperature was low. Further, we noticed that the particle size was also reduced if pressure pulses by an exploding wire were applied to the boiling film around the molten metal drop, and that the fragmented particles became more spherical when the water temperature was higher. Finally we confirmed that amorphous phase was formed by steam explosion in the case of palladium-silicon alloy with 20 atomic percent silicon.

INTRODUCTION

Some alloys obtained by solidifying the melt very rapidly are known to be possessed of certain special characteristics. For example, they have fine crystalline structure, segregation in microscopic scale is reduced or disappeared in them, supersaturated solid solutions are formed in some cases, or metastable or amorphous phases are formed in the other. Owing to those characteristics, rapidly solidified alloys are expected to exhibit desirable qualities, for instance, in mechanical strength, plasticity, resistance to corrosion, density, or electromagnetic properties.

Rapidly solidified alloys have been manufactured in the forms of thin films and wires or of fine powder. Illustrated in Fig.1 are typical examples of the techniques employed to manufacture amorphous alloys[1]. In this figure, (a)-(e) illustrate the techniques for manufacturing thin ribbons, (f)-(h) show the methods for producing thin wires, and (i)-(k) represent the way how to obtain fine particles. Common to all of the techniques, the solid phases available are small in dimensions. Thickness of the foil or ribbon, or diameter of the wire or powder is smaller than the order of 100μm. It is impossible to obtain products with larger thickness or diameter because the rate of cooling is limited.

Recently the powder of rapidly solidified alloy can be formed to any shape by hot, cold or dynamic press without losing the original characteristics[2]. The demand for the powder produced by rapid solidification is expected to increase.

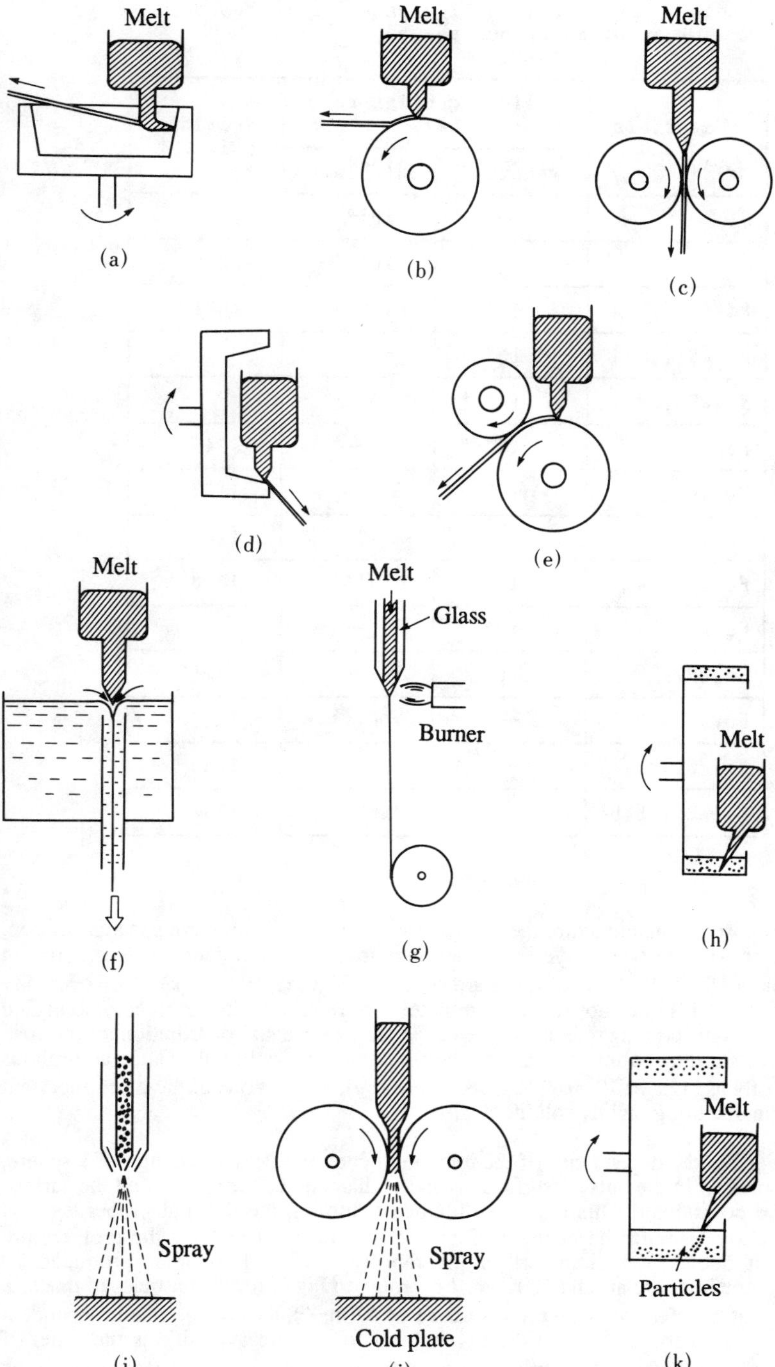

Fig.1 Various techniques of manufacturing amorphous materials

Table 1 Critical cooling rate for formation of amorphous phase(1)

Materials	Critical cooling rate, K/s		Melting point, K
	Measured	Calculated	
Ni	—	$10^{10.5}$	1725
Ge	—	$10^{5.7}$	1210
$Pd_{82}Si_{18}$	10^{4}	10^{4}	1071
$Pd_{78}Si_{16}Cu_{6}$	$10^{2.8}$	10^{2}	
$Ni_{40}Pd_{40}P_{20}$	$10^{2.8}$	$10^{2.1}$	
$Fe_{80}P_{20}$	$10^{4.9}$	—	945
$Fe_{80}B_{20}$	$10^{5.4}$	—	1174
$Fe_{83}B_{17}$	$10^{6.0}$	$\sim 10^{6}$	1448
$Fe_{80}P_{13}C_{7}$	$10^{4.8}$	10^{4}	1258
$Fe_{80}P_{13}B_{7}$	$10^{5.5}$	—	
$Fe_{80}B_{13}P_{7}$	$10^{4.9}$	—	
$Fe_{79}Si_{10}B_{11}$	—	$10^{5.3}$	1419
$Ni_{75}Si_{8}B_{17}$	—	$10^{5.0}$	1340
$Co_{75}Si_{15}B_{10}$	—	$10^{5.5}$	1393

Methods proposed to manufacture the powder by rapid solidification are gas atomization, atomization in rotating water layer, pulse atomization, centrifuge method, spark erosion method, and so forth(3). Some of these are shown in Fig.1 (i),(j) and (k). However, any one of these methods does not satisfy completely the requirements for rapidly solidified particles such as the cooling rate, the shape of the particle, the size distribution, or the cost. Especially, as for the cooling rate, since the critical rate of cooling to realize amorphous phase is mostly as high as 10^4-10^6 K/s [see Table 1(1)], it raises one of the most important issues in manufacturing rapidly solidified alloys.

Shown in Fig.2 is the dependence of the maximum cooling rate at the center of a sphere, which is cooled from the outside surface, upon the diameter of the sphere and the surface heat transfer coefficient. This result is obtained assuming the thermal properties of a duralumin alloy (4), and the heat transfer coefficient is defined for the temperature difference of 600K. The diameters of the rapidly solidified particles produced by conventional methods are around 100 μm. According to Fig.2, for the particles of this size and for the heat transfer coefficient lower than 1 MW/(m^2K), the cooling rate is dependent upon the rate of heat transfer and the maximum cooling rate available is the order of 10^7K/s. Thus, developing the technique of manufacturing rapidly solidified alloys in the form of small particles and attaining at the same time a high heat transfer rate are required.

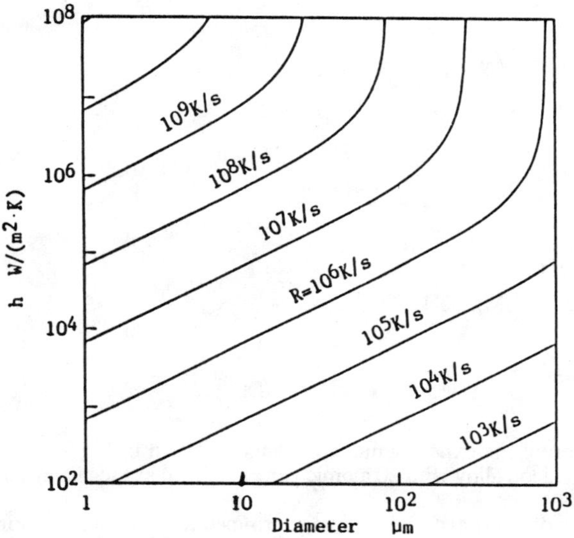

Fig.2 Maximum rate of cooling at the center of a sphere (Duralumin A-7075, ΔT=600K)

It has been known that a molten metal can be fragmented into small particles by a violent thermal interaction or steam explosion between the molten metal and water. The heat flow rate during such a thermal interaction is considered to be very high because of direct contact between the two liquids. According to an experiment simulating the process of steam explosion by applying a shock wave to film boiling, a maximum heat flux up to 160 MW/m^2 has been observed for a short period(5). Converted into the heat transfer coefficient, this value corresponds to 270 $kW/(m^2K)$, being a very high coefficient compared to the conventional heat transfer processes.

If an appropriate use of the steam explosion is possible, a rather simple and inexpensive technique of manufacturing rapidly solidified alloy particles will be established. In a preceding work(6) we have found that the probability of occurrence of the steam explosion is greatly increased by replacing the air in the space above the water surface with steam. In this paper, we confirm further the factors influencing fragmentation of the molten metal by thermal interaction. Then, we attempt to manufacture particles of amorphous alloy making use of the knowledge obtained by the preliminary experiment.

NOMENCLATURE

D	:	particle diameter, m
d	:	distance between an exploding wire and a pressure transducer, m
h	:	heat transfer coefficient, $W/(m^2K)$
P	:	pressure change, Pa
R	:	maximum cooling rate at the center of a particle, K/s
T	:	temperature, °C or K
w	:	mass of a particle, kg
θ	:	angle of diffraction of X-ray, deg

Subscripts

h : molten metal
i : initial
l : water
m : mean
max : maximum

PRELIMINARY EXPERIMENT USING TIN

Method of Experiment

A schematic diagram of the experimental apparatus is shown in Fig,3. We used tin first and then palladium-silicon alloy with 20 atomic percent silicon as the test materials.

At first we heated 2g of a test material to a predetermined temperature in a nozzle ①(6mm inside diameter) made of stainless steel, which was placed in an electric furnace②filled with argon gas, and let it melt. By opening a shutter ⑤ we made the molten alloy drop into the water in a vessel ⑦ (400 mm wide, 200 mm thick and 350 mm deep water pool) and let thermal interaction with water take place. After each of the runs we collected the fragments of solidified metal with a paper filter placed at the bottom of the vessel and measured the distribution of the particle size by drying and classifying the fragments with sieves. Following the result of the preceding study(6) we filled the space above the water pool with steam in order to promote fragmentation by steam explosion. In this case the steam condensed on the surface of water, raising the temperature there up to the saturation temperature. We made circulate the water in the vessel through a constant temperature bath ⑨to restrict the thickness of the thermal boundary layer within a few millimeters, thus kept the temperature around the depth where the thermal interaction took pace almost uniform. The temperature T_1 appearing hereafter corresponds to this bulk temperature. We reduced the rate of condensation of steam by placing on the water surface an insulating plate with a hole, through which a molten metal drop passed.

Effect of Atmosphere and Water Temperature on Fragmentation of Molten Metal

In order to reconfirm and supplement the results of the previous experiment(6), we investigated the effects of the water temperature and the atmosphere above the water surface upon the fragmentation of molten metal by thermal interaction. We filled the space above the pool of water either by air or by steam and varied the bulk temperature of water T_1 as 20 °C (only for air atmosphere), 40 °C, 60 °C and 80 °C. The initial temperature of the drop of molten tin, T_{hi}, was kept 1000 °C. Shown in Fig.4 is the mean diameter of the fragmented particles measured in this experiment. The mean diameter D_m is defined as,

$$D_m = \Sigma wD / \Sigma w \tag{1}$$

where the diameter D is represented by the size of the mesh which the particles can pass through. As for the particle that cannot pass through the largest mesh(2.8mm), we have regarded the length of the second axis of the particle, assuming the particle to be an ellipsoid, as the equivalent diameter. w in Eq.(1) denotes the mass of the particle, the equivalent diameter of which is D. It is known from Fig.4 that smaller particles were obtained when the atmosphere above the water surface was steam and when the water temperature was lower. Even when the atmosphere was air, smaller particles, comparable to the case of steam atmosphere, were sometimes available at lower water temperatures. However, dispersion of the mean diameter in this case was very great. It was often the case that fragmentation hardly took place. Therefore, a condition desirable to produce rapidly solidified particles is to fill the atmosphere with steam and keep water temperature low.

Effect of Temperature of Molten Metal on the Mean Particle Diameter

We examined the effect of the initial temperature of molten metal on the mean particle diameter. We kept constant the water temperature, T_1, as 40 °C and filled the atmosphere

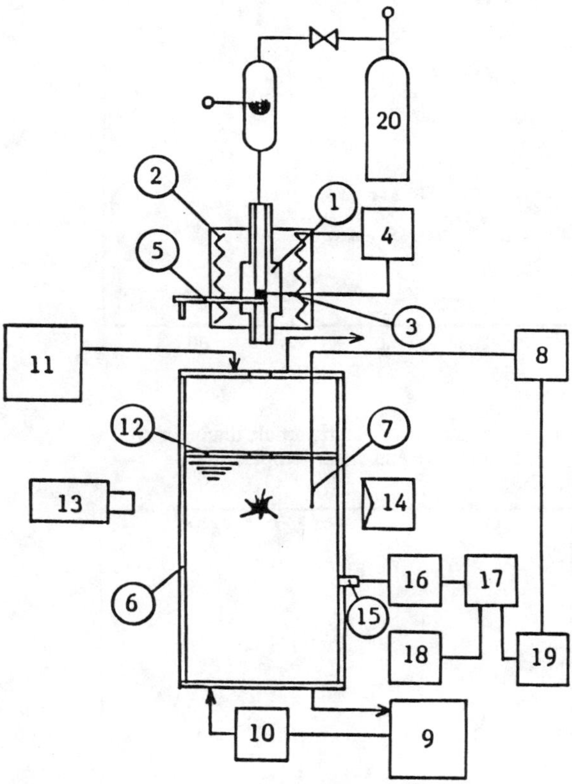

1.Nozzle	11.Boiler
2.Heater	12.Insulator
3.CA Thermocouple	13.Video or Camera
4.Temp. Controller	14.Stroboscope
5.Shutter	15.Pressure Transducer
6.Water Bath	16.Amplifier
7.CA Thermocouple	17.Transient Memory
8.Cold Junction	18.Oscilloscope
9.Water Bath	19.Pen Recorder
10.Pump	20.Ar Bomb

Fig.3 Schematic diagram of experimental apparatus

with steam. The result shown in Fig. 5 indicates that the mean diameters do not depend on T_{hi} and are around 200 - 500 μm.

An observation using a high-speed video facility revealed that the thermal interaction occurred at a depth 2-3 cm from the water surface (which was about 10 ms after the metal

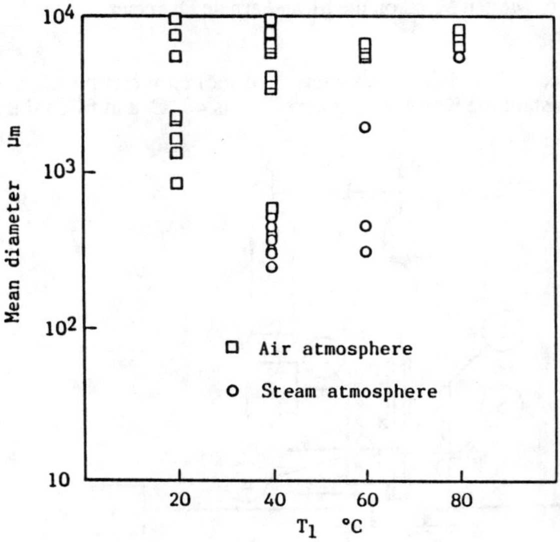

Fig.4 Dependence of mean particle diameter on atmosphere and water temperature

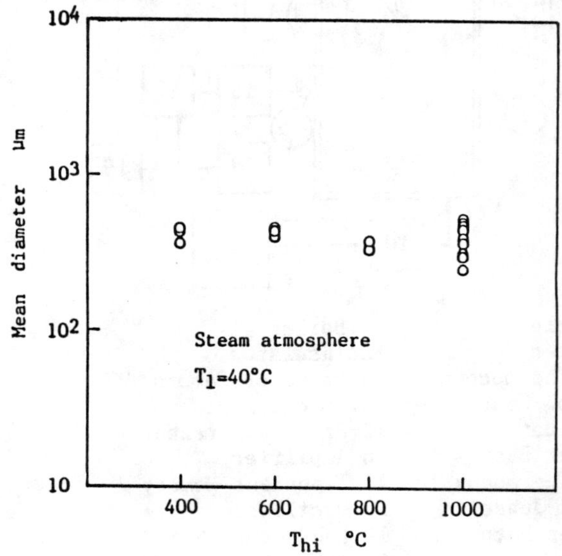

Fig.5 Dependence of mean particle diameter on initial drop temperature

drop plunged into the water pool). This means that the temperature fall of the metal drop before the thermal interaction occurred did not depend so much on the initial metal temperature. It does not seem necessary that the molten metal is cooled to a definite temperature before the thermal interaction takes place. We have considered that the result of Fig.5 shows that the mean particle diameter is not affected by the temperature of the molten metal at the moment when the thermal interaction takes place.

The result mentioned above suggests that the mean diameter of the particles produced by the steam explosion which occurs spontaneously may not be so small as those produced by conventional techniques.

Effect of External Pressure Pulses on the Mean Particle Diameter

The next step we took was to enhance the fragmentation of particles by applying external pressure pulses on the molten metal-water system during film boiling. Such a method had been adopted by several investigators(7)-(13) in studying the basic mechanism of thermal interaction.

A circuit to operate an exploding wire for generating the pressure pulses is shown in Fig.6. The exploding wire ㉑ 0.2mm in diameter and 5mm long, made of copper, is melted very fast by a big discharge current from a capacitor ㉒ (1000 μF, 1 kV). Pressure waves with several peaks are generated due to rapid expansion and condensation of vapors of copper and water formed by explosion.

In order to initiate the discharge from the capacitor with an appropriate timing, we placed a junction of chromel-alumel thermocouple ㉔ in the water pool just at the location which the drop of molten metal passed through. When the molten metal drop touched the junction, the thermal emf, amplified by an amplifier ㉕, was compared with a standard voltage ㉗ by a comparator ㉖. When the molten metal hit the junction and the amplified emf exceeded the standard voltage, then an electric pulse generator ㉘ was activated and SCR(thyrister) was triggered to initiate the discharge.

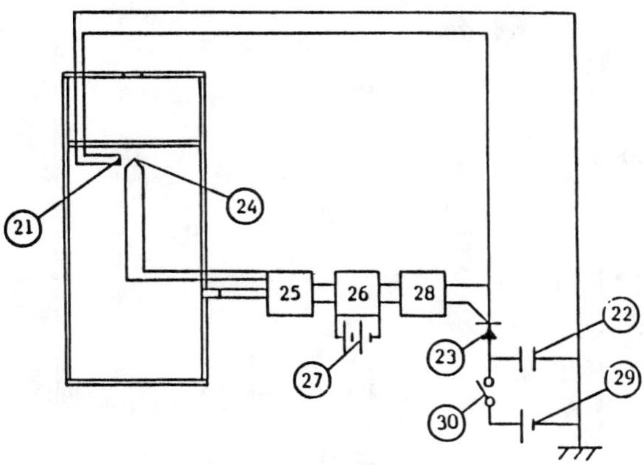

21.Copper Wire
22.Capacitor(1000μF)
23.SCR (Thyristor)
24.CA Thermocouple
25.Amplifier
26.Comparator
27.Battery
28.Pulse Generator
29.D.C. High Voltage Power Supply
30.Switch

Fig.6 Circuit to operate exploding wire

Plotted in Fig.7 are the peak pressures P_{max} generated by exploding wires at different distances from the pressure transducer. The distance we employed in the experiment was about 3 cm, indicating that the peak pressure was about 2 MPa. The pressure waves shown also in Fig.7 represent a typical pattern generated by the exploding wire.

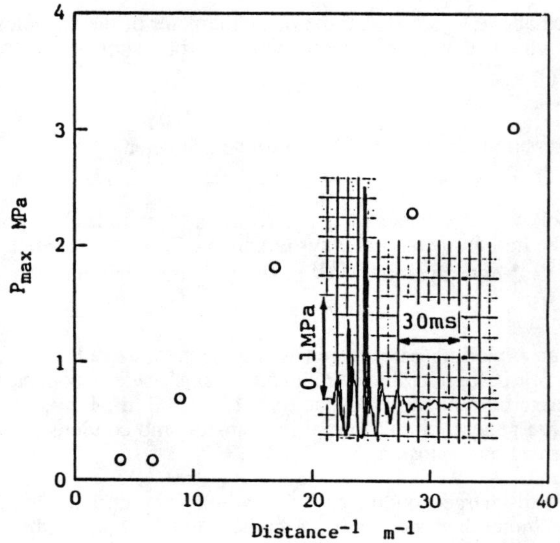

Fig.7 Maximum pressure generated by exploding wire

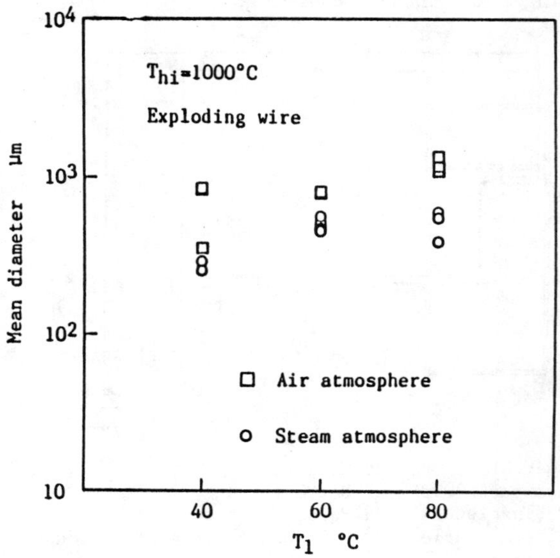

Fig.8 Mean particle diameter obtained by using exploding wire

Figure 8 shows the mean diameters of the fragmented particles when the thermal interaction was triggered by external pressure pulses. Comparison with Fig.4 suggests that the mean diameters here are not so much different from those obtained under the conditions of T_1= 40 °C, steam atmosphere, and without external pressure waves. However, the external pressure pulses seem especially effective in enhancing fragmentation under the conditions that the molten metal drop cannot be fragmented to small particles unless the external pressure waves are applied.

Shapes of the Fragmented Particles

The pictures of the fragmented particles taken through an electron microscope are shown in Fig.9. Here (a) is taken at T_1= 40 °C and with no external pressure pulses, and (b) is taken at T_1 = 80 °C and with the external pressure pulses applied. For both of the cases the atmosphere is steam and T_{hi}=1000 °C. It is said that the particles of spherical shape are preferable when forming them to a certain shape, since the porosity could be reduced. Comparing both of the pictures in Fig.9, the shapes of particles in (a), which is for T_1 = 40 °C, are irregular, whereas those in (b), for T_1 = 80 °C with the external pressure pulses applied, are more spherical. It may be because the fragmented molten metal particles are solidified more slowly in the warmer water, approaching spherical shape by the action of the surface tension. An important point here is that steam exposition with fragmentation of

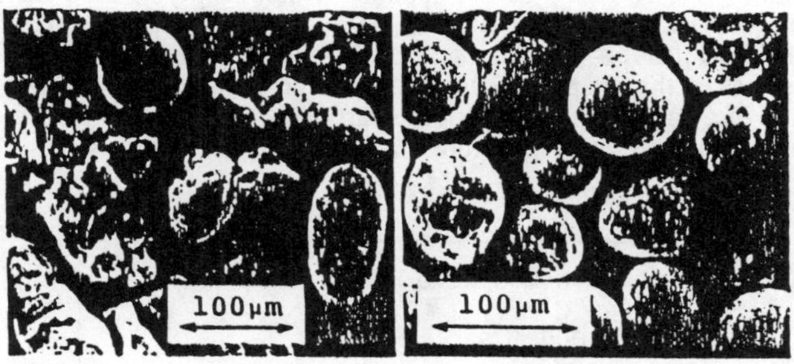

Fig.9 Electron microscope pictures of fragmented tin
(a) T_l =40°C
(b) T_l =80°C, with exploding wire

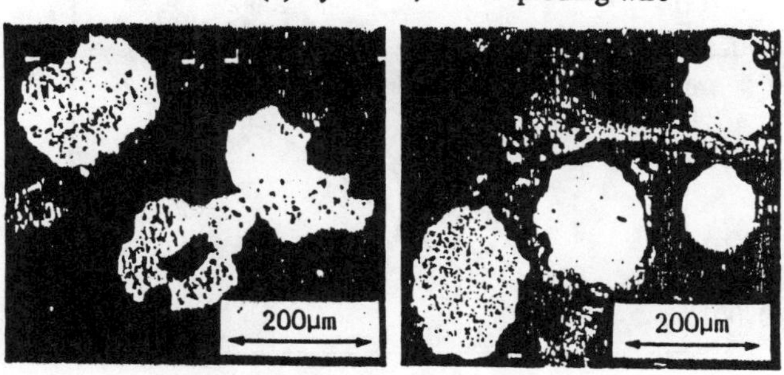

Fig.10 Electron microscope pictures of crosssections of fragmented tin
(a) T_l =40°C
(b) T_l =80°C, with exploding wire

the molten metal does not occur in the water of 80°C, unless external pressure pulses are applied. The use of exploding wire or another means of generating external pressure pulses is effective in obtaining spherical particles.

The electron microscopic pictures of the crosssections of the particles are shown in Fig.10. The pictures (a) and (b) correspond to (a) and (b) in Fig.9, respectively.

PRODUCTION OF AMORPHOUS PARTICLES OF PALLADIUM-SILICON ALLOY

The results mentioned so far indicate that metal powder by rapid solidification can be produced by making use of the thermal interaction between a molten metal and water. We attempted to manufacture the particles of amorphous alloy using $Pd_{80}Si_{20}$. The reasons why we chose this material were as follows: (1) The critical quenching rate for obtaining the amorphous phase is estimated to be around 10^4 K/s [see the value for $Pd_{82}Si_{18}$ in Table 1], (2) the melting point of this alloy is 760 °C which is lower than 1000 °C, the temperature attainable with our furnace, and (3) the alloy is more resistant to oxidation than tin because it contains palladium, a noble metal. The experiments have been carried out under the conditions of steam atmosphere, T_1 = 40 °C and T_{hi} = 1000 °C.

The size distribution of the fragmented particles is shown in Fig.11. Smaller particles are obtained by applying the pressure pulses by the exploding wire, though the mean diameter is a little larger than that of tin. Electron microscopic pictures of the particles are shown in Fig.12. The picture (a) shows the products obtained without applying the external pressure pulses, whereas (b) shows those with the external pressure pulses. The equivalent diameters range 45-90 μm for the both pictures. It seems that application of the external pressure pulses makes the shapes of the particles more spherical. Figure 12(c) is a magnified picture of a part of the surface of the particle at the center of Fig.12(b). A very smooth surface such as is seen in this picture is a characteristic of the amorphous phase.

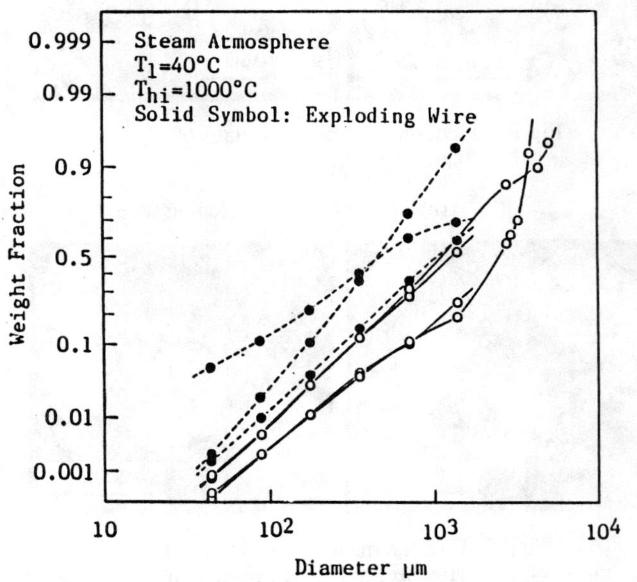

Fig.11 Mass distribution of fragmented Pd-Si alloy

The results of the X-ray diffraction analysis of the surfaces of the particles with different sizes are shown in Fig.13. A small peak of diffraction intensity from $Pd_{80}Si_{20}$, suggesting formation of crystalline structure, appears at $2\theta = 18$ deg for larger particles, but it disappears for smaller particles. Peaks at $2\theta = 13.7$ deg and 20 deg observable for the particles smaller than 45 μm [Fig.13(b)] are possibly due to foreign materials such as copper, originated from the exploding wire. We have judged that most of the smaller fragmented particles are of amorphous phase, the upper limit of the size being 350μm when the external pressure pulses are applied, and 180μm without pressure pulses.

CONCLUSION

We investigated first the factors influencing the fragmentation of a molten metal by thermal explosion. Particles with smaller sizes were obtained when the atmosphere above the water surface was filled with steam and the water temperature was low. We confirmed that fragmentation was promoted by applying external pressure pulses on the molten metal during film boiling, even under such conditions that only rather coarse particles could be obtained by spontaneous thermal interaction. In this case more spherical particles were formed when the bulk water temperature was higher. Furthermore, we found that the rate of cooling available by steam explosion was high enough for obtaining amorphous phase of a palladium-silicon alloy.

These results suggest that the method of fragmenting molten metal with the use of steam explosion is promising as a technique for producing rapidly solidified metals, though there still remains an issue of reducing the size of particles still more.

Finally we would like to thank Mr. Katsumi Goto, Mr. Hiroyasu Itamura and Mr. Masotoshi Fujino for their kind help and instruction for the X-ray diffraction analysis and for the electron microscopic photography.

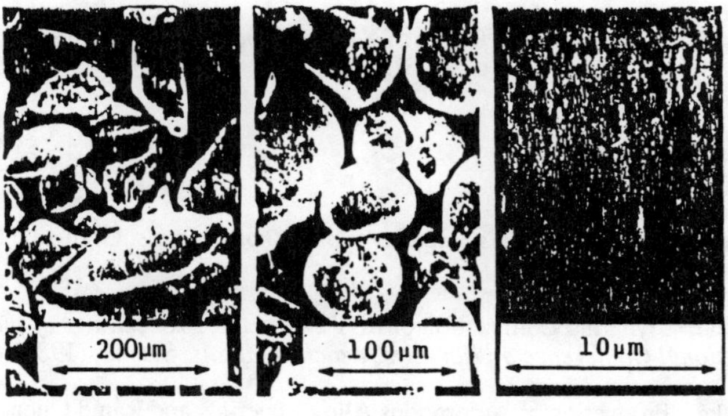

Fig.12 Electron microscope pictures
of fragmented Pd-Si alloy
(a) without using exploding wire
(b) with exploding wire
(c) magnified picture of surface(b)

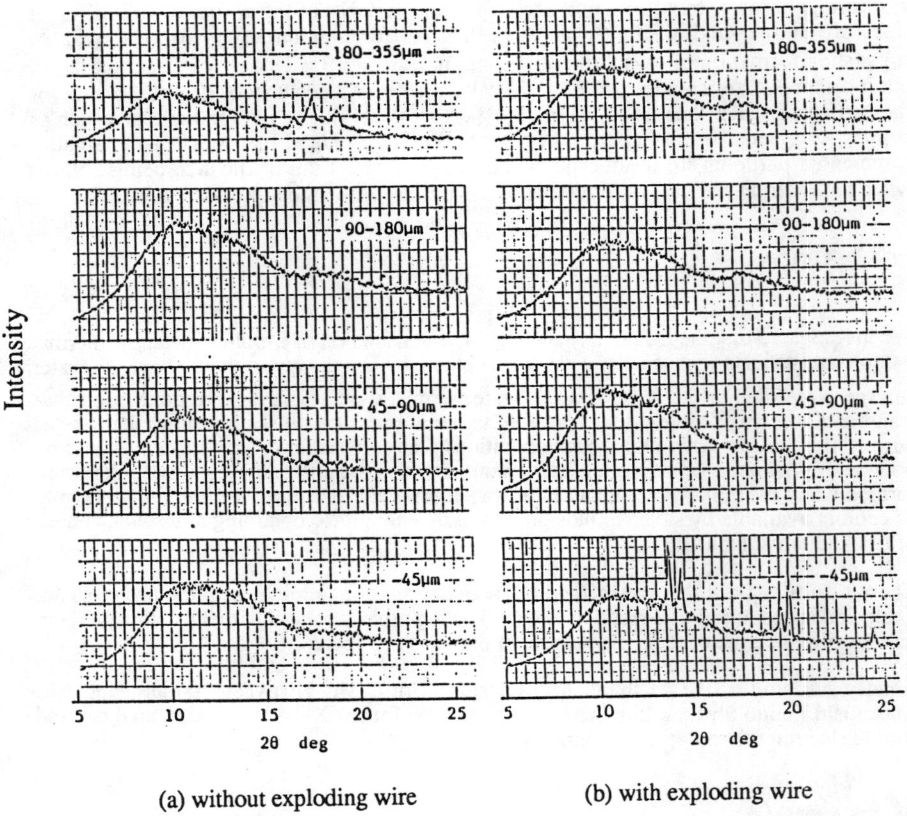

Fig.13 Result of X-ray diffraction analysis of Pd-Si alloy

REFERENCES

1. T. Masumoto(ed.), *Materials Science of Amorphous Metals*, Ohm-sha(1985), (in Japanese).

2. C.F.Cline, "Dynamic Compaction of RST Powders", *Proceedings of the 4th International Conference on Rapidly Quenched Metals,* Sendai, (1981), pp.129-132.

3. I. Ohnaka, "Production of Amorphous Alloy Materials and Rapid Quenching Techniques", *Journal of the Japan Society of Mechanical Engineers,* Vol.88(1985), pp.1060-1064 (in Japanese).

4. Japan Society of Mechanical Engineers(ed.), *Heat Transfer Data Book*(4th Edition), JSME(1986), p.318, (in Japanese).

5. A. Inoue, "Vapor Explosion", *Journal of the Heat Transfer Society of Japan,* Vol.19(1980), pp.22-28 (in Japanese).

6. R. Akiyoshi, S. Nishio and I. Tanasawa, "A Study on the Effect of Non-condensable Gas in the Vapor Film on Vapor Explosion", *Trans. JSME,* Series B, Vol.54(1988), pp.630-635, (in Japanese).

7. L.S. Nelson and L.D. Buxton, "The Thermal Interaction of Molten LWR Core Materials with Water", *Transactions of American Nuclear Society,* Vol.26(1977), pp.397-398.

8. L.S. Nelson and L.D. Buxton, "Steam Explosion Triggering Experiments with Oxide Corium-E Simulants", *Transactions of American Nuclear Society,* Vol.27(1977), p.663.

9. L.S. Nelson and L.D. Buxton, "Effects of Pressure on Steam Explosion Triggering in Corium-E Simulants", *Transactions of American Nuclear Society,* Vol.28(1978), pp.448-449.

10. L.S. Nelson, L.D. Buxton and M.N. Planner, "Steam Explosions with Molten Iron Oxides", *Transactions of American Nuclear Society,* Vol.30(1978), pp.365-366.

11. L.S. Nelson and P.M. Duda, "Steam Explosions of Molten Iron Oxide Drops: Easier Initiation at Small Pressurizations", *Nature,* Vol.296(1982), pp.844-846.

12. V.H. Arakeri, I. Catton and W.E. Kastenberg, "An Experimental Study of the Molten Metal/Water Thermal Interaction Under Free and Forced Conditions", *Nuclear Science and Engineering,* Vol. 66(1978), pp.153-166.

13. A. Inoue and S.G. Bankoff, "Destabilization of Film Boiling Due to Arrival of a Pressure Shock: Part 1 Experimental", *Trans. ASME ,Journal of Heat Transfer* , Vol.103(1981), pp.459-464.

Summary of General Discussion of Session on Solidification and New Materials by Rapid Quenching

R. Viskanta, I. Ohnaka, and Y. Hayashi

The papers presented in these two sessions have demonstrated that mathematical modeling is a powerful tool, especially if supplemented by experimental modeling, to aid in the understanding and control of complex solidification systems. A wide range of processes involving solidification of multicomponent systems are being simulated mathematically with different levels of detail and sophistication to understand the process and to establish optimal operating conditions and control. Recent macroscopic transport models have been quite successful in predicting important features of solidification in binary systems. However, despite some successes, difficulties have been experienced in achieving good quantitative agreement between predictions and measurables, such as solidus and liquids front locations and temperature and composition fields. Such difficulties may be a consequence of several model assumptions as well as deficiencies in the requisite property data base.

The key research questions in solidification of mixtures which have been identified include:

- Refinement of macroscopic models to relax some of the assumptions
- Development of improved descriptions of transport in the mushy region to better account for anisotropic features and parameters
- Development of capability to model break-up of dendrites and floating of crystals
- Interaction of micro- and macro- processes during solidification may be important, and an ability to model is essential in those physical situations where it is important
- Three-dimensional models of solidification with turbulent flow in the melt are needed for large ingots and castings
- Improved numerical algorithms are needed to solve the macroscopic transport equations numerically, because time-dependent two- and three- dimensional solidification problems require very large computer resources
- Ability to accommodate greatly different time and length scales during solidification needs to be developed
- Modeling of nonequilibrium phenomena in some situations (i.e., rapid solidification) is very important but has not been incorporated in macroscopic models
- Implementation of both macroscopic and micro-macroscopic models requires development of improved property data base
- Description of peripheral heat transfer from the solidifying system are difficult to find in the literature
- Most of all, validation of mathematical models requires reliable experimental data, and innovative experimental techniques need to be developed to acquire the needed fundamental data in solidifying systems, but particularly in the mushy region
- **Coupled thermomechanical models need to be developed for solidification processes to enable prediction of combined heat and stress-strain.**
- **Modeling of solidification during filling time in case of thin shape casting**
- **Modeling of atomization and heat transfer including solidification in powder production**

GENERAL TOPICS (1)

Production and Heat Transfer Characteristics of Liquid Ice as New PCM

SHOICHIRO FUKUSAKO and MASAHIKO YAMADA
Department of Mechanical Engineering
Hokkaido University
N-13, W-8, Kita-ku, Sapporo 060, Japan

ABSTRACT

A brief review concerning some of investigations for freezing of the aqueous binary solution to produce the liquid ice and for melting of the liquid ice to release the cold thermal energy is presented. Attention is initially focused on the morphological studies of the frozen layer obtained using an aqueous binary solution. It is demonstrated that the solid-gas-liquid three-phase fluidized bed system is one of the best heat-transfer systems for the liquid ice.

INTRODUCTION

Ice thermal storage is recently becoming a widely accepted form of electrical load management. Air conditioning when treated in summer as instantaneous load is one of diverse loads which violate sound economic practice for the electric utility. One way of addressing load management for the air conditioning load is through the ice thermal storage, which is greatly available as a load management device.

A number of studies of ice thermal storage systems [1, 2], in which common ice is produced using extra electric power supply in the midnight and is used for air conditioning load during the daytime, have been performed. Common ice has been adopted as the phase change medium (PCM) in such a system. The common ice has an usefulness of being able to make the reservoir smaller than that of water due to its large latent heat of fusion.

Recently, a great attention has been paid to liquid ice as PCM instead of the common ice, because the liquid ice is practically transportable by use of a piping and because the melting heat-transfer performance is favorably controllable to be highly great. Then, utilization of the thermal storage system using the liquid ice [3] has been extensively examined. However, there remains a variety of unknown characteristics of both producing the liquid ice by freezing the aqueous binary solution and releasing the cold thermal energy from the liquid ice.

The objective of the present paper is to review briefly some of the studies on freezing of the aqueous binary solution to produce the liquid ice and on melting of the liquid ice to release the cold thermal energy. Attention is first focused on the morphological investigations of the frozen layer which is made of the aqueous binary solution. Three-phase fluidized bed system of the ice-air-solution is revealed to be quite available heat-transfer system for the liquid ice.

PREVIOUS WORKS

Freezing and Melting of Phase Change Medium

Previous works on freezing and melting problems of phase change medium along horizontal or vertical surface, around horizontal cylinder, and inside horizontal tube or sphere have been extensively reviewed in [4].

Recently, Hayashi et al. [5] proposed latent heat thermal energy storage using an aqueous solution as PCM and studied freezing heat-transfer characteristics along the frozen layer within a shell tube, both experimentally and analytically. Himeno et al. [6] investigated the characteristics of solidification of a binary mixture of p-$C_6H_4Cl_2$ and p-$C_6H_4Br_2$ and found that non-uniform concentration profiles were formed in both solid and liquid phases due to segregation. Matsuoka et al. [7] examined a composite thermal energy storage material which consists of organic and copper fiber and claimed that supercooling on freezing was greatly reduced.

Himeno et al. [6] studied the characteristics of melting of a binary mixture of p-$C_6H_4Cl_2$ and p-$C_6H_4Br_2$ in a horizontal cylindrical capsule. They found that non-uniform concentration profiles were formed in both solid and liquid phases since density stratification occurred and that these non-uniform density profiles generated several horizontal cells by free convection in the melting process. The melting process of an unified n-octadecane between two isothermal concentric horizontal cylinders was determined by Betzel and Beer[8] using experimental techniques and using a combined analytical and numerical method.

Freezing of Binary Solution

A number of investigations have been reported in the literature pertaining to the freezing of binary solutions. O'Callaghan et al. [9] adopted a model which accounts for the mushy zone to investigate theoretically heat and mass transfer during solidification of an eutectic binary solution. In their analysis, the velocity of propagation of the liquid-mush interface was taken to be identical to that of the solid-liquid interface. An experimental and theoretical study of the selective freezing of a dilute solution on a cold ice surface was reported by Fang et al.[10]. The comparison between theory and experiment showed good agreement. Huppert and Worster [11] studied the solidification of Na_2CO_3-water solutions and compared their prediction with experimental data. In their model, the solid fraction in the mushy zone was assumed to be constant and latent heat of fusion was released at the interface. Worster[12] determined theoretically the effect of the mushy zone on the solidification of an alloy cooled from below. His model accounted for the simultaneous action of heat and mass diffusion.

Bennon and Incropera [13] derived a set of continuum conservation equations for binary solid-liquid phase change systems by integrating semi-empirical laws and microscopic descriptions of transport behavior with principles of classical mixture theory. They [14] also used these conservation equations to model and solve numerically the problem of solidification of a binary alloy in a cavity cooled through one of its side wall. Comparisons between experimental and predicted results for the rectangular cavity by Beckermann and Viskanta [15] and Christenson et al. [16] agree with respect to important qualitative features, such as the nature of flows and solute redistribution, but fail to achieve close quantitative agreement. Christenson et al. [17] also conducted an experiment of solidification for an aqueous sodium carbonate solution in a horizontal cylindrical annulus with varying initial concentration. The time evolution of the mean properties of a binary system as one component crystallizes at a constant temperature boundary was studied by Leitch [18].

Recently, microscopic study of coupled heat and mass transport phenomena during the unidirectonal solidification of a binary solution were carried out by Kourosh et al.[19, 20] who used a specially designed freezing stage for transmitted light microscopy. Nelson et al. [21] studied numerically solidification of a binary Na_2CO_3 solution in a horizontal cylindrical annulus using a continuum formulation with a control-volume based, finite-difference scheme. Simple and effective numerical and analytical solution of one-dimensional solidification of dilute binary alloys was proposed by Gupta [22]. Cao and Poulikakos [23] reported an experimental investigation on the solidification of an ammonium chloride aqueous solution in a rectangular cavity cooled through its top wall. Freezing of water-salt solutions on a horizontal wall was studied experimentally and analytically by Braga and Viskanta [24], who proposed a model for the temperature distribution during solidification from below of subeutectic aqueous solutions.

FREEZING OF AQUEOUS BINARY SOLUTION WITHOUT MAIN FLOW

Freezing Characteristics of Ethylene Glycol Solution [25]

Experimental apparatus and procedures. The test section is made of transparent lucite box with inner dimensions of 50x40x100 mm except both the front side of pair glass and the bottom side of copper plate with 3 mm thickness as the cooled plate. The temperature of the plate was favorably controlled by spraying the temperature-controlled brine from below. Ethylene glycol solutions ranging from 5 to 50 wt% in concentration were adopted as the testing fluids. Cooling temperature T_w, initial concentration C_0, and uniform initial temperature T_∞ were set as parameters. Observations of the freezing characteristics such as the shape of frozen layer and the freezing rate, and the temperature-distribution variation were extensively carried out.

Morphology of frozen layer and characteristics of freezing behavior. Three types of frozen layer, such as white opaque, dendritic, and slant dendritic ice layer were in general observed. The dendritic ice layer (Fig. 1) was generally obtained, while the opaque ice layer (Fig. 2) for low concentration and quite lower cooling temperature than fusion temperature T_f, and one (Fig. 3) for high concentration. In addition to the above ice layers, the slant dendritic ice layer (Fig. 4) was observed. In this case, there may be also a quite interesting fact that the ice crystal is easy to come off the cooled plate.

Though the mechanism how the ice crystal separates from the cooled surface has not yet been known well, both the buoyancy force acting to the ice crystal and the non-equilibrium melting at the interface between the ice crystal and cooled surface may mainly cause the separation. This characteristics may be favorably utilized for the process of making the liquid ice.

To determine the effects of the parameters on the freezing characteristics, following non-dimensional temperatures θ_w and θ_∞ were defined as follows.

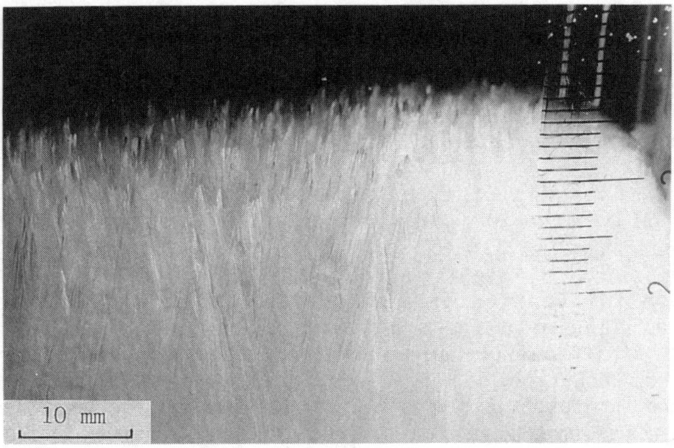

FIGURE 1. Dendritic frozen layer, C_0=30 wt%, T_w=-28 °C, T_∞=-13 °C

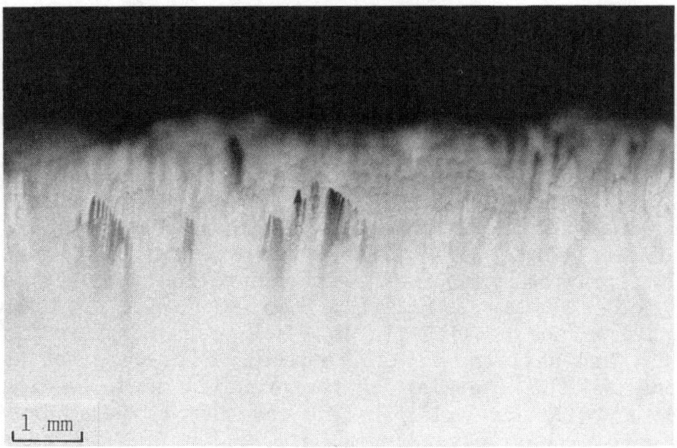

FIGURE 2. White opaque frozen layer, C_0=30 wt%, T_w=-53 °C, T_∞=-2.5 °C

$$\theta_w = (T_f - T_w)/(T_0 - T_f) \quad (1) \qquad \theta_\infty = (T_\infty - T_f)/(T_0 - T_f) \quad (2)$$

The large θ_w and small θ_∞ correspond to quick freezing, while the small θ_w and large θ_∞ correspond to slow freezing. The observed results are plotted in Fig. 5.

The resultant figure is divided into three regions as denoted by the dashed-dotted lines. Region A in Fig. 5 denotes the case of non-freeze or slant dendritic ice-layer formation, region C denotes the case of white opaque ice layer, and in the region B dendritic ice layer forms except for the concentration of 50 wt%. It appears from the figure that the white opaque and the slant dendritic ice layers may be formed for the quick and the slow cooling, respectively.

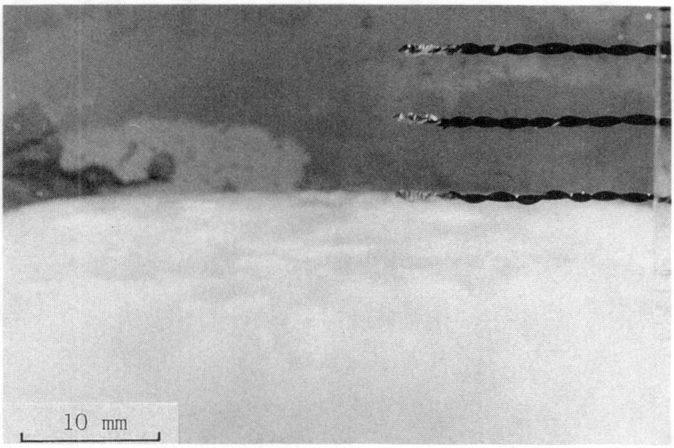

FIGURE 3. White opaque frozen layer, C_0=50 wt%, T_w=-56 °C, T_∞=-28 °C

FIGURE 4. Slant dendritic frozen layer, C_0=10 wt%, T_w=-6.5 °C, T_∞=-2.0 °C

FREEZING OF AQUEOUS BINARY SOLUTION WITH MAIN FLOW

Freezing along a Vertical Wall [27, 28]

Experimental apparatus and procedures. The experimental apparatus consists essentially of the test section, testing-solution circulating loop, cooled-brine circulating loop, and associated instrumentations. The vertical duct has a rectangular cross section of 70 mm x 50 mm, which is made of both transparent lucite plate and pair glass to observe the frozen layer. A copper plate of 50 mm in width, 500 mm in length, and 3 mm in thickness was utilized for the cooled plate.

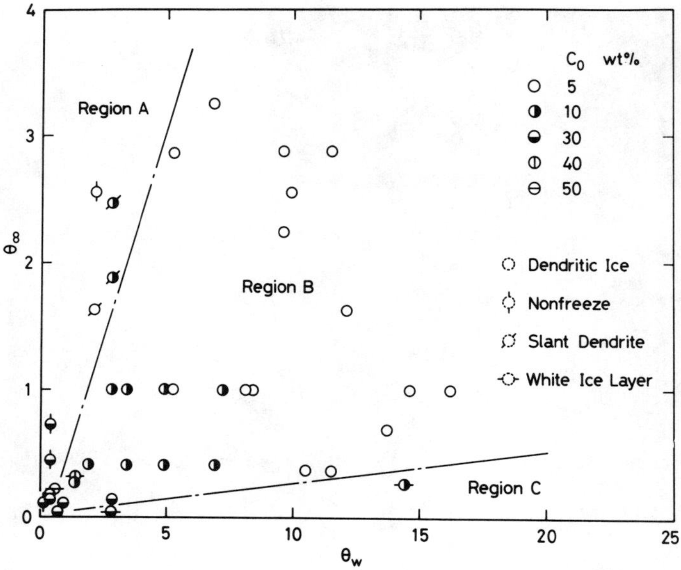

FIGURE 5. Data arrangement by non dimensional temperature

Ethylene glycol solutions of 10 and 30 wt% in concentration were utilized as testing fluids. Observations of the onset and the growing process of the frozen layer and measurements of the frozen-layer thickness at the steady state were successively performed under a variety of parameters of fluid concentration C_∞, fluid temperature T_∞, cooled-wall temperature T_w, and fluid velocity U_∞. The state, in which any change in both the frozen-layer thickness and its surface behavior has not been observed over an hour, was defined as a steady state.

Morphology of frozen layer at steady state. Three kinds of the frozen layers were observed depending on the fluid velocity and freezing velocity, as shown in Fig. 6. As the effect of main fluid flow, the followings are considered; (1) the heat transfer at the interface between flow liquid and frozen layer is promoted; (2) the solute, which is contained among the ice crystals of the frozen layer, is washed away, and then concentration distribution near the interface is disturbed; and (3) the needle-shaped ice crystals are appreciably destroyed.

The cause for the formation of the dendritic ice crystals is considered to be the constitutional supercooling which may be based on the precipitation of the solute to the freezing front. As the solute at the interface is washed away, the concentration distribution is forced to vary, which results in somewhat different dendritic frozen layer, as shown in Fig. 6 (a) and (b). It seems that the fine needle-like ice crystals may not fully grow because the tip of the dendritic ice crystals is destroyed by the fluid flow (see Fig. 6(a)). Furthermore, as the solute near the interface is washed away by the flow, the frozen layer becomes to be more dense and eventually to hold completely the solute within it, and the fine-needle ice crystals progressively tends to disappear as shown in Fig. 6(b).

The line grooves were observed to take place on the frozen-layer

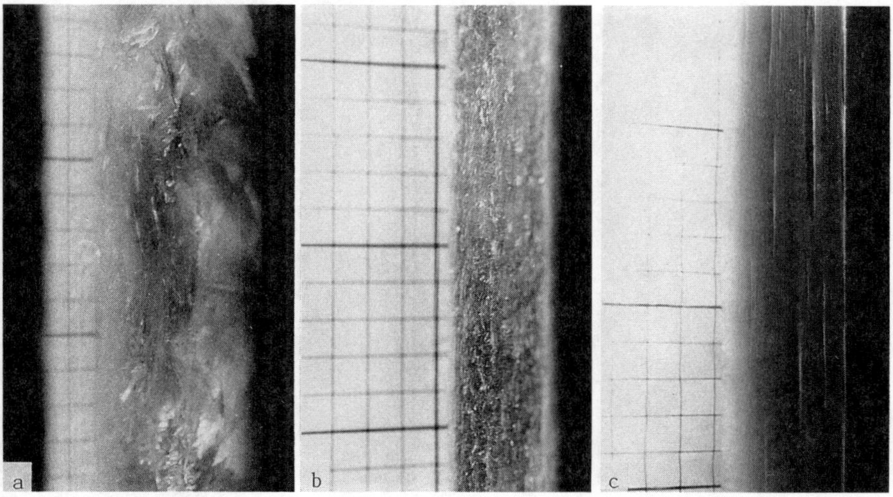

FIGURE 6. Morphology of ice layer, (a) C_0=30 wt%, T_w=-19 °C, T_∞=-12.5 °C, U_∞=0.02 m/s, (b) C_0=30 wt%, T_w=-18.5 °C, T_∞=0.0 °C, U_∞=0.02 m/s, (c) C_0=10 wt%, T_w=-12 °C, T_∞=-2.0 °C, U_∞=0.3 m/s

surface when the fluid velocity is high, as shown in Fig. 6(c). It appears that the grooves may be formed since the high concentration solute, which is washed away from the frozen layer, flows along the surface and dissolves the frozen layer (high concentration causes depression of freezing point).

Another interesting effect due to fluid flow is an occurrence of the separations of the frozen layer and of the ice flake, as shown in Fig. 7. When the fluid temperature and the cooled-surface temperature are high, the separations of both the ice layer and the ice flakes were in general observed.

The morphologies of the frozen layer observed are divided into four regions as shown in Fig. 8. Region A of the figure corresponds to the condition under which the frozen layer of Fig. 6(a) is formed. In region B the frozen layer shown in Fig. 6(b) is in general observed. On the other hand, the frozen layer with vertical line grooves along the surface (see Fig. 6(c)) is formed in region C, and for greater Reynolds number Re and smaller cooling temperature ratio θ_c $(=(T_f-T_w)/(T_\infty-T_f)$, T_f:fusion temperature, T_w:cooled wall temperature, T_∞:inlet fluid temperature), there takes place no frozen layer (region D). It was also found that the separation of the frozen layer occurred in region B adjacent to region D.

Freezing along a Horizontal Tube [29]

Experimental apparatus and procedures. The experimental apparatus essentially is quite similar to that for freezing along a vertical wall. The duct has a cross section of 200 mm x 120 mm, which is made of transparent lucite plate and pair glass to observe the freezing characteristics along the circular tube. The test tube, which is a copper concentric tube (41.2 mm o.d.), was placed horizontally at the center of the test section. To maintain the uniform temperature of the tube surface, the temperature-controlled coolant was intensively spouted from many holes drilled through inner tube wall. Forty C-A thermocouples of 0.1 mm in diameter were peripherally distributed in order to measure the

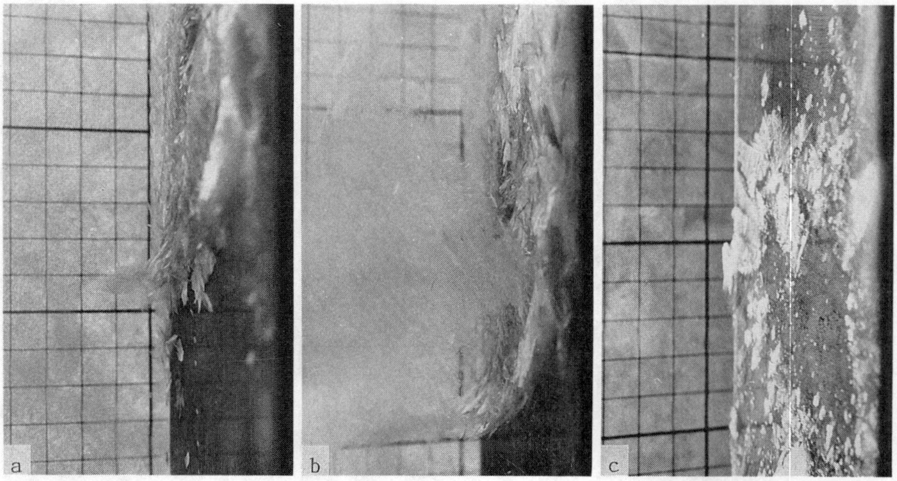

FIGURE 7. Morpholoy of ice layer, C_0=30 wt%, T_w=-18.5 °C, T_∞=0.0 °C, U_∞=0.1 m/s

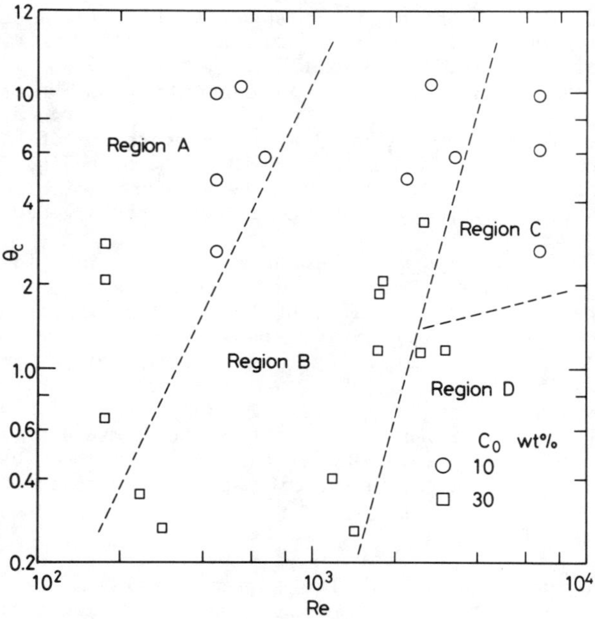

FIGURE 8. Data arrangement by nondimensional temperature

temperature distribution along the tube. The parameters covered were solution concentration, flow-fluid temperature, cooled-surface temperature, and flow velocity. Attention was focused on observations of both the onset and the growing processes of the frozen layer and also quantitative measurements of the frozen rate at the steady state were carried out.

Morphology of frozen layer at steady state. The effects of the cooled-surface temperature, fluid temperature, fluid concentration, and fluid velocity on the freezing characteristics were extensively determined. Figure 9 demonstrates the effect of the fluid velocity on the freezing behavior along a horizontal cooled tube for fluid concentration in 5 wt%, cooled surface temperature -15 °C, and fluid temperature 0 °C. When there is no main flow (see Fig. 9(a)), dendritic frozen layer normal to the cooled surface forms. It appears that the solute segregated due to the freezing may fall down along the frozen surface, which tends to melt the downward frozen surface, thus resulting the vertical line grooves, where typical dendritic shape of ice crystals gradually disappears. As the fluid velocity increases, the location of the step which was formed around the separation points tends to advance upward as shown in Figs. 9(b) and 9(c). For U_∞ = 0.07 m/s (see Fig. 9(d)) there appears no step and no clear dendritic frozen surface but somewhat mirror-like frozen surface due to melting action.

Mapping of the frozen-layer features observed under the parameters covered is shown in Fig. 10. In region A a trace of dendritic ice crystals along the upper part of the frozen layer on the tube (see Fig. 9(b)) is in general observed. In this region where the flow velocity is

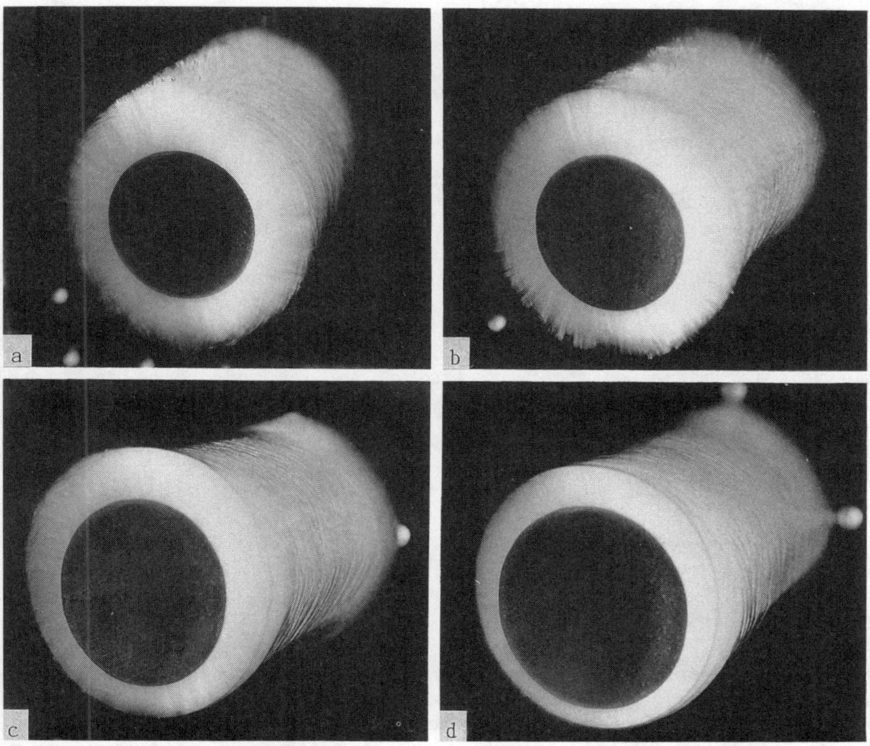

FIGURE 9. Effect of velocity on freezing, C_0=15 wt%, T_w=-12 °C, T_∞=0.0 °C, (a) U_∞=0.0 m/s, (b) U_∞=0.01 m/s, (c) U_∞=0.03 m/s, (d) U_∞=0.07 m/s

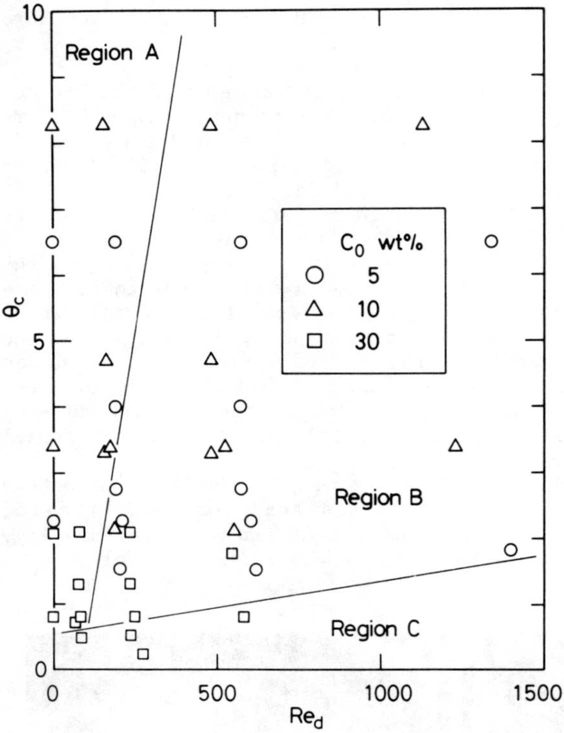

FIGURE 10. Data arrangement by nondimensional temperature

quite small, the effect of melting due to fluid flow is to be neglected and also the vortices in the wake is quite weak. Therefore, it appears that the upper part of the frozen layer on the tube may be scarcely influenced by the fluid flow. In region B the frozen layer as shown in Fig. 9(d) is observed. Surface of the frozen layer is rather smooth due to the melting action and there is no trace of the dendritic ice crystals on the surface. This is interpreted in terms of the fact that as the fluid velocity increases the dendritic ice needles tend to be destroyed and also there take place the vortices downstream the tube, which causes the effect of the melting on the all of the frozen surface around the tube. Region C denotes the case of non-freeze or non-two-dimensional frozen-layer formation.

MELTING OF QUIESCENT LIQUID ICE

Melting Along a Horizontal Tube [31]

Experimental apparatus and procedures. The experimental apparatus fundamentally consists of the test section, test-solution circulating loop, air-circulating loop, and measurement instrumentations. The testing solution, which was cooled in advance in the reservoir, was derived by a circulating pump. The test section to determine the melting heat-transfer rate was made up of transparent lucite plate, with inside dimensions of

240 mm width, 90 mm depth, and 270 mm height.

Three kind tubes (D=30, 20, and 15 mm) were heated electrically by stainless foil (20 mm in thickness) being wound up on lucite tube. The test tube was horizontally set at 100 mm distance from the distributor plate. Mixture of fine ice particles with average 1 mm in diameter and ethylene glycol solution was adopted as a testing liquid ice. Ice packing factor (I.P.F.) is 62 % through all the experiments.

Melting characteristics along a cylinder. Figure 11 shows the melting characteristics of the liquid ice along a horizontal tube. It is to be noted that the melting interface mainly tends to advance to horizontal direction, which is quite different from the normal melting behavior well known when P.C.M. or pure ice around a horizontal tube melts. In the figure, characteristics of the stratified layers are distinctly indicated along with schematic drawings. Formation of the stratified-layer interface may be considered as follows. As the melting progresses, there forms the potentially stable concentration distribution after some of interaction between melt water and solution. The liquid near the bottom of the tube, whose density is greatest within the melt zone, is driven upwards by buoyancy forces along the heated tube; but it is gradually retarded by minus buoyancy force because the density of the environmental melt liquid decreases vertically; eventually upward flow is forced to change course to move horizontally or downwards; this fact appears to cause the stratified-layer interface. Thus, the interface is the double-diffusive interface with a couple of heat and concentration diffusion.

Heat-transfer characteristics. Figure 12 presents the effect of the tube diameter on the average heat-transfer coefficient, h, as the time, t, elapses. It reveals in general that the heat-transfer coefficient initially decreases sharply, passes through a minimum, then increases gradually, and eventually levels off or somewhat decreases. Sharp decrease in heat-transfer coefficient at first stage suggests that heat

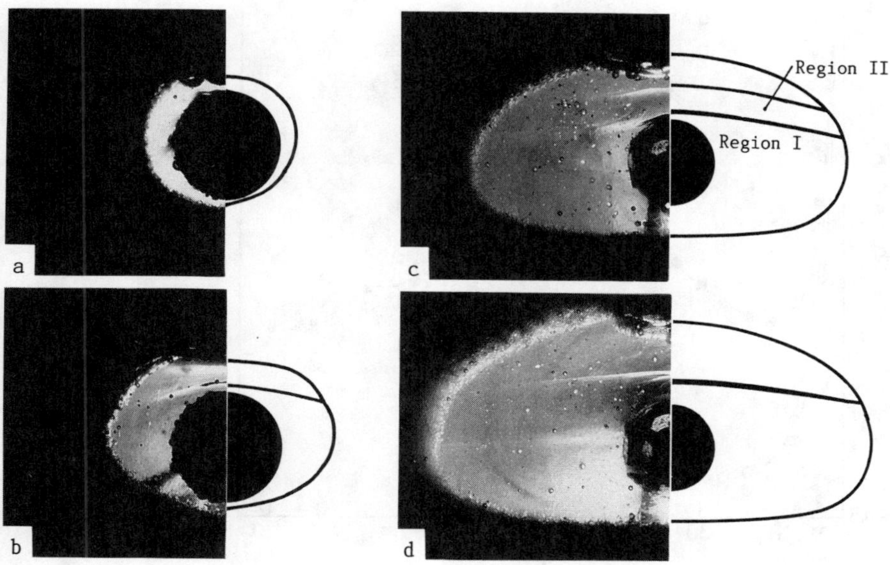

FIGURE 11. Melting behavior and characteristics of straitified layer, (a) t=10 min., (b) t=60 min., (c) t=130 min., (d) t=145 min.

transfer is mainly performed by conduction. Increasing heat transfer after taking a minimum corresponds both to onset of free convection motion within the melt zone and to increase in its motion based on the development of the melt zone. Variation in heat-transfer coefficient in the free convection region denotes the delicate effect of the appearance or disappearance of the stratified-layer interface on the melting heat-transfer characteristics in such a system.

Melting Along a Bundle of Tubes [31]

Experimental apparatus and procedures. Experimental apparatus and procedures are quite same as those for previous experiments of melting along a tube except that eleven heated tubes of 20 mm in diameter were arranged in staggered array with 50 mm in pitch in the test section.

Melting characteristics along a bundle of tubes. Figure 13 indicates the melting characteristics along a bundle of heated tubes immersed in the liquid ice. As the time elapses, the melting interface is first observed to mainly progresses to horizontal direction, which is quite identical to that around a horizontal heated tube (see Fig. 11). When 40 min. elapsed, there happen the combinations of the melt zones as shown in Fig. 13(b). Eventually, the melt zone shown in Fig. 13(d) is attained.

Heat-transfer characteristics along a bundle of tubes. Figure 14 shows the local heat-transfer coefficient distributions against the angle from the bottom of the tube. The distribution is identical to that for single tube when 30 min. elapsed. It is, however, of great interest that the

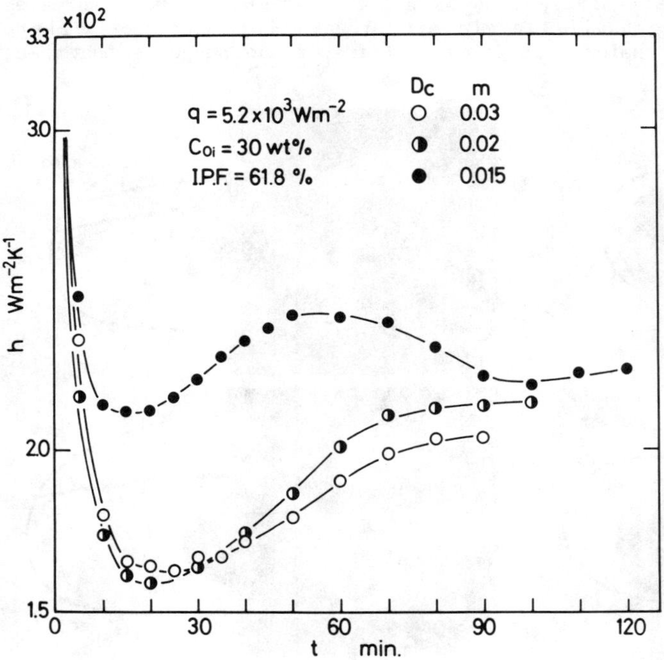

FIGURE 12. Average heat-transfer coefficient behavior

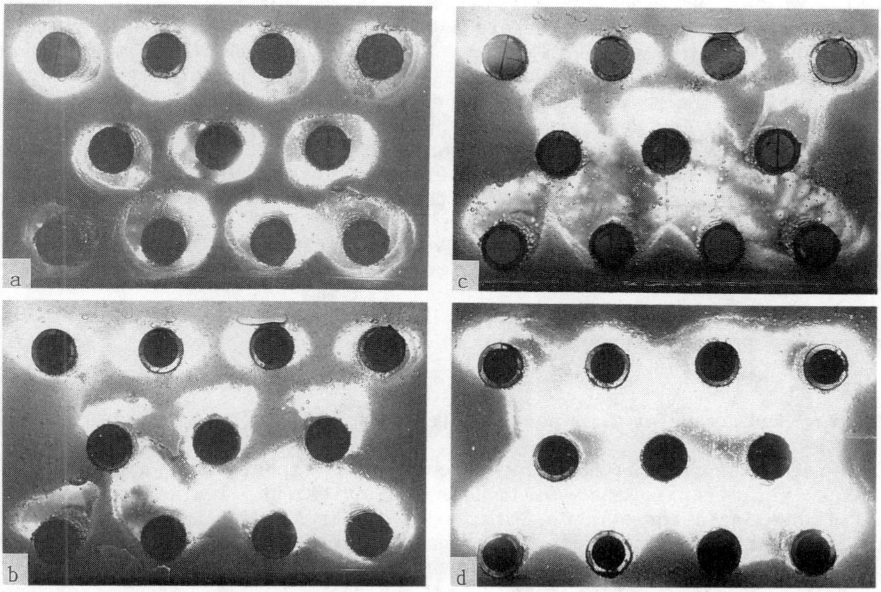

FIGURE 13. Melting behavior in quiescent bed, (a) t=30 min., (b) t=50 min., (c) t=40 min., (d) t=60 min.

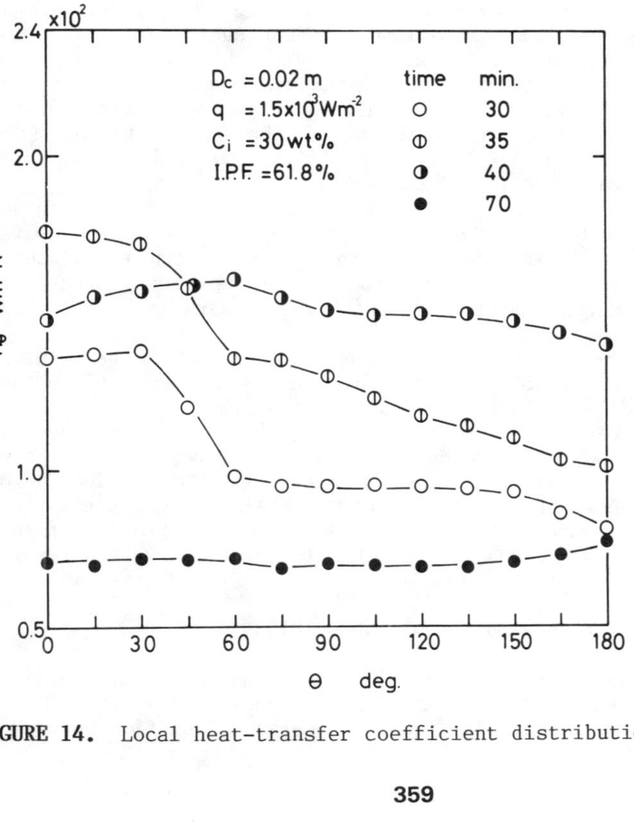

FIGURE 14. Local heat-transfer coefficient distribution

heat transfer increases greatly between elapsed times of 35 min. and 40 min., which exhibits quite different characteristics from those of single tube. This is interpreted in terms of the combination of the melt zones indicated in Figs. 13(b) and 13(c). As pointed out previously, there is the stratified layer within the respectively independent melt zone shown in Fig. 13(a); the density of the upper layer in the melt zone is in general greater than that of the lower layer. According to combination of the melt zones, the fluid of the lower layer in the melt zone around the upper row tube changes places with the fluid of the upper layer in the melt zone around the lower row tube; the fluid with low temperature and low concentration flows into the melt zone of the upper row tube, which activates the convection, thus resulting in an increase in heat transfer.

MELTING OF FLUIDIZED LIQUID ICE

Melting Heat Transfer from a Horizontal Tube [32]

Experimental apparatus and procedures. The experimental apparatus is basically similar to that for melting of quiescent liquid ice. To make solid-gas-liquid three phase fluidized bed, air was utilized as an acting gas. The air was introduced to the test section after passing through the temperature controlling section and orifice. In order to get uniform air flow, the calming section made up of honeycomb box was located at the bottom of the test section. The distributor plate consisted of a piece of 200 mesh brass wire screen was sandwiched between two perforated lucite plates with holes of 2 mm in diameter spaced 5 mm center-to-center.

Melting heat-transfer characteristics. In Fig. 15, average heat-transfer coefficient and bulk temperature of test section for solid-gas-liquid three-phase fluidized bed are presented as a function of the elapsed time. Except a period of just starting operation, the heat-transfer coefficient provides uniform value, while the bulk temperature increase gradually based on the melting of the ice particles (this corresponds to decrease in I.P.F.). After about 120 min. elapsed, the heat-transfer coefficient decreases gradually.

This is attributed to the fact that the frequency of the contacts between the ice particles and the heated tube is lowered by the melting of the ice particles and also the heat capacity of the bed is reduced as the time elapses. In the present experiments, the uniform heat-transfer coefficients were virtually attained for the ice packing factors (I.P.F.) ranging from 62 to about 25 %.

Comparison of the average heat-transfer coefficient between the solid-liquid two-phase fixed bed and the solid-gas-liquid three-phase fluidized bed is indicated in Fig. 16. It is clear that the values of the average heat-transfer coefficient for three-phase fluidized bed are apparently over twenty times as large as those for two-phase quiescent bed, It appears that increasing air-flow rate into the liquid ice may cause increase in heat transfer. The characteristics of the Fig. 16 suggest that the heat-transfer system of the ice-air-liquid three-phase fluidized bed system is enormously available for heat-removal system from the liquid ice.

Melting Heat Transfer from a Bundle of Tubes [31]

From a technological point of view, it is of great importance to determine the melting heat-transfer characteristics from a bundle of tubes immersed within the ice-air-liquid three-phase fluidized bed.

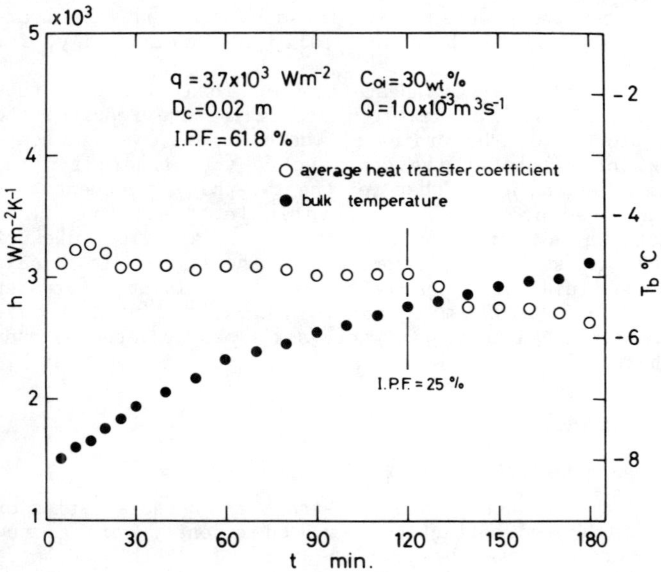

FIGURE 15. Average heat-transfer coefficient behavior

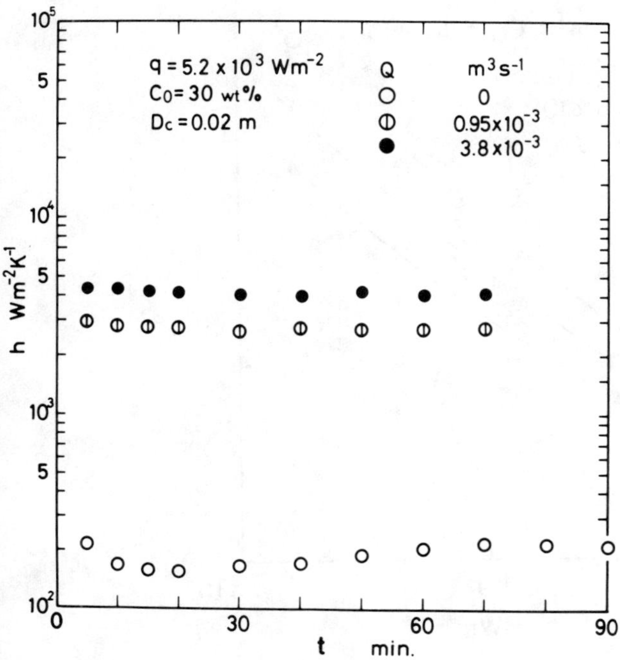

FIGURE 16. Comparison of average heat-transfer coefficient

Vertical and horizontal distributions of heat-transfer coefficient in a tube bundle were extensively determined. It was found that the heat-transfer coefficient was uniform, both vertically and horizontally, for the parameters covered.

The presentation of the melting heat-transfer results for a tube bundle is made in Fig. 17. With a view to generalizing the results, the quantity h_{fb}/h_c is plotted on the ordinate, where h_{fb} is average heat-transfer coefficient of a tube at the center of a tube bundle for the three-phase fluidized bed and h_c is that for the two-phase quiescent bed. The heat flux is adopted as the abscissa variable. There are three groupings of the data which correspond respectively to air-flow rates of 1.0, 2.0, and 3.8×10^{-3} m^3/s. From an inspection of the figure, it reveals that the effect of the fluidization on the heat transfer is quite greater than expected. For $Q = 3.8 \times 10^{-3}$ m^3/s and $q = 1.5 \times 10^3$ W/m^2 the heat transfer for the fluidized bed is about twenty-six times as large as that for the quiescent bed.

CONCLUDING REMARKS

This paper considered the production and heat-transfer characteristics of the liquid ice, to which great attention has been recently paid as new PCM. The article reported first the previous works on both the heat-transfer problems of the phase change medium and freezing problems of the aqueous binary solutions. Consideration was mainly focused on melting heat-transfer characteristics of the liquid ice. It was demonstrated that

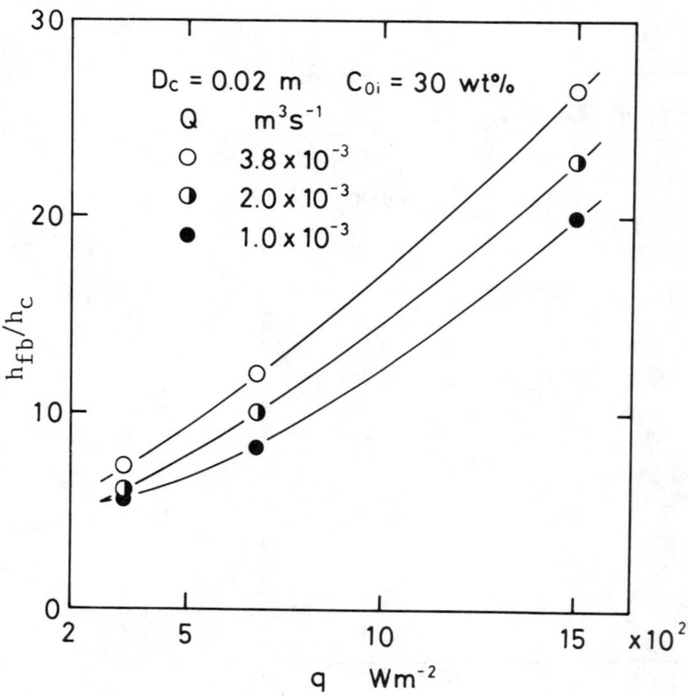

FIGURE 17. Heat-transfer characteristics in three-phase fluidized bed

the liquid ice was greatly available when used as the solid-gas-liquid three-phase fluidized bed.

REFERENCES

1. Miyasaka, A., Current Trends of Ice Storage System in USA, Refrig. J., vol.714, pp.32-42, 1987.
2. Sagara, N., Performance of Glycol Ice Storage System With Ice-On-Coil Type Heat Exchanger, Refrig. J., vol.714, pp.43-51, 1987.
3. Endo, M., and Hoshino, M., Crystal Liquid Ice Thermal Storage System, Refrig. J., vol.715, pp.25-30, 1987.
4. Fukusako, S., and Seki, N., Heat Transfer Problems on Ice and Snow, Cheng, K.C., and Seki, N., eds., Freezing and Melting Heat Transfer in Engineering - Selected Topics on Ice-Water Systems, and Welding and Casting Processes, Hemisphere Pub. Corp. (in print).
5. Hayashi, Y., Kunimine, K., and Yamaguchi, K., Study on Latent Heat Thermal Energy Storage Using Aqueous Solution as PCM, Trans. JSME, vol.54, pp.452-458, 1988.
6. Himeno, N., Hijikata, K., and Sekikawa, A., Latent Heat Thermal Energy Storage of a Binary Mixture-Flow and Heat Transfer Characteristics in a Horizontal Cylinder, Int. J. Heat Mass Transfer, vol.31, pp.359-366, 1988.
7. Matsuoka, M., Tsubohara, K., and Yasukawa, T., Behavior of Phase Transformation and Heat Transfer Characteristics in a Latent Heat Storage Unit With Composite Material, J. Chem. Engng. Japan, vol.21, pp.238-243, 1988.
8. Betzel, T., and Beer, H., Solidification and Melting Heat Transfer to an Unifixed Phase Change Material (PCM) Encapsulated in a Horizontal Concentric Annulus, Wärme- und Stoffübertr., vol.22, pp.335-344, 1988.
9. O'Callaghan, M.G., Cravalho, E.G., and Huggins, C.E., An Analysis of Heat and Solute Transport During Solidification of an Aqueous Binary Solution, Int. J. Heat Mass Transfer, vol.25, pp.553-573, 1982.
10. Fang, L.J., Cheung, F.B., Linehan, J.H., and Pederson, D.R., Selective Freezing of a Dilute Salt Solution on a Cold Ice Surface, J. Heat Transfer, vol.106, pp.385-393, 1984.
11. Huppert, H.E., and Worster, M.G., Dynamic Solidification of a Binary Melt, Nature, vol.314, pp.703-707, 1985.
12. Wroster, M.G., Solidification of an Alloy from a Cooled Boundary, J. Fluid Mech., vol.167, pp.481-501, 1986.
13. Bennon, W.D., and Incropera, F.P., A Continuum Model for Momentum Heat and Species Transport in Binary Solid-Liquid Phase Change Systems - I. Model Formulation, Int. J. Heat Mass Transfer, vol.30, pp.2165-2170, 1987.
14. Bennon, W.D., and Incropera, F.P., A Continuum Model for Momentum, Heat and Species Transport in Binary Solid-Liquid Phase Change Systems - II. Application to Solidification in a Rectangular Cavity, Int. J. Heat Mass Transfer, vol.30, pp.2171-2187, 1987.
15. Beckermann, C., and Viskanta, R., Double-Diffusive Convection During Dendritic Solidification of a Binary Mixture, Physic. Chem. Hydrodyn., vol.10, pp.195-213, 1988.
16. Christenson, M.S., Bennon, W.D., and Incropera, F.P., Solidification of an Aqueous Ammonium Chloride Solution in a Rectangular Cavity-II. Comparison of Predicted and Measured Results, Int. J. Heat Mass Transfer, vol.32, pp.69-79, 1989.
17. Cristenson, M.S., and Incropera, F.P., Experiments on Solidification of an Aqueous Sodium Carbonate Solution in a Horizontal Cylindrical Annulus, J. Heat Transfer, vol. 111, pp.998-1005, 1989.

18. Leitch, A.M., Evolution of a Binary System Crystallizing in a Confined Region, Int. J. Heat Mass Transfer, vol.32, pp.2087-2098, 1989.
19. Kourosh, S., Crawford, M.E., and Diller, K.R., Microscopic Study of Coupled Heat and Mass Transport During Unidirectional Solidification of Binary Solutions - I. Thermal Analysis, Int. J. Heat Transfer, vol.33, pp.29-38, 1990.
20. Kourosh, S., Diller, K.R., and Crawford, M.E., Microscopic Study of Coupled Heat and Mass Transport During Unidirectional Solidification of Binary Solutions - II. Mass Transfer Analysis, Int. J. Heat Mass Transfer, vol.33 pp.39-53, 1990.
21. Nelson, D.G., Incropera, F.P., and Bennon, W.D., Numerical Simulation of Solidification in a Horizontal Cylindrical Annulus Charged With an Aqueous Salt Solution, Int. J. Heat Mass Transfer, vol.33, pp.367-380, 1990.
22. Gupta, S.C., Numerical and Analytical Solutions of One-Dimensional Freezing of Dilute Binary Alloys With Coupled Heat and Mass Transfer, Int. J. Heat Mass Transfer, vol.33 pp.593-602, 1990.
23. Cao, W.Z., and Poulikakos, D., Solidification of an Alloy in a Cavity Cooled Through its Top Surface, Int. J. Heat Mass Transfer, vol.33, pp.427-434, 1990.
24. Braga, S.L., and Viskanta, R., Solidification of a Binary Solution on a Cold Isothermal Surface, Int. J. Heat Mass Transfer, vol.33, pp.745-754, 1990.
25. Fukusako, S., and Yamada, M., Freezing Characteristics of Ethylene Glycol Solution, Wärme- und Stoffübertr., vol.24, pp.303-309, 1989.
27. Fukusako, S., Yamada, M., and Morizane, H., Freezing Characteristics of Ethylene-Glycol Solution, Proc. 1st KSME-JSME Thermal and Fluids Engng. Conf., vol.1, pp.55-60, 1988.
28. Fukusako, S., Yamada, M., and Morizane, H., Freezing Characteristics of Ethylene-Glycol Solution on the Vertical Cooled Plate, Trans. JAR, vol.7, pp.57-65, 1990.
29. Fukusako, S., and Yamada, M., Freezing Characteristics Along a Horizontal Cooled Tube Immersed in Aqueous Binary Solution, Trans. JSME, vol.56, 1990 (in print).
30. Fukusako, S., and Morizane, H., Melting Heat Transfer along a Horizontal Heated Cylinder Immersed in Liquid Ice, Proc. JSME Thermal Engng. Conf., pp.72-73, 1989.
31. Fukusako, S., Yamada, M., and Morizane, H., Heat-Transfer Control in Melting Along a Bundle of Tubes Immersed in Liquid Ice, Proc. 1990 JSME Annual Meeting, vol. C, pp.58-60, 1990.
32. Fukusako, S., and Morizane, H., Melting Heat-transfer Characteristics from a Horizontal Heated Cylinder in Liquid Ice, Proc. 27th Nat. Heat Transfer Symp. Japan, vol.3, pp.928-930, 1990.

Generics Structure of Flow and Temperature Fields in Welding and High Energy Beam Processing

MICHAEL M. CHEN
Department of Mechanical and Industrial Engineering
University of Illinois at Urbana-Champaign
Urbana, Illinois 61801, USA

ABSTRACT

Thermocapillary convection plays a dominant role in the fluid flow and heat transfer of a number of materials processing techniques, especially welding and other processing modalities involving melting due to concentrated energy beams, such as laser and electron beam processing and plasma processing. Typically, these processes produce a pool of molten material with intense temperature gradients along the surface, and it is the surface tension gradient associated with the surface temperature variation which drives the flow. The dimensionless parameter for thermocapillary convection is the Marangoni number, which can be viewed as the Peclet number based on the elementary thermocapillary velocity scale. For most of the processes cited above, the Marangoni number is of the order of 10^4 or more. Thus the flow is expected to possess singular regions, including thermal and viscous boundary layers and a unique thermocapillary singularity at the corners. These singular regions render numerical computations extremely challenging. In addition, common numerical approximations, such as upwind differencing, can be particularly damaging because of the phenomenon's strong dependence on accurate temperature gradients in these singular regions. Failure to appreciate these difficulties often leads to inaccurate or erroneous numerical results and hence incorrect conclusions concerning processing. In this paper the physics of thermocapillary convection under common materials processing conditions will be described. The structure of temperature and velocity distributions, the scalings of the boundary layers and corner singular regions, and their known implications on materials processing will be discussed. In addition, common pitfalls for numerical computations of such flows will be reviewed.

NOMENCLATURE

The nomenclature used here should be self-evident. Only selected symbols are listed below. The dimensionless parameters are defined and discussed in Eqs. (6) through (12), (24), and (25).

English Symbols

- D characteristic pool dimension; length scale for surface temperature gradient
- L_x, L_y length scales for the singular cold corner region
- x direction tangential to free surface
- y direction normal to free surface
- T temperature
- u velocity

Greek symbols

α heat diffusivity
δ thicknesses of the boundary layers along free surface
λ thicknesses of the boundary layers along hot and cold walls
σ surface tension
γ temperature coefficient of surface tension
μ viscosity
ν kinematic viscosity
ρ density

Superscripts

o denotes "true" Peclet, Reynolds, and Weber numbers

Subscripts

c denotes core flow
f denotes reference scales and dimensionless parameters based on a prescribed heat flux
s denotes a scale of appropriate order of magnitude
r denotes a reference quantity, not necessarily of the appropriate order of magnitude
ν denotes viscous boundary layer thicknesses
α denotes thermal boundary layer thickness

INTRODUCTION

Thermocapillary convection refers to the coupled heat transfer/fluid mechanics phenomenon driven by surface tension gradients associated with non-uniform surface temperatures. A closely related phenomenon is soluto-capillary convection, associated with concentration gradients. Combinations of thermocapillarity and one or more forms of solutal capillarity may also be found, leading to significantly greater complexity. Due to the high temperatures often encountered in the melting processes associated with materials processing employing concedntrated heat sources, thermocapillary convection is frequently the dominant mode of convective motion. Even when other forms of convection, including those due to buoyancy, hydromagnetics and rotation, are dominant or important over the major portion of the flow domain, it is still possible for thermocapillary convection to play controlling roles in small regions, due to the different asymptotic behaviors of these driving forces. Major consequences of thermocapillary convection are: the alteration of temperature distribution and hence the melt pool shape, solidification and cooling rates, and in turn the product microstructure; the influence on solute mixing and hence the composition distribution; the distortion of free surface shape and in turn the degradation of surface finish. Last but not least, thermocapillary convection may generate hydrodynamic instabilities, with associated oscillations and variations in product quality.

A number of reviews of thermocapillary convection exist in the literature. General discussions have been given by Ostrach (1977, 1982, Schwabe (1981), Schwabe and Scharmann (1981) and others. Recently, thermocapillary instability have been reviewed by Davis (1987). An important feature of the thermocapillary flow encountered in materials processing is that it is usually of high intensity (i.e. of high Marangoni number, see discussion below). The present review will focus on the qualitative and quantitative aspects of steady, intense thermocapillary flows relevant to materials processing, employing concentrated heat sources,with particular attention directed to the scaling laws, so that existing results and plans for future analyses can be put in proper perspective. Computational difficulties unique to thermocapillary flows will be pointed out. Impact of the existing results on materials processing will also be discussed.

SAMPLE FLOW CONFIGURATIONS IN MATERIALS PROCESSING

In the presence of surface tension non-uniformities associated with non-uniform temperature distributions, the fluid at the surface is pulled in the direction of increasing surface tension. This motion creates a normal gradient of the tangential velocity, accompanied by a shear stress which balances the force due to the surface tension gradient. The shear stress in turn acts on the fluid in the interior, thus setting up a bulk convective motion. For most materials, the temperature coefficient of the surface tension is negative, hence the thermocapillary surface velocity is usually in the direction of decreasing temperature. For conciseness such a material will serve as the model for our discussions in this paper. The discussions can be easily reinterpreted for materials with positive temperature coefficients or for soluto-capillary convection.

Thermocapillary convection can not exist without two other attributes of capillarity: that the curvature of the free surface is proportional to the interfacial pressure difference, and that the tri-state line between the solid, liquid and gas phases is governed by contact angle considerations. Although these aspects of the problem are often neglected in current efforts to understand the already complex problem of thermocapillary flow with flat surfaces, their potential importance to materials processing should not be overlooked.

Figure 1.a represents the typical flow configuration encountered in arc welding as well as other processes employing intense, concentrated energy beams, such as laser and electron beam processing (Szekelly 1986, Giedt 1987, Mazumder 1991). The pool size can be slightly larger or significantly larger than the size of the energy beam, with a width-to-depth ratio of the order of unity. The pool diameters vary from sub-millimeters to the order of 10 mm. For a typical material, thermocapillary flow thus creates a toroidal vortex, flowing outward toward the rim at the surface and returning below. The pressure distribution associated with this driven vortex may cause a small dip of the free surface at the center and an elevation near the rim. Under certain arc welding conditions, the flow field and the free surface shape may also be influenced by the electromagnetic forces of the arc current and the shear stress and fluid pressure of the arc plasma, though thermo-buoyancy effects are likely to be small due to the small dimensions involved.

Under more intense irradiation, the dip in the free surface, either due to thermocapillarity or other forces, may become so deep that a "keyhole" is formed, shown in Fig. 1.b. This is of considerable practical importance since the keyhole permits the welding of thick sections using laser or electron beam, or plasma arc processes. The pattern of thermocapillary flow can be quite complex. In the presence of scanning, when welding a seam there is also a strong component of the flow around the hole, from the hot upstream side to the cold downstream side.

In all cases the cold surface represents a phase change surface where melting or solidification takes place. Hence it is expected that there is strong coupling between the heat transfer characteristics of the thermocapillary convection, heat conduction in the solid, and the shape of the pool. There is likely to be some distortion of the shape of the free surface, though the extent to which the free surface shape influences the convection pattern for the cases of Figs. 1.a and 1.b is unclear. Free surface shape clearly plays dominant roles in the formation of the key hole, illustrated in Fig. 1.b.

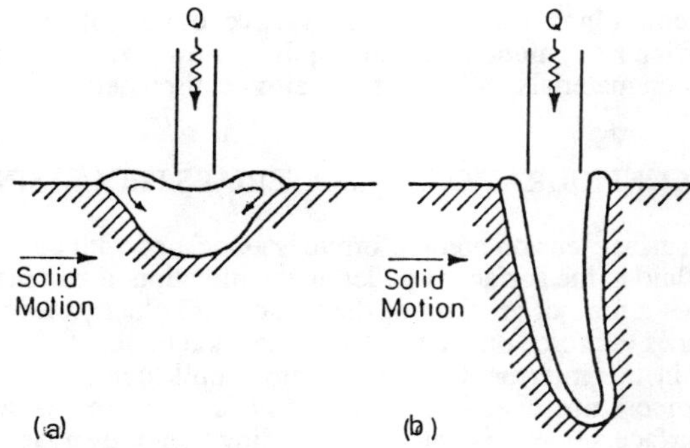

FIGURE 1. Typical configurations for thermocapillary flow in high energy beam processing materials. (a) shallow pool processes, including welding; (b) key hole formation.

GOVERNING EQUATIONS AND QUALITATIVE DIMENSIONAL CONSIDERATIONS

The thermocapillary effect is described by the expression

$$\mu \frac{\partial u}{\partial y} = \frac{\partial \sigma}{\partial x} \tag{1}$$

where for simplicity, we have assumed two dimensionality and x and y in the tangential and normal directions, respectively. For dimensional and scaling considerations, it is convenient to define a negative temperature coefficient

$$\gamma \equiv - \frac{d\sigma}{dT} \tag{2}$$

The surface tension gradient can now be expressed in terms of the temperature gradient:

$$\mu \frac{\partial u}{\partial y} = - \gamma \frac{\partial \sigma}{\partial x} \tag{3}$$

This expression serves as the boundary condition for the flow equations. As a rule, for non-trivial problems the surface temperature distribution depends strongly on the flow field. Hence the temperature and flow fields are strongly coupled.

Equation (3) yields a thermocapillary velocity scale:

$$u_s = \frac{\gamma \Delta T}{\mu} \frac{\delta_v}{D} \tag{4}$$

This is not useful for defining primary dimensionless parameters, since the length scale δ_v is not known a priori. Instead, for dimensional considerations the following reference velocity, which can be evaluated solely from given parameters but is not necessarily representative of the magnitude of the actual thermocapillary velocity, can be used.

$$u_r = \equiv - \frac{\gamma \Delta T}{\mu} \tag{5}$$

In a heat transfer system, the measure of the intensity of convection relative to conduction is the Peclet number, defined as the product of characteristic velocity and characteristic length divided by the heat diffusivity. The Peclet number based on the thermocapillary reference velocity is called the Marangoni number, which is the basic dimensionless parameter of thermocapillary convection:

$$Ma = Pe = \frac{u_r D}{\alpha} = \frac{\gamma \Delta T D}{\mu \alpha} \tag{6}$$

There is no special name for the Reynolds number associated with u_r, so it is customary to retain the name Reynolds number:

$$Re = \frac{u_V D}{\nu} = \frac{\gamma \Delta T D}{\mu \nu} \tag{7}$$

Typical Ma's for many materials processing conditions are of the order of 10^4 or above. It should be pointed out that the Ma and Re defined here can not be compared directly to conventional Pe and Re, due to the fact that u_r does not represent the magnitude of the real velocities encountered. Typical values of u/u_r from current literature are in the range 0.01 to 0.1. Hence the true Peclet and Reynolds numbers are somewhat smaller than Ma, but still greatly exceed unity. These values are high enough for the flow to exhibit the well known singularities of high Peclet number and high Reynolds number flows in the form of thermal and viscous boundary layers. We shall later point out that these boundary layers, being driven from the boundary rather than the fluid interior, behave differently from conventional boundary layers. Furthermore there is also a singularity at the corner which is rather unique for the thermocapillary flow.

Since for high Reynolds number flow the pressure scale is ρu^2 the dimensionless parameter governing the surface curvature is the Weber number, which is a measure of the reference pressure of the flow field versus the capillary pressure difference:

$$We = \frac{\rho u_r^2 D}{\sigma} = \frac{\rho \gamma^2 \Delta T^2 D}{\mu^2 \sigma} \tag{8}$$

These three parameters, Ma, Re, and We, are the basic dimensionless parameters appearing in a simple thermocapillary flow problem. These must be supplemented by contact angle conditions at the three phase line. The roles of the latter on thermocapillary flow in materials processing is poorly understood at present, and will not be dealt with in this discussion. The readers are referred to review articles on the subject (Davis 1983).

In addition to these parameters, other dimensionless parameters can be obtained as combinations of these parameters.

$$Pr = Ma/Re \text{ and } Ca = \frac{\mu u_V}{\sigma} = \frac{\gamma \Delta T}{\sigma} = We/Re \tag{9,10}$$

Capillary number Ca replaces the Weber number as the parameter governing curvature in low Reynolds number flows.

Where gravity is important, the Bond number may also be relevant:

$$Bo = \frac{g\rho D^2}{\sigma} \tag{11}$$

When a surface heat flux distribution is prescribed instead of a temperature difference, an implicit reference temperature difference can be defined from a reference heat flux q, leading to the definition of flux Marangoni, Reynolds and Weber numbers according to Eqs. (6), (7), and (8):

$$\Delta T_f = \frac{qD}{k}, \; Ma_f = \frac{\rho c_p \gamma q D^2}{\mu k^2} \tag{12,13}$$

$$Re_f = \frac{\rho \gamma q D^2}{\mu^2 k}, \text{ and } We_f = \frac{\rho \gamma^2 q^2 D^2}{\mu^2 k^2 \sigma} \tag{14,15}$$

Again, since D is not representative of the thermal boundary layer thickness, ΔT_f is not a realistic estimate of the actual temperature difference. Hence the flux Marangoni number and flux Reynold number cannot be equated to the Marangoni and Reynolds numbers in scaling considerations.

HIGH Ma BEHAVIOR AND BOUNDARY LAYER SCALING

A qualitative appreciation of the structure of high Marangoni number convection can be obtained by examining isotherms shown in Fig. 2, taken from Zehr, et al. (1987). They are for a square cavity with differentially maintained wall temperatures, with a taut, adiabatic free surface. It is seen that as Ma increases, the temperature distribution gradually changes from the linear distribution characteristic of conduction solutions to one characteristic of high Peclet number convection, with thermal boundary layers along both isothermal walls as well as along the adiabatic free surface. A unique feature not found in conventional high Peclet number convection is the concentration of the temperature gradient near the cold corner. This is brought out more dramatically by the surface temperature and velocity distributions shown in Fig. 3.a, taken from Zebib, et al. (1985). It is seen that along the free surface, thermocapillary effects at high Ma appear to be concentrated near the hot and cold corners. The velocity distribution in the vicinity of the cold corner is shown in greater detail in Fig. 6, from Zehr et al. (1987). This is apparently due to the fact that the intense convection associated with the surface motion toward the cold wall has brought the warm fluid forward, significantly compressing the region of temperature variation. In midspan, between the hot and cold walls, there is a region with essentially constant surface temperature. The flow therefore can be viewed as a half jet, maintained by inertia, issuing from the hot corner.

The surface temperature and velocity distribution for a cavity maintained by a given surface heat flux is shown in Fig. 3.b, taken from Kamotani and Ostrach (1987). It is seen that the velocity curve for the highest Marangoni number also shows a peak at the cold corner. It is not clear why this velocity peak is less pronounced than the ones in Fig. 3.a. On the other hand, pronounced peaks were also found in a two-dimensional weld pool simulation at comparable Marangoni numbers, shown in Fig. 3.c.

These figures suggest that the structure of thermocapillary flow at high Marangoni numbers is as shown schematically in Figs. 4.a and 4.b. To avoid clutter, only the thermal

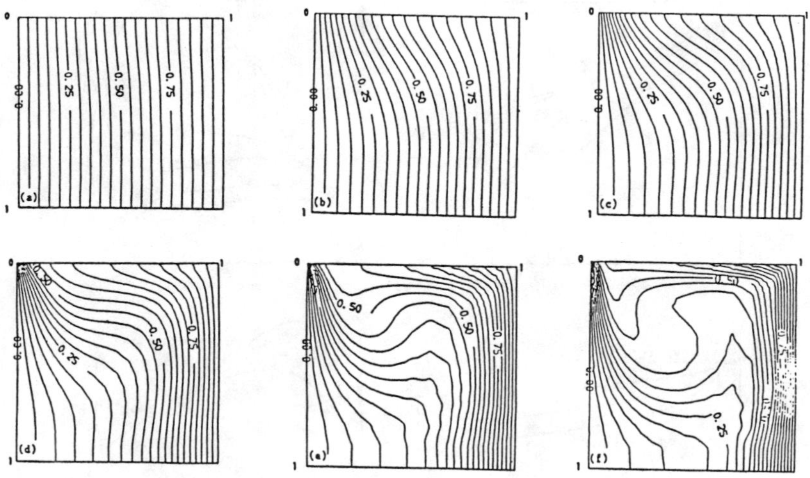

FIGURE 2. Isotherms in a square cavity with differential wall temperatures showing the formation of boundary layers and singular corner regions at high Marangoni numbers. (a) Ma = 1; (b) Ma = 100; (c) Ma = 300; (d) Ma = 1000; (e) Ma = 3000; (f) Ma = 10,000. (From Zehr, et al. 1987). Cold wall is to the left.

boundary layer is shown, with the viscous boundary layer being thicker or thinner according to whether the Prandtl number is greater or smaller than unity. Note that the only difference between these two figures is found in the hot end of the free surface. In Fig. 4.a, corresponding to the case when the heating is caused by a hot surface, the flow originates at the hot wall, the thermocapillary motion forming a boundary layer, which is labeled as region Ia. This is followed by region II, where the thermocapillary driving force has diminished due to a reduced temperature gradient. In the limit of vanishing thermocapillarity, the frictionless free surface acts just like a symmetry plane, and the boundary layer approaches a half jet. Region III is the singular region near the cold corner. Region IV is a conventional rigid wall boundary layer along the cold wall, where additional heat loss takes place. Finally, the recirculating flow impinges on the hot wall, forming again a conventional rigid wall boundary layer, labeled as region V. Region VI is the core flow.

In Fig. 4.b, corresponding to the case when heat flux is delivered directly to the free surface, the thermocapillary boundary layer originates in a stagnation flow region formed by the fluid drawn from the interior. This region is labeled Ib. This is followed by regions II, III, and IV as in Fig. 4.a. Region V does not exist due to the absence of a hot wall. If a relatively diffuse heat flux distribution is employed, region II may be much smaller or completely eliminated.

The following discussion of the scalings of the different regions is based on the papers of Ostrach (1982), Cowley and Davis (1983), Zebib, et al. (1985), and Chan, et al. (1985,6,8) as well as heuristic considerationsby the present author.

Of the different regions shown in Fig. 4, the one easiest to analyze is probably Ib, since in the limit of infinite Marangoni number, the stagnation flow draws from a quiescent isothermal core flow instead of another boundary layer. Near the center line, the heat flux distribution and in turn the temperature distribution can be approximated by a parabola. The problem can then be solved by techniques closely related to that used to solve the classical

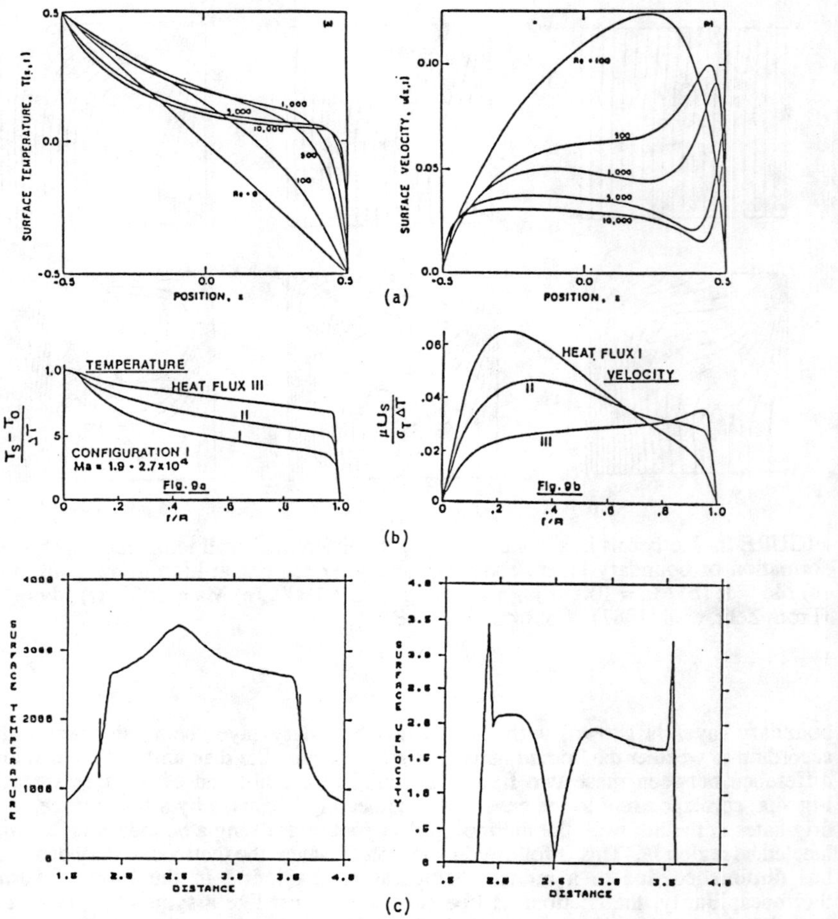

FIGURE 3. Typical surface temperature and velocity distributions at various Marangoni numbers for (a) a square cavity with an imposed temperature difference (from Zebib, et al. (1985); (b) a cavity with an imposed uniform heat flux (from Kamotani and Ostrach, 1987); (c) a two-dimensional weld pool with a concentrated heat flux (Zehr, 1987). In (a) and (b), cold wall is to the right.

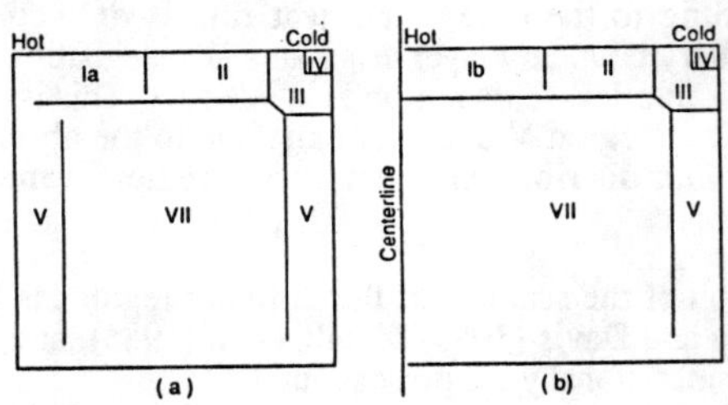

FIGURE 4. Structure of thermocapillary flow at high Ma. (a) A cavity with differential wall temperatures; (b) half of a cavity with a centrally applied non-uniform heat flux distribution.

Hiemenz stagnation flow problem (Schlicthing, 1979). This was done by Chan, et al. (1985) for a pool of infinite depth and width, and hence infinite Marangoni number. In particular the results for the asymptotic cases of low and high Prandtl numbers (Chan, et al. (1987), yield universal velocity and temperature distributions, shown in Fig. 5. Their results are presented in terms of variables defined for infinite Marangoni number. The original paper should be consulted for the definition of these variables. For the case of large finite Marangoni number surface heating concentrated in a region of horizontal extent D, the scalings for the surface velocity, the thermal and velocity boundary layer thicknesses can be obtained easily by invoking the thermocapillary boundary condition, Eq. (4), and the balance of the advective and diffusive terms for the u-momentum and energy equations. Care must be exercised for the low Prandtl number case, when the thermal boundary layer is in a region of vanishing tangential velocity, and the dominant advective effect is due to the normal velocities. The results are:

$$\text{For } Pr << 1, \frac{u_s}{u_r} \sim Re^{-1/3}; \frac{\delta_v}{D} \sim Re^{-1/3}; \frac{\delta_\alpha}{D} \sim Ma^{-1/3}\,Pr^{-2/3} \tag{16}$$

$$\text{For } Pr >> 1, \frac{u_s}{u_r} \sim Re^{-1/3}; \frac{\delta_v}{D} \sim Re^{-1/3}; \frac{\delta_\alpha}{D} \sim Ma^{-1/3}\,Pr^{-1/6} \tag{17}$$

As is the case for all heat transfer problems, both the high Pr and low Pr scalings lead to the same result for Pr = 1. Hence Eqs. (16) and (17) are expected to apply to the entire Pr range.

It is instructive to use Eqs. (16) and (17) to evaluate the real Peclet number and Reynolds number, based on the more realistic velocity scale. We shall distinguish these from the nominal Peclet (i.e. Marangoni) and Reynolds numbers used before by the superscript o, since they will be used in later discussions.

$$Pe^o \sim Ma^{2/3}Pr^{1/3},\ Re^o \sim Re^{2/3} \tag{18}$$

Similarly, relationships between the flux Marangoni number and the regular Marangoni number are also instructive:

$$\text{For } Pr << 1,\ Ma = Ma_f^{3/4}\,Pr^{-1/2}\ ;\ \text{For } Pr >> 1,\ Ma = Ma_f^{3/4}\,Pr^{-1/8} \tag{19,20}$$

The dimensionless variables of Chan, et al. (1985b), as shown in Fig. 5 are consistent with this scaling. Readers are referred to the original paper for the definitions of the variables in Fig. 5.

The analogous problem for region Ia, near the hot wall of a cavity with differentially wall temperatures, is considerably more difficult. The structure of the flow for Pr = 1 was considered by Zebib, et al. (1985), in conjunction with their numerical computations. They concluded that

$$\text{For } Pr \sim 1, \frac{u_s}{u_v} \sim \frac{\delta_v}{D} \sim \frac{\delta_\alpha}{D} \sim Re^{-1/3} \tag{21}$$

This is in agreement with the scaling suggested by Ostrach (1982), also for Pr = 1. The extension of their result to Pr << 1 would appear to be straightforward, following the same procedures for, and leading to the same results as in Eq. (16). However, for high Prandtl number fluids, the surface velocity is much more closely tied to the viscous core flow, hence a simple extension of Eq. (21) to higher Prandtl numbers is not possible. A very detailed analysis for the hot corner region for Ma >> 1, Pr >> 1 was carried out by Cowley and Davis (1983), who showed that

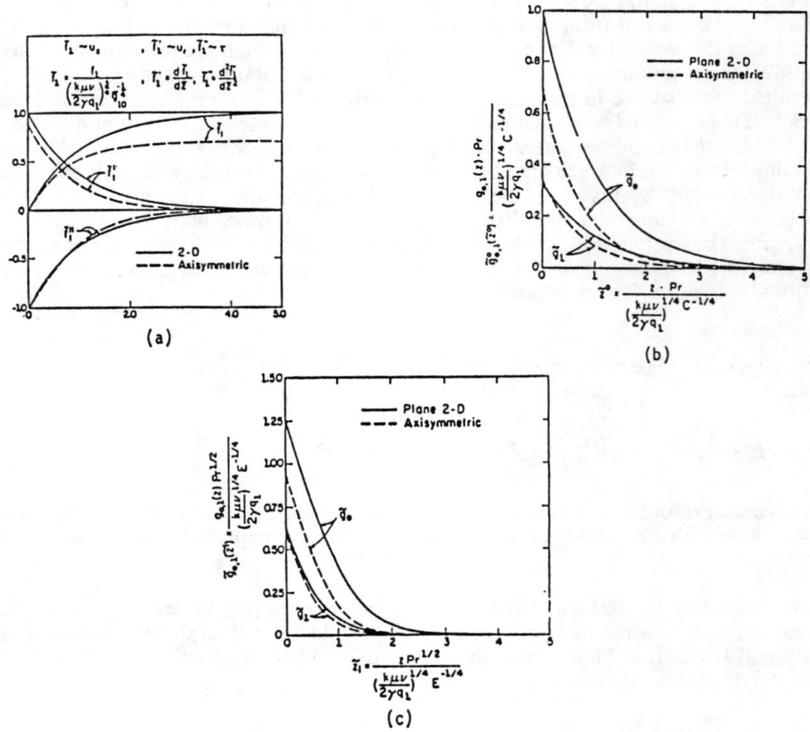

FIGURE 5. Velocity and temperature distributions at the stagnation point under a heat source at infinite Ma. (a) velocity distribution; (b) temperature distribution for Pr << 1; (c) temperature distribution for Pr >> 1 (from Chan, et al. 1987).

$$\text{For Pr} \gg \text{Ma}^{6/7} \gg 1, \frac{u_s}{u_v} \sim \text{Ma}^{-1/7}; \frac{\delta_\alpha}{D} \sim \text{Ma}^{-3/7} \tag{22}$$

The scaling for 1 << Pr << $Ma^{6/7}$ is less straight forward. Readers are referred to the original paper for details. Furthermore, Cowley and Davis predicted a high surface velocity region at a point $DM^{-2/7}$ from the hot wall, in agreement with the computed results of Zebib, et al. (1985), for high Prandtl numbers.

Cowley and Davis (1983) continued their analysis to region II, and presented a solution which is essentially half of the similarity solution for a jet.

For the cold corner, Zebib, et al. gave the scaling

$$\frac{u}{u_v} \sim \text{Re}^{-1/3}; \frac{L_x}{D} \sim \text{Re}^{-2/3}; \frac{L_y}{D} \sim \text{Re}^{-1/3} \tag{23}$$

This, however, does not appear to agree with the velocity distributions presented by themselves and by Zehr, et al. (1987). Recent results from Zehr (1991) showing the details of the velocity distribution in the vicinity of the cold corner is shown in Fig. 6.

The scaling of the vertical extent of this corner singularity is important in determining the non-uniform local heat flux on the vertical wall. In typical materials processing applications, the cold wall represents the solid-liquid phase boundary. Hence any concentrated heat flux would have important consequences on the shape of the interface. The heat flux distribution from Zehr (1991), is shown in Fig. 7. The reader's attention is drawn to the fact that the heat flux is presented in log scale. Hence the non-uniformity of the heat flux is extremely severe. It would seem then that the structure of the cold corner flow is one of the most critical issues to be studied in the future.

Despite the potential importance of the cold corner region, there does not seem to a definitive study of its structure in the literature. Unpublished computations by Zehr (1990) for effective Marangoni numbers exceeding 10^6 suggest that the horizontal and vertical dimensions of the singular region appear to vary with $1/Ma^a$, with the exponent a is close to unity. However, precise scaling has yet to be established.

The boundary layer on the lower part of the cold wall, region IV, should be a conventional rigid wall boundary layer, as pointed out by Zebib, et al. (1985). However, they gave the scaling of the boundary thickness for Pr = 1 as $Re^{-1/2}$, which is probably a careless error. As indicated earlier, the real Peclet and Reynolds numbers are not equal to the nominal Marangoni and Reynolds numbers. They are typically proportional to Ma raised to an exponent less than unity. Hence, what they intended was probably $R_e^{o-1/2}$. Since the Prandtl number dependence of convectional boundary layers is well known, and since the Prandtl numbers relevant to materials processing tend to be either much greater or much smaller than unity, it would be more useful to give the asymptotic relationship for large and small Pr:

$$\text{For Pr} \ll 1, \frac{\lambda_v}{D} \sim R_e^{o-1/2}\,;\, \frac{\lambda_\alpha}{D} \sim P_e^{o-1/2} \tag{24}$$

$$\text{For Pr} \gg 1, \frac{\lambda_v}{D} \sim R_e^{o-1/2}\,;\, \frac{\lambda_\alpha}{D} \sim P_e^{o-1/2}\, Pr^{-1/3} \tag{25}$$

The boundary layer on the hot wall, region V, is again a conventional boundary layer. Hence, Eqs. (24) and (25) apply. These are also in agreement with the hot wall scalings of Cowley and Davis (1983).

The core region, VI, is essentially isothermal and does not play an important role in heat transfer. However, for very high Prandtl number flows the core flow determines the characteristic surface velocity. Cowley and Davis (1983) thus gave

$$\frac{u_c}{u_v} \sim Ma^{-1/7}\ (Pr \gg 1) \tag{26}$$

As a tangible interpretation of the above discussion, the orders-of-magnitude of the dimensions of the different regions are estimated and shown in Fig. 8, for two typical weld pools in steel and aluminum. One of the most striking results is the smallness of the cold corner region. It should be noted that this dimension is nevertheless large enough for the Navier-Stokes equations, which had been used as the basis for the theoretical deductions, to be valid. However, the effects of impurities and their influence on thermocapillar flow were not taken into consideration.

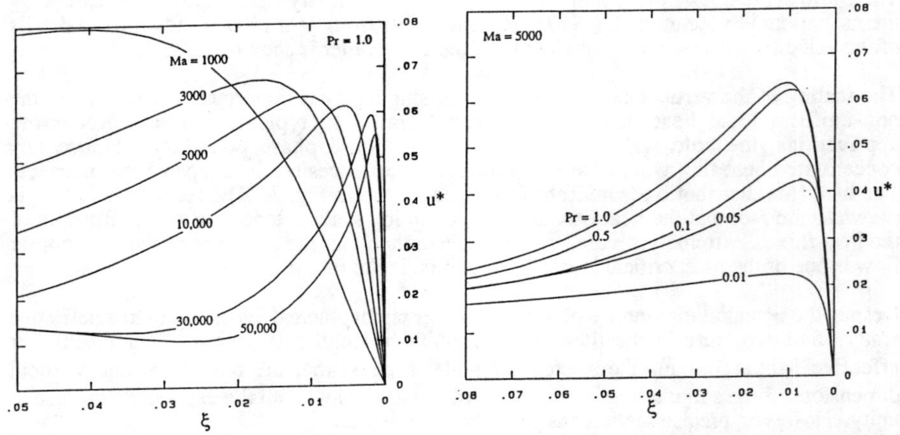

FIGURE 6. Expanded curves of the velocity distribution at the cold corner, Zehr (1991).

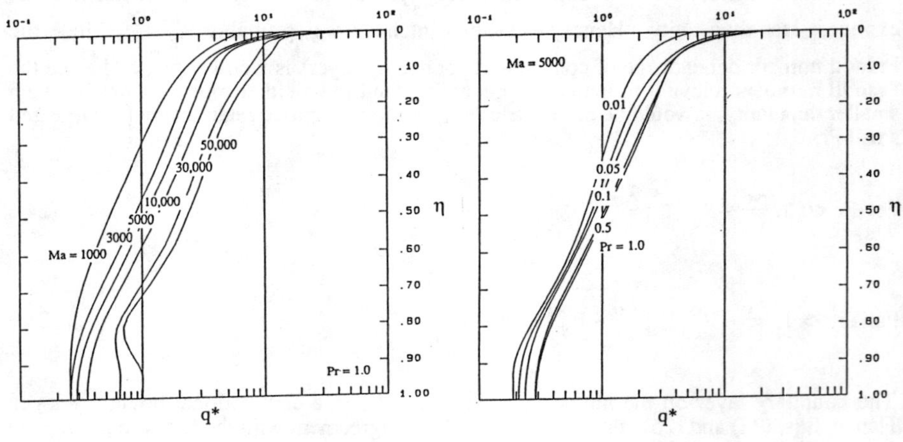

FIGURE 7. Heat flux distribution on the cold wall, Zehr (19).

The above review indicates that while the general structure of high Marangoni number convection is known, the scalings of some key regions are still not understood. Since scalings can be useful for assessing computation techniques and organizing computational results, continued effort in this area is essential. We mention in closing here that Marangoni boundary layers have also been examined by, among others, Napolitano and Golia (1981) and Gill (1983), who discussed similarity and other solutions for boundary layers in media of infinite extent. The scalings are generally consistent with the above discussion for the configuration involved.

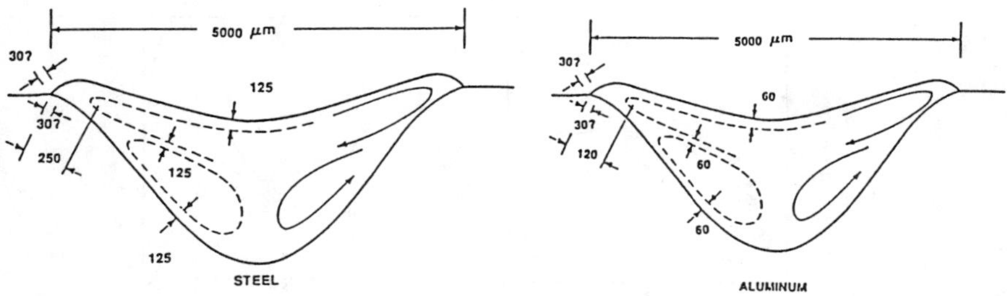

FIGURE 8. Estimates of the dimensions of difference flow regions in typical weld pools.

SHAPE OF THE FREE SURFACE

All the calculations cited above are for flat free surfaces. Physically this assumption implies that the superficial pressure variations generated by fluid mechanical forces are small relative to the pressure difference which can be generated by the surface tension in conjunction with surface curvatures. The parameters which govern this phenomenon is the Weber number or the capillary number, depending on whether the flow is of high or low Reynolds number. Since Ma is usually large in materials processing, the Weber number is probably the more relevant. Typical Weber numbers, as defined in Eq. (8) on the basis of the reference velocity u_r, are of the order of 10^4. In other words, the assumption of flat surfaces would require considerable scrutiny.

There are actually several different issues here. The first issue is whether the conclusions we have reached in the last section concerning boundary layers are still valid. This issue was addressed by Cowley and Davis (1983). They pointed out that while the quantitative details of the flow field is intimately dependent on the free surface shape, the qualitative and semi-quantitative aspects of the problem, particularly the structure of the flow and the scalings of the boundary layers and other singular regions, should not be affected.

The second issue is the quantitative issue of computing the free surface shape numerically. This has apparently not been done for realistic Weber numbers. Zebib et al. (1987) presented some first order results, which are in effect surface shape based on the pressure distribution obtained from flat surface computations. Their low and high Prandtl number results are shown in Figs. 9.a and 9.b. Similar first order computations of the free surface for a three-dimensional weld pool was also presented by Chan et al. (1986), reproduced here as Fig. 9.c. While these calculations serve a useful qualitative purpose, one should refrain from drawing important conclusions from them. Due to the pressure field's strong dependence on boundary shape, a self-consistent calculation for realistically large Weber numbers may be significantly different from these first order results.

The third issue to be considered is whether the large Weber number would cause its own singular behaviors, just as the large Marangoni number. This issue has not been seriously investigated, but the answer is probably yes. Since a radius of curvature of the order of D is insufficient to balance the flow field pressure variations, it is likely that the pressure variations responsible for turning the flow at the corner must occur within a very small region, where the local Weber number is of the order unity.

The large Weber numbers encountered in welding could be one of the factors which facilitates the formation of the keyhole. To this author's knowledge, no self-consistent

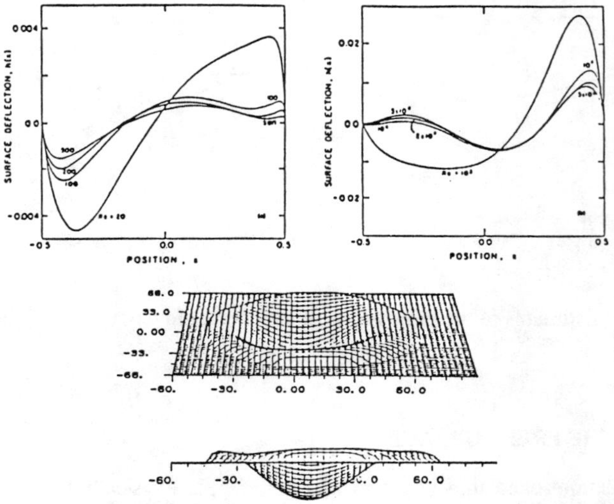

FIGURE 9. Shape of the free surface based on linearized analyses. (a) and (b) cavity with differential wall temperatures, Pr = 0.1 and 50, respectively; (c) three-dimensional weld pool at Ma ~ 10^4, Pr = 0.15.

analysis of keyhole formation exists in the literature. A discussion of several factors involved and model calculation of the flow around the hole in the horizontal plane has been presented by Giedt and Wei (1982) and Wei and Giedt (1985).

IMPACT ON MATERIALS PROCESSING

Heat Flux Non-uniformity At The Cold Corner

Perhaps the most striking feature of high Ma thermocapillary flow is the singularity at the cold corner. Among its many consequences is the intense concentration of the heat flux at the upper edge, as shown in Fig. 7. Since in materials processing the cold wall is usually the melting or solidification interface, this intense concentration of the heat flux at the corner can have significant effects on the shape of the melt pool. Very likely a small "continental shelf" may form at the corner, as suggested faintly by the sketch in Fig. 1.a. A shelf is clearly seen in the recent calculation of Kanouf (1987). Such a shelf may have important consequences on the uniformity of the depth of penetration, and on the cooling rate of the solid. Unfortunately, most of the published computations of the melt pool probably did not have the fine mesh required to resolve this point.

Instabilities and Oscillations

Many weld seams are characterized by the appearance of "chevrons," or frozen waves formed at an angle to the direction of travel. Crystals grown under the Czochralski process have also been known to show ripples or periodic composition variations. These are signs of oscillations, possibly due to one of the thermocapillary instabilities. Observations by

Kamotani, et al. (1984), (1989), Ostrach et al. (1985), have shown that the instability tended to start near the cold corner, a possible consequence of the high velocity and high gradients found there. For a review of thermocapillary instability, the readers are referred to Davis (1987).

Free Surface Ripple at Corner

As discussed in Section 5, at large Weber numbers a free surface singularity may also develop near the corner. This problem has not yet been investigated, but a likely consequence is the development of a ripple, of small radius of curvature, at the edge of the free surface. Since the edge of the free surface is the first area to solidify under cooling, the ripple can influence the surface finish of the solidified work. This could be responsible for the formation of undercut and raised ridges during welding processes.

Buoyancy and Other Driving Forces

The above discussion showed that the cold corner singularity is unique to the thermocapillary flow. Hence even in buoyancy, JxB or other driving forces exist, a region of intense thermocapillary motion and localized heat transfer may exist. The processing consequence of this have not been seriously examined.

COMPUTATIONAL CHALLENGES

The structure of high Marangoni number convection discussed above is particularly challenging for computational modeling. The most obvious is the mesh fineness requirement. This is brought out dramatically by Figs. 3 and 6. At Ma = 10,000, region III is less than 0.01 of the width of the cavity, but it contains the most dramatic variations of both the velocity and temperature. This clearly rules out uniform mesh. The problem remains challenging even with variable mesh. Assuming that a minimum of 20 mesh increments is needed for each region of Fig. 4 for marginal accuracy, it is clear that at least a 80 x 60 mesh is required. This is so because regions I, II, III, and IV require different mesh sizes in the x-direction, and regions II, III, and IV require different mesh sizes in the y-direction. The effect of mesh size on the computed surface velocity distribution for Ma = 3,000 is shown in Fig. 9. It is seen that with uniform 26 x 26 and 51 x 51 meshes, the results were clearly unsatisfactory. The variable 26 x 26 mesh was not satisfactory. The variable 51 x 51 mesh gave results which were qualitatively satisfactory, but was too coarse for quantitative accuracy, at least near $x = 1$. Clearly, the problem would be progressively more severe for Ma = 10,000 or higher. The mesh fineness requirement for high Marangoni number flows could be one of the reasons why the many early modeling efforts have completely missed the intense corner flow shown in Figs. 3 and 6.

Most common finite difference algorithms employ three-point differencing schemes which at best can attain second order accuracy for uniform mesh, but revert to first order accuracy for variable mesh. Usually this is of no significant concern since the discretization error can be kept low by using gradually changing mesh sizes. When the mesh size requirement in different regions of the flow is drastically different, as in the case of high Marangoni number thermocapillary flow, it may no longer be possible to keep mesh size increments low enough to maintain effective second order accuracy.

A more serious problem of discretization error, in this case shared by both the finite difference and finite element methods, is the false diffusion caused by the popular upwind-differencing scheme. In conventional problems with boundaries parallel to coordinate directions this is not a serious problem, since the false diffusion is mainly introduced in the direction of motion, which is typically parallel to the boundary and typically has only a small gradient component. Even with inclined boundaries much of the problem has been alleviated by skew upwinding. Unfortunately, for thermocapillary flow, the flow is primarily driven by the temperature gradient component in the direction of flow. Hence the false diffusion can potentially lead to significant errors. It is essential that sufficiently fine mesh is used to reduce the false diffusion. The use of hybrid upwinding schemes so that false diffusion can be completely eliminated for small mesh Peclet numbers is recommended. Second order schemes, such as Quick, would be particularly desirable.

Maintaining an adequately fine mesh for each region is particularly difficult for real pool shapes encountered in materials processing problems, in comparison with the prototype calculations for rectangular cavities cited in this paper. Clearly some type of form-fitting non-orthogonal mesh generation or coordinate transformation scheme would be highly desirable if not essential in some instances.

CONCLUDING REMARKS

In this review we have summarized current knowledge on steady thermocapillary flow under conditions relevant to materials processing. Special attention has been directed to the singular nature and the structure of the flow as Marangoni number and Weber number approach infinity. The potential impact of such a structure on materials processing has been illustrated by a few sample problems. Computational difficulties in modeling such flows have also been pointed out. It is suggested that unless vigilance is maintained in the choice of mesh size and computational scheme, significant error may result and important flow features may be lost. Much of the details of such flows and their materials processing impact are not yet thoroughly understood. Continued research effort by scientists of both materials and thermofluids communities are required.

ACKNOWLEDGMENT

The author wishes to acknowledge his indebtedness to C. L. Chan, J. Mazumder, and R. Zehr for valuable discussions, and the Office of Naval Research for financial support. The extraordinary assistance by R. Zehr and the Mechanical and Industrial Engineering Publications Office in preparing the manuscript on very short notice is particularly appreciated.

REFERENCES

Chan, C. L., M. M. Chen, and J. Mazumder, 1985, "Thermocapillary Convection in the Central Region of a Deep Melt Pool due to Intense, Non-uniform Surface Heating," Paper No. 85-HT-21, ASME - AIChE National Heat Transfer Conference, Denver, CO,

Chan, C. L., R. Zehr, J. Mazumder, and M. M. Chen, 1986, "Three-Dimensional Model for Convection in Laser Weld Pool," Modeling and Control of Casting and Welding Processes, S. Kou and R. Mehrabian, Eds., The Metallurgical Society, Inc., Publ., pp. 229-246.

Chan, C. L., J. Mazumder, and M. M. Chen, 1988, "Effect of surface tension driven convection in a laser melted pool: 3-dimensional perturbation model," J. App. Phys., 64:6166-6174.

Chan, C. L., M. M. Chen, and J. Mazumder, 1988, "Asymptotic solution for thermocapillary flow at high and low Prandtl numbers due to concentrated surface heating," J. Heat Transfer, 110:140-146.

Cowley, S. J., and S. H. Davis, 1983, "Viscous Thermocapillary Convection at High Marangoni Number," J. Fluid Mech., Vol. 135, pp. 175-188.

Davis, S. H., 1983, "Contact Line Problems in Fluid Mechanics," J. Appl. Mech., Vol. 50, pp. 977-982.

Davis, S. H., 1987, "Thermocapillary Instabilities," Ann. Rev. Fluid Mech., Vol. 19, pp. 403-435.

Giedt, W. H., and P. S. Wei, 1982, "Temperature and Velocity Distribution in the Liquid Flowing around the Front of an Electron Beam Welding Cavity," Proc. Seventh International Heat Transfer Conference, Hemisphere Pub., New York, NY.

Giedt, W., 1987, paper in this symposium.

Gill, W. N., 1983, "Surface Tension Driven Boundary Layer in Floating Zones," Chem. Eng. Commun., Vol. 23, pp. 57-61.

Kamotani, Y. and Lee, J. J., 1989, "Oscillatory Thermocapillary Flow in a Liquid Column Heated by a Ring Heater, PCH PhysicoChemical Hydrodynamics, Vol. 11, No. 5/6, pp. 729-736.

Kamotani, Y., S. Ostrach, and M. Vargas, 1984, J. Crystal Growth, Vol. 66, pp. 83.

Kanouff, M. P., 1987, "Numerical Modeling of Momentum and Energy Transport in a Weld Pool Using an Adapting, Non-orthogonal Grid," Paper AIAA-87-1443, AIAA 19th Fluid Dynamics, Plasma Dynamics and Laser Conference, June 8-10, Honolulu, HI.

Mazumder, J., 1990, paper in this volume.

Napolitano, L. G., and C. Golia, 1981, "Coupled Marangoni Boundary Layers," Acta Astronautica, Vol. 8, pp. 417-434.

Ostrach, S., 1977, "Motion Induced by Capillarity," Physichochemici-hydrodynamics-V. G. Levich Festschrift, Advanced Pub. Ltd.

Ostrach, S., 1982, "Low Gravity Fluid Flows," Ann. Rev. Fluid Mech., Vol. 14, pp. 313.

Ostrach, S., Y. Kamotani, and Lai, C. L., 1985, "Oscillatory Thermocapillary Flows," PCH PhysicoChemical Hydrodynamics, Vol. 6, No. 516, pp. 585-599.

Schlichting, H., 1979, Boundary Layer Theory, 7th Edition McGraw-Hill Book Company, New York.

Schwabe, D., 1981, "Marangoni Effects in Crystal Growth Melts," Physico Chemical Hydrodynamics, Vol. 2, pp 263-280.

Schwabe, D., and A. Scharmann, 1981, "Marangoni Convection in Open Boats and Crucibles," J. Crystal Growth, Vol. 52, pp. 435-449.

Szekelly, J., 1986, "The Mathematical Modeling of Arc Welding Processes," Advances in Welding Science and Technology, S. A. David, Ed., Pub. Amer. Soc. Met.

Wei, P. S., and W. H. Giedt, 1985, "Surface Tension Gradient Driven Flow around an Electron Beam Welding Cavity," The Welding Journal, Vol. 64, pp. 251s-259s.

Zebib, A., G. M. Homsy, and E. Meiburg, 1985, "High Marangoni Number Convection in a Square Cavity," Phys. Fluid, Vol. 28, pp. 3467-3476.

Zehr, R., M. M. Chen, and J. Mazumder, 1987, "Thermocapillary Convection of a Differentially Heated Cavity at High Marangoni Numbers," ASME Paper No. 87-HT-229, National Heat Transfer Conference, Pittsburgh, PA, Aug. 10-12.

Zehr, R. L. 1990, unpublished computations.

Zehr, R., L., 1991, Dissertation, Department of Mechanical and Industrial Engineering, University of Illinois at Urbana-Champaign, Urbana, IL USA.

Thermal Oscillations in Materials Processing

SIMON OSTRACH and YASUHIRO KAMOTANI
Department of Mechanical and Aerospace Engineering
Case Western Reserve University
Cleveland, Ohio, USA

INTRODUCTION

With the increasing need for more perfect single crystals for the electronics industry, microwave devices, infrared detectors, and many other electromechanical and magneto- and electro-optical devices an awareness developed nearly 20 years ago that what happens in the fluid state during the processing of crystals significantly impacts the solid state. In almost all of the commonly used methods the parent phase is a liquid or a gas and the growth is essentially a mass transfer process coupled with fluid flows, heat transfer, phase change, and, in some cases chemical reactions. The fluid flows in closed reactors (such as in Bridgman techniques) are generated by density gradients in a gravitational field (natural convection). In open-reactor vapor deposition and some solution growth methods mixed forced and free convection occurs. In melt growth systems where free surfaces exist, like Czochralski and floating-zone methods, surface tension gradients can influence or cause fluid flows (thermo- and diffuso-capillary flows).

In practice it was observed that uncontrolled convection, as inferred from temperature oscillations, often resulted in poor crystal morphology. Thus, many crystal deficiencies were attributed to natural convection, where that term was used in a generic sense to refer to unstable, unsteady, untameable, and undesirable flows, Ref. 1. For the most part early attempts at elucidating fluid motions in crystal growth were based on ad hoc models or analogies to similar (but not identical) flow problems. Critical evaluations of that work were made in Refs. 2 and 3.

With the opportunity to grow crystals aboard spacecraft in reduced (but not zero) gravitational environments it was believed that the presumed deleterious effects of natural convection would be eliminated. A number of crystal growth experiments were, therefore, conducted aboard Skylab and other spacecraft but the results have been much less conclusive than expected. It thus became evident that the complex physicochemical transport phenomena that occur in crystal growth warrant more systematic and intensive study than had been made. Therefore, the nature and present state of knowledge of natural convection were described and its relation, as well as that of other flow phenomena, to materials processing was identified in Ref. 4. The various other possible driving forces for fluid flows in low-gravity environments were also identified therein.

Representative crystal growth processes were then examined in Ref. 5 to indicate gaps in knowledge that need to be filled by research in order to obtain the understanding to answer such questions as:

(i) What are the conditions for the occurrence of the fluctuations and what is their nature?

(ii) What effects do they have on the crystal interface and solute distribution?

(iii) How can the deleterious effects be minimized or eliminated?

It is the purpose of the present paper to focus on two types of flows (indicated in Ref. 5) viz., thermosolutal and thermocapillary, which ensuing research has shown can lead to fluctuations in configurations similar to those in crystal growth. These phenomena which had not previously been studied also can occur in other materials processing techniques.

THERMOSOLUTAL CONVECTION

In considering closed reactor systems, such as closed tube chemical vapor and horizontal Bridgman growth, it was pointed out in Ref. 5 that density gradients are established in the fluid by the simultaneous occurrence of both temperature and concentration gradients. Flows induced in a fluid system by both temperature and concentration gradients in the presence of a body force field are referred to as thermosolutal convection. It was indicated in Ref. 6 that no information existed on thermosolutal convection in enclosures with the two gradients normal to gravity. Since such configurations correspond to the closed reactor systems an extensive research effort was initiated to obtain an understanding of them.

Since it was shown in Ref. 5 and numerous references cited therein that even the simpler cases of purely thermally driven confined buoyancy flows are extremely sensitive to the configuration and imposed boundary conditions and that the flows may contain sub-flow regions (cells and layers) which are difficult to resolve purely numerically, it was decided to obtain guidance from experiments to determine the more complex thermosolutal flows.

Experimental Design

Dimensionless parameters. In order to design an experimental model that can be related to the prototype situation and to have the data be most general it is necessary to use the appropriate dimensionless parameters on which the related phenomena depend. It was shown in Ref. 6 that the fundamental parameters that describe thermosolutal convection are the:

thermal Grashof number, $Gr_T = \beta g \Delta T H^3/\nu^2$

Prandtl number, $Pr = \nu/\alpha$

Schmidt number, $Sc = \nu/D$

buoyancy ratio, $N = \bar{\beta}\Delta C/\beta\Delta T$

aspect ratio, $A = H/L$

where g is the gravitational acceleration, ν the fluid kinematic viscosity, α the thermal diffusivity, and D the diffusion coefficient. The height and width of the enclosure are H and L, respectively. The imposed temperature and concentration differences are ΔT and ΔC, respectively between the two vertical walls separated by the distance L. Density variation due to temperature is represented by the volumetric expansion coefficient, β, and that due to concentration by the volumetric solutal expansion coefficient, $\bar{\beta}$. Instead of the (Gr_T, N) combination the (Gr_T, Gr_s) is sometimes used, where Gr_s is defined as

$$Gr_s = \bar{\beta} g \Delta C H^3/\nu^2 = N Gr_T, \text{ solutal Grashof number.}$$

Among actual crystal growth systems the above parameters vary widely. For liquid metals Pr is about 0.01, and for aqueous solutions it is about 10. Sc is generally larger than Pr. For crystal growth using the horizontal Bridgman, open boat or closed-tube vapor deposition techniques, A ranges approximately from 0.1 to 1. Gr is generally in the range from 10^4 to 10^6.

Test apparatus. After exploring numerous ways to obtain even quasi-steady confined buoyancy flows driven by the simultaneous action of both horizontal temperature and concentration gradients it was decided to use an electrochemical method based on a diffusion-controlled electrode reaction to create the concentration gradients. Horizontal temperature gradients were imposed by heating or cooling the electrodes in such a manner that the two gradients were either augmenting or opposing. The ranges of parametric values covered in the experiment are essentially limited by the electrochemical system and are: $Gr_T = -1.7 \times 10^6$ to 1.7×10^6, Sc = 2583 to 2701, N = -45 to 45, Pr = 7, and A = 0.13 to 0.55. It should be noted that the Lewis number, Le = Sc/Pr = α/D ranges from 369 to 386, which indicates that the diffusion of heat is much larger than the diffusion of species so that double diffusion effects are to be expected. Clearly, the experimental system does not cover the entire range of parametric values that are associated with all crystal growth methods but it is representative of some. Since the main objective of the research was to obtain basic information on confined thermosolutal convection and the electrochemical system enabled flow visualization and temperature and mass transfer measurements it was deemed satisfactory for those purposes.

A schematic sketch of the experimental system is presented in Fig. 1. A copper sulphate-acid solution ($CuSO_4 + H_2SO_4 + H_2O$) is used as the electrolyte test fluid. When a voltage is applied to the electrodes, copper dissolves into the solution at the anode and is deposited on the cathode. As a result the fluid density near cathode (anode) becomes lower (higher) than that of the bulk of the solution. The migration of the cupric ions in the electrical field is eliminated by the sulphuric acid in the solution, which acts as a supporting electrolyte. The transport of the cupric ions in the cell is, thereby, controlled only by diffusion and convection.

The auxiliary system consists of a d.c. power supply, instruments to measure the total current and potential in the cell, and a variable resistance to control the current. More complete details on the apparatus and test procedures can be obtained from Refs. 7 and 8.

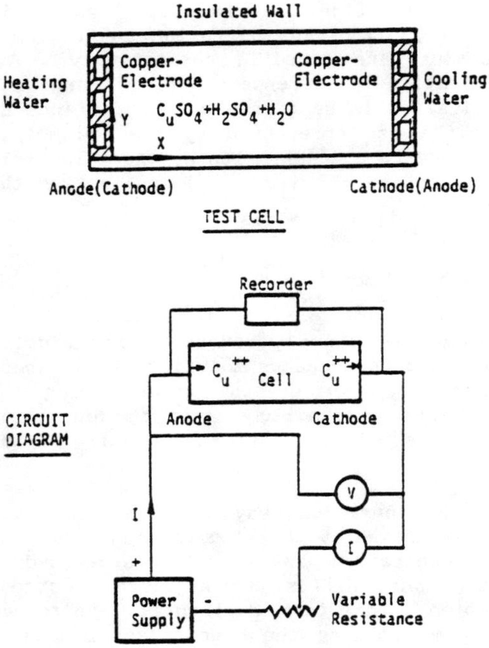

FIGURE 1. Illustration of test cell and electric circuit

Test procedure. Although the temperature of the copper end walls can easily be measured by thermocouples the concentration levels at the walls cannot be so easily determined. The limiting current technique in the last two decades has proved to be a convenient and accurate means for the determination of local and average mass transfer rates in electrochemical flow cells. The concepts of mass transfer associated with the limiting current measurement and the related techniques for determining the concentrations at the walls are outlined in Refs. 7 and 8.

When the limiting current is obtained the temperature distribution at the midsection is measured. The flow pattern and the concentration distributions are visually observed. Since the color of the solution changes with cupric ion concentration a qualitative picture of the flow pattern is obtained by the color distribution. Small particles and dye streaks are also used to visualize local flow structures.

Three types of situations were studied. The first type of experiments was done under isothermal conditions (pure solutal convection). In the second type the anode temperature is maintained below that at the cathode so that the thermal and solutal buoyancy forces augment each other. The anode temperature is higher than the cathode temperature so that the buoyancy forces were opposed in the third type, see Fig. 2. In the last two cases the electrochemical system is started after the thermal convection becomes steady. If the system is started in other ways the concentration boundary conditions become less well defined because the limiting current changes significantly with time as thermal convection develops in the cell. Pure thermal convection in enclosures similar to the present one is described in Ref. 10.

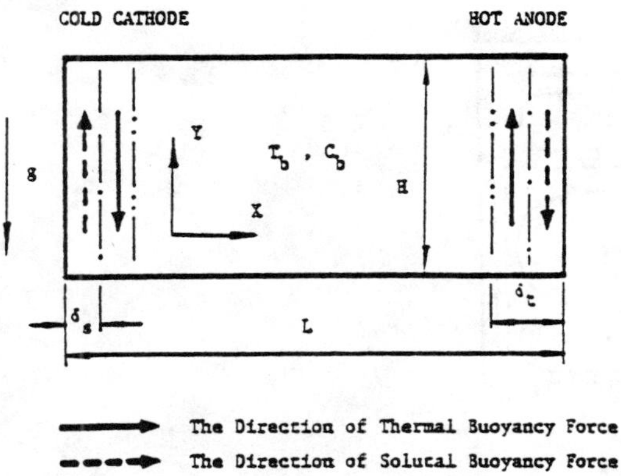

FIGURE 2. Sketch of test enclosure and orientation of thermal and solutal buoyancy forces

Experiments similar to the present ones were performed with an osmotic system, Ref. 11, in a salt-water solution. The two buoyancy forces were imposed simultaneously in that system and the observed flows differed somewhat from those obtained by the electrochemical system. In the work of Ref. 11 the boundary conditions are not maintained as well as in the electrochemical system.

A shadowgraph system, described in Ref. 12, was used for qualitative observations of the flow patterns for the opposed force cases.

Results and Discussion

Flow patterns. For the experimental conditions the solutal boundary layers on the vertical end walls are much smaller than the thermal boundary layers. Various complex flow patterns are obtained due to double-diffusion effects as described in Refs. 7 and 8. The flow patterns were found to depend significantly on the buoyancy ratio, N. In cases for which the buoyancy forces are augmenting if N is greater than about 6 a three-layer pattern appears. Within each layer the flow direction is determined by the thermal convection. The top and bottom layers grow with time. The aspect ratio does not influence the pattern significantly. For cases in which the buoyancy forces are opposed two different types of multilayer flows are observed when N is large enough ($N > 10$) depending on the aspect ratio. At moderately low A ($A = 0.66$) zonal multilayers appear (Fig. 3) and at very low A ($A = 0.13$) multilayers are tilted.

For N less than about 6 in augmenting cases and less than 10 in opposing cases the flow forms a unicellular pattern with secondary cells, Fig. 4, near the end walls which is similar to the patterns obtained in pure thermal convection, Ref. 10.

Temperature and velocity distributions as well as the mass transfer rate are profoundly influenced by the different flow patterns. Over the range of parameters studied for the augmenting cases the mass transfer correlates well with that for a corresponding vertical flat plate. The correlations of mass transfer rate for the opposing cases is much more complex, Refs. 7 and 8.

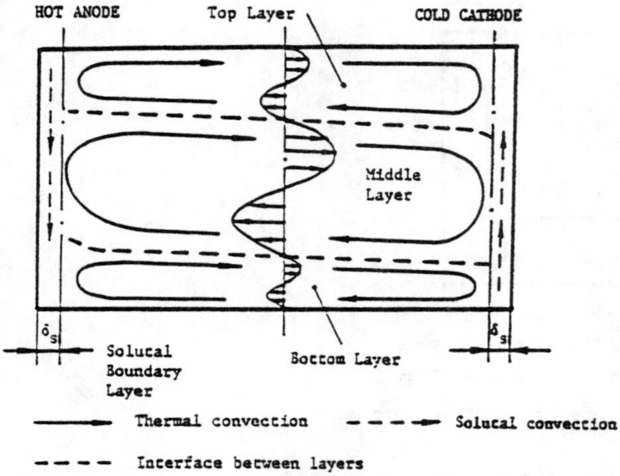

FIGURE 3. Sketch of typical multilayer flow structure

Under certain conditions the flow in the solutal boundary layer for opposing forces oscillates or fluctuates randomly, as was observed, at first, from oscillations in the electrochemical cell electrical potential. Since the potential fluctuations observed for the opposing buoyancy forces cases are directly associated with unsteady flows it was thought that these unsteady characteristics of thermosolutal flows could be important in understanding crystallographic imperfections during the growth of crystals. Therefore, more research was directed to determine the conditions for the fluctuations, to describe in more detail the nature of the transport phenomena under those conditions, and to determine the mechanism for their occurrence, Refs. 9, 13, and 14. It is these aspects that will be given emphasis in the present paper.

In the experiments the solutal Rayleigh number, Ra_s, ranges from 10^8 to 10^{10} when Gr_s is of the order of 10^5 to 10^6 with A = 0.13 to 0.5. Therefore near the vertical end walls thin solutal boundary layers are expected. Their thicknesses are of the order of 0.1 to 0.2 mm. As in most aqueous electrochemical systems the Lewis number, Le, is large (about 400) which indicates that the thermal boundary layers, if they exist, are much thicker than those of the solutal boundary layers. Therefore, two different boundary layers are defined, viz., the inner solutal layer and the outer thermal layer. The inner layer always refers to the boundary layer which is closer to the walls and the outer layer refers to the thicker thermal boundary layer. Because the Prandtl number is 7 for the present system the thermal and flow boundary layers are of the same order of magnitude. For opposing buoyancy forces, $N < 0$, but for simplicity of presentation absolute values of N will be used throughout. With an improved visualization technique both multi-layer flows and secondary flow cells are carefully observed and measured. In order to classify the flow regimes two special values of N are defined: N_s represents the value of the buoyancy ratio for which secondary flow cells just appear and N_c the value for which the multi-layer flow pattern just disappears. This enables three flow regimes to be identified: (1) multi-layer flows, $N \geq N_s$, (2) secondary cell flows, $N \leq N_c$ and (3) mixed flows, $N_c < N < N_s$.

For opposing buoyancy forces counter flows occur which lead to viscous shear forces and entrainment phenomena. These coupled with double diffusion result in the multi-layered flow patterns which are quasi-steady, and are depicted schematically in Fig. 3. A shadowgraph is shown in Ref. 13. The mechanism of formation of the multi-layer flows is described in Refs. 7 and 8. For fixed A multi-layer flows are observed at relatively large N, i.e., $N \geq N_s$. Changes in the concentration of the bulk solution, which affects Gr_s also changes N_s. The measured values of N_s can be correlated by:

$$N_s = 0.39\ Gr_s^{1/4} A^{-1/2} \tag{1}$$

The multi-layer flows completely disappear when $N \leq N_c$ and then an unicellular core flow with secondary cells near the vertical walls are clearly identified, which is referred to as the secondary cell flow. A shadowgraph of a typical secondary flow pattern is presented in Ref. 13 and a schematic sketch is shown in Fig. 4. Over some of the region the flow is unsteady. The skewed arrows indicate the unsteady flow sub-stream, which consists of solutal blobs or plumes generated by the breakup of the solutal boundary layer. The unsteadiness of the flow field is believed to be induced by a thermosolutal instability, which is explained in Ref. 13. From the experimental data it is found that the value of N_c which defines the secondary flow regime can be correlated by:

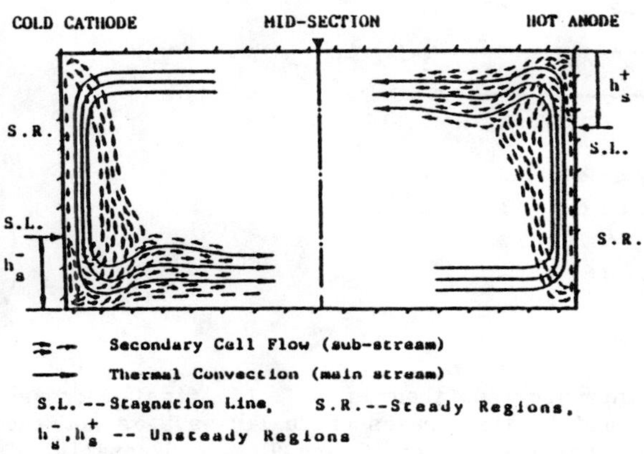

FIGURE 4. Sketch of a typical unicellular core flow with secondary flow cells near the vertical end walls

$$N_c = 0.13\ Gr_s^{1/4} A^{-1/2} \tag{2}$$

<u>Unsteady mass transfer</u>. For constant A and Gr_s, as the thermal Grashof

number Gr_T is increased (N decreased) it is found that the total limiting current undergoes a steady-unsteady-steady process, as shown in Fig. 5 in which the current variations with time are depicted for various values of N. Based on this, two particular values of N can be defined to distinguish the steady current region from the fluctuating current region. The values of N_{CU} and N_{CL} represent the upper and lower bounds for the current transition from steady to fluctuating states, respectively. N_{CU} are represented by case (b) in Fig. 5 whereas N_{CL} are represented by case (j). For either $N \geq N_{CU}$ or $N \leq N_{CL}$ the mass transfer (proportional to the limiting current) is steady as shown in cases (a), (b), (j) and (k) in Fig. 5. Otherwise ($N_{CU} > N > N_{CL}$) the limiting current fluctuates, as depicted by cases (c) to (i).

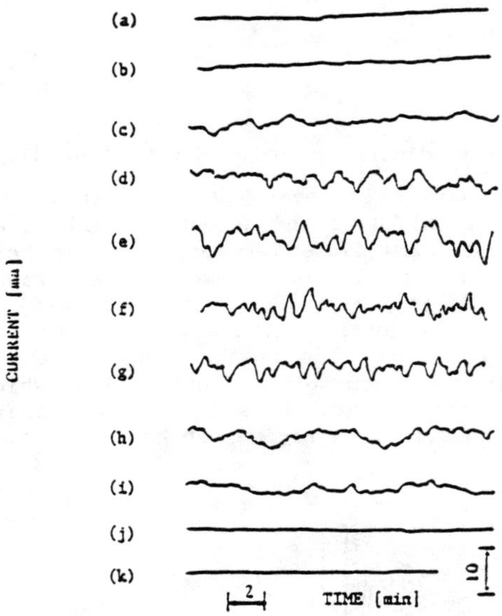

FIGURE 5. Influence of buoyancy ratio N on the current variations for A = .5 and Gr_s = 9.6x10^6 with (a) N = 22.4, (b) N_{CU} = 18.4, (c) N = 15.3, (d) N = 13.1, (e) N_C = 10.2, (f) N = 9.2, (g) N = 8.3, (h) N = 6.0, (i) N = 5.0, (j) N_{CL} = 4.6, (k) N = 3.7

The current fluctuation intensity, defined in Ref. 14, first increases, passes through a maximum, and then decreases gradually as N is decreased from N_{CU} to N_{CL} for each combination of A and Gr_s. Generally, for moderately low A (0.5) the current fluctuations occur over wider ranges of N and the fluctuations are relatively weak. For those cases with very low A (0.13) the fluctuations occur over smaller ranges of N but are relatively stronger. This indicates that the fluctuations are more easily suppressed for the lower aspect ratios but the fluctuations, once they occur, are stronger. From many tests covering various A, Gr_s, and N ranges it was found that the fluctuations are influenced not only by N but also by Gr_s and A. The N_{CU} correlate well by

$$N_{CU} = 0.25\ Gr_s^{1/4} A^{-2/5} \tag{3}$$

and the data for N_{CL} are fitted well by

$$N_{CL} = 0.06\ Gr_s^{1/4} A^{-.45} \tag{4}$$

For N either greater than N_{CU} or less than N_{CL} a steady current reading is recorded. In the former case the solutal buoyancy dominates the flow field and the solutal flow in the inner solutal layer is steady and laminar (Ref. 13). In the latter case it is surprising to find no current fluctuation at all even though a vigorous secondary cell flow exists near the vertical electrodes (Ref. 13). This indicates that the unsteady flow generated by the secondary cells in the far-field plays no role in the unsteady mass transfer at the solid boundary. The unsteady mass transfer occurs only when the solutal layers are unsteady which, under the present conditions, is attributed to a thermosolutal instability, Ref. 13.

In Refs. 7 and 8 it was indicated that the mass transfer rate is profoundly influenced by the countercurrent flow in the opposing buoyancy cases. The dimensionless mass transfer rate, Sherwood number, first decreases as N is reduced, passes through a minimum, and then increases.

Unsteady temperature variations. The temperature field near the end walls is measured by a thin thermocouple probe which is positioned 2 mm away from the cathode surface in order not to disturb the solutal boundary layers which are on the order of 0.1 to 0.2 mm thick. It is observed that in the secondary cell flow regime the maximum amplitude of temperature fluctuations occurs in the vicinity of the mid-height of the cathode so that location was chosen for the probe position.

For fixed A and Gr_s the influence of N on the temperature field is shown in Fig. 6.

In the multilayer regime ($N \geq N_s$) steady temperature readings were recorded as shown by cases (a) and (b). For $N_s > N \geq N_{CU}$, within the mixed flow regime, weak temperature fluctuations associated with weak secondary cell flows are detected, as indicated by case (c) and (d) but the current is still quite steady, as shown in Fig. 5 cases (a) and (b). For the range $N_{CU} > N > N_C$ (within the mixed flow regime) the secondary cell flow becomes fairly strong and, thereby, more vigorous temperature fluctuations are observed, cases (e) and (g) and the corresponding current fluctuations are also stronger, cases (c) to (e) Fig. 5. At $N = N_C$ the flow changes from the mixed flow to the secondary cell regime and the amplitude of the current fluctuations reaches a maximum. The temperature fluctuations also reach their maximum value when $N = N_C$ as shown in cases (g) and (h) of Fig. 6. As N is further reduced from N_C to N_{CL} the current fluctuation intensity decreases from its maximum value to zero and the temperature fluctuations follow a similar trend as indicated by cases (h) to (l) in Fig. 6. However, when N equals N_{CL} or is lower weak temperature fluctuations are still observed, cases (l) and (m)

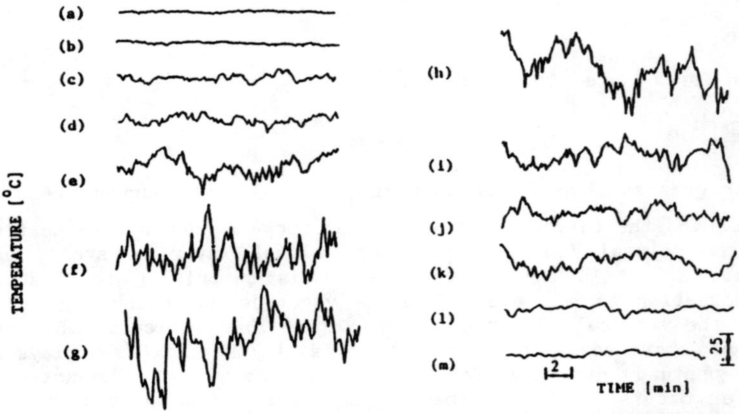

FIGURE 6. Influence of buoyancy ratio N on the temperature variations measured at mid-height and 2 mm from the cathode surface for A = .5 with (a) N = 102, $Gr_s = 1.73 \times 10^5$, (b) N = 28.1, $Gr_s = 6.0 \times 10^6$ and for A = .5, $Gr_s = 9.6 \times 10^6$ with (c) N = 22.4, (d) $N_{CU} = 18.4$, (e) N = 13.1, (f) N = 11.5, (g) $N_C = 10.2$, (h) N = 9.2, (i) N = 8.3, (j) N = 6.0, (k) N = 5.0, (l) $N_{CL} = 4.6$, (m) N = 3.7

whereas the current fluctuations have disappeared, cases (j) and (k) Fig. 5. The differences between the current and temperature fluctuations are due to the fact that the former are a boundary-layer phenomenon which is only related to the unsteady solutal boundary layer, whereas the temperature fluctuations reflect any local unsteady flow field which may occur either near the wall or elsewhere in the enclosure. For $N \leq N_{CL}$ the secondary flow cell is stretched and pushed away from the cathode wall. As a result the effect of the unsteady cell flow is weakened only near the surface but it is enhanced in the far-field. This indicates that temperature fluctuations are detected anywhere in the enclosure where the unsteady cell flow passes over the probe, regardless of any current fluctuations.

Flow classifications. From the discussion above it is obvious that thermosolutal convection in enclosures is a very complex process. In order to provide an overall view of the process and to summarize the transport phenomena the correlation equations (1) to (4) are plotted in Fig. 7 together with two other important observed quantities, $N_{I,MIN}$ and $\hat{N}_{I,MAX}$ which are the values of N at which the minimum mass transport rate (current) and maximum current fluctuation intensity are obtained. In this figure I and T represent the steady current (mass transfer rate) and temperature and $\tilde{I}$ and $\tilde{T}$ indicate fluctuating current and temperature, respectively. The abbreviations M. L. and S. C. represent multilayer and secondary cell flow regimes, respectively and M. L. + S. C. denotes the mixed flow regime. Four solid lines represented by N_s, N_C, N_{CU}, and N_{CL} divide the flow field into three regimes with five regions. Region I ($N \geq N_s$) is the multilayer flow regime, discussed in Ref. 13, wherein both the temperature and current are steady and the convective flow near the cold cathode is laminar and bidirectional. Region II, defined by $N_s > N \geq N_{CU}$, covers part of the mixed flow regime, in which the current is

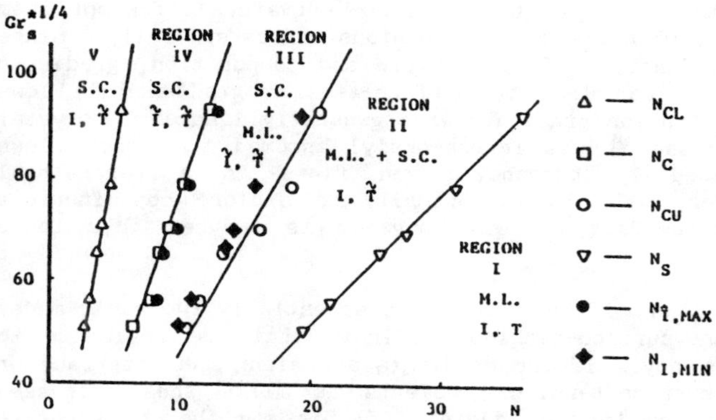

FIGURE 7. Relation between the flow regimes and the fluctuations of temperature and current (including the minimum mass transfer rate I_{MIN} and the maximum current fluctuation intensity $\hat{I}_{MAX}$) in the $(Gr_s^{*1/4}, N)$ -plane

steady but the temperature is not. As discussed in Ref. 13 near the vertical end walls the flow is steady and bidirectional and a weak secondary flow cell exists in the end region. Region III includes the remainder of the mixed flow regime ($N_{CU} > N > N_C$). Near the end walls the flow field consists of both steady and unsteady sections with bidirectional flows in both, Ref. 13. In this region both the temperature and current fluctuate. Region IV covers part of the secondary cell flow regime ($N_C \geq N > N_{CL}$). The flow field near the end walls consists of a steady section with unidirectional flow and an unsteady one with bidirectional flow, Ref. 13. Fluctuation of both temperature and current occur in that Region. Region V is defined by $N \geq N_{CL}$. Very strong secondary cell flow occurs and the major portion of the cathode is covered by a unidirectional steady flow. Therefore, the current is steady even though temperature fluctuations in the far-field are still observed.

THERMOCAPILLARY CONVECTION

The access of a low-gravity environment aboard an orbiting spacecraft opened new possibilities for the growth of crystals. Since the effects of gravity and natural convection (two of the main sources of imperfections in crystal growth on earth) are reduced in space the chances of obtaining better quality crystals are enhanced. Certainly, the deleterious effects of thermosolutal convection, as described above, would be favorably affected by reduced gravity.

The low-gravity environment also seems to offer the opportunity for the containerless processing of materials, wherein crystal contamination by container impurities could be minimized or eliminated, because surface tension becomes a dominant force. The floating-zone method is an example of such a containerless process. Inherent to such configurations and to pulling techniques, such as the Czochralski method, are free surfaces along which are temperature and/or concentration gradients. The presence of an interface between two fluid phases can influence the motion of fluids when either the interface has a finite curvature different from that at equilibrium or when the surface-tensions varies spatially. Since surface tension is a function of temperature and composition, gradients of either one or both can lead to surface-tension gradients. Flows generated by surface-tension gradients were generally ignored in crystal growth processes because it was (erroneously) assumed that such flows would be overshadowed by buoyancy-driven flows in a terrestrial environment. However, with the opportunity to explore containerless processing aboard spacecraft wherein buoyancy is reduced this topic assumed new importance.

It was pointed out in Ref. 15 that there are essentially two basic modes of flow generated by surface-tension gradients that are analogous to those induced by buoyancy. If the gradients are along the interface or free surface, the surface-tension gradients act like shear stresses applied by the interface to the adjoining fluids and thereby generate flows or affect existing ones. These are analogous to conventional convection and are called thermocapillary or diffusocapillary flows according to whether they are caused by temperature or concentration gradients, respectively. These flows can, of course, affect transport processes.

If one of the aforementioned gradients is perpendicular to the interface a "Marangoni instability" can occur under proper conditions that can lead to cellular flows. This situation is analogous to the thermally unstable type of buoyancy flows. This type of flow had received more attention than those with gradients along the free surface, Refs. 16 and 17.

Since flows due to gradients along the interface are the more important type for containerless processing of materials and have received relatively less consideration than the other type, an extensive research program was initiated to gain an understanding of them. This will now be described below.

Dimensionless Parameters

In dealing with new problems it was indicated in Ref. 18 that dimensionless parameters are most useful for obtaining qualitative features of the flow and for estimating flow velocities. The parameters also indicate the dominant physical factors, mathematical simplifications, data correlations, and proper theoretical and experimental models. The first explicit derivation of the fundamental dimensionless parameters was presented in Ref. 19. It was shown therein that, in analogy to buoyancy flows, there are different reference velocities obtained from the tangential stress balance, e.g., for viscous-type thermocapillary flows the characteristic velocity, U_R, is

$$U_R = (|\partial\sigma/\partial T|\Delta T/\mu)(D/L)$$

where σ is the surface tension, T temperature, ΔT a characteristic temperature difference, μ the absolute viscosity, and D and L are two different characteristic lengths. For boundary-layer flows

$$U_R = [(\partial\sigma/\partial T)^2 \Delta T^2 \nu/\mu^2 L]^{1/3}$$

where ν is the kinematic viscosity, see Ref. 4 for details. These velocities are limited to Prandtl numbers of unit order of magnitude and related expressions for limiting values of Prandtl number are derived in Refs. 19 and 20. Note that the reference velocities are given in terms of the physically important variables for establishing thermocapillary flows. The relevant dimensionless parameters based on the reference velocities that appear in the basic equations are the surface-tension Reynolds number, $Re_\sigma = [(\partial\sigma/\partial T)\Delta T D]/\mu\nu$, the Marangoni number, $Ma = PrRe_\sigma$, an aspect ratio, $A = D/L$, a dynamic Bond number $Bd = \rho g L^2/(\partial\sigma/\partial T)\Delta T$, and the relative temperature variation, $\beta\Delta T$, where ρ is the fluid density. The surface-tension Reynolds number appears in the equations of motion and plays the same role as the conventional Reynolds number. The Marangoni number appears in the energy equation and plays the same role as the conventional Peclet number. The relative importance of buoyancy to thermocapillary forces is determined by the coefficient of the buoyancy term which for viscous-type flows is $Bd\ (\beta\Delta T)$ and for boundary-layer flows is $Bd(\beta\Delta T)(A/Re_\sigma)^{2/3}$. It is thus evident that the dynamic Bond number and not the static one $Bo = \rho g L^2/\sigma$ is the correct parameter for that purpose and it indicates that in a low-gravity environment thermocapillary flows will be significant.

In most of the work the interface has been assumed to be flat (fixed). However, in general, the interface can be curved or deformed as a result of the forces acting on it and, therefore, its shape must be determined as part of the solution. Proper formulation of the problem requires balances of tangential and normal stresses at the interfaces. Other dimensionless parameters can appear in the boundary conditions depending on the forces that act on the fluid. In the existing literature a capillary number, Ca, is often cited as another parameter for free surface problems, where it is obtained from a balance of viscous and curvature terms in the normal stress equations as $Ca = (Bd/Bo)A$ so that no new parameters are obtained from the boundary conditions except for the contact angle.

Because surface tension is the driving force for flows the condition of the free surface is most important. Oxide formation or other surface reactions and contamination (due to dust and impurities in the apparatus or environment) can seriously inhibit or even eliminate surface-tension driven flows. This is not necessarily undesirable for practical applications but it makes thermocapillary experiments difficult.

Early Work

Most of the early research focused on the floating-zone method of growing crystals, Fig. 8, in which a non-uniform heat flux is imposed along a melt free surface that induces a thermocapillary flow in the melt. Based on the assumption that the free surface was fixed (nondeformable) numerical studies were made of thermocapillary flows in liquid cylindrical bridges between two solid rods that were maintained at different temperatures, Refs. 22 and 23. This configuration, Fig. 8, is called a half zone and is intended to simulate floating-zone processing. It has also been used in many experimental studies. The main feature of the numerical work is the prediction of steady flow cells (toroidal vortices) that encompass the fluid. Numerical studies in rectangular slots, Refs. 24 and 25, gave similar results.

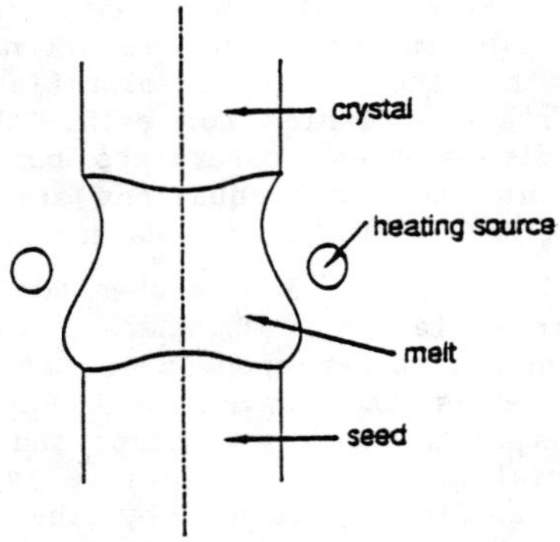

Floating Zone Crystal Growth

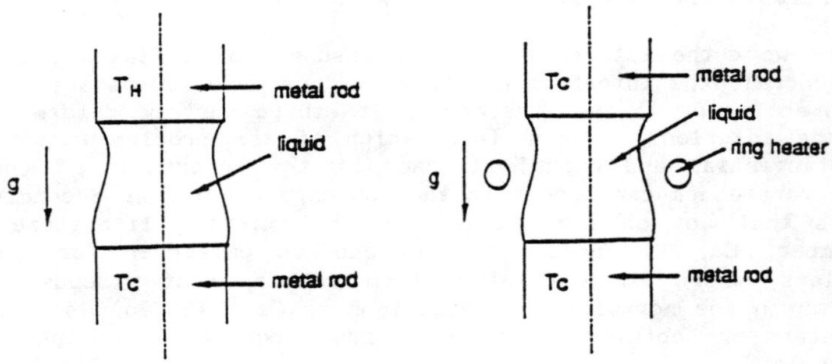

Half-Zone Simulation

Full-Zone Simulation

FIGURE 8. Floating-zone simulation

In numerous experimental studies, Refs. 26 to 30, steady toroidal flows were observed. In addition, a particularly significant finding in that work was that the flow undergoes a transition from the steady flow to an oscillatory state under certain conditions. The reason for this transition is not fully understood, but it is believed that the oscillations would be detrimental for containerless processing. Thus, understanding of the physical mechanism of oscillatory thermocapillary convection is important and much of the recent research is directed to determine the onset conditions for the oscillations, their nature, and their causes.

Recent Work

To gain insight into the physical mechanism responsible for the transition from steady to oscillatory thermocapillary flows a systematic, quantitative, and more extensive experimental study was made as reported in Ref. 31. A number of fluids different from those in the other works were used and the aspect ratio (liquid column length to diameter ratio) was varied so that the range of parametric values was greater than previously. No description of the apparatus, test procedures, and instrumentation will be given herein because they are presented in Ref. 31. More complete experimental data can be found in Refs. 31 and 32.

The value of Ma at which the oscillations start is referred to as the critical Marangoni number, Ma_{cr}. If the oscillations are considered to be characterized by the primary dimensionless parameters Ma, Pr, and A, then Ma_{cr} should depend on A and Pr only. A plot of Ma_{cr} determined under various conditions is given in Fig. 9 and it is obvious that the onset conditions cannot be defined by the two basic parameters Ma and A for a given fluid (fixed Pr). This is contrary to what is commonly believed and it was the first clue that perhaps surface deformations were important because another parameter, at least, is required to correlate the data and the only other basic parameters in the problem formulation are related to the surface deformations, e.g., the capillary number.

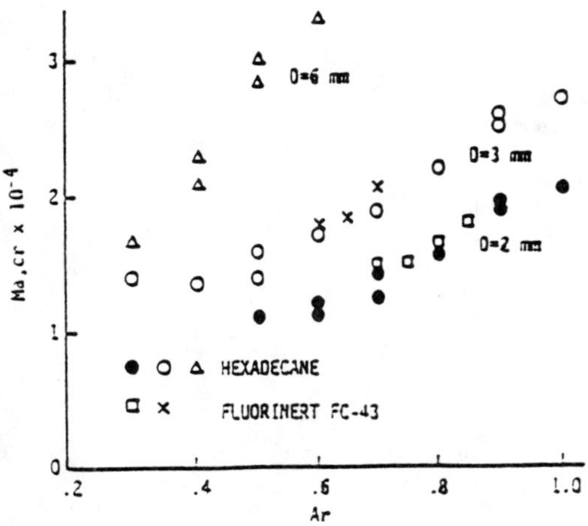

FIGURE 9. Critical Marangoni number versus aspect ratio

The temperature distribution along the free surface is a possible source of information because it is directly related to the flow driving force. Some typical surface temperature profiles are presented in Fig. 10. When Ma is below the critical value the convection is relatively weak and the temperature varies smoothly along the free surface. As Ma is increased the convection alters the surface temperature distribution (and, hence, the driving force) and results in large temperature gradients near the top and bottom ends of the zone. In the region between, the temperature is relatively uniform and the driving force is diminished there. Such surface temperature distributions were also obtained numerically in Ref. 23 wherein that distribution was determined from the solution and are also observed experimentally in Ref. 29 wherein, however, the reason for their occurrence was incorrectly interpreted, see Ref. 33.

On the basis of this evidence it was suggested in Ref. 31 that a coupling among the velocity and temperature fields and the surface deformation is responsible for the oscillations, with the last causing a time lag between the surface and bulk (return) flows. Because of the time lag alternate weak and strong convection periods ensue that lead to the oscillations. The entire process will be described in greater detail subsequently. From observations of the flow it was noted that the cyclic motion does not have to be in phase at all radial planes, Fig. 11. In fact, the oscillatory recirculating motions at different radial planes interact in such a way that the entire flow structure becomes non-symmetric at any instant, i.e., a strong period in one radial plane coincides with a weak period in the diametrically opposed plane. Such behavior was also observed in Ref. 34.

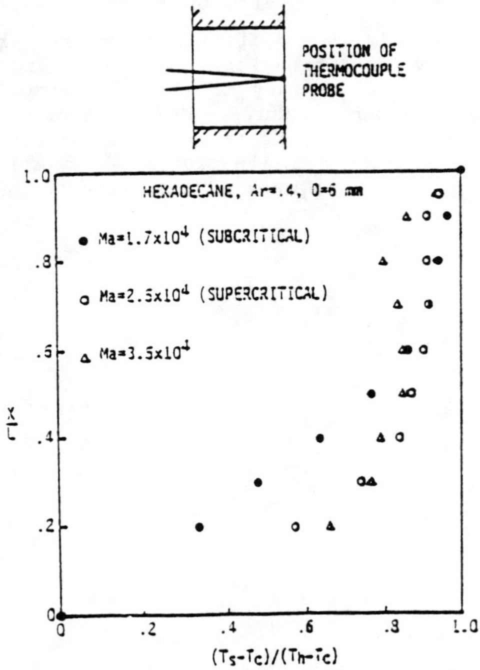

FIGURE 10. Surface temperature distribution

With the insights provided by the experimental work theoretical studies of the coupling effects on thermocapillary convection were made in Ref. 35. In the first part of that work three idealized time-dependent thermocapillary flows were analytically solved to illustrate explicitly the time lag between the velocity and temperature fields due to surface deformation. It was shown thereby in Ref. 36 that with a deformable free surface, fluid can accumulate near the free surface until the pressure rises sufficiently to generate a return flow and surface waves are generated in this way. Such a mass accumulation seems to delay the "message" transmission of the flow, i.e., leads to a time lag.

To develop the concept further, the dimensionless parameters in a floating-zone configuration are obtained by a scaling analysis which includes deformable surfaces, takes into account that the appropriate length scales are not necessarily the physical dimensions but may be boundary-layer thicknesses and/or the lengths over which the driving force truly acts (the end regions for the S-shaped surface temperature distribution), and that different characteristic velocities are associated with the various flow regimes. In this way the normal stress balance is truly normalized so that the coefficients of the various terms (dimensionless parameters) do, in fact, indicate relative magnitudes. The relevant parameters of surface deformation under zero-gravity conditions are derived because of their possible importance in defining the onset of oscillations through the establishment of a time lag between the velocity and temperature fields. These parameters are presented in Ref. 33 for all possible flow regimes and for large Ma the relevant deformation parameter is $S = |\sigma_T|\Delta T/\sigma_o$, which is the capillary number based on the appropriate velocity. For different fluids, S is modified by the Pr.

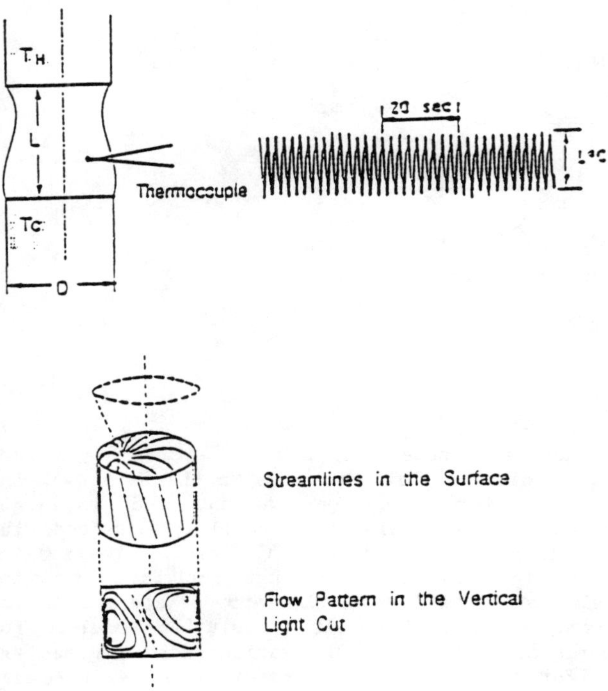

Important Parameters

Ma

$S = \frac{1}{Pr}\frac{\sigma_T \Delta T}{\sigma}$ (in the range Ma = 5×10^3 - 2×10^4)

FIGURE 11. Oscillatory thermocapillary flow in half-zone configuration

On the basis of that work the experimental data for the onset of oscillations in half zones for various Pr and A are correlated by using the deformation parameters, S, in Fig. 12. (See Ref. 13 for details.)

The seemingly anomalous behavior for liquid $NaNo_3$ is due to radiation heat loss from the zone because the liquid temperature in those experiments was approximately 300°C whereas the other experiments were performed at room temperature.

Fig. 12

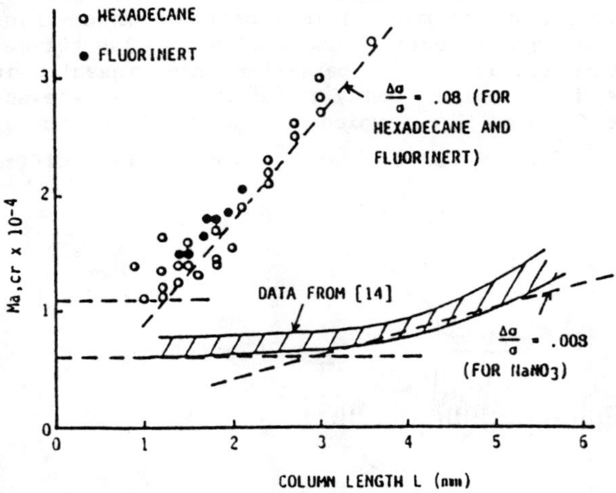

FIGURE 12. Conditions for onset of oscillations

More careful scrutiny of the numerical and experimental surface temperature distributions indicated that for large Ma the extent of the region over which the driving force acts near the hot "end" is larger than that at the cold, which is actually quite small. Therefore, the hot-end region is the dominant one for driving the flow. The scaling analysis was reconsidered in Ref. 21 to account for this, but the free-surfaces were considered to be fixed. The appropriate scales for the variables are derived, which indicate the resolution required for both numerical and experimental studies. The various flow regimes are identified and criteria for their existence are determined as are also their general flow characteristics. The scaling analysis of Ref. 35 with deformable surfaces will have to be modified to include this greater generality of the end wall conditions.

<u>Other configurations</u>. All the above work on thermocapillary oscillations was limited to the half-zone configuration with an imposed temperature difference. The question then arises as to how the geometric configuration and imposed thermal signature influences the oscillations. To this end experiments were conducted on a liquid (silicone oil) column suspended between two rods maintained at the same temperature, Ref. 37. The thermal signature is imposed at the free surface by a nichrome ring heater around the column at mid-height, Fig. 8. This imposes a heat flux

on the surface and more closely simulates actual floating-zones. Alumina particles illuminated by a light sheet indicated flow patterns and very fine (0.025 mm) thermocouples yielded temperatures. The distance of the ring heater from the liquid column (4 mm in diameter) was varied from 0.3 to 0.8 mm. Experimental conditions are Pr = 27, A varied from 0.5 to 0.75 and Ma ranged up to 2×10^4.

The steady flow pattern was one of two counter-rotating toroidal cells even though the liquid column had a non-cylindrical shape. As the heater temperature is increased the flow becomes more active and eventually oscillatory flow appears. Apparently because of buoyancy the oscillatory flow occurs first in the lower half of the zone and with further increases in the heater temperature the top half also becomes oscillatory. During oscillations the flow pattern in the bottom half is similar to that in the half-zone configuration, depicted in Fig. 13, and explained in detail in Ref. 38.

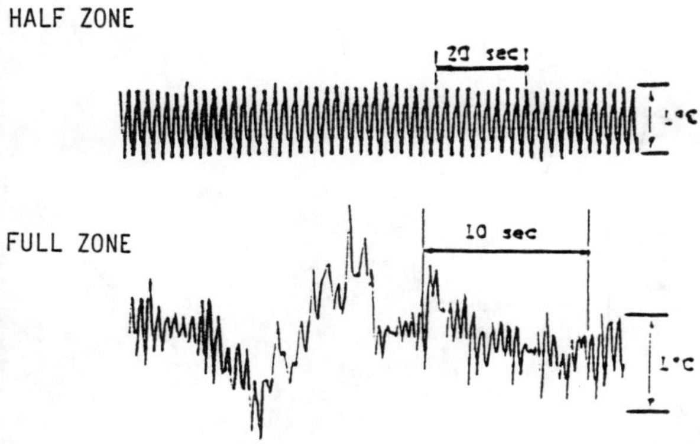

FIGURE 13. Oscillation patterns

Temperature fluctuations are measured in various locations in the fluid and their patterns are different at different locations. In the half-zone the toroidal cell flow structure is bounded by solid rods and the driving-force temperature difference is fixed. In the present configuration (full-zone) each cell is less confined and the overall temperature difference can change. As shown in Fig. 13 the oscillation pattern is more orderly in the half-zone.

The temperature difference at which the oscillations appear is called the critical temperature difference, ΔT_c. The ΔT_c measured at various A and heat fluxes (varied by heater diameter and gap between the heater and liquid column) show considerable scatter and do not show any systematic dependence on the liquid column geometry or imposed heat flux distribution. However, it is observed that with increasingly diffuse heat flux (larger gap and/or larger heater diameter) larger ΔT are required to obtain strongly oscillatory flow. A more controlled set of experiments is needed to obtain better ΔT_c data and to determine the parameters that define the onset of oscillation.

To supplement the experiment the flow and temperature fields were numerically determined from the SIMPLER algorithm with a 41 x 41 non-uniform grid. The computations were made only for steady flows with a fixed cylindrical surface and includes buoyancy effects. A comparison of the free-surface temperature distribution shows reasonable agreement, Fig. 14. However, slightly lower temperatures are measured at the zone midheight where the liquid is heated. The reason for this discrepancy is that the thermal boundary layer along the free surface becomes very thin near the mid-height for large Ma because relatively cool fluid from the bulk fluid impinges on the surface at mid-height just where the fluid is heated so that a thin thermal boundary layer is formed which cannot be resolved even by the thin thermocouple. The thinness of the thermal boundary layer plays an important role in the oscillation mechanism as will be indicated later.

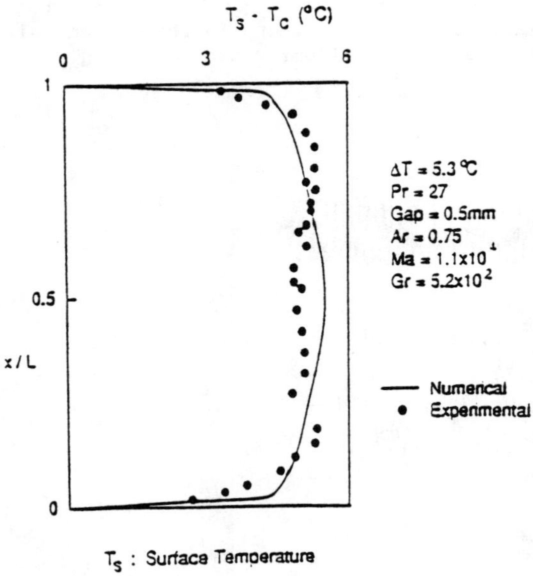

FIGURE 14. Comparison between numerical and experimental surface temperature distribution in oscillatory flow

The surface shape is computed by a domain perturbation method from the velocity and pressure fields for a fixed free surface. Due to the flow of fluid away from the mid-height to the ends along the surface, fluid accumulates near the ends and the surface bulges there and is depressed near mid-height and this plays a part in causing the oscillations.

The oscillation mechanism based on these experiments seems to be the following: During the oscillations the surface flow in a given radial plane is observed to move alternately fast and slow which moves the free surface in and out. During the fast period, fluid is removed from the mid-height region and accumulates near the ends. The opposite occurs during the slow period. Although the associated surface deformations are small relative to the zone size, they are not small compared to the thermal boundary-layer thickness. Thus, when fluid is removed from the surface it deforms inward (eliminating the thermal boundary layer) and the surface temperature decreases; similarly the temperature is increased where fluid accumulates (surface bulges). As a result the fast period leads to decreased surface temperature gradients which slow the motion and the slow period produces a gradient to accelerate the flow. In this way the process repeats.

Another configuration to receive attention is a cylindrical container with a center-post heater, Fig. 15. By cooling the container walls a fixed temperature difference can be imposed on a fluid layer. This configuration differs significantly from the others studied. To minimize the effects of buoyancy the test section diameter is 4.78 mm. The test fluid is 2 cs silicone oil. Temperature and frequency measurements are made with thermocouples, a RMS meter, an oscilloscope, and a potentiometer. Flow visualization is obtained with a microscope and video system. See Ref. 39 for details. The ranges of parameters covered in the experiment are: Pr = 18 to 27, A = 0.62 to 1.95, $Re_\sigma = 1.30 \times 10^3$ to 1.62×10^4, $Ma = 3.10 \times 10^4$ to 3.04×10^5, and Bd = 0.57. The steady flow pattern is shown in Fig. 16. When the flow becomes oscillatory the pattern becomes three-dimensional as indicated in Fig. 17. The pathline of a particle at the free surface and down to a depth just before it turns upward is similar to that in steady flows. However, as it flows upward to resurface it follows a wavy path, which becomes more pronounced with increasing ΔT, and is unlike any steady flow observations. After resurfacing the particle takes a new path. The pattern repeats and the flow rotates in the direction first taken by the particle. The temperature fluctuations are similar to those in the half zone and details of the frequency and intensity of the oscillations as functions of the parameters were presented in Ref. 36.

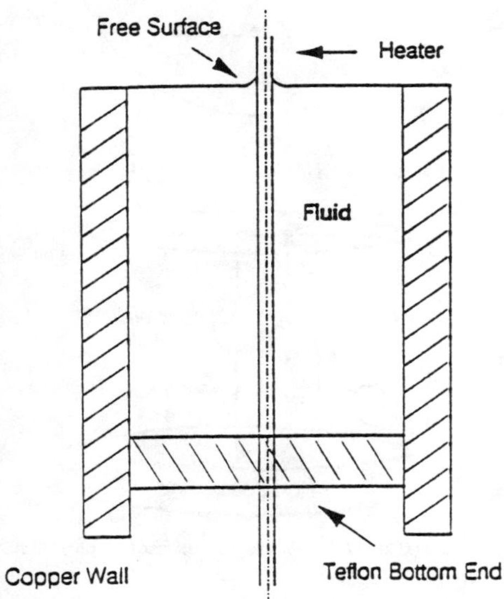

FIGURE 15. Test section

A thermal infrared imager was used to obtain almost instantaneous thermal images of the entire free surface. As sketched in Fig. 18 the image indicates a rotating pattern during oscillations. The rotation frequency agrees with the oscillation frequency determined by thermocouples.

From these observations it would appear that the oscillation mechanism is as follows, see Fig. 19. As fluid is pushed away from the high temperature region a standing wave is formed at the radial plane which lowers the temperature there because of the thin thermal boundary layer. The return flow is enhanced because of the surface curvature. The back flow moves toward the heater and accumulates there (because the flow was retarded in that region) which reverses the free surface curvature. The return flow is thereby retarded and the fluid near the heater becomes hotter and drives fluid away so that wave changes again and the process is repeated. The phase changes linearly in the azimuthal direction so that the wave pattern appears to rotate. The wave patterns at different radial planes are depicted in Fig. 20. The thermal image rotation is due to the rotation of the wave pattern. The wave motion is not necessarily axisymmetric.

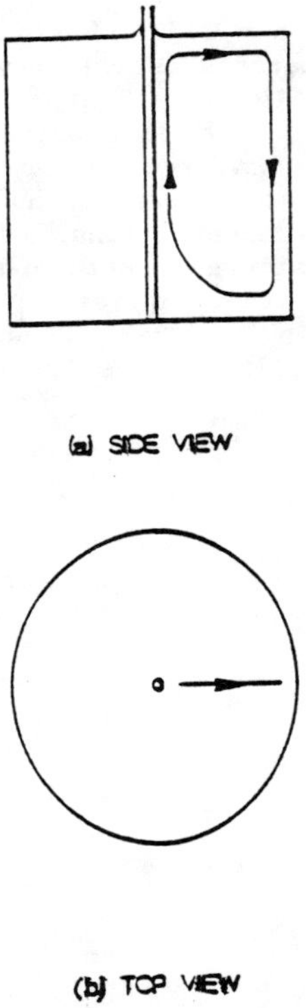

FIGURE 16. Typical flow pattern by flow visualization (non-oscillatory)

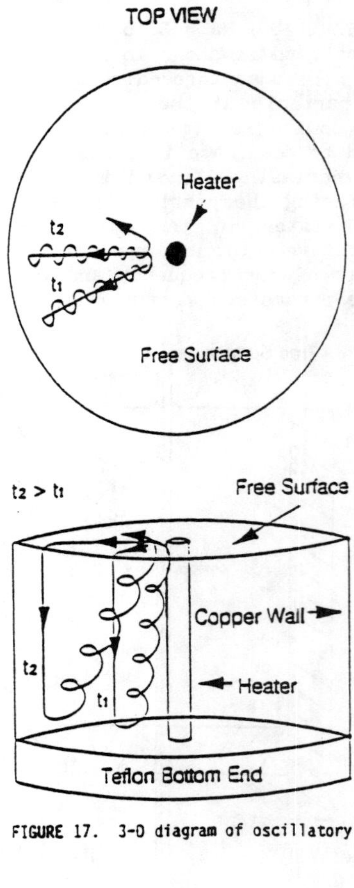

FIGURE 17. 3-D diagram of oscillatory flow

FUTURE NEEDS AND DIRECTIONS

It should be clear from the above that transport oscillations in materials processing are due to extremely complex phenomena. In thermosolutal flows there is the interaction among opposing buoyancy forces, counter flows and double diffusive effects. Coupling among the imposed thermal signature, the surface flow, and the surface deformation cause the oscillations in thermocapillary flows. The work done to date indicates design options for avoiding the oscillations. For example, from Fig. 7 conditions under which closed growth reactors should be operated to avoid oscillations are indicated. It would appear from the work done to date that the imposed thermal signature may be one design option to avoid oscillations in thermocapillary convection. However, as helpful as this work is in giving insights to the oscillatory phenomena and details of how they are affected by various parameters, it is limited to fluids with large Prandtl numbers. This was necessitated by experimental expediency but clearly this research must be extended to include low Pr fluids which are realistic for many materials processes. The oscillation mechanisms may be quite different because the differences in the diffusivity of viscosity and heat for low Pr fluids can alter the couplings described herein for large Pr fluids. The experimental challenges in doing similar research with low Pr fluids are formidable, but it will be worth the effort by indicating how the deleterious effects of transport oscillations can be eliminated in meaningful ways.

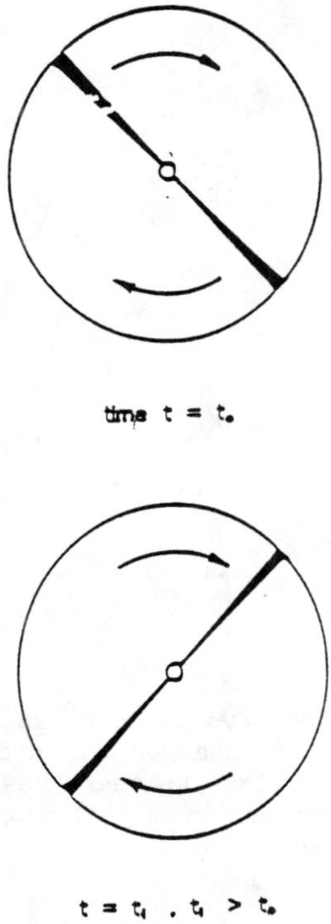

FIGURE 18. Thermal image pattern by infrared scanning

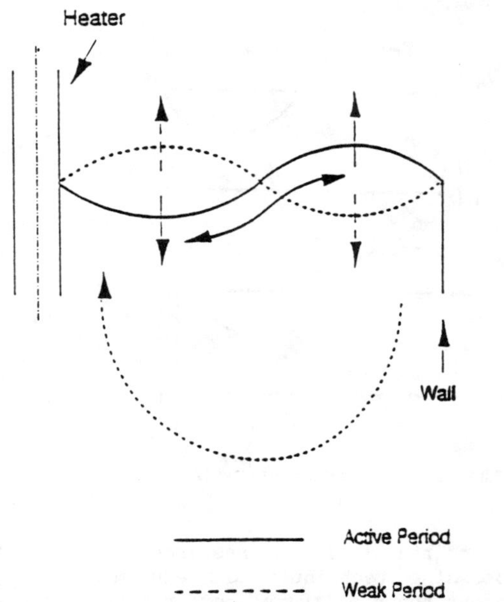

FIGURE 19. Concept of standing waves during oscillation

Even though the thermocapillary experiments were performed on tiny liquid bridges to eliminate buoyancy effects, the buoyancy effects were still of the same order as the thermocapillary convection (Bd unit order). Therefore, it is necessary to study thermocapillary flows in a low-gravity environment not only to reduce buoyancy further but also to reduce the static Bond number so that the zones will be more cylindrical and larger which will yield a larger range of Ma, and the surface damping and deformation characteristics will be different due to the reduced hydrostatic pressure. Such an experiment is manifested for the USML-1 mission and is described in Ref. 40.

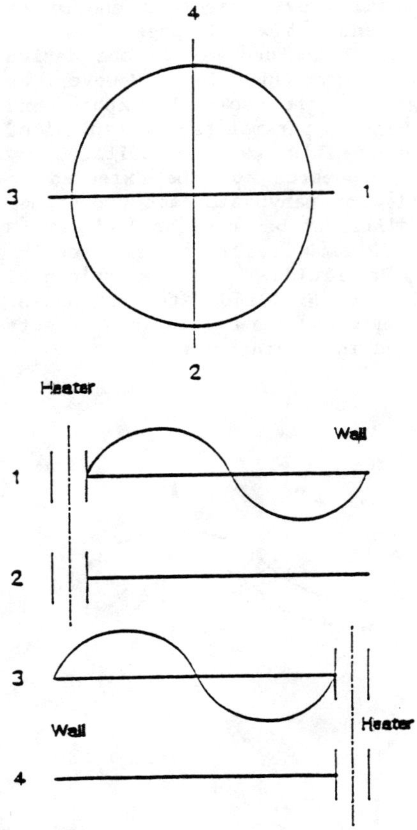

FIGURE 20. Different wave patterns

The end result of such research will indicate the ways in which materials processing techniques and equipment need to be changed in order to improve the quality and performance of the materials. Whether these will be accepted by industry, which has a penchant for staying with empirically developed methods, remains to be seen.

ACKNOWLEDGEMENT

This work was supported by the National Science Foundation under Grant MEA 84-01883, by the National Aeronautics and Space Administration under Grant NAG 3-570, and by the Battelle Advanced Materials Center for the Commercial Development of Space.

REFERENCES

1. Carruthers, J. R., Thermal Convection Instabilities Relevant to Crystal Growth From Liquids, in Preparation and Properties of Solid State Materials, ed. W. R. Wilcox, Vol. 3,pp. 1-121, Marcel Dekker, New York, 1973.

2. Solan, A. and Ostrach, S., Convection Effects in Crystal Growth by Closed-Tube Chemical Vapor Transport, in Preparation and Properties of Solid State Materials, ed. W. R. Wilcox, Vol. 4, pp. 63-110, Marcel Dekker, New York, 1974.

3. Pimputkar, S. M. and Ostrach, S., Convective Effects in Crystals Grown from Melt, Jour. Crystal Growth, Vol. 55, pp. 614-646, North-Holland Publishing Co., 1981.

4. Ostrach, S., Low-Gravity Fluid Flows, Ann. Rev. Fluid Mech., Vol 14, pp. 313-345, 1982.

5. Ostrach, S., Fluid Mechanics in Crystal Growth - The 1982 Freeman Scholar Lecture, Jour. Fluids Eng., Vol. 105, pp. 5-20, 1983.

6. Ostrach, S., Natural Convection with Combined Driving Forces, PhysicoChemical Hydrodynamics, Vol. 1, No. 4, pp. 233-247, Pergamon Press, 1980.

7. Kamotani, Y., Wang, L. W., Ostrach, S., and Jiang, H. D., Experimental Study of Natural Convection in Shallow Enclosures with Horizontal Temperature and Concentration Gradients, Int. Jour. Heat Mass Transfer, Vol. 28, No. 1, pp. 165-173, 1985.

8. Ostrach, S., Jiang, H. D., and Kamotani, Y., Thermosolutal Convection in Shallow Enclosures, in Proc. ASME/JSME Thermal Engineering Joint Conf., Vol. 2, eds., P. J. Marto and I. Tanasawa, pp. 159-168, 1987.

9. Jiang, H. D., Ostrach, S., and Kamotani, Y., Thermosolutal Convection with Opposed Buoyancy Forces in Shallow Enclosures, in Natural Convection in Enclosures-1988, eds. R. S. Figliona and P. G. Simpkins, HTD-Vol. 94, ASME Book No. G00476, 1988.

10. Kamotani, Y., Wang, L. W., and Ostrach, S., Experiments on Natural Convection in Low Aspect Ratio Enclosures, AIAA Jour., Vol. 21, pp. 290-294, 1983.

11. Lee, J. and Hyun, M. T., Experimental Study of Natural Convection due to Combined Buoyancy in a Low Aspect Ratio Enclosure, in Proc. Joint ASME/JSME Thermal Engineering Conf., eds., P. J. Marto and I. Tanasawa, Honolulu, 1987.

12. Jiang, H. D., Ostrach, S., and Kamotani, Y., Thermosolutal Convection with Opposed Buoyancy Forces in Shallow Enclosures, Ph.D. Thesis, Case Western Reserve University, Cleveland, Ohio, FTAS/TR-87-192, 1988.

13. Jiang, H. D., Ostrach, S., and Kamotani, Y., Thermosolutal Convection Flow Regimes with Opposed Buoyancy Forces in Shallow Enclosures, PCH PhysicoChemical Hydrodynamics, Vol. 10, No. 516, pp. 599-613, Pergamon Press, 1988.

14. Jiang, H. D., Ostrach, S., and Kamotani, Y., Unsteady Thermosolutal Transport Phenomena due to Opposed Buoyancy Forces in Shallow Enclosures, Accepted for Publication, Jour. Heat Transfer, 1990.

15. Ostrach, S., Motion Induced by Capillarity in Physicochemical Hydrodynamics: V. G. Levich Festschrift, ed., D. B. Spalding, Vol. 2, Advance Publ. Ltd., pp. 571-589, 1977.

16. Sternling, C. V. and Scriven, L. E., Interfacial Turbulence: Hydrodynamic Instability and the Marangoni Effect, AIChE Jour., Vol. 5, pp.514-523, 1959.

17. Kenning, D. B. R., Two-Phase Flow with Nonuniform Surface Tension, Appl. Mech. Rev., Vol. 21, pp. 1101-1111, 1968.

18. Ostrach, S., Role of Analysis in the solution of Complex Physical Problems, Proc. 3rd Intntl. Heat Transfer Conf., Vol. 6, pp. 32-43, 1966.

19. Ostrach, S., Convection Due to Surface-Tension Gradients, in (COSPAR) Space Research, ed., M. J. Rycroft, Vol. 19, pp. 563-570, Pergammon Press, 1979.

20. Rivas, D. and Ostrach, S., Thermocapillary Flows of Liquid Metals in Rectangular Enclosures, Ph.D. Thesis (Also FTAS/TR-87-190), Department of Mechanical and Aerospace Engineering, Case Western Reserve University, Cleveland, Ohio, 1987.

21. Hu, D., Ostrach, S., and Kamotani, Y., Thermocapillary Flows in an Enclosed Unit Order Aspect Ratio, Ph.D. Thesis (Also FTAS/TR-90-196), Department of Mechanical and Aerospace Engineering, Case Western Reserve University, Cleveland, Ohio, 1990.

22. Clark, P. A. and Wilcox, W. R., Influence of Gravity on Thermocapillary Convection in Floating Zone Melting of Silicon, Jour. Crystal Growth, Vol. 50, pp. 461-469, 1980.

23. Fu, B.-I. and Ostrach, S., Numerical Solutions of Floating-Zone Thermocapillary Flows, in Transport Phenomena in Materials Processing, ASME PED-Vol. 10, HTD-Vol. 29, 1983 (Also Proc. 4th European Symp. on Materials Sciences under Microgravity, ESA SP-191, pp. 239-245, June 1983).

24. Sen, A. K. and Davis, S. H., Steady Thermocapillary Convection in Two-Dimensional Slots, Jour. Fl. Mechs., Vol. 121, pp. 163-186, 1982.

25. Strani, M., Piva, R., and Graziani, G., Thermocapillary Convection in a Rectangular Cavity: Asymptotic Theory and Numerical Simulation, Jour. Fl. Mechs., Vol. 130, pp. 347-376, 1983.

26. Chun, Ch.-H. and Wuest, W., A Microgravity Simulation of Marangoni Convection, Acta Astronautica, Vol. 6, pp. 681-686, 1978.

27. Schwabe, Scharmann, A., Preisser, F., and Oeder, R., Experiments on Surface Tension Driven Flow in Floating Zone Melting, Jour. Crystal Growth, Vol. 43, pp. 305-315, 1978.

28. Chun, Ch.-H. and Wuest, W., Experiments on the Transition from the Steady to the Oscillatory Marangoni Convection of a Floating Zone under Reduced Gravity Effect, Acta Astronautica, Vol. 6, pp. 1073-1082, 1979.

29. Chun, Ch.-H. and Wuest, W., Experiments on Steady and Oscillatory Temperature Distribution in a Floating Zone due to Marangoni Convection, Acta Astronautica, Vol. 7, pp. 479-488, 1980.

30. Schwabe, D. and Scharmann, A., Some Evidence for the Existence and Magnitude of a Critical Marangoni Number for the Onset of Oscillatory Flow in Crystal Growth Melts, Jour. Crystal Growth, Vol. 46, pp. 125-131, 1979.

31. Kamotani, Y., Ostrach, S., and Vargas, M., Oscillatory Thermocapillary Convection in a Simulated Floating-Zone Configuration, Jour. Crystal Growth, Vol. 66, pp. 83-90, 1984.

32. Vargas, M., Ostrach, S., and Kamotani, Y., Surface-Tension Driven Convection in a Simulated Floating-Zone Configuration, M.S. Thesis (Also FTAS/TR-82-159), Department of Mechanical and Aerospace Engineering, Case Western Reserve University, 1982.

33. Ostrach, S., Kamotani, Y., and Lai, C. L., Oscillatory Thermocapillary Flows, PCH PhysicoChemical Hydrodynamics, Vol. 6, No. 516, pp.585-599, 1985.

34. Schwabe, D., Marangoni Effects in Crystal Growth Melts, PCH PhysicoChemical Hydrodynamics, Vol. 2, No. 4, pp. 263-280, 1981.

35. Lai, C. L., Ostrach, S., and Kamotani, Y., Studies of Thermocapillary Oscillation Phenomena, Ph.D. Thesis (Also FTAS/TR-84-177), Department of Mechanical and Aerospace Engineering, Case Western Reserve University, Cleveland, Ohio, 1984.

36. Lai, C. L., Ostrach, S., and Kamoani, Y., The Role of Surface Deformation in Unsteady Thermocapillary Flow, in Heat Transfer in High Technology and Power Engineering, ed., W. J. Yang and Y. Mori, Hemisphere Publ. Co., 1987.

37. Kamotani, Y., and Lee, K. J., Oscillatory Thermocapillary Flow in a Liquid Column Heated by a Ring Heater, PCH PhysicoChemical Hydrodynamics, Vol. 11, No. 5/6, pp. 729-736, 1989.

38. Preisser, F., Schwabe, D., and Scharmann, A., Steady and Oscillatory Thermocapillary Convection in Liquid Columns with Free Cylindrical Surface, Jour. Fl. Mech., Vol. 126, pp. 545, 1983.

39. Lee, J. H., An Experimental Study of Thermocapillary Convection in a Cylindrical Container, M.S. Thesis, Department of Mechanical and Aerospace Engineering, Case Western Reserve University.

40. Kamotani, Y., and Ostrach, S., Design of a Thermocapillary Flow Experiment in Reduced Gravity, Jour. Thermophysics, AIAA, Vol. 1, No. 1, Jan., 1987.

Summary of General Discussion of Session on General Topics (1)

T. Aihara, Tohoku University
J. Mazumder, University of Illinois

Since the session on General Topics (1) covers various areas, the general discussion was divided into two parts. The first half was concerned with Professor S. Fukusako's paper on production and heat-transfer characteristics of liquid ice; the latter half was concerned with Professor M.M. Chen's paper on thermo-capillary flow in materials processing and Professor S. Ostrach's paper on thermal oscillations.

In the first half, co-chairman T. Aihara made brief comments and suggestions showing examples of agitation systems and solid diffusion at high temperatures, which are summarized as follows.

Professor Fukusako developed a new technique for controlling a phase-change process; his technique may be expanded into a remote-control agitation technique for material refining.

Transient solid diffusion of carbon atoms in a Ti-TiC functional gradient material at high temperatures seriously affects the physical properties of the material. Hence, great attention should be concentrated not only upon the phase change, but also upon the transient solid diffusion.

In the latter half, co-chairman J. Mazumder made brief comments and suggestions on thermo-capillary flow and thermal oscillations, summarizing the papers by Professor M. M. Chen and Professor S. Ostrach.

Both Professors Chen and Ostrach stressed that careful "scaling" is essential for understanding of the thermo-capillary flow. Both concluded that thermo-capillary flow is complex, oscillatory and sometimes very singular. In addition, Professor Chen has drawn our attention to the following problem areas.

1. Full computation and numerical analysis for therm-capillary flow is expensive and difficult. Sometimes naive computation without guidance from proper scaling may even be misleading.

2. Physics for contact angle and boundary conditions for time dependent solidification at the surface are unknown.

Professor Chen *recommended* problem-based analysis and computations with the application of the "asymptotics" technique whenever possible.

Professor Ostrach has carried out extensive studies on thermo-capillary flow using high Prandtl number fluids. Based on this study he concluded that improved thermal signature may be used to suppress oscillatory behavior in thermo-capillary flow. In addition,

Professor Ostrach identified the following "needs" for better understanding of the thermo-capillary flow.

1. Future research with low Prandtl number fluids.

2. Careful examination of interaction between:
(a) Opposing buoyancy forces and double diffusive flows in thermo-solutal flows, and
(b)Imposed thermal signature, surface flow and surface deformation in thermo-capillary flow.

3. Micro-gravity experiments:
(a) To separate buoyancy effects from thermo-capillary and thermo-solutal flows, and
(b) To provide different damping conditions for free surface.

Another area needing attention from this community, not covered by the authors of this session, is the effect of vaporization and subsequent recoil pressure on thermo-capillary convection. These are important phenomena for many high energy beam processing techniques (e.g. laser, electron beam, etc.).

GENERAL TOPICS (2)

Heat and Mass Transport in the Photonics Manufacturing Technology

F. T. GEYLING
SEMATECH
Austin, Texas, USA

Abstract

The photonics technology comprises the design and manufacture of optical fibers, as well as the associated electronic signal processing equipment. Since the latter manufacturing technology is already represented at this seminar, in terms of bulk and epitaxial crystal growth, this talk will concentrate on glass fibers.

Optical fiber manufacture involves the fabrication of preforms, the drawing of fibers and subsequent coating. Preforms can be produced by the MCVD (modified chemical vapor deposition), the OVD (outside vapor deposition), or the VAD (vapor axial deposition) processes. Different photonic applications and manufacturers may favor different processes. Hybrid techniques are also possible; e.g. 'overcladding' an OVD or VAD preform by collapsing a tube onto it, akin to the final stages of MCVD. We illustrate MCVD by discussing the internal deposition and sintering process by which an optically graded material is created and the subsequent, thermal collapse of the tubular preform into a solid rod. Relatively simple representations of mass and heat transport, as well as chemistry, proved effective for optimizing this fabrication sequence in the 1980's. More stringent demands on the finished product have motivated more recent modeling efforts to represent 3D, transient gas dynamics with variable fluid properties, as well as radiative heat transfer effects and more sophisticated chemical kinetics. At the same time, more refined, optical diagnostics were used for process control. A similar evolution, from simple to more realistic models, could be observed in the OVD and VAD technologies.

Fiber drawing is a central step in the photonics manufacturing sequence and plays a dominate role in the quality control of long-distance optical cables. The manner of preform melting and ambient control, together with subsequent coating, govern the mechanical strength, optical waveguide properties and uniformity of the resulting fibers. The draw-down phenomenon and the coating flow may be subject to instabilities, including possible thermal feed-back. Recent endeavours to create fibers with strong polarization by locking (thermal or mechanical) stresses into their cross-sections raise questions of long-term integrity, i.e. the possibility of creep relaxation and/or static fatigue.

Early work will be described in each of these areas, with pointers to subsequent refinements in the late 1980's.

Modeling and Simulation of Manufacturing Processes of Advanced Composite Materials

WOO IL LEE
Seoul National University
Seoul 151-742, Korea

GEORGE S. SPRINGER
Stanford University
Stanford, California 94305, USA

I. INTRODUCTION

Fiber reinforced organic matrix composites are manufactured by many different methods, such as autoclave curing, filament winding, compression molding, resin transfer molding, and pultrusion. Regardless of the methods used, care must be taken to employ the correct procedures during manufacture. Improper manufacturing procedures may result in parts which are of unacceptable quality or costs. To ensure that both the quality and the cost are acceptable, the manufacturing process must be tailored for each particular application and thus it is important to select the appropriate manufacturing conditions. The required conditions are established through variables which can be manipulated during the processing. Applied temperature and pressure are two of the most important variables, and are common to the many different types of manufacturing processes. There are additional variables specific to different processes. For example, in filament winding the winding tension and winding speed, in pultrusion the pulling speed of the material into the die must carefully be selected and controlled. The aforementioned process variables can be chosen empirically. However, the empirical, trial and error methods are expensive and time consuming. Moreover unless tried out exhaustively, which is almost impossible due to the cost and time, they do not guarantee that the resulting manufacturing conditions are optimum. It is far more advantageous and convenient to establish the required process conditions and process variables by models based on fundamental engineering principles. Some models had been proposed for autoclave curing [1,2], filament winding [3] and pultrusion [4,5] processes. Some investigators did numerical simulation of the mold filling during resin transfer molding [6,7]. Recent developments of new thermoplastic resins initiated research in manufacturing processes for composites with these materials. Process models were first proposed by Lee and Springer [8] and currently being studied by many other researchers.

In this paper a brief description is given of a) the existing models for autoclave curing of thermosetting matrix composites, and b) modeling of manufacturing process for thermoplastic matrix composites.

II. MODEL FOR THERMOSETTING COMPOSITES

One of the most common ways of manufacturing parts made of thermosetting matrix composites is by autoclave curing. In the autoclave curing process, the part to be cured along with other necessary materials is placed in an autoclave (Fig.1). Inside the autoclave heat and pressure are applied to the part . Elevated temperature accelerates the chemical reactions occurring during the cure, while the applied pressure squeezes out excess resin, compacts the plies, and reduces the void content. It is obvious that the durations and magnitudes of the

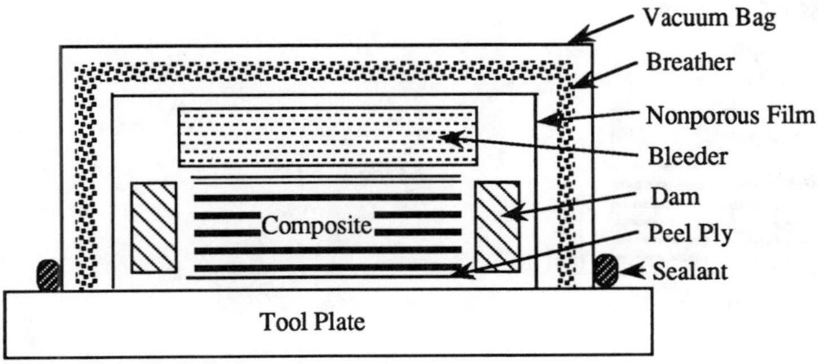

Figure1. Components of autoclave curing assembly.

applied temperatures and pressures (generally referred to as the cure cycle) affect significantly the quality of the part. The requirements which must be satisfied during autoclave cure can be summarized as follows:

a) The temperature inside the material must not exceed a preset value at any time during cure;

b) At the end of cure the resin content must be uniform and must have the desired value;

c) The material must be cured uniformly and completely;

d) The cured composite must have the lowest possible void content;

e) The curing must be achieved in the shortest time.

Consider a fiber-reinforced epoxy resin composite of thickness L (Fig. 2). The composite is constructed of continuous fiber unidirectional prepreg tape. An absorbent material (referred to as the "bleeder")is placed on one or both sides of the composite. As the curing is started, the composite-bleeder system is exposed to a known temperature. At some time, pressure is applied to the system, squeezing excess resin from the composite into the bleeder. Both the cure temperature T and the cure pressure P may vary with time in an arbitrary manner. The objective is to determine the following parameters during cure:

a) The temperature T inside the composite as a function of position and time;

b) The degree of cure of the resin α as a function of position and time;

c) The resin viscosity μ as a function of position and time;

d) The amount of resin in the composite (resin volume fraction) and in the bleeder as a function of time.

A model providing the aforementioned information is described below in two parts. The first part of the model results in the temperature, viscosity, and degree of cure. The second part yields the resin flow into the bleeder and the resin content of the composite. The model is presented for one-dimensional geometry; that is, for unidirectional composites in which the properties vary only in the x-direction (Fig.2) and in which the resin flow is only in the x-direction. Resin flow parallel to the fibers (a phenomenon that may arise in practice) is not considered.

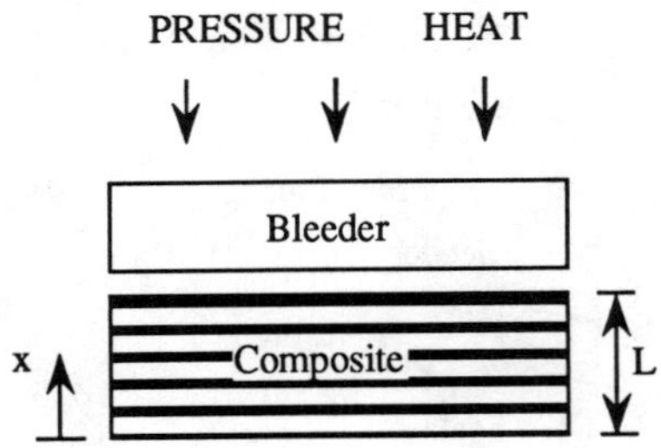

Figure2. Composite-bleeder system.

Thermo-Chemical Submodel

The temperature distribution inside the laminate can be described by one-dimensional unsteady heat conduction equation

$$\rho C \frac{\partial T}{\partial t} = \frac{\partial}{\partial x}\left(K \frac{\partial T}{\partial x}\right) + \rho \dot{H}_R \quad (1)$$

where ρ and C are the density and specific heat of the composite, respectively, K is the thermal conductivity in the direction perpendicular to the plate, and T is the temperature. The parameter H_R is the rate of heat generated by chemical reactions and is defined by

$$\dot{H}_R = H_{ult} \frac{d\alpha}{dt} \quad (2)$$

where H_{ult} is the total or ultimate heat of reaction during cure, and $d\alpha/dt$ is the cure rate. The rate of heat generation during cure is proportional to the rate of cure. The degree of cure α is related to time and temperature by the following form of differential equation

$$\frac{d\alpha}{dt} = f(\alpha, T) \quad (3)$$

Many different expressions for the dependence of the cure rate on the temperature and on the degree of cure have been proposed in the past. The expression most appropriate for the material under consideration should be employed in the calculations. Values for the constants that appear in these expressions, as well as the value of the heat of reaction H_{ult}, can be determined experimentally by differential scanning calorimetry [9,10]. The following form is used frequently for epoxy resins [9,10]

$$\frac{d\alpha}{dt} = [A_1 \exp(-\frac{\Delta E_1}{RT}) + A_2 \exp(-\frac{\Delta E_2}{RT})\alpha^m](1-\alpha)^n \quad (4)$$

where A_1, A_2, ΔE_1, ΔE_2, m and n are constants and R is the universal gas constant.

Solutions to Eqs.(1) to (4) along with applicable initial and boundary conditions yield the temperature T, and the degree of cure α as functions of position and time inside the material. Once these variables are known the resin viscosity μ can be calculated from an appropriate empirical expression relating μ to both T and a (or to t). One form of such a relation is given as follows [9,10]

$$\mu = \mu_\infty \exp(\frac{\Delta E_\mu}{RT} + K\alpha) \tag{5}$$

where μ_∞, ΔE_μ and K are constants.

Resin Flow Submodel

Upon application of pressure, resin may be squeezed out from spaces between the individual fibers and from the space between adjacent plies. In fact, owing to the proximity of the fibers, relatively small amount of resin is removed from the space between the fibers. The bulk of the resin originates form the space between adjacent plies. Due to the edge constraints, resin may flow only in the direction normal to the plane of the laminate. When the resin flow is only in the normal direction, then first only the top layer moves while resin is being squeezed out from space between the first and second layers (Fig.3) [11]. When the fibers reach the second layer, the first two layers move in unison towards the third layer, squeezing resin out from the space between the second and third layers. This sequence is repeated for the subsequent layers. Thus, the layers are "compacted" in a wavelike, cascading manner. Therefore a pressure differential exists only across the region where resin flow occurs and in the remaining region the pressure is constant. Recently, there have been some data reported which suggest that, as the cure progresses, the pressure in the resin at the bottom center of the laminate becomes lower than the applied pressure [12,13]. Such decrease in resin pressure could occur only if the fibers were to "touch" and hence carry load [12,13]. This situation does not arise as long as all the fibers are surrounded by resin, i,e., until there is at least one thin resin layer in the prepreg across which there is no fiber-to-fiber contact.

The resin flow through the composite and through the bleeder is considered to be flow through porous media; at any instant of time, the velocity V in the composite and in the bleeder are represented by Darcy's law

$$V = -\frac{S}{\mu}\frac{dP}{dx} \tag{6}$$

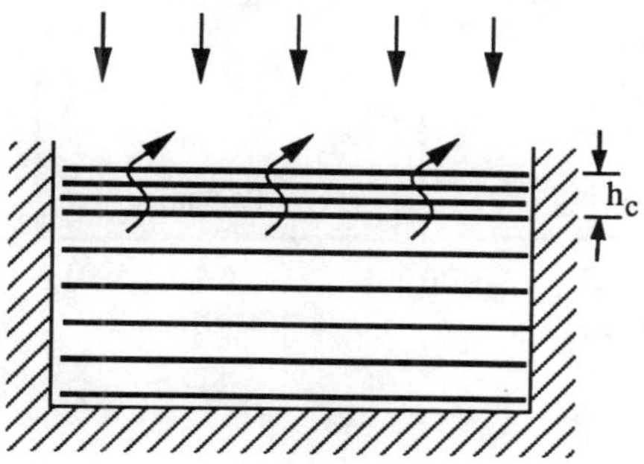

Figure 3. Illustration of the resin flow and the compaction process.

where S is the apparent permeability, μ is the resin viscosity, and dP/dx is the pressure gradient. The rate of change of mass M in the composite thus becomes

$$\dot{M} = -\rho AV = -\rho AS \frac{\Delta P}{\int_0^{h_c} \mu dx} \qquad (7)$$

where ρ is the resin density, A is the cross-sectional area, and h_c is the thickness of the layer through which resin flow takes place (Fig.3). The parameter ΔP is the difference between the applied pressure and the pressure of air in the bleeder. If the mass of excess resin in one ply is ΔM, the number of compacted plies, n is

$$n = \frac{1}{\Delta M} \int_0^t \dot{M} dt \qquad (8)$$

Equations (7) and (8) provide the relations for calculating the resin flow.

In order to illustrate the type of informations which can be generated by the model, sample calculations were done for Hercules AS4/3501-6 graphite/epoxy composite laminates. The results are shown from Fig.4 for temperature and degree of cure as functions of position and time for different heat-up rates. In Fig.5, the number of compacted plies as a function of time is presented for different heat-up rates. The information as in Fig.4 and 5 can be used in selecting optimum cure cycle which satisfies the requirement stated previously.

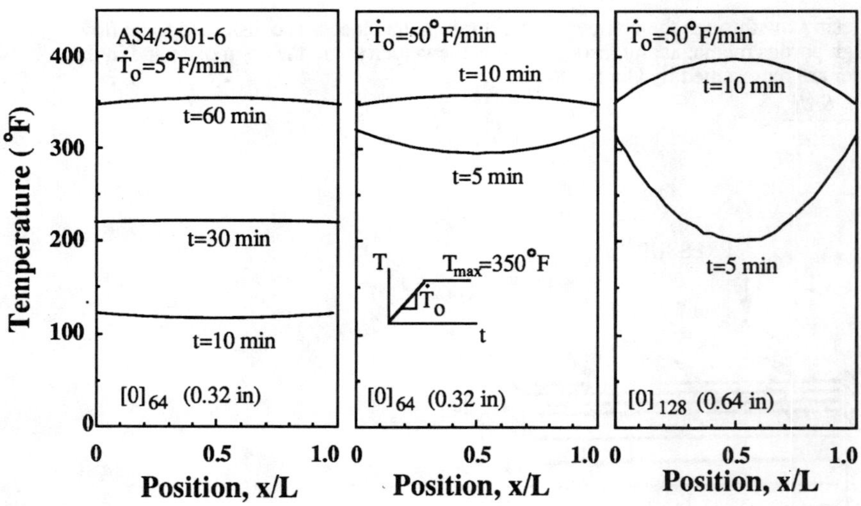

Figure 4. Temperature distribution as a function of time for different heat-up rates and different thicknesses. Cooling rates indicated are the surface cooling rates.

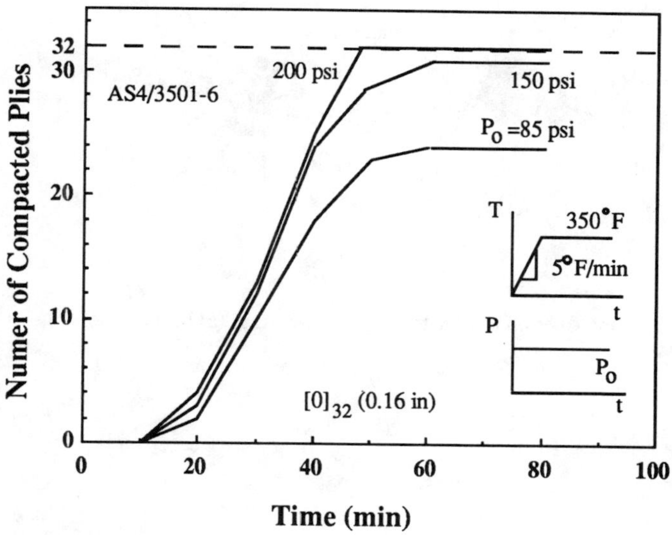

Figure 5. Number of compacted plies as a function of time for different pressure. The laminate thickness is 32 plies (0.16 in).

III. MODEL FOR THERMOPLASTIC COMPOSITES

Processing of thermoplastic matrix composites consists of three major steps. First, matrix is introduced in to space between the fibers in each tow, impregnating the tow. Second, individual plies are "consolidated" such that there is good contact as well as good bond between adjacent plies. Third, the composite is heated and then cooled at a rate which provides the desired "crystallinity" in the matrix. The requirements which must be met during processing thermoplastic matrix composites are as follows:

a) The tows must be fully impregnated;

b) The bond between individual plies must be complete;

c) The crystallinity must have the desired value throughout the composite.

In this section a model is presented describing each of the above three steps. The model is divided into the three submodels designated as "impregnation," "consolidation," and "crystallinity." Each of these submodels is developed separately. The three submodels are then combined, so as to be applicable in situations where some or all of the phenomena described by the different submodels occur simultaneously.

<u>Impregnation Submodel</u>

Impregnation is accomplished by introducing resin into fiber tows. For the purpose of impregnating effectively, process temperature is kept as high as possible to reduce the

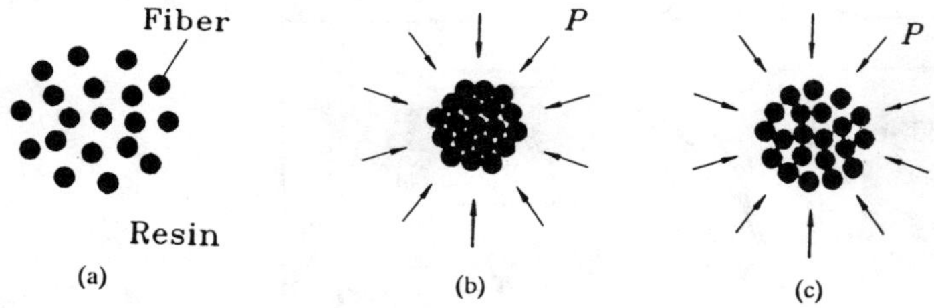

Figure 6. Deformation of the fiber bundle by the applied pressure: (a) initial state; (b) ideal case; (c) real case.

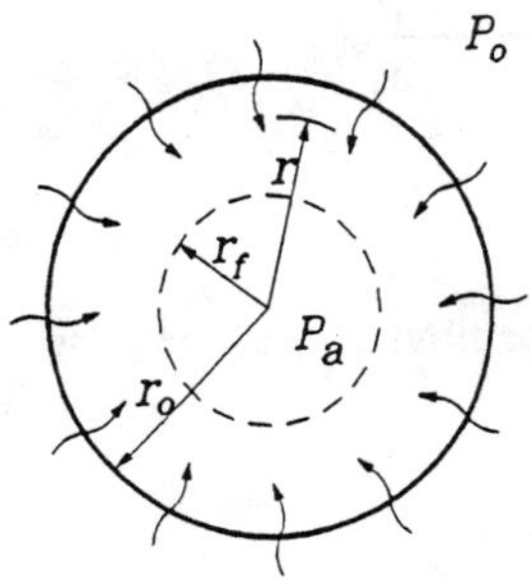

Figure 7. Illustration of the radial resin flow in a fiber tow

viscosity of the resin. At the same time pressure is applied to accelerate the impregnation process. In the ideal case, where the individual fibers are perfectly straight and parallel to each other, as pressure is applied, the fiber bundle collapses inward leaving no room for the resin to penetrate (see Fig.6). In the real situation, however, each of the fibers is not perfectly straight but rather wavy in nature. Therefore as the fiber bundle collapses inward by the applied pressure, physical contacts between the fibers take place at some points. Between these contact points there exists room for the resin to pass through (see Fig.6). It is assumed that the radial flow of the resin is a flow which passes through a porous medium as in Fig.7 and is governed by Darcy's law. Therefore the radial velocity of the resin can be expressed as

$$V_r = -\frac{S}{\mu}\frac{dP}{dr} \tag{9}$$

where V_r is the radial velocity of the resin, S is the permeability of the fiber bundle, P is the applied pressure, r is the radial position. The mass flow rate in the radial direction may be expressed by the continuity condition as

$$\dot{M} = -2\pi\rho V_r r\varepsilon \tag{10}$$

where ρ is the density of the molten resin, and ε is the porosity of the fiber bundle. Equations (9) and (10) along with boundary conditions ($P=P_o$ at $r=r_o$, $P=P_a$ at $r=r_f$) yields

$$\frac{\dot{M}}{2\pi\rho\varepsilon}\ln(\frac{r_f}{r_o}) = -\frac{S}{\mu}(P_0 - P_a) \tag{11}$$

where r_o is the outer radius of the fiber bundle, r_f is the location of the resin front, P_o is the applied pressure, and P_a is the atmospheric pressure. The mass flow rate can be expressed in terms of dr_f/dt as

$$\dot{M} = 2\pi\rho r_f \varepsilon(-\frac{dr_f}{dt}) \tag{12}$$

Equations (11) and (12) yield

$$t = \frac{\mu r_o^2}{4S(P_0 - P_a)}[2(\frac{r_f}{r_o})^2 \ln(\frac{r_f}{r_o}) + 1 - (\frac{r_f}{r_o})^2] \tag{13}$$

The degree of impregnation D_{imp} can be defined as the ratio of the number of impregnated fibers to the total number of fibers of a tow. Assuming that the fibers are uniformly distributed, the degree of impregnation can be expressed as

$$D_{imp} = \frac{\text{area impregnated}}{\text{tow area}} = 1 - (\frac{r_f}{r_o})^2 \tag{14}$$

Substituting Eq.(14) into Eq.(13), we finally obtain the degree of impregnation as a function of the process time, the applied pressure, the viscosity, the initial radius and the permeability of the fiber bundle

$$t = \frac{\mu r_o^2}{4S(P_0 - P_a)}[(1 - D_{imp})\ln(1 - D_{imp}) + D_{imp}] \tag{15}$$

From Eq.(15) given the values of μ of the molten resin, r_o and S of the fiber bundle, we can estimate the degree of impregnation as a function of time and the processing variables, namely applied pressure, temperature, and tow size.

Consolidation Submodel

When thermoplastic composites are being processed individual plies consolidate into a laminate by bonding at the interfaces [14]. This bonding consists of two phenomena. First, two adjacent ply surfaces coalesce and come into "intimate contact." Second, bond forms at the ply interface by a process called autohesion. In the following, models are presented which describe the effects of the processing variables (temperature, pressure, time) on the intimate contact and autohesion processes. These processes occur simultaneously. However, for the sake of clarity, the models of these processes are developed separately.

a) Intimate Contact: Laminates are formed by " laying up" plies. Since the ply surfaces are uneven, spatial gaps exist between the plies prior to the application of heat and pressure. Therefore, thermoplastic matrix composites must actually be deformed to produce intimate contact between adjacent surfaces. The irregular ply surface can be approximated by a surface consisting of a series of identical rectangles (Fig.8). Due to the applied force (or pressure) the rectangular elements spread along the interface. We are interested in the extent of this spread (degree of intimate contact) as a function of applied temperature, force (pressure), and time. With reference to Fig.8, the degree of intimate contact is defined as

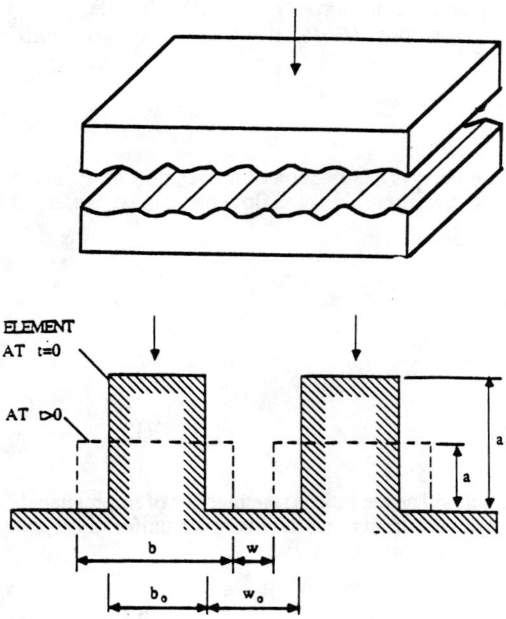

Figure 8. Illustration of the idealized ply surface used in the intimate contact submodel

$$D_{ic} = \frac{b}{b_0 + w_0} \tag{16}$$

where b_o and b are the initial and instantaneous (at time t) widths of each rectangular element, respectively, and w_o is the initial distance between two adjacent element. The volume of the rectangle is assumed to be constant throughout the deformation (i.e., $a_o b_o = ab$). For the squeezing flow of a rectangle between parallel plates, the applied pressure P is related to the height of the rectangle by [8]

$$\frac{a_0}{a} = [1 + 5\frac{P}{\mu_{mf}}(1 + \frac{w_0}{b_0})(\frac{a_0}{b_0})^2 t]^{1/5} \tag{17}$$

where a_o and a are the initial and instantaneous heights of each rectangular element. Here μ_{mf} is the viscosity of the fiber-matrix mixture. Therefore the degree of intimate contact can be expressed in terms of applied pressure and time as

$$D_{ic} = \frac{1}{1 + b_0 / w_0}[1 + 5\frac{P}{\mu_{mf}}(1 + \frac{w_0}{b_0})(\frac{a_0}{b_0})^2 t]^{1/5} \tag{18}$$

b) Autohesion: Once two adjacent interfaces come into intimate contact, a bonding process between the interfaces starts. In the case of two similar thermoplastic interfaces the bond is formed mostly by autohesion. During autohesion segments of the chain like molecules diffuse across the interface (Fig.9). The extent of the molecular diffusion and, hence, the bond strength increase with time. A convenient way of characterizing the extent of autohesion is through the bond strength. Thus a degree of autohesion D_{au} may be defined as [14]

$$D_{au} = \frac{B}{B_{\infty}} \tag{19}$$

where B is the bond strength at time t and B_{∞} is the ultimate bond strength, i.e., the strength of a completely bonded interface. The degree of autohesion can be approximated by the following expression proposed by previous investigators (e.g., see [14]) for amorphous polymers

$$D_{au} = \kappa t^{1/4} \tag{20}$$

where t is the time elapsed from the start of the autohesion process (i.e., the time when the interfaces come into intimate contact). κ is a constant which is related to the temperature through the Arrhenius relation

$$\kappa = \kappa_0 \exp(-\frac{E}{RT}) \tag{21}$$

κ_o and E are constants. It is noted that the crystallinity of the material resulting from cooling may affect the accuracy of Eq.(20).

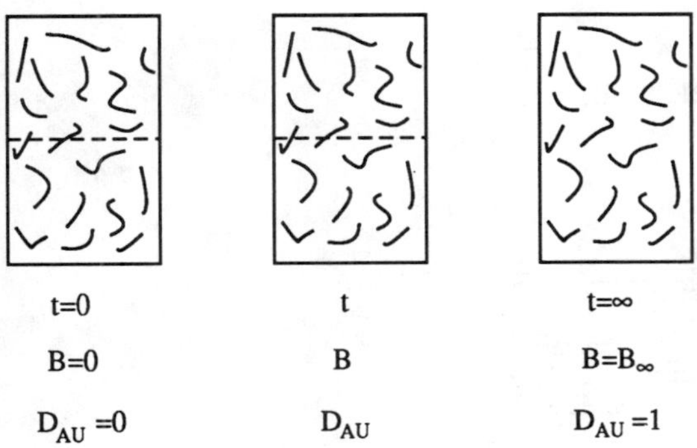

Figure 9. Illustration of autohesion process

Crystallinity Submodel

For semicrystalline thermoplastic matrix composites the degree of crystallinity affects significantly the mechanical properties of the composite [15]. The desired crystallinity is achieved during the processing by cooling the composite from the melt temperature at an appropriate cooling rate. The objective here is to establish a model which relates the cooling rate applied during processing to the crystallinity of the material. For clarity, the model is developed for a flat plate in which the temperature varies only across the plate but not along the plate (one dimensional problem, Fig.10). During cooling the crystallinity of the polymer changes. The instantaneous degree of crystallinity depends on the temperature and on the rate of change of the temperature. Therefore, to determine the degree of crystallinity as a function of position and time, the temperature distribution inside the laminate must be known at all times. For a flat plate the temperature distribution can be calculated by the following differential equation

$$\rho C\frac{\partial T}{\partial t}=\frac{\partial}{\partial x}(K\frac{\partial T}{\partial x})+\rho\dot{H}_c \tag{22}$$

The second term on the right-hand side of Eq.(22) represents the heat generated due to crystallization. The heat generated due to crystallization is related to the rate of crystallinity change as

$$\dot{H}_c=H_T\frac{dc}{dt} \tag{23}$$

where c is the crystallinity of the matrix and H_T is the theoretical ultimate heat of crystallization. The crystallinity is related to the cooling rate and the temperature

$$c=g(\frac{dT}{dt},T) \tag{24}$$

A typical form of the equation can be found elsewhere [8,16].

The temperature and the crystallinity as functions of position and time can be calculated by Eqs.(22) and (24) along with appropriate initial and boundary conditions.

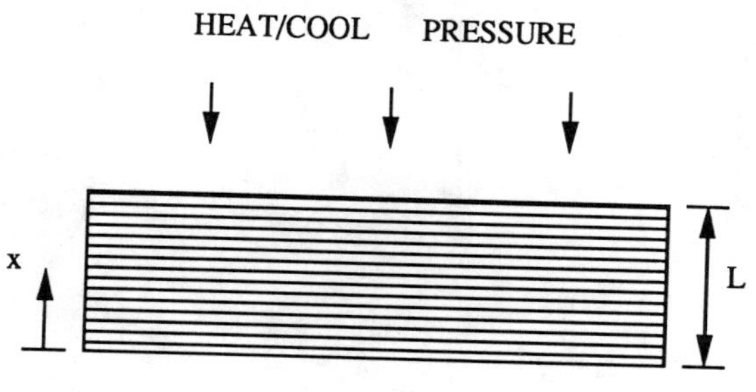

Figure 10. Geometry used in the crystallinity submodel

To illustrate the type of information which can be generated by the aforementioned model, numerical results for the graphite/PEEK composite are presented in Fig.11-13. In Fig.11 the degree of impregnation as a function of time for different pressure is shown. The change in degree of intimate contact is illustrated for different temperatures in Fig.12. In Fig.13, temperature and crystallinity distributions as functions of time, as well as the average crystallinity as a function of cooling rate were presented. By generating results such as shown here, the process variables can be selected which result in a composite of satisfactory quality.

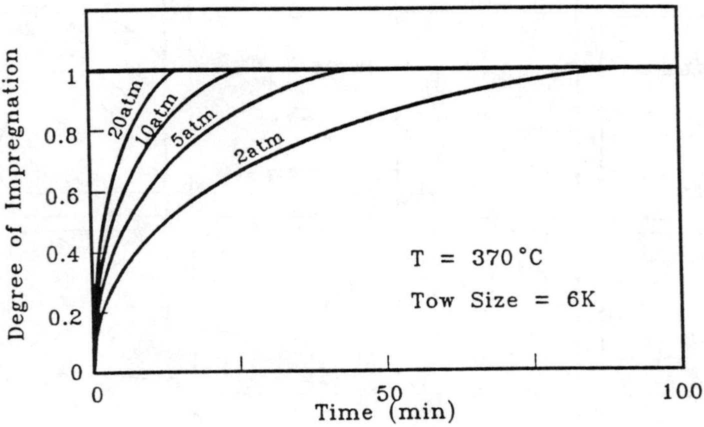

Figure 11. Degree of impregnation as a function of time for different pressures. Temperature is 370°C and tow size is 6K.

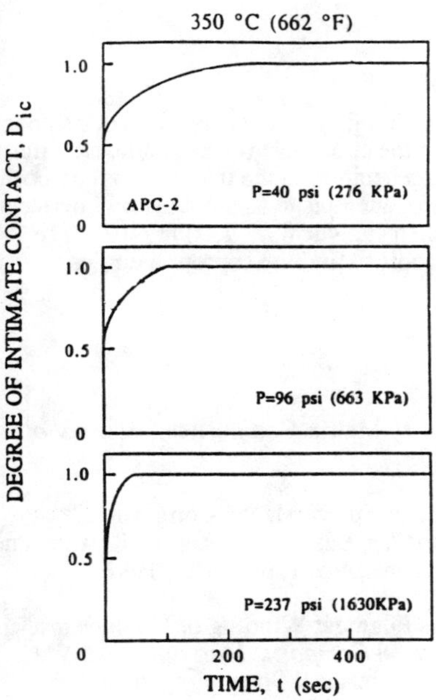

Figure 12. Degree of intimate contact as a function of time for different pressures. Temperature is 350°C.

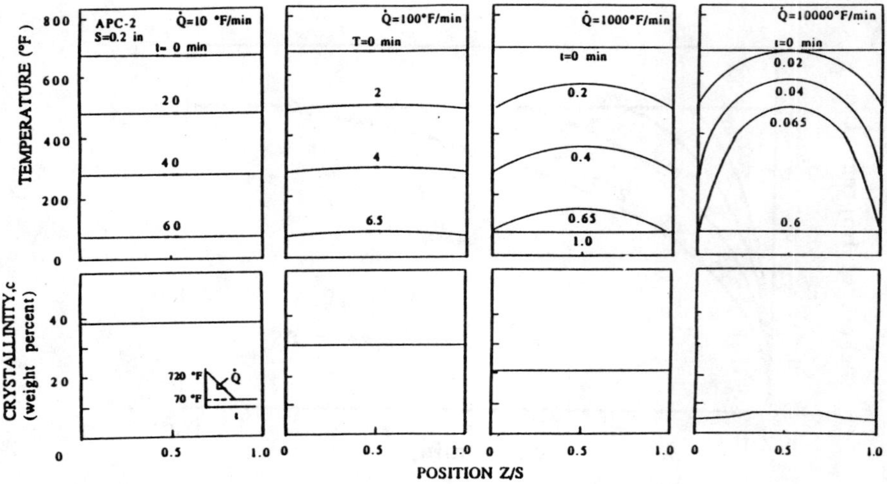

Figure 13. Top: Temperature distributions a function of time. Bottom: Crystallinity distribution after cooling to 70°F. The thickness of composite is 0.2 in. Cooling rates indicated are the surface cooling rates.

IV. CONCLUDING REMARKS

The models presented in this paper can be used in obtaining optimum cure cycle of composite materials. These models are particularly useful since the cure cycle (heat-up rate, etc.) must be modified to account for the effect of internal heat generation as the thickness of composite laminate changes. Recently, there are some successful attempts to find out the optimum cure cycle via an expert system (e.g., using rules rather than analytical models) [17-19]. The models described above can serve as indispensable tools to find the appropriate rules.

V. REFERENCES

1. Loos, A. C. and Springer, G. S., "Curing of Epoxy Matrix Composites," *Journal of Composite Materials*, Vol.17, pp.135-169, 1983.

2. Springer, G. S., "A Model of the Curing Process of Epoxy Matrix Composites," *Progress in Science and Engineering for Composites*, Eds. T. Hayashi, K. Kawata, and S. Umekawa, Japan Society for Composite Materials, Vol.1, pp.23-35, 1982.

3. Calius, E. P. and Springer, G. S., "Modeling the Filament Winding of Composites," *Proceedings of the Fifth International Conference on Composite Materials*, Eds.W.C. Harrigan, Jr., J. Strife and A.K. Dhingra, The Metallurgical Society, pp.1071-1088, 1985.

4. Han, C. D., Lee, D. S. and Chin, H. B., "Development of a Mathematical Model for the Pultrusion Process," *Polymer Engineering and Science*, Vol.26, pp.393-404, 1986.

5. Batch, G. L. and Macosko, C. W., "Pultrusion Modeling," in *the Manufacturing Science of Composites*, Proceedings of Manufacturing International '88, Ed. T. G. Gutowsky, the American Society of Mechanical Engineers, 1988.

6. Coulter, J. P. and Guceri, S. I., Resin Impregnation During the Manufacturing of Composite Materials, CCM Report No.88-07, University of Delaware, 1988.

7. Um, M.-K. and Lee, W. I., "A Study on the Mold Filling Process in Resin Transfer Molding," to appear in *Polymer Engineering and Science*.

8. Lee, W. I. and Springer, G. S., "A Model of the Manufacturing Process of Thermoplastic Matrix Composites," *Journal of Composite Materials*, Vol.21, pp.1017-1055, 1987.

9. Lee, W. I., Loos, A. and Springer, G. S., "Heat of Reaction, Degree of Cure and Viscosity of Hercules 3501-6 Resin," *Journal of Composite Materials*, Vol.16, pp.510-520, 1982.

10. Dusi, M. R., Lee, W. I., Ciriscioli, P. R. and Springer, G. S., "Cure Kinetics and Viscosity of Fiberite 976 Resin," *Journal of Composite Materials*, Vol.21, pp.243-261, 1987.

11. Springer, G. S., "Resin Flow During the Cure of Fiber Reinforced Composites," *Journal of Composite Materials*, Vol.16, pp.400-410, 1982.

12. Gutowsky, T. G., Cai, Z., Bauer, S., Boucher, D., Kingery, J. and Wineman, S., "Consolidation Experiments for Laminate Composites, " *Journal of Composite Materials*, Vol.21, pp.650-669, 1987.

13. Gutowsky, T. G., Morigaki, T. and Cai, Z., "The Consolidation of Laminate Composites, " *Journal of Composite Materials*, Vol.21, pp.172-188, 1987.

14. Dara, P. H. and Loos, A. C., "Thermoplastic Matrix Composite Processing Model," Report CCMS-85-10, VPI-E-85-21, Virginia Polytechnic Institute and State University, 1985.

15. Lee, W. I., Talbott, M. F., Springer, G. S. and Berglund, L. A., "Effect of Cooling Rate on the Crystallinity and Mechanical Properties of Thermoplastic Composites," *Journal of Reinforced Plastics and Composites*, Vol.6, pp.2-12, 1987.

16. Ozawa, T., "Kinetics of Nonisothermal Crystallization," *Polymer*, Vol.12, pp.150-158, 1971.

17. Servais, R. A., Lee, C. W. and Browning, C. E., "Intelligent Processing of Composite Materials," in *Materials for the Future*, Society for the Advancement of Material and Process Engineering, Vol.31, pp.764-775, 1986.

18. Ciriscioli, P. R. and Springer, G. S., "Dielectric Cure Monitoring-A Critical Review," *SAMPE Journal*, Vol.25, pp.35-42, 1989.

19. Ciriscioli, P. R., "An Expert System for Autoclave Curing of Composites," Ph. D. Thesis, Department of Aeronautics and Astronautics, Stanford University, 1990.

Reaction Engineering Principles of Combustion Synthesis of Advanced Materials

JORGE E. GATICA, HEMANT W. DANDEKAR, HENDRIK J. VILJOEN, and VLADIMIR HLAVACEK
Laboratory for Ceramic and Reaction Engineering
Department of Chemical Engineering
State University of New York at Buffalo
Buffalo, New York 14260, USA

INTRODUCTION

Several solid-solid, gas-solid or liquid-solid synthesis reactions are accompanied by a large release of energy, becoming self-sustaining after ignition without additional energy input. These processes require only a short-term service to initiate the reaction, a combustion front appears at the ignition point in the form of a sharp interface which propagates towards the unreacted material. The temperature raises abruptly at the reaction front reaching typically 2000 K, and up to 4000 K in some cases. This process, sometimes referred to as Self-propagating High-temperature Synthesis (SHS), is certainly a unique technology for preparing advanced ceramic materials.

In the last two decades the synthesis of materials by combustion reactions has been gaining popularity, becoming an important process as a part of modern technology .The history of Combustion Synthesis, however, can be traced back to the past century. Indeed, the first recorded combustion synthesis was performed early in the nineteenth century by Berzelius (1843) who ignited amorphous zirconium powders to form the oxide. By the end of the century, Goldschmidt (1895) was successful in reducing metal oxides by aluminum powder, thus discovering the thermite process. He can also be credited as the first person who had used combustion synthesis for technical applications.

The modern history of the application combustion synthesis principles dates back to 1967 when Shkiro, Borovinskaya and Merzhanov (Merzhanov, 1990) synthesized TiB_2 from the elemental precursors. Since then a large number of materials has been synthesized and the list is still growing. The wide range of applications of Combustion Synthesis principles includes the manufacturing of certain gases (e.g., BCl_3 from chlorination of B_4C), metals (aluminothermic operations), refractory high temperature materials (borides or carbides), coatings, lubricants (molybdenum and tungsten sulfides), pesticides (magnesium, aluminum and zinc phosphides), superconductors, isolation of high-level nuclear waste, fabrication of advanced nuclear fuel (uranium-zirconium carbonitride), and metallic hydrides as shield materials.

Besides its simplicity, the synthesis by Combustion reactions holds several other advantages over conventional methods like carbothermal reduction or wet chemical methods. Methods for the synthesis of pure elements have evolved into maturity. As a result, one of the key steps in the production of advanced ceramic parts, the precursors for the synthesis reactions are readily available with high purity degree. The large amount of energy that is released by the reaction also vaporizes volatile impurities which may be adsorbed on reactants. In the case of a solid product the elevated temperature after the reaction has been completed can induce plastic behavior, if post-processing is performed in this state a dense body becomes a very realistic goal. This feature becomes attractive for materials which conventionally required several steps, like milling, classification and hot-pressing to obtain a final dense shape.

Present research concentrates on problems such as scale-up of combustion synthesis reactors, the synthesis of tailored materials, the better utilization of the heat of reaction for the processing of products and the numerical and analytical analysis of the combustion process.

DESCRIPTION OF PROCESS

A Combustion Synthesis process can be categorized as either a solid-solid or a solid-fluid reaction. This classification relates to the states of the precursors and it is tacitly assumed that a system of three components is not considered. However, the appearance of a second phase at the front (e.g. melting of one component) is well-known (Margolis, 1983; Dimitriou, 1989). The preferred mode for industrial operation is the planar wave. The reaction front keeps the form of a flat sheet and it transverses the sample, converting all reactants in a homogeneous way. Scale-up of sample dimensions leads to spatiotemporal instabilities which have undesirable effects like local melting, incomplete conversion and damage to reactor linings.

Solid-Solid Combustion Synthesis

In the case of synthesis reactions from solid precursors, the reactants are intimately mixed and pressed into a preform, usually in the form of a cylindrical pellet, or the powder mixture is contained in an inert boat. An external heat source is required to ignite the sample. For this purpose electrical igniters, lasers, chemical igniters or resistively heated wires can be used. A reaction front appears in the form of a propagating sharp interface between reactants and products (cf. Fig. 1).

The propagation of a combustion front may be stable or unstable depending on the kinetic and thermophysical characteristics of the reacting system. The instability of the reaction front can be one or multi-dimensional depending on the geometric characteristics of the combustion system (Merzhanov, 1981; Sivashinsky, 1981; Dvoryankin *et al.*, 1982). In cylindrical geometries, different combustion modes have been observed, such as

- Planar, stable.
- Planar, pulsating.
- Non-planar, one-and multi-head spinning.

The stability of a planar front is lost by the occurrence of a Hopf-type bifurcation and the system behaves in oscillatory fashion. The oscillatory patterns become increasingly complex away from the oscillatory limit .Taking melting effects into account in their model, Bayliss and Matkowsky (1987) observed similar dynamic behavior of the front propagation phenomena in condensed phase combustion. They showed that the pulsating front developed into oscillations whose peaks became progressively sharper and steeper as their bifurcation parameter (a combination of the dimensionless heat of reaction and melting parameters) moved away from the primary bifurcation point. They demonstrated that as a secondary bifurcation point was exceeded, a period doubling bifurcation occurred. More recently, Dimitriou *et al.* (1989) showed that as the system departs from the first bifurcation point a series of period doubling bifurcation occurs and condensed phase combustion evolves to non-deterministic behavior .Similar results were reported by Bayliss and Matkowsky (1990) who performed a numerical study of the problem. These authors found that the inclusion of melting effects in the model changed the bifurcation behavior of the system: periodic windows appeared in the parameter space and finally an intermittency behaviour was observed.

These phenomena have serious implications for the scale-up of any combustion synthesis process and they must be properly analyzed by analytical and numerical methods.

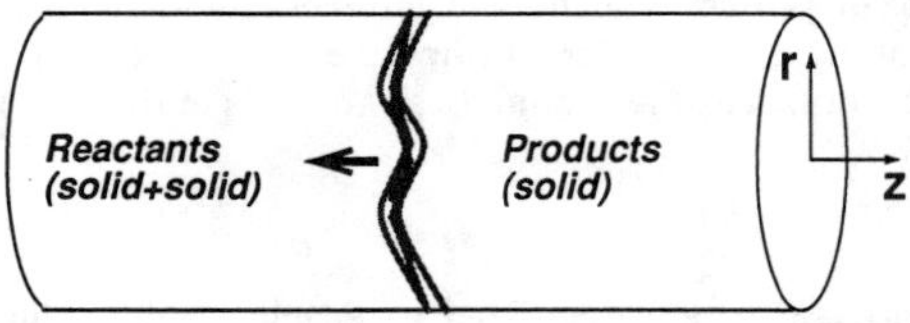

FIGURE 1. Schematic of Solid-Solid Combustion in a cylindrical preform.

Gas-Solid Combustion Synthesis

In these systems the combustion takes place between a gaseous phase and a porous solid phase. A very important factor in these cases is the flow of gas through porous solid by means of filtration or permeation. Hence, this process is often termed as *filtration combustion.* This process finds use in producing nitrides and hydrides. Various configurations have been proposed in the past, one of the most commonly studied is the counter-flow configuration shown in Fig. 2a. Metal powder is packed in a cylindrical reactor closed at one end, whose walls are capable of withstanding very high temperature. The open end of the tube is exposed to gaseous reactant at very high pressure. The reaction is initiated at the sealed end by a short term service heat source, such as as resistively heated graphite or tungsten strip. The gaseous oxidizer necessary for the reaction is transported by filtration through the permeable powder. The flow is driven by the pressure gradient caused by the consumption of gaseous reactant at the front near the sealed end. Once initiated, under appropriate conditions, the front propagates down the length of the tube. However, there are certain disadvantages with the counter-flow configuration amongst them: i) limitations on the length of the tube which will lengthen the filtration path requiring higher operating pressure; ii) difficulty in product removal especially for fused solids iii) for the case of low pressure operation if pore closures take place due to sintering/volumetric expansion the inner layers of solid will remain unconverted. Taking these factors into consideration, a new configuration called the *cross-flow configuration* was proposed recently by Kumar *et al.*(1988). This configuration is shown in Fig. 2b. This configuration has been studied in detail by Dandekar *et al.* (1990b). Gas access to the front is considerably improved in this configuration, the gaseous reactant being transported not only by longitudinal flow from above, but also, by transverse (counter-flow) flow from regions ahead of the front. As a result nitridation is completed at relatively lower pressures. Product removal is easier since the reactor is open at the top.

As in solid-solid systems non-linear phenomena such as spin-combustion, and pulsating combustion fronts have been observed in the past (Merzhanov et al, 1973). However, the analysis of these systems is considerably more involved since, gas-flow to the front is another complication which has to be accounted for. Besides avoiding regions of operation where such non-linear phenomena occur, additional parameters need to be considered. At low pressures, the bed height has to be limited in order to maintain a sufficient gradient in pressure for the flow to be driven to the front, in order to sustain the reaction. Effect of structural changes due to melting, sintering and volumetric expansion also need to be taken into consideration. Structural changes strongly affect the thermal conductivity and permeability of the medium.

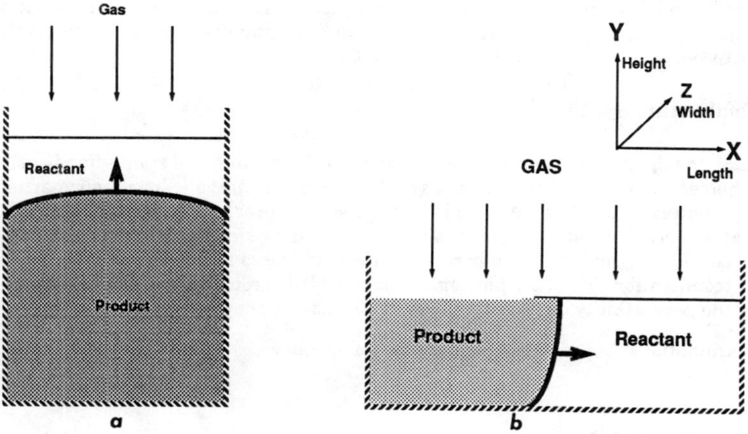

FIGURE 2. Configurations for gas-solid combustion synthesis; a) Counter-flow, b) Cross-flow configuration.

FORMULATION OF THE MATHEMATICAL MODEL

Solid-solid Combustion Synthesis

Let us consider the system represented schematically in Fig. 1. It represents a cylindrical preform of well-mixed reactants. The reaction between two solid reactants A and B is assumed of first order with respect to the limiting reactant A. The thermophysical properties (density, thermal conductivity, heat capacity) are assumed constant, diffusion of reactants or products is neglected and the reaction rate temperature dependence is of the Arrhenius form. Based on the above assumptions the gasless burning of a condensed system is described by the following set of dimensionless equations :

Mass Balance

$$\frac{\partial \eta}{\partial \tau} = -R(\eta, \theta) \tag{1}$$

Energy Balance

$$\frac{\partial \theta}{\partial \tau} = \underline{\nabla} \cdot \underline{\underline{\lambda}} \cdot \underline{\nabla} \theta + R(\eta, \theta) \tag{2}$$

subject to,

$$\begin{array}{ll} -\underline{n} \cdot \underline{\underline{\lambda}} \cdot \underline{\nabla}\theta = \mathrm{Bi}\ \theta & \text{at the lateral boundaries, i.e. } \rho = a \\ \theta(\rho, \phi, \xi, \tau) = 0 & \text{at } \xi = \pm\infty \\ \eta(\rho, \phi, \xi, \tau) = 0 & \text{at } \xi = -\infty \\ \eta(\rho, \phi, \xi, \tau) = 1 & \text{at } \xi = +\infty \end{array} \tag{3}$$

These equations have been rendered non-dimensional by defining the characteristic length as the thermal thickness of the reaction front (i.e. $l^\star = k_e/(\omega \overline{\rho C_p})$) and the time scale as $t^\star = l^\star/\omega$, where ω represents the reaction front propagation velocity for a stable combustion mode. The variables η and θ are the dimensionless concentration and temperature, $R(\eta, \theta)$ stands for the reaction rate expression, $\underline{\underline{\lambda}}$ is the thermal conductivity tensor, $\underline{n}$ is the outwards normal vector, and Bi is the Biot number. The dimensionless axial, radial and angular coordinates are ξ, ρ and ϕ, respectively while τ stands for the dimensionless time.

Even though this model is based on a simple one-step global Arrhenius kinetics, it predicts qualitatively the behavior of gasless reacting systems and its simplicity makes it amenable to analytical treatment.

Gas-Solid Combustion Synthesis

In order to study the dynamics of gas-solid systems detailed modeling and simulation is necessary. A mathematical model in the form of a system of coupled parabolic-hyperbolic partial differential equation is formulated. A pseudo-homogeneous representation is assumed, the thermophysical properties of the solid and gas are considered constant, a irreversible chemical reaction is considered, and local bed porosity variations due to sintering and volumetric expansion is accounted for. The flow phenomena is modeled through Darcy's law with the dependence of the permeability on porosity given by Carman-Kozeny equation.

Under these assumptions, the governing equations are as follows:

Mass balance on the gas phase

$$\frac{\partial \alpha \bar{\varepsilon} \bar{\rho}_g}{\partial \tau} = -\underline{\nabla} \cdot \underline{v} \alpha \bar{\varepsilon} \bar{\rho}_g - \gamma^{-1} R(\eta, \theta) \tag{4}$$

Energy balance on the system

$$\frac{\partial H}{\partial \tau} = \underline{\nabla} \cdot \underline{\underline{\lambda}} \cdot \underline{\nabla} \theta - \underline{\nabla} \cdot \underline{v} \bar{\varepsilon} H_g + \left[(1 + \delta_1 - \delta_2)^{-1} + \gamma^{-2} \right] R(\eta, \theta) \tag{5}$$

Momentum Balance

$$\underline{v} = -\kappa \underline{\nabla} \pi \tag{6}$$

These equations are completed with a reaction rate expression, $R(\eta, \theta) = f(\eta) S(\eta) exp\left(\frac{\theta}{1+\beta\theta}\right) \pi^{\nu}$, and a equation of state for the gas (e.g. ideal gas).

The boundary conditions for the system involve impermeable boundaries at all sides except the top which is open to the gaseous atmosphere. Convective heat losses are assumed from all sides, and radiation heat losses are accounted for only at the top. For a certain time the boundary at left (see Fig. 2) is heated to the adiabatic temperature, to ignite the system, after which normal heat losses are assumed from this boundary.

NUMERICAL SOLUTION OF THE GOVERNING EQUATIONS

Due to the strongly nonlinear nature exhibited by the equations describing combustion processes, analytical techniques as of yet have failed to produce solutions for the full form of the governing equations. Nevertheless, considerable progress in understanding reaction systems has been made possible through the application of analytical techniques to some limiting situations. For instance, a wealth of information and has been accumulated over the past decade showing results based on the use of large activation energy asymptotics. Although credit should be given to similar approaches for providing physical insight in many reaction problems, the models described are simplified and limitations on the techniques clearly exist. To overcome these limitations the systems of governing equations, together with the appropriate boundary conditions, must be normally solved by numerical techniques.

Solution with Adaptive Meshes

For the application of numerical methods, a number of grid points is distributed over the physical domain or, equivalently, the domain is cut up into small pieces, called cells. The discretization of the governing equations yields a system of algebraic equations expressing relationships between values of the different variables at neighboring grid points.

Because of finite computer resources, it will not be possible to monitor the solution at every point of the region in space on which the problem has been defined. A finite number of grid points is therefore distributed over the field. Each of these grid points acts as an observer, monitoring the solution locally and relaying information about it to its neighbors. Efficiency in distributing these observers should be the guiding principle; a limited, fixed number of grid points or grid cells should be arranged in such a way that the largest possible amount of information can be extracted from their use.

In combustion-type problems, as in a host of others, most of the important physics are often restricted to small parts (or, "boundary layers,") inside the domain. To be able to give a sufficiently accurate representation of the solution on the grid, the grid point distribution should be influenced by the behavior of the very solution it is monitoring. By concentrating part of the available grid points inside these boundary layers, efficient use is made of the resources at hand. Certainly, for realistic simulation of combustion problems is rarely feasible to implement a numerical solution method that is reliable and accurate at a reasonable cost and yet does not utilize some form of grid adaption.

Among the regriding approaches available, the technique suggested by Degrève *et al.* (1987) has been followed for the results presented next. It consists in a variational approach which has been derived following the original idea of Saltzmann (1982). The number of grid points remain fixed, a weighting function is computed along with the solution and the grid points are dynamically distributed in order to minimize a given objective function. The adaption is done in terms of a time and space dependent transformation which proportions grid size on the derivative. Thus we transform a non-equidistant physical domain (x,y) to an equidistant computational domain (ξ,η) with the transformation functions $\xi(x,y,t)$ and $\eta(x,y,t)$. The mesh generation is governed by three functionals representing smoothness, orthogonality and adaption. The adaption functional represents a measure of how well the mesh adapts to a given weight function w = w(x,y) based on the gradients and curvature associated with temperature and conversion. Minimization of this functional causes small cell sizes in regions of high gradients. However, rapid variations in cell size and lack of grid orthogonality can have a deleterious effect on the accuracy of calculations by introducing substantial errors in finite difference approximations. The need to control properties of smoothness and orthogonality therefore, becomes imperative. For this a weighted average of these functional in an integral form is minimized over the whole domain. This leads to a system of elliptic partial differential equations which are discretized over the whole domain and a solution is obtained by iteratively by an ADI scheme for the movement of grid points. An important aspect of grid generation is to prevent the grid from becoming singular, called grid point limiting. Grids become singular whenever the interval size (one dimension), area (two dimensions) or volume (three dimensions) become zero or negative. Since any simulation done on a singular grid will terminate catastrophically, such a situation has to be avoided. Furthermore, a small cell size restricts the size of time step especially for explicit schemes. Details of the mesh generation and grid point limiting algorithm are available in Degrève (1990).

The adaptive grid algorithm used is for a two dimensional adaptive grid. For the three-dimensional simulations, an intermediate level of adaptive grid, is generated, where two-dimensional adaption is performed in 2D planes and the position of the 2D planes is determined by a one dimensional grid generation. Thus a 2-D mesh is used in the x-y plane and the position of these planes in the z direction is determined by one dimensional adaption.

Discretization of the Governing Equations

Once the physical domain has been discretized appropriately and a way has been chosen to change the grid points locations in response to the evolving solution, the governing equations can be approximated on the so constructed generalized mesh.

The use of a generalized mesh makes the approximation of the governing partial differential equations more complicated than it would be on a orthogonal cartesian or polar grid. Two different approaches have been followed in the past to accomplish the discretization of the equations in a generalized coordinate system, each one having its own merits (Anderson *et al.*, 1984).

The first approach is based on the transformation of the governing equations to the generalized curvilinear coordinates. Since the transformation originates additional terms ("cross derivatives,") and variable coefficients ("metrics,"), the resulting equations become more complicated. The additional terms appear because of the non-orthogonality of the curvilinear coordinates in the physical space. Depending on the rearrangement of the different terms after substitution of the transformation relations, the resulting transformed equations can be obtained in a conservative or a non-conservative form. For more information on the subject we refer the reader to Hindman (1982).

The second approach is known by the name of "Control Volume Formulation," (CVF). This approach takes into account the fact that the governing equations frequently represent the conservation of a transported quantity for an infinitesimally small control volume; thus, it is easily realizable that the conservation equations could be written for any arbitrary control volume of finite size. Precisely, the approach consists in dividing the physical domain into smaller "cells," and rewriting the governing equations for these finite-size control volumes. A typical arrangement of 27 grid points and the corresponding control volume for the central point is depicted in Fig. 3.

The discretization of the governing equations now consists in approximating the integral form on the generalized mesh formed by quadrilateral cells. These last forms of the equations are extremely useful since continuity of fluxes the boundaries of neighboring control volumes can be easily ensured. Conservation of the transported quantity, even for the discrete numerical approximation, will thus be maintained which is a desirable property to have for any numerical method used for the simulation of physical conservation laws.

Treatment of the Transport Terms

The convective and diffusive/conductive transport terms could be rewritten by resorting to the Gauss-divergence theorem. When implementing this equation numerically care has to be taken such as conservation of the transported property is ensured. At first glance, the transport provided by the global movement of the fluid ($\underline{n} \cdot \underline{v}$), the convective flux resulting from the control volume displacement in time ($\underline{n} \cdot \underline{v}_s$), and the diffusive transport ($\underline{n} \cdot \underline{\underline{\lambda}} \cdot \underline{\nabla}$) do not seem to present any particular problem. The properties , the velocities and the gradients can be evaluated at a given time and assumed constant during the time integration step .This is certainly true, however, as the control volume V moves in time its boundary surface S deforms and the outwards normal becomes time dependent (i.e., $\underline{n} = \underline{n}(\tau)$). It is this particular fact which calls for a careful treatment of the numerical implementation of these transport terms. Thus, if only transport due to the control volume displacement exist. Then, for a constant field (e.g., $\theta(x, y, z, t) = \theta_o \forall\, x, y, z, t$), the integral form of the conservation law reduces to

$$\frac{d}{d\tau} \int_V dV = \oint_S \underline{n} \cdot \underline{v}_s \, dS \tag{7}$$

This is the so called Geometric Conservation Law (GCL) and it simply states that the change in the control volume is due to the movement of the control volume sides. Although this sounds straightforward, the GCL has important repercussion on how the discrete approximation to the governing equations has to be carried out. Lack of satisfaction of this law in a numerical sense will cause a uniform field (for instance) to become non-uniform just due to the movement of the grid points and not due to any physical phenomena. It is therefore extremely important to find out exactly how the product ($\underline{n} \cdot \underline{v}_s$) is to be approximated.

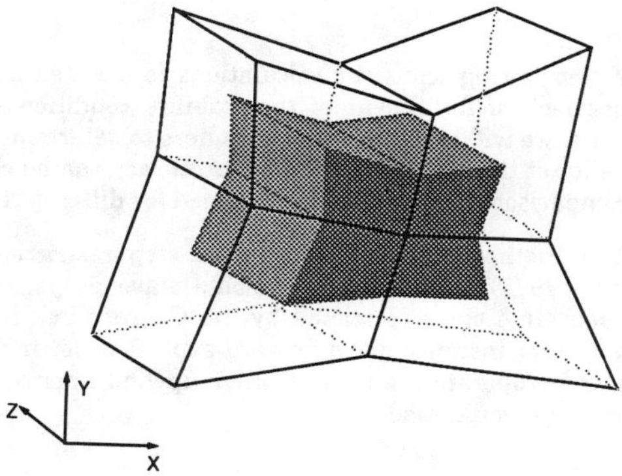

FIGURE 3. Arrangement of grid points and control volume for a three-dimensional grid.

Treatment of the Convective Transport Terms

Precise calculation of practical problems where chemical reactions and fluid flow phenomena interact relies on the accuracy and stability of the numerical schemes adopted for the space-time discretization. Several studies have been devoted to modify and asses existing numerical algorithms. The quest for an optimal discretization, however, still continues.

One of the main difficulties arises from the convective transport terms. This transport has some peculiar features, at difference of the diffusive/conductive transport the propagation of information occurs only in the flow direction. A perturbation at a particular position in a purely convective field it is not supposed to influence regions located upstream. It is this property of the convective transport that a numerical approximation should posses. Centered approximations, although formally accurate, do not mimic this property. Thus the use of central differences in convective transport has shown some undesirable stability properties: at low Reynolds numbers it provides a satisfactory representation for the convective transport, as the flow speed increases, spurious oscillations appear, the unphysical oscillatory responses are augmented and disastrous non-convergence is the final result (Roache, 1972).

The usual remedy has been to replace the central differences approximation with a first order upwind approximation, or even a hybrid between central differences and first order upwind (Spalding, 1972). These methods have been shown to introduce a substantial amount of artificial dispersion which might obscure the accurate modeling of a given physical process. Two second order algorithms have been suggested as alternates to the first order upwind: the skew-upwind approach and quadratic upstream interpolation. The skew-upwind scheme (Raithby, 1976) constitutes a modification to the first order upwind in the sense that the scheme is applied along a skewed streamline, thus the upwind scheme is used in a vector sense rather that along the main directions. The second alternate has been suggested by Leonard (1979) and is known by the acronym QUICK (Quadratic Upstream Interpolation for Convective Kinematics), this method proposes a local quadratic interpolating surface on a cell basis.

A number of studies have provided a comparison between these different methods (cf. Braaten and Shyy, 1986). These studies have shown that the two alternate approaches are indeed more accurate than the first order schemes. In this paper we will adopt the QUICK scheme; it has shown to be a reliable approximation for convection-dominated problems and its generalization to non-regular grids can be efficiently accomplished.

Choice of the Time Step

The choice of a proper time step during transient calculations is dictated by two criteria: stability and accuracy. Since lack of fulfillment of the stability condition would lead to inaccurate unstable calculations, we will limit the discussion here to determine the time step which would ensure the stability of the numerical method. Accuracy can be checked by the more heuristic approach of comparison between results obtained for different time steps.

It is well known that explicit methods are subject to time step restriction for reasons of numerical stability (Roache, 1972). The one-dimensional wave propagation and heat conduction problems have their time steps restricted by the Courant-Friedrich-Lewy and Fourier conditions, respectively. For instance, for a forward explicit Euler in time, centered finite differences for second order operators and first order upwind approximation of the convective terms, these restrictions would read

$$\Delta\tau_D \leq \frac{\Delta\xi^2}{2\tilde{\lambda}}$$
$$\Delta\tau_C \leq \frac{\Delta\xi}{|\underline{v}|}$$

Here $\underline{v}$ is the flow velocity and $\tilde{\lambda}$ is the maximum of the coefficients for the diffusive terms. Even though no diffusive terms are present in the mass balance, the way the velocity is calculated (using Darcy's Law) and substituted in the mass balance, the convective terms in fact translate into diffusive terms in pressure. Thus $\tilde{\lambda}$ should be taken as the maximum of the permeability and the thermal conductivity of the medium. These criteria simply establish that in one time step changes at one particular grid point can only be "transported," to its immediate neighbors. In the case of non-equidistant grids, these criteria will impose different restrictions for every grid point; $\Delta\tau$ should be small enough such that even the most restrictive condition is met.

To obtain a time step stability criterion for a particular transport equation the most widely used approach is the stability analysis developed by Von Neuman. This analysis considers the decay or amplification of each Fourier component of the solution (Roache, 1972). Although elegant for equidistant grids, the method becomes quite cumbersome for general discretizations (non-orthogonal and non-equidistant). A similar approach is followed.

For the particular problem under analysis, the physical property (in the example, enthalpy) is being transported by both mechanisms simultaneously. In such event, a restriction to the time step can be found by combining the restriction for every individual mechanism, i.e.

$$\frac{1}{\Delta\tau_{DC}} \leq \frac{1}{\Delta\tau_C} + \frac{1}{\Delta\tau_D}$$

the formula can be applied to every subelement in each cell and select the local minimum. The final time step restriction will be dictated by the global minimum after all cells have been scanned.

STABILITY OF SOLID-SOLID COMBUSTION

Steady planar combustion mode of propagation implies a constant propagation velocity (or burning rate) ω. In the case of non-adiabatic conditions the combustion front curves and its propagation speed decreases. If the heat losses become excessive the front may even extinguish near the cylinder surface reducing the effective dimensions of the final product. Expressions for ω can be derived from match asymptotic analysis (Kaper *et al.*, 1986; Kovalenko, 1987).

Since planar combustion mode of operation is characterized by a constant burning rate ω, a new coordinate system attached to the reaction front can be defined as: $z = x + \omega t$.

The length/radius ratio will be considered large enough such that the combustion front will resemble a freely propagating combustion wave, i.e. independent of the initiation (end) effects. In a very recent work, Varma *et al.* (1990) analyzed numerically combustion in finite length rods. These authors showed that truly constant-pattern propagation is not realized in finite size pellets. Hence, the analysis presented next can be considered strictly valid for infinitely long cylinders only.

Since reactions of the SHS-type are characterized by large activation energy and heat of reaction, the reaction front will be considered confined to a very thin region ("sheet,") (cf. Fig. 1). Using a similar approach as outlined by Buckmaster and Ludford (1983) a mathematical model for the stability analysis can be derived. In this model, the reaction term is approximated as (Thiart *et al.*, 1990a),

$$R(\eta, \theta) \approx \delta_F \exp\left[\alpha(\theta - \theta_m)\right]$$

where, $\delta_F = \sqrt{1 + \left(\frac{\partial F}{\partial \rho}\right)^2 + \frac{1}{\rho^2}\frac{\partial^2 F}{\partial \phi^2}}\ \delta(\xi - F(\rho, \phi, \tau))$, is the Dirac delta function, where $F(\rho, \phi, \tau)$ indicates the instantaneous location of the reaction front. A similar model has been already used by Matkowsky and Sivashinsky (1978) in the analysis of pulsating fronts in condensed phase combustion.

The adiabatic case was studied by Sivashinsky (1981). This author solved for the basic state (planar propagation mode) in a closed form and found the conditions for the onset of spinning wave propagation mode. In practice, ideal adiabatic conditions will never be realized and the stability analysis should include heat losses from the reaction system to the surrounding medium. Once heat losses are introduced, the maximum temperature $\theta_m < 1$, the "flame sheet," (front) curves (i.e., from the reacted zone, the reaction surface will resemble a convex disc) and the basic temperature depends on the radial coordinate as well. For weakly non-adiabatic conditions the Biot number (the parameter characterizing the heat losses) will be small. In such a case, the Biot number can be used as an expansion parameter, and the solution can be sought as an asymptotic series in terms of the Biot number. Correct to $O(\mathrm{Bi})$, the basic solution is given by:

$$\theta_0^{\pm} = \theta_0^{\pm(h)} + \theta_0^{\pm(p)} \qquad (8)$$

with,

$$\theta_0^{+(p)} = -\frac{1}{2a}\left[\rho^2 + 4\xi\right] \qquad (9)$$

$$\theta_0^{-(p)} = -\frac{1}{2a}\left[\rho^2 - 4\xi\right]\exp(\xi) \qquad (10)$$

$$\theta_0^{\pm(h)} = \frac{1}{2a}\sum_{j=1}^{\infty} A_j^{\pm} J_0(k_j\rho)\exp\left[\left(1 \mp \sqrt{1 + 4k_j^2}\right)\xi/2\right] \qquad (11)$$

where k_j denotes the j-th zero of

$$0 = -k_j J_1(k_j a) + \mathrm{Bi}\, J_0(k_j a) \qquad (12)$$

where the symbols + and − refer to positions behind and ahead of the front, respectively, and the coefficients $A_j^{\pm}$ with $j = 1, \ldots, N$ can be determined by substituting the basic solution into the matching conditions at the front (cf. Thiart *et al.*, 1990a).

If this basic solution is now perturbed, the following dispersion relation is obtained :

$$0 = 4\sigma^3 + (1 + 4\kappa_m^2 + 4\alpha - \alpha^2)\,\sigma^2 + \left[\alpha + 4\alpha\kappa_m^2 - 2\alpha^2\left(\widetilde{C} - 1\right)\right]\sigma$$
$$+\alpha^2\left[\kappa_m^2 - \widetilde{C}\left(\widetilde{C} - 1\right)\right] \quad (13)$$

where,

$$\widetilde{C} = 1 + \frac{4\mathrm{Bi}}{a} + \frac{\mathrm{Bi}}{2a}\sum_{j=1}^{\infty}\{A_j^-\,[(_j^-)^2 - 1] - A_j^+(_j^+)^2\}\left[\frac{\int_0^a J_0(k_j\rho)\,J_n^2(\kappa_m\rho)\,\rho\,d\rho}{\int_0^a J_n^2(\kappa_m\rho)\,\rho\,d\rho}\right]$$

A wealth of information on the different modes is locked up in this relation. In Fig. 4 few of the critical modes (i.e. for Re(σ) = 0) are shown.

The values of n and m in Fig. 4 relate to the eigenfunctions of the perturbed problem, i.e.,

$$\theta_1^{\pm} = T^{\pm} J_n(\kappa_m\rho)\exp\left[i\sigma\tau + in\phi + \left(2 \mp \sqrt{1 + 4(\sigma + \kappa_m^2)}\right)(\xi - F_0)/2\right] \quad (14)$$
$$F_1 = P\,J_n(\kappa_m\rho)\exp\left[i\sigma\tau + in\phi\right] \quad (15)$$

where κ_m is the m-th root of,

$$0 = \frac{dJ_n(\kappa_m\rho)}{d\rho} + \mathrm{Bi}\,J_n(\kappa_m\rho) = 0 \quad (16)$$

while $T^{\pm}$ and P are arbitrary constants which can be related through, $T^- = -P\,(1 + 1/\alpha)$ and $T^+ = -P\,\sigma/\alpha$.

If the eigenfunctions are super-imposed on the basic solution, the different modes become apparent. In Fig. 5 the oscillation of the mode $n = 0$, $m = 2$ is shown. At the beginning of the period, the center of the sample had a hot spot and a hot ring formed near the surface, half a period later the center and the ring had cooled down and a hot ring formed near the middle. The solution oscillates between these two extremes with a frequency

$$4\sigma^2 = \alpha + 4\alpha(\kappa_m^{\star})^2 - 2\alpha^2(\widetilde{C} - 1) \quad (17)$$

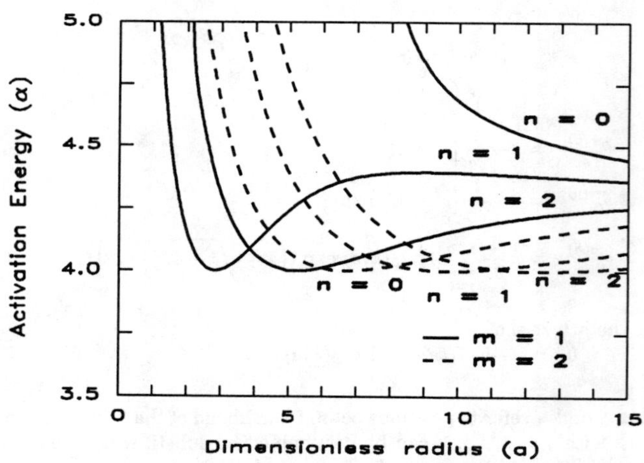

FIGURE 4. Neutral stability curves for non-adiabatic Solid-Solid Combustion of cylindrical preforms.

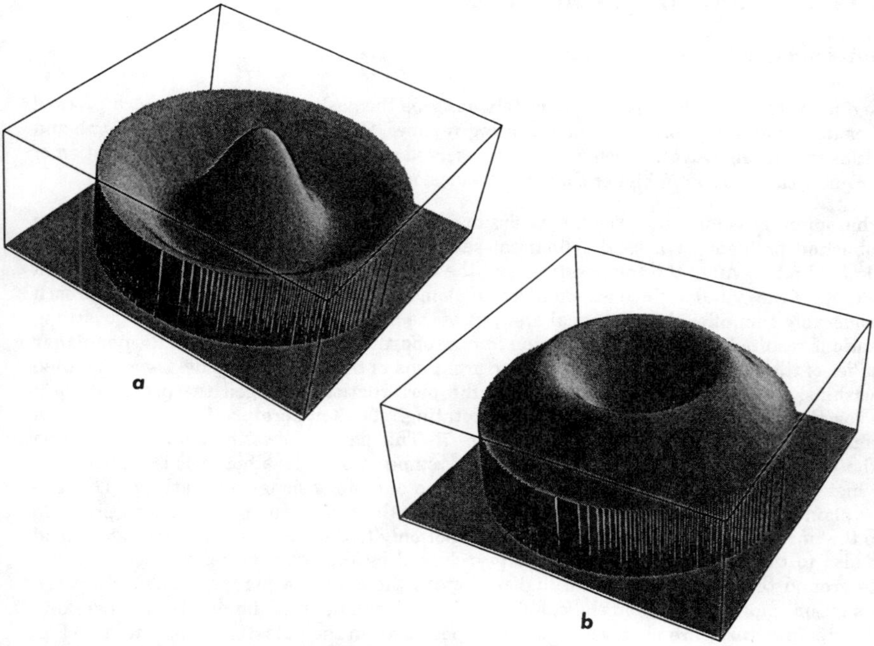

FIGURE 5. Shape of the reaction front for a perturbation with wave-numbers $n = 0$, $m = 2$. (a) $\sigma\tau = 2k\pi$, (b) $\sigma\tau = (2k-1)\pi$.

In Fig. 6 the mode $n = 1, m = 1$ is shown. The front has a warm spot, submerged under the surface, that propagates in a helical way along the transversal axis of cylinder.

The neutral stability curves in Fig. 4 converge to the value $2 + \sqrt{5}$ as $a \longrightarrow \infty$ and form a dense set. This indicates that, at large aspect ratios, the planar front can lose stability to any of a large number of other modes. The stability of these fronts can be determined by a non-linear analysis. This result might serve as an explanation for the experimental observation of chaotic combustion of large systems.

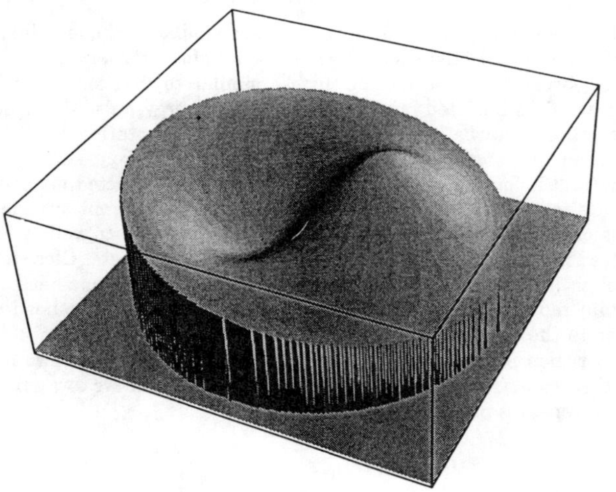

FIGURE 6. Shape of the reaction front for a perturbation with wave numbers $n = 1$, $m = 1$.

RESULTS OF NUMERICAL SIMULATION

Solid-Solid Combustion Synthesis

One dimensional simulations can accurately describe the gasless combustion, which proceed either in planar stable or in planar pulsating regime. However, the description of one-and multihead rotating waves, which have been observed experimentally , requires simulation of the combustion process in higher than one dimensions.

If the spinning combustion regime is assumed as a variety of surface combustion, the combustion problem for a solid cylindrical sample is replaced by the same problem for a hollow cylinder. After the temperature and the concentration are averaged over the small thickness of the cylinder, the mathematical problem becomes two dimensional. This approach considerably simplifies the numerical analysis of the problem. The solution to the governing equations resulted into a planar pulsating front propagation. In order to locate the nonplanar families of the solution small sinusoidal perturbations in the temperature field were imposed (Sivashinsky (1981)). The introduction of the perturbations changed the propagation of the reaction front from planar to two-head rotating. The temperature field on the lateral surface of a thin annulus is presented in Fig. 7. This figure shows the dynamic evolution of the two-head rotating front. The maximum temperature at the hot spot is higher than the highest temperature that would be observed in one dimensional calculations. This can be explained as follows : the hot spots generated by the introduced perturbations protrude into the unreacted mixture. They have access not only to the fresh material straight ahead but also to the left and right. The energy released by the reaction per unit volume and time around the hot spots is much higher than in the case of a planar reaction front (an assumption implicitly made in the one-dimensional formulation of the problem). Therefore, higher temperatures are observed in the hot spots than in the pulsating planar fronts. Fig. 7 shows also the mesh point distribution used in the simulation of the front propagation. The grid points concentrate around the two rotating heads where the steepest temperature and concentration gradients occur. therefore, serving as a characteristic example of the usefulness of an adaptively moving grid. The total number of grid points used in these calculations was eight times smaller than the number of equispaced points required to obtain the same degree of resolution of the front propagating phenomena.

As it has been shown theoretically (Sivashinsky, 1981; Margolis et al., 1985) as well as experimentally (Maksimov et al., 1985) the propagation of the combustion front in cylindrical pellets is, inherently, three dimensional. Spinning foci form not only on the surface of the sample but also inside it.

For the sake of comparison, let us consider a situation similar to the one illustrated by Fig. 7. Therefore, the diameter of the specimen is chosen sufficiently large, so that a two-head rotating wave is observed on the lateral surface similar to that shown in Fig. 7. The reaction is considered as initiated by a short time high energy pulse (laser beam) imposed on a small area (the perturbation necessary to generate a rotating wave is imposed by choosing an ignition spot sligthly away from the center) of the front surface of the sample. The combustion front is formed and propagates towards the unreacted material. Some very interesting results regarding the propagation of the combustion front are shown in Fig. 8. The simulations predict the simultaneous existence of corrugated as well as fully developed one - and two-head rotating waves depending on the radial coordinate. Close to the specimen center only a slightly corrugated pulsating front is observed. A one-head rotating wave appears at certain radial positions ($0.1 < r/R < 0.3$) (Fig. 8a). A two-head rotating wave is observed close to the lateral surface of the cylindrical sample ($0.8 < \xi < 1.0$) (Fig. 8b). In the transition region between one-and two-head rotating waves the two hot spots move towards each other, collapse into one, and again reappear and move away from each other. This phenomenon repeates periodically.

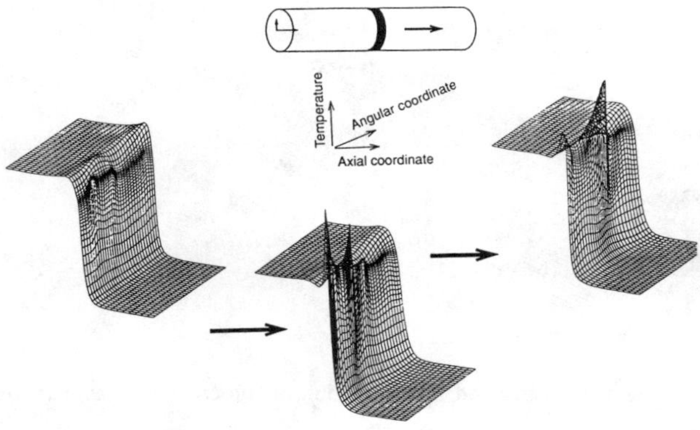

FIGURE 7. Sequence of snapshots for a spinning wave in adiabatic combustion of cylindrical preforms.

Summarizing, numerical simulations of solid-solid combustion process in one, two and three spatial dimensions reveal that different combustion modes might occur: planar constant pattern, planar pulsating, one- and multihead rotating. It should be noted, however, that the occurrence of phase transformations (melting) might change the propagation characteristics, i.e transition between unstable and stable combustion.

Gas-Solid Combustion Synthesis

Early analytical studies in filtration combustion can be found in works by Aldushin and co-workers (cf. Aldushin *et al.*, 1980). Degrève *et al.*(1987) have presented two dimensional numerical simulations for counter-flow filtration combustion, in which they predict the presence of non-planar fronts with moving hot spots for highly exothermic systems with high activation energies. Counter-flow filtration has also been studied by Dandekar *et al.* (1990a), for adiabatic and non-adiabatic systems. It has been shown that at high pressures, the stability criteria for solid-solid systems can be extended to gas-solid systems. At very low pressures when the filtration of gas is a limiting factor, all systems propagate with a constant front propagation velocity.

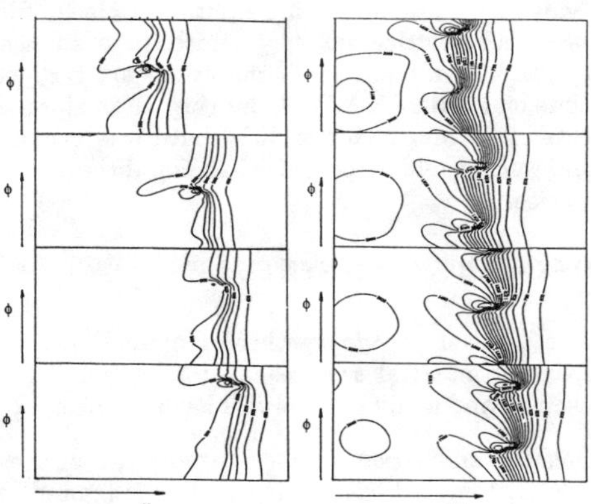

FIGURE 8. Sequence of snapshots for a spinning wave in non-adiabatic combustion of cylindrical preforms. (a) Near the center of the sample ($r = 0.1R$), (b) Outermost surface ($r = R$).

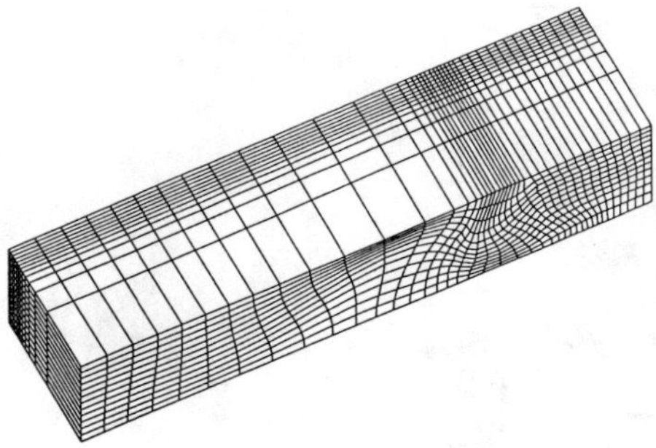

FIGURE 9. Three-dimensional mesh used in simulation of the cross-flow configuration for gas-solid combustion.

Cross-flow filtration combustion shows some unique front propagation phenomena. At high pressure hot spots appear and oscillate in the vertical direction as the front propagates horizontally down the length of packed powder bed (Dandekar *et al.*, 1990b). They have also shown the importance of limiting bed height in achieving high final conversion, at low pressures. High initial porosity is recommended for systems which show large volumetric expansion on reaction, due to difference in molar densities of reactant and product. A quantitative comparison between numerical predictions and experimental observations has been presented by Dandekar et al. (1990c), for the system $Ta - N_2$.

However, past studies are at most two dimensional. A full three dimensional study will show if a two dimensional approximation for cross-flow is indeed accurate, neglecting the variations of temperature, conversion and density in the direction of the width. Preliminary results from three dimensional study show some surprising results. In Fig. 9 a typical adaptive mesh used at a particular instant in the simulation, is shown. Fig. 10 shows temperature contours of planes at different positions along the width. As is clearly visible there are hot spots formed at high pressures. However, a closer examination reveals that these hot spots are not uniform across the width. Indeed at Z=0.0 no hot spots are visible, on the other hand at Z=3.47 one hot spot is visible and at Z=4.21, one can see 2 hot spots. This clearly points out to the fact that just as there are vertical hot spots as the front moves horizontally , one can also observe transverse hot spots in the direction of the width. Unfortunately, the present medium does not lend itself well to represent on of four dimensional data (three dimensional data varying in time). A video tape produced with 102 time frames of isothermal surfaces in color showed the presence of a cylindrical hot spot, which unfurls much like a roll of cloth and then coalesces again. The three dimensional simulations are very expensive, requiring almost 3-4 hours of CPU time on a CRAY X-MP/48, per run. Data storage is also a problem. For the animation an excess of 64 Megabytes of disk space was necessary for storing the temperature profiles. Future work would concentrate in using three dimensional simulations to make quantitative comparisons.

NEW APPLICATIONS OF COMBUSTION SYNTHESIS

Combustion synthesis offers a novel way to synthesize materials and it has acquired a permanent place in the fields of material science and combustion. There are still many questions which are not resolved and many possibilities have not been exploited.

One of the new applications of Combustion synthesis is the use of a external centrifugal field. When a mixture of aluminium and iron-oxide powders is ignited in the presence of a centrifugal field, separation will occur. If the system is properly controlled (cf. Merzhanov, 1980) a metallic pipe with a dense inner ceramic lining can be synthesized. Several patents on metal-ceramic composite pipes produced by thermite operation in a centrifugal field, are held by Odawara and co-workers (Odawara *et al.*, 1984; Odawara *et al.*, 1986a; Odawara and Yamazaki, 1986; Odawara *et al.*, 1986b).

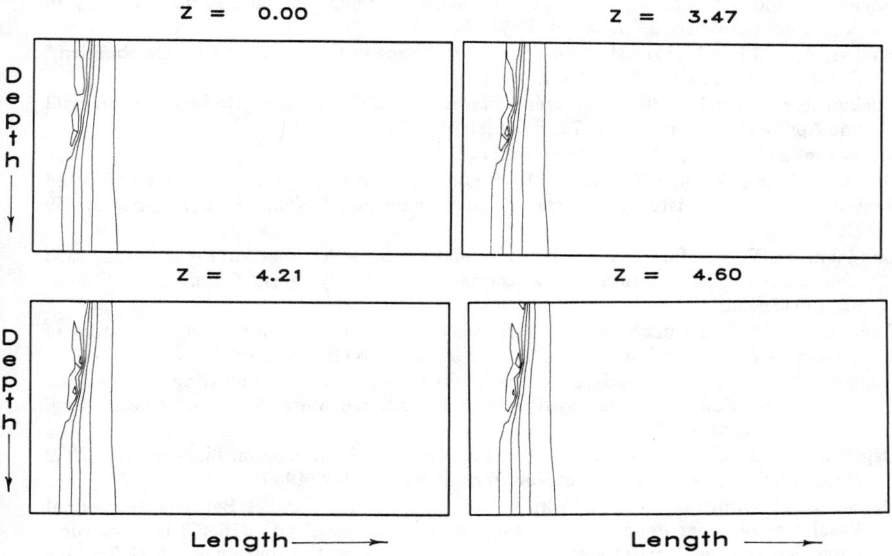

FIGURE 10. Contours showing isotherms across z planes in filtration combustion, $\tau = 15$, $\beta = 0.08$, $\gamma = 0.05$, $L_x = 20$, and $L_y = 5.0$, $L_z = 5.0$

Another application of CSIM is the direct densification of refractory materials. Niiler *et al.* (1989) studied the compaction of TiB_2 and TiC tiles by a shock wave. Once the reaction was complete, an explosive charge was set off that propelled a flyer plate against the sample. They obtained 90%+ densification in this manner, but the fracture toughness was not comparable to the hot-pressed product. The explosive compaction relies on a shock wave that propagates through the sample and the time that elapses before the system is unloading, is 30msec. The densification will exhibit some inertia since gas emission and material flow can occur on slower time scales. Therefore it is sensible to look at methods that will apply pressure for longer times.

One alternative is to react the system in a die and to apply a uniaxial load once the reaction is complete. This arrangement is similar to a hot-press. An additional feature that can be added to this system is to apply radial cooling simultaneously. The large temperature gradients which develop, cause large compressive thermal stresses in the sample. To analyze the effect of thermal stress Thiart *et al.* (1990b) adopted a simplified approach of elastic behaviour and solved the thermo-elastic equations for different values of the Biot number. Stresses at the center of the preform can reach 1500 MPa for a short period. There exists an interesting pay-off between maximum stress and time of exposure and the cooling rate must be controlled to leave ample time for outgassing and densification and still maintain the required stress level. Another application of the rapid quenching is to capture certain metastable phases.

REFERENCES

Aldushin, A.P., B.S. Seplyarskii and K.G. Shkadinskii "Theory of Filtration Combustion," *Fiz. Gor. Vzryva* **16**, 36-45 (1980).

Anderson, D.A., J.C. Tannehill and R.H. Pletcher *Computational Fluid Mechanics and Heat Transfer*, Mc Graw-Hill, New York (1984).

Bayliss, A. and B.J. Matkowsky "Fronts, Relaxation Oscillations, and Period Doubling in Solid Fuel Combustion," *J. Comp. Phys.* **71**, 147 (1987).

Bayliss, A. and B.J. Matkowsky "Two Routes to Chaos in Condensed Phase Combustion," *SIAM J. Appl. Math.* **50**, 2 (1990).

Buckmaster, J. and G. Ludford, *Lectures of Mathematical Combustion*, Society for Industrial and Applied Mathematics, p. 73, Philadelphia (1983).

J. J.Berzelius *Lechrbuch Der Chemie*, **3**, 1242 (1843).

Braaten, M. and W. Shyy "A Study of Recirculating Flow Computation Using Body-Fitted Coordinates: Consistency Aspects and Mesh Skewness," *Num. Heat Transfer* **9**, 559 (1986).

Dandekar, H. W., J. A. Puszynski, J. Degrève, and V. Hlavacek, "Reaction Front Propagation Characteristics in Non-catalytic Exothermic Gas-solid Systems," *Chem. Eng. Comm.* **92**, 199 (1990a).

Dandekar H. W., J. A. Puszynski and V. Hlavacek, "A Two-Dimensional Numerical Study of Cross-flow Filtration Combustion," *AIChE J.*, in press (1990b).

Dandekar H. W., C. C. Agrafiotis, J. A. Puszynski and V. Hlavacek, "Modeling and Analysis of Filtration Combustion for Synthesis of Transitional Metal Nitrides," *Chem. Eng. Sci.*, **45**(8), 2499 (1990c).

Degrève, J. "Use of Adaptive Meshes in the Simulation of Combustion Phenomena," *Ph D Dissertation*, State University of New York at Buffalo, NY (1989).

Degrève, J., P. Dimitriou, J. Puszynski, V. Hlavacek, S. Valone and R. Behrens "Numerical Resolution of Front Phenomena by Regriding Techniques," pp. 376-403 in *Supercomputer Research in Chemistry and Chemical Engineering* (K.F. Jensen and D.G. Truhlar eds.), ACS, Washington, DC (1987).

Dimitriou, P.A. "Gassless Combustion in Condensed Systems. Modeling, Analysis and Simulation," *Ph.D. Dissertation*, State University of New York at Buffalo, NY (1989).

Dimitriou, P.A., J. Puszynski and V. Hlavacek "On the Dynamics of the Equations describing Gasless Combustion in Condensed Systems," *Comb. Sci. and Tech.* **68**, 101 (1989).

Dvoryankin, A.V., A.G. Strunina and A.G. Merzhanov " Trends in the Spin Combustion of Thermites," *Fiz. Gor. Vzryva* **18**, 10 (1982).

Goldschmidt, H. *German Patent 96317* (1895).

Hindman, R.G. "Generalized Coordinate Forms of Governing Fluid Equations and Associated Geometrically Induced Errors," *AIAA J.* **16**, 1359-1367 (1982).

Kaper, H.G., G.K. Leaf, S.B. Margolis and B.J. Matkowsky "On Nonadiabatic Condensed Phase Combustion," *Sandia Report SAND86-8666*, Sandia National Laboratories, Alburquerque, NM (1986).

Kovalenko, Yu. A. "Limiting Conditions for Steady Gasless Combustion with Heat Transfer at the Wall," *Fiz. Gor. Vzryva* **23**, 6 (1987).

Kumar S., C. A. Agrafiotis, J. A. Puszynski and V. Hlavacek, "Heat Transfer Characteristics in Combustion Synthesis of Ceramics," *AIChE Symp. Ser.*, **263**, 50 (1988).

Kuczynski G. C., "Self-Diffusion of Sintering of Metallic Particles," *Trans. AIME*, **185**, 169 (1949).

Leonard, B.P. "A stable and Accurate Convective Modeling Procedure Based on Quadratic Upstream Interpolation," *Comp. Meth. Appl. Mech. Engng.* **19**, 59 (1979).

Maksimov, E.I., and K. G. Shkadinksii, "Steady-State Combustion Stability of Gasless Compositions," *Fiz. Gor. Vzryva* **7**, 454, (1971).

Margolis, S.B. "An Asymptotic Theory of Condensed Two-Phase Flame Propagation," *SIAM J. Appl. Math.* **43**, 351 (1983).

Matkowsky, B. and G. Sivashinsky "Propagation of a Pulsating Reaction Front in Solid Fuel Combustion," *SIAM J. Appl. Math.* **35**, 465 (1978)

Merzhanov, A., A. Filonenko, and I.P. Borovinskaya "New Phenomena in Combustion of Condensed Systems," *Dokl. Akad. Nauk SSSR* **208**, 892 (1973).

Merzhanov, A.G. and I.P. Borovinskaya "A New Class of Combustion Processes," *Comb. Sci. and Tech.* **10**, 195 (1975).

Merzhanov A.G., A.R. Kachin, B.I. Yuchvid, I.P. Borovinskaya and G.A. Vishnyakova *U.S. patent 4,217,948* (1980).

Merzhanov, A.G. "SHS Process: Combustion Theory and Practice," *Archivum Combustionis* **1**, 23 (1981).

Merzhanov, A.G. "Self-Propagating High-Temperature Synthesis: Twenty Years of Search and Findings," in *Combustion and Plasma Synthesis of High-Temperature Materials*, eds. Z.A. Munir and J.B. Holt, pp. 1-53, VCH Publishers, Inc, New York, NY (1990).

Munir, Z.A. "Synthesis of High Temperature Materials by Self-Propagating Combustion Methods," *Ceramic Bulletin* **67**, 342 (1988).

Niller A., L. J. Kecskes , T. Kottke, P. H. Netherwood and R. F. Benck, "Explosive Consolidation of Combustion Synthesisized Ceramics: TiC and TiB_2," *Ballistic Research Laboratory*, Aberdeen MD, December (1988).

Odawara O., Y. Nakajima and M. Sato *Jpn. Kokai Tokkyo Koho JP 60,228,679 (85,228,679)* (Apr. 1984).

Odawara O., Y. Ishii and H. Yamazaki *Jpn. Kokai Tokkyo Koho JP 61,238,975 (86,238,975)* (Oct. 1986).

Odawara O. and H. Yamazaki *Jpn., Kokai Tokkyo Koho JP 61,238,976 (86,238,976)* (Oct 1986).

Odawara O., I. Yasumasa and Y. Hicoshi *Jpn. Kokai Tokkyo Koho JP 61,238,974 (86,238,974)* (Oct. 1986).

Puszynski, J., J. Degrève and V. Hlavacek "Modeling of Exothermic Solid-Solid Non-catalytic Reactions," *Ind. Eng. Chem. Res.* **26**, 1424 (1987).

Puszynski, J., J. Degrève, S. Kumar and V. Hlavacek "Propagation of Reaction Fronts in Exothermic Heterogeneous Noncatalytic Systems Solid-Solid and Solid-Gas," *Lectures in Applied Mathematics*, 24 (1986).

Puszynski, J., J. Degreve and V. Hlavacek "Modeling of Exothermic Solid-Solid Noncatalytic Reactions," *Ind. Eng. Chem. Res.* **26**, 1424 (1987).

Puszynski, J., S. Kumar, P.A. Dimitriou and V. Hlavacek "A Numerical and Experimental Study of Reaction Front Propagation in Condensed Phase Systems," *Z. Naturforsch* **43a** 1017 (1988).

Raithby, G.D. "Skew Upstream Differencing for Problems Involving Fluid Flow," *Comp. Meth. Appl. Mech. Engng.* **9**, 153 (1976).

Roache, P.J. *Computational Fluid Dynamics*, Hermosa Publishers, Alburquerque, NM (1972).

Saltzmann, J. "A Variational Method for Generating Multidimensional Adaptive Grids," *Report DOE/ER/03077-174*, Courant Mathematics and Computing Laboratory, New York University (1982).

Sivashinsky, G. "On Spinning Propagation of Combustion Waves," *SIAM J Appl. Math.* **40**, 432 (1981).

Spalding, D.B. "A Novel Finite Difference Formulation for Differential Expressions involving both First and Second Derivatives," *Int. J. Num. Meth. Appl. Mech. Engng.* **4**, 551 (1972).

Thiart, J.J., H.J. Viljoen, N.F.J. van Rensburg, J.E. Gatica and V. Hlavacek "Stability of Non-Adiabatic Solid-Solid Combustion," submitted (1990a).

Thiart, J.J., H.J. Viljoen, J.E. Gatica and V. Hlavacek "Direct Densification in Combustion Synthesis of Inorganic Materials,", submitted (1990b)

Varma, A., G. Cao and M. Morbidelli "Self-Propagating Solid-solid Noncatalytic Reactions in Finite Pellets," *AIChE J.* **36**(7), 1032 (1990).

ACKNOWLEDGMENTS

This work was supported in part by NSF grant CTS 89-15787. The computations were performed using a CRAY-XMP/48 at the National Center for Supercomputing Applications through NSF grant CBT 89-0018N. Finally the authors want to acknowledge results provided by Drs. Jan Degrève and Panagiotis A. Dimitriou.

Rapporteur's Reports of Session on General Topics (2)

M. E. Glicksman, Rensselaer Polytechnic Institute

Modeling and Simulation of Manufacturing Processes of Advanced Composite Materials, Woo Il Lee, Soul National University, and George S. Springer, Stanford University

Fiber reinforced composites represent a broad class of "engineered" materials, where stiffness, strength, toughness, and density can be manipulated by careful design and selection of the polymer matrix and the fiber phase. The fiber and matrix are joined together in an autoclave curing process, where heat and pressure are permitted to act over time to redistribute and modify the resin matrix.

Two distinct classes of composites are considered in this paper: 1) *thermosets,* which have cross–linking reactions occur in the polymer matrix, and 2) *thermoplastics,* in which the matrix must be redistributed through the fibers and partially crystallized.

Autoclaves provide the temperature–pressure environment in which cross–linking reactions occur, resin redistribution occurs by flow and and absorption into "bleeders", and voids are squeezed out from the fiber/matrix interface by resin flow and pressure effects. The autoclave temperature must not exceed critical values, in order to obtain optimum composite properties, and the cure cycle must be as short as practicable to reduce costs.

The authors described models which permit estimation of the internal temperature distribution within the composite, accounting for the volume heat production due to chemical reaction, resin viscosity versus time and temperature, using constitutive relations which depend explicitly on temperature and degree of curing, and finally, resin volume fraction in the composite and "bleeder" as functions of time. The temperature, time, reaction extent were related through a thermo–chemical submodel using a one–dimensional energy/reaction equation and a two–term Arrhenius relation as the ODE for the extent of reaction. Resin flow and redistribution were calculated on the basis of a D'Arcy flow model for the fibers and the "bleeder". It was pointed out that the compaction within the autoclave occurs as a kinematic wave, wherein each fiber layer advances sequentially toward the middle of the composite as the resin flows past the fibers into the surrounding "bleeder". Calculations showed how the rate of heating affected the thermal distribution within the composite, and how the compaction wave advanced as a function of time at different autoclave pressures.

The topic of thermoplastic processing was explained as occurring in three essential steps: 1) *impregnation*, 2) *consolidation*, and 3) *crystallization.* The flow of resin through the fiber tows is calculated with a radial infiltration model,whereas, in reality, such flows occur at fiber locations that have slight misalignments and waviness. Again, a D'Arcy's law model is used, and the autoclave temperature at this stage is kept high enough to assure a relatively low viscosity of the polymer resin. The so–called lay–up of the plies occurs by a complicated deformation process, wherein the plies reach intimate contact and autoadhesion with the resin occurs. Bond strength between fiber and matrix develops via a diffusion–limited precess with sublinear time kinetics, and an Arrhenius temperature dependence.

Crystallization of the polymer matrix is essential for the development of mechanical properties. The time–temperature history as the autoclave is cooled has a strong effect on the degree of crystallinity achieved. Crystallinity is calculated by solving a one–dimensional energy equation, accounting for the latent heat of crystallization. The low thermal conductivity of most polymers makes this calculation extremely sensitive to the kinetics of crystallization assumed in the model.

The authors demonstrated their process model by giving graphical results of ply impregnation versus time, contact versus time at various pressures, temperature distributions, and, finally, the degree of crystallinity versus composite thickness. These models are of considerable help to polymer composite engineers in achieving efficient autoclave operations and producing quality fiber reinforced composites. The problems associated with forming complex composite structures, particularly when they achieve large size remains one of the key pacing problems in this industry. Indeed, ongoing research on photo–induced polymerization is an attempt to build up composites layer by layer, and avoid the autoclave environment altogether. Such processes are still a long way off, and the understanding of conventional autoclave processing remains extremely important to the composite materials industry.

Heat and Mass Transport in the Photonics Manufacturing Technology, F.T. Geyling, SEMATECH

The author gave an historical overview of the process design problems encountered in the 1980's when photonic technology began its first serious inroads into the communications industry in the form of optical wave guides, known as optical fibers. The author recounted the successful approaches taken by Bell Telephone Laboratories as it attempted to move graded refractive index glass fibers, invented in the 1970's, out of the research laboratory and into pilot production of long–distance optical cables. The problems of preforms, drawing, and coating were described with various heat and mass transport models that provided much needed insight and direction for continuing studies of these processes.

Modified chemical vapor deposition, MCVD, is a typical horizontal, open reactor system operating under Poiseuille flow with radial mass transport. Temperature effects and boundary–layer solutions were discussed, along with thermophoretic "trajectory" modeling of the deposition of Si, Ge, and boron oxides along the surrounding quartz reactor walls. Such modeling provided estimates of the deposition profiles along the wall. Deposition efficiency was then modified to maximize production by optimizing the oxyacetylene torch distribution and the wall cooling. Current extensions of this work now involve numerical simulation of the gas dynamics within the quartz tube, accounting for tube rotation, buoyancy–induced convection, and more realistic chemical kinetics of the oxidation process.

Process modeling for *particle deposition* accomplished through outside vapor deposition, OVD, and vapor axial deposition, VAD, involve transport studies of flame flow around cylindrical tubes. Particle deposition rates are controlled by the rate of thermophoresis and ballistic trajectories. Early studies used boundary layer approximations, and recently employ numerical simulation. The author briefly described optimization now used by manufacturers of these fibers, e.g., Corning Glass, Sumitomo, Hitachi, and Fujitsu.

To achieve monolithic graded glass bodies consolidation is needed. Each successive pass of the torches produces a layer of consolidated particles. The studies of layer consolidation is modeled by studies of transport in three interacting phases: 1) gas stream, 2) the porous oxide deposit, 3) the supporting quartz wall. Sintering was described by a constitutive equation based on surface tension–driven sintering. These models explained the influence on consolidation rates of dopants, trapping of bubbles, and approaches to accelerate the rate of consolidation. Improvements might include pore network coarsening, as described earlier in the Conference by Professor Glicksman.

Collapse of the pre-form was described as a one–dimensional extensional flow phenomenon. Cross-sectional irregularity was often due to necking instability in the section. Flow visualization experiments proved very useful in elucidating this phenomenon. Other instabilities such as "draw resonance" occurred, which is a well known effect in the textile industry. Enhance cooling below the drawing neck tends to suppress these instabilities. Residual stresses resulting from flow close to the softening point are also of interest, both from the standpoint of optical polarization, which can enhance fiber performance, and the deleterious occurrences of so-called static fatigue and stress relaxation.

The final process is *coating,* to protect the fiber from the external environment. This is accomplished by passing the fiber through an open coating cup. The author showed that high throughputs, *viz*., > 1.5 m/sec, led to excessive meniscus deformation of the free surface of the coating fluid, which could lead to "skips" and loss of wettability. Dynamic wetting, originally studied by Deryagin in the 1930's becomes a significant issue at high coating speeds. The use of pressure–assisted coating to enhance the dynamic wetting was described as effective to speeds up to 20 m/sec. Self–excited transverse fiber oscillations that lead to coating anomalies also occur at high coating speeds, and an intense shear layer can develop from the localized heating of the coating fluid by the fiber.

In summary, the processes underlying modern optical fiber production were comprehensively reviewed, and the audience gained a full appreciation of the broad sweep of thermal, fluid, and materials science needed to do process engineering in the photonic industry of the 1990's.

Summary of General Discussion of Session on General Topics (2)

Ken Okazaki, Toyohashi University of Technology

One of the very interesting subjects which was presented in this session is the application of combustion to material processings. In the combustion synthesis, the heat of reaction or high temperature itself, radical species formed from the decomposition of parent molecules, or the combination of the both could be used to synthesize the materials. Here, the comments given by Dr. H. J. Viljoen including the present status and futere views of combustion synthesis are summarized.

Combustion synthesis of inorganic materials is now widely accepted as a viable alternative route for the production of certain ceramic materials. To put it in the proper perspective, one can categorize combustion synthesis processes as homogeneous flame reactions and heterogeneous flame reactions.

Flame reactors are operated at the lower end of the plasma scale and they are suitable for reactions which do not require large amounts of energy. The big advantage of the flame reactors is the control over particle size distribution; however collection of the powder and scale-up pose problems. The high temperature gas-phase chemistry is also very complex and in most of the cases it is unchartered territory; especially when ternary or higher systems are considered. Progressing to glow discharges, this technology is already well-established in the electronics industry. Although the sustenance of the two processes (flame and glow discharge) differs markedly, they create very similar chemical environments. The glow discharge will find application over a much wider field. If the system is properly exploited, film deposition on temperature sensitive substrates, stress-free films and enhanced rates of deposition can be achieved. The plasma, however, is very sensitive to external perturbations and a proper understanding of the physics and the interaction with its surrounding medium is necessary.

Heterogeneous flame synthesis includes the solid-solid and the solid-gas type of self-propagating reactions. Conditions for these reactions are a high activation energy and a high adiabatic temperature rise. Reactions which do not meet these criteria are not necessarily ruled out, and chemical furnaces, or external heating in furnaces can be used. It becomes obvious that the system is now forfeiting its simplicity. Another complication one has to deal with, is the heat removal/shielding when the process is scaled up. In its present form, the process is of batch type, a major advance will be the design of a continuous system. Fluidized bed operations have a lot of good features and is one possible way to operate a combustion synthesis process continuously.

Combustion synthesis offers a lot of imaginative post processing applications. Direct densification, and the capture of metastable phases by rapid quenching are examples where the heat of reaction was used to process the product material. The method also offers the possiblity to manufacture graded materials and it can be used for welding ceramic and metal parts.

Overall, a lot of research is still required to realise the full potential of the process, but it is certainly one of the most exciting areas to explore.

STEEL PROCESSING

Quenching with Liquid Jet Impingement

R. VISKANTA and F. P. INCROPERA
Heat Transfer Laboratory
School of Mechanical Engineering
Purdue University
West Lafayette, Indiana 47907, USA

ABSTRACT

A state-of-the-art review of liquid jet impingment heat transfer is presented in this article. This topic has many fundamental features of interest and is also pertinent to numerous technological applications, most notably cooling of metals during processing. Single phase convection and boiling heat transfer to impinging jets are reviewed, and the authors' current research on planar water jets is emphasized. Methods for enhancing heat transfer are discussed, and future research needs are identified.

INTRODUCTION

Production of steel, aluminum and other metal alloys having desired mechanical and metallurgical properties requires accurate temperature control during processing. Applications include strip investment, plate mill, and bar mill cooling [1-3]; quenching of extrusions, forgings and continuous castings [4,5]; and cooling of rolls in metal rolling [6]. For example, accelerated cooling is often a major objective in steel production, since it improves the mechanical properties of the final product by providing desired structural changes with respect to grain size and the ratios of ferrite, pearlite and laminite. Optimum strength and toughness properties of hot-rolled steel, for example, can be achieved by refining the ferrite grain size and precipitation conditions through accelerated cooling at approximately 15 K/s from the initial post-roll temperature. An important process parameter used to control strength and toughness is the finish coiling temperature, and to obtain good properties, accelerated cooling is terminated at approximately 500 °C. Accelerated cooling also yields a relatively low carbon equivalent, thus improving the weldability of steel.

To achieve desired cooling rates and temperature control for strip metals, several methods have been developed (Fig. 1), and cooling efficiencies (heat removal rate per unit volume of

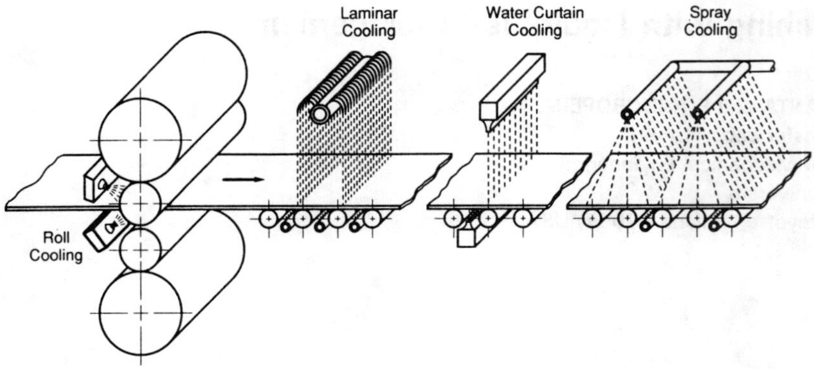

FIGURE 1. Schematic of cooling systems along a runout table for cooling strip steel.

water consumed) have been compared [1,7,8]. The results show that planar water jets (waterwalls, water curtains) which span the entire strip are about 50% more efficient than spray nozzles in removing heat from a strip at 900 °C and traveling at 10 m/s. In addition to providing the highest specific rate of cooling, waterwalls also provide more uniform cooling with respect to temperature variations across the strip width [7]. Although spray cooling is also used extensively, as, for example, to cool continuous castings and strip metals on a runout table, space limitations do not permit discussion of this option. The interested reader is referred to existing reviews of the subject [9,10].

The primary purpose of this paper is to assess the relationship of single- and- multi-phase convection heat transfer using liquid jets to metals processing. In sequence, the following sections deal with the fundamentals of convection heat transfer encountered when metals are cooled by impinging liquid jets and with specific results obtained for rectangular and axisymmetric jets.

FUNDAMENTAL HEAT TRANSFER PROCESSES

Several fundamental heat transfer processes are associated with liquid jet impingement cooling. Slot (rectangular) and axisymmetric water jets are typically used in metallurgical applications, many of which involve moving surfaces which must be cooled. A good example is the hot rolling process in which a series of counter-rotating cylindrical rollers reduce a steel strip to its final thickness [11]. For low carbon steels, rolling usually begins at about 1200 °C and is completed well above 725 °C. After rolling, the strip is typically cooled along a runout table by jets of water to about 500 °C and is subsequently directed to a coiler.

To understand the importance of plate motion, it is useful to consider the heat transfer regimes occurring on a steel strip as it is cooled (Fig. 2). The following description is based on visual observations at a hot rolling mill during the manufacture of steel strips. It is consistent with observations made by Kokado et al. [12] for a stationary plate cooled from 900 °C by a circular water jet. Beneath the jet and extending perhaps several jet widths, heat transfer is by single phase forced convection. In this zone (regime I), the cooling effectiveness of the jet is very high and the surface temperature is too low for boiling. Away from the stagnation streamline, the water near the plate is heated and the cooling effectiveness diminishes. The plate's surface temperature increases, and the onset of boiling

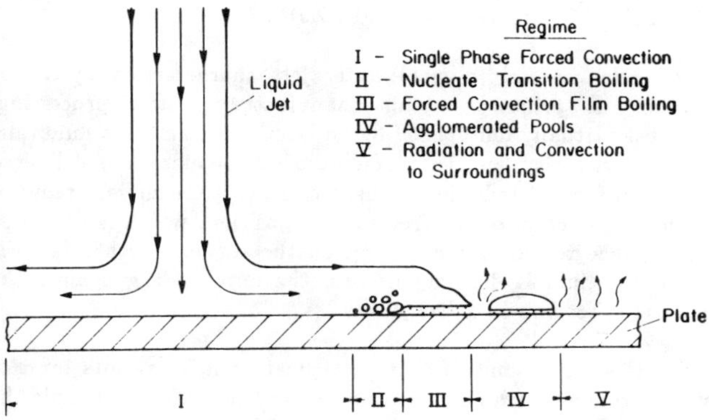

FIGURE 2. Heat transfer regimes adjacent to an impinging jet on a stationary plate.

is eventually reached. A comparatively narrow region (regime II) of nucleate and transition boiling may separate regime I from regime III, where forced convection film boiling occurs. Eventually, water overriding the vapor layer may agglomerate into pools (regime IV) due to surface tension effects, and a significant fraction of the surface may be void of liquid and vapor. Heat transfer occurs by convection and radiation beneath the pools and principally by radiation to the surroundings from the unwetted surface. The pools migrate in a random-walk fashion until they either evaporate or are lost at the edge of the plate. For regime V, heat transfer occurs by radiation and convection from the dry plate to the surroundings.

In many applications, surface motion is an important consideration. If the plate in Fig. 2 moves to the right, the thickness of the vapor layers beneath the pools and in the film boiling regime decreases since vapor flow is promoted by the plate motion. However, if the plate moves to the left with sufficient velocity, the vapor layer thickness may increase since vapor can be drawn toward the jet by the plate motion. Of course, both cases occur simultaneously on a moving plate since an impinging jet divides on its surface. Locations for transitions between the regimes depicted in Fig. 2 may also depend on surface motion and in turn affect the local surface temperature. In film boiling, plate motion influences the thicknesses of the vapor layer and the boundary layer in the liquid [13]. Consequently, the importance of radiation relative to conduction across the vapor layer also depends on the plate speed.

This brief discussion of metal cooling during manufacture using liquid jets has revealed an array of complex single and two-phase convective transport phenomena. Heat transfer to an impinging liquid jet from either a stationary or moving plate is by single phase convection, nucleate boiling, transition and film boiling, and by radiation. Conditions governing transition from one mode of heat transfer to another are needed, along with expressions for the heat transfer itself. Impingement heat transfer depends on a large number of variables, such as the free stream turbulence, partial pressure of noncondensibles, and surface characteristics which cannot be evaluated under mill conditions.

Knowledge of the local heat transfer coefficient or the local heat flux at the surface of a quenched material is a necessary boundary condition if the temperature distribution in the material and the rate of cooling are to be predicted from the transient energy equation in the stationary or moving metal. The state of this knowledge and recent developments are discussed. Emphasis is placed on experimental approaches, because convection phenomena are not sufficiently well understood to permit accurate modeling. Many details, including a discussion of radiation transfer, cannot be included in this brief account.

CONVECTIVE HEAT TRANSFER TO IMPINGING LIQUID JETS

Convective heat transfer to circular or slot (planar) water jets characterized by a free surface has found application in primary metals manufacturing and processing. Propagation of a free surface jet is virtually unimpeded by an immiscible ambient fluid (air) of substantially lower density and viscosity, and jet momentum can be efficiently delivered to and redirected along a solid surface. Prediction of heat transfer to such jets requires analysis of the full flow field and prediction of the free surface position, which is governed by a balance of pressure and surface tension forces acting on the surface. Other factors, such as stress relaxation when the fluid is discharged from the nozzle, drag against the ambient gas, and gravity can also affect flow in a free jet.

In this subsection we highlight theoretical and experimental heat transfer results for slot and circular free surface jets. Parameters which affect convective heat transfer include the Reynolds number of the jet, the Prandtl number of the fluid, turbulence and pressure gradient in the free stream, the velocity profile of the jet, the distance of the nozzle from the cooled surface, interference between adjacent jets if more than one jet is present, and other effects [14]. For the sake of completeness, reference is made to a large body of literature on heat and mass transfer to gaseous jets in gaseous surroundings [15-17].

Slot Jets

Because of the ability to provide uniform cooling along its width, planar jets are preferred in many applications. Within the single-phase forced convection region, there are three distinct boundary layer flow regimes. Immediately beneath the jet ($x/w_j \cong 0.5$), a stagnation zone is characterized by a nearly constant free-stream velocity gradient, du_δ/dx, in which case the free-stream velocity is proportional to the streamwise coordinate, $u_\delta = Cx$. In the impingement zone, ($0.5 \cong x/w_j \cong 3$), du_δ/dx approaches zero and u_δ approaches the jet impingement velocity V_j. Because of the favorable pressure gradient and fluid acceleration in the stagnation and impingement zones, turbulence is abated and the boundary layer flow is laminar. However, even low level turbulence will persist in the free-stream and interact with the laminar boundary layer. Free-stream turbulence, as well as surface roughness or other sources, provides disturbances which eventually induce transition to a turbulent boundary layer on the plate. In fact free-stream turbulence strongly influences development and transition of the laminar boundary layer and hence local heat transfer from the plate. In the uniform parallel flow regime, $x/w_j \gtrsim 3$, and $u_\delta/V_j \cong 1$.

Whether the flow in a jet is truly laminar or turbulent, at and near the stagnation point and in the boundary layer that develops along the impingement surface, the flow is laminar. Local heat transfer to an impinging liquid jet has been modeled under the simplifying assumptions of a laminar and uniform flow in the free stream to obtain a first order estimate of momentum and energy transfer using a similarity analysis. The Hiemenz or Falkner-Skan class of uniform flows past a wedge of included angle $\pi\beta$ typify such flows [18]. In the laminar stagnation and impingement zones, the predictions [19-21] yield an expression of the form

$$Nu_x = A\,[Re_{x^*}\,(u_\delta/V_j)]^{1/2}\,Pr^n \tag{1}$$

where (u_δ/V_j) is obtained from an inviscid flow solution. The Hiemenz and Falkner-Skan analyses assume $u_\delta = Cx$, where C is the velocity gradient du_δ/dx. It should be pointed out that both $Nu_x = f\,(Re_{x^*}, Pr)$ and $Nu_j = g\,(Re_j, Pr, x/w_j)$ can be used as correlating functions for the local heat transfer coefficient, and correlating equation (1) can be recast as

$$Nu_j = B\,[Re_j\,f\,(x/w_j)]^{1/2}\,Pr^n \tag{2}$$

For a prescribed fluid (prescribed k, ν, and Pr), heat transfer at the stagnation line is determined by the velocity gradient, $C = du_\delta/dx$. With increasing velocity, the increased advection and thinning of the hydrodynamic and thermal boundary layers enhance heat transfer. Moreover, since complete acceleration of the fluid along the impingement surface normally occurs in less that three jet widths, a decrease in the jet width causes the fluid to accelerate over a shorter distance, thereby increasing the velocity gradient and heat transfer coefficient. Conversely, effects of the jet velocity profile, which have received only limited consideration, are not so apparent.

Recently, Zumbrunnen [22] used a similarity solution for the two-dimensional stagnation region of a laminar, planar jet impinging on a moving surface to calculate the velocity field, which, in turn, was used to determine the heat transfer distribution. The results show that heat transfer decreases at downstream positions (relative to the direction of surface motion) and increases with increasing surface velocity and surface temperature gradients.

The first experimental study of single-phase heat transfer to a planar, liquid jet appears to have been performed by McMurray et al. [23], who measured local convection coefficients h on an electrically heated nichrome foil. A secondary maximum in h was observed downstream of the maximum at the stagnation line and was attributed to boundary layer transition. In the laminar stagnation and impingement zones, data were well correlated by an expression of the form given by equation (1). Assuming that the local convection coefficient at a distance x/w_j is equivalent to that for uniform free-stream flow over a wedge, Inada et al. [21] performed a laminar boundary layer analysis of flow and heat transfer in the impingement and parallel flow regions. The analysis successfully predicted data for a jet Reynolds number of $Re_j = 940$, but substantially underpredicted the data of McMurray et al. [23], which corresponded to much larger Reynolds numbers, $Re_j \approx 10^5$. Inada et al. attributed this difference to the effects of free-stream turbulence. The result was confirmed by Miyasaka and Inada [24].

Recently, Vader et al. [14] performed convective heat transfer experiments from a heated surface to an impinging, planar jet of water having a near-uniform velocity profile and Reynolds and Prandtl numbers in the ranges $2 \times 10^4 \leq Re_j \leq 9 \times 10^4$ and $2.7 \leq Pr \leq 4.5$. They correlated the stagnation line heat transfer data by the equation

$$Nu_j = 0.28\, Re_j^{0.58}\, Pr^{0.4} \tag{3}$$

with thermophysical properties evaluated at the film temperature and use of a Prandtl number exponent, which is consistent with reported results [15,16]. Nusselt numbers computed from equation (3) are up to 50% greater than the analytical result for a laminar, impinging jet. The discrepancy was attributed to the effect of free stream turbulence on the measured results.

Stagnation heat transfer coefficients from a heated surface to a planar jet of water with a nonuniform velocity profile have been measured by Wolf et al. [25]. The stagnation flow Nusselt numbers were correlated by

$$Nu_j = 0.116\, Re_j^{0.71}\, Pr^{0.4} \tag{4}$$

The correlation is based on Reynolds and Prandtl number ranges of $1.7 \times 10^4 \leq Re_j \leq 7.9 \times 10^4$ and $2.8 \leq Pr \leq 5.0$. Figure 3 shows the data of Wolf et al. and equation (4), as well as the correlation of Vader et al. [equation (3)] for a nearly uniform velocity profile. Figure 3 shows enhancement due to the nonuniform velocity profile and a corresponding increase in the Reynolds number exponent of 22 percent. Such an increase is consistent with increased levels of free-stream turbulence in the nonuniform profile jet.

Local heat transfer downstream of the stagnation region has been predicted by laminar boundary layer analyses [21,26]. The boundary layer equations can readily be solved once the streamwise pressure variation is known. The analysis of Inada et al. [21], which encompasses both the impingement and parallel flow regions, assumes that the local heat

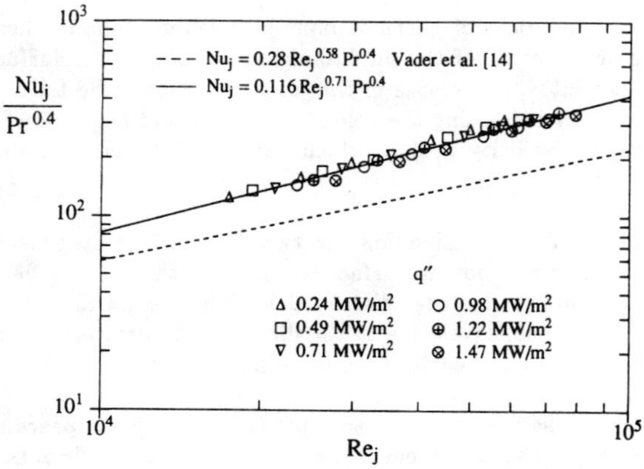

FIGURE 3. Correlation of stagnation line heat transfer data for $1.8 \leq \overline{V}_j \leq 4.5$ m/s, from Wolf et al. [25].

transfer coefficient at a distance x/w_j is equivalent to that for a wedge flow having the same velocity gradient (du_δ/dx) at the same distance from the stagnation line. By using results of published analyses, Inada et al. [21] predicted local heat transfer in the laminar boundary layer. The model shows that heat transfer is affected strongly as the nozzle is moved in close proximity to the plate. Values of H/w_n less than 1.5 caused the maximum heat transfer coefficient to occur near x/w_n=0.5, instead of at the stagnation line. Movement of the nozzle beyond H/w_n=1.5 had little effect on the local heat transfer coefficient.

Measured local convective heat transfer coefficients downstream of the stagnation line for an impinging jet of water have been reported [14,23,25,27]. The effect of the impinging jet velocity on the local coefficient distribution is shown in Fig. 4. Since the velocity is zero at the stagnation line, the boundary layer immediately downstream of that line will be laminar. Heat is added to the fluid as it moves away from the stagnation line, causing the thermal boundary layer thickness and thermal resistance to to increase. Hence, the heat transfer coefficient decreases as long as the boundary layer remains laminar. Boundary layer turbulence is initiated by free-stream flow disturbances, surface roughness, acoustic noise, wire trips or other sources, and transition consists of a sequence of events which are distributed in the streamwise direction. Incipience of transition to turbulence is marked by an increase in the heat transfer coefficient. When the turbulent mixing near the wall becomes sufficiently vigorous to reduce the resistance to heat transfer across the thermal boundary layer, the heat transfer coefficient increases from a local minimum. Coefficients increase between the onset and completion of transition to fully turbulent boundary layer flow.

Vader et al. [14] correlated their local data for $100 \leq Re_{x^*} \leq 10^5$ by the equation,

$$Nu_x = 0.89\ Re_{x^*}^{0.48}\ Pr^{0.4} \tag{5}$$

with the thermophysical properties evaluated at the film temperature. Equation (3) should be used to compute stagnation heat transfer ($Re_{x^*} \lesssim 100$), while equation (5) may be used for Reynolds numbers up to the critical value ($Re_{x,c^*} = 3.6 \times 10^5$). Equation (5) is compared with the empirical correlation of McMurray et al. [23] and the data of Vader et al. [14] in Fig. 5. The McMurray et al. correlation is also valid in the impingement and parallel flow region and is in good agreement with equation (5). Comparisons of local heat transfer coefficient data for nonuniform and near-uniform velocity profiles at the nozzle exit show that enhancement is considerable for the nonuniform case at all locations on the impingement surface [25]. Enhancement (up to 79% at the stagnation line) is postulated to be due primarily to higher levels of turbulence in the nonuniform jet.

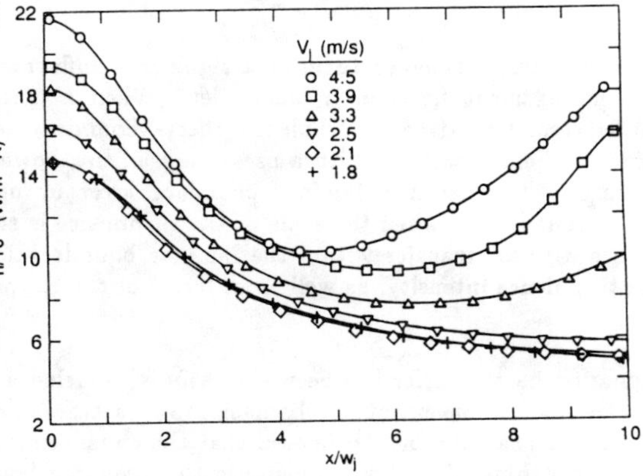

FIGURE 4. Effect of jet velocity on the local heat transfer coefficient to water, $q''_w = 0.25\ MW/m^2$ and $T_w = 30\,^\circ C$, from Vader et al. [14].

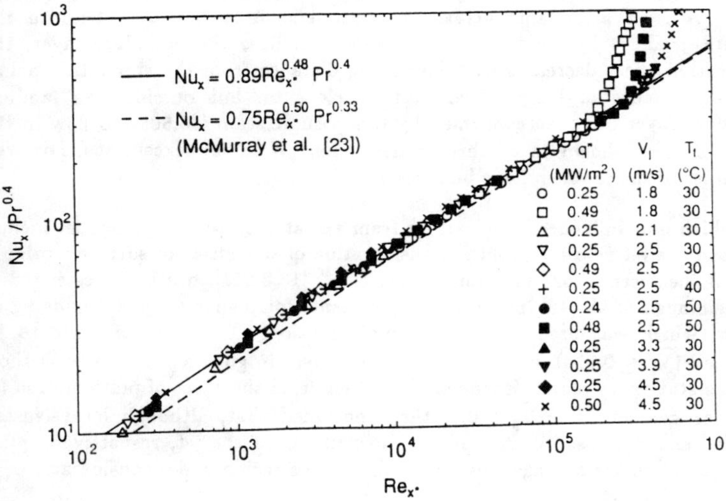

FIGURE 5. Correlation of single-phase convection data downstream of the stagnation line, from Vader et al. [14].

Axisymmetric Jets

The flow field of an axisymmetric, free liquid jet impinging on a horizontal plate is conventionally divided into potential and viscous regions. The viscous region before the hydraulic jump may include stagnation, boundary layer, transition and similarity regions. Previous theoretical approaches have used similarity [28,29,32] and integral [28-31] methods to predict local heat transfer coefficients by studying the different regions separately. However, no theoretical predictions have been reported for turbulent free jets impinging on a stationary or a moving surface. Experimental heat transfer results recently have been reported by several investigators [33-36]. However, with the exception of the work by Barasanti et al. [35], most of the authors used relatively small diameter nozzles, rendering the range of dimensionless parameters nonprototypic of materials processing applications.

Heat Transfer Enhancement

As noted previously, free-stream turbulence is believed to have a significant influence on conditions in the stagnation and impingement flows of a planar jet. When a laminar boundary layer pressure gradient is zero, free stream turbulence affects boundary layer transition but has no other influence on heat transfer. With a pressure gradient, however, free stream turbulence can enhance heat transfer in a laminar boundary layer by up to 100% [37-46]. Moreover, it has been found that, when the scale of the turbulence is small relative to the boundary layer thickness, heat transfer within the laminar boundary layer must be correlated in terms of the turbulence intensity, as well as in terms of the Reynolds and Prandtl numbers.

Although the enhancement of stagnation heat transfer has been successfully modeled as an eddy diffusion process [47-49], the actual enhancement mechanism may be traceable to coherent turbulent structures. It has, for example, been proposed that the enhancement of stagnation heat transfer is due to a stretching of vortex filaments in the diverging flow of the stagnation region. If a vortex in the jet approaches the stagnation point with its axis parallel to the flow direction, it will be elongated by the favorable pressure gradient in the stagnation and impingement regions causing a reduction in its diameter and amplification of its speed and the magnitude of its vorticity. Such vortices disrupt the developing laminar boundary layer, with an attendant increase in heat transfer. However, because the favorable pressure gradient in the impingement zone stabilizes the boundary layer, the vorticity amplification effect decreases with increasing distance from the stagnation point. Nevertheless, the existence of longitudinal vortices close to, but outside, the laminar stagnation boundary layer has been confirmed by flow visualization [46,50] and flow in the boundary layer has been shown to be three-dimensional. With few exceptions, however, experimental studies of the problem have been restricted to air.

Other features which may influence heat transfer from a heated plate to a planar jet include surface motion and the jet velocity profile. Consideration of the effects of surface motion is limited and, until the recent work of Zumbrunnen et al. [13,22,51], has been restricted to gas jets. Van Heiningen et al. [42] performed experiments for planar air jets impinging on a rotating drum, which was used to simulate plate motion. The ratio of plate to jet velocities was small ($\overline{V}_p \leq 0.014$), however, and results were in good agreement with those corresponding to a stationary plate. Huang et al. [52] modeled the effect of plate motion for a turbulent planar gas jet. For $\overline{V}_p \leq 0.5$, they concluded that, although local Nusselt numbers increased and decreased upstream and downstream of the jet, respectively, plate motion had little effect on the average Nusselt number in the region under consideration.

With respect to the effects of the velocity profile on liquid jet impingement heat transfer, turbulent and laminar nozzle flows differ in two fundamental ways. First, the fully developed velocity profile for turbulent flow is much more uniform than that for laminar flow. Hence, large differences between the centerline velocity associated with fully developed and uniform velocity profiles would not be expected for turbulent nozzle flows. A turbulent velocity profile would therefore be expected to yield only moderate enhancements in heat transfer relative to a uniform profile for a fixed flow rate. Second, equation (1) predicts heat transfer in a laminar boundary layer without free-stream turbulence. As discussed, stagnation flow provides ideal conditions for turbulence enhancement due to a strong pressure gradient that accelerates the fluid along the surface. For a cylinder in crossflow, it has been shown that an increase in the free-stream turbulence intensity from 0 to 3 percent can yield enhancements in the local heat transfer coefficient by as much as 80 percent [40,47]. Similar trends have been reported for impinging jets [14,25,26,42,53]. Clearly, altering the shape of a velocity profile in a high Reynolds number flow will also modify its turbulence characteristics, suggesting that for turbulent jets of differing velocity profile, the enhancement due to the profile would be superimposed on that due to free-stream turbulence.

Kestin [38] makes a dimensional argument that, in addition to the Reynolds number, turbulence intensity and scale must be matched in order to achieve dynamic similitude in a turbulent flow. When the turbulence scale is small with respect to an appropriate length scale, such as boundary layer thickness, he argues that the time averaged effect of turbulence scale on the local transport of mass, momentum, energy and species is negligible. For such cases Kestin concludes that the Nusselt number is a function of the Reynolds number, Prandtl number and free-stream turbulence intensity, $Nu = f(Re, Pr, Tu)$. This result has been widely affirmed by experimentalists, who have shown that functions of the form $(Nu/Nu_o) = 1 + f(Re, Tu)$ and $Nu/Re^{1/2} = f(Tu\ Re^{1/2})$ are suitable for correlating data [47,54].

Measured convective heat transfer results [14,25] show that inherent and induced turbulence in the jet can significantly enhance the local heat transfer. Theoretical studies of the turbulence phenomena suggest that the stretching of vortex lines in the impingement region induces three-dimensional flow disturbances just outside of the velocity boundary layer. The available vorticity amplification models indicate that the Prandtl number Pr and velocity gradient C in the stagnation region affect heat transfer in addition to the free-stream turbulence intensity Tu. Although the free-stream turbulence intensity in a free-surface, liquid jet has not been measured, qualitative analysis of the data corroborates the dependence of heat transfer enhancement on Tu, Pr and C. However, correlations that treat Tu as a variable have not been developed for liquid jets.

BOILING HEAT TRANSFER WITH LIQUID JET IMPINGEMENT

Introduction

Nucleate, transition and film boiling heat transfer regimes are expected to occur during impingement quenching of stationary metal parts, moving hot rolled strip, and continuous castings. All three regimes may occur, for example, on a moving strip at some distance from an impinging jet or at the same location on a stationary metal part, but at different times during the quenching process. Consider a typical temperature history for metal as it is quenched in a stationary liquid bath (Fig. 6). Four stages of heat transfer are revealed [55]. Regime I depicts the temperature response from the moment of immersion to formation of a stable vapor film. During regime II stable film boiling occurs. Stable film break-up begins, goes through transition boiling, and ends in nucleate boiling in regime III. Regime IV denotes non-boiling heat transfer. Although regime III corresponds to the largest cooling rate, it is the least understood, even for conventional saturated boiling conditions, unrelated to metallurgical applications involving subcooled boiling with forced internal or external flow of the coolant. The extensive literature related to conventional boiling processes is described in representative reviews [56-58], as is the mechanism of quenching surfaces, particularly uncovered nuclear reactor cores during postulated loss-of-coolant accidents [59]. However, because of unique geometrical and flow conditions, little in the way of existing data is applicable to the quenching of metals.

Factors which influence boiling heat transfer in metallurgical applications include the unsteady nature of the boiling process due to the continuously changing surface temperature, the Leidenfrost temperature which demarcates the film and transition boiling regions, and the subcooling, velocity and purity of the coolant. The effect of subcooling is clearly evident from the experimental results reported by Mitsutsuka and Fukuda on boiling phenomena during immersion of a silver rod [55] and steel plate [60] in water. Photographs were taken of a 28 x 220 x 220 mm low-carbon steel plate immersed in water at either 27 °C or 65 °C. The initial temperature of the steel was 930 °C. For the low water temperature, film boiling began immediately and 2-3 seconds later nucleate boiling commenced on all four

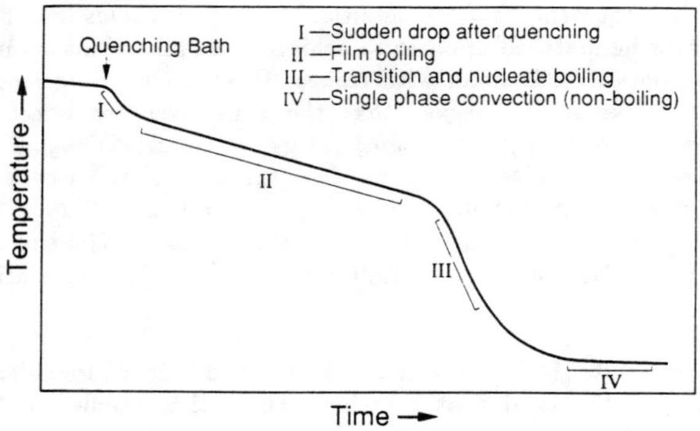

FIGURE 6. Schematic representation of a typical quenching curve illustrating surface temperature variation with time, after Mitsutsuka and Fukuda [55].

edges. At the higher water temperature, film boiling lasted 10 s, after which nucleate boiling began at the edges and moved towards the upper-middle portion of the plate. No simultaneous occurrence of nucleate boiling over the entire plate was observed. The results suggest the importance of water subcooling for higher and more uniform cooling rates. Water subcooling also affects the transition boiling regime, as reported by Mitsutsuka and Fukuda [55]. They used a silver cylinder 10 mm in diameter and 30 mm long to study transition boiling during a quench. The cylinder was heated to 800 °C and quenched in water at different temperatures. They observed that, when the difference between the initial metal temperature and the water temperature was high enough, the shift from stable film to nucleate boiling was instantaneous. In other words, there was no noticeable transition boiling region. This phenomenon occurred when the water temperature was below 50 °C. When the water temperature was above 50 °C, there was a noticeable transition boiling regime. The transition from film to nucleate boiling is important in quenching but has received little attention [56-58].

The heat transfer coefficient during quenching is a strong function of the surface temperature. Urbanovich et al. [61] measured the heat transfer coefficient on a 30 x 120 x 160 mm steel plate immersed in flowing water. The steel was initially at a temperature of 800-900 °C and the water temperature was less than 45 °C. The results show that in the stable film boiling region, $T_s > 300\,^\circ C$, the water velocity has little effect on the heat transfer coefficient. However, this result may be due to the relatively small range of velocities (0 to 0.74 m/s) considered. The heat transfer coefficient is a strong function of water velocity in nucleate and transition boiling ($110 < T_s < 300\,^\circ C$), as well as in the single phase convection ($T_s < 110\,^\circ C$) regimes. The peak heat transfer coefficient occurs at about $T_s = 110\,^\circ C$, which does not agree with the description provided by Urbanovich et al. [61]. They claim that nucleate boiling starts at approximately 300 °C, which should correspond to the maximum heat flux, rather than the boiling-single phase (112 °C) transition. Mitsutsuka and Fukuda [60] found the maximum heat transfer coefficient in the 200 to 350 °C surface temperature range, which is consistent with expectations. A possible explanation of differences in the shapes of the h vs. T_s curves reported by the two groups of investigators may be in the existence of two transition boiling curves, as suggested by Witte and Lienhard [62].

Prediction and optimization of the cooling rate to achieve desirable metallurgial and mechanical properties requires knowledge of the boiling curve (Fig. 7) for the liquid and the hydrodynamic conditions employed. Since the initial metal temperature (Point 1) is usually sufficiently high to induce film boiling, continued cooling will result in achievement of the

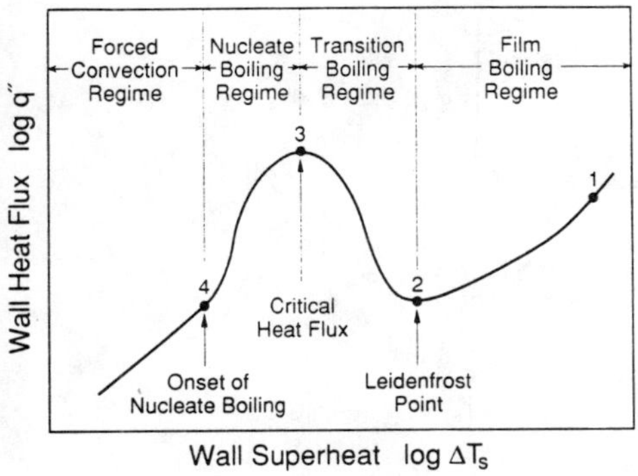

FIGURE 7. Typical boiling curve.

Leidenfrost temperature (Point 2). At this point the vapor film begins to breakup, liquid droplets in the jet penetrate to the metal surface, and transition to nucleate boiling commences. Simulation of the quenching process requires modeling of the boiling curve. Particularly, careful characterization of the minimum heat flux (Point 2) and of the critical heat flux (Point 3) is needed. In the remainder of this section discussion is focused on the definition of the minimum and maximum heat fluxes and the corresponding wall superheats at these points.

The study of Ishigai et al. [63], who measured heat transfer to a plane water jet impinging on a high-temperature flat surface at its stagnation point using a transient method, is pertinent to the foregoing discussion. The boiling curve determined by this method is shown in Fig. 8. Film boiling prevails from the beginning of the process to point A. During this period a stable vapor film is formed between the cooled surface and the water. The vapor interface is smooth and the surface is visible through the water. At point A, the stable film collapses and boiling noise is generated. The heat flux is a minimum, and the temperature corresponds to the Leidenfrost point. As the temperature decreases between points A and B, the heat flux sharply increases, but then remains nearly constant between points B and C. The boiling curve has a "shoulder" in this region. The vapor interface appears white due to frequent and instantaneous liquid/solid contacts from point A to point C. The surface becomes "wetted" when the temperature exceeds that of point C. The heat flux increases and the wetted zone spreads over the entire surface in the transition boiling regime (between points C and D). As the surface temperature falls, boiling passes through the critical heat flux (point D), and with a continued decrease in the surface temperature, boiling ceases (point E). To the left of point E, heat is transferred by single-phase convection.

Nucleate Boiling

Until the recent work of Vader [64], previous studies of boiling in planar jet impingement have been restricted to the stagnation region. Ishigai et al. [63] performed steady-state and quenching experiments, respectively, for 0.1 mm and 2 mm thick stainless steel strips. In the quenching experiment the heat flux passed through a minimum at the collapse of stable film boiling and achieved a maximum at the onset of stable nucleate boiling. In the steady-state experiments, data fell on a single, fully developed nucleate boiling curve, irrespective of jet velocity and subcooling. Miyasaka and Inada [24] also obtained forced convection nucleate boiling data which collapsed to a single boiling curve. Vader [64] was able to

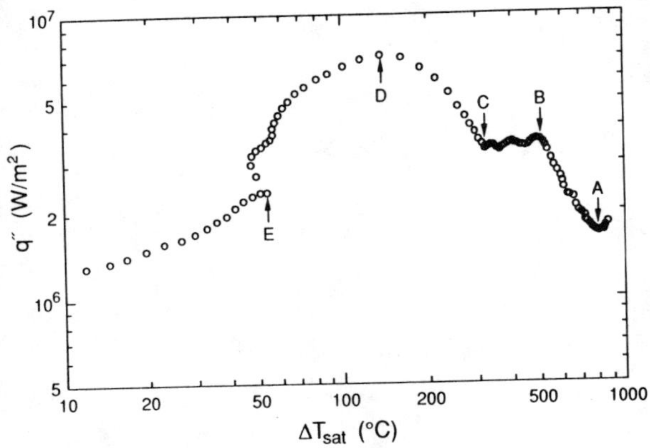

FIGURE 8. Boiling curve for a slot jet of water on a polished stainless steel strip heated electrically, after Ishigai et al. [63].

obtain partial and fully developed nucleate boiling heat transfer data and flow visualization in the laminar region beneath the jet, as well as in turbulent regions extending as much as ten jet widths from the stagnation line. Interactions between the onset of turbulence and nucleate boiling were delinated, and the importance of nucleation on transition was determined. Thus far, however, the onset of nucleation has not been correlated, its effect on transition has not been quantified, and heat transfer correlations have not been determined for other than fully developed nucleate boiling.

Partial and fully developed nucleate boiling heat transfer to an impinging planar jet of water have been studied by Vader [64]. Since a rising jet temperature or heat flux enhances nucleate boiling, partial boiling convection coefficients increase substantially as q'', T_f, and V_j are increased. Figure 9 shows some typical local heat transfer coefficients along the impingement surface for a jet velocity of $V_j = 4.5$ m/s. The results reveal that for $q'' = 1.23$ MW/m^2 partial boiling starts at $x/w_j \sim 6$, and the location moves upstream as the heat flux q'' is increased. Vader's results have shown that the convection coefficient becomes increasingly sensitive to the heat flux, while the sensitivity to the jet velocity and subcooling is diminished as boiling approaches the fully developed condition. Both Ruch and Holman [65] and Miyasaka and Inada [24] report that $q'' \sim \Delta T_{sat}^3$ for fully developed boiling in the stagnation region of an impinging liquid jet. Hence, it follows that $h = q''/\Delta T_{sat} \sim \Delta T_{sat}^2$. The fully developed boiling data of Vader do not span a sufficient range of heat fluxes and superheats to verify this result.

Boiling heat transfer literature for circular and slot jets of water and other liquids impinging on a hot surface has been reviewed by Ma and Bergles [66]. The general conclusion is that the fully developed boiling curve depends on fluid properties and is independent of the jet parameters and the test section dimensions. On the other hand, the critical (burnout) heat flux depends on the jet diameter, jet velocity and the dimension of the heated surface. Most of the studies used relatively small diameter jets, which are not prototypic of metallurgical applications. The works of Miyasaka et al. [24,67] are exceptions, involving slot jets 10 mm wide. Results of fundamental boiling heat transfer studies for water jets impinging on hot surfaces will be highlighted below.

Extensive experiments for burnout in the stagnation region of free, circular jets have been performed by Katto and co-workers [68-70], and critical heat flux (CHF) data have been correlated. Improved correlations of the data have also been developed by Lienhard et al.

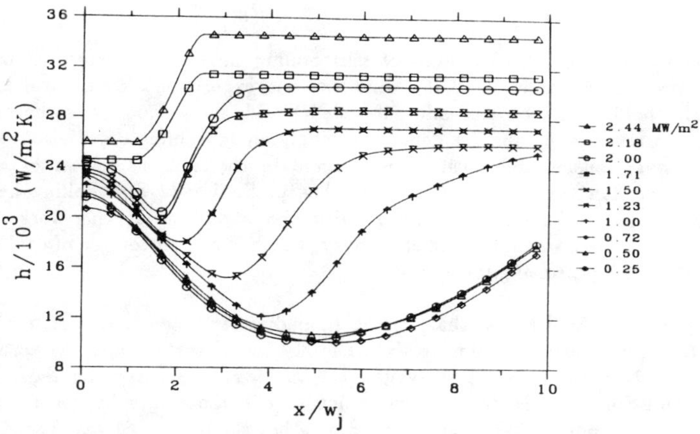

FIGURE 9. Effect of heat flux on the local heat transfer coefficient to an impinging planar jet of water, $V_j = 4.5$ m/s and $T_f = 30\,^\circ$C, from Vader [64].

[71-73]. Experiments for a planar jet have been performed by Miyasaka et al. [67], who determined the effects of jet velocity and subcooling on CHF. They correlated the critical heat flux data in the stagnation region ($x/w_j \cong 0$) of an impinging planar jet by the equation

$$q''_{crit} = 0.16 h_{fg}\, \rho_v\, (1 + 0.86\, V_j^{0.38}) \left(1 + 0.112 \left(\frac{\rho_f}{\rho_v} \right)^{0.8} \left[\frac{c_{pf} \Delta T_{sub}}{h_{fg}} \right] \right) \left[\frac{\sigma g (\rho_f - \rho_v)}{\rho_v^2} \right]^{1/4} \tag{6}$$

They also noted that heat fluxes in excess of that required for departure from nucleate boiling caused the boiling curve slope to be greater than the slope of the fully developed nucleate boiling curve. Stable film boiling was not established until the heat flux was substantially greater than q_{crit}. The value of q_{crit} predicted by equation (6) is of interest, however, because it identifies the upper heat flux limit for fully developed boiling at the heated surface. The dependence of q_{crit} on the velocity revealed by equation (6) is not as strong as for water at atmospheric pressure in parallel flow on an electronically heated flat plate, but the strong linear dependence on subcooling is retained [74].

Transition Boiling

In the transition boiling region, the heat flux at end of the shoulder (point C) in the boiling curve of Fig. 8 increases with the impinging jet velocity and subcooling [63], but the wall temperature at the other end of the shoulder (point B) does not depend on either the jet velocity or the subcooling. In their work, Miyasaka et al. [67] identified two transition boiling regimes, both for pool boiling and impinging jet boiling on a platinum surface. The heat flux was found to increase with the impinging jet velocity, and there was no film boiling even for $\Delta T_{sat} = 500\,^\circ$C.

The experimental results of Miyasaka et al. [67] clearly show that high jet impingement velocities and subcoolings can completely suppress film boiling, since the liquid jet reaches the hot surface and prevents formation of a stable vapor film. In the transition regime violent nucleate boiling is evident, and the boiling curves appear to be continuous extensions of the boiling curves for ordinary saturated pool boiling.

Film Boiling

Numerous experimental and analytical studies of film boiling have been performed and state-of-the-art reviews are available [56-58,75]. Analyses and experiments on natural and forced convection film boiling in saturated and subcooled liquids on stationary isothermal bodies have shown that the Nusselt number depends strongly on subcooling. Radiation is a significant mode of heat transfer and must be considered in the analysis. Most of the analyses are for laminar flow, and only recently [76] has prediction of film boiling heat transfer been reported for turbulent flow. Also, with the exception of the work by Zumbrunnen et al. [13], surfaces were assumed stationary, which is not prototypic of cooling applications in the materials processing industries.

Experimental observations [24,63] reveal that, in the immediate impingement region of a subcooled water jet, there is no vapor film between the hot metal surface and the water, which is in direct contact with the surface, even if the surface temperature exceeds the Leidenfrost point. This condition is due to penetration of the vapor film by the liquid. When subcooling is reduced, boiling has been observed on a hot steel plate. Away from the jet impingement zone, film boiling is observed. As the plate is cooled, the surface begins to be wetted with water and a stable vapor film is not formed.

Wetting is a critical factor which controls cooling of a hot metal being quenched. A relationship between the wetting (Leidenfrost) temperature and the cooling water temperature for a circular jet of water impinging on a 18 Cr-8 Ni stainless steel plate has been established by Kokado et al. [12]. The measured wetting (Leidenfrost) temperature which separates the wetted and nonwetted zones is represented by an empirical equation,

$$T_L = 1150 - 8\, T_w \qquad (^\circ C) \tag{7}$$

where T_w is the water temperature. This relation is valid in the range $T_w = T_{w,crit} = 68\,^\circ C$, because at temperatures lower than $68\,^\circ C$, wetting occurs as soon as the water impinges on the hot plate. In a very recent review, Klimenko and Snytin [75] have shown that for a stationary surface the Leidenfrost temperature depends strongly on the material and coolant property combinations; however, experiments have not been performed and q''_{min} and T_L have not been measured when the surface is in motion. Experimental data [10,77-79] show that the Leidenfrost temperature depends not only on the water subcooling but also on the fluid motion and on the plate thermophysical properties through the thermal-penetration-coefficient, $a = \sqrt{k\,\rho\,c}$. The Leidenfrost temperature was found to decrease with increasing a [79]. The results demonstrate conclusively the importance of the material properties and fluid motion on the Leidenfrost temperature and the cooling rate during stable film boiling. The effects of surface condition (scale and surface roughness) and plate motion on the Leidenfrost temperature do not appear to have been investigated.

The experimental results of Ishigai et al. [63] clearly show the shift to higher wall superheats and heat fluxes with the increase in subcooling. For low subcooling, the heat flux increases with the impinging jet velocity, but the temperature at which the minimum heat flux occurs is insensitive to this velocity. For high subcooling, the boiling curve shifts to higher superheats and higher heat fluxes, and the shoulder width increases with increasing jet velocity. Ishigai et al. correlated their experimental minimum heat flux data by an empirical equation,

$$q''_{min} = 5.40 \times 10^4\,(1 + 0.527\,\Delta T_{sub})\,V_j^{0.607} \qquad (W/m^2) \tag{8}$$

The impinging jet velocity V_j and the subcooling ΔT_{sub} are seen to have a strong effect on the minimum heat flux.

The effects of subcooling and mass flux on transition boiling of subcooled water flowing in a tube [80] are similar to those of Ishigai et al. [63] for a plane water jet impinging on a hot surface. There is, however, a major difference between the results of the two studies. Heat fluxes in the transition region measured by Ragheb et al. [80] did not reveal as strong a dependence on subcooling as did the measurements of Ishigai et al. Although differences may be due to the different geometry, the results suggest that upstream history can influence the boiling curve. The presence of a vapor blanket just upstream of a transition boiling section tends to reduce the transition boiling heat flux considerably.

Lamvik and Iden [81] measured the average convective heat transfer coefficient from a heated aluminum surface ($\sim 500\,^{\circ}$C) to a single jet and two adjacent circular jets of water. Average coefficients are reported for a single horizontal jet impinging on a vertical surface, for a vertical jet impinging on a horizontal surface from below, and for a vertical jet impinging on a horizontal surface from above. It was found that the average coefficient depends strongly on the orientation of the jet with respect to the gravitational field. For a given jet diameter and impingement velocity the coefficient is largest for a vertical jet impinging upward on a horizontal surface from below. Measurements of local surface temperatures during quenching were not reported, but based on the experimental conditions, it is presumed that both film and transition boiling occurred simultaneously on the test plate during the experiments. The hydrodynamic characteristics of the jet were not included as parameters in the average heat transfer coefficient correlations reported.

Based on a comparison of data with predictions, Kokado et al. [12] have developed the following empirical fit for the convective film boiling heat transfer coefficient,

$$h = 200\,(2420 - 21.7\,T_w)\,(T_s - T_{sat})^{-0.8} \qquad (W/m^2\,K) \tag{9}$$

The equation demonstrates that the heat transfer coefficient is a strong function of both T_w and the surface superheat, $T_s - T_{sat}$.

The effect of surface motion on forced convection film boiling heat transfer has been analyzed by Zumbrunnen et al. [13]. Since experiments can not be performed in the laboratory to acquire the test data corresponding to mill conditions, analysis provides the only means for obtaining the needed scaling relations. In an integral analysis of vapor and liquid boundary layer flows for film boiling and $\overline{V}_p \leq 20$, they determined the extent to which plate motion can enhance and suppress heat transfer, respectively, downstream and upstream of the impinging jet. A five-fold increase in convection heat transfer was predicted for a dimensionless plate speed of $\overline{V}_p = 20$. Experiments performed for planar water jet impingement on a moving plate with $\overline{V}_p \leq 0.41$ revealed a measureable increase in single-phase convection heat transfer upstream of the jet with increasing $\overline{V}_p$ but no definitive changes downstream of the jet [51]. In forced convection boiling, maximum heat transfer was found to occur upstream of the jet.

SUMMARY

Simulation of a quenching process is difficult because of the need to predict local heat transfer coefficients in single-phase forced convection, convective nucleate boiling and convective film boiling from a moving metal to an impinging liquid jet. Also needed are the conditions for transitions from one mode of heat transfer to another. Liquid jet impingement cooling is difficult to analyze because of the large number of complicated flow regimes and variables controling the process. The effects of some important parameters such as the free-stream turbulence intensity in the jet, the partial pressure of noncondensibles in the nucleation gas embryo, and the presence of scale and/or oxide layers are not known or readily controllable under mill conditions. Furthermore, single phase and nucleate boiling heat transfer can not be evaluated independently. The minimum heat flux and the Leidenfrost temperature depend on the metal and on the scale or oxide film at the surface, which can seldom be established. Boiling incipience, for example, promotes the onset of turbulence in the boundary layer on the moving metal but does not appear to have been studied and is not understood. As with any complex physical process, there are a large number of factors which influence the local heat transfer rate.

Scaling from small laboratory experiments performed on stationary or slowly moving test specimens to prototypic mill conditions where the strip may be moving as fast as 30 m/s is a key issue. It is not possible to simulate these speeds in the laboratory, and it is not practical to perform tests in a mill because the conditions cannot be controlled or it is too costly to interrupt production. Models need to be developed and validated so that the results of small laboratory experiments can be scaled to prototypic mill conditions.

Enhancement of heat transfer to the jet can reduce water consumption and hence energy requirements for pumping and water purification. Fundamental work on the effects of free stream turbulence for single phase convection heat transfer in the impingement region is needed for both slot and axisymmetric nozzles. Validated mathematical-numerical model results are needed to identify nozzle designs, arrangements and hydrodynamic conditions for heat transfer augmentation and cooling system design optimization.

For most metallurgical applications film and transition boiling regimes are more important than the nucleate boiling regime. However, the effects of thermal and hydrodynamic conditions, surface and thermophysical properties of materials, and the motion of the quenched material on film and transition boiling are poorly understood. For example, the effects of thin oxide films on the minimum heat flux and the Leidenfrost temperature have not been studied and are not known.

The extent of the wetted region, as measured from the stagnation line or point of an impinging jet, is an important parameter which demarcates the single-phase convection-transition boiling region, where heat transfer is very effective, from the film boiling region. The wetting mechanism and its dependence on jet hydrodynamic conditions, thermal conditions in the material, and other relevant factors must be established. Knowledge of wetting is also needed to identify quenchants, other than water, and forced convection conditions to achieve desirable hardness through the control of wetting kinematics [82].

ACKNOWLEDGEMENTS

The authors wish to acknowledge support of their research on liquid jet impingement cooling by the National Science Foundation under Grants CPE-8414613 and CTS-8912831.

NOMENCLATURE

c	=	specific heat of fluid (J/kg °C)
g	=	gravitaional acceleration (m/s^2)
H	=	distance between the nozzle discharge and the impingement surface (m)
h_{fg}	=	latent heat of vaporization (J/kg)
h	=	heat transfer coefficient (W/mK)
k	=	thermal conductivity of fluid (W/mK)
Nu_j	=	Nusselt number (hw_j/k)
Nu_x	=	Nusselt number (hx/k)
Pr	=	Prandtl number ($\mu c/k$)
q	=	heat flux (W/m^2)
Re_j	=	jet Reynolds number ($V_j w_j/\nu$)

Re_{x*} = plate Reynolds number based on local velocity ($u_\delta x/\nu$)
Tu = turbulence intensity
T_s = surface temperature (°C)
T_w = water temperature (°C)
ΔT_{sat} = surface minus saturation temperature difference (°C)
u_δ = local free stream value of the x-component of velocity (m/s)
V_j = jet velocity at impingement (m/s)
$\bar{V}_p$ = dimensionless plate velocity (V_p/V_j)
w_j = width of jet at impingement surface with origin at the stagnation line (m)
w_n = discharge width of the nozzle (m)
σ = surface tension (N/m)
x = coordinate along impingement surface (m)
ν = kinematic viscosity of fluid
ρ = density of fluid

Subscripts

f = refers to fluid
j = refers to jet
ν = refers to vapor
sat = refers to saturation
sub = refers to subcooling

REFERENCES

1. Roberts, W.L., *Hot Rolling of Steel*, Marcel Dekker, New York, 1983.
2. Southwick, P.D. (ed.), *Accelerated Cooling of Steel*, The Metallurgical Society, Warrendale, PA, 1985.
3. Ruddle, G.E., and Crawley, A.F. (eds.), *Accelerated Cooling of Rolled Steel*, Pergamon Press, New York, 1987.
4. Chevier, J., Simon, A., and Beck, G., Optimal Cooling Rate and Process Control in Metallic Parts Heat Treatment, *Heat and Mass Transfer in Metallurgical Systems*, eds. D.B. Spalding and N.H. Afgan, pp. 535-544, Hemisphere, Washington, D.C., 1981.
5. Brimacombe, J.K., Agarwal, P.K., Baptista, L.A., Hibbins, S., and Prabhar, B., Spray Cooling in Continuous Casting of Steel, *63rd National Open Hearth and Basic Oxygen Steel Conference Proceedings*, Vol. 63, pp. 235-252, Washington, DC, 1980.
6. Tseng, A.A., Gunderia, A.S., and Ni, D.S., Thermal Behavior of Roll and Strip Steel Rolling with Non-Conventional Roll Designs, *Energy Optimization in Manufacturing and Materials Processing*, pp. 89-99, Advanced Study Institute, Indian Institute of Technology, Madras, India, 1989.
7. Kohring, F.C., Waterwall Water Cooling Systems, *Iron and Steel Engineer*, Vol. 62, pp. 30-36, June 1985.
8. Tacke, G., Litzke, H., and Raquet, E., "Investigations into the Efficiency of Cooling Systems for Wide-Strip Hot Rolling Mills and Computer Aided Control of Strip Cooling, in *Accelerated Cooling of Steel*, ed. P.D. Southwick, pp. 35-54, The Metallurgical Society, Inc., Warrendale, PA, 1986.

9. Brimacombe, J.K., Agarwal, P.K., Baptista, L.A., Hobbins, S., and Prabhakar, B., "Spray Cooling in the Continuous Casting of Steel", in the *Proceedings of the 63rd National Open Hearth and Basic Oxygen Steel Conference*, vol. 63, pp. 235-252, The Iron and Steel Society of the American Institute of Mining, Metallurgical and Petroleum Engineers, Washington, D.C., 1980.

10. Jeschar, R., Scholz, R., Reiners, U., and Maass, R., 1987, Kühltechniken zur thermischen Behandlung von Werkstoffen, *Stahl und Eisen*, vol. 107, no. 6, pp. 251-258.

11. McGannon, H.E. (ed.), *The Making, Shaping and Treating of Steel*, 9th Edition, United States Steel Corp., Herbick and Held, Pittsburgh, 1971.

12. Kokado, J., Hatta, N., Tokuda, H., Harada, J., and Yasuhira, N., An Analysis of Film Boiling Phenomena of Subcooled Water Spreading Radially on a Hot Steel Plate, *Archiv für Eisenhüttenwesen*, vol. 55, pp. 113-118, 1984.

13. Zumbrunnen, D.A., Viskanta, R., and Incropera, F.P., Analysis of Forced Convection Film Boiling on Moving Metallic Strips and Plates During Manufacture, *Journal of Heat Transfer*, vol. 110, pp. 889-896, 1989.

14. Vader, D.T., Incropera, F.P., and Viskanta, R., Local Convective Heat Transfer from a Heated Surface to an Impinging, Planar Jet of Water, *International Journal of Heat and Mass Transfer* (in press).

15. Martin, H., Heat and Mass Transfer Between Impinging Gas Jets and Solid Surfaces, *Advances in Heat Transfer*, eds. T.F. Irvine, Jr. and J.P. Hartnett, vol. 13, pp. 1-60, Academic Press, New York, 1977.

16. Downs, S.J., and Jones, E.H., Jet Impingement Heat Transfer - A Literature Survey, ASME Paper No. 87-HT-35, 1987.

17. Polat, S., Huang, B., Mujumder, A.S., and Douglas, W.J.M., Numerical Flow and Heat Transfer Under Impinging Jets: A Review, *Annual Review of Fluid Mechanics and Heat Transfer*, eds. C.L. Tien and T.C. Chawla, vol. 2, pp. 157-197, Hemisphere, Washington, D.C., 1989.

18. Schlichting, H., *Boundary Layer Theory*, 7th Edition, pp. 95-99, pp. 163-166, McGraw-Hill, New York, 1979.

19. Levy, S., Heat Transfer to Constant-Property Laminar Boundary-Layer Flows with Power Function Free-Stream Velocity and Wall-Temperature Variation, *Journal of the Aeronautical Sciences*, vol. 19, no. 5, pp. 341-348, 1952.

20. Evans, H.L., Mass Transfer Through Laminar Boundary Layers - Further Similar Solutions to the B-Equation for the Case B = 0, *International Journal of Heat and Mass Transfer*, vol. 5, pp. 35-57, 1962.

21. Inada, S., Miyasaka, Y., and Izumi, R., A Study of Laminar-Flow Heat Transfer Between a Two-Dimensional Laminar Water Jet and a Flat Plate, *Bulletin of the Japan Society of Mechanical Engineers*, vol. 24, no. 196, pp. 1803-1810, 1981.

22. Zumbrunnen, D.A., Convective Heat and Mass Transfer in the Stagnation Region of a Planar Jet Impinging on a Moving Surface, *Proceedings of the 5th AIAA/ASME Thermophysics and Heat Transfer Conference*, eds. K. Vafai et al., HTD-Vol. 129, pp. 11-20, ASME, New York, 1990.

23. McMurray, D.C., Myers, P.S., and Uyehara, O.A., Influence of Impinging Jet Variables on Local Heat Transfer Coefficients Along a Flat Plate Surface with Constant Heat Flux, *Proceedings of the Third International Heat Transfer Conference*, Vol. II, pp. 292-299, AIChE, New York, 1966.

24. Miyasaka, Y., and Inada, S., The Effect of Pure Forced Convection on the Boiling Heat Transfer Between a Two-Dimensional Subcooled Water Jet and a Heated Surface", *Journal of Chemical Engineering of Japan*, vol. 13, no. 1, pp. 22-28, 1980.

25. Wolf, D.H., Viskanta, R., and Incropera, F.P., Local Convective Heat Transfer from a Heated Surface to a Planar Jet of Water with Nonuniform Velocity Profile, *Journal of Heat Transfer* (in press).

26. Miyasaki, H., and Silberman, E., Flow and Heat Transfer on Flat Plate Normal to a Two-Dimensional Laminar Jet Issuing from a Nozzle of Finite Height, *International Journal of Heat and Mass Transfer*, vol. 15, pp. 2097-2107, 1972.

27. Zumbrunnen, D.A., Incropera, F.P., and Viskanta, R., Convective Heat Transfer Distributions on a Plate Cooled by Planar Water Jets, *Journal of Heat Transfer*, vol. 111, pp. 889-896, 1989.

28. Wang, W.S., Dagan, Z., and Jiji, L.M., Heat Transfer Between a Solid Surface with Non-Uniform Wall Temperature or Wall Heat Flux - 1: Solution for the Stagnation Region, *International Journal of Heat and Mass Transfer*, vol. 32, pp. 1351-1360, 1989.

29. Wang, X.S., Dagan, Z., and Jiji, L.M., Heat Transfer Between a Circular Free Impinging Jet and a Solid Surface with Non-Uniform Wall Temperature or Heat Flux - 2: Solution for the Boundary Layer Region, *International Journal of Heat and Mass Transfer*, vol. 32, pp. 1361-1371, 1989.

30. Carper, H.J. Jr., Impingement Cooling of Liquid Jets, *Proceedings of ASME 1989 Winter Annual Meeting*, pp. 23-30, ASME, New York, 1989.

31. Lin, X., and Lienhard, J.H. V, Liquid Jet Impingement Heat Transfer on a Uniform Flux Surface, *Heat Transfer Phenomena in Radiation, Combustion and Fires*, HTD-Vol. 106, pp. 523-530, ASME, New York, 1989.

32. Nakoryakov, V.E., Pokusaev, B.A., and Troyan, E.N., Impingement of an Axisymmetric Liquid Jet on a Barrier, *International Journal of Heat and Mass Transfer*, vol. 21, pp. 1175-1184, 1978.

33. Jiji, L.M., and Dagan, Z., Experimental Investigation of Single Phase Multi-Jet Impingment Cooling of an Array of Electronic Heat Sources, *Proceedings of the International Symposium on Cooling Technology for Electronic Equipment*, pp. 265-283, Hemisphere, Washington, D.C., 1987.

34. Stevens, J., and Webb, B.W., Local Heat Transfer for an Axisymmetric, Single-Phase Liquid Jet," *Heat Transfer in Electronics*, pp. 113-119, ASME, New York, 1989.

35. Barasanti, G., Faggiani, S., and Grassi, W., Single-Phase Forced Convection Cooling of Heating Surfaces by Liquid Jet-Impingement, *Heat and Technology*, vol. 7, no. 2, pp.1-11, 1989.

36. Womac, D.J., Aharoni, A., Ramadhyani, S., and Incropera, F.P., Single Phase Liquid Jet Impingement Cooling of Small Heat Sources, *Heat Transfer - 90*, vol. 4, pp. 149-154, Hemisphere, Washington, D.C., 1990.

37. Kestin, J., Meader, P.E., and Wang, H.E., Influence of Turbulence on the Transfer of Heat from Plates With and Without a Pressure Gradient, *International Journal of Heat and Mass Transfer*, vol. 3, pp. 133-154, 1961.

38. Kestin, J., The Effect of Free Stream Turbulence on Heat Transfer Rates, *Advances in Heat Transfer*, eds. T.F. Irvine, Jr. and J.P. Hartnett, Vol. 3, pp. 1-32, Academic Press, New York, 1966.

39. Donaldson, C.D., Snededeker, R.S., and Margolis, D.P., A Study of Free Jet Impingement. Part 2. Free Jet Turbulent Structure and Impingement Heat Transfer, *Journal of Fluid Mechanics*, vol. 45, pp. 477-512, 1971.

40. Sparrow, E.M., and Wong, T.C., Impingement Heat Transfer Coefficients due to Initially Laminar Slot Jets, *International Journal of Heat and Mass Transfer*, vol. 18, pp. 597-605, 1975.

41. Sparrow, E.M., and Lee, L., Analysis of Flow Field and Impingement Heat/Mass Transfer Due to a Nonuniform Slot Jet, *Journal of Heat Transfer*, vol. 97, pp. 191-197, 1975.

42. van Heiningen, A.R.P., Mujumdar, A.S., and Douglas, W.J.M., Numerical Prediction of the Flow Field and Impingement Heat Transfer Caused by a Laminar Slot Jet, *Journal of Heat Transfer*, vol. 98, pp. 654-658, 1976.

43. Kataoka, K., and Mizushina, T., Local Enhancement of the Rate of Heat-Transfer in an Impinging Round Jet by Free-Stream Turbulence, *Proceedings of the 5th International Heat Transfer Conference*, vol. II, pp. 305-309, 1974.

44. Kataoka, K., Komai, T., and Nakamura, G., Enhancement Mechanism of Mass Transfer in a Turbulent Impinging Jet for High Schmidt Numbers, ASME Paper No. 78-HT-5, 1978.

45. Lyell, M.H., and Huerra, P., Linear and Nonlinear Stability of Plane Stagnation Flow, *Journal of Fluid Mechanics*, vol. 161, pp. 295-312, 1985.

46. VanFossen, G.J., Jr., and Simoneau, R.J., A Study of the Relationship Between Free-Stream Turbulence and Stagnation Region Heat Transfer, *Journal of Heat Transfer*, vol. 109, pp. 10-15, 1987.

47. Smith, M.C., and Kuethe, A.M., Effects of Turbulence on Laminar Skin Friction and Heat Transfer, *Physics of Fluids*, vol. 9, pp. 2337-2344, 1966.

48. Miyazaki, H., and Sparrow, E.M., Analysis of Effects of Free-Stream Turbulence on Heat Transfer and Skin Friction, *Journal of Heat Transfer*, vol. 99, pp. 614-619, 1977.

49. Torii, K., and Yoshida, M., Free-Stream Turbulence Effects on Heat and Mass Transfer from Spheres, in *Structure of Turbulence in Heat and Mass Transfer*, pp. 245-264, Hemisphere, Washington, D.C., 1980.

50. Morkovin, M.V., On the Question of Instabilities Upstream of Cylindrical Bodies, NASA CR-3231, pp. 1-106, 1979.

51. Zumbrunnen, D.A., F.P. Incropera, and Viskanta, R., A Method and Apparatus for Measuring Heat Transfer Distributions on Moving and Stationary Plates Cooled by a Planar Liquid Jet, *Experimental Thermal and Fluid Science*, vol. 3, no. 2, pp. 202-213, 1990.

52. Huang, P.G., Mujumdar, A.S., and Douglas, M.J.M., Numerical Prediction of Fluid Flow and Heat Transfer under Turbulent Impinging Slot Jet With Surface Motion and Crossflow, ASME Paper No. 84-WA/HT-33.

53. Ishigaki, H., Heat Transfer in a Periodic Boundary Layer Near a Two-Dimensional Stagnation Point, *Journal of Fluid Mechanics*, vol. 56, pp. 619-627, 1972.

54. Lowery, G.W., and Vachon, R.I., The Effect of Turbulence on Heat Transfer from Heated Cylinders, *International Journal of Heat and Mass Transfer*, vol. 18, pp. 1229-1242, 1975.

55. Mitsutsuka, M., and Fukuda, K., Transition Boiling and Characteristic Temperature in Cooling Curve During Water Quenching of Heated Metal, *Transactions of the Iron and Steel Institute of Japan*, vol. 16, pp. 46-50, 1976.

56. Jordan, D.P., Film and Transition Boiling, *Advances in Heat Transfer*, eds. T.F. Irvine, Jr. and J.P. Hartnett, vol. 5, pp. 55-128, Academic Press, New York, 1968.

57. Kalinin, E.K., Berlin, I.I., and Kostiouk, V.V., Transition Boiling Heat Transfer, *Advances in Heat Transfer*, eds. J.P. Hartnett and T.F. Irvine, Jr., vol. 18, pp. 241-323, Academic Press, New York, 1987.

58. Lienhard, J.H., Things We Don't Know About Boiling Heat Transfer: 1988, *International Communications in Heat and Mass Transfer*, vol. 15, pp. 401-428, 1988.

59. Nelson, R.A., Mechanisms of Quenching Surfaces, *Handbook of Heat and Mass Transfer*, ed. N.P. Cheremisinoff, pp. 1103-1153, Gulf, Houston, 1986.

60. Mitsutsuka, M., and Fukuda, K., Boiling Phenomena and Effects of Water Temperature on Heat Transfer in the Process of Immersion Cooling of a Heated Steel Plate, *Transactions of the Iron and Steel Institute of Japan*, vol. 19, pp. 162-169, 1979.

61. Urbanovich, L.I., Niskovskikh, V.M., Gur'ev, V.S., Gyraya, A.P., and Fedorov, L.A., Heat Engineering Research Into Effectiveness of Slab Cooling After Continuous Billet Caster, *Steel in the USSR*, vol. 12, pp. 254-257, 1982.

62. Witte, L.C., and Lienhard, J.H., On the Existence of 2 'Transition' Boiling Curves, *International Journal of Heat and Mass Transfer*, vol. 25, pp. 771-779, 1982.

63. Ishigai, S., Nakanishi, S., and Ochi, T., Boiling Heat Transfer for a Plane Water Jet Impinging on a Hot Surface, in *Heat Transfer - 1978*, Hemisphere, vol. 1, pp. 445-450, Washington, D.C., 1978.

64. Vader, D.T., Convective and Boiling Heat Transfer from a Heated Surface to an Impinging, Planar Jet of Water, Ph.D. Thesis, Purdue University, 1988.

65. Ruch, M.A., and Holman, J.P., Boiling Heat Transfer to a Freon-113 Jet Impinging Upward Onto a Flat, Heated Surface, *International Journal of Heat and Mass Transfer*, vol. 18, pp. 51-60, 1975.

66. Ma, C.F., and Bergles, A.E., Boiling Jet Impingement Cooling of Simulated Microelectronics, *Heat Transfer in Electronic Equipment - 1983*, eds. S. Oktay and A. Bar-Cohen, pp. 5-12, ASME, New York, 1983.

67. Miyasaka, Y., Inada, S., and Owase, Y., Critical Heat Flux and Subcooled Nucleate Boiling in Transient Region Between a Two-Dimensional Water Jet and a Heated Surface, *Journal of Chemical Engineering of Japan*, vol. 13, pp. 29-35, 1980.

68. Katto, Y., and Monde, M., Study of Burn-Out in a High Heat Flux Boiling System with an Impinging Jet, in *Proceedings of the Fifth International Heat Transfer Conference*, vol. IV, pp. 245-249, JSME, Tokyo, 1974.

69. Monde, M., and Katto, Y., Burnout in a High-Heat Flux Boiling System with an Impinging Jet, *International Journal of Heat and Mass Transfer*, vol. 21, pp. 295-305, 1978.

70. Katto, Y., and Shimazu, M., Upper Limit of CHF in the Saturated Forced Convection Boiling on a Heated Dish with a Small Impinging Jet, *Journal of Heat Transfer*, vol. 101, pp. 265-269, 1979.

71. Lienhard, J.H., and Eichhorn, R., On Predicting Boiling Burnout for Heaters Cooled by Liquid Jets, *International Journal of Heat and Mass Transfer*, vol. 22, pp. 774-776, 1979.

72. Lienhard, J.H., and Hasan, M.Z., Correlation of Burnout Data for Disk Heaters Cooled by Liquid Jets, *Journal of Heat Transfer*, vol. 101, pp. 383-384, 1979.

73. Sharan, A., and Lienhard, J.H., On Predicting Burnout in the Jet Disk Configuration, *Journal of Heat Transfer*, vol. 107, pp. 398-401, 1985.

74. Gunther, F., Photographic Study of Surface Boiling Heat Transfer to Water with Forced Convection, *Transactions of the American Society of Mechanical Engineers*, vol. 73, pp. 115-123, 1951.

75. Klimenko, V.V., and Snytin, S. Yu., Film Boiling Crisis on a Submerged Heating Surface, *Experimental Thermal and Fluid Science*, vol. 3, no. 5, pp. 467-479, 1990.

76. Wang, B.-X., Shi, D.-H., and Peng, X.-F., An Advance on the Theory of Forced Turbulent-Flow Film Boiling Heat Transfer for Subcooled Liquid Flowing Along a Horizontal Flat Plate, *International Journal of Heat and Mass Transfer*, vol. 30, pp. 137-141, 1987.

77. Owen, R.A., and Pulling, D.J., Wetting Delay: Film Boiling of Water Jets Impinging on Hot Flat Metal Surfaces, *Multiphase Transport*, ed. T.N. Veziroglu, vol. 2, pp. 639-669, Hemisphere, Washington, D.C., 1979.

78. Müller, H.R., and Jeschar, R., Heat Transfer During Water Spray Cooling of Nonferrous Metals, *Zeitschrift für Metallkunde*, vol. 74, pp. 257-264, 1983.

79. Jeschar, R., Reiners, U., and Scholz, R., Heat Transfer During Water and Water-Air Spray Cooling Zone of Continuous Casting Plants, *Steelmaking Proceedings-Fifth Iron and Steel Congress*, vol. 69, pp. 511-521, Washington, D.C., 1986.

80. Ragheb, H.S., Cheng, S.C., and Groenveld, D.C., Observations in Transition Boiling of Subcooled Water Under Forced Convective Conditions, *International Journal of Heat and Mass Transfer*, vol. 24, pp. 1127-1137, 1981.

81. Lamvik, M., and Iden, B.-A., Heat Transfer Coefficient by Water Jets on a Hot Surface, *Heat Transfer 1982*, eds. U. Grigull, E. Hahne, K. Stephan, and J. Straub, vol. 1, pp. 369-375, Hemisphere, Washington, DC, 1982.

82. Tensi, H.M., and Stitzelberger-Jakob, P., Evaluation of Apparatus for Assessing Effect of Forced Convection on Quenching Characteristics, *Materials Science and Engineering*, vol. 5, pp. 718-724, 1989.

Stability of Mist Cooling in Thermo-Mechanical Control Process of Steel

SHIGEFUMI NISHIO and HIDETOSHI OHKUBO
Institute of Industrial Science
University of Tokyo
7-22-1, Roppongi, Minato-ku, Tokyo 106, Japan

ABSTRACT

Mist cooling has found many applications in industry, but its heat transfer mechanism is still not clear. The reason for this is that mist cooling heat transfer depends on many factors. Understanding the effects of these factors has become important because of the advent of the so-called Thermo-Mechanical Control (TMC) process in the steel industry and other new applications. The aim of the TMC process is the on-line control of metallurgical structure of steel. From the metallurgical and mechanical viewpoints, this process requires that steel plates are uniformly cooled down to a prescribed temperature. Yet, such a stable operation is not possible because of surface oxidizing and fouling. This paper presents an overview of the mist-cooling heat transfer and the results of authors' recent experiments on stabilizing the TMC process.

1. INTRODUCTION

The terms like 'spray cooling', 'mist cooling', and 'fog cooling' are commonly used when describing the cooling systems in which a hot surface is cooled by impinging droplets. To describe such a system the term, mist cooling is used in this paper.

The mist-cooling heat transfer is a kind of dispersed-flow heat transfer that can provide uniform object cooling and cover a wide range of cooling rates by simply changing the droplet mass flux. Thus, the mist-cooling heat transfer has been widely used in many practical applications, particularly in the metallurgical industry and nuclear reactor safety devices. Other areas that mist-cooling has been tried are: thermal tempering of glass plates(1), embrittling of waste materials in cryogenic recycling(2), and cooling of electronic devices(3).

One recent application of mist-cooling is the so-called "Thermo-Mechanical Control (TMC) process," which aims at on-line control of metallurgical structure of steel to produce value-added steels. Conventionally, heat-treatment process of steel has been an off-line process that consists of the sub-processes, rolling, heating, quenching and annealing. These processes are now being replaced by new on-line processes to produce value-added steels, while saving energy. In the TMC process, the rolling and the cooling processes are controlled and coupled, and the former precedes the latter. We will call these processes CC and CR processes meaning the contolled-cooling and the controlled-rolling processes respectively.

In a conventionally cooled process, the hardness of steel in general depends on both the cooling rate and the plate thickness. Coarsening the grain size intensifies the hardenability of steel and higher cooling rates increase the hardness. Thus in the TMC process, to homogenize the cross-sectional hardness distribution, the controlled-rolling process adequately controls the cross-sectional grain-size distribution of the steel plate. During this controlled rolling, the grains at the surface of the plate will become smaller than that at the center. When the steel plate is next cooled in the cooling process, the cross-sectional hardness distribution of the plate becomes uniform because the enhanced hardenability at the center of the plate compensates the slower cooling rate. This effect is called as the "self-control effect of steel." The main problem to be overcome in this cooling process is the undesirable non-uniformity of cooling that leads to shape and size-defects.

In recent years, the TMC process has been used to produce value-added steels in a sophisticated system, in which the cooling process is divided into two stages. During the first stage($0<\tau<\tau_o$, in Fig.1), the steel plate is cooled down roughly to 700K at a relatively high cooling rate. During the second stage that follows the first stage ($\tau_o<\tau$), the plate is gradually cooled down to room temperature as shown in Fig.1 (line A). The metallurgical structures such as martensite, bainite and ferrite are controlled by selecting the cooling rate of the first stage, which usually uses mist cooling.

Steel plates produced in the TMC process excel in hardness, toughness and weldability. Generally, the high-strength steels are produced by adding alloying elements and increasing the carbon concentration. Therefore, the TMC process can reduce alloying elements and the carbon concentration to the minimum. The main problem to be overcome in this system is to stabilize the final temperature of steel in the first stage of the cooling process. Controlling this temperature is important because its uniformity determines the metallurgical and mechanical characteristics of steel plates produced.

Yet, stabilizing the final temperature in the first stage is difficult because mist-cooling systems have a quench temperature that is roughly equal to the desired final-temperature. Also, the quench temperature strongly depends on the extent of surface oxidizing or fouling. For example, supposing the line A in Fig.1 represents the cooling curve of a weakly oxidized surface, this curve will move to line B when the surface is fully oxidized and to line C when the surface is fresh. Such a behavior of the cooling curve implies that, to secure a stable operation for the TMC process, it is needed to understand the heat transfer characteristics of mist cooling, particularly the effects of surface oxidizing and fouling. The surface oxidizing and fouling definitely change the surface roughness, wettability, and thermal characteristics, but their effects are very complicated and not clear.

This paper presents an experimental study of the independent effects on mist-cooling heat transfer of the surface parameters: roughness, wettability, and thermal characteristics. We shall first discuss the effects of the mist-flow parameters on mist-cooling heat transfer in sec.2 and next, the effects of surface parameters in the sec.3.

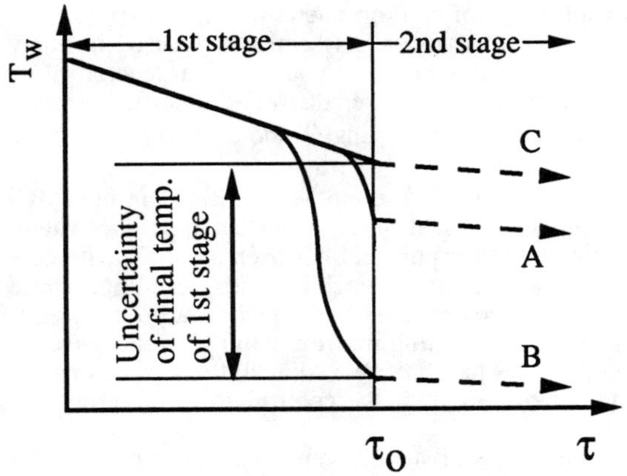

FIGURE 1. Variation of cooling curve in TMC process

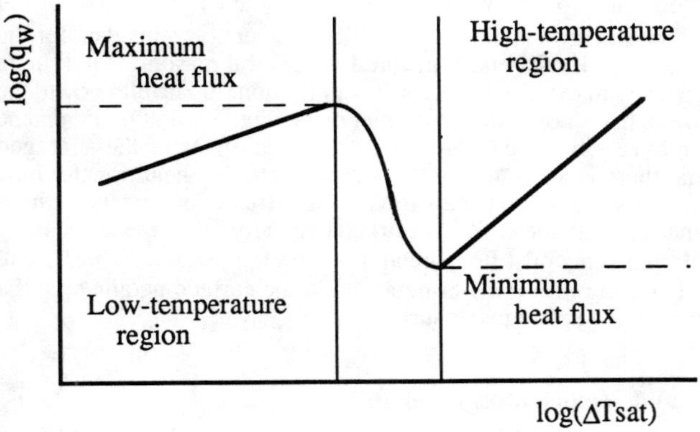

FIGURE 2. Schematic diagram of characteristics of mist-cooling heat transfer

2. EFFECTS OF MIST-FLOW PARAMETERS

Figure 2 schematically shows the mist-cooling heat-transfer characteristic, which has a maximum and a minimum heat flux point similar to boiling heat transfer characteristic. These points divide the temperature axis into three temperature regions: the 'low-temperature region', 'transition region' and 'high-temperature region' in the order, low to high temperature.

The existing experimental data of mist-cooling heat transfer(4)-(7) on upward-facing horizontal flat plates are shown in Fig. 3(8), where symbol "D" denotes the droplet mass

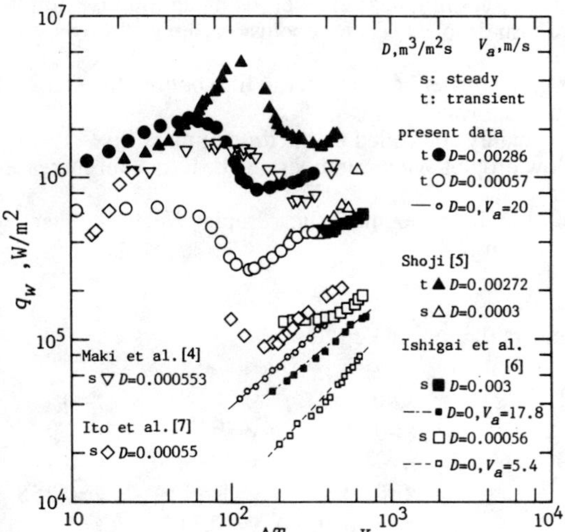

FIGURE 3. Diversity of experimental data obtained at the same droplet mass flux

flux of mist flow, and open and closed symbols denote the data for D=0.0006 and $0.003 m^3/m^2s$ respectively. One can see that there is significant diversity in the data for the same droplet mass flux. Such diversity may have occured for several reasons. First, there is a possibility of an inconsistency in the droplet-mass-flux definition. It is well known, the mist flow generated by a nozzle has a non-uniform droplet mass-flux distribution in a plane perpendicular to the nozzle axis. Obviously, the reported values of D are the averaged droplet mass-fluxes. Second, there is a possibility that the mist-cooling heat transfer may depend on mist-flow parameters other than the droplet mass flux. For instance, these parameters may be the mean droplet diameter "d", the droplet velocity "V_d", the air velocity "V_a" because, mostly, the droplets in a mist flow are at the room temperature. Third, there is a possibility that the mist-cooling heat transfer depends on the surface parameters: the thermal and surface conditions of the heat transfer surface.

2.1 Effects of Diameter and Velocity of Droplets

In investigating the effects of mist-flow parameters on mist-cooling heat transfer, the simplest approach would be to correlate statistically the data from actual mist-cooling systems. Customarily, the mist cooling heat transfer coefficient, h, is correlated to the mist-flow parameters, the diameter, velocity, and mass flux of droplets by the following equation:

$$h=const.d^x V_d^y D^z \tag{1}$$

Table 1 lists the values of the exponents x, y and z obtained from experiments(2),(4)-(6),(9)-(11) and analysis(7). The respective values of x, y and z are seen to be so different, showing that the statistical approach so far has not established a universal relationship between h and the mist-flow parameters.
Therefore, to understand the independent effects of mist-flow parameters, other approaches have been tried. The first is the study of the dynamics of a single droplet impinging on a heated surface(12). The second is the study of the dynamics and heat transfer of a single stream of impinging droplets(13)-(15). The last is the complex case of parallel streams of impinging droplets(16). For this case, with independent control of droplet size, velocity, and mass flux, Choi and Yao(16) measured mist-cooling heat transfer, using an impulsed multi-orifice mist-generator that produces uniform droplets. Their conclusions are as follows:

1. Horizontally impinging sprays give a lower heat transfer at film boiling than the corresponding vertical impinging sprays.
2. The film boiling heat transfer is mainly controlled by the droplet mass flux.
3. At low droplet mass flux and low droplet Weber number, the heat transfer increases with the droplet Weber number.
4. At high droplet Weber number or high droplet mass flux, droplet Weber number does not significantly affect heat transfer.

TABLE 1. Effects of mist-flow parameters ($h=const.d^x V_d^y D^z$)

Tanaka et al.(9) [experimental]	x=0,	y=0,	z=0.7
Maki et al.(4) [experimental]	x=0,	y=7/4,	z=1/9
Yanagi et al.(10) [experimental]	x=---,	y=0.75,	z=0.75
Shoji(5) [experimental]	x=0.717,	y=1.55,	z=0.255
Ishigai et al.(6) [experimental]	x=-0.259,	y=-0.319,	z=1
Fujibayashi et sl.(11) [experimental]	x=0,	y=0.38,	z=0.64
Awonorin(2) [experimental]	x=-0.5,	y=0.5,	z=---
Ito et al.(7) [analytical]	x=-0.5,	y=-0.5,	z=1

Following these results, we assume in this paper that the droplet diameter and velocity have much weaker influences on heat transfer than the droplet mass flux.

2.2 Effect of Droplet Mass Flux

For a single stream of impinging droplets, Shoji et al.(15) showed that the heat transfered to a droplet, in the high-temperature region can be estimated by the following equation.

$$Q=C(\pi d^3/6)\rho_l c_{pl}\Delta T_{sub}, \tag{2}$$

where ρ_l, c_{pl}, ΔT_{sub}, and C are the liquid density, the liquid specific heat, the droplet subcooling, and a constant respectively; their recommended value of C is 0.5.
For a mist flow, as stated above, Choi and Yao showed that the heat flux in the high temperature-region of dilute mists is mainly controlled by the droplet mass flux. Thus, the heat flux in the high-temperature region can be estimated by Eq.(3), provided the following assumptions are valid:

1. The effects of d and V_d on mist-cooling heat transfer is not dominant.
2. The effect of interaction between impinging droplets is negligible.
3. The total heat flux to mist flow can be estimated as a summation of droplet heat flux q_d, convective heat flux q_a, and radiative heat flux q_r.
4. Equation (2) gives the heat transfer to a single droplet.

$$q_w=q_d+q_a+q_r=C(\rho_l c_{pl}\Delta T_{sub})D+q_a+q_r \tag{3}$$

The mist-cooling heat-transfer coefficient predicted by Eq.(3) is plotted in Fig.4 together with the experimental data of mist-cooling heat transfer(7), (8),(16). Here, C was taken as 1.0 for upward-facing horizontal flat plates and as 0.5 for vertical plates because, for upward-facing horizontal flat plates, individual droplets can impinge repeatedly on the plate. Figure 4 shows that Eq.(3) works well as a first approximation in predicting the mist-cooling heat transfer in the high temperature region. A detailed discussion of this predicting method is

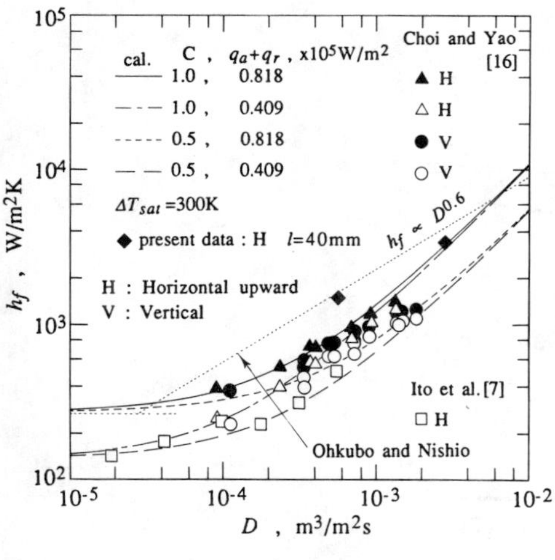

FIGURE 4. Prediction of heat transfer of dilute mist flow in high-temperature region

presented in ref.(8).
From Fig.4, the following conclusions can be drawn. First, in the high-temperature region, the droplet heat transfer q_d is dominated by sensible heat transfer to droplets and not by latent heat transfer. This is true because Eq.(3) does not include the latent heat of vaporization h_{lv}. Second, in dilute mist cooling, each droplet contributes almost independently to the droplet heat transfer because q_d is proportional to D.
The first conclusion has been confirmed by studies of highly-subcooled film-boiling heat transfer. The second conclusion does not seem to hold for dense sprays: droplets tend to form a liquid film as the droplet mass flux increases. For example, as shown in Fig.5, Monde(17) showed that the maximum heat flux for dilute mists at the liquid saturation temperature, $q_{max,dm}$ is roughly equal to $0.8\rho_l h_{lv} D$. But for dense sprays, the maximum heat flux is much less than the $q_{max,dm}$.
Finally, a set of statistical correlating equations, which may be useful is given below. They were obtained by The Iron and Steel Institute of Japan(18).

$$
\begin{aligned}
h &= 10^{10.74} Tw^{-3.56} D^{0.622} V_d^{0.0378}, && \text{for } 200^{\circ}C < Tw < 450^{\circ}C \\
h &= 10^{3.33} Tw^{-0.857} D^{0.662} V_d^{0.308}, && \text{for } 450^{\circ}C < Tw < 600^{\circ}C \\
h &= 10^{1.40} Tw^{-0.136} D^{0.629} V_d^{0.273}, && \text{for } 600^{\circ}C < Tw
\end{aligned}
\qquad (4)
$$

where the units of h, T_w, D, V_d are kcal/m^{2}hK, $^{\circ}$C, l/m^{2}min, m/s respectively.

3. EFFECTS OF SURFACE PARAMETERS

Baumeister and Hamill(19) and Nishio and Hirata(20) experimentally showed that the Leidenfrost temperature was a function of thermal properties of the surface material; the poorer the surface thermal conductance the higher the temperature. For boiling, Nishio(21) showed for liquid nitrogen that a thin layer of insulating material on a surface would raise the minimum heat flux temperature; the thicker the coating-layer the higher the temperature rise. Thus, one would expect the mist-cooling heat transfer to depend on the thermal properties of a heat transfer surface. In addition, the surface parameter, wettability is expected to have strong effects on mist-cooling heat transfer, because the minimum heat flux temperature T_M of mist cooling is affected by surface oxidizing. This section summarizes our experimental results on the effects of these surface parameters on mist-cooling heat transfer.

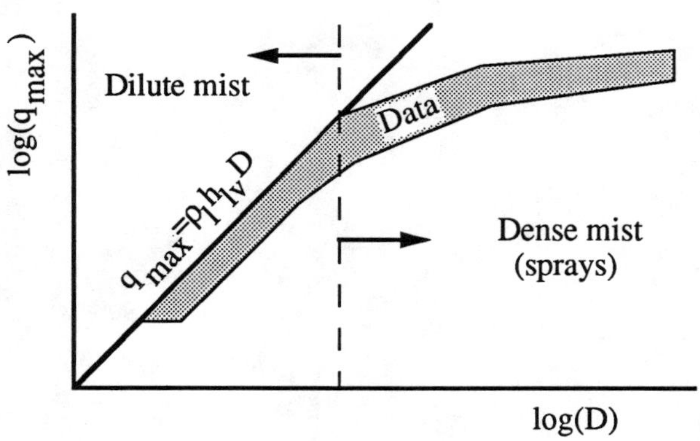

FIGURE 5. Relation between maximum heat flux and droplet mass flux

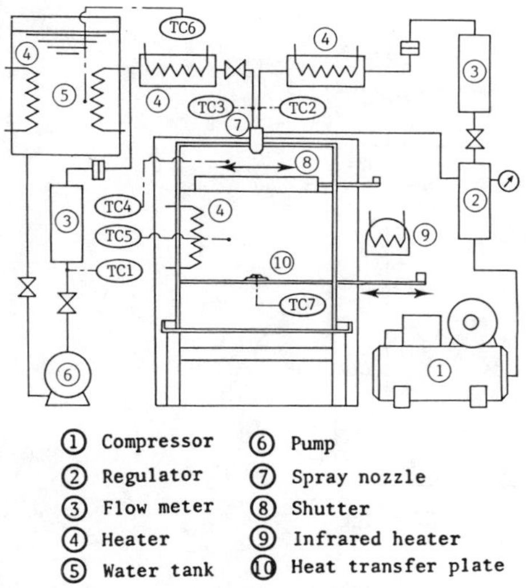

FIGURE 6. Experimental apparatus

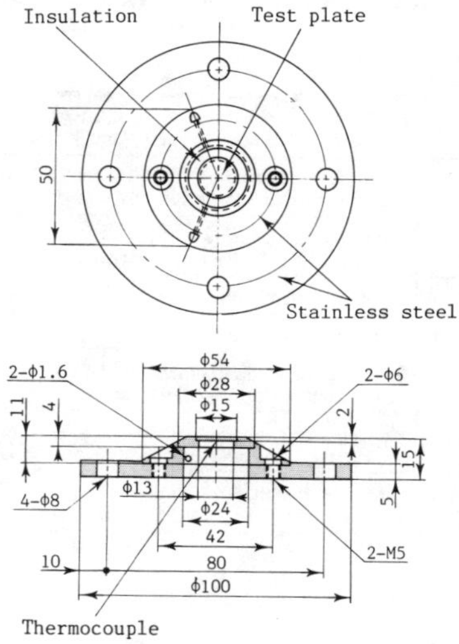

FIGURE 7. Heat transfer surface

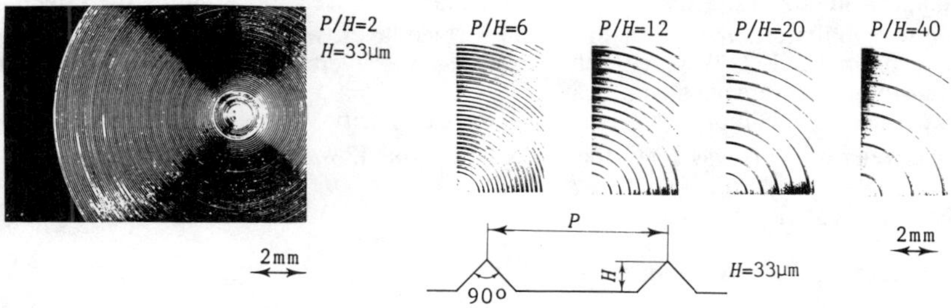

FIGURE 8. Spiral grooves simulating surface roughness

3.1 Experimental Apparatus and Procedure

In this experiment, the effects of surface parameters on mist-cooling heat transfer were experimentally investigated by independently changing the surface roughness, the surface wettability, and surface thermal-resistance-layer thickness.

Figure 6 shows a schematic diagram of the experimental apparatus, which uses a full cone pneumatic atomizing nozzle (Fuso-Seiki, Lumina PR-10f1.5) to produce a water-air mixture as a mist flow. The hydrodynamic aspects of the mist flow are measured as described in ref.(1). In the apparatus, the mist-flow impinges on a heat transfer plate 300mm away from the nozzle. This heat transfer plate was fitted into a diatomite holder with a ceramic bond as shown in Fig.7. We used plates made of silver (99.9% pure), aluminum, nickel, stainless steel (SUS303) and fused quartz. The typical dimensions of a plate were 15mm in diameter and 1.5mm in thickness. Mist cooling experiments were conducted by a quenching method. The temperature of the heat transfer plate was measured by a CA thermocouple attached to

TABLE 2. Methods of surface treatment

No.	method of surface treatment	θ_s
1	Surface treated by chromate process ($Cr(OH)_3 \cdot Cr(OH) \cdot CrO_4$); surface layer thickness is below 0.7μm.	70
2	Etched surface (etched with 5% NaOH at 50°C for 2min., and after cleaning, neutralized with 30% HNO_3 at 15°C for 1min.); *Ra*=0.50μm.	62
3	Mirror-finished surface (polished with polishing cloth and γ-Al_2O_3(ϕ40nm)); *Ra*=0.01μm.	60
4	Smoothed surface (polished with buff); *Ra*=0.23μm.	55
5	Machined surface; *Ra*=0.45μm.	48
6	Sulphuric acid anodized surface (and boiled with distilled water); surface layer thickness is below 0.7μm.	46
7	Phosphoric acid anodized surface (and boiled with distilled water); surface layer thickness is below 0.7μm.	37
8	Phosphoric acid anodized surface; surface layer thickness is below 0.7μm.	37

the back side of the plate. The heat transfer characteristics were calculated by the lumped heat-capacity approximation in the cases of silver, aluminum, and nickel plates, but they were calculated by an inverse heat conduction method in the cases of stainless steel and fuzed quartz plates.

To examine the effect of surface roughness, regular spiral grooves (Fig.8) were made on the stainless steel and the sliver heat transfer plates(22). Next, to examine the effects of surface wettability and the thermal resistant surface layer, we used(23) treated aluminum plates listed in Table 2. To prepare thermal resistance layers of different thicknesses, the heat transfer plates were anodized.

The experimental conditions were as follows: the air velocity V_a was 20m/s, the temperature T_l of the water droplets was 21°C, the droplet mass flux D was 0.0003-0.005m^3/m^2s.

3.2 Experimental Results

3.2.1 Effect of surface roughness

The effect of surface roughness on mist-cooling heat transfer is shown in Fig.9, where H and P denote the roughness height and pitch. The figure shows that the roughness height and pitch-to-height ratio(P/H) do not have strong effects on the mist-cooling heat transfer. Yet, in the cooling process, small differences in the mist-cooling heat flux add and cause a significant change in the temperature-time curve. This effect is shown in Fig.10, where the cooldown time from 700°C to 150°C is plotted against the roughness height and pitch-to-height ratio. The cooldown time is scaled by that of the smooth plates. When either the roughness pitch or the pitch-to-height ratio increases, the scaled cooling time reaches a maximum at a certain value of the respective variable and becomes nearly constant after that value.

3.2.2 Effect of surface wettability

It is difficult to describe the surface wettability quantitatively. A measure of the surface wettability is the contact angle, which can be measured in many ways: equilibrium, static advancing or receding, dynamic advancing or receding, saturated contact angles. According to Tamai et al.(24), the saturated contact angle is a measure of adsorption energy of the surface; the saturated contact angle θ_s is defined as the contact angle for which the surface contamination by readsorption reaches the state of saturation. Thus, θ_s was chosen as the measure of the surface wettability in this paper.

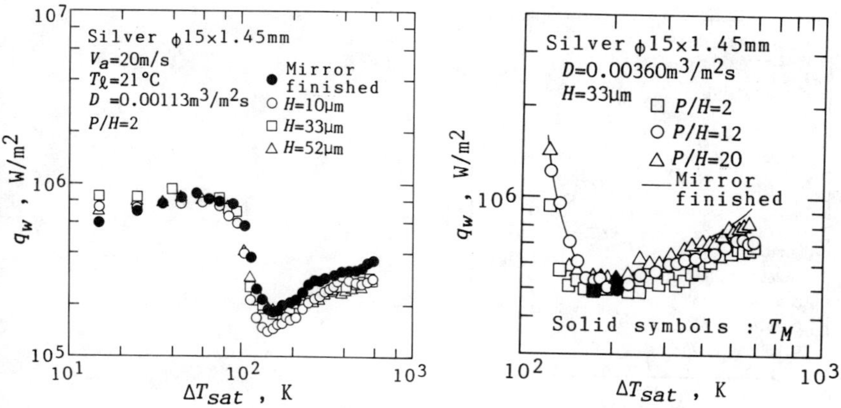

FIGURE 9. Effect of surface roughness on mist-cooling heat transfer

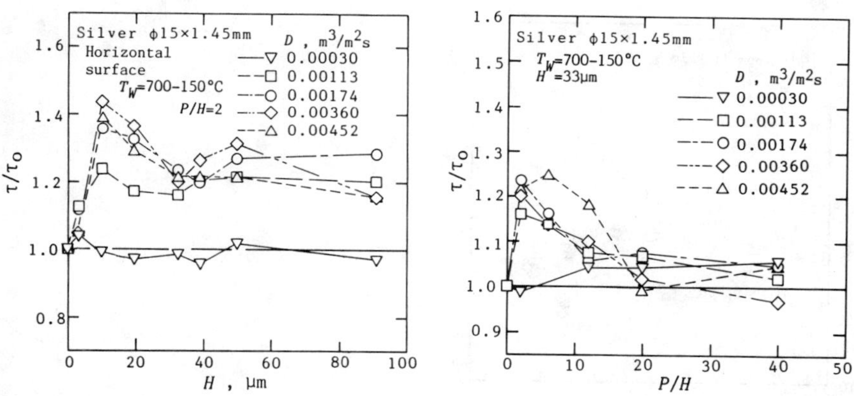

FIGURE 10. Effect of surface roughness on cooldown time

The effects of surface wettability are presented in Fig.11. The figure shows that the effect of surface wettability primarily appears in the minimum heat flux superheat ΔT_M and in the transition region, where the heat flux increases with increasing surface wettability.

3.2.3 Effect of thermal characteristics of surface layer

The thermal characteristics of the heat-transfer surface layer were varied by using aluminum plates having different anodized layer thicknesses. Since the anodized layer has a much lower thermal conductivity than that of aluminum, it offers a thermal resistance that increases with increasing the anodized layer thickness.

Figure 12 shows our experimental results. The surface layer thickness has an effect similar to the surface wettability: both the minimum-heat-flux temperature and the heat flux in the transition region increase markedly when the surface-resistance-layer thickness increases.

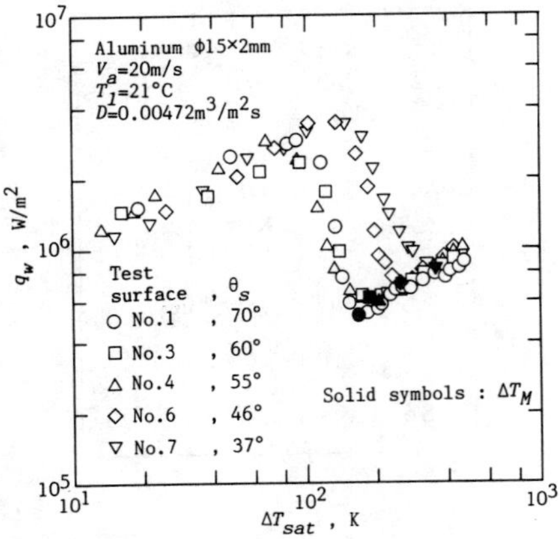

FIGURE 11. Effect of surface wettability on mist-cooling heat transfer

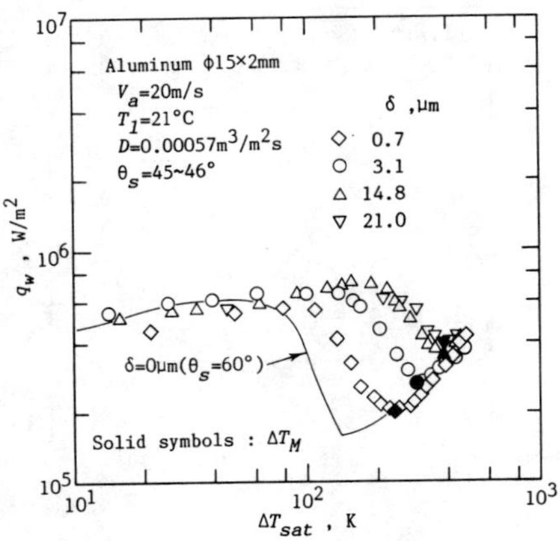

FIGURE 12. Effect of surface thermal-resistance-layer on mist-cooling heat transfer

4. CONCLUDING REMARKS

In this paper, we presented a brief review of the parametric effects on mist-cooling heat transfer. The mist-flow parameters we considered were droplet diameter, velocity, and droplet mass flux. Of these mist-flow parameters, mainly the droplet mass flux (or the sensible heat transfer to droplets) controls the droplet heat transfer in dilute mist-cooling, in the high-temperature region. On the other hand, both of the two surface-parameters, wettability and thermal-resistance of surface layer, considerably affect the minimum heat flux temperature and heat fluxes in the transition region. The surface roughness height and pattern do not have strong effects on mist-cooling heat transfer. But a discrete distribution of high surface protrusions will stabilize the cooling curve against changes in the roughness height and pattern. These conclusions suggest that in stabilizing the cooling process of the TMC process, it is effective to roughen surfaces of steel plates and control the degree of surface oxidizing which changes wettability and thermal conductance of the surface layer.

REFERENCES

1. Ohkubo, H. and Nishio, S., Mist Cooling for Thermal Tempering of Glass, *JSME International Journal, Series II*, vol.31, no.3, pp.444-450, 1988.

2. Awonorin, S.O., Film Boiling Characteristics of Liquid Nitrogen Sprays on a Heated Plate, *International Journal of Heat & Mass Transfer*, vol.32. no.10, pp.1853-1864, 1989.

3. Chyu, M.-C. and Bergles, A.E., An Analytical and Experimental Study of Falling-Film Evaporation on a Horizontal Tube, *Journal of Heat Transfer*, vol.109, no.4, pp.983-990, 1987.

4. Maki, H., Tanaka, Y. and Maezawa, M., Study of Spray Cooling, *Proceedings of the 57th JSME Fall Annual Meeting*, no.790-16, pp.295-302, 1979, (in Japanese).

5. Shoji, M., Experimental Study of Spray Cooling, *Proceedings of the 15th National Heat Transfer Symposium of Japan*, vol.15, pp.187-189, 1978, (in Japanese).

6. Ishigai, S., Nakanishi, S. and Ochi, T., Spray Cooling of a Hot Surface, *Proceedings of the 16th National Heat Transfer Symposium of Japan*, vol.16, pp.316-318, 1979, (in Japanese).

7. Ito, T., Takata, Y. and Liu, Z.-H., On the Water Cooling of Hot Surfaces(Analysis of Fog Cooling in the Region Equivalent to Film Boiling), *Transaction of JSME, Series B*, vol.55, no.511, pp.805-813, 1989, (in Japanese).

8. Ohkubo, H. and Nishio, S., Study of Transient Heat Transfer Characteristics of Mist Cooling from Horizontal Upward Facing Surface , *Transaction of JSME, Series B*, to be published, (in Japanese).

9. Tanaka, S., Ohmori, S. and Sugimoto, M., Study on Water Cooling of Hot Steel Sheet, *Technical Report of Mitsubishi Heavy Industry*, vol.2, no.2, pp.192-199, 1965, (in Japanese).

10. Yanagi, K., Setoguchi, K., and Hayashi, K., Study on the Characteristics of Fog Made by Newly Developed Nozzle, *Technical Report of Mitsubishi Heavy Industry*, vol.9, no.6, pp.792-798, 1972, (in Japanese).

11. Fujibayashi, A., Teramoto, T., Fukuda, S. and Hirata, M., Cooling of Steel by a Superheated Liquid Jet, *Proceedings of the 22nd National Heat Transfer Symposium of Japan*, vol.22, pp.70-72, 1985, (in Japanese).

12. Wachters, L.H.J, and Westerling, N.A.J., The Heat Transfer from a Hot Wall to Impinging Water Drops in the Spheroidal State, *Chemical Engineering Science*, vol.21, pp.1047-1056, 1966.

13. Pedersen, C.O., An Experimental Study of the Dynamic Behavior and Heat Transfer Characteristics of Water Droplets Impinging upon a Heated Surface, *International Journal of Heat & Mass Transfer*, vol.13, pp.369-381, 1970.

14. Ueda, T., Enomoto, T. and Kanetsuki, M., Heat Transfer Characteristics and Dynamic Behavior of Saturated Droplets Impinging on a Heated Vertical Surface, *Bulletin of JSME*, vol.22, no.167, pp.724-732, (1979).

15. Shoji,M., Wakunaga, T. and Kodama, K., Heat Transfer from a Heated Surface to Impinging Subcooled Droplets(Heat Transfer Characteristics in Non-Wetting Region), *Transaction of JSME, Series B*, vol.50, no.451, pp.716-723, 1984, (in Japanese).

16. Choi, K.J. and Yao, S.C., Mechanisms of Film Boiling Heat Transfer of Normally Impacting Spray, *International Journal of Heat & Mass Transfer*, vol.30, no.2, pp.311-318, 1987.

17. Monde.M, Study of Critical Heat Flux in a High Heat-Flux System with Water Sprays, *Transaction of JSME*, *Sires B*, vol.45, no.394, pp.849-858, 1979, (in Japanese).

18. *Forced Cooling of Steel*, pp.98-101, The Iron and Steel Institute of Japan, (1988), (in Japanese).

19. Baumeister, K.J. and Simon, F.F., Leidenfrost Temperature-Its Correlation for Liquid Metals, Cryogens, Hydrocarbons, and Water, *Journal of Heat Transfer*, vol.95. no.2, pp.166-173.

20. Nishio, S. and Hirata, M., Direct Contact Phenomena between a Liquid Droplet and High Temperature Solid Wall, *Proceedings of the 6th International Heat Transfer Conference*, vol.1, pp.245-250, 1978.

21. Nishio, S., Cooldown of Insulated Metal Plates, *Proceedings of 1983 ASME-JSME Thermal Engineering Joint Conference*, vol.1, pp.103-109, 1983.

22. Ohkubo, H. and Nishio, S., Study on Accurate Prediction on Mist Cooling Characteristics (Part I, Effects of Surface Roughness), *Heat Transfer-Japanese Research*, vol.18, no.5, pp.43-50, 1989.

23. Ohkubo, H. and Nishio, S., Study on Accurate Prediction of Heat Transfer Characteristics of Mist Cooling(Effects of Surface Wettability), *JSME International Journal*, *Series B*, vol.33, no.2, pp.326-332, 1990.

24. Tamai, Y., Matsunaga, T., and Suzuki, K., The Effect of Water Vapor on Contamination of Metallic Oxide Surface, *Bulletin of Chemical Society of Japan*, vol.50, no.7, pp.1181-1182, (1977).

Behavior of Minimum Heat Flux (MHF) Point and Surface Effect in Transient Cooling Process of Water-Carbon Steel System

KYOUHEI MURATA
Heat and Energy Technology Lab.
R&D Laboratories-III
Nippon Steel Corporation
20-1 Shintomi Futtsu, Chiba-Ken 299-12, Japan

1. INTRODUCTION

Cooling of steel during heat treatment is mainly a transient cool ing process of an atmospheric subcooled water-carbon steel system. Quench-and-temper was a principal method of heat treatment for steel in the past. In recent years, an on-line thermomechanical treatment process, called the thermomechanical control process(TMC P), has become predominant, resulting in a sharp increase in the production of heat treated steels.

The heat transfer surface conditions have drastically changed and become diverse with the TMCP. In the conventional quenching process, the heat transfer surface is normally covered by an oxide film. In the TMCP, a fresh surface revealed by rolling works as the heat transfer surface. There, the oxide film is very thin and reveals the surface roughness transferred from the work roll.

When we take a look at an overall picture of heat treating processes currently used for steel products, we can point out that heat transfer surface conditions have greatly diversified, mainly with respect to oxidation and surface roughness.

In the cooling process during heat treatment, a temperature dispersion, called the unstable boiling cooling phenomenon, often occurs as shown in Fig.1. This phenomenon may cause shape and size-defects(due to uneven thermal stress) in heat-treated products after cooling and may bring about variations in material properties in some steel products.

Fig.1 shows typical dispersions of cooling curve groups obtained when 220-mmOD and 12.7-mm wall seamless steel pipe was fit with eight thermocouples at equal spacings on the inside circumference and was cooled under uniform heating temperature and uniform water spray conditions(ΔT_{sub}=86.5K). The mild cooling portions of curve groups ② and ③ in the initial stage of cooling correspond to film boiling. Curve group ① involves little film boiling, and cooling is considered to start with transition boiling. A maximum temperature deviation tends to occur when film boiling is coexistent with non-film boiling in the cooling curve groups.

The following research themes on heat transfer engineering may be noted from Fig.1:

1) In the transient high-subcool water cooling process from high temperature, the temperature dispersion originates in the ini-

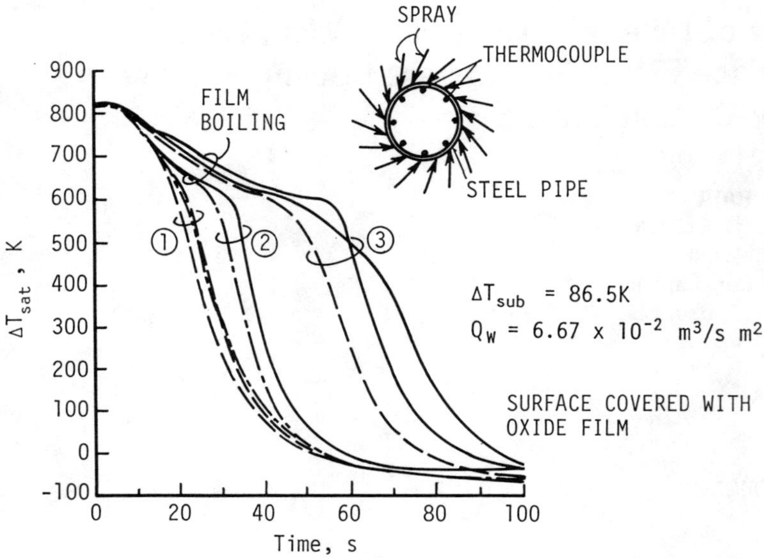

FIGURE 1. Cooling curves at inside surface of seamless steel pipe illustrating unstable boiling cooling phenomenon

tial stage of cooling. There is a possibility that the surface effect may be present in the boiling heat transfer phenomenon under industrial surface conditions.

2) There is a possibility that the initiation of film boiling and minimum heat flux(MHF) point are influenced by the surface condition.

3) There is a possibility that film boiling heat transfer is influenced by the surface condition. If the slope of the portion of the cooling curve that corresponds to film boiling varies with the surface condition as shown in Fig.1,it is expected that the time of occurrence of the quench point varies to cause the unstable boiling cooling phenomenon.

To verify the above possibilities,it is necessary to clarify the behavior of the MHF point on the boiling curve of the water-carbon steel system and the associated surface effect.
The following findings are obtained about film boiling heat transfer and MHF point behavior for various saturated liquids, including water:

1) Heat transfer by film boiling is not influenced by the oxidation,cleanliness,or roughness of the heat transfer surface[1-3].

2) The MHF point shifts towards the high wall superheat and high heat flux side as the oxidation,fouling,and wettability of the heat transfer surface improve[1,2].

Empirical equation (1) of Dhir and Purohit is famous for the minimum film boiling temperature Tmin of subcooled water[4]:

$$\Delta Tmin=101+8\Delta Tsub \quad (1)$$

A problem with this empirical equation is that the minimum film boiling temperature Tmin exceeds the hydrodynamic Tmin value of Berenson,thermodynamic limiting liquid superheat of Spiegler et. al.and Lienhard,and critical temperature of water[5-7].
Nishio found that Dhir and Purohit-type dependence holds when liquid-solid contact occurs first at the peripheral edge of the heat transfer surface,called the preferential collapse of the vapor film,and then such contact zone propagates from the periphery to the center[8].
Kikuchi et.al. clarified that ΔTmin is the initial superheat of the heat transfer surface when the solid-liquid contact surface temperature reaches the thermodynamic limiting liquid superheat due to the heat-resistant layer effect of the heat transfer surface[9]. The phenomenon of boiling heat transfer in the cooling process during heat treatment as shown in Fig.1 cannot be fully explained by the past findings.
The purpose of this study was to experimentally clarify the behavior of the MHF point and the surface effect in the forced flow and transient cooling process of the atmospheric subcooled water-carbon steel system. The subcool of water and the oxidation and surface roughness conditions of the heat transfer surface were selected as experimental parameters.

2. EXPERIMENTAL APPARATUS AND PROCEDURE

The boiling curves of an atmospheric subcooled tap water-carbon steel(0.14%C,0.19%Si,1.0%Mn,0.018%P,0.006%S)system were measured under impinging jet and bottom surface cooling conditions.

2.1 Experimental Apparatus

Fig.2 schematically shows the apparatus used to measure the boiling curves under the conditions described above. The sample used for temperature measurement in the cooling process was 25.0mm in outside diameter and 3.00±0.015mm in thickness. To investigate the effect of oxidation of the heat transfer surface,the heating chamber was maintained in a nitrogen gas atmosphere,and the oxygen concentration in the heating chamber was maintained at three levels of 0.01,0.4 and 1.0%. Another experiment was conducted in an atmosphere of 0.002% oxygen concentration in a different apparatus of the sealed type. Its details are omitted here.
The thermocouple welded to the insulated top surface of the sample was a sheathed chromel-alumel thermocouple and measured 1.0 mm in outside diameter and 0.18 mm in wire diameter. The accuracy of temperature measurement was±6K,including temperature error due to sample thickness error.
The sample was heated in the upper heating chamber to 1188K for specified time,was lowered out of the heating chamber and left to cool in air for 4 s,and its water cooling was started at 1153K.

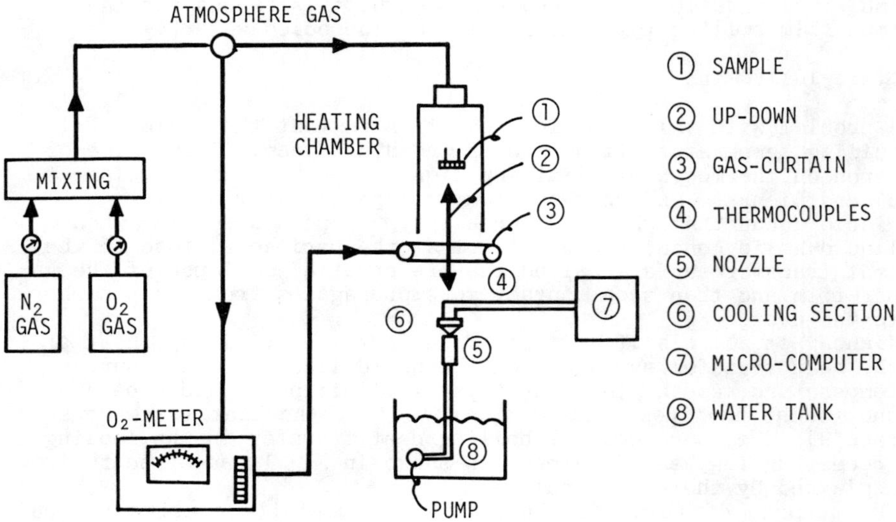

FIGURE 2. Schematic diagram of experimental apparatus for measuring heat transfer

The heat transfer surface of the sample faced downward and cooled by an upward impinging water jet(average water flux of 0.395 $m^3/s\ m^2$ for the total heating surface). The subcool of the cooling water was 10 to 80K.

2.2 Surface Condition of Sample

The sample had their surfaces artificially conditioned to a roughness range from mirror finish to extreme roughness of Rmax=50μm. The conditioning methods employed were buffing,shotblasting, abrasive wheel grinding,and NC lathe turning. The surface roughness produced by shotblasting consists of pits,while that produced by the other methods except buffing consists of linear V-shaped grooves.

2.3 Method for Calculating Back Boiling Curves

The boiling curves were calculated back from the cooling curves using the program of Ariyoshi and Fukuda for analysis of inverse heat conduction problems by the finite element-least square method[10]. The inverse calculation error of heat flux,including the error resulting from the temperature measurement error,is estimated at ±5%.
Whether or not film boiling occured in the initial stage of cooling was determined by reffering to whether or not any film boiling portion existed in the boiling curve,as well as to whether or not any mild cooling portion existed in the cooling curve.

3. EXPERIMENTAL RESULTS

3.1 Minimum Heat Flux

Typical boiling curves of the water-carbon steel system are shown in Fig.3. In Fig.4, the minimum heat flux qmin is shown as a function of water subcool with surface roughness taken as a parameter. Under all surface conditions except Rmax=20μm, the MHF point does not occur for highly subcooled water, and qmin exhibits water subcool dependence of the exponential type. Under the surface condition of Rmax=20μm, the MHF point occurs for highly subcooled water, and qmin exhibits water subcool dependence of the linear type. Approximate empirical equations for these conditions are given in Table 1.

Fig.5 shows the effect of oxygen concentration in the heating atmosphere on the minimum heat flux qmin. Where the oxygen concentration is not less than 0.01%, qmin moderately increases with the oxidation of heat transfer surface. The above experimental results confirmed that qmin in the subcooled water-carbon steel system is influenced by surface roughness and oxidation.

3.2 Minimum Film Boiling Temperature

Fig.6 shows the minimum film boiling temperature △Tmin as a function of water subcool with the surface condition of the sample taken as a parameter. Where film boiling occurs, △Tmin is practi-

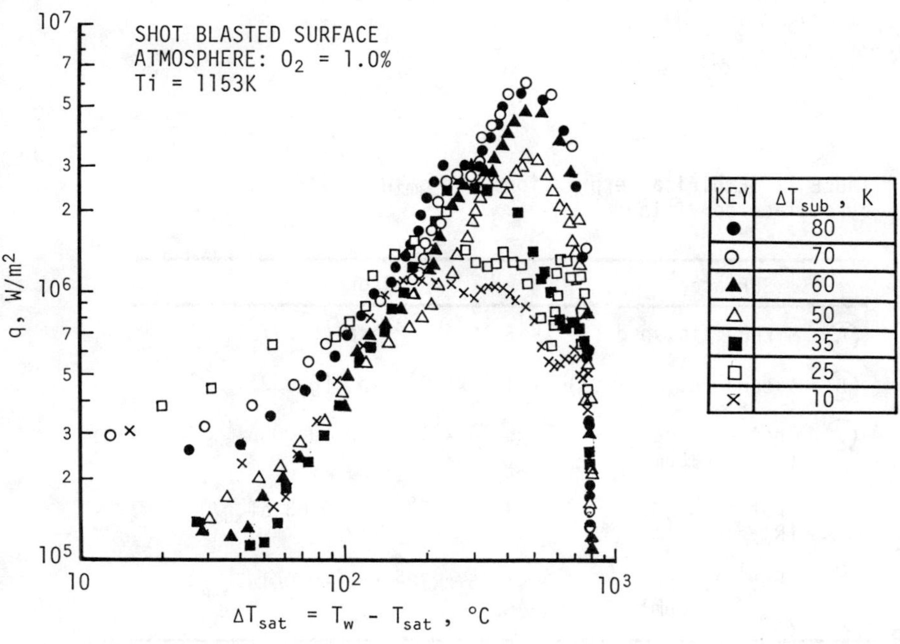

FIGURE 3. Typical boiling curves of subcooled tap water-carbon steel combination system

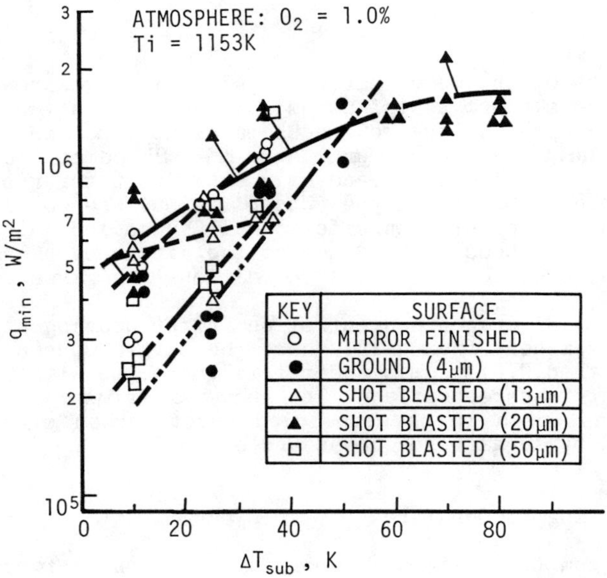

FIGURE 4. Effects of subcooling and surface condition on minimum heat flux of tap water-carbon steel combination system

TABLE 1. Empirical expressions for $q_{min} = f(\Delta T_{sub})$ by surface condition

	Surface	Expression
Ⓐ	Mirror finished	$3.12 \times 10^5 \exp(0.037\Delta T_{sub})$
Ⓑ	Ground ($R_{max} \sim 4\mu m$)	$1.0 \times 10^5 \exp(0.053\Delta T_{sub})$
Ⓒ	Shot blasted ($R_{max} \sim 13\mu m$)	$5.14 \times 10^5 \exp(0.009\Delta T_{sub})$
Ⓓ	Shot blasted ($R_{max} \sim 20\mu m$)	$4.4 \times 10^5 (1 + 0.036\Delta T_{sub})$
Ⓔ	Shot blasted ($R_{max} \sim 50\mu m$)	$1.66 \times 10^5 \exp(0.041\Delta T_{sub})$

For applicable scope of ΔT_{sub}, see Fig. 4

cally independent of water subcool and sample surface condition, and is about 500 to 630K.

Fig.7 shows the effect of oxygen concentration in the heating atmosphere on the minimum film boiling temperature ΔTmin. the oxygen concentration dependence of ΔTmin was found to change drastically as the oxygen concentration crosses the boundary of about 0.01%. when the oxygen concentration in the heating atmosphere is 0.002%, ΔTmin falls to an intermediate level between the limiting water superheat and hydrodynamic ΔTmin.

3.3 Condition for Initiation of Film Boiling

Table 2 summarizes the effects of water subcool and surface condition on the occurrence of film boiling. When the water has a subcool of 35K or lower, film boiling occurs irrespective of the surface condition of the sample. When the water has a subcool of 60K or over, film boiling hardly occurs on the whole. It should be noted that film boiling selectively and stably occurs even with high-subcool water under the sample surface condition of Rmax=20 μm(pits).

Table 3 shows that the surface roughness range permitting the occurrence of film boiling tends to expand when the oxygen concentration in the heating atmosphere is reduced to 0.01%.

Table 4 shows that film boiling does not occur on a linear-grooved surface even when the oxygen concentration in the heating atmosphere is low at 0.01%. There are cases of film boiling occurring when the oxygen concentration in the heating atmosphere is 0.4 to 1.0%. This film boiling occurred evidently because the oxide scale formed on the sample during heating was removed in

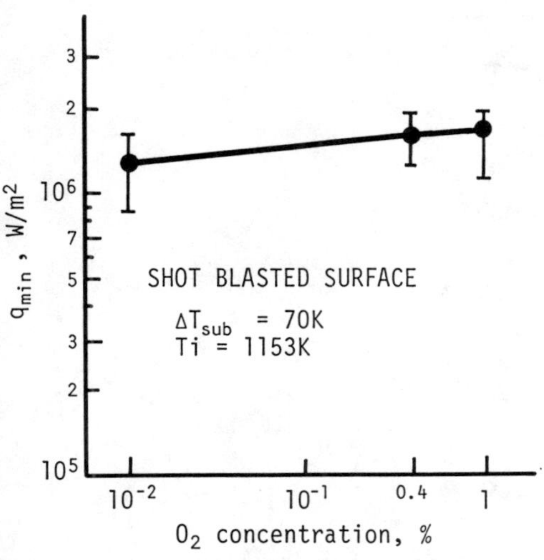

FIGURE 5. Effect of oxygen concentration in heating atmosphere on q_{min} of tap water-carbon steel combination system

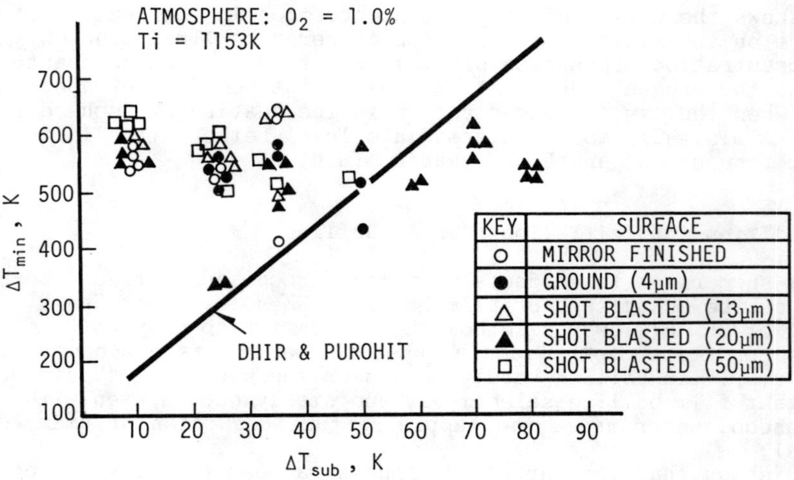

FIGURE 6. Effects of subcooling and surface condition on minimum film boiling temperature of tap water-carbon steel combination system

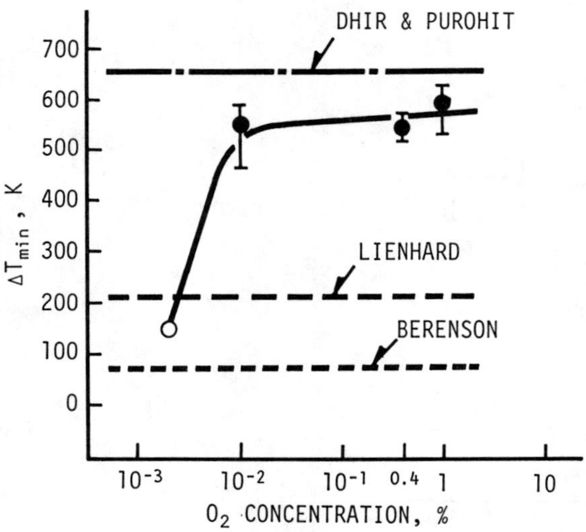

FIGURE 7. Effect of oxygen concentration in heating atmosphere on minimum film boiling temperature of tap water-carbon steel combination system

TABLE 2. Effects of water temperature and surface condition on occurrence of film boiling (O_2 concentration: 1.0%)

Surface	Subcooling						
	10K	25K	35K	50K	60K	70K	80K
Mirror finished	×	×	×	○	○	○	○
Ground (4μm)	×	×	×	×	○	○	○
Shot blasted (13μm)	×	×	×	○	○	○	○
Shot blasted (20μm)	×	×	×	×	×	×	×
Shot blasted (50μm)	×	×	×	○	○	○	○

Legend

○ : No film boiling

× : 50% or more occurrence of film boiling

TABLE 3. Effects of oxygen concentration in heating atmosphere and surface condition on occurrence of film boiling (ΔT_{sub} = 70K)

Surface	O_2 concentration		
	0.01%	0.4%	1.0%
Mirror finished	—	○	○
Ground (4μm)	△	○	○
Shot blasted (13μm)	×	○	○
Shot blasted (20μm)	×	△	×
Shot blasted (50μm)	×	○	○

Legend

○ : No film boiling

△ : ∿ 25% occurrence of film boiling

× : 50% or more occurrence of film boiling

TABLE 4. Effects of oxygen concentration in heating atmosphere and surface condition on occurrence of film boiling (ΔT_{sub} = 70K, linear V-shaped grooves)

	O_2 concentration		
Surface	0.01%	0.4%	1.0%
A (4μm)	○	○	○
B (6μm)	○	○	○
C (12μm)	○	×	△
D (25μm)	○	○	×
E (35μm)	○	○	△
F (50μm)	○	○	×

Legend

○: No film boiling

△: ~ 25% occurrence of film boiling

×: 50% or more occurrence of film boiling

the subsequent cooling process.

These experimental results confirmed that the initiation of film boiling in the transient cooling process depends on the subcool of water and the geometry of surface roughness and oxidation(wettability and oxide film) of the sample.

4. DISCUSSIONS

4.1 Surface Condition Dependence of Film Boiling Heat Transfer

The minimum heat flux qmin clearly changes with the surface roughness condition of the sample as shown in Fig.4,while the minimum film boiling temperature Tmin depends little on the surface roughness condition of the sample as shown in Fig.6. It is inferred from these findings that film boiling heat transfer changes with the surface condition of the sample. For,the emissivity of engineering heat transfer surfaces is presumed to change with their conditions.

In Fig.1,the cooling curve groups of steel pipe differ in the slope of the portion corresponding to film boiling. The above discussion indicates that this phenomenon can actually occur under engineering surface conditions.

4.2 Rise in Minimum Film Boiling Temperature with Oxidation

Fig.7 clearly shows that the minimum film boiling temperature exceeds the limiting liquid superheat and critical temperature of water when the heat transfer surface is oxidized.
When the oxygen concentration in the heating atmosphere is 0.01% or higher, an oxide film of such thickness as to act as an heat-resistant layer is formed, and $\Delta Tmin$ rises by the aforementioned mechanism proposed by Kikuchi et.al.[9].
When the oxygen concentration in the heating atmosphere is below 0.01%, an oxide film of such thickness as to act as a heat-resistant layer is not formed, and the wettability of the heat transfer surface is expected to change drastically according to the oxidized condition of the sample.
An outstanding question for future study is whether or not the minimum film boiling temperature of a completely oxidation-free heat transfer surface agrees with the hydrodynamic $\Delta Tmin$ of Berenson, or the determination of the lower limit of $\Delta Tmin$.

4.3 Condition for Initiation of Film Boiling

Kokado et.al. found in their transient impinging jet cooling experiment of a tap water-stainless steel(18Cr-8Ni) system that no film-boiling vapor film is formed when the subcool of water is 32 K or over[11]. This temperature agrees fairly well with the experimental results of the present study.
In the cooling of steel during actual heat treatment, however, film boiling often occurs even with high-subcool water, as shown in Fig. 1. It was made clear that this phenomenon is due to the combined effect of the geometry of dents characterizing the surface roughness and the wettability of the heat transfer surface.
It was also found that the unstable boiling cooling phenomenon where film boiling is co-existent can occur on actual engineering surfaces and that this phenomenon is attributable to a surface effect mainly comprising the roughness and oxidation of the heat transfer surface.

5. CONCLUSIONS

The behavior of the minimum heat flux(MHF) point and the initiation of film boiling have been found to be heavily dependent on the oxidation and roughness of the heat transfer surface.
The mechanism by which the unstable boiling cooling phenomenon occurs on the cooling of steel during heat treatment can be explained by these results.
It will assume increasing importance in the future to study the surface effect in boiling heat transfer under various subcooled water-solid combinations, and to deepen research and development on the methods of controlling the boiling curve shape by use of various surface effects.

6. NOMENCLATURE

Tsat=Saturation temperature of water,K
Tmin=Minimum film boiling temperature,K
Ti=Cooling start temperature,K
ΔTsub=Subcool of water,K
ΔTsat=Wall superheat,K
ΔTmin=Tmin-Tsat,K
qmin=Minimum heat flux,W/m^2
Qw=Average water flux density,m^3/s m^2

REFERENCES

1. Bergles, A. E. and Thompson, Jr., W. G., The Relationship of Quench Data to Steady-State Pool Boiling Data, Int. J. Heat Mass Transfer, vol.13, no. 1, pp. 55-68, 1970.

2. Veres, D. R. and Florschuetz, L. W. A Comparison of Transient and Steady-State Pool-Boiling Data Obtained Using the Same Heating Surface, Transacions of ASME,Journal of Heat Transfer, vol.93, no. 2, pp. 229-232, 1971.

3. Berenson, P. J., Experiments on Pool-Boiling Heat Transfer, Int. J. Heat Mass Transfer, vol.5, pp. 985-999, 1962.

4. Dhir, V. K. and Purohit, G. P., Subcooled Film-Boiling Heat Transfer from Spheres, Nuclear Engineering Design, vol.47, no. 1, pp. 49-66, 1978.

5. Berenson, P. J., Film-Boiling Heat Transfer from a Horizontal Surface, Transacions of ASME, Journal of Heat Transfer, vol. 83, no. 3, pp. 351-358, 1961.

6. Spiegler, P., Hopenfeld, J., Silberberg,M., Bumpus, Jr., C.F. and Norman, A., Onset of Stable Film Boiling and the Foam Limit, Int. J. Heat Mass Transfer, vol.6, pp. 987-989, 1963.

7. Lienhard, J. H. Correlation for the Limiting Liquid Superheat, Chemical Engineering Science, vol.31, no. 9, pp. 847-849, 1976.

8. Nishio, S.,Uemura, M. and Sakaguchi, K., Film Boiling Heat Transfer and Minimum-Heat-Flux (MHF)-Point Condition in Sub-cooled Pool Boiling, JSME International Journal, vol.30, no. 226, pp. 1274-1281, 1987.

9. Kikuchi, Y., Hori, T. and Michiyoshi, I., Minimum Film Boil-ing Temperature for Cooldown of Insulated Metals in Saturated Liquid, Int. J. Heat Mass Transfer, vol.28, no. 6, pp. 1105-1114, 1985.

10. Fukuda, K. and Ariyoshi, T., Analysis of Inverse Heat Condu-ction Problem by Finite Element-Least Squares Method, Testu to Hagane, vol.70, no. 8, pp. 869-874, 1984.

11. Kokado, J., Hatta, N., Takuda, H., Harada, J. and Yasuhira, N., An Analysis of Film Boiling Phenomena of Subcooled Water Spreading Radially on a Hot Steel Plate, Archiv für Eisen-hüttenwesen, vol.55, no. 3, pp. 113-118, 1984.

Heat Transfer Characteristics of Roller Quench System in Continuous Annealing Line

SHUZO FUKUDA and YUTAKA OHKUBO
Research and Development Division
NKK Corporation
Kawasaki 210, Japan

Abstract

A new quenching system has been developed in the continuous annealing Line of steel strip. The system consists of cooling the steel strip by contact with internally water-cooled rolls and called Roller Quench System(RQ).

For the development of RQ process, contact thermal conductance between the surface of strip and roll were investigated. Thermal behaviors of roll shell were analyzed through theoretical simulation.

Finally, mathematical model of the RQ process was established, which is useful for both the design and operation of RQ process.

1. Introduction

In continuous annealing line(abbr. CAL), which produces cold rolled steel sheet used for the manufacture of automobile parts, electric appliances etc., cooling(quenching) process is one of the most important parts which decide the mechanical property(ex. drawing quality) of steel sheet.[1)]

Some types of quenching processes has been developed and actually used in industry.

- gas jet cooling (GJ)
- water quench (WQ)
- mist jet cooling (MJ)

These processes have advantages and disadvantages respectively.

WQ and MJ processes use water as coolant, so the strip is oxidized and descaling process is necessary. GJ and MJ processes require huge amount of compressed gas, so the cost of electric power becomes high.

To overcome these disadvantges, NKK has developed the Roll Quenching system(RQ).

As shown in Fig. 1, the RQ process consists of cooling the steel strip by contact with internally water-cooled rolls.

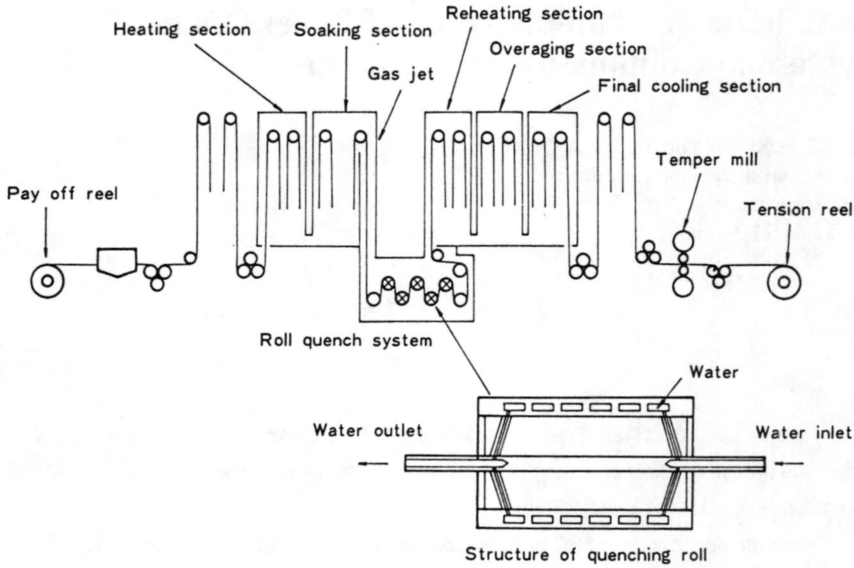

Fig. 1. Line arrangement of the NKK-CAL RQ system.

For the development of RQ process, basic heat transfer data concerned to the contact heat transfer were requested for the design of system.

Contact thermal conductance between the surface of strip and roll were investigated.

Furthermore, because the cooling process is an unsteady process, thermal behaviors of roll shell were analyzed through theoretical simulation in order to predict the final cooling ability of the RQ process. These basic data of contact thermal conductance and the mathematical model of the RQ process are useful for both the design and operation of RQ process.

2. Basic Experiments for Contact Thermal Conductance

2-1. Experimental Procedure

To measure the contact thermal conductance hc, the experimental apparatus shown in Fig. 2 was prepared. The apparatus consists of a pair of axisymmetric metal blocks with the contact surfaces of 10 mm^{ϕ} for heat transfer.

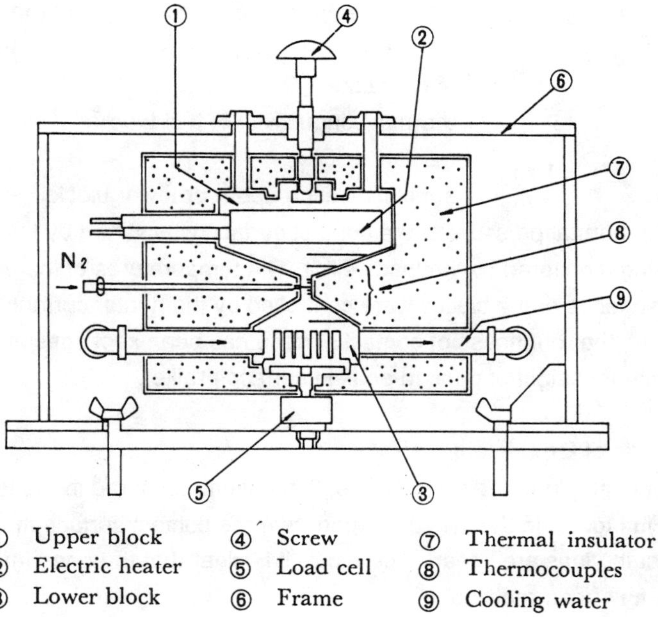

Fig. 2. Experimental apparatus for contact thermal conductance.

The upper block is made of steel and has an electric heater inside, corresponding to the high temperature strip. The lower ones with water-cooled fins are made of steel and copper to correspond to the water-cooled roll shell.

In the cylindrical parts of both blocks with 10 mm diameter, one-dimensional steady heat flux is generated, and the temperature gradients of both cylinders are measured by C-A thermocouples which are inserted into the center of the test blocks.

Further, the heat flux $\dot{q}$ and the contact thermal conductance hc are calculated from the following equations.

$$\dot{q} = -\lambda_u \left(\frac{\Delta T}{\Delta x}\right)_u = -\lambda_l \left(\frac{\Delta T}{\Delta x}\right)_l \qquad \text{.............................(1)}$$

$$h_c = \frac{\dot{q}}{(\theta_u - \theta_l)} \qquad \text{......................................(2)}$$

where, ($\Delta T / \Delta x$): vertical temperature gradient in the cylindrical part (°C/m)

θ : temperature (°C)

λ : thermal conductivity of test blocks (kcal/m·h·°C)

u, l : suffix indicating upper or lower block.

The normal pressure at the contact surface was varied by the screw④in Fig. 2 being monitored by the load cell⑤. The temperatures of the contact heat transfer surfaces of the blocks were controlled by the heater current. In this experiment, the roughness of contact surface has been kept constant; 5 μm as Rz (maximum height of peak to peak in surface profile).

2-2. Results of Experiment

An example of the temperature distribution measured in the test blocks shown in Fig. 3. In the cylindrical part near the contact surface,the temperature distributions are linear. Therefore, it is clear that the one-dimensional heat flux has been realized.

When the materials of the two blocks are different, the temperature gradient is not the same. The values of the heat fluxes calculated from this data coincided well with each other, and reliability of the experiments were confirmed.

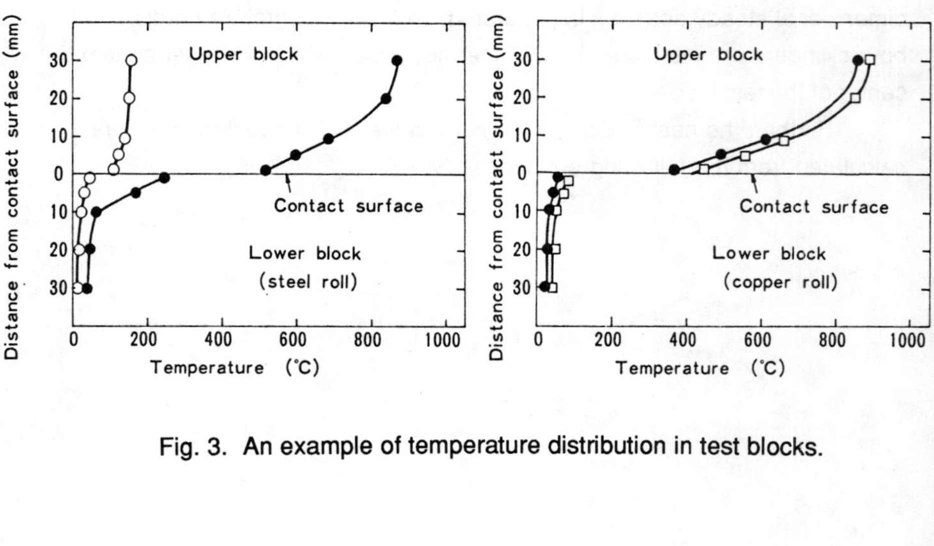

Fig. 3. An example of temperature distribution in test blocks.

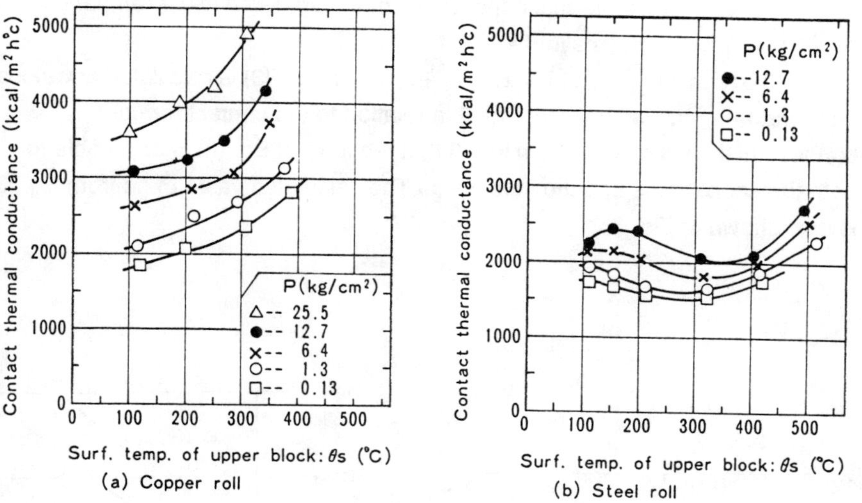

Fig. 4. Relation between contact thermal conductance and surface temperature of the upper block.

The contact thermal conductance is higher in copper than in steel, and is dependent much on the normal pressure. This should be explained by the difference of hardness between copper and steel.

The relation between contact thermal conductance and normal pressure is shown in Fig. 5. Contact thermal conductance was expressed by Tachibana as follows when the surface roughness of both materials are same. 2)

$$h_c = \frac{10^4}{(\delta+23)} \cdot \frac{P}{H_V}\left(\frac{\lambda_H \cdot \lambda_L}{\lambda_H+\lambda_L}\right)\frac{10^6}{2\delta}\lambda_g \qquad \text{..............................(3)}$$

where, Hv : Vicker's hardness of smaller one (kg/mm^2)

P : normal pressure (kg/cm^2)

δ : peak-to-peak roughness of contact surface; Rz (μm)

λ_g : thermal conductivity of ambient gas (kcal/m·h·°C).

λ_H : thermal conductivity of material with higher temperature (kcal/m•h•°C).

λ_L : thermal conductivity of material with lower temperature (kcal/m•h•°C)

The term $(P/H_V)\ [\lambda_H \cdot \lambda_L/(\lambda_H+\lambda_L)]$ in Eq. (3) is calculated at various temperatures, and is shown in Fig. 6 in relation to the contact termal conductance. It is evident from Fig. 6 that the form of Eq. (3) is applicable to both the copper rolls and the steel rolls. The calculated value of hc from Eq. (3) is also shown in Fig. 6.

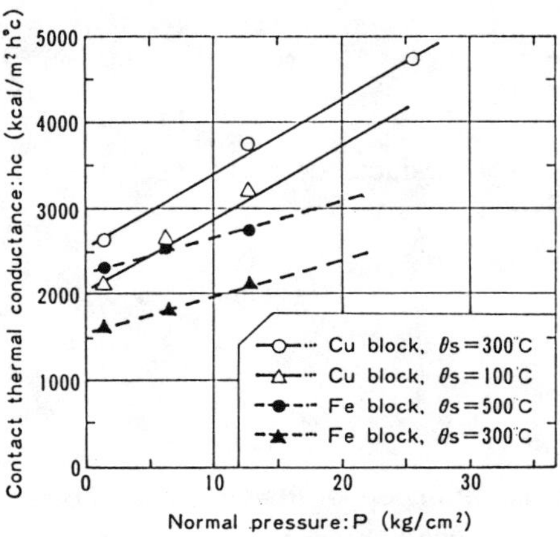

Fig. 5. Contact thermal conductance and normal pressure.

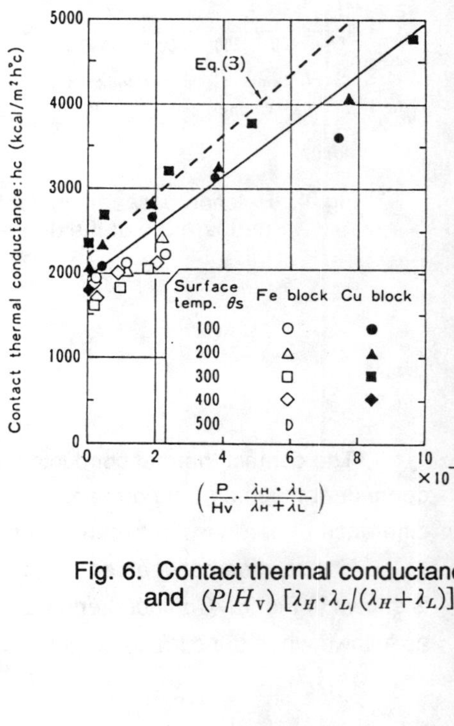

Fig. 6. Contact thermal conductanc and $(P/H_V)\ [\lambda_H \cdot \lambda_L/(\lambda_H+\lambda_L)]$

Tachibana has obtained Eq. (3) from the data of over 15 kg/cm^2 normal pressure. Nevertheless, the results of our experiments showed a tendency to agree very well with it. Most research works on contact thermal conductance reported hitherto are concerned with the case of fairly large normal pressures. In continuous annealing line of steel strip, where the unit tension in the strip is about 3 kg/mm^2 at most, the normal pressure (P kg/mm^2) is lower than about 0.6kg/cm^2.

There is little data of hc with such a low normal pressure.

From the results described above, the following Eq. (4) is obtained. It predicts the contact thermal conductance at low normal pressure which has not been reported.

$$h_c = K_1\left(\frac{P}{H_V}\right)\left(\frac{\lambda_H \cdot \lambda_L}{\lambda_H + \lambda_L}\right) + K_2 \qquad \text{.............................(4)}$$

where, K1, K2: constants determined by the surface roughness of the roll and the strip(—).

Our experimental results and Eq. (4) are very useful in predicting the cooling rate of the strip while designing the RQ process.

3. Overall Heat Transfer Coefficient of the Water-cooled Roll

3-1. Analysis of Unsteady Heat Transfer with the Rotation of the Roll

In the cooling process of RQ system, the overall heat transfer coefficient (U_O) between the cooling water and the strip can be defined using their actual measured temperatures. (c·f Fig. 7)

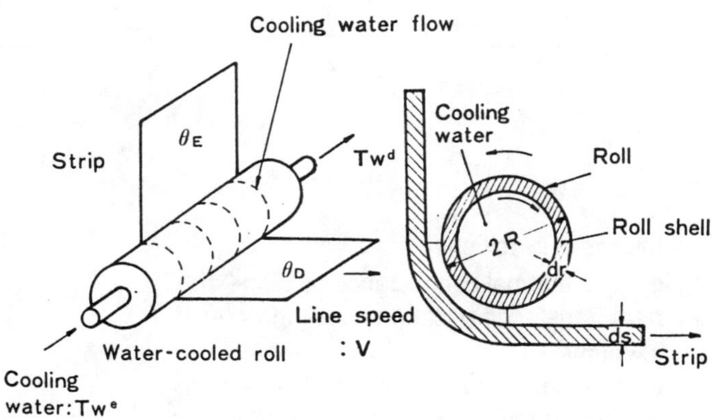

Fig. 7. Heat transfer model of Roll Quench system.

U_O is defined by the following equation (5):

$$U_o = \frac{\rho C_p d_s V}{L_s} \ln \frac{\theta_e - \theta_w}{\theta_d - \theta_w} \qquad \text{..........................(5)}$$

where, U_o : overall heat transfer coefficient (kcal/m$^2\cdot$h$\cdot$°C)

C_p : specific heat of strip (kcal/kg$\cdot$°C)

d_s : thickness of strip (m)

L_s : contact length of roll with strip (m)

V : line speed (m/h)

θ_d : outlet strip temperature (°C)

θ_e : inlet strip temperature (°C)

θ_w : mean temperature of cooling water (°C)

ρ : density of strip (kg/m^3).

In Eq. (5), θ_d and θ_e are the mean temperature in strip thickness. U_o in Eq. (5) is the signficant parameter in designing the RQ process and it would be very profitable if, in advance, U_o could be estimated from the conditions of the facilities and the operation. When thermal resistances exist as layer-by-layer, equation for the overall heat transfer coefficient U_o is known in general. To apply this equation for the shell of the water-cooled roll and the strip in Fig.7, the following equation is obtained.

$$U'_o = \frac{1}{\frac{1}{h_w} + \frac{d_r}{\lambda_r} + \frac{1}{h_c} + \frac{0.5 d_s}{\lambda_s}} \qquad \text{..........................(6)}$$

where, d_r : thickness of roll shell (m)

h_c : contact thermal conductance (kcal/m$^2\cdot$h$\cdot$°C)

h_w : heat transfer coefficient of cooling water (kcal/m$^2\cdot$h$\cdot$°C)

λ_r : thermal conductivity of roll shell (kcal/m$\cdot$h$\cdot$°C)

λ_s : thermal conductivity of strip (kcal/m$\cdot$h$\cdot$°C).

On the other hand, the roll shell is heated by the strip at the contact area and cooled down by the cooling water at the non-contact area. This heating and cooling process is repeated periodically, so that U_o, which shows the cooling ability of the actual RQ system, is obtained at unsteady thermal states. Contrary to this, U'_o of Eq. (6) is in steady thermal states. Therefore, U_o and U'_o are generally very different from each other, and it is impossible to predict the cooling ability of the actual process from Eq. (6).

Therefore, the theoretical calculation of the unsteady heat transfer process of a single water-cooled roll was carried out, and the overall heat transfer coefficient U_O, which should be obtained in actual process, was estimated. The calculations have been performed for the heat transfer in the direction of the roll radius, by the finite difference method of the one-dimensional differential equation with a cylindrical coordinate.

Figure 8 shows an example of the calculated results. In Fig. 8, the temperature distribution profiles of the radial direction in the roll shell at this stational thermal cycle are shown when the contact angle with strip is 45 and 90 deg, at the starting point of contact (inlet) and the end point of contact (outlet), respectively. The temperature of the outer surface of the roll shell varies about 30°C between inlet and outlet in this case.

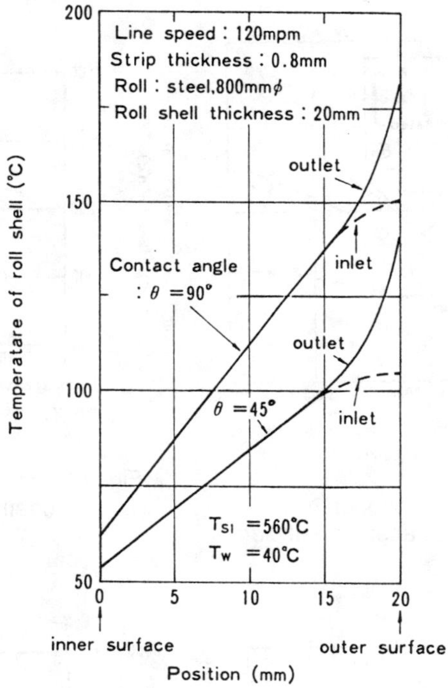

Fig. 8. Radial temperature distribution of roll shell.

3-2. Influence of Conditions of Installation and Operation on U_O.

From the simulation results mentioned above and Eq. (5), the overall heat transfer coefficient U_O of both the single roll and the plural rolls were obtained. This should correspond to that of actual process.

Hence, we studied the degree of influence which should be affected by each parameter of the actual process on U_O.

(a) Material of roll shell : Fig. 9

In Fig. 9, U'_O calculated by Eq. (6) are also shown which are much different from the simulation results, indicating that Eq. (6) is not proper to predict U_O.

(b)Heat transfer coefficient of colling water :Fig. 10

(c)Thickness of roll shell :FIg. 11

(d)Roll contact angle of strip :Fig. 12

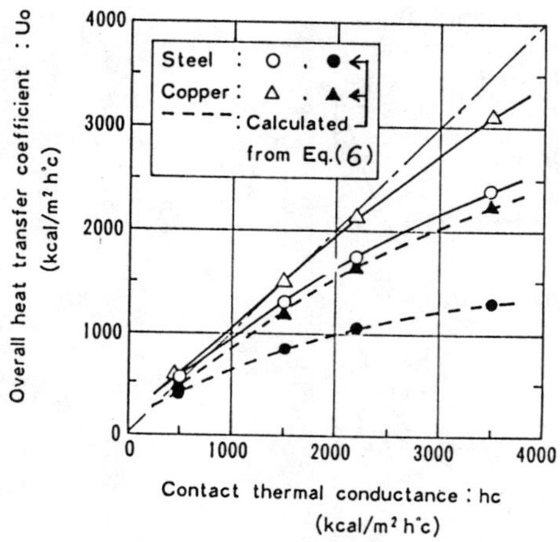

Fig. 9. Contact thermal conductance and overall heat transfer coefficient.

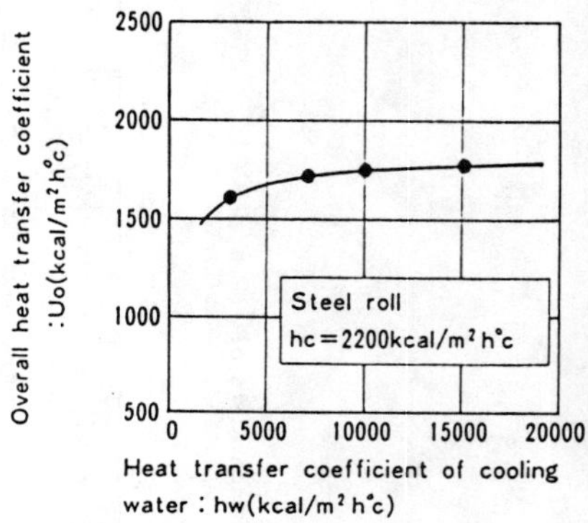

Fig. 10. Influence of heat transfer coefficient of cooling water

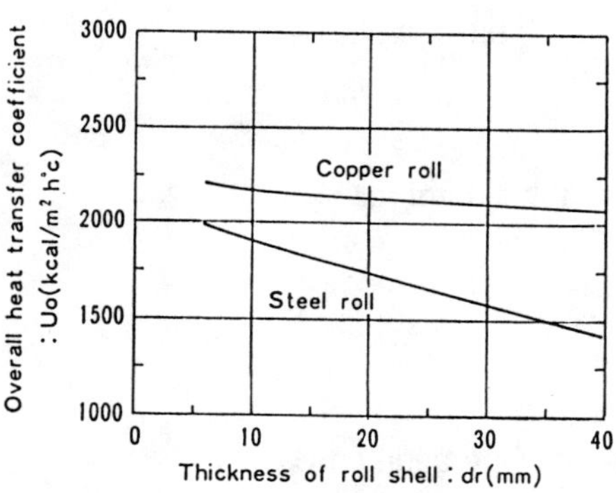

Fig. 11. Influence of thickness of roll shell.

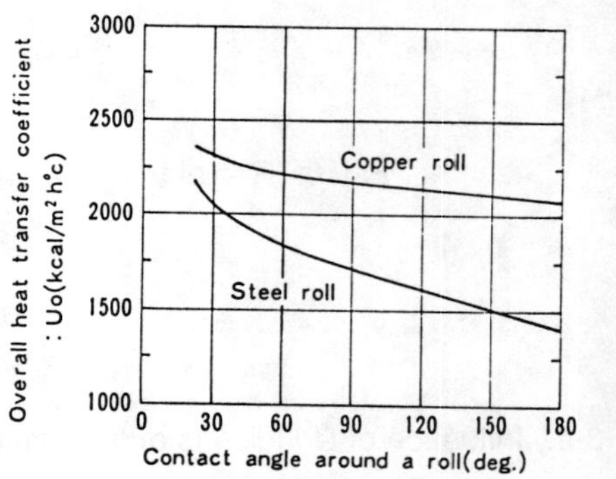

Fig. 12. Influence of contact angle.

The larger the contact angle becomes, the lower U_o decreases as shown in Fig. 12, and the degree of change is larger in the steel roll than in the copper roll. This is because the ratio of the cooling time for the roll shell decreases whith the increase of contact angle, and the mean temperature of the roll shell becomes higher. Thermal behavior like this is due to the unsteady cooling mechanism of the Roll Quench process.

From these results, we summarized as conclusions:

(1) The major factors of U_o are as follows:

i) material of the roll shell,
ii) thickness of the roll shell,
iii) contact angle, and
iv) starting temperature of cooling

(2) If the contact angle is constant, U_o is not so much changed by the roll diameter and the line speed. Consequently, in actual process, as the contact angle becomes smaller when the roll diameter is large and the line speed is low, U becomes larger.

(3) In the major factor described above, the contact angle is variable in opration, and much attention shall be paid to the apportionments of the contact angle. From this information, we have simplified the model equations of U_o for practicall use as Eqs. (7) and (8).

For steel roll; $$U_o = 20.1 h_c^{0.80} d_r^{-0.22} \theta^{-0.23} \qquad \text{.....(7)}$$

For copper roll; $$U_o = 3.15 h_c^{0.90} d_r^{-0.052} \theta^{-0.052} \qquad \text{.....(8)}$$

where, d_r : thickness of roll shell(mm)

h_c: contact thermal conductance(kcal/m^2•h•°C)

U_o: overall heat transfer coefficient(kcal/m^2•h•°C)

θ : contact angle (deg)

4. Outline of Roll Quench Process in Actual Production Line

The knowledge about cotact thermal conductance and the model for the cooling ability of water-cooled roll which are shown in this report are very profitable for design and operation of our RQ process.

A full commercial scale RQ system has been installed in No. 2 CAL of the Fukuyama Works in 1982.

Table 1 shows the specifications of the water-cooled roll equipment of the actual production line. The strip thermometers have been equipped at the inlet and outlet of the Roll Quench section, and the water-cooled roll is fitted with a lifting device to vary the length of its contact with the strip. This makes automatic control of the outlet strip temperature possible.

The overall heat transfer coefficients U_o obtained in the actual line are shown in Fig. 13 in comparison with those predicted by Eq. (7). Both values coincide very well, showing the reliability of Eq. (7). In the thickness range of 0.4 to 1.6mm, the cooling rates are obtained 100 to 450°C/s.

Table 1. Specification of RQ.

Line specification	Production line		No. 2 CAL, Fukuyama
	Line speed (mpm)		Max. 180 mpm
	Strip size	Thickness (mm)	0.4 to 1.6
		Width (mm)	610 to 1 300
Specification of water-cooled rolls	Number of rolls		5 rolls
	Material of rolls		steel
	Overall heat transfer coefficient (kcal·m²·h·°C)		1 500 to 1 700
	Contacting length (m)		1.0 to 5.2
	Flow rate of cooling water (t/h, each roll)		3.5 to 20.00

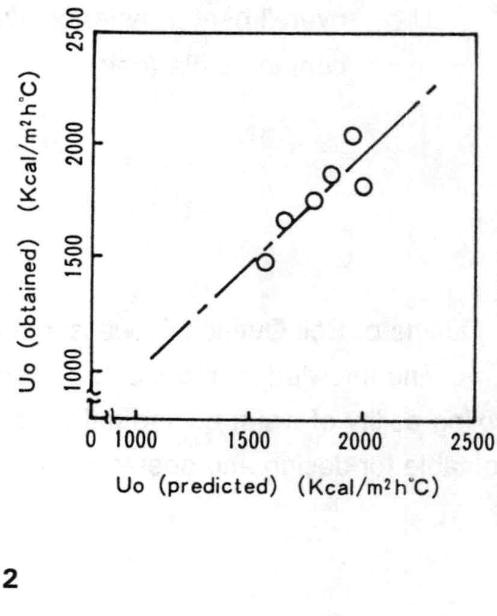

Fig. 13. Overall heat transfer coefficient obtained in CAL and predicted value.

5. Conclusions

New quenching system with water-cooled rolls has been developed.

In the first stage of the basic research, contact thermal conductance was studied by experiment. Further, the caluculation of the unsteady heat transfer with the rotation of the water-cooled roll was performed, and the thermal behavior of the roll shell has been cleared up. The model equation which predicts the overall heat transfer coefficient U_0 of the actual process has been obtained also. A lot of phenomena observed in the actual process can be explained from this model of U_0 , which is useful for designing and controlling the system.

A full commercial scale RQ system installed in the No. 2 CAL of Fukuyama Works has been operating successfully.

REFERENCES

1) S. Osaka, N. Tanaka, K. Matsudo, Y. Fukuoka and H. Ishioka :Iron Steel Eng., (1980), Nov., 67.
2) F. Tachibana: J. Japan soc. Mech. Eng., 55 (1952), 102.

Temperature Equalization by the Radiative Converter and the Impinging Jet for a Slab in Continuous Casting-Direct Rolling Process

KOICHI ICHIMIYA
Department of Mechanical System Engineering
Yamanashi University
Takeda-4, Kofu, Yamanashi 400, Japan

RYOZO ECHIGO
Department of Mechanical Engineering
Tokyo Institute of Technology
Ohokayama-2-12-1, Meguro-ku, Tokyo 152, Japan

ABSTRACT

An effective method was proposed to reduce the thermal energy at the edge heating of high temperature steel slab in continuous casting-direct rolling process and flatten the temperature distribution. This method was composed of the porous medium converting gas enthalpy into thermal radiant energy and the impinging jet removing the heat to the edge of the slab. At the first stage, the heat transfer characteristics of the porous medium and the impinging jet were described in detail. At the next stage, the transient process on equalizing the temperature distribution was analyzed numerically at thermal conductive-convective field by using the basic model. In addition, to simulate the continuous casting-direct rolling process in the experiment, a preheated slab was cooled by thermal radiation and fed into the proposed temperature equalizer. It was proved that the radiative converter (porous medium) equalized the slab temperature quickly owing to the heating of the slab edge.

INTRODUCTION

In order to produce the high temperature steel slab, many processes were included ; melting, production ingot, heating, cutting and rolling process, as shown in Table 1. Much energy more than $2x10^3$ MJ was supplied to produce a slab of 1 ton . Much energy was lost and should be compensated in such processes. Continuous casting process [1], which stepped the producing ingot and heating , was proposed to save the thermal energy. In the process, the total energy reduced to third-four. In addition, the thermal energy approached to one-third in the continuous casting-direct rolling (CC-DR) process [2,3,4] , in which continuous casting process was connected directly to the rolling process. This process is very effective to save not only the cost, but also the energy.

TABLE 1. Production system of the slab.

IC-CCR Process	Melting - Ingot Casting - Heating - Blooming - Slab - Heating - Rolling
CC-DR Process	Melting - Continuous Casting - Rolling

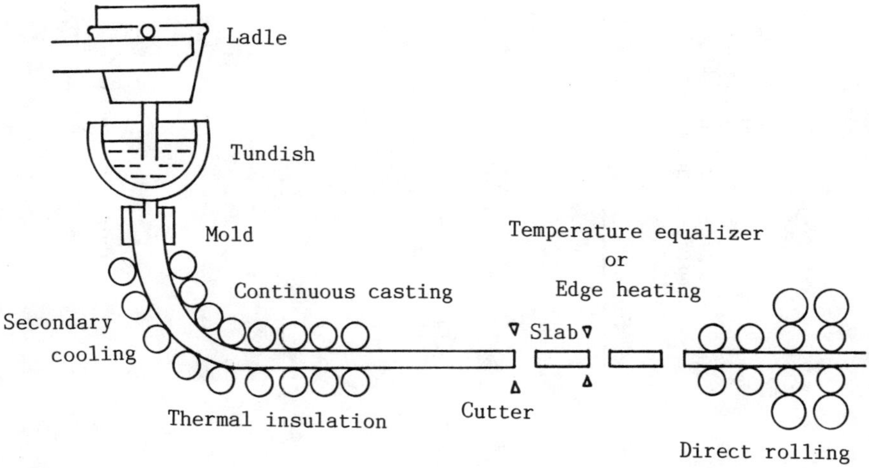

FIGURE 1. Continuous casting-direct rolling process.

FIGURE 1 represents the outline of the CC-DR process. In the continuous casting process, slabs are cooled usually by spray impingement and thermal radiation [5,6]. Particularly, temperature within the slab is distributed nonuniformly since the edge temperature drops rapidly. Maximum temperature difference within the slab is required to be less than 100 °C because of the degradation of the quality. Several methods [4] are considered to heat the edge of the slab in the process as shown in Table 2. The heating by the combustion gas or the electric magnetic induction method is adopted mainly as the temperature equalizer. However, the former method is not sufficiently established in view of the heating efficiency. On the other hand, the latter one has the good controllability and the high efficiency, but has the high construction cost. In this situation, an effective method is introduced in the next chapter.

BASIC MODEL FOR EQUALIZING TEMPERATURE DISTRIBUTION

FIGURE 2 shows a basic model of an effective method as a temperature equalizer, which is composed of the injection of high temperature gas and the porous medium. The high temperature energy is kept at the center of the slab for heating the cooled edge. If the gas temperature is set at the value between the temperature at the center and at the edge of the slab, the thermal energy would be transported from the center to the edge within the slab by thermal conduction and the forced convection of the impinging jet. If the exhaust energy be applied as a working gas , it would be also effective in the production system . In addition, the porous medium is set at the downstream region and absorbs the thermal energy. It converts the gas enthalpy into the thermal radiation, which is emitted to the edge of the slab. It is expected that the temperature distribution within the slab is flattened in high efficiency by the thermal radiation shielding effect of the porous medium and the impinging jets effect and the cost is lower since the structure is simple. Heat transfer system is very complex owing to the conjugation of transient conduction,convection and thermal radiative heat transfer.

TABLE 2. Edge heating of the slab.

	Covering by insulator	Edge heating by combustion gas	Edge heating by electro magnetic induction	Temperature equalizing by porous medium and impinging jets
Structure	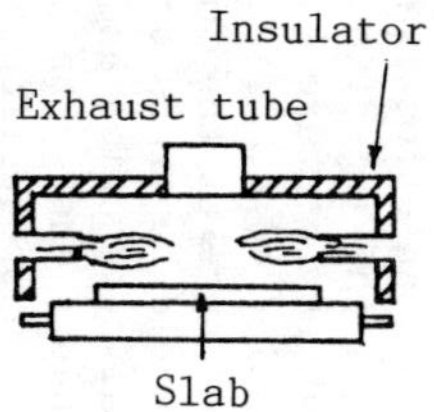	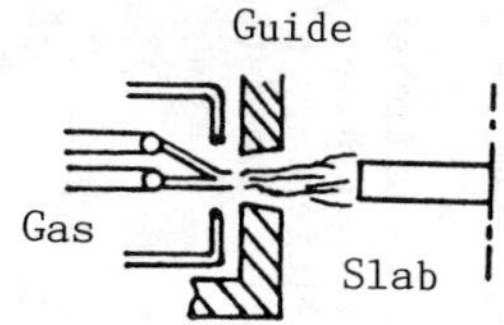	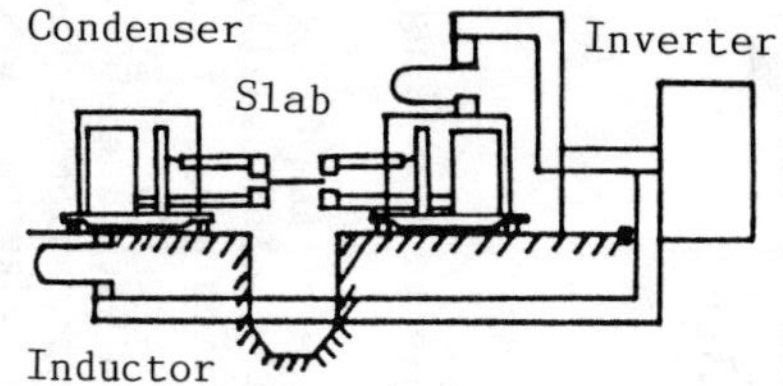	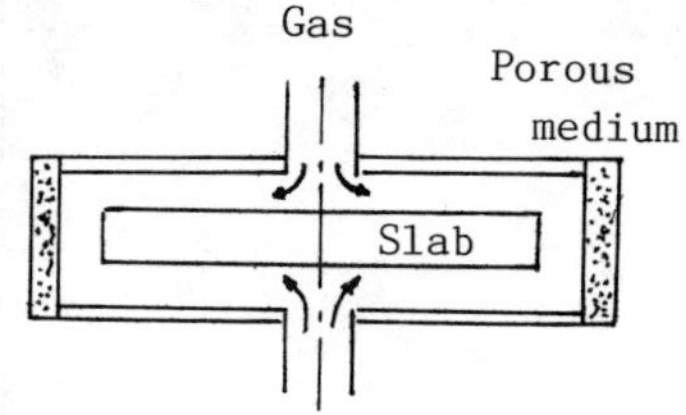
Equipment cost	Low	Low	High	Low
Characteristics	•Low efficiency at slab corners •Heating over the whole surface	•Higher efficiency at slab corners	•Higher efficiency at slab corners •Thermal control	•Higher efficiency at slab corners

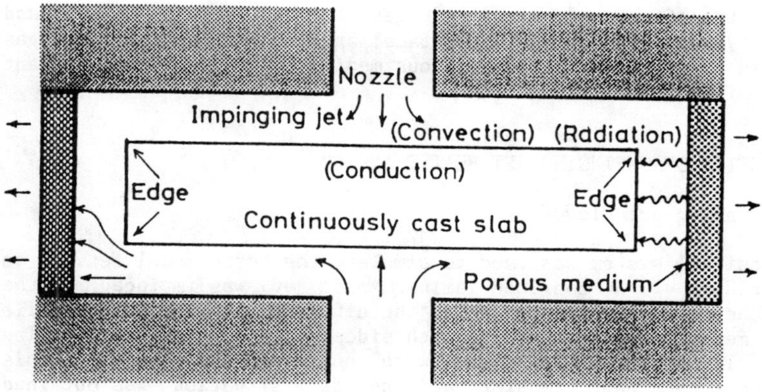

FIGURE 2. Basic model of the temperature equalizer.

HEAT TRANSFER CHARACTERISTICS OF POROUS MEDIUM

The heat transfer characteristics of the porous medium was examined numerically and experimentally as the effective energy conversion method between gas enthalpy and thermal radiation by Echigo [7]. When a porous medium is installed in a duct or a combustion furnace, the gas enthalpy is converted into radiant energy propagating to the upstream region and a large temperature difference takes place between up-and down-stream regions of the porous medium. FIGURE 3 shows the gas temperature profiles along the flowing direction with the porous medium being set in a parallel plate duct. The letter Θm is the dimensionless temperature, X dimensionless length, and X1 the location of porous medium . The conditions are as follows. Reynolds number Re is 150, Prandtl number is 0.7, the dimensionless parameter determined by thermal radiation and heat conduction N_R is 0.05 and the dimensionless parameter pertinent to the overall coefficient of heat transmission through the duct wall P is 0.114 and 0.0114 . According to the figure, the enthalpy drops drastically at the position of the porous medium and the magnitude is larger in the upstream region due to the radiant contribution of higher temperature. The basic principle is attributed not only to high emissive power of solid body compared with that of gas, but also to the extremely

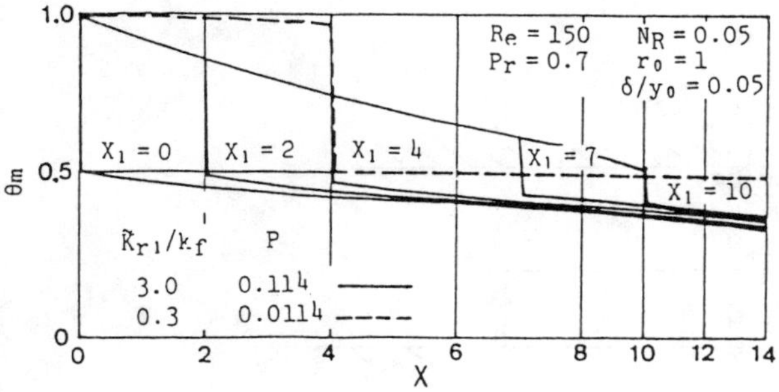

FIGURE 3. Gas temperature profiles through porous medium placed in a parallel plate duct .

high heat transfer coefficient of gas to the porous medium and substancially large heat transfer surface area. Consequently, the gas enthalpy is transfered to the porous medium and the strong radiant energy is emitted from it.

CHARACTERISTICS OF IMPINGING JET HEAT TRANSFER

Laminar Impinging Jet Flow

A two-dimensional system was used to simulate the basic model set up, as shown in FIGURE 4. A confined laminar slot jet was impinged to the center of the heated surface by using the air as a working fluid, while the porous media were set near the both sides of the heated slab. The the upper wall and the l_1-l_2 surface were insulated thermally. In this system, the local heat transfer and the flow situation was obtained numerically and experimentally [8]. Basic equations governing the flow and the temperature field were formed by introducing the stream function and the vorticity . They were reduced to the conservative form and were solved by applying the upwind iteration method.

Flow situation was represented in FIGURE 5 (a) and (b) for H(=(h2-h1)/d, d;half width of a nozzle) =1.0, $C(=(l_2-l_1)/d)=5.0$ and Re=600. After the fluid impinges to the upper surface of the slab, it is accelerated to the mainstream direction and the position of the maximum velocity deviates from the impinged surface. The flow is separated at the end of the slab. A big recirculating flow is generated between the side surface of the slab and the porous medium, and a small recirculating flow occurs near the bottom of the slab.

FIGURE 6 show the numerical and experimental results of the local Nusselt numbers. The origin is the center of the slab surface and the both edges of the slab are X=10 and X=-10. The side of the slab is shown in Y section. The figure denotes the effect of the distance between the nozzle and the impinged surface for Re=1000 and C=2.5. The local heat transfer coefficient takes a peak at stagnant point and decreases along the flowing direction. In the experiment, Nusselt number on the impinging point is 2.5 times in magnitude than that on the edge for H=1.5, and 3 times for H=0.5. Heat transfer at the bottom of the side wall increases locally for a large recirculating flow between the side wall and the porous medium.

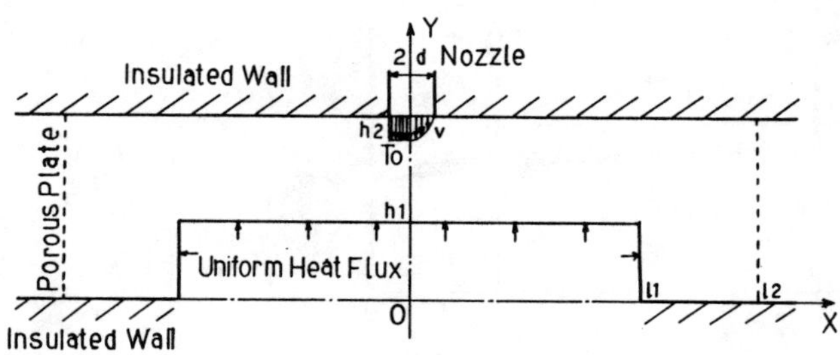

FIGURE 4. Co-ordinate system.

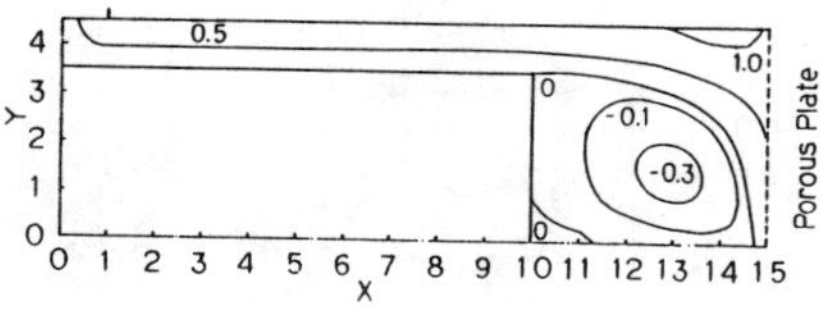

(a) Dimensionless stream function.

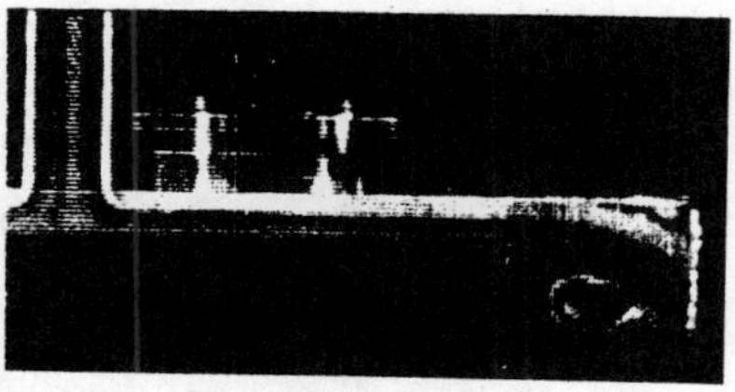

(b) Flow visualization.

FIGURE 5. Flow situation.

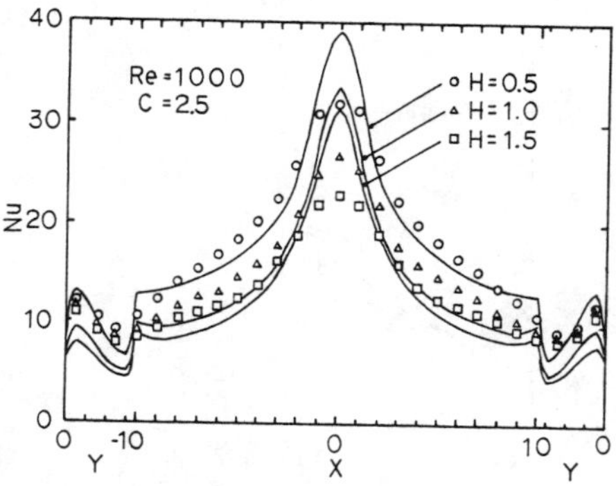

solid line : calculated value

symbols : experimental value

FIGURE 6. Local Nusselt number (laminar slot jet).

Turbulent Impinging Jet Flow

Characteristics of the impinging jet heat transfer were examined for the turbulent slot jet. Gardon and Akfirat[9,10] obtained experimentally the effects of turbulence on the local impinging heat transfer of a two-dimensional slit and a circular jet without a confined plate. They also expressed the correlation of the mean heat transfer characteristics. Kumada and Mabuchi[11] examined experimentally the mass transfer at the stagnation point, the mean Sherwood number and the two-dimensional approximate theory. Martin [12] arranged the heat and mass transfer between impinging gas jets and solid surfaces for comparatively wide space.

We determined experimentally the characteristics of the local heat transfer due to impingement of a confined turbulent air jet on a heated surface for comparatively narrow space (refer to FIGURE.4)[13]. FIGURES 7(a),(b) and (c) show the local Nusselt number profiles for the dimensionless distance between the nozzle and the heated surface H=1.5, 1.0 and 0.5 . Pictures of the oil film are simultaneously presented. In FIGURE 7(a), the local Nusselt numbers peak at the stagnation point and at $X=\pm 7$. The position of former peaks do not depend on Reynolds number. The latter peaks tend to become unclear with decrease of Re. As a result of the correspondence between the heat transfer and the picture of the oil film, the flow region was divided into four sections. The first is the impinged point where the color is white. The second is black where the flow changes abruptly the direction. The third is white where the oil film remains and the local Nusselt number reaches a minimum . The fourth is where the local Nusselt number decreases after the second peak. In the case of H=1.0 (FIGURE 7(b)),the local Nusselt number

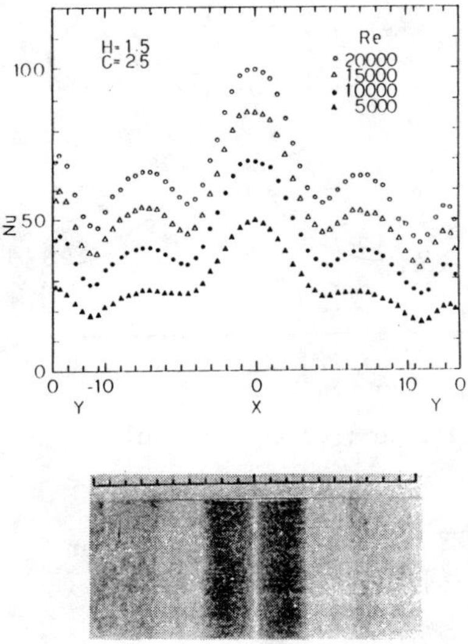

(a) H=1.5.

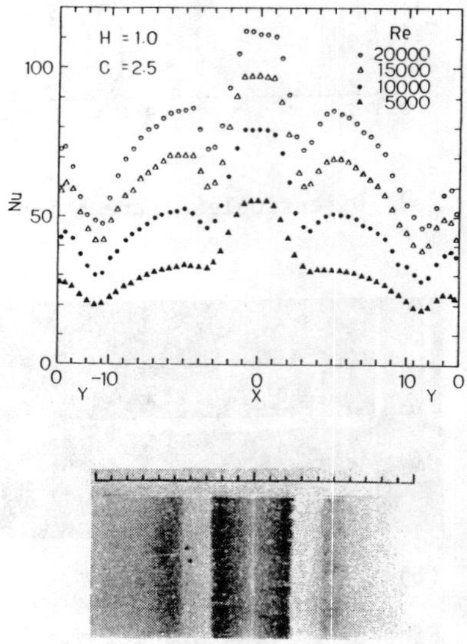

(b) H=1.0.

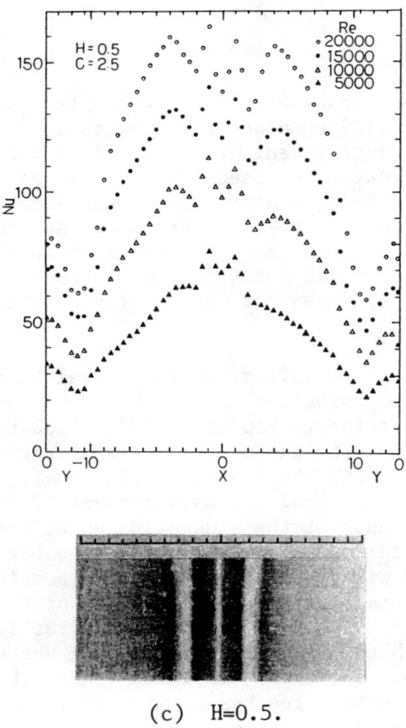

(c) H=0.5.

FIGURE 7. Local Nusselt number (turbulent slot jet).

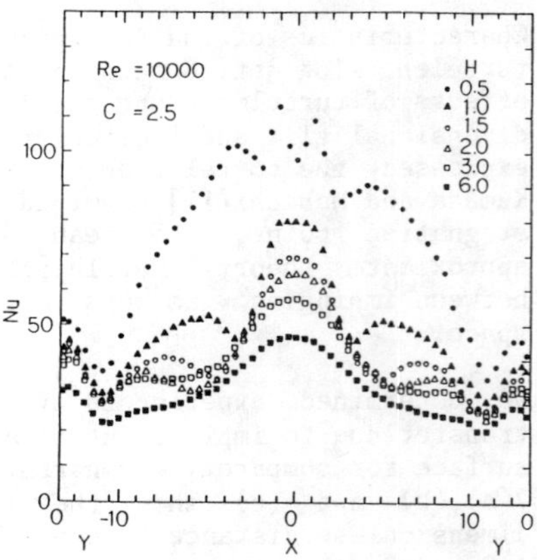

FIGURE 8. Local Nusselt number (effect of H).

becomes flat near the stagnant point the position of the peak moves to the upstream region X=±5. In FIGURE 7(c)(H=0.5), the heat transfer situation of H<1 is different from that of H>1. Peaks of the local Nusselt number appear at both sides of a stagnant point X=±1. It is estimated that the high turbulence intensity at the nozzle exit causes the direct improvement of the local heat transfer at X=±1. The local heat transfer is improved by the impinged effect and the acceleration effect.

FIGURE 8 displays the heat transfer augmentation in the whole region and the movement of the position of the peak to an upstream region with decrease of H for Re=10000 and C=2.5. Local Nusselt number at the stagnant point is expressed in the functions of Re and H, as follows,

$$Nu=0.78Re^{0.5}\,H^{-0.3}\ .$$

UNSTEADY HEAT TRANSFER WITHIN A SLAB IN THERMAL CONDUCTIVE-CONVECTIVE FIELD

Unsteady flow and heat transfer process were examined numerically in the system (FIGURE 9) based on the basic model(refer to FIGURE 2), where the porous medium was set on the position l_2 near the slab and the laminar slot jet was impinged to the center of the slab surface [14].

Governing Equations and Numerical Analysis

Governing differential equations are expressed in eqs. (2),(3),(4) and (5) by introducing the stream function Ψ, vorticity ω and following

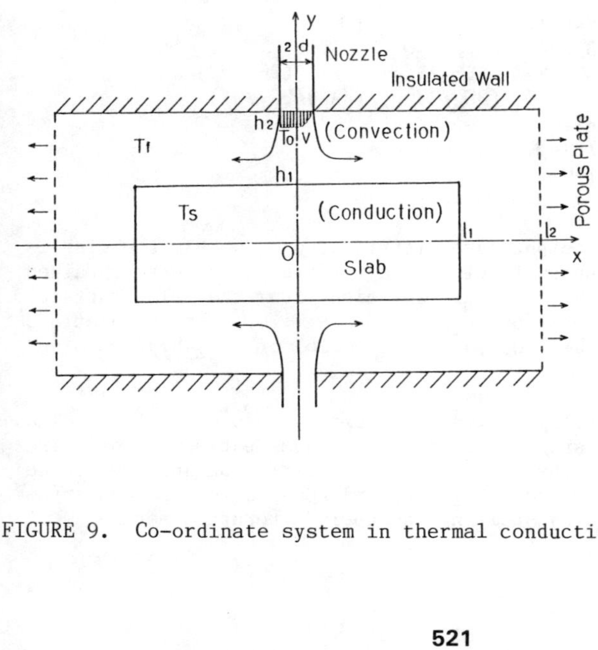

FIGURE 9. Co-ordinate system in thermal conductive-convective field .

dimensionless parameters (1),

$$u=\frac{\partial\psi}{\partial y},\quad v=-\frac{\partial\psi}{\partial x},\quad \omega=\frac{\partial v}{\partial x}-\frac{\partial u}{\partial y}$$
$$U=\frac{u}{v_o},\quad V=\frac{v}{v_o},\quad \Psi=\frac{\psi}{v_o d},\quad \Omega=\frac{\omega d}{v_o},\quad \tau=\frac{v_o}{d}t \tag{1}$$

The letters u, v and t are velocities along x and y direction and time, respectively.

Dimensionless equations for the flow are as follows,

$$\frac{\partial^2\Psi}{\partial X^2}+\frac{\partial^2\Psi}{\partial Y^2}=-\Omega \tag{2}$$

$$\frac{\partial\Omega}{\partial\tau}+\frac{\partial\Psi\partial\Omega}{\partial Y\partial X}-\frac{\partial\Psi\partial\Omega}{\partial X\partial Y}=\frac{2}{Re}\left(\frac{\partial^2\Omega}{\partial X^2}+\frac{\partial^2\Omega}{\partial Y^2}\right) \tag{3}$$

Dimensionless energy equations are expressed as follows,
for the working gas,

$$\frac{\partial\Theta_f}{\partial\tau}+\frac{\partial\Psi\partial\Theta_f}{\partial Y\partial X}-\frac{\partial\Psi\partial\Theta_f}{\partial X\partial Y}=\frac{2}{Re\cdot Pr}\left(\frac{\partial^2\Theta_f}{\partial X^2}+\frac{\partial^2\Theta_f}{\partial Y^2}\right) \tag{4}$$

for the slab,

$$\frac{Pr\cdot Re}{2}\frac{\partial\Theta_s}{\partial\tau}=\frac{\alpha_s}{\alpha_f}\left(\frac{\partial^2\Theta_s}{\partial X^2}+\frac{\partial^2\Theta_s}{\partial Y^2}\right) \tag{5}$$

A Greek letter α is thermal diffusivity and the suffixes s and f mean slab and working gas, respectively.

Dimensionless boundary conditions of the flow and the temperature are as follows,

at $0\leq X\leq 1$, $Y=h_2/d$: $U=0$, $V=(3/2)(X^2-1)$, $\Theta_f=1$
at $1\leq X\leq l_2/d$, $Y=h_2/d$: $U=0$, $V=0$, $\partial\Theta_f/\partial Y=0$
at $0\leq X\leq l_1/d$, $Y=h_1/d$: $U=0$, $V=0$, $(\partial\Theta_s/\partial Y)=(\lambda_f/\lambda_s)(\partial\Theta_f/\partial Y)$
at $X=l_1/d$, $0\leq Y\leq h_1/d$: $U=0$, $V=0$, $(\partial\Theta_s/\partial Y)=(\lambda_f/\lambda_s)(\partial\Theta_f/\partial Y)$
at $X=0$, $0\leq Y<h_1/d$: $\partial\Theta_s/\partial X=0$
at $l_1/d\leq X\leq l_2/d$, $Y=0$: $\partial U/\partial Y=0$, $V=0$, $\partial\Theta_f/\partial Y=0$
at $0\leq X<l_1/d$, $Y=0$: $\partial\Theta_s/\partial Y=0$
at $X=l_2/d$, $0\leq Y\leq h_2/d$: $\partial U/\partial Y=0$, $V=0$, $\partial\Theta_f/\partial X=0$

At $\tau=0$, the dimensionless temperature distribution Θ_s within the slab is assumed to be the values shown in Fig. 11(a). These values were obtained previously by assuming the thermal radiation from the slab surface. Initial temperatures in the flow passage Θ_f is 1.0. The calculations were performed on $Pr=1$, $Re=1000$, and $\alpha_s/\alpha_f=0.16$ for $(h_2-h_1)/d=0.5$ and $(l_2-l_1)/d=2.5$.

Numerical method is generally obeyed by Gosman's procedure [15]. The flow and energy equations are expressed in the same patterns which are composed of the unsteady, the convection , the diffusion and the source terms. Governing equations are integrated and reduced to the conservative form. They are solved by applying the upwind iteration method.

Numerical Results

Dimensionless stream functions are represented in FIGURES 10 (a),(b) and (c). At $\tau=1$, eddies are not generated at the end of the nozzzle exit and the flow direction changes abruptly since the distance between nozzle and slab surface is narrow. At $\tau=5$, eddies take place clearly between the side surface and the porous medium and at the end of the upper wall. At $\tau=61.54$, the flow situation becomes almost steady.

FIGURE 11(a),(b) and (c) show the dimensionless isothermal lines in the system. FIGURE 11(a) represents the initial temperature distribution in a slab and the initial fluid temperature is $\Theta_f=1.0$. While the dimensionless slab surface temperature is $\Theta_w=1.04$ at the stagnation point, Θ_w reduces to 0.854 at the end. FIGURE 11(b)($\tau=700$) shows that heat is transported with gas flow from the slab surface since the isothermal line of $\Theta_f=1.001$ moves to the upper downstream region . The temperature within the slab decreases for the heat conduction. For example, while the region of $\Theta_s=1.4$ is within X=6.15 at $\tau=0$, it moves within X=5.0 at $\tau=700$. The dimensionless edge temperature increases to $\Theta_s=1.008$ at $\tau=1600$ corresponding to real time about 2 minutes (FIGURE 11(c)).

Dimensionless heat flux on the slab surface $Q(=q\cdot d/(\lambda\cdot T_o))$is represented as a parameter of τ in FIGURE 12. Upper surface of the slab is from X=0 to X=10, and Y corresponds to the side surface. Local heat flux is the highest at the stagnation point X=0, and decreases appreciably near $X\simeq2$. In addition, it becomes zero between X=7 and X=8. The negative values near the end of the slab mean that heat flows from the fluid to the slab.

The edge temperature of the slab is denoted as a parameter of time τ in FIGURE 13. The temperatures are compared for three different boundary conditions ; (A) the present model with impinging jet (solid lines), (B) the heat transfer by thermal conduction in quiet gas (chain lines) and (C) the insulated conditionon the slab surface (dotted lines). Near the stagnation point, Θ_w is the lowest in the case (A) because of impingement, and is the highest in the case (C). This situation is maintained till $X\simeq7$. On the other hand, at the end of the slab, Θ_w is the lowest in the case (B), and increases in order (A) and (C) at $\tau=100$.However, the edge temperature of the slab is the highest in the case (A) and decreases in order (C) and (B) at $\tau=700$. This indicates that the heat transportation by the impingement is superior to the thermal conductive effects of working gas and slab and contributes to the temperature increase of the slab edge. While the ratio of temperature at the stagnation point and that at the slab edge is 1.218(=1.04/0.854) at $\tau=0$, it reduces to 1,162(=1.172/1.008) at $\tau=1600$. This reveals that the surface temperature tends to be uniformlized with increase of the edge temperature.

EXPERIMENT OF TEMPERATURE EQUALIZATION BY RADIATIVE CONVERTER

Experiments were carried out on heat transfer performances of the temperature equalizer with the radiative converter (the porous medium) for a slab [16].

Experimental Facility and Procedure

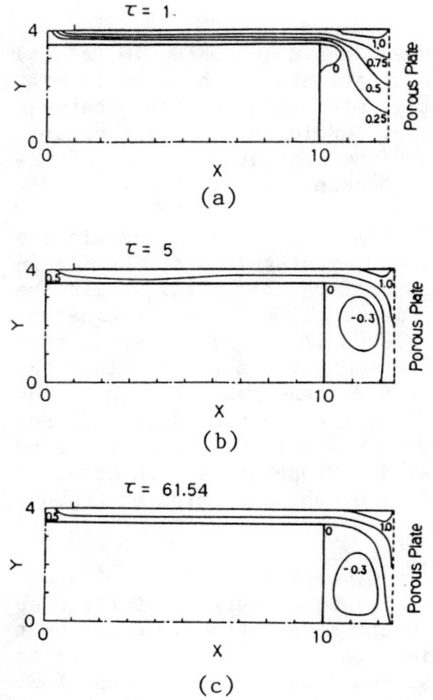

FIGUTRE 10. Dimensionless stream function.

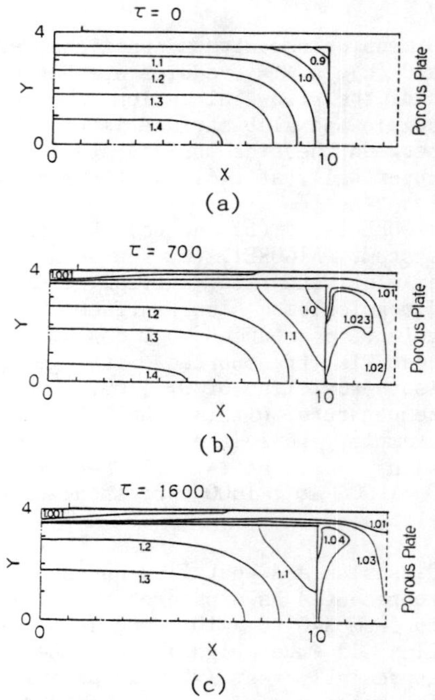

FIGURE 11. Dimensionless isothermal lines.

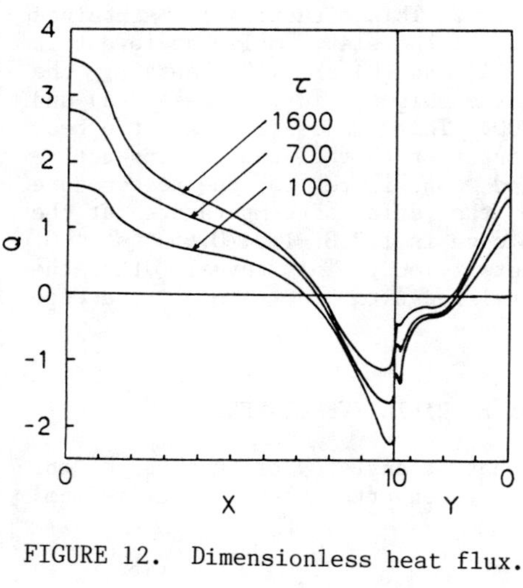

FIGURE 12. Dimensionless heat flux.

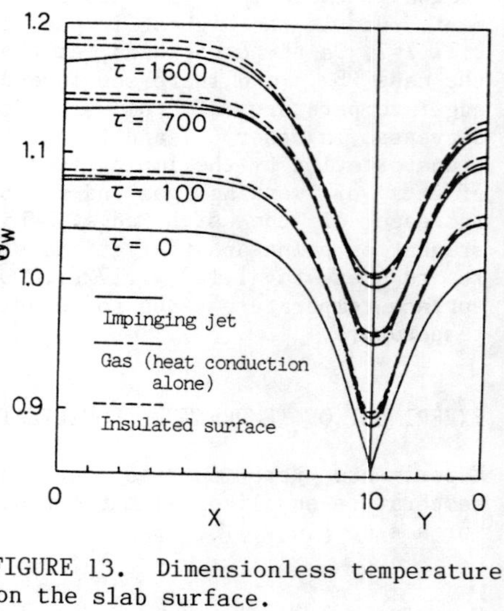

FIGURE 13. Dimensionless temperature on the slab surface.

FIGURE 14 shows a schematic diagram of the experimental facility. The facility was composed of three parts, i.e. the slab, the preheating furnace and the temperature equalizer. The slab was arranged to traverse on rails between the preheating furnace and the temperature equalizer. The slab consisted of a heat-resisting stainless steel (type 310) ingot of about 24.4 kg with dimensions 40x200x400 mm. In the preheating furnace, four radiant cup-burners of 5 kw output were embedded on the upper and lower walls. The temperature equalizer was constructed by lining the stainless steel frame with ceramic fiber boards; its inside dimensions are 80x240x400 mm. The slot nozzles of 40 mm width were fixed on the upper and lower walls. The radiative converters made of porous media were installed at the side walls of the temperature equalizer. The porous media were piles of fifteen stainless-steel-wired mesh screens of nominal mesh length 1/80 inch, whose optical thickness τ_o is 7.5. Ten thermocouples were embedded within the slab and eleven were set in the temperature equalizer.

To simulate exactly the continuous casting process is difficult in a laboratory experiment. Therefore, the alternative way in which an uniformly preheated slab was cooled in an open space , was adopted in the present experiment. The jet temperature was set to be higher than that at the edge and lower than that at the stagnation point. The experiments were conducted on the conditions of the preheated slab temperature To=1350 K, the cooling period t=200 sec and the inflow-gas temperature T_{gin}=1100 K. Reynolds number Re at nozzle exit was changed from 660 to 1600. The heat transfer in the CC-DR process is the transient thermal conduction problem subjected to the thermal radiation boundary condition . The parameter governing the problem is attributed to the radiation number defined by $Nr=\lambda_s/(\varepsilon\sigma T_o{}^3\ 1)$ [17,18], where σ is the Stefan-Boltzmann constant, ε is the emissivity, T_o is the temperature λ_s is the thermal conductivity and 1 is the characteristic length (half thickness of the slab). The radiation number Nr in the present experiment is 9.32.

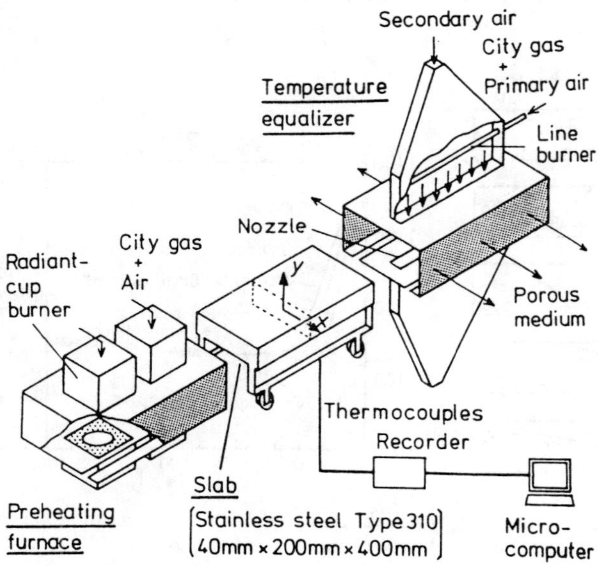

FIGURE 14. Fxperimental facility.

This is about seven times in magnitude than Nr in the CC-DR process. Therefore, the contribution of the radiative converter to the slab edge heating is expected to be relatively minor compared with that in the CC-DR operation.

Experimental Results

The slab was fed into the temperature equalizer subsequent to the cooling period of 200 sec. FIGURE 15 shows the contours of isotherms in the right-upper quarter of the slab. In the figure the center of the slab is taken as the coordinate-system origin. At t=0 sec, the edge temperature is the lowest in the slab as a result of the radiative cooling from the surfaces. At 32 sec, the surface temperature near the stagnation point recovers rapidly . In the initial stage, temperature equalization by thermal conduction is dominant. As the time progresses, the edge temperature increases owing to the local heating of the slab edge by the radiative converter.

FIGURE 16 shows the temperature variation with time at the stagnation point, the center and the edge. The temperatures at the stagnation point and the edge increases quickly as soon as the slab are fed into the temperature equalizer. On the contrary, the temperature of the center decreases at the initial stage and becomes lower than that of the stagnation point t=200 sec.

Energy balance between the radiative converter and the the slab side is considered in this section. One-dimensional radiative heat transfer is assumed for simplicity, as shown in FIGURE 17. The temperature distribution of the porous medium Tp is approximately linear. In the porous medium used in the present experiment, the scattering is so small that the radiative heat flux q_r^- , q_r^+ in the backward and forward directions are obtained by using the exponential integral function.

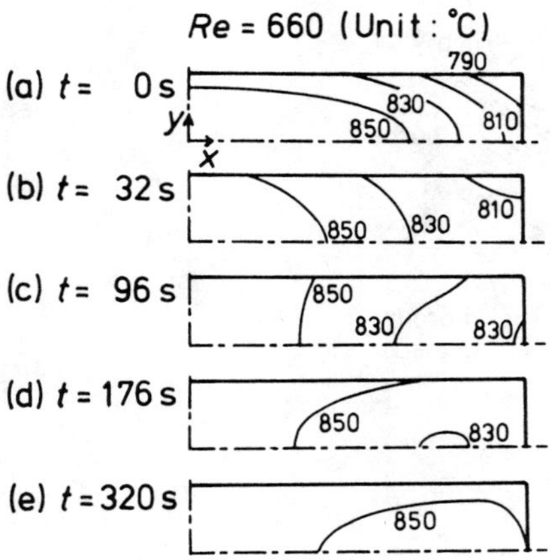

FIGURE 15. Isothermal lines in the right-upper quarter of the slab.

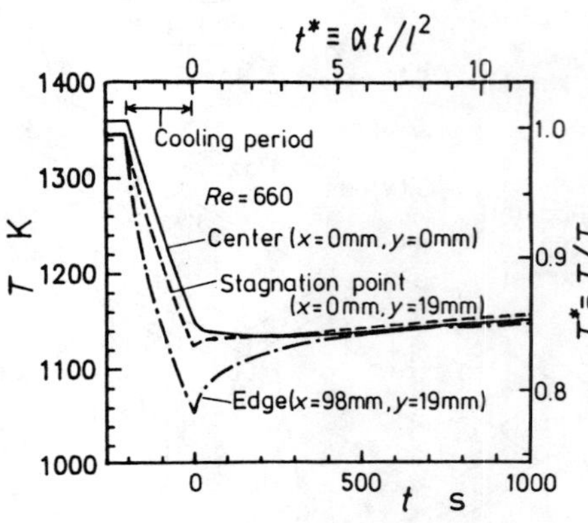

FIGURE 16. Variation of slab temperatures with time.

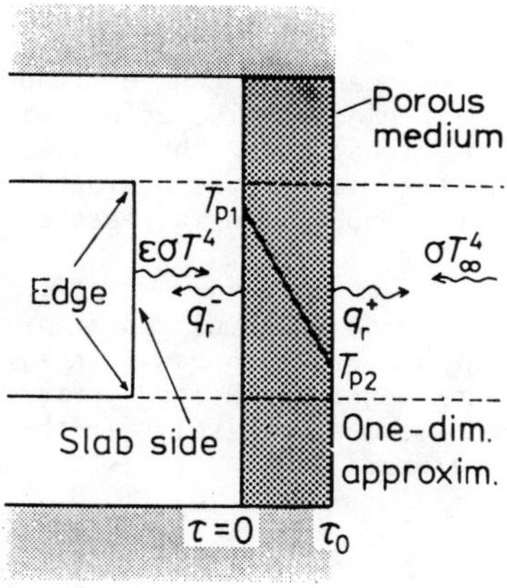

FIGURE 17. Energy balance between radiative converter and slab surface.

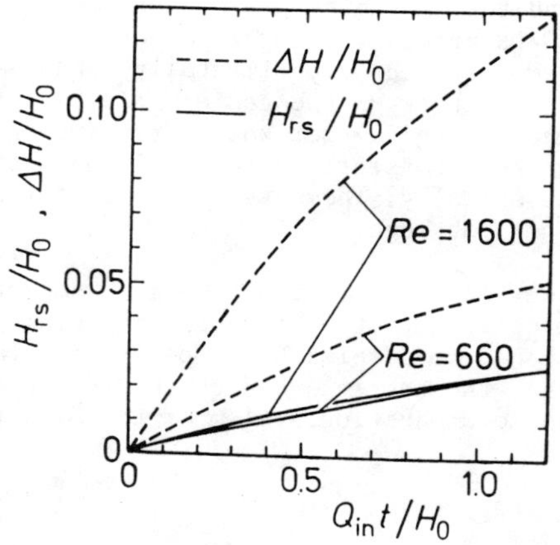

FIGURE 18. Variation of Hrs and ΔH with energy in flow.

$$q_r^-(\tau) = 2\sigma T_\infty^4 E_3(\tau_0-\tau)+2\int_\tau^{\tau_0}\sigma T_p^4(\tau')E_2(\tau'-\tau)d\tau' \quad (6)$$

$$q_r^+(\tau) = 2[\varepsilon\sigma T^4+(1-\varepsilon)q_r^-(0)]E_3(\tau)+2\int_0^\tau\sigma T_p^4(\tau')E_2(\tau-\tau')d\tau' \quad (7)$$

An amount of energy transfer from the radiative converter to the slab side Hrs is obtained by integrating the net heat flux $q_{rs}\cdot A$ with the time, where A is the area of the slab side.

$$q_{rs} = \varepsilon q_r^-(0) - \varepsilon\sigma T^4 \quad (8)$$

$$H_{rs}(t) = \int_0^t Q_{rs}\,dt'' = \int_0^t q_{rs}A\,dt'' \quad (9)$$

The increase of ΔH (=H(t)- Ho, where Ho is an amount of energy storage in a slab at t=0 sec) in the energy storage in the slab H is obtained by using the mean slab- temperature which is calculated on the basis of the temperature distribution.

FIGURE 18 shows the variation of Hrs and ΔH with the amount of energy inflow Q_{in}. The ratio of Hrs and ΔH arroaches about 0.5. It proves that the radiative converter effectively heats the slab edge. In addition, ΔH increases markedly with Reynolds number. This implies that the radiative and convective heat transfer on the slab surface makes a major contribution to the slab heating for high Reynolds number.

CONCLUSIONS

An effective temperature equalizer composed of the radiative converter and the impinging jet was proposed simulating the edge heating in the CC-DR process. The effect of the radiative converter (porous media) was demonstrated experimentally and was more remarkable to convert the energy between the gas enthalpy and the thermal radiation. The effect of the impinging jet which the impingement flow transported the thermal energy transfered by thermal conduction within the slab, was evaluated numerically. These two effects contribute to the temperature increase at the slab edge.

We hope the porous medium, which has the good thermal radiative properties, the high heat resistace and the high corrosion resistance, should be developed to operate the radiative converter for a long period and the thermal energy should be controlled more effectively in the whole production system of the high temperature steel slab.

REFERENCES

1. Tsubakihara,, O., Kusano, A., Terada, T., Yamamoto, T., Shirabe, K., and Ohhashi, W., Progress of the Iron and Steel Technologies in Japan in the Past Decade : Continuous Casting,(in Japanese), Journal of the Iron and Steel Institute of Japan, vol.71, no.3, pp.425-446, 1985.

2. Hachiya, S., Yamaguchi, T., Tanaka, I., Iso, H., and Tsubakihara, O., A Continuous Casting Machine Designed for CC-DR and its Operation Performance, Iron & Steelmaker, vol.10, no.9, pp.18-28, 1983.

3. Hachiya, S., Asano, S., and Uekawa, S., CC-DR in Sakai Works,(in Japanese), Tekkou-kai , vol.33, no.7, pp.52-60, 1983.

4. Kawasaki, S., Arita, H., Kikunaga, M., Chida, Y., and Kajita, Z., Secondary Cooling Control Model for Continuous Casting-Direct Rolling Process,(in Japanese), Seitetsu Kenkyu, no.313, pp.13-19, 1984.

5. Schrewe, H., Developments in Continuous Casting of Steel,Ironmaking and Steelmaking, vol.8, no.2, pp.85-90, 1981.

6. Iida, Y., Kodama, M., Suzuki, K., Yamazaki, J., and Kojima, S., Development of Surface Temperature Control in Slab Caster,(in Japanese), Kawasaki Steel Technical Report, vol.11, no.3, pp.121-133, 1979.

7. Echigo, R., Effective Energy Conversion Method between Gas Enthalpy and Thermal Radiation and Application to Industrial Furnaces,Proceedings of 7 th Int. Heat transfer Conf.,Munchen, vol.6, pp.361- 366, 1982.

8. Ichimiya, K., Kobayashi, K., and Echigo, R., Fundamental Study on the Flattening of Temperature Distribution of a High Temperature Steel Slab (1st Report, Impingement Heat transfer and Flow Situation around a Heated Slab Caused by a Laminar Slot Jet),(in Japanese), Trans. of JSME, Ser.B, vol.52, no.474, pp.904-909, 1986.

9. Gardon, R., and Akfirat, J.C., The Role of Turbulence in Determining the Heat Transfer Characteristics of Impinging Jets, Int. J. Heat and Mass transfer, vol.8, pp.1261-1272, 1965.

10. Gardon, R., and Akfirat, J.C., Heat Transfer Characteristics of Impinging Two-Dimensional Jets, Trans. ASME., Ser. C, J. of Heat Transfer, vol.83, no. 1, pp.101-107, 1966.

11. Kumada, M., and Mabuchi, I., Study on the Impingement Heat Transfer,(in Japanese), Trans. of JSME, vol.35, no.273, pp.1053-1061,

1969.

12. Martin, H., Heat and Mass Transfer between Impinging Gas Jets and Solid Surfaces, Advances in Heat Transfer, ed. Hartnett, J.P. and Irvine, T.F. , vol.13, pp.1-60, Academic Press, 1977.

13. Ichimiya, K., Kobayashi, K., and Echigo, R., Heat Transfer Characteristics due to a Confined Turbulent Impinging Jet Flow, Proceedings of 1st KSME/JSME Thermal and Fluids Engineering Conf., Seoul, vol.2, pp.393-397, 1988.

14. Ichimiya, K., Kobayashi, K., and Echigo, R., Basic Study on the Flattening of Temperature Distribution of Steel Slab for Direct Rolling (Unsteady-State Heat transfer in Heat Conductive-Convective Field on the Flattening of Temperature Distribution) , Proceedings of 2nd ASME/JSME Thermal Engineering Joint Conf., Honolulu, vol.5, pp.79-84, 1987.

15. Gosman, A.D., Pun, W.M., Runchal, A.K., Spalding, D.B., and Wolfshtein, M., Heat and Mass Transfer in Recirculating Flow, Academic Press, London, 1969.

16. Echigo, R., Yoshida, H., and Mochizuki, T., Temperature Equalization by Radiative Converter for a Slab in Continuous Casting- Direct Rolling, Proceedings of 2nd ASME-JSME Thermal Engineering Joint Conf., Honolulu, vol.5, pp.63-69, 1987.

17. Zerkle, R.D., and Sunderland, J.E., The Transient Temperature Distribution in a Slab Subject to Thermal Radiation, Trans. of ASME, Ser.C, J. of Heat Transfer, vol.87, no.1, pp.453-459, 1984.

18. Crosbie, A.L., and Viskanta, R., Transient Heating or Cooling of One-Dimensional Solids by Thermal Radiation, Proceedings of 3rd Int. Heat Transfer Conf., Chicago, vol.5, pp.146-153, 1966.

A Summary of General Discussion of Session on Water Cooling in Steel Processing

W. J. Yang, University of Michigan
S. Nishio, University of Tokyo

Water jet-, spray- and film-cooling are very popular in steel processing, and it is indispensable to get good understanding of their heat transfer characteristics for control of the metallurgical and mechanical structures of steel. However, heat transfer characteristics in such cooling systems are still not clear because they depend on many parameters listed in the following:

water jet & water film	water spray
(1)liquid velocity (2)liquid temperature	(1')droplet & air velocities (1")droplet mass flux (2') droplet & air temperatures (3) droplet diameter
(4)surface size (5)surface roughness (6)thermal resistance of surface layer(oxide layer or coating layer) (7)thermal capacity of surface (8)surface material (9)transient or steady-state cooling	

However, this complicated aspect of the phenomena seems to indicate that we have many controllable parameters to secure desirable cooling curves.

In the session entitled "Steel Processing", Viakanta and Incropera presented a model of water-jet cooling heat-transfer on a moving steel plate. In the model, heat transfer area is divided into three regions;the impinging region dominated by single-phase heat transfer, the liquid film region by boiling heat transfer, and the dry region by the Leidenfrost heat transfer. They presented a correlation of single-phase heat transfer in the impinging region and also pointed out the importance of the effects of the parameters on boiling heat transfer in the liquid film region. Nishio and Ohkubo presented a review of the effects of flow and surface parameters on spray cooling heat transfer. They proposed a simple model of heat transfer in high-temperature region of dilute sprays and showed that heat flux in the high-temperature region of dilute sprays can be estimated as the summation of the sensible heat transfer heating up the droplets to the saturation temperature, convective heat transfer by air flow, and radiative heat transfer. They presented also experimental data of the independent effects on spray cooling heat transfer of surface roughness, wettability and thermal conductance of surface layer. Murata dealt with cooling problems in the thermo-mechanical control process(TMCP) of steel. In the paper, the effects on relatively-dense-spray heat transfer of spray subcooling, surface oxidization and roughness parameters (height and pattern) were discussed. He reported that a certain pattern of high surface protrusions brought about a stable cooling curve in the TMCP.

In the session entitled "Casting and Molding", Specht and Jescher presented experimental results of the effects of spray subcooling, surface roughness and surface material on heat transfer of relatively dense spray. Heidt, Specht and Jescher presented an analytical model of film-boiling heat transfer of water film-cooling and also presented the experimental data of the effects of concentration of impurities in cooling water on film cooling.

In summary, the heat transfer of water jet-, spray- and film-cooling strongly depends on wettability and thermal characteristics of the surface. In general, single-phase heat transfer and film-boiling heat transfer can be predicted by boundary layer theory. The incipience of nucleate boiling can be estimated by the so-called nucleation criteria if the probability density function of pre-existing nucleus is given. In addition, we have many correlations for nucleate boiling heat transfer, and the critical heat flux can be estimated by droplet dryout or macrolayer dryout models. However, as for the transition-boiling heat transfer and the minimum heat flux condition, we have no effective models to estimate the effects of flow and surface parameters.

Here again, cooling heat transfer by water changes depending on many surface parameters, but the effects of these parameters should be clarified to search for how to control the cooling curve to obtain the prescribed metallurgical and mechanical structures of steel.

CASTING AND MOLDING (1)

Heat Transfer in Continuous Casting During Spray-Water Cooling

E. SPECHT and R. JESCHAR
Institute of Energy Process Technology
Technical University of Clausthal
Germany

INTRODUCTION

For the quenching of hot metals different techniques are employed taking advantage of the high heat transfer of vaporising liquids.

Essentially, three techniques can be differentiated as illustrated in Fig. 1: the film, the immersion and the spray quenching [1,2]. For film quenching water is poured over the metals. For immersion quenching the workpieces are dipped into baths of different liquids, and for spray quenching they are impinged by water coming from jets. Some examples for film quenching shall be mentioned now: the direct quenching for thermal–mechanical rolling in order to obtain defined structural properties, the quenching of e.g. inductively heated axles or forging metals for surface hardening, and the quenching during the continuous casting of non–ferrous metals. The main scope of immersion quenching is hardening, and that of spray quenching the continuous casting of steel.

To obtain defined properties of workpieces with these quenching techniques, it is necessary to know the standard heat transfer conditions and the possibilities to influence the heat transfer. This article only deals with spray quenching. In the following, the mechanism of the heat transfer will be explained first, and after that, the difference to the other two techniques will be clear.

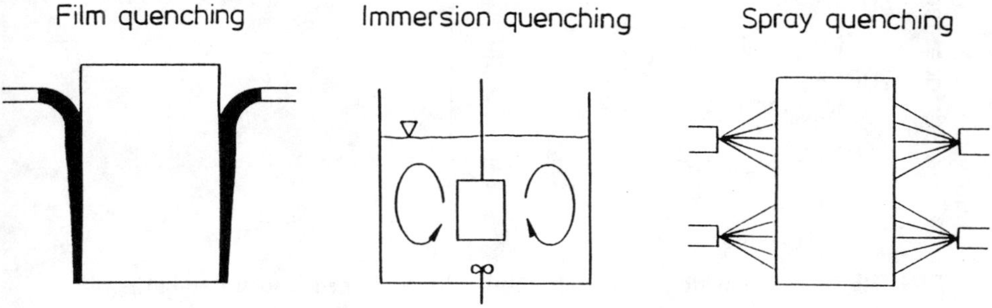

FIGURE 1: Quenching techniques

MECHANISM OF HEAT TRANSFER

Independent of the type of quenching a stable vapour film forms over the hot surface. This has an insulating effect so that the cooling during this film boiling is relatively low. The vapour film collapses as soon as the temperature of the surface falls below the Leidenfrost temperature. In the subsequent ranges of partial film boiling and nucleate boiling a considerably larger heat flow is transferred, allowing the solid to cool rapidly.

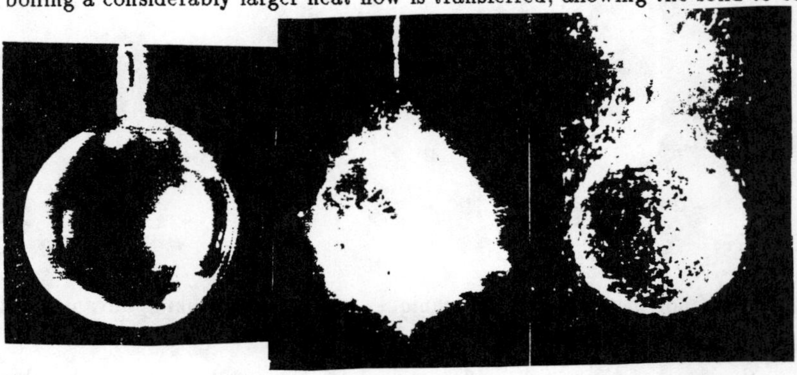

FIGURE 2: Photographs of film and nucleate boiling for a sphere [3,4]

In Fig. 2 the various phases of heat transfer are illustrated with a sphere dipped into a water bath [3,4]. One can recognize the compact vapour film formed around the sphere, as well as the phase a short time after the collaps of the film, and the nucleate boiling. The quenching curves associated with it are shown in Fig. 3. The temperature of the liquid serves as parameter. In the range of stable film boiling, which is passed through first, the temperature of the sphere falls slowly, almost independently of the bath temperature and

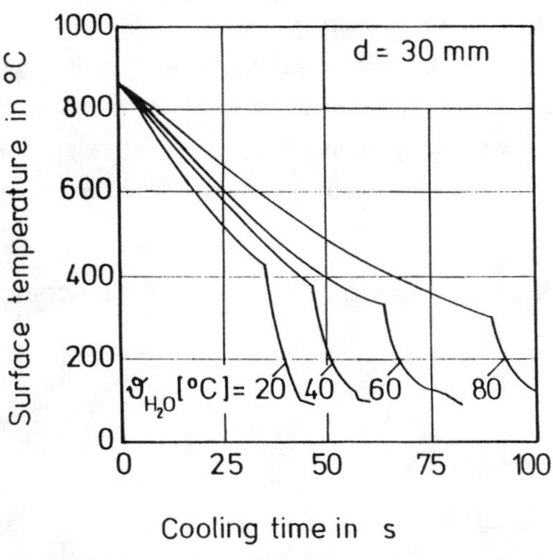

FIGURE 3: Quenching curves of nickel spheres dipped into water baths of different temperatures [3,4]

of the liquid velocity. As soon as the temperature falls below the Leidenfrost temperature and the film collapses, the surface temperature drops rapidly. The higher the Leidenfrost temperature is, i.e. the sooner the film collapses, the shorter is the total quenching time. An increase in the Leidenfrost temperature of a few degrees Kelvin is sufficient to reduce the quenching time of the solid considerably. These correlations show clearly that for a defined quenching of workpieces it is necessary to influence not only the heat transfer in the range of stable film boiling, but also the position of the Leidenfrost point. The heat loss during film boiling and the possibilities to influence the Leidenfrost temperature depend both on the type of the quenching technique, as will be explained further down below.

Fig. 4 shows the principle mechanism of heat transfer for the three cooling techniques described above. In all three cases the heat flow $\dot{q}$ emitted from the surface is transported through the vapour film by conduction ($\dot{q}_\lambda$) and by radiation ($\dot{q}_\epsilon$). The heat flux by radiation is much lower (generally less than 10%) than is the heat flux conducted through the vapour film. At the vapour–liquid interface the heat flow $\dot{q}$ is partly given to the liquid by convection ($\dot{q}_\alpha$), the remaining part being combined as latent heat of vaporization ($\dot{q}_v$). The temperature curves are also given in Fig. 4. Within the film the vapour temperature falls from the surface temperature ϑ_w to the saturation temperature ϑ_s of the liquid. From this temperature of the vapour–liquid interface the temperature decreases further to the ambient temperature ϑ_L of the liquid. The thickness δ of the vapour film and, therefore, the transferred heat flow are greatly influenced by the extent of the heat flow $\dot{q}_\alpha$ given

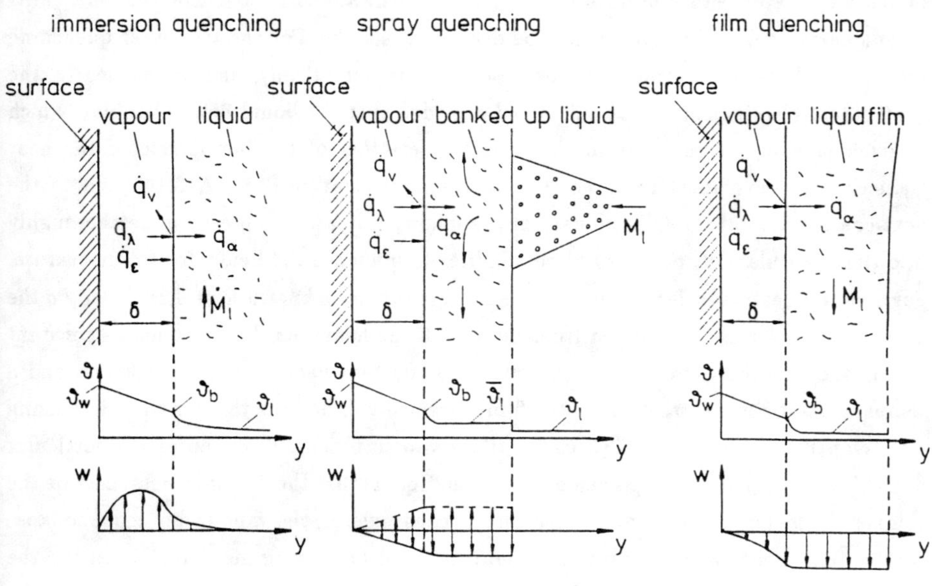

FIGURE 4: Mechanism of heat transfer during film boiling

into the liquid [2,5,6], if the liquid temperature lies below the saturation temperature. This is the case for instance for water with liquid temperatures below 60 °C. For technical quenching processes the latent heat of vaporization can therefore be neglected. The heat loss of the solid can be calculated approximately with the equations

$$\dot{q} = \frac{\lambda}{\delta} \cdot (\vartheta_w - \vartheta_s) \tag{1}$$

or

$$\dot{q} = \alpha \cdot (\vartheta_s - \vartheta_L) \tag{2}$$

if either the film thickness δ or the heat-transfer coefficient was known. The convective heat transfer is determined by the velocity profile in the liquid. Hence, in the lower part of Fig. 4 the velocity profiles are also given qualitatively. It can be seen that these profiles differ for all three quenching techniques. For film quenching, an approximately uniform velocity exists in the draining liquid over the cross section. The vapour flow is impressed by the liquid and, as a result, an approximately linear velocity course adjusts. In immersion quenching the flow in the liquid is formed by the vapour flow, in so far as the bath was initially inactive. As a result of the buoyancy the vapour flows in an upward direction, taking the liquid with it, as can be seen from the velocity curve. In spray quenching the flow of the vapour is also impressed by that of the liquid. The impinged liquid can flow off either in an upward or in a downward direction. For this quenching technique the velocity profile cannot be described exactly. For the two other quenching techniques the velocity profile can be described mathematically, and, consequently, the heat-transfer coefficient can be calculated provided that the liquid flow is laminar, which is given in many cases. The mathematical description of the mechanism of the heat transfer is not explained here; we refer to the nomenclature [2,5,6,7,13,14], and to the next contribution of the Oji seminar. In the following it will be entered more thoroughly into the particulars of the essential possibilities to influence the Leidenfrost temperature. For a known heat-transfer coefficient α and, therefore, for a known heat-flux density $\dot{q}$ the film thickness δ can be calulated from Eq. (1). If the heat-flux density remains constant during the quenching process, e.g. because of a constant heat- transfer coefficient and a temperature difference $\vartheta_w - \vartheta_s$, so the film thickness will become thinner with decreasing wall temperature according to Eq. (1), until the vapour film finally collapses at a particular film thickness. Furthermore, it can be seen from Eq. (1) that the film thickness must be the thinner the higher the heat-flux density is. In consequence, the vapour film must collapse earlier, that means at higher wall temperatures. All processing measures increasing the heat transfer also shift the Leidenfrost temperature to higher values. As the heat-transfer coefficient is also dependent on the flow length along with the flow velocity the Leidenfrost temperature is dependent on this flow length too. So, the Leidenfrost temperature lies the higher the smaller the overflowed solid is and the shorter the flow length is, that means, the vapour film collapses always at the edges first and the wetting fornt moves from this point into the direction of the flow. At the end of these reflections it should be emphasized that from Eq. (1) the vapour film thickness can be calculated in dependence on the wall temperature, the thickness or the wall temperature, however, which is responsible for the collapsing of the vapour film, is not calculable yet. At this time there is still no criterion in order to predict the Leidenfrost temperature. One has to rely on experimental results. Those experiments show that besides parameters influencing the convective heat transfer there are others which have an influence, such as different kinds of solid material, surface roughness, surface stress, impurities of the liquid and so on. In the following the heat transfer for spray quenching will be dealt with exclusively, which so far has been researched in detail only experimentally, because as mentioned above a theoretical description of the velocity profile is not possible yet.

EXPERIMENTAL ARRANGEMENT

The heat transfer was determined by an insteady and a steady experimental technique. For the insteady method a cylindrical metal sample, which has been heated in a tube furnace first, is sprayed with water on its front side. With thermocouples, which are placed in different positions in the middle of the sample, the quenching curves are registrated with a fast working measuring equipment. From the measured temperature-time-curves the local temperature distribution is numerically calculated with the Fourier's differential equation. From the temperature gradient at the quenched surface one finally gets the heat-flux density and the heat-transfer coefficient belonging to it.

For the steady method an electric current is passed through a metal sample. Because of Ohm's resistance the sample heats itself up. In order to limit the temperature of the sample the electrical power is regulated. The result is an equilibrium between the electrical power supplied, and the heat loss caused

by spraying with water. The steady temperature profile is calculated from the electrical power and from the core temperature of the sample. With that the heat-flux density can also be determined.

Compared with the insteady method the steady measuring method offers the advantage to reduce the time of heating up the samples. The steady method, however, is only suitable for the range of the stable film boiling.

The heat-transfer coefficient determined with the above mentioned methods is defined as

$$\alpha = \frac{\dot{q}}{\vartheta_w - \vartheta_s} \quad , \tag{3}$$

- α = total heat–transfer coefficient,
- $\dot{q}$ = heat–flux density,
- ϑ_w = surface temperature, and
- ϑ_s = saturation temperature.

The heat–transfer coefficient defined by the above equation contains a heat flux due to radiation. Neither the radiative nor the convective portion of the heat loss can be separated. Owing to this the total heat–transfer coefficient will be used below.

A description of the measuring methods is given in [6,8,9,10] in more detail.

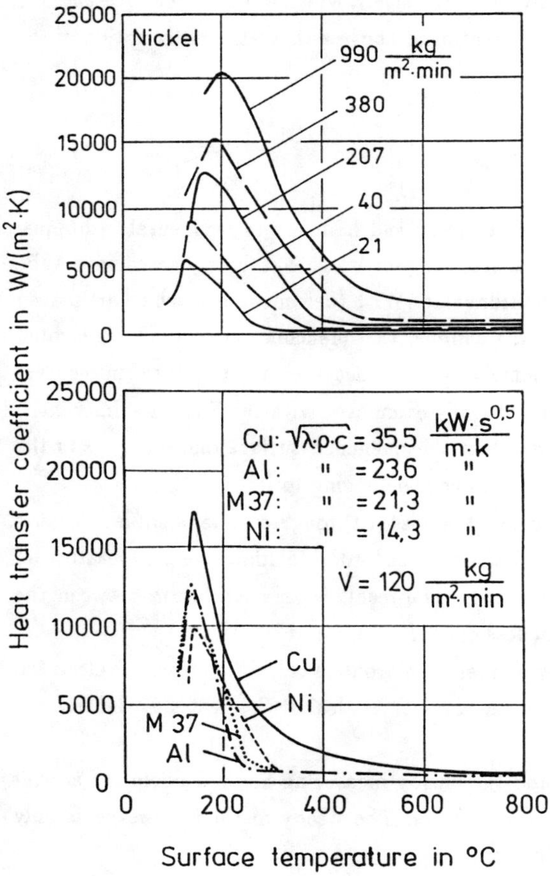

FIGURE 5: Total heat–transfer coefficient for spray quenching as a function of the surface temperature. Above: influence of the impingement density. Below: influence of the type of metal [11]

EXPERIMENTAL RESULTS

Fig. 5 shows heat–transfer coefficients obtained in this way in dependence on the surface temperature for water used as spray liquid [11]. In the upper figure the parameter is the impingement density, which indicates how much liquid reaches the hot solid per unit of time and area. An increase in the impingement density causes the heat transfer to improve, and the Leidenfrost temperature to shift to higher values. The lower figure indicates that the heat transfer is also dependent on the type of metal, albeit only below the Leidenfrost temperature.

Furthermore, from both figures it can be seen that in the range of the stable film boiling the heat–transfer coefficient is independent of the surface temperature. For continuous casting of steel, for which spray quenching is used most frequently, the surface temperatures are above 1000 °C. Therefore, the heat transfer always occurs in the range of stable boiling. Hence, it will be dealt with the possibilities of influencing the heat transfer in this range particularly. It was investigated, if there would exist further parameters in addition to the impingement density. For this reason, some conditions were changed:

- momentum flow of the water at the nozzle exit,
- adjustments, which are specific for the nozzle, as for instance spray angle, diameter of the nozzle, distance between nozzle and surface, and drop velocity,
- water spray quenching (water pressure atomization), and water–air–spray quenching (water–air–atomization)
- size of the sample
- roughness of the surface
- temperature of the water

The result was that the influences of the momentum flow, of the adjustments specific for the nozzle, and of the water pressure — such as water–air–atomization — can be combined in a single parameter, namely in the impingement density, and therefore they do not have to be considered separately. Furthermore, this parameter can be measured very simply. A dependence of the measuring results on the chosen size of the sample could not be found.

The influence of the roughness on the heat transfer is shown in Fig. 6 [12]. The average roughness of the surface was determined by DIN 4768. The impingement density was fixed in these experiments. One can recognize in this plot that there does not exist any influence of the surface roughness in the range of stable film boiling. On the other hand, the Leidenfrost temperature shifts to higher values with increasing roughness.

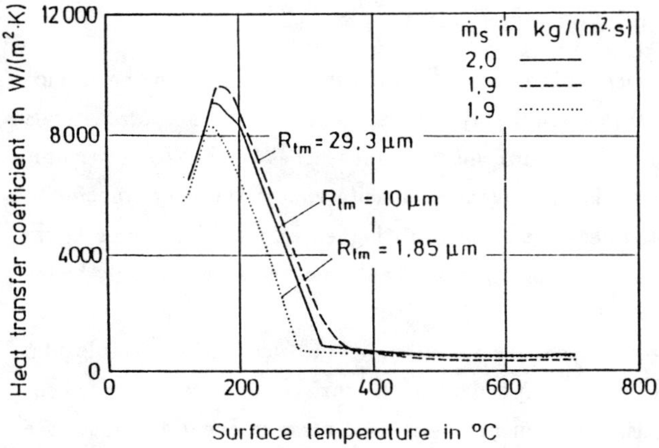

FIGURE 6: Total heat transfer coefficient in dependence on the the surface temperature with the mean roughness of the metal surface as parameter [6,12]

The present results are for water temperatures of 20 °C. In Figure 7 the heat-transfer coefficient is plotted in dependence of the impingement density with the water temperature as parameter for a surface temperature of 700 °C. For high values of the impingement densities the heat-transfer coefficient, and together with it the heat-flux density decrease with rising water temperature. As in practical operation the water temperatures are about 20 °C, their influence can therefore be neglected.

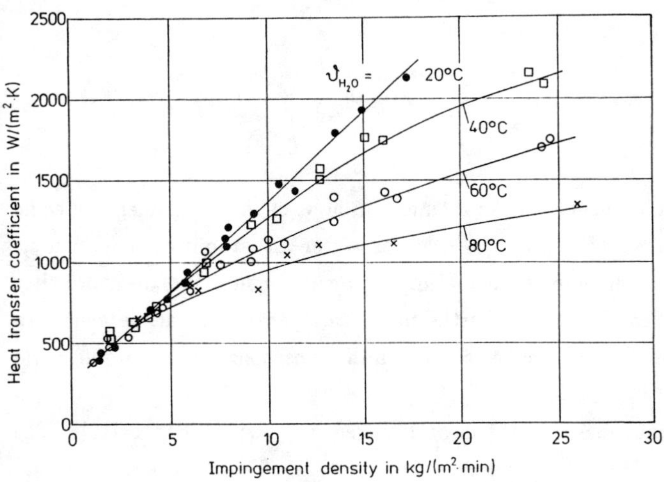

FIGURE 7: Total heat transfer coefficient in dependence on the impingement density with the liquid temperature as parameter [6,12]

Finally, in Figure 8 the heat-transfer coefficient is shown in dependence on the impingement density. The measuring points are for different nozzle geometries (cone nozzle, fan-jet nozzle) and water pressure-atomization, such as for instance water-air-atomization. The straight line can be reproduced by the equation

$$\frac{\alpha_{ges}}{W/(m^2 \cdot k)} = 200 + 1.82 \cdot \frac{\dot{m}}{\mathrm{kg/(m^2 \cdot min)}} \tag{4}$$

which is valid for the range of 100 to 2000 $\mathrm{kg/(m^2 \cdot min)}$.

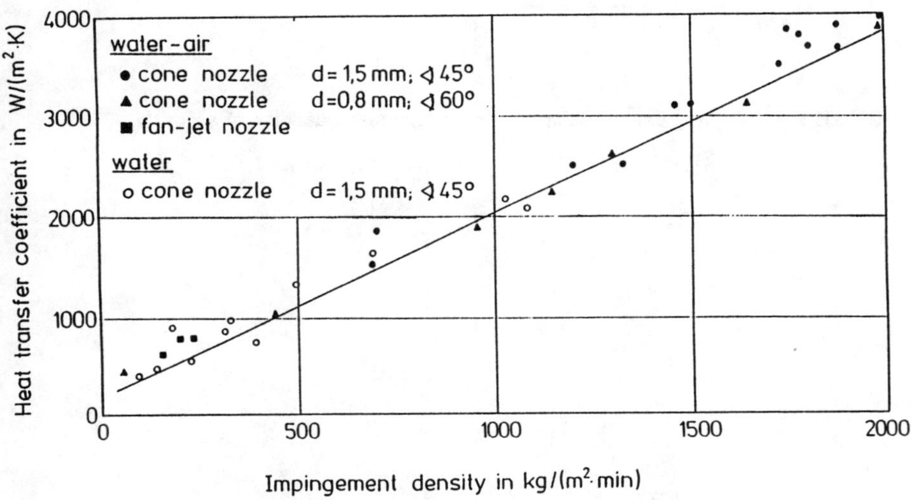

FIGURE 8: Heat transfer coefficient in the range of film boiling for spray quenching with water as a function of the impingement density [6,11,12]

As the impingement density is the only variable controlling the heat flow in the range of the stable film boiling, the local distribution of the impingement density is of particular importance. Figure 9 shows an example of a typical radial distribution of the impingement density. Figure 10 shows the field of the heat-transfer coefficient for a slot jet as a result of such a measured distribution of the impingement density.

It is evident that the very large gradients for the heat-transfer coefficient with the high absolute maximum can generate large temperature differences at the metal surface and, therefore, material defects (e.g. cracks). These can be avoided by producing the most uniform impingement density distribution possible. This can be achieved, e.g. by cooling with several nozzles and arranging these in such a way that their water distributions overlap accordingly.

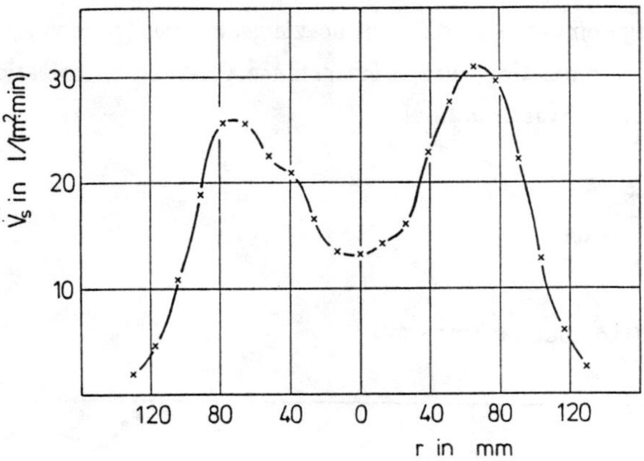

FIGURE 9: Typical distribution of the impingement density of a nozzle

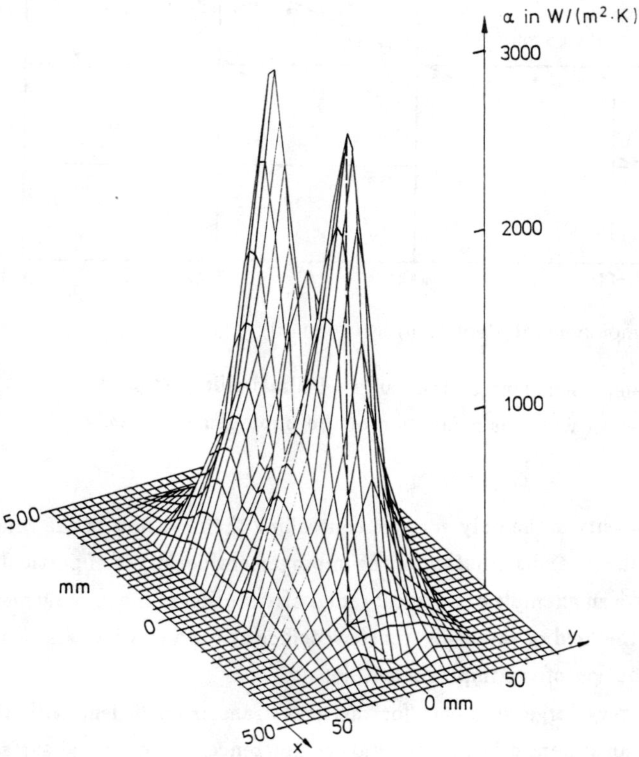

FIGURE 10: Characteristics of a nozzle according to the heat-transfer coefficient for a slot nozzle

In industrial practice, two-component nozzles, i.e. water-air nozzles, are being increasingly employed. The addition of air as second quenching medium mainly serves to influence the nozzle characteristics. The cooling effect of these nozzles is, however, only determined by the magnitude of the impingement density.

According to Figure 5 the heat flow in the range of partial film boiling and nucleate boiling depends not only on the impingement density but also on the type of metal. This influence can be described by the coefficient of heat penetration $\sqrt{\lambda \cdot \rho \cdot c}$. Figure 11 shows the dependence of the Leidenfrost temperature ϑ_{Le}, the burn-out-temperature ϑ_{bo}(the temperature of maximum heat flux density) and the maximum heat transfer coefficient α_{max} on the impingement density for four metals. All three variables approximately increase exponentially with the impingement density. Hence, the Leidenfrost temperature, the burn-out-temperature and the maximum heat transfer coefficient can be described by the relation

$$\alpha_{max}, \vartheta = B \cdot \left[\frac{\dot{m}}{\mathrm{kg/(m^2 \cdot min)}}\right]^n \tag{5}$$

The variables B and n are compiled in Table 1 for the four given metals.

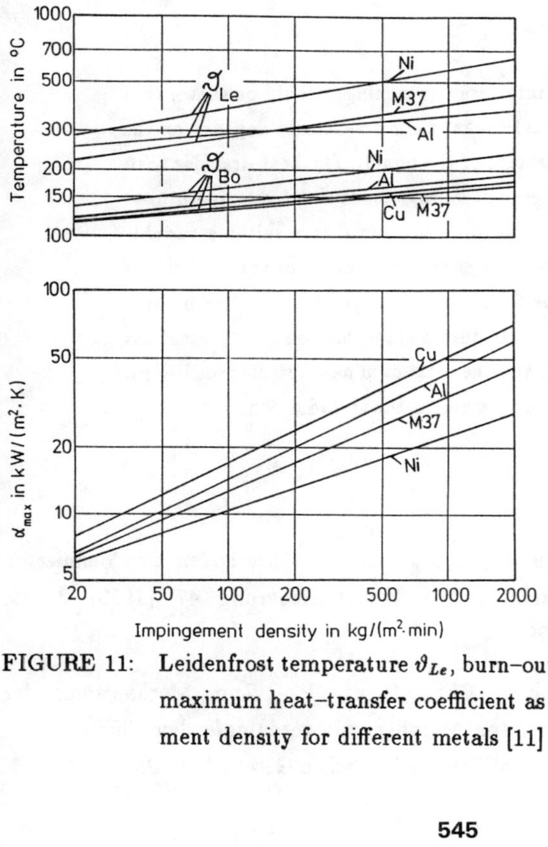

FIGURE 11: Leidenfrost temperature ϑ_{Le}, burn-out-temperature ϑ_{bo}, and the maximum heat-transfer coefficient as a function of the impingement density for different metals [11]

	$\sqrt{\lambda \cdot \rho \cdot c}$	ϑ_{Le}		ϑ_{Bo}		α_{max}	
		B	n	B	n	B	n
Metal	$\frac{kW \cdot \sqrt{s}}{m^2 \cdot K}$	°C	-	°C	-	$\frac{W}{m^2 \cdot K}$	-
Copper	35,5	-	-	97	0,08	1840	0,48
Aluminium	23,6	190	0,09	83	0,11	1530	0,48
Brass	21,3	126	0,17	80	0,11	1890	0,41
Nickel	14,3	156	0,19	72	0,16	2360	0,32

TABLE 1: Data to calculate the Leidenfrost temperature/ and the burn-out-(DNB)-temperature for spray quenching

To calculate the coefficient of heat penetration it must be taken into account that the material values at the surface are decisive. For copper, in particular, an oxide layer always forms at the surface in which $\sqrt{\lambda \cdot \rho \cdot c}$ contains considerably smaller values than in the metal itself.

ABSTRACT

Because of numerous investigations into spray quenching it could be shown that in the range of stable film boiling the heat-transfer coefficient only depends on the impingement density and on the temperature of the liquid. The heat-transfer coefficient increases linearly with the impingement density for low water temperatures. The drop impulse increasing with added air does not serve as a suitable variable to describe the heat transfer. Neither an influence of roughness of the surface nor an influence of the sample size could be found in the range of stable film boiling. Equations were given for the heat flux density during stable film boiling, Leidenfrost temperature, the burn-out-temperature and the maximum heat transfer coefficient. The mechanism of the heat transfer was explained in the introduction.

REFERENCES

[1] Jeschar, R.; Köhler, Chr.; Specht, E.; Heidt, V.: Methoden zur definierten Abkühlung metallischer Werkstoffe. Die Industriefeuerung 47 (1989) 34-39, Gaswärme int. 38 (1989) 223-229.

[2] Jeschar, R.; Specht, E.; Köhler, Chr.; Heidt, V.: Zum Mechanismus der Wärmeübertragung beim Kühlen von Metallen mit verdampfenden Flüssigkeiten. Abhandlung der Braunschweigischen Wissenschaftlichen Gesellschaft, Jahrbuch 1989.

[3] Maaß, R.; Jeschar, R.: Einflußgrößen des Wärmeübergangs beim Abschrecken von Metallen in Wasser, Gaswärme int. 38 (1989) 142–150.

[4] Maaß, R.: Untersuchung zur Wärmeübertragung beim Abschreckhärten. Thesis, Technical University of Clausthal, Germany (1988).

[5] Köhler, Chr.; Specht. E.; Jeschar, R.: Heat transfer with film quenching of vaporising liquids. Steel research (1990) Nr. 11.

[6] Reiners, U.: Wärmeübertragung durch Spritzwasserkühlung heißer Oberflächen im Bereich der stabilen Filmverdampfung. Thesis, Technical University of Clausthal (Germany), (1987).

[7] Köhler, Chr.: Wärmeübertragung von heißen Oberflächen durch Wasserfilmkühlung im Bereich der stabilen Filmverdampfung. Thesis, Technical University of Clausthal, Germany (1990).

[8] Reiners, U.; Jeschar, R.; Scholz, R.; Zebrowski, D.; Reichelt, W.: A measuring method for quick determination of local heat transfer coefficients in spray water cooling within the range of stable film boiling. Steel research 56 (1985) 239–246.

[9] Jeschar, R.; Scholz, R.; Reiners, U.: Heat transfer during one and two–phase water–spray cooling in hot metals. The Sixth Japan–Germany Seminar. Proceedings. (1984) 146–159.

[10] Jeschar, R.; Reiners, U.; Scholz, R.; Heat Transfer During Water And Water–Air Spray Cooling in the Secondary Cooling Zone of Continuous Casting Plants. Proceedings of the 69th Steelmaking Conference in Washington, Vol. 69 (1986) 511–521.

[11] Müller, H.R.; Jeschar, R.: Wärmeübergang bei der Spritzwasserkühlung von Nichteisenmetallen. Metallkunde 74 (1983) 257–264.

[12] Reiners, U.; Jeschar, R.; Scholz, R.: Wärmeübertragung bei der Stranggußkühlung durch Spritzwasser. Steel research 60 (1989) 442–450.

[13] Shigechi, T.; Ito, T.; Nishikawa, K.: Forced Convection Film Boiling Heat Transfer from a Horizontal Cylinder to Liquid Cross–Flowing Upwards. Bulletin of the JSME 26 (1983) No. 214–10. 554.

[14] Nakayama, A.; Koyama, H.: Two phase boundary layers in laminar film boiling. Wärme– und Stoffübertragung 19 (1985) 211–216.

Heat Transfer in Continuous Casting During Film Cooling

V. HEIDT, E. SPECHT, and R. JESCHAR
Institute of Energy Process Technology
Technical University of Clausthal
Germany

INTRODUCTION

In steel industry and in nonferrous metal industry liquid metal is mainly cast in continuous casting plants. For this process the metal passes through a water cooled mould first, where it sets on the surface. After, the strand, which is still fluid inside the core, is further cooled down in the so-called secondary cooling zone by water [1]. For nonferrous metals the cooling water is, as illustrated in figure 1, mostly given onto the surface as gush of water by way of a distribution system, from where it runs down along the strand as closed water film.

For water film cooling other possible applications are the thermal-mechanical rolling and the surface-hardening of inductively heated engineering components out of steel. In order to obtain defined properties of workpieces and to avoid surface irregularities, it is necessary to know the standard heat-transfer conditions and the possibilities to influence them. In this paper a mathematical model, which was developed for the film cooling, is presented, and the results gained by it are compared with experimentally determined values. Furthermore, the qualitative influence of the cooling water quality on the quenching progress is shown by experimental results.

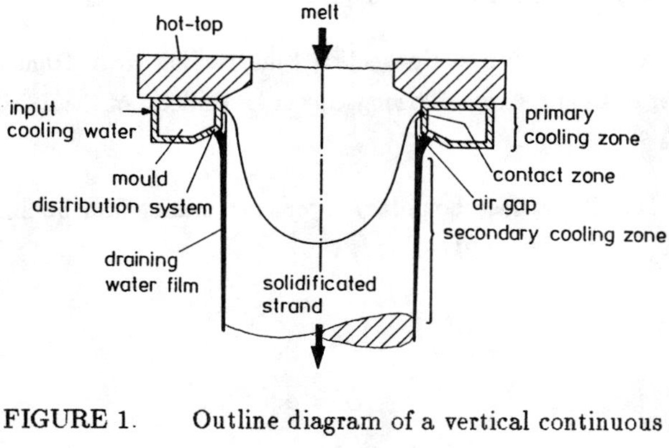

FIGURE 1. Outline diagram of a vertical continuous caster for nonferrous metals

MECHANISM OF HEAT TRANSFER

If for the cooling of hot substances the surface temperature is far above the saturation temperature of the cooling water, a stable closed vapour film will form, of which the thermal conductivity is far less than that of the liquid. This process is called stable film boiling. After reaching the Leidenfrost temperature at the surface the insulating vapour film collapses. In the following range of the partial film boiling the emitted heat flow increases very strongly, as shown in figure 2 for a typical profile. With this the cooling speed also rises many times over. After the burn-out-temperature has passed through, which is characterised by the maximum appearing heat-flow, the area of the nucleate boiling is entered.

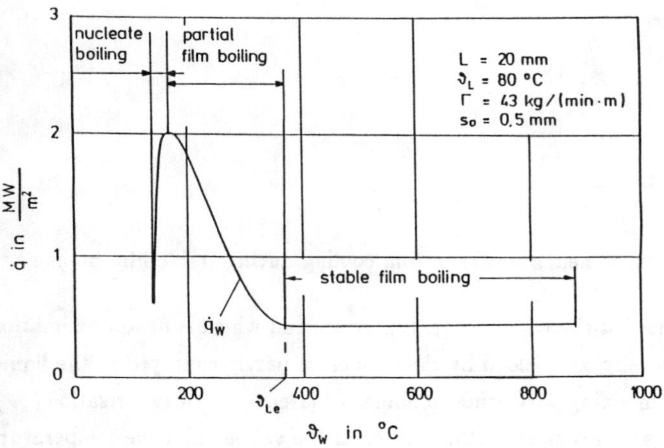

FIGURE 2. Typical cooling curve for the film cooling

For concrete calculations of the cooling process the range of the stable film boiling is of particular interest, as the period of this phase essentially determines the total cooling time. After falling below the Leidenfrost temperature the body temperature decreases very quickly because of the heat flow rising rapidly. Figure 3 presents the basic mechanism of the heat transfer in the range of stable film boiling taking as an example film cooling. The heat-flux density $\dot{q}_W$ emitted from the surface is transported to the vapour-liquid interface by conduction ($\dot{q}_\lambda$) and by radiation ($\dot{q}_\varepsilon$). With that the conduction fraction is reduced by the heat flow $\dot{q}_{SP}$, which serves as superheating the vapour. As it has turned out, the vapour mass flow, however, is so small in the technically important cooling water temperatures below 60 °C that compared to the conduction fraction the superheating fraction may be neglected.

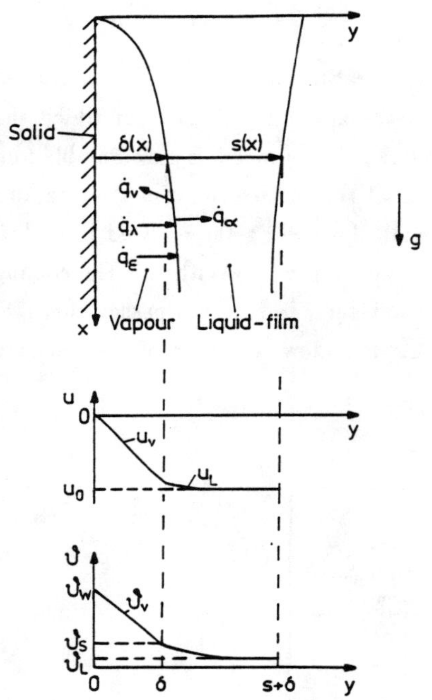

FIGURE 3. Mechanism of heat transfer in film cooling during stable film boiling

The heat flow which arrives at the vapour–liquid interface and which is by approximation equal to the heat–flux density $\dot{q}_W$ yielded by the surface, is partly conveyed to the liquid by convection ($\dot{q}_\alpha$), the remaining part being combined as latent heat of vaporization ($\dot{q}_V$). The temperature curves are also given in this figure. In the vapour film the temperature goes down from the surface temperature ϑ_W to the saturation temperature ϑ_S of the cooling liquid at the vapour–liquid interface. In the liquid the temperature decreases further to the bulk liquid temperature. The thickness δ of the vapour film and, therefore, the transferred heat flow are greatly influenced by the extent of the heat flow $\dot{q}_\alpha$ given into the liquid, as will be explained below. On the other hand, this convective heat transfer is determined from the velocity profile in the liquid. An approximately uniform velocity exists in the draining liquid over the cross section, as illustrated in figure 3. The vapour flow is impressed by the liquid and as a result, an approximately linear velocity course adjusts.

The hitherto qualitatively explained mechanism of the heat transfer is described in the following mathematically.

MATHEMATICAL MODEL OF HEAT TRANSFER FOR FILM COOLING

For the heat flow conducted through the vapour film the following equation is valid

$$\dot{q}_\lambda = \frac{\lambda_V}{\delta} \cdot (\vartheta_W - \vartheta_S) \tag{1}$$

where λ_V is the mean thermal conduction coefficient of the vapour, δ the thickness of the vapour film, ϑ_W the wall temperature and ϑ_S the saturation temperature. In most cases the radiation superposing this heat flow can be neglected. For water temperatures below $\vartheta_L = 60$ °C and surface temperatures up to 800 °C, the radiation part is lower than 10% of complete heat flow from the wall. At the vapour–liquid interface the heat flow is partly transported further to the liquid by convection and the remaining part is combined as latent heat of vaporization

$$\dot{q}_\lambda = \dot{q}_\alpha + \dot{m}_V \cdot h_{LV} \tag{2}$$

where $\dot{m}_V$ is the evaporating mass flow and h_{LV} the latent heat of vaporization. With these equations the heat flow $\dot{q}_\lambda$ transported by the wall and the film thickness δ can be calculated, if $\dot{q}_\alpha$ and $\dot{m}_V$ are known.

To calculate the heat flow $\dot{q}_\alpha$ it is assumed that the velocity u is uniform over the cross section of the liquid. If u_0 is the initial velocity at which the liquid reaches the workpiece, then the velocity along the workpiece due to gravity is

$$u = \sqrt{u_o^2 + 2gx} \tag{3}$$

where g is the acceleration due to gravity and x the flow route. From the mass balance

$$\frac{\partial u}{\partial x} + \frac{\partial v}{\partial y} = 0 \tag{4}$$

the following is obtained for the velocity component v transverse to the direction of flow

$$v = -\frac{g \cdot y}{\sqrt{u_o^2 + 2gx}} \tag{5}$$

From the Fourier differential equation

$$u\,\frac{\partial \vartheta}{\partial x} + v\,\frac{\partial \vartheta}{\partial y} = a_L\,\frac{\partial^2 \vartheta}{\partial y^2} \tag{6}$$

(a_L thermal diffusivity of the liquid) results for the temperature distribution in the liquid

$$\frac{\vartheta - \vartheta_s}{\vartheta_L - \vartheta_s} = \text{erf}\left(\sqrt{\frac{3}{4}} \cdot \sqrt{\frac{\frac{g}{a_L} \cdot y^2}{\sqrt{u_o^2 + 2gx} - \frac{u_o^3}{u_o^2 + 2gx}}}\right) \tag{7}$$

where ϑ_S is the saturation temperature at $y = \delta$ and ϑ_L the liquid temperature at $y = \infty$ and $x = 0$. The heat flow density transferred by convection at the liquid–vapour interface has to be as high as the heat flow density which is transported by conduction from the vapour–liquid interface to the liquid.

$$\dot{q}_\alpha = -\lambda_L \left.\frac{\partial \vartheta}{\partial y}\right|_{y=\delta} \tag{8}$$

Using equation (7) yields

$$\dot{q}_\alpha = \sqrt{\frac{3}{\pi}} \cdot \sqrt{\frac{Gn_x \cdot (Pe_x^2 + 2 \cdot Gn_x)}{(Pe_x^2 + 2 \cdot Gn_x)^{3/2} - Pe_x^3}} \cdot \frac{\lambda_L}{x} (\vartheta_S - \vartheta_L) \tag{9}$$

In this equation means

$$Pe_x = \frac{u_o \cdot x}{a_L} \tag{10}$$

the Peclet number, and

$$Gn_x = \frac{g \cdot x^3}{a_L^2} \tag{11}$$

a dimensionless number, here called Gravity number. To calculate the evaporating mass flow $\dot{m}_V$ it is assumed that the velocity distribution in the vapour film is linear

$$u_V = \sqrt{u_o^2 + 2gx} \cdot \frac{y}{\delta} \tag{12}$$

As such, the vapour mass flow in x–direction is

$$\dot{M}_x = \int_0^\delta \rho_V \cdot u_V \cdot b \cdot dy = \frac{1}{2} \rho_V \sqrt{u_o^2 + 2gx} \cdot b \cdot \delta \tag{13}$$

where ρ_V is the mean vapour density, b the width of the vapour film and y the coordinate vertical to the flow route. The evaporating mass flow is equal to the change in mass of the vapour in the direction of flow

$$\dot{m}_V = \frac{1}{b} \cdot \frac{d\dot{M}_x}{dx} \tag{14}$$

With this, from the equations (1), (2) and (4) one obtains a differential equation for the local change of the vapour film thickness δ. The solution of the differential equation is

$$\frac{\delta}{x} = \frac{2}{\sqrt{3\cdot\pi}} \cdot \frac{\rho_L}{\rho_V} \cdot Ph \cdot \left(\sqrt{1 + \frac{\pi}{Ph\cdot\theta} \cdot \frac{\lambda_V}{\lambda_L} \cdot \frac{\rho_V}{\rho_L}} - 1 \right) \cdot \sqrt{\frac{(Pe_x^{\,2} + 2\cdot Gn_x)^{3/2} - Pe_x^{\,3}}{Gn_x \cdot (Pe_x^{\,2} + 2\cdot Gn_x)}} \tag{15}$$

with

$$Ph = \frac{c_L \cdot (\vartheta_S - \vartheta_L)}{h_{LV}} \tag{16}$$

the phase change number, and

$$\theta = \frac{\vartheta_S - \vartheta_L}{\vartheta_W - \vartheta_S} \tag{17}$$

a dimensionless temperature.

The equation for the heat-flux density conducted through the vapour layer can subsequently be derived from equation (1)

$$\dot{q}_\lambda = \frac{\sqrt{3\cdot\pi}}{2} \cdot \frac{\rho_V}{\rho_L} \cdot \frac{\lambda_V}{x} \cdot \frac{1}{Ph} \cdot \frac{\vartheta_W - \vartheta_S}{\sqrt{1 + \frac{\pi}{Ph\cdot\theta} \cdot \frac{\lambda_V}{\lambda_L} \cdot \frac{\rho_V}{\rho_L}} - 1} \cdot \sqrt{\frac{Gn_x \cdot (Pe_x^{\,2} + 2\cdot Gn_x)}{(Pe_x^{\,2} + 2\cdot Gn_x)^{2/3} - Pe_x^{\,3}}} \tag{18}$$

With above equations it can now be calculated which portion of the heat conducted through the film is transferred into the liquid

$$\frac{\dot{q}_\alpha}{\dot{q}_\lambda} = \frac{2}{\pi} \cdot \frac{\lambda_L}{\lambda_V} \cdot \frac{\rho_L}{\rho_V} \cdot Ph \cdot \theta \cdot \left(\sqrt{1 + \pi \cdot \frac{\lambda_V}{\lambda_L} \cdot \frac{\rho_V}{\rho_L} \cdot \frac{1}{Ph\cdot\theta}} - 1 \right) \tag{19}$$

The remaining part is used for the vaporization of the liquid. Figure 4 shows this ratio for water as a function of the liquid temperature with the surface temperature of the solid as parameter.

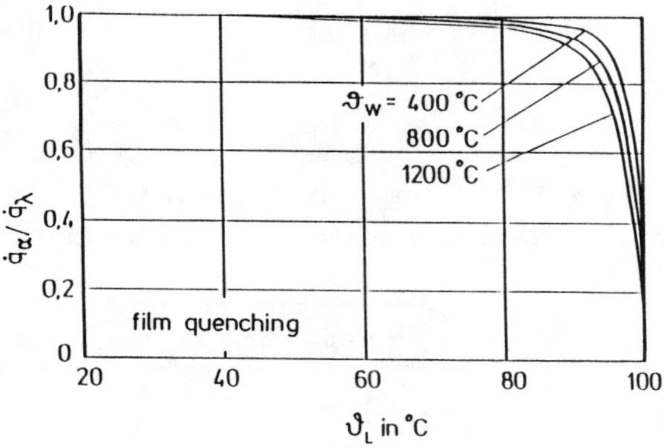

FIGURE 4. Portion of the total heat flux transported into the liquid

Hence, for water temperatures up to 80 °C the heat–flux density $\dot{q}_\lambda$ is almost entirely transferred into the liquid. As can be displayed mathematically also, equation (17) changes into equation (9), and disregarding the initial velocity u_0 ($Pe = 0$) one obtains from equation (15)

$$\frac{\delta}{x}\left(\vartheta_L << \vartheta_S;\ Pe = 0\right) = \sqrt[4]{\frac{2\cdot\pi^2}{9}} \cdot \frac{\lambda_V}{\lambda_L} \cdot \frac{1}{\sqrt[4]{Gn_x}} \cdot \frac{\vartheta_W - \vartheta_S}{\vartheta_S - \vartheta_L} \tag{20}$$

e.g. neglecting the acceleration due to gravity ($Gn = 0$)

$$\frac{\delta}{x}\left(\vartheta_L << \vartheta_S;\ Gn = 0\right) = \sqrt{\pi} \cdot \frac{\lambda_V}{\lambda_L} \cdot \frac{1}{\sqrt{Pe_x}} \cdot \frac{\vartheta_W - \vartheta_S}{\vartheta_S - \vartheta_L} \tag{21}$$

From this equation it is evident that the thickness of the vapour film is proportional to the temperature difference $\vartheta_W - \vartheta_S$. Hence, the thickness of the film adjusts such that for a decreasing surface temperature the transferred heat–flux density remains constant according to equation (1). An influence of the surface temperature is observed only at very high water temperatures, when the heat is no longer transported predominantly to the liquid and as such causes the liquid to vaporize [2].
Figure 5 shows the thickness of the vapour film for water of 20 °C as a function of the surface temperature. The curves are calculated with equation (21), which is valid for high initial velocities u_0 and short flow routes x. It can be seen that the film thicknesses δ are in the μm–range and are as such extremely thin when compared to the liquid film. Thus the thickness of the vapour film is of the same order of magnitude as the temperature boundary layer in the liquid film.

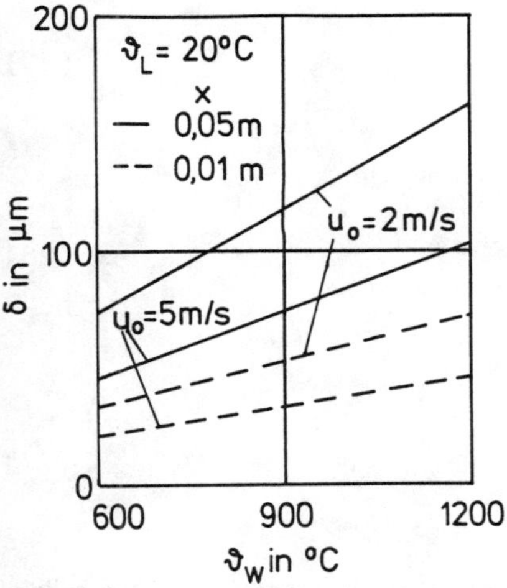

FIGURE 5. Indications for the magnitude of the vapour film thickness calculated by neglecting the acceleration due to gravity ($Gn = 0$)

The given results are confirmed by further model calculations which are based on the boundary layer equations for the momentum and energy transport and which, furthermore, take the temperature dependency of the material values into account [3].
In many technical cases, it is not so much the local heat–flux density which is of interest rather than the mean heat–flux density which is transported along the length of the solid L

$$\bar{\dot{q}}_\alpha = \frac{1}{L} \int_0^L \dot{q}_\alpha \cdot dx \tag{22}$$

for which the following equation can be derived from equation (9)

$$\bar{\dot{q}}_\alpha = \frac{2}{\sqrt{3 \cdot \pi}} \cdot \frac{\lambda_L}{L} \cdot (\vartheta_S - \vartheta_L) \sqrt{\frac{(Pe^2 + 2 \cdot Gn)^{3/2} - Pe^3}{Gn}} \tag{23}$$

The Peclet and Gravity numbers in this equation are to be formed using the length of the solid L, as according to equations (10) and (11), in place of the flow route x. If a heat–transfer coefficient is defined according to

$$\bar{\dot{q}}_\alpha = \alpha \cdot (\vartheta_S - \vartheta_L) \quad , \tag{24}$$

then equation (23) results in the following as Nusselt function for the range of film boiling

$$Nu = \frac{2}{\sqrt{3 \cdot \pi}} \sqrt{\frac{(Pe^2 + 2 \cdot Gn)^{3/2} - Pe^3}{Gn}} \qquad (25)$$

If the acceleration of the liquid can be neglected, this results in

$$Nu \; (u_o^2 \gg g \cdot L) = \frac{2}{\pi} \cdot \sqrt{Pe} \qquad (26)$$

if, on the other hand, the initial velocity can be neglected,

$$Nu \cdot (u_o^2 \ll g \cdot L) = \sqrt[4]{\frac{128}{9 \cdot \pi^2}} \; \sqrt[4]{Gn} \qquad (27)$$

is valid. In the first case the heat–flux density decreases with dimensions of the solid proportional to $L^{1/2}$, in the second case only proportional to $L^{1/4}$.
The cooling speed in the range of film boiling can technically only be raised by an increase in the initial velocity u_0, as the overflowed length L is specified by the dimensions of the workpiece. Thereby, the film thickness will be reduced too. As a consequence, the vapour film collapses sooner, so that the Leidenfrost temperature will shift to higher values.

EXPERIMENTAL RESULTS

Influence of Process Parameters

In the following, the results gained from the analytical model are compared with experimental results. Employing the experimental apparatus a nickel sample with a quadratic cross–section is heated in an electrical furnace to 850° C. Subsequently, the sample, which is placed flush into a vertical wall, is quenched at its frontal surface by a water film. That film emerges about 3 mm above the upper edge of the sample from a slot nozzle with an adjustable slot width s_0. The temporal temperature profile is measured in three places on the center line of the sample, that is 0.5 mm, 2.5 mm, and 10.5 mm away from the quenched frontal surface.
The following parameters were varied in the experiments:

- the water temperature (20 °C, 40 °C, 60 °C)
- the slot width of the nozzle ($s_0 = 0.5$ mm, 1 mm, 1.5 mm)

- the pressure in the nozzle and with it the exit velocity ($0.7\ m/s \leq u_0 \leq 4\ m/s$), which is equivalent to an irrigation density Γ of 20 to 120 kg/(min · m)
- and the edge lengths of the samples L = 8, 15, 20, 30, 40 mm.

In figure 6 the transferred heat-flux density during stable film boiling is plotted against the jet velocity u_0 for the three adjusted slot widths. An influence of the slot width and, therefore, of the water film thickness is not recognizable. From the temperature curve

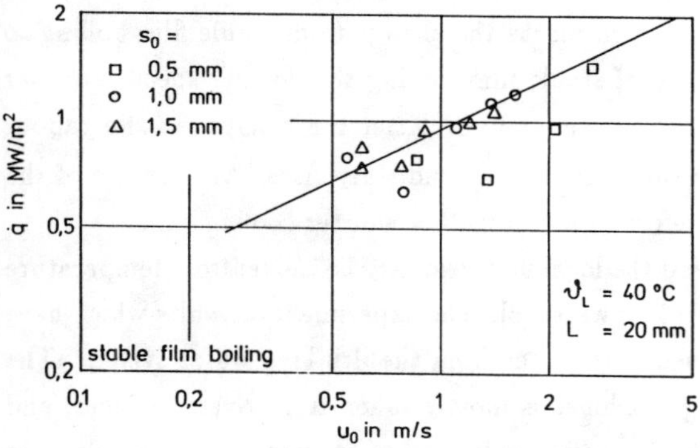

FIGURE 6. Influence of the nozzle slot width s_0 and of the jet velocity u_0 on the heat-flux density

according to equation (7) it follows that the thickness of the boundary layer in the liquid lies in the magnitude of 100 μm, and is consequently much thinner than the thickness of the water film s_0. As such, this water film thickness cannot influence the temperature curve and with that the heat transfer. The influence of the water temperature is given in figure 7. In addition, the heat-transfer coefficient calculated from equation (24) is plotted as a function of the jet withdrawal velocity. According to the Nusselt function (25) the

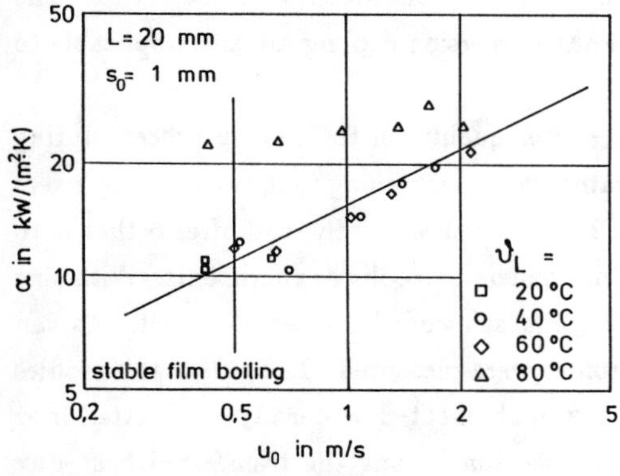

FIGURE 7. Influence of the water temperature on the heat-transfer coefficient

heat–transfer coefficient must be independent of water temperatures below 80° C, which is confirmed by the experimental results. The dependence of the heat–flux density and of the heat–transfer coefficient on the square root of the jet velocity u_o according to figures 6 and 7 is also confirmed by equation (24) or equation (26). The same is valid also for the dependence of the sample size L, which, however, is not shown separately in a figure.

Influence of Cooling Water Quality

The most important influence quantity for cooling processes with vaporizing liquids is the level of the Leidenfrost temperature, which marks the change from stable film boiling to partial film boiling [4,5]. In the range of stable film boiling the cooling speed is rather low because of the isolating effect of the vapour film. After the collapse of the vapour film at the Leidenfrost temperature the heat transfer suddenly rises. An increase of the Leidenfrost temperature results therefore in an altogether shorter cooling time.
Apart from the water temperature and the local flow velocity the Leidenfrost temperature depends also on the quality of the cooling water [6]. The experimental results which have been documented so far were obtained with water from the drinking water system. The cooling water for continuous casting machines is mostly taken from rivers or lakes, and can therefore be polluted with suspended solids, salts and microorganisms depending on the location of the plant. There are additional pollutions during the production process such as casting powder, for instance. The following results, which show the influence of the cooling water quality on the cooling progress, were obtained from immersion cooling experiments. A nickel sphere with 30 mm in diameter was heated to 850 °C in an electrical furnace, and was subsequently immersed into a calm water bath. With a thermocouple, which had been fixed 0.5 mm below the surface at the south pole of the sphere, the cooling profile was measured. On the assumption that the temperature profiles are symmetrical to the centre dot of the sphere during the cooling, the surface temperature and the emitted heat–flux density were calculated [7]. In the presented experiments different cooling water qualities were used, while all other test conditions were not varied. The experimental results demonstrated here for the immersion cooling are also applicable to the film cooling for broad predictions.
Figure 8 shows the influence of the cooling–water quality on the cooling process during immersion cooling experiments. Distilled water and water coming from two different rivers were tested. The cooling water from river 2 was stored smoothly, and after 6 that is to say 11 days the water was used again for the immersion cooling experiments. With this method the water could change biologically, and suspended matter could set. As can be noticed, completely different cooling profiles were measured. In figure 9 the profiles calculated from the cooling curves in figure 8 of the heat–flow density are plotted over the surface temperature. In the range of stable film boiling the transferred heat–flux

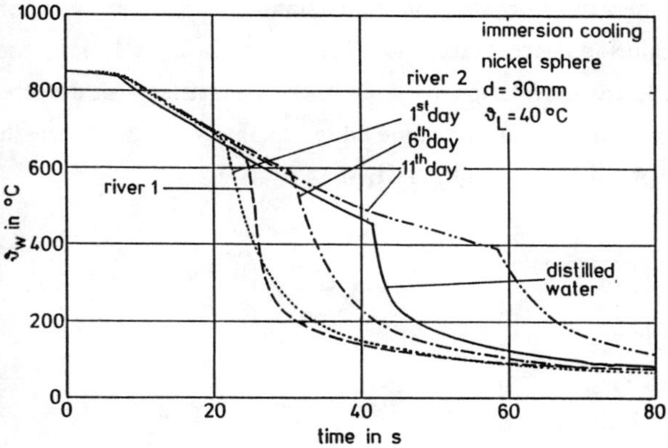

FIGURE 8. Quenching as a function of time for different cooling water qualities

density is the same for all experiments. On the other hand, the Leidenfrost temperatures are about 100° C higher with the fresh water from both rivers than is the Leidenfrost temperature of the distilled water. With increasing storing time the cooling effect of the water from river 2 declines strongly. After the eleventh day the Leidenfrost temperature is even below that of the distilled water. One of the most important pollutions of the cooling water, which is employed in technical installations, are dissolved salts. Within these salts causing scale, especially calcium and calcium carbonate as well as chlorides appear most frequently.

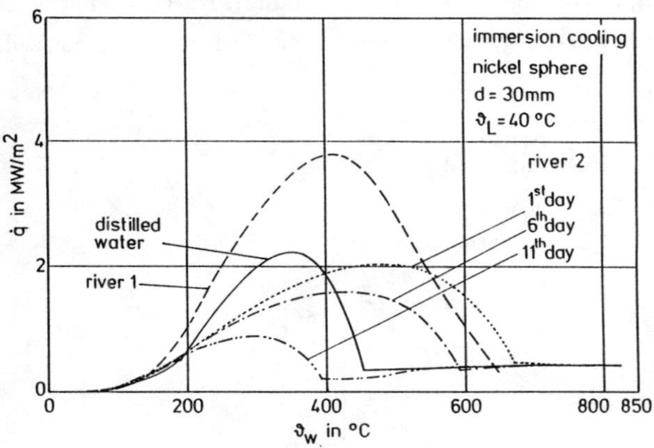

FIGURE 9. Heat–flux density as a function of surface temperature for different cooling water qualities

In figure 10 the Leidenfrost temperatures obtained in immersion cooling experiments are plotted in dependence on the chloride concentration for NaCl– and $CaCl_2$–solutions produced with distilled water. A linear correlation between these quantities was detected by approximation for both salts. Referring to the same added quantity of Cl^-–Ionen the Leidenfrost temperature is influenced stronger by $CaCl_2$ than by NaCl.

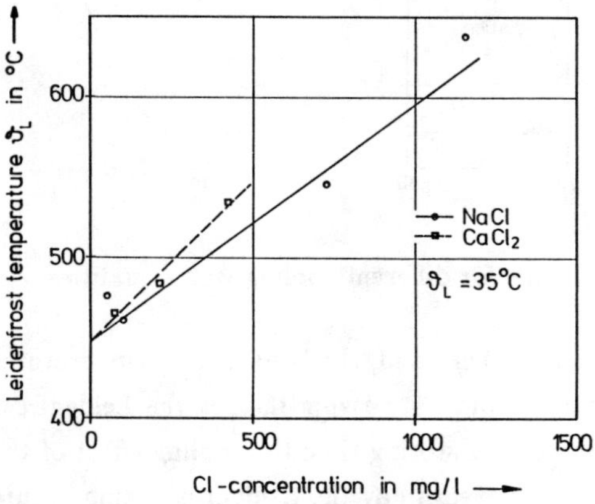

FIGURE 10. Leidenfrost temperature as a function of chloride concentration for NaCl– and $CaCl_2$–solutions

In figure 11 the profile of the heat–flow density over the surface temperature is presented with a mixed solution consisting of 660 mg/l $CaCl_2$ and of 150 mg/l $NaHCO_3$. For comparison the curves for a solution of 660 mg/l $CaCl_2$ and for distilled water are marked.

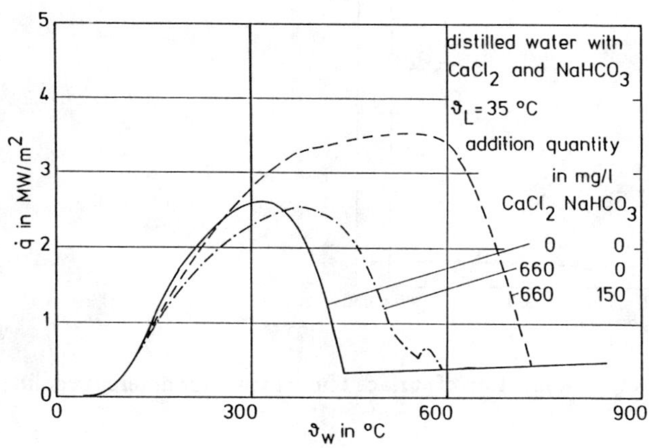

FIGURE 11. Heat–flux density as a function of surface temperature for different cooling water hardnesses

In the range of stable film boiling the same heat-flux density was measured for these experiments. But it becomes evident that the addition of an only small amount of sodium hydrogen carbonate in combination with calcium chloride causes a great shifting of the Leidenfrost temperature towards higher values compared with distilled water and a pure calcium chloride solution. Besides, the maximum appearing heat-flux density rises considerably in the burn-out-temperature.
The investigations show that the cooling water quality is of great importance for technical cooling processes. But it should be taken into consideration that the quality of the cooling water not only depends on the place the water is obtained from, but can also be dependent on seasonal fluctuations. Generally, an increase of the salt concentration leads to a higher Leidenfrost temperature, and therefore to a faster cooling.

References

[1] Jeschar, R., Köhler, C., Specht, E., Heidt, V., Methoden zur definierten Abkühlung metallischer Werkstoffe, *Gas wärme internat. 38*, pp. 223–229, 1989, and *Die Industriefeuerung 47*, pp. 34–39, 1989.

[2] Köhler, C., Specht, E., Jeschar, R., Heat Transfer with Film Quenching of Vaporizing Liquids, *steel research 61*, 1990, forthcoming.

[3] Köhler, C., *Wärmeübertragung von heißen Oberflächen im Bereich der stabilen Filmverdampfung*, thesis TU Clausthal, 1990.

[4] Maaß, R., *Untersuchung zur Wärmeübertragung beim Abschreckhärten*, thesis TU Clausthal, 1988.

[5] Maaß, R., Jeschar, R., Einflußgrößen des Wärmeübergangs beim Abschrecken von Metallen in Wasser, *Gas wärme internat. 38*, pp. 142–150, 1989.

[6] Yu, H., The Effect of Cooling Water Quality on Aluminium Ingot Casting, *Light Metals 1985*, AIME, pp. 1331–1347.

[7] Jeschar, R., Maaß, R., Wärmebehandlung beim Abschrecken kugelförmiger Körper in Wasser, *Gas wärme internat. 35*, pp. 370–375, 1986.

CASTING AND MOLDING (2)

Visualization of Flow and Solidification of Polymer Melt in the Injection Molding Process

YASUO KUROSAKI and ISAO SATOH
Department of Mechanical Engineering for Production
Tokyo Institute of Technology
Oohkayama, Meguro-ku, Tokyo 152, Japan

ABSTRACT

Visualization of both the flow and the solidification of polymer melt in a mold plays an important role for understanding the mechanisms of residual stress and deformation generation in molded polymer during the injection molding process. This paper first reviews the studies conducted so far concerning visualization of polymer melt behavior in the mold cavity. Next, an experimental investigation is described in which real-time visualization of the polymer melt in an injection molding process is achieved by utilizing birefringence changes in the polymer melt.

1. INTRODUCTION

The injection molding process is one of the most popular methods for manufacturing polymer plastic products. The main process of injection molding is the injection of molten polymer into cold cavities, where the polymer is solidified to form fabricated products. The injection molding process involves non-isothermal fluid mechanics with complex rheological properties and solidification processes.

The mechanical properties of injection-molded plastics are influenced by many injection conditions: the injection pressure, injection melt temperatures, mold wall temperatures, and so forth. It has been known that these conditions can control the residual stresses in the molded product. It is accordingly important to predict the stress field remaining in the molded plastic to obtain products with desirable mechanical properties.

In recent years, efforts have been made to use the injection molding process to produce optical parts such as camera lenses, compact disks, and opto-magnetic memory disks. For such optical parts it is important to reduce not only shape deformation, but also optical distortion. It is well known that the optical distortion, i.e., the nonuniformity of birefringence, is strongly connected to the magnitude of residual stress in the product. A better understanding of the mechanisms of stress development and residual stress generation during the injection molding process is required so that the magnitude of residual stress can be minimized by optimized molding conditions.

There have been many studies concerning the relationship between the molding conditions and the mechanical properties of injection-molded polymer [1-22]. These studies can be classified into the following three groups according to their targets and methodologies:

(1) Experimental studies concerned with the mold filling process and/or with stress generation and residual stress in the molded polymer [1-8],
(2) Studies of the relationship between the residual stress and mechanical properties of the molded product, such as elongation, strength, elastic modulus, etc. [9-15],
(3) Numerical simulation of the flow, temperature and stress fields in an injection-molded polymer based upon a rheological constitutive equation [16-22].

The studies in the first group have contributed to the production of polymer plastic precision devices such as opto-magnetic memory disks. It is very important to observe the flow and solidification processes in the mold so as to clarify the mechanism of residual stress generation within the plastic products. However, since the mold die is usually made from opaque metallic materials, it is not easy to observe the phenomena within the mold without some sophisticated procedures for visualization.

A few visualization techniques have been reported so far for the investigation of flow and stress fields in molded polymer. These visualization studies can be categorized as follows:

1. Observation without Windows ("Short-shot" method)
   Visualization of an advancing melt front of polymer in the mold cavity employing the "short-shot" procedure.
2. Observation with Windows
   (1) Visualization of the filling process.
   Direct observation of the filling process in a mold through a transparent window.
   (2) Visualization of the development of stress birefringence due to cooling.
   Observation of flow patterns and visualization of the development of stress birefringence during mold filling and cooling.
   (3) Visualization of the stress birefringence due to flowing and cooling.
   Visualization of the real time stress birefringence patterns to determine the mechanisms of both stress development and of residual stress generation in injection-molded polymer.

The first three works above are briefly reviewed, and then the visualization of stress generation, as well as the solidification process, in injection-molded polymer is reported in this paper.

2. REVIEW OF OBSERVATION STUDIES OF FLOW AND STRESS FIELDS IN MOLDED POLYMER

2.1 Visualization of the Advancing Melt Front of a Polymer with the "Short-Shot" Procedure

The advancing speed and shape of a melt front in a mold cavity affects the weld line appearing in plastic products and the molecule orientation distribution. The weld line can be caused by inadequate mold filling conditions. Flow visualization experiments are indispensable for studying the location and shape of an advancing melt front, i.e., for the overall flow process of the polymer melt in the mold cavity. Krueger and Tadmor [1] elucidated the overall flow pattern in a complex configuration cavity with various insert shapes by employing the "short-shot" procedure. The overall flow patterns measured in their experiment were compared to the simulated ones obtained with the Flow Analysis Network computation method by Broyer, et al. [2]. In addition to the shape of the advancing front, they examined weld line formation and location, and concluded that weld lines were formed whenever two advancing fronts met each other.

The "short-shot" method consists of the following steps:

(1) The initial ram position - the injection volume of polymer melt - of the injection molding machine is set to fill the whole mold cavity with a small "cushion" (full-shot). After the molded polymer is completely cooled, the specimen is removed from the mold cavity.
(2) Injection molding processes are repeated under the same molding conditions, except for the initial ram position. Successive shots are made with progressively shorter ram positions resulting in a series of "short-shots".
(3) The series of short-shots show the shapes and locations of the flow front at various stages of the cavity filling process, plus any subsequent relaxation.

The "short-shot" procedure, which is usually adopted in practical manufacturing, has the advantage of visualizing the filling process even in a mold constructed with opaque walls. It has been pointed out, however, that the shapes and locations of the advancing flow front simulated with this procedure do not coincide with the actual ones, because the obtained flow front is given by the solidified polymer [5].

2.1.1 The overall flow pattern observed with the "Short-Shot" procedure. Krueger and Tadmor experimentally observed advancing flow fronts in five mold configurations. The overall flow pattern and the advancing front profiles were determined through a series of short shots. One of these superimposed photographs of the short shots for each configuration and the traces of the corresponding shapes of the advancing fronts are shown in Fig. 1, which also shows the shape of the inserts used and the location of the visually observable weld lines (broken curves).

This figure shows a characteristic, initially radial flow pattern which is rapidly converted into a flat advancing front. The front exhibits a sharp curvature at the side wall due to wall effects, and a mild curvature over the whole profile due to thickness nonuniformities. The insert shaped like a *T* splits the front into two advancing fronts, creating a long weld line downstream at the insert which continues all the way to the end of the cavity. The weld line is curved exactly following the "meeting line" of the two advancing fronts.

Analyzing the weld lines in the various samples, Krueger and Tadmor concluded that the weld lines are formed wherever advancing fronts "meet".

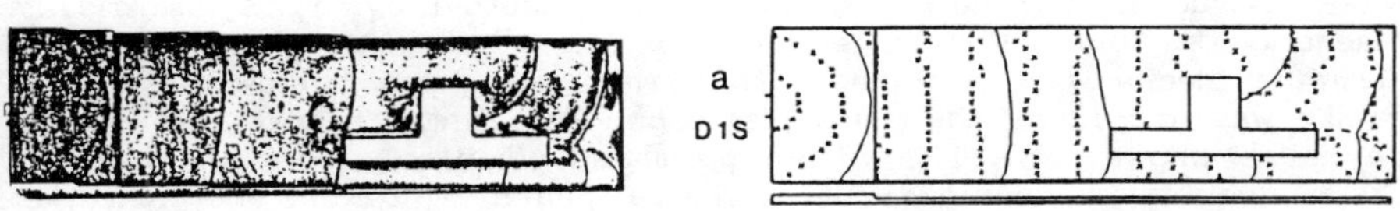

FIGURE 1. Photographs of superimposed short shots [1]: experimental (solid lines) and simulated (×) advancing front profiles.

2.1.2 Comparisons of predicted and measured front profiles. Krueger and Tadmor also conducted numerical simulations of the filling process. The overall flow patterns obtained experimentally were compared to theoretical predictions using the Flow-Analysis-Network computational method. In their theoretical prediction, the polymer flow was assumed to be isothermal, and the effects of the side walls on the polymer flow were neglected. The details of this method and computational conditions are given in Refs.[1] and [2].

Simulation results are also plotted in Fig. 1. The calculated advancing front profiles are marked with × symbols. The front profiles obtained by the short-shot experiments are indicated by solid lines, and the experimental weld line locations are shown by dotted lines. The agreement between the predicted and measured front profiles is sufficiently good considering the restrictive isothermal flow assumption in the model and the possible front distortion in short shots. The reasonable prediction of the advancing front profiles indicates that no large melt temperature drop occurred during the filling process under the given conditions.

2.2 Direct Visualization of the Filling Process using a Mold with Transparent Windows

The "short-shot" procedure is not completely capable of visualizing the time-dependent shape and location of a melt front. A few studies [3-6] have tried to observe the motion of a flow front by using molds with transparent windows. White and Dee [3] experimentally observed the flow pattern during the injection molding of rheologically low-density polyethylene and polystyrene melts under various molding conditions.

The utilization of a mold which has transparent windows is the optimal way to observe the time-dependent polymer flow during injection molding. However, the installation of the transparent windows on the mold involves great difficulties in construction because the windows, which are normally made from glass, are in direct contact with the polymer melt, which is under high temperature and pressure conditions. White and Dee, therefore, utilized capillary rheometers as a polymer melt feed system, instead of a practical injection molding machine.

Yokoi and co-workers [5] have recently reported results of jetting phenomena obtained by using a metal mold with a glass window installed upon a normal injection molding machine under normal injection pressures. They pointed out that the conventional observation method using a mold with two glass windows was not completely satisfactory to simulate actual melt flows within a mold, because the allowable injection pressure was quite low.

2.2.1 Low injection pressure case. White and Dee [3] used a modified rheometer as an injection molding apparatus. They attached a combined nozzle-mold assembly, instead of a capillary die, to the lower end of the barrel. The apparatus used is schematically illustrated in Fig. 2. The mold was constructed in three parts, and clamped together by four symmetrically placed bolts. This mold was designed for flow visualization, and the outer clamping blocks were provided with Pyrex plate glass windows. The mold cavity was formed by the glass plates and the aluminum plate, which was partially hollowed out. The design permitted simple changing of the mold cavity geometry by using different central plates. White and Dee used four different mold shapes. The mold temperature was controlled with four electric heaters installed in the outer aluminum clamping blocks, which permitted the observation of the flow patterns under the conditions of both isothermal and cold cavity walls.

Typical observed results are shown in Fig. 3, where the mold walls were held at the same temperature (200°C) as that of the molten polymer at the inlet. White and Dee observed two basic flow regimes depending upon the injection rate of the molten polymer into the mold, that is, filling at a

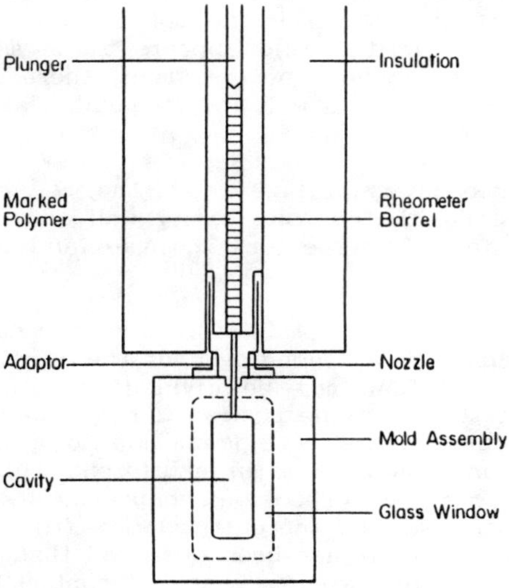

FIGURE 2. Injection molding apparatus used by White and Dee [3].

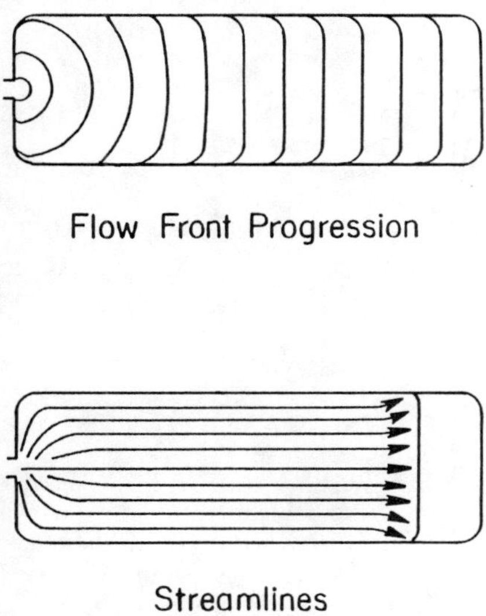

FIGURE 3. Polymer melt filling a mold in isothermal flow [3].

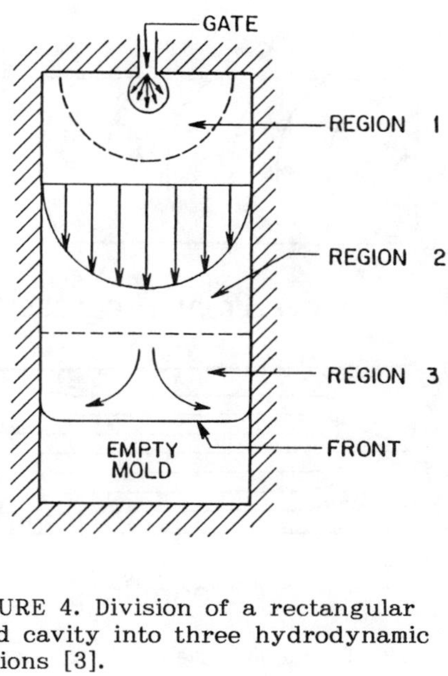

FIGURE 4. Division of a rectangular mold cavity into three hydrodynamic regions [3].

slow injection rate and at a high injection rate.

The mold filling process for slow injection is divided into three stages (see Fig. 4). First, the melt spreads into the mold through the narrow constriction of the gate in an approximately radial manner. Second, after the corners are filled, there is a transition region in which the front changes from a circular shape to an almost flat profile. Finally the nearly flat front continues to move forward until the mold is filled. The front shape is flat except for some convex curvature near the mold wall. In the final region the velocity of the center of the mold appears to be more rapid than that near the walls. In the vicinity of the front, there are significant velocity components towards the mold wall and perpendicular to the direction of movement of the front.

When the melt is injected into a mold cavity at high rates, the melt may enter as a jet. The jet may then strike the opposite cavity wall and pile up on itself. Later, a front of uniform filling enters the gate and fills the entire mold.

The filling process in a cold mold (80°C) by a hot (200°C) polymer melt is shown in Fig. 5. There are two regions of flow depending upon the injection rate. At low injection rates, the mold flow may be divided into three stages like the hot mold case. First there is a nearly radial flow entering the mold. After the corners are filled, there is a transition to a new steady front shape. However, the front flow behaves in a different manner from the one observed in the isothermal case. The front shape has a distinct curvature in the non-isothermal case, and the outward flow from the core to the mold walls is more pronounced. It has been observed that the polymer on the wall comes from the melt front moving around to contact the wall (the "fountain" flow). It has also been observed that stagnant regions develop in the corners and near the mold wall and that the melt flow field behind the advancing front has a sharper velocity profile than in the isothermal case. On the other hand, White and Dee concluded that the mold wall temperature does not play a significant role in the melt flow at high

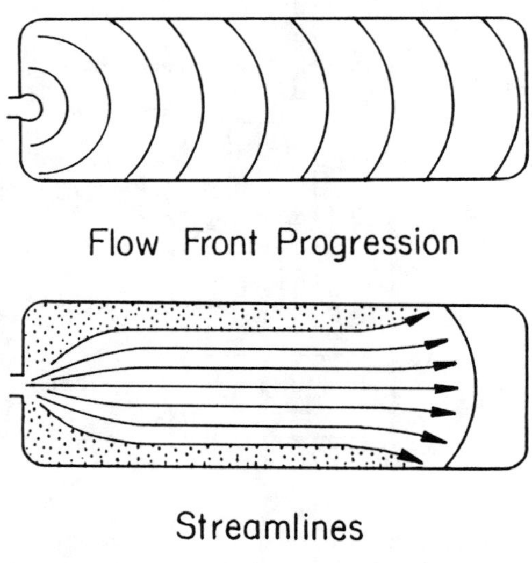

FIGURE 5. Polymer melt filling a mold with cold walls [3].

injection rates since the "jetting" melt does not stick on the cold wall.

2.2.2 Practical injection pressure case. Figure 6 shows a side view of the glass-insert metal mold developed by Yokoi and co-workers [5, 6]. A glass prism was inserted into a hole in a metal mold, and one face of this prism faced the cavity. The oblique face of the prism reflected the optical image from the cavity, and the reflected image transmitted out through the window. The glass prism was fixed by two sets of wedges and spacers which were used to apply an initial compression. The high hydrostatic pressure during the injection process prevented the brittle glass prism from breaking.

Using this test mold, Yokoi et. al. conducted an observation of both simple uniform filling processes and jetting phenomena under high pressure. Their molding conditions are listed in Table 1.

They also observed the behavior of the flow fronts of short shots in order to verify the accuracy of the "short-shot" procedure. Typical observed results are shown in Fig. 7. As shown in this figure, after the short shot the flow front gradually moves 3.7 mm forward during cooling due to creep relaxation. From this result, they concluded that the "short-shot" procedure does not always simulate precisely the filling process in injection molding.

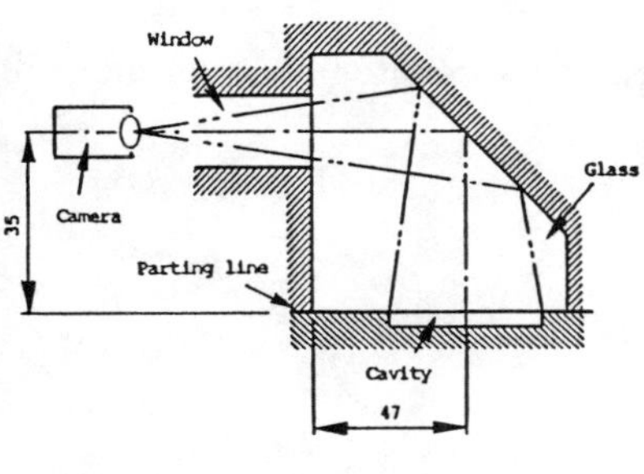

FIGURE 6. Glass-insert mold [5].

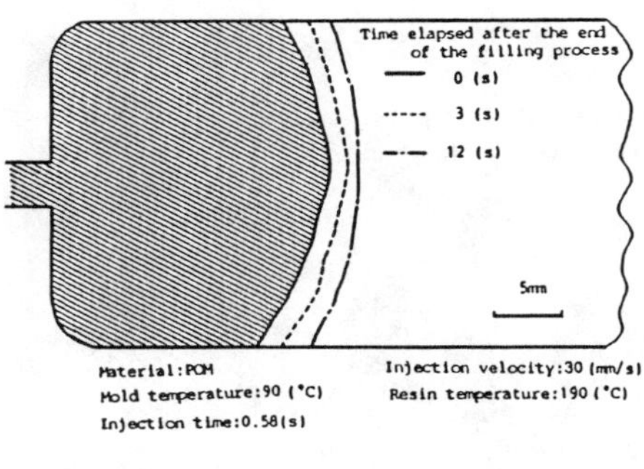

FIGURE 7. Flow front behavior during the cooling process [5].

TABLE 1. Experimental conditions [5].

	A Jetting	B Simple uniform filling
Material	POM	POM
Resin temperature (°C)	190	190
Mold temperature (°C)	80	80
Injection velocity (mm/s)	30	15
Injection pressure (MPa)	56	33
Injection time (s)	0.9	2.4

2.3 Flow Patterns and Development of Stress Birefringence due to Mold Cooling

The visualization techniques described in the previous two examples enable us to obtain information about the filling process of injection molding. However, it is also necessary to get a better understanding of the mechanisms of stress generation and residual stress in a molded polymer in order to obtain molded optical precision devices such as camera lenses and

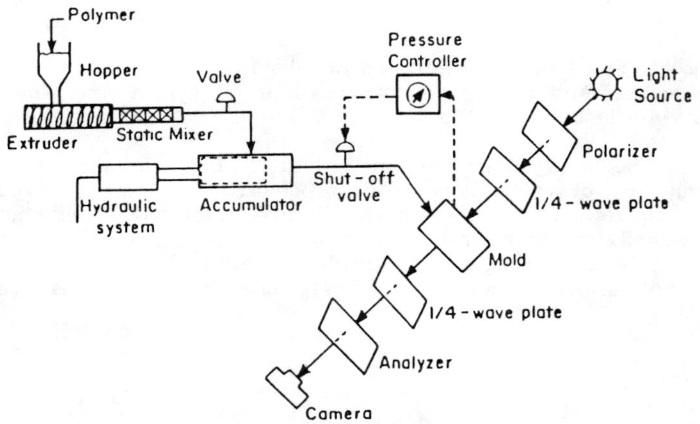

FIGURE 8. Schematic diagram of the experimental apparatus used by Han and Villamizar [7].

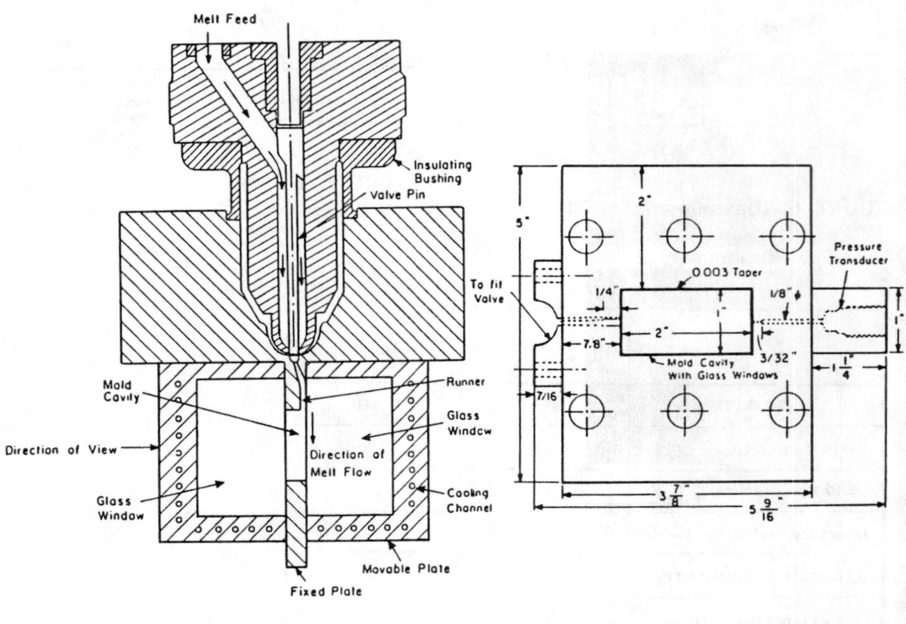

FIGURE 9. Schematic diagram of the mold employed by Han and Villamizar [7].

opto-magnetic memory disks.

There have been only a few papers concerning visualization of the stress fields in molded polymer during injection molding. Han and Villamizar [7] carried out an observation of the mold filling and stress birefringence pattern development during mold cooling operations. They constructed a rectangular mold cavity with glass windows on both sides, and the development of stress birefringence was visualized by using the technique of photoelasticity [23].

The experimental apparatus used by Han and Villamizar is schematically shown in Fig. 8. The apparatus consisted of three major parts: (a) the polymer melt feed system, (b) the mold with transparent windows made of quartz glass, and (c) the optical system. A rectangular mold cavity with glass windows on both sides was constructed in order to observe the stress birefringence patterns. Figure 9 shows a plane view and a side elevation of the mold. In their experiment, the mold was placed on an optical bench equipped with a polarizer, an analyzer, quarter-wave plates, and a light source, as shown in Fig. 8. The injection pressure, the injection melt temperature, and the mold temperature were variable.

Figure 10 shows a series of photographs indicating stress birefringence patterns during the mold filling and packing processes as polystyrene was injected into a hot mold cavity (i.e., under an isothermal condition). It is observed that no stress is seen in the mold cavity even in the packing process. This is explained by the fact that, in their experimental configuration, the melt flow has no effect on the observed stress field, that is, the shear stress generated in the molded polymer due to the melt flow is dominated by the gap-wise stress distribution in the mold cavity.

Figure 11 is a series of photographs of stress birefringence patterns showing the development of stresses after the mold cavity had started to

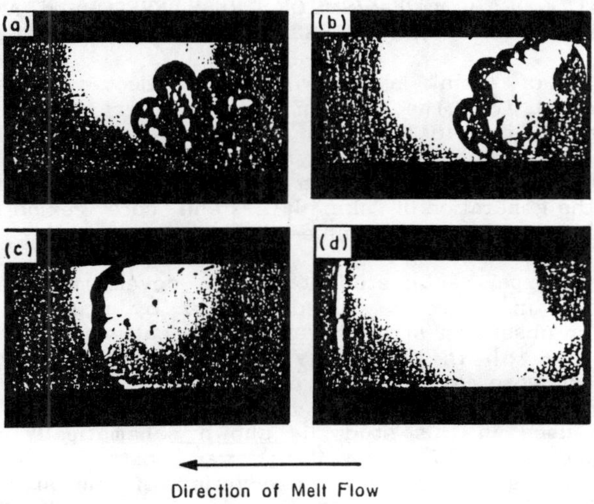

FIGURE 10. Photographs showing the mold filling characteristics and stress birefringence patterns as polystyrene is injected into a hot mold cavity [7]: (a) after 10 sec, (b) after 20 sec, (c) after 30 sec, (d) after 40 sec.

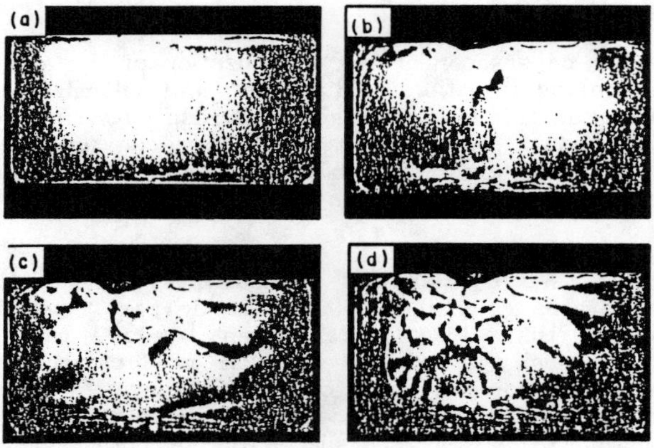

FIGURE 11. Photographs showing the development of stress birefringence patterns after the mold cavity has started to cool [7]: (a) t=0, (b) after 16 min, (c) after 24 min, (d) after 60 min.

cool. The stress birefringence observed during the cooling process is considered to be due to thermal shrinking of the polymer. This figure indicates that the stresses start to develop from all four sides, and move towards the center of the cavity.

As seen from their visualization results, the residual stress in the molded polymer is partially caused by thermal shrinkage of the molten polymer. The mechanisms of the residual stress generation, however, can not be satisfactorily clarified by their results because the effects of melt flow on the residual stress distributions have not been investigated.

3. A REAL-TIME VISUALIZATION OF THE MECHANISMS OF STRESS DEVELOPMENT AND RESIDUAL STRESS GENERATION USING STRESS BIREFRINGENCE CHANGE

The mechanical properties of "skin" layers within the injected molded polymer are quite different from the ones of the "core" region. The difference is believed to be due to both the molecular orientation and residual stress distribution in the molded polymer. However, so far the details of the generation process have not been clear. It is important to know the mechanisms of the generation of "skin" layers and "core" region in order to obtain a molded product with desirable properties.

The authors performed an experimental study on stress development and residual stress generation in injection molded polymer by utilizing a visualization technique for observing stress birefringence patterns during the entire molding process [8]. In this study, the mechanisms of the generation of "skin" layers were clearly displayed.

The experimental setup used in this study is shown schematically in Fig. 12. A rectangular mold cavity (65 x 12 x 4 mm) was constructed with Plexiglas windows on both sides to allow observation of the stress birefringence patterns. The details of the test mold are shown in Fig. 13. The mold used in this study was maintained at room temperature by air cooling because the injection molding process was repeated with a very low frequency (a few cycles/hour).

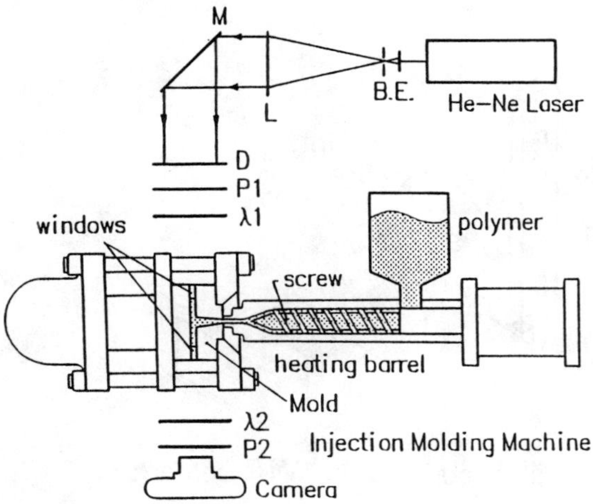

Injection Molding Machine	—	Minimat 5/7
Diameter of Screw	—	16 mm
Max. Injection Pressure	—	223 MPa
Max. Injection Volume	—	3.8 cm^3
Max. Mold Clamping Force	—	68.6 kN

FIGURE 12. Experimental apparatus used by the present authors [8].

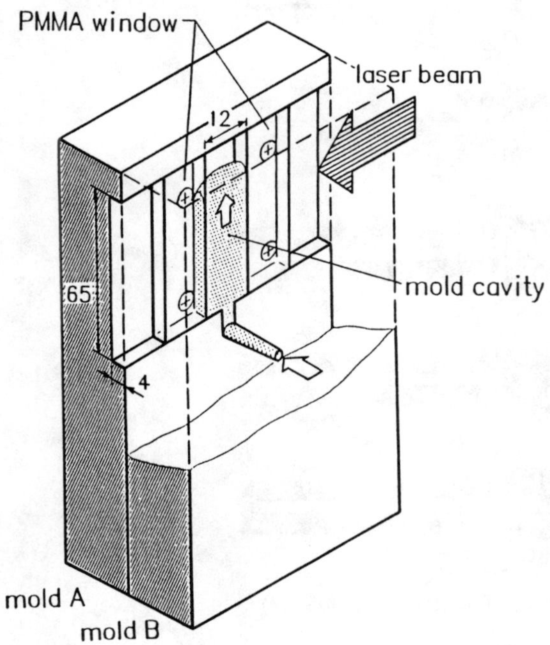

FIGURE 13. Test mold with transparent windows [8].

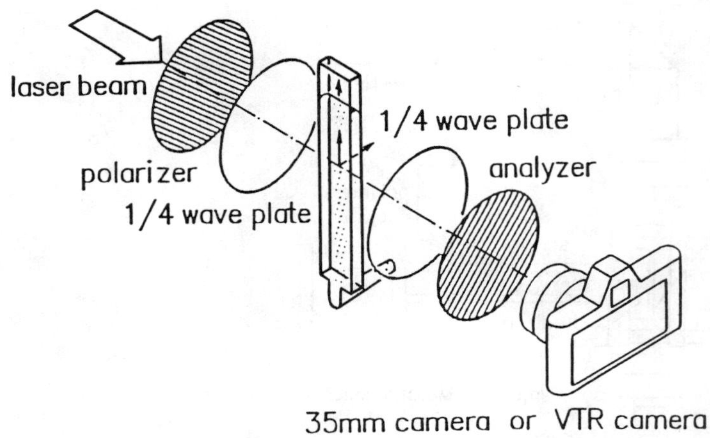

FIGURE 14. Optical setup for visualization of birefringence patterns [8].

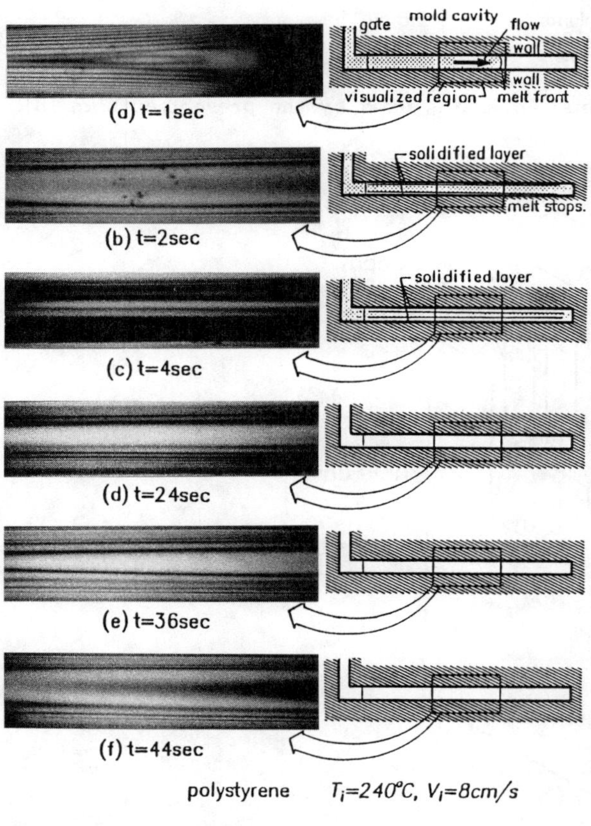

FIGURE 15. Birefringence patterns observed in molded polymer during the injection molding process [8].

The test mold was installed in an optical setup consisting of a polarizer, an analyzer, 1/4-wave plates and a light source (He-Ne laser, λ=632.8nm) so that the stress birefringence patterns within the polymer melt could be observed. The optical setup for the visualization is schematically shown in Fig. 14. The isochromatic fringe patterns due to the stress birefringence in the X–Y plane were visualized by this means, and the fringe patterns were recorded with photographs or with a high speed VTR.

Figure 15 shows a series of photographs of birefringence patterns taken during the mold filling and cooling processes. This figure clearly shows that stress birefringence appears in the mold cavity during the filling process but disappears immediately, except in the thin solidified layer near the walls, after the polymer melt stops, when the mold cavity is filled up with the polymer melt. During the cooling process, birefringence again appears in the core region and develops slowly for about 40 seconds. After the molded specimen is completely cooled (t=44sec), birefringence corresponding to the residual stress remains.

The real-time observation of the development of stress birefringence in the mold cavity enables us to classify the behavior of stress birefringence during the injection molding process into three successive stages: (1) the flowing stage, (2) the relaxation stage, and (3) the solidifying stage. Therefore, an attempt was made to explain the behavior of birefringence in the injected polymer in each stage.

3.1 Flowing Stage

Figure 16 shows a typical birefringence pattern observed in the flowing stage and its schematic diagram; the molten polymer injected into the mold cavity flows from left to right in the photograph. The rounded head shape of the melt front can be observed in the right end of the photograph, and dark, built-up layers can be observed in the polymer adjacent to the mold walls. These are solidified layers where the polymer melt is cooled by the

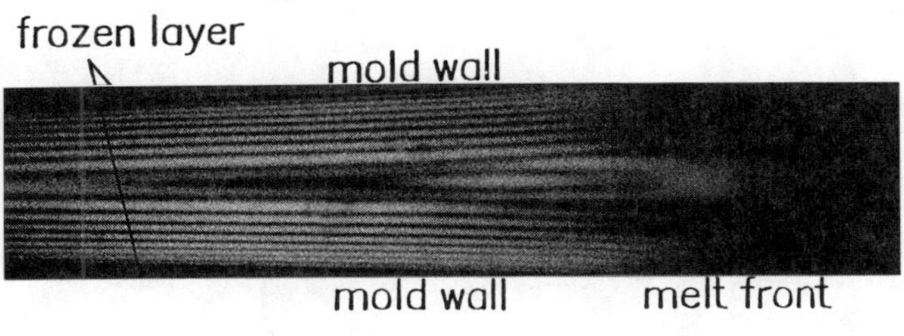

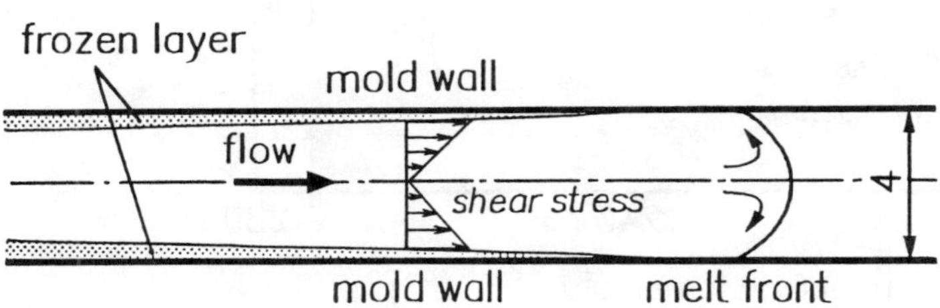

FIGURE 16. Birefringence pattern observed in the flowing stage [8].

mold wall. These frozen (solidified) layers originate just behind the melt front, and their thickness increases with downstream distance from the melt front.

In the melt flowing region between the frozen layers, the shear stress in the flow adds to the observed birefringence pattern. The birefringence pattern in this region is symmetric with respect to the middle plane of the mold cavity, and the fringes are almost parallel to the mold wall.

The stress birefringence due to flowing is largest at the boundary between the flowing region and frozen layers, and its maximum value is affected by the injection processing conditions: the temperature and velocity of the injected melt. Figure 17 shows the maximum birefringence during the flowing stage for various conditions of injection processing. As shown in this figure, the maximum birefringence due to flowing decreases with increasing melt temperature and decreasing melt velocity.

In the frozen layers near the walls, the birefringence due to shear stress must be "frozen." However, in the present experiments, the "frozen-in" residual stress in the frozen layers is invisible because the stepwise change of the refractive index of the polymer melt due to solidification deflected the light so much that it did not pass through the test mold.

3.2 Relaxation Stage

The birefringence due to shear stress observed in the flowing stage disappears rapidly after the cavity is filled up. Figure 18 presents a series of photographs showing a typical disappearance of birefringence after the melt flow stops; in this figure, schematic diagrams of stress birefringence distributions are also shown. The disappearance of birefringence, i.e., the relaxation of shear stress, does not occur instantaneously, but takes 0.1 - 1.0 seconds. This is because the polymer melt is a viscoelastic fluid.

The relaxation time t_R (the period required for the disappearance of stress

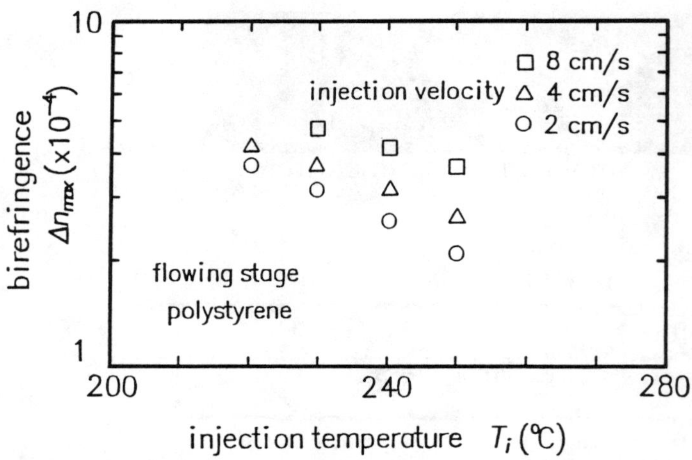

FIGURE 17. Relation between the maximum birefringence during the flowing stage and the injection melt temperature [8].

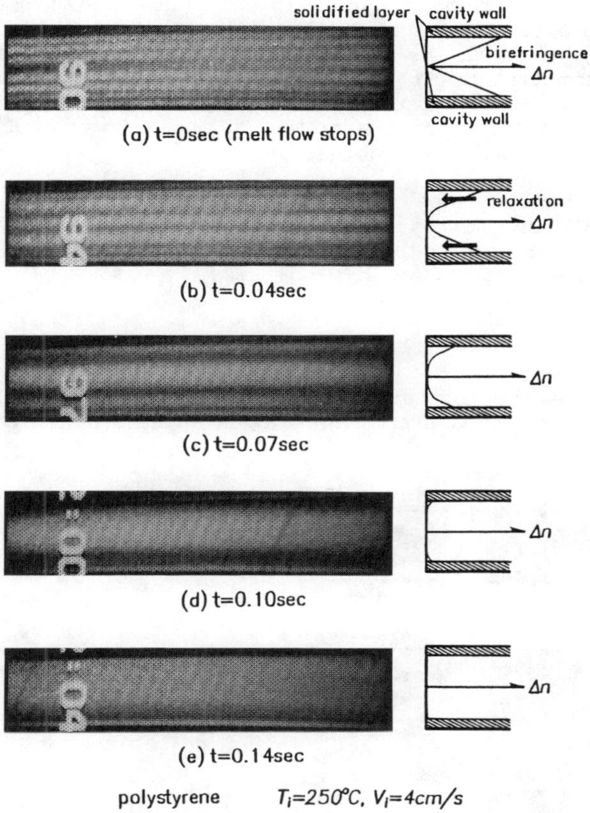

FIGURE 18. Disappearance of birefringence during the relaxation stage [8].

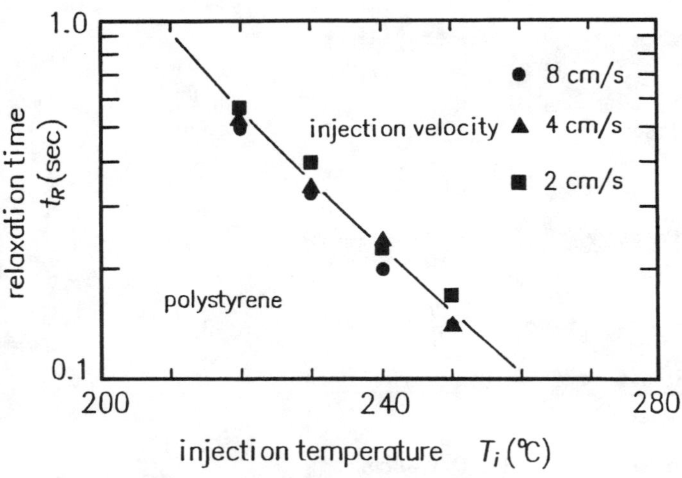

FIGURE 19. Relation between the relaxation time and the injection melt temperature [8].

birefringence in the flowing region) under different injection process conditions is plotted as a function of injected melt temperature T_i in Fig. 19. This figure clearly shows that the relaxation time t_R rapidly decreases with increasing melt temperature T_i, but t_R is hardly affected by the injection velocity. This is explained by the fact that, from a rheological point of view, the viscosity for stress relaxation of the polymer melt is reduced with an increase in melt temperature.

As shown in Fig. 18, no birefringence can be observed in the polymer melt except in the frozen ("skin") layers after relaxation. This reveals that the final residual stress frozen in the core of the molded specimen is mainly induced in the later cooling stage.

3.3 Cooling Stage

After the relaxation stage, the polymer melt in the mold cavity is cooled by the cavity walls. In this stage, birefringence again appears in the core

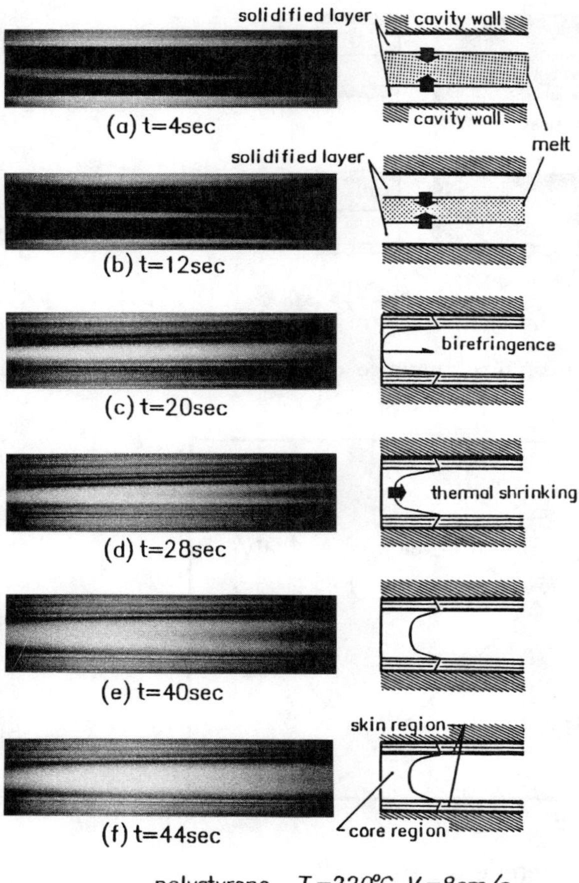

FIGURE 20. Birefringence patterns observed in the molded polymer during the cooling stage [8].

region of the injected polymer. A series of photographs of birefringence patterns observed in the solidifying stage is shown in Fig. 20. This series of photographs clearly shows that the thickness of the solidified layer gradually increases; the solidification process (glass-transition) of the whole specimen takes 20 - 25 seconds.

The birefringence observed in the unsolidified core region is mainly caused by thermal shrinking of the polymer due to cooling. From the practical point of view, therefore, it is worth noting that a high-pressure packing process will be effective in reducing the thermal stress occurring in the core region. The tensile stress due to thermal shrinking in the unsolidified core region is consequently reduced by the high pressure of the packing process. In the present study, however, the effectiveness of a high-pressure packing process was not examined because the transparent windows had to be protected from excessive pressure.

Thermal shrinking of the polymer proceeds slowly even if the temperature of the entire region of the molded specimen becomes lower than the glass-transition temperature of the polymer (t>20 sec). Then an immovable birefringence pattern, which corresponds to the residual stress frozen in the specimen, can be observed after the molded polymer is completely cooled. A careful inspection of the photograph for t=44sec shows that the residual stress region in the specimen can be classified into two regions: (1) the "skin" layers adjacent to the mold walls, where the unrelaxed shear stress remains, and (2) the "core" region, where the thermal stress due to cooling is frozen. Therefore, it can be concluded that the amount of residual stress remaining in molded parts is determined by both the injection velocity of the polymer melt and the cooling rate of the mold.

4. CONCLUDING REMARKS

This paper presented a brief review of three earlier works concerning the observation of flow and stress fields in injection-molded thermo-plastic polymer, as well as a description of our research on the visualization of stress development, residual stress generation, and solidification processes in the molded polymer. As seen from these studies, visualization techniques are powerful tools to reveal phenomena occurring in injection mold cavities. The results obtained through visualization are promising for improving the quality of injection-molded products.

REFERENCES

1. W.L. Krueger and Z. Tadmor: Injection Molding into a Rectangular Cavity with Inserts, Polymer Eng. Sci., 20-6 (1980) pp.426-431
2. E. Broyer, C. Gutfinger and Z. Tadmor: Trans. Soc. Rheol., 19 (1975) p.423
3. J.L. White and H.B. Dee: Flow Visualization for Injection Molding of Polyethylene and Polystyrene Melts, Polymer Eng. Sci., 14-3 (1974) pp.212-222
4. L.R. Schmidt: A Special Mold and Tracer Technique for Studying Shear and Extensional Flows in a Mold Cavity During Injection Molding, Polymer Eng. Sci., 14-11 (1974) pp.797-800
5. H. Yokoi, T. Hayashi, K. Toda and N. Morita: Direct Observation of Jetting Phenomena under a High Injection Pressure by Using a Prismatic-Glass Inserted Mold, ANTEC '88 (1988) pp.329-333
6. H. Yokoi, T. Hayashi, Y. Murata and K. Yonezawa: Direct Observation of the Process of the Plastic Flow by Using a Glass-Inserted Mold

(Comparison with Numerical Simulations), Polymer Preprints, Japan, 37-4 (1988) p.1215
7. C.D. Han and C.A. Villamizar: Development of Stress Birefringence and Flow Patterns During Mold Filling and Cooling, Polymer Eng. Sci., 18-3 (1978) pp.173-179
8. Y. Kurosaki, I. Satoh and K. Ishii: Visualization of Flow and Solidification of Polymer Melt in the Injection Molding Processes, Heat Transfer in Manufacturing and Material Processing (1989 National Heat Transfer Conf.), HTD-113 (1989) pp.63-70
9. R.S. Stein and A.V. Tobolsky: An Investigation of the Relationship Between Polymer Structure and Mechanical Properties (Part I: Relationship Between Structure, Mechanical Properties, and Birefringence), Textile Res. J., 18-4 (1948) pp.201-223
10. R.S. Stein and A.V. Tobolsky: An Investigation of the Relationship Between Polymer Structure and Mechanical Properties (Part II: An Experimental Study of the Stress and Birefringence Properties of Several Relaxing Polymeric Systems), Textile Res. J., 18-5 (1948) pp.302-314
11. R.S. Stein, S. Krimm and A.V. Tobolsky: An Investigation of the Relationship Between Polymer Structure and Mechanical Properties (Part III: A Thermodynamic and Optical Analysis of the Behavior of Polyethylene, Polyvinyl Chloride, and Lactoprene), Textile Res. J., 19-1 (1949) pp.8-22
12. J.R. Saffell and A.H. Windle: The Influence of Thermal History on Internal Stress Distributions in Sheets of PMMA and Polycarbonate, J. of Applied Polymer Sci., 25 (1980) pp1117-1129
13. M.M. Qayyum and J.R. White: Time-Dependent Birefringence in Glassy Polymers during Stress-Relaxation and Recovery, Polymer, 23 (1982) pp.129-136
14. J. Bowman: The Structure and Mechanical Properties of an Injection-Moulded Acetal Copolymer, J. of Materials Sci., 16 (1981) pp.1151-1166
15. A. Siegmann, S. Kenig and A. Buchman: Residual Stresses in Injection-Molded Amorphous Polymers, Polymer Eng. Sci., 27-14 (1987) pp.1069-1078
16. C.A. Hieber and S.F. Shen: A Finite-Element/Finite Difference Simulation of the Injection-Molding Filling Process, J. of Non-Newtonian Fluid Mech., 7 (1980) pp.1-32
17. H. Mavridis, A.N. Hrymak and J. Vlachopoulos: Finite Element Simulation of Fountain Flow in Injection Molding, Polymer Eng. Sci., 26-7 (1986) pp.449-454
18. C.G. Gogos, C. Huang and L.R. Schmidt: The Process of Cavity Filling Including the Fountain Flow in Injection Molding, Polymer Eng. Sci., 26-20 (1986) pp.1457-1466
19. C.A. Hieber, L.S. Socha, S.F. Shen, K.K. Wang and A.I. Isayev: Filling Thin Cavities of Variable Gap Thickness: A Numerical and Experimental Investigation, Polymer Eng. Sci., 23-1 (1983) pp.20-26
20. P.G. Lafleur and M.R. Kamal: A Structure-Oriented Computer Simulation of the Injection Molding of Viscoelastic Crystalline Polymers Part I: Model with Fountain Flow, Packing, Solidification, Polymer Eng. Sci., 26-1 (1986) pp.92-102
21. M.R. Kamal and P.G. Lafleur: A Structure-Oriented Computer Simulation of the Injection Molding of Viscoelastic Crystalline Polymer Part II: Model Predictions and Experimental Results, Polymer Eng. Sci., 26-1 (1986) pp.103-110
22. A.I. Isayev and C.A. Hieber: Toward a Viscoelastic Modelling of the Injection Molding of Polymers, Rheol. Acta, 19 (1980) pp.168-182
23. M.M. Frocht: Photoelasticity, 1, Willy, New York (1954)

Modeling and Numerical Simulation of Heat Transfer During the Filling Stage of Injection Molding

F. DUPRET and L. DHEUR
Unité de Mécanique Appliquée
Université Catholique de Louvain
Place du Levant, 2
B-1348 Louvain-la-Neuve, Belgium

1. INTRODUCTION

As long as design of injection molded parts is based on experience, it remains difficult to predict moldability and to reduce the level of residual stresses and resulting deformation of the cooled part. Hence, selecting suitable processing conditions requires a long and costly tuning of molding tools, unless the predictive power of numerical simulation is used.

Injection molding is a cyclic process. Each cycle consists of the following sequence of operations : (i) preplasticization of the polymer; (ii) filling of the cavity; (iii) packing of the material within the mold; (iv) cooling of the injected polymer; (v) ejection of the solidified part. In this paper, we present a model and a general algorithm for simulating the nonisothermal filling of thin planar or three-dimensional parts. We extend the methods of Vanderschuren and Dupret (1986), Dupret and Vanderschuren (1988), and Couniot et al (1989). Among previous investigations on the same topic, the works of Williams and Lord (1975), Lord and Williams (1975), Hieber and Shen (1980), Shen (1986), and Hieber (1987) must also be referred to.

Numerical simulation of the filling stage of injection molding represents however a challenging objective, since one is facing a three-dimensional, transient and variable domain system of partial differential equations. This is a consequence of the complexity of the melt front motion during filling and of the coupling between momentum and energy equations in the system. Consequently, following Schlichting (1968) and all the previously mentioned authors, the model must be simplified by introducing the Hele-Shaw approximation. The resulting system of equations is two-dimensional for the pressure field, and three-dimensional for the temperature field. However, following Dupret and Vanderschuren (1988), a simplified modelling of fountain flow must then be used in the form of a suitable front thermal boundary condition.

We here present a theoretical proof of the validity of this simplified model. We also focus on the influence of thermal exchanges on the kinematics of the filling stage. This influence appears to be very important in case of a variable mold thickness. In addition, we analyze solidification of the fluid on the cold walls during filling. In some cases, complete filling of the part is not possible. These results are successfully compared with experiments, and so validate our model and our numerical tool.

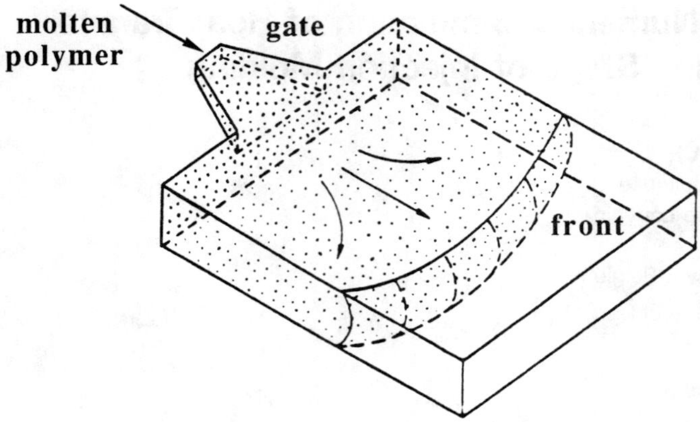

FIGURE 1. Schematic view of the injection molding problem.

2. DESCRIPTION OF THE PROBLEM

2.1 Geometrical and material hypotheses

We analyze the nonisothermal filling of thin cavities which are characterized by a midsurface S_m of continuous thickness 2h. The midsurface is composed by a set of planar facets, and the thickness is assumed to be small compared with the characteristic dimensions of the mold and also to vary slowly on the midsurface; hence, thickness derivatives in the midsurface directions are small compared to one.

We consider the creeping flow of a generalized Newtonian fluid, with a viscosity dependence on the absolute temperature T and the shear rate $\dot{\gamma}$. We assume that elastic effects are negligible. The viscosity is given by a power law of the form :

$$\mu(\dot{\gamma}, T) = m(T)\, \dot{\gamma}^{n-1} \quad , \tag{1}$$

where n is the power index and m(T) is a material function. For many materials, a fairly good approximation for m(T) is the exponential law :

$$m(T) = m_0 \exp(-aT) \quad , \tag{2}$$

as long as the temperature is far from the melting temperature T_m. In order to take into account solidification on the walls during filling, we also use the modified law :

$$m(T) = m_0 \exp(-aT) + m_1 \exp(-(T/T_m)^{n'}) \quad , \tag{3}$$

which was introduced in Vanderschuren and Dupret (1986) as a first approach of the solidification phenomenon : choosing sufficiently high values for n' and m_1 induces an important increase of the viscosity under T_m. The graph of the material function m(T) is shown in Figure 2 for both laws (2) and (3).

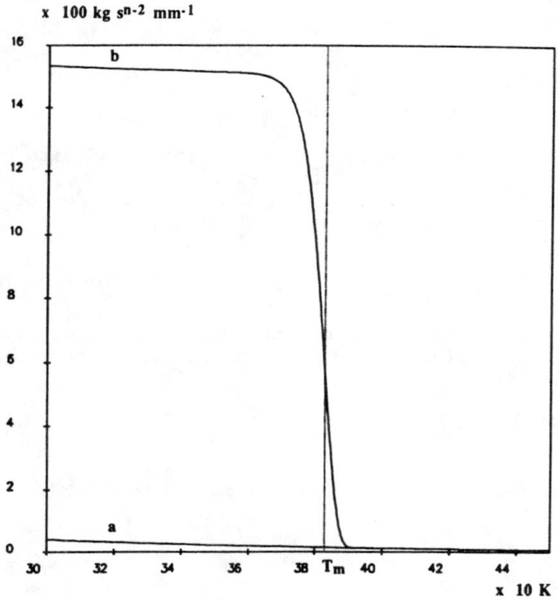

FIGURE 2. Viscosity as a function of temperature : a. exponential law; b. modified law.

Remaining material hypotheses consist in assuming that the polymer thermal conductivity k, specific heat c, and specific mass ρ do not depend on temperature.

2.2 Mass and momentum equations

We must solve the mass, momentum, and energy equations. The mathematical system is closed by suitable front, inlet, and wall boundary conditions, which are supposed to be symmetric with respect to the midsurface. In order to obtain a manageable system of equations, the problem must be simplified by means of dimensional analysis, following the method of Hele-Shaw (Schlichting 1968). The latter was applied to the injection molding problem in the case of a flat midsurface by Lord and Williams (1975) and Hieber and Shen (1980), and extended to the case of nonplanar midsurfaces by Vanderschuren and Dupret (1986). However, a definite proof taking nonisothermal effects into account was still missing. This important issue is examined in Section 3.

Consider any particular facet of the mold; let $Oxyz$ denote a corresponding system of Cartesian axes, where (x, y) and z stand for the midsurface and gap coordinates, respectively. From the Hele-Shaw theory, pressure variation in the thickness of the mold is negligible. The pressure field $p(x, y, t)$ obeys a second order nonlinear partial differential equation associated with mass conservation :

$$\frac{\partial}{\partial x}\left(S\,\frac{\partial p}{\partial x}\right) + \frac{\partial}{\partial y}\left(S\,\frac{\partial p}{\partial y}\right) = 0 \quad , \tag{4}$$

with

$$S = ||\nabla p||^{(1-n)/n} \int_0^h z^{1/n+1} \, m(T)^{-1/n} \, dz \quad . \tag{5}$$

A vanishing value of the pressure is specified as a boundary condition on the flow front, since capillary effects are not taken into account; a vanishing pressure normal derivative is imposed on the lateral sides of the mold, where only a tangential velocity must be obtained.

2.3 Energy equation

The energy equation represents a balance among heat advection (mainly in the midsurface direction), heat diffusion (in the gapwise direction), and viscous heating. The simplified form is :

$$\frac{DT}{Dt} = \frac{k}{\rho c} \frac{\partial^2 T}{\partial z^2} + \frac{m(T)}{\rho c} \left(\frac{S}{h}\right)^{n+1} ||\nabla p||^{n+1} \quad , \tag{6}$$

where DT/Dt stands for the temperature material derivative, which may be decomposed as follows :

$$\frac{DT}{Dt} = \frac{\partial T}{\partial t} + v_x \frac{\partial T}{\partial x} + v_y \frac{\partial T}{\partial y} + v_z \frac{\partial T}{\partial z} \quad . \tag{7}$$

When the mold thickness is constant, thermal convection in the gapwise direction is neglected in our computations. When the mold thickness is variable, the velocity component v_z is approximated by considering only a locally convergent, divergent, or parallel velocity field (Dupret and Vanderschuren 1988).

Boundary conditions consist in imposing the inlet temperature and suitable relations on the walls and the front. Following Dupret and Vanderschuren (1988), these equations are introduced in order to take into account the transient thermal behaviour inside the walls and the fountain effect on the front. Wall conditions are written as :

$$\frac{1}{3} \rho_s \, c_s \, e \frac{\partial T}{\partial t} + \frac{k_s}{e} (T - T_e) + k \frac{\partial T}{\partial n} = 0 \quad , \tag{8}$$

where ρ_s, c_s, k_s are the wall specific mass, specific heat, and thermal conductivity, respectively, T_e is the coolant temperature, and e is the thermal thickness (which is lower or equal to the wall-coolant distance). Further analysis shows that heat penetration from the hot polymer to the cold wall is progressive, and that contact thermal resistance may be observed between wall and polymer. Hence, estimating e is difficult, as will be verified in Section 5.

Fountain effect is schematically represented in Figure 3. The simplified model of Dupret and Vanderschuren (1988) is derived from the following assumptions :

1. The front region is infinitely thin and made of straight segments perpendicular to the midsurface;
2. Heat diffusion and viscous heating are neglected in the front region;
3. The residence time of material points in the front is small compared to the characteristic time scale of the filling.

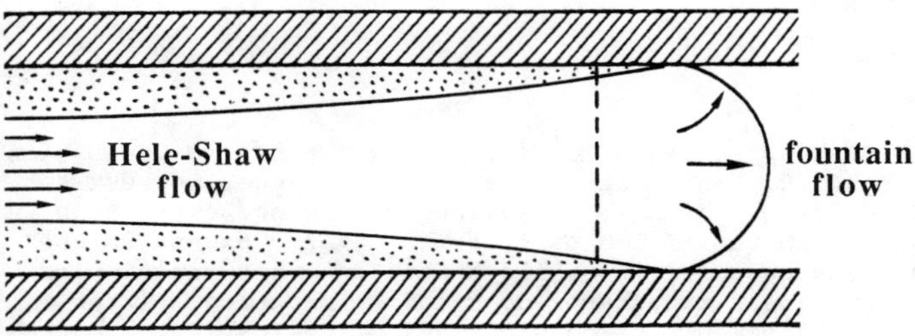

FIGURE 3. Schematic cross section of the flow.

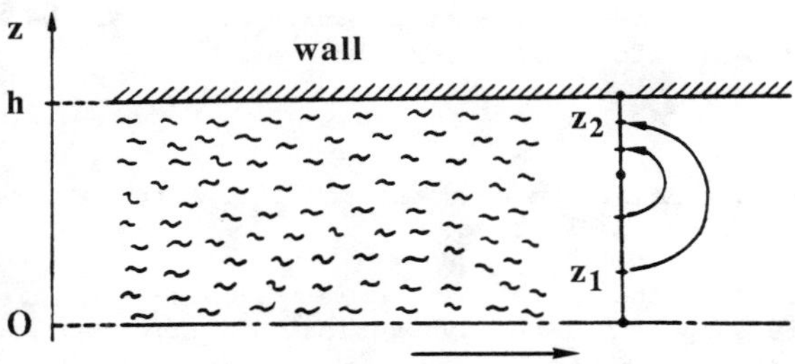

FIGURE 4. Fountain flow modelling.

Hence, the temperature of any material point entering the front at level z_1 and leaving the latter at level z_2 (Fig. 4) will remain unchanged. The front condition is, therefore,

$$T(z_2) = T(z_1) \quad . \tag{9}$$

As will be shown in Section 3, this model is the one which is in agreement with the Hele-Shaw approximate form of the equations.

2.4 Runner system

A simplified sprue and runner model is used (Couniot et al. 1989). The inlet temperature T_i and the flow rate Q_i or the pressure p_i are imposed at the sprue. A runner is represented by an equivalent cylindrical channel where a specific relationship is assumed between the flow rate Q and the loss of head Δp,

$$Q = K \, \Delta p^{1/n} \quad , \tag{10}$$

where K is an appropriate constant. Nonisothermal effects and front motion are not taken into account. Therefore, initial fronts are assumed to be located at a close distance from the gates.

3. MODELLING

3.1 Mathematical method

Obtaining the Hele-Shaw approximate model yields a singular perturbation problem (Schlichting 1968, Bender and Orszag 1978). Hence, we must set apart outer and inner flow regions : in injection molding, outer flow takes place everywhere in the filled part of the cavity, except in the following regions where peculiar inner flows develop : the vicinity of the fronts, the inlet gates, the sharp edges and bifurcations of the midsurface, the sudden enlargements or contractions of the cavity, and the lateral sides of the mold. However, in this paper, only the outer and front inner flows will be analyzed. Details will be published in Dupret and Vanderschuren (1990).

The method consists in considering the asymptotical behavior of the mass, momentum, and energy equations for a one-parameter family of flows. We let tend towards zero the dimensionless number ε defined by :

$$\varepsilon = \delta/L \quad , \tag{11}$$

where δ is the order of magnitude of the mold thickness $2h$ and L is any characteristic dimension of the midsurface.

We assume dynamical similitude of the flows when ε tends to zero. This means that the pathlines are supposed to be asymptotically independent of ε so long as all variables are written in a suitable nondimensional form. The latter differs depending on whether the outer or inner flow region is considered.

3.2 Hele-Shaw flow

Within the outer (or main region), coordinates are scaled using the characteristic dimensions L in the midsurface directions and δ in the gapwise direction, respectively. It is clear that dynamical similitude involves that dimensionless mold geometries are independent of ε. Scaling of the independent time variable is performed in the following way : throughout this section, we may suppose without loss of generality that injection is performed at a constant flow rate Q_i. Hence, we take the total filling period τ as time scale.

All the remaining scalar reduced variables are then obtained by using L as unit of length, τ as unit of time, and a suitable unit of pressure :

$$P = m_0 \, \varepsilon^{-n-1} \, \tau^{-n} \quad . \tag{12}$$

The latter selection solely yields a nondimensional pressure field which asymptotically does not vanish nor tends to infinity for a fixed power index n. Hence, P provides the order of magnitude of the pressure field for low values of ε.

In order to achieve dynamical similitude, it is also necessary that reduced temperatures T' tend towards a given limit when ε tends to zero. These are defined by the ratio :

$$T' = (T-T_e) / (T_i-T_e) \quad , \tag{13}$$

where T_e and T_i are the imposed wall and gate temperatures (for the sake of simplicity, in this section, the runner system is not taken into account and the thermal thickness is supposed to vanish).

However, the asymptotical reduced temperature field which is provided by the energy equation is not always nonvanishing and bounded. These conditions will be satisfied if the reduced thermal diffusivity $k'/\rho'c'$, volume heat capacity $\rho'c'$, and viscosity thermal exponent a' are of the orders $O(\varepsilon^{-2})$, $O(1)$, and $O(1)$, respectively. Equivalently, the latter conditions are written :

$$k/\rho c = O(\delta^2/\tau) \quad , \tag{14}$$

$$\rho c = O(P/(T_i-T_e)) \quad , \tag{15}$$

$$a = O(1/(T_i-T_e)) \quad . \tag{16}$$

This means that molding thin parts requires both a low polymer thermal diffusivity (otherwise, cooling down effect would be too strong and viscosities too high) and a sufficient heat capacity (for maintaining viscous heating effect under an acceptable level).

It is easy to show from the assumptions of this subsection that field equations asymptotically yield the Hele-Shaw simplified flow model.

3.3 Front flow

Turning back our attention to the front region (which is of the same order of magnitude δ as the gap thickness), we will first observe that associated inner flows are separate, in the sense that a different asymptotical system of equations must be solved at any time and for any point of this region. It may indeed be shown that front inner flows are asymptotical enlargements of infinitely thin flow domains observed during an infinitely short time interval. Therefore, within the front region, coordinates are scaled using the characteristic thickness δ in any direction, since the orders of magnitude of the corresponding velocity components are equivalent (and equal to L/τ). This also indicates that $\varepsilon\tau$ must be taken as scale for the independent time variable.

On the other hand, reduced scalar variables are then obtained by using δ as unit of length and $\varepsilon\tau$ as unit of time, while the previous unit of pressure P must still be selected, since the orders of magnitude of the outer and inner pressure fields cannot differ. Pressure and temperature fields are scaled exactly as within the main flow region.

Consider now the asymptotical forms of the field equations within the front region. Resulting from the selected inner scaling and from conditions (14) to (16), it is possible to show that three asymptotical properties characterize the front inner flows :

1. The reduced pressure vanishes;
2. The reduced temperature is constant along the pathlines;
3. The flow obeys the Stokes system of equations, with zero heat diffusion and viscous heating.

This is exactly what is required for obtaining the front model of Dupret and Vanderschuren(1988), as described in Subsection 2.3. We observe that the importance of inner flows is to provide the correct boundary and jump conditions for calculating the simplified main flow. In particular, with respect to the region of Hele-Shaw flow, simplified fronts must be straight segments perpendicular to the midsurface, since the asymptotical front region is infinitely thin with respect to the characteristic dimension L.

4. NUMERICAL METHOD

In order to obtain an approximate solution, we use the numerical algorithm which is developed in Dupret and Vanderschuren (1988) and Couniot et al. (1989). Calculations are performed using successive finite element meshes, which are generated on the filled part of the midsurface.

For describing the algorithm, it is sufficient to analyze a single time step. Let F_n and R_n denote the fronts and the temporary mesh at time t_n, and let $[p^n]$ and $[T^n]$ stand for the associated vectors of nodal pressures and temperatures. Lagrangian (bi)quadratic elements are used for discretizing the pressure field on the midsurface, while Lagrangian (bi)linear elements combined with Chebyshev collocation in the gap are used for the temperature field. We wish to calculate F_{n+1}, R_{n+1}, $[p^{n+1}]$, and $[T^{n+1}]$.

4.1 First step : predicting temperatures on the old mesh

Provisional new nodal temperatures $[\hat{T}^{n+1}]$ are obtained on the old mesh R_n by performing Eulerian time integration using a suitable transient scheme.

4.2 Second step : locating the new fronts

From the Hele-Shaw approximation, average velocity components are proportional to the pressure gradient. The latter is calculated on the front by using a modified Gresho method. Therefore, any front node may be relocated at the end of the considered time interval $[t_n, t_{n+1}]$ according to Euler explicit integration. Front conditioning is then performed in order to avoid ill-behaved temporary meshes.

4.3 Third step : remeshing

A new mesh R_{n+1} must be generated on the filled part of the midsurface. We select at the outset a well-behaved fixed mesh which covers the whole midsurface. Temporary meshes are generated by recovering as many elements as possible from this fixed mesh.

Several remeshing procedures have been investigated for obtaining a submesh in the region which is contiguous to the fronts. The definite algorithm is based on Delaunay triangulation. In the case of multi-faceted midsurfaces, remeshing is performed facet by facet in such a manner that conformity of the global mesh is ensured (Couniot et al. 1990).

4.4 Fourth step : recalculating temperatures on the new mesh

Extrapolation of the provisional temperature field $[\hat{T}^{n+1}]$ on the new mesh R_{n+1} provides the final nodal temperatures $[T^{n+1}]$. These calculations are performed taking into account the front thermal condition (9). When complete solidification occurs during filling, a particular underrelaxation technique is necessary for getting rid of unwanted oscillations in the vicinity of the front. Details will be published in Dheur et al. (1990).

Temperature extrapolation is obtained through the technique of Dupret and Vanderschuren (1988), as modified in Couniot et al. (1989). The method consists in defining on the midsurface a mesh formed by a set of convex quadrilaterals which cover the inter-front region (i.e. the domain enclosed by successive fronts F_n and F_{n+1}). Each quadrilateral is associated with a segment of the front F_n. Quadrilaterals may be folded around the edges of the midsurface, and may slightly overlap in case of front meeting. Extrapolation requires to project on the old front F_n the integration nodes of the inter-front region. The correspondence is easily established by using the local coordinates of the quadrilaterals forming the inter-front mesh.

4.5 Fifth step : calculating the new pressure field

The pressure equation (4) is solved and provides the new pressure field $[p^{n+1}]$ on R_{n+1}.

5. RESULTS AND DISCUSSION

Three problems are analyzed. Injected polymers are either a polycarbonate (1st and 2nd cases) or a polyethylene (3rd case). Material, geometrical and kinematical data are given in Table 1.

5.1 First problem

This example (Figs. 5 to 6) depicts the filling of a nonplanar part at two different flow rates. We may observe that gate pressure history is almost identical in both cases, since cooling down effect is more important and viscosity is higher at a low flow rate. The strong pressure increase which is observed at the end of the filling is a consequence of assuming polymer incompressibility in the simplified model.

5.2 Second problem

The second example (Figs. 7 to 9) deals with the molding of a rectangular plate of nonconstant (but continuous) thickness. Results obtained at two different flow rates are compared with isothermal calculations. The influence of thermal exchanges on the geometry of the pathlines and the

TABLE 1. Material data and injection conditions

		problem 1	problem 2	problem 3
polymer		polycarbonate	polycarbonate	polyethylene
m_0	kg s^{n-2} m^{-1}	1.44 10^9	1.44 10^9	1.48 10^6
a	K^{-1}	2.26 10^{-2}	2.26 10^{-2}	1.2 10^{-2}
n	-	0.665	0.665	0.45
n'	-	-	-	100
m_1	kg s^{n-2} m^{-1}	-	-	1.49 10^6
T_m	K	-	-	383
ρ	kg/m^3	1200	1200	950
c	J/kg K	2200	2200	2230
k	W/m K	0.215	0.215	0.33
ρ_s	kg/m^3	7800	7800	7800
c_s	J/kg K	460	460	500
k_s	W/m K	46	46	40
2h	mm	2	1 to 4	2
e	mm	3	5	a) 5 b) 8
T_e	K	383	383	313
T_i	K	533	533	435
Q_i	cm^3/s	a) 10 b) 20	a) 5 b) 10	-

filling history is here very strong. Viscous heating is indeed more important within the thick upper rib, where velocities and shear rate are high, than within the remaining of the part, where pressure gradient is low. Kinematic effects result from the dependence of velocity profiles on temperature profiles in the gap, through viscosity law (1).

5.3 Third problem

The third example (Figs. 10 to 12) shows a comparison between numerical and experimental results obtained in similar conditions. In order to validate our thermal model, we analyze complete solidification during the filling of a rectangular part. To this end, injection is performed by imposing a low pressure p_i at the sprue. Loss of head inside the runner is calculated from (10) with $K = 3.79\ 10^{-27}$ (in SI units). As solidification occurs, complete filling of the part is not possible. Hence, experimental and numerical last fronts may be compared, together with flow rates (as functions of time). For this very stringent test, comparison shows a good agreement, but indicates that influence of the wall thermal thickness is important. Hence, a better modelling of heat transfer inside the walls appears to be necessary. On the other hand, thermal shrinkage should be taken into account in the mass balance.

6. CONCLUSIONS

We have presented a general and accurate method for calculating the non-isothermal filling of thin planar or three-dimensional parts. Several examples have been investigated. Comparison with experiments has shown a very good agreement. The results illustrate how thermal exchanges may have an important kinematic influence in the process.

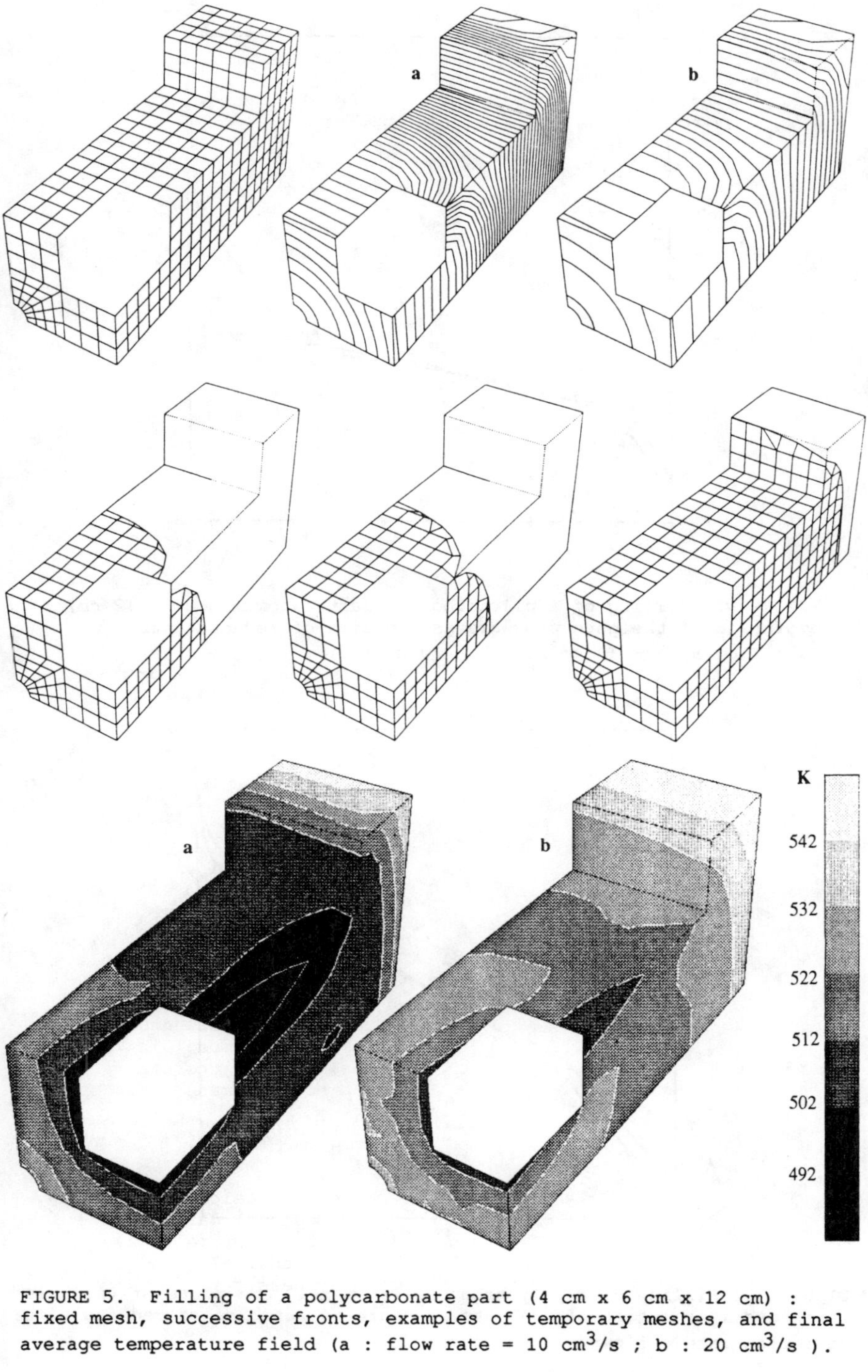

FIGURE 5. Filling of a polycarbonate part (4 cm x 6 cm x 12 cm) : fixed mesh, successive fronts, examples of temporary meshes, and final average temperature field (a : flow rate = 10 cm^3/s ; b : 20 cm^3/s).

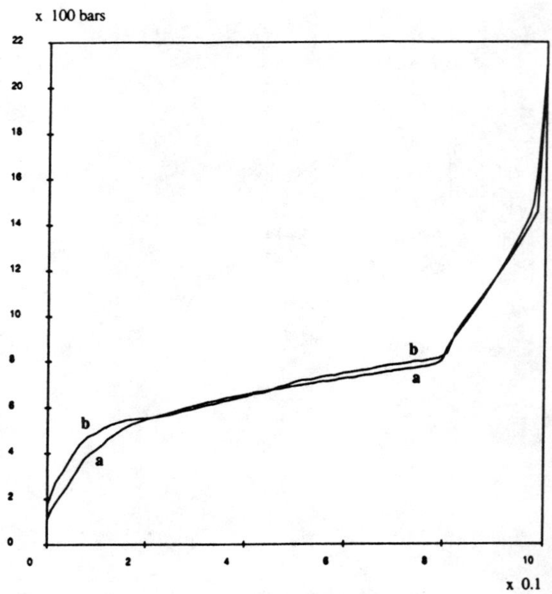

FIGURE 6. Filling of a polycarbonate part (4 cm x 6 cm x 12 cm) : evolution of the gate pressure as function of reduced time (a : flow rate = 10 cm^3/s ; b : 20 cm^3/s).

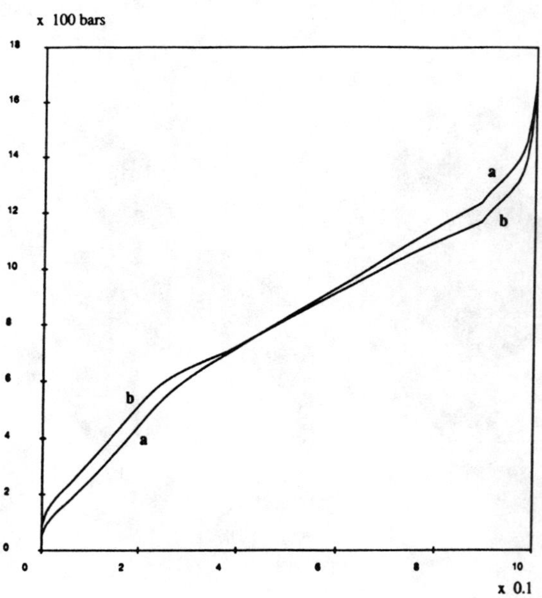

FIGURE 7. Filling of a polycarbonate 6 cm x 12 cm plate of variable thickness : evolution of the gate pressure as function of reduced time (a : flow rate = 5 cm^3/s ; b : 10 cm^3/s).

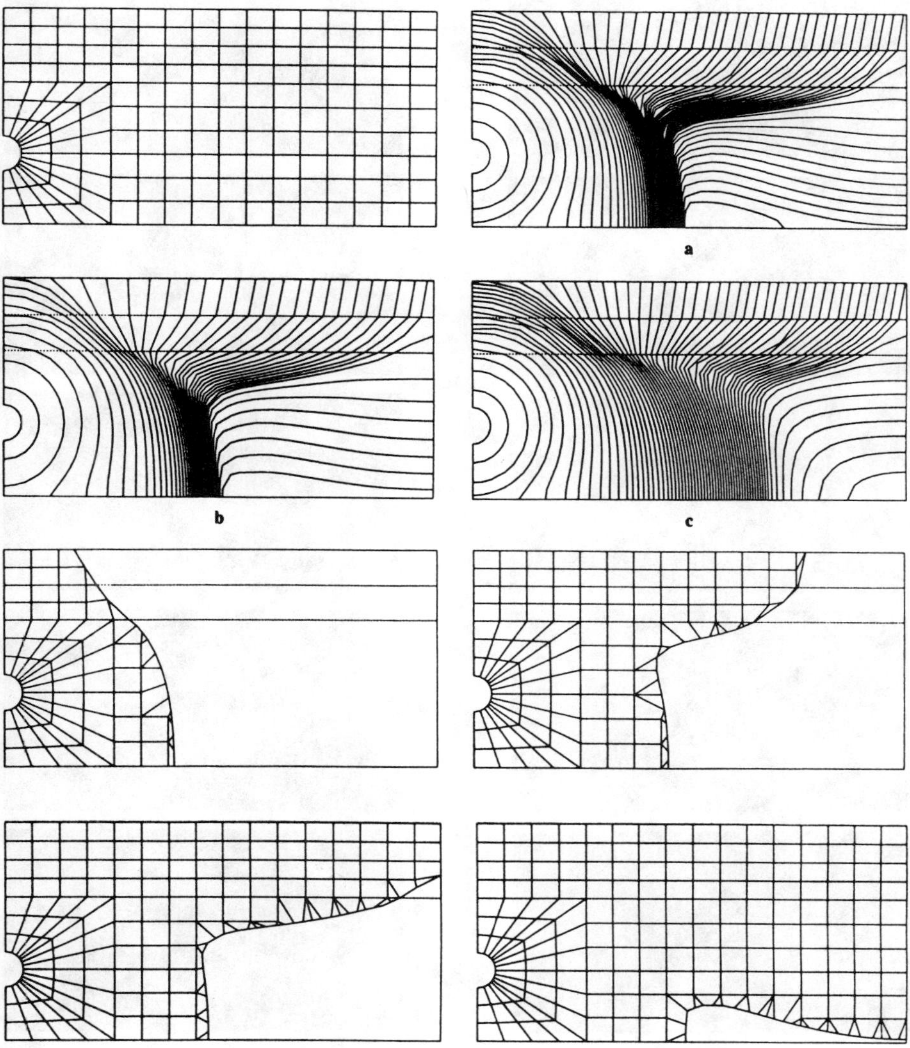

FIGURE 8. Filling of a polycarbonate 6 cm x 12 cm plate of variable thickness (1 mm to 4 mm) : fixed mesh, successive fronts, and examples of temporary meshes
(a : flow rate = 5 cm^3/s ; b : 10 cm^3/s ; c : 5 cm^3/s , isothermal).

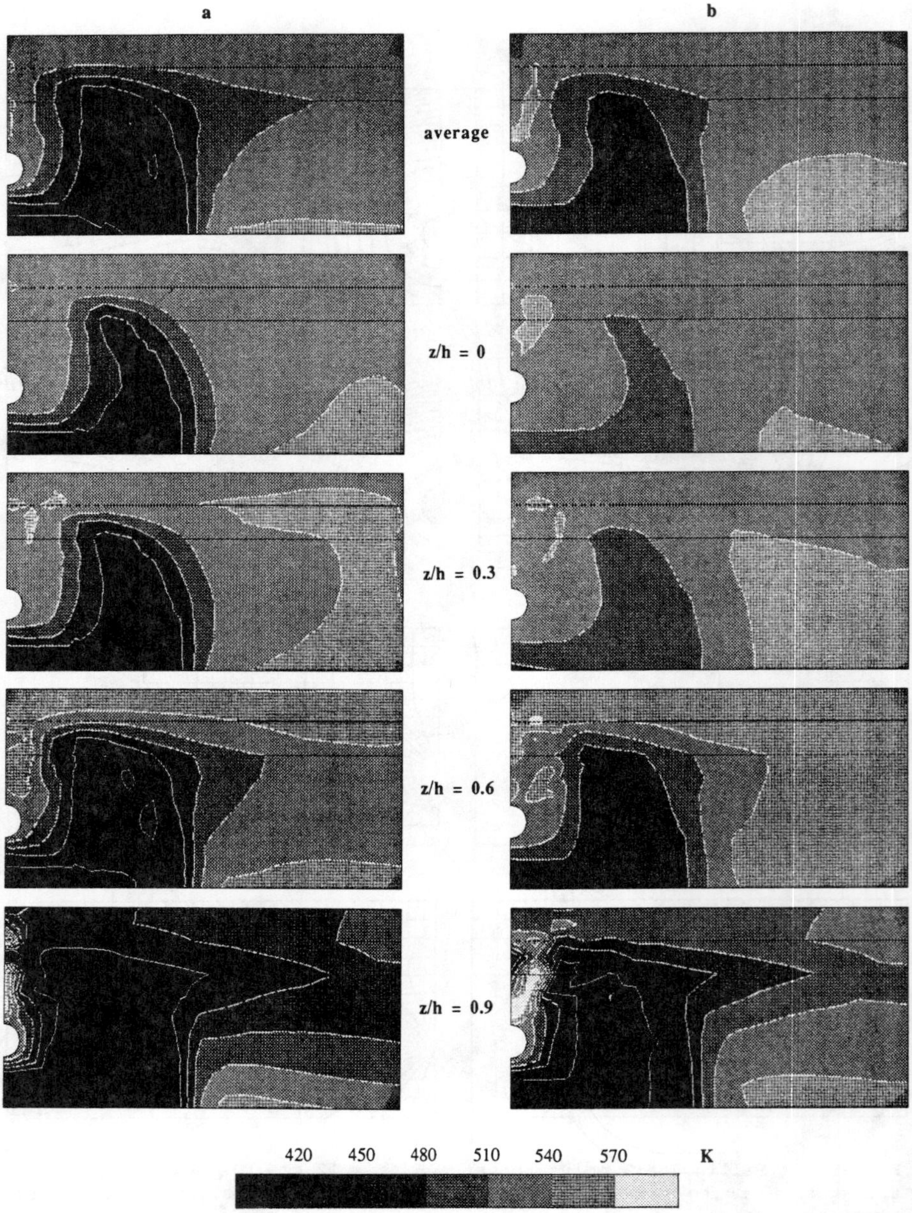

FIGURE 9. Filling of a polycarbonate 6 cm x 12 cm plate of variable thickness (1 mm to 4 mm) : final temperature field (a : flow rate = 5 cm^3/s ; b : 10 cm^3/s).

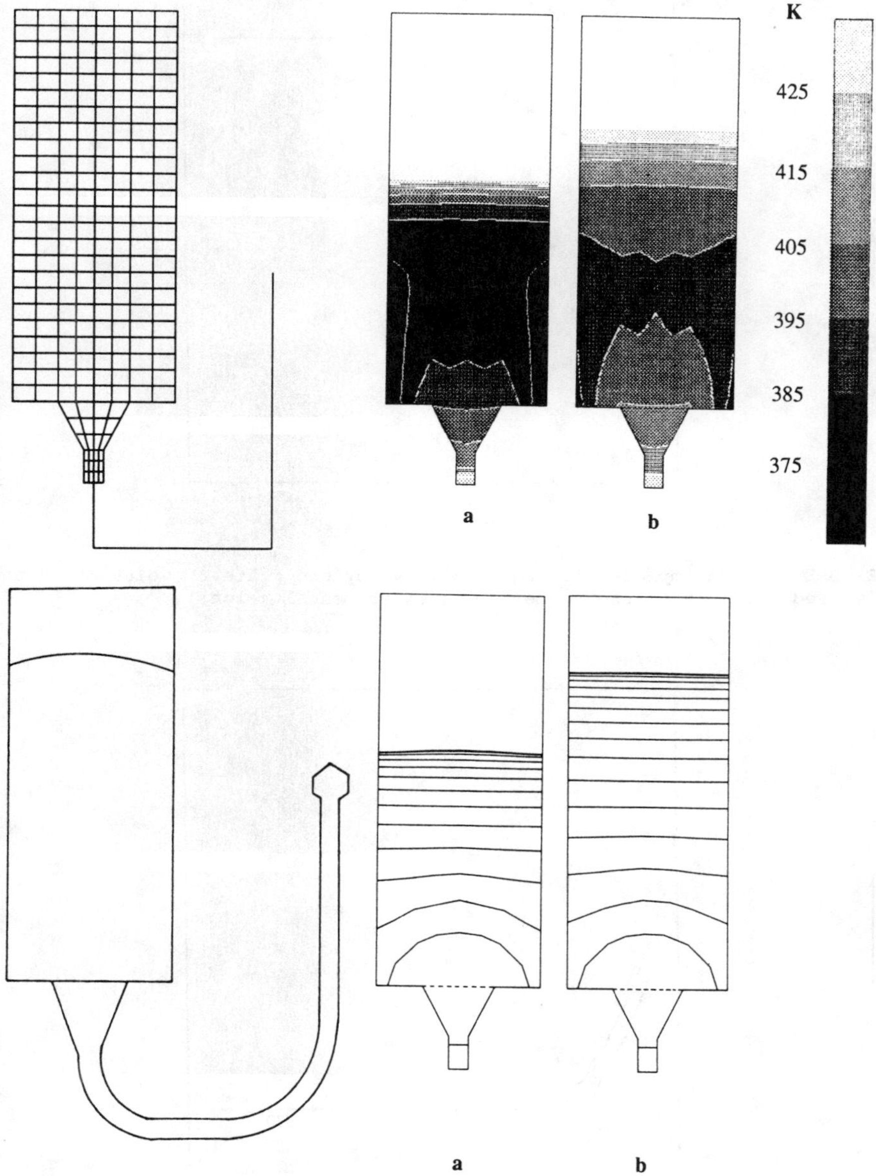

FIGURE 10. Incomplete filling of a polyethylene plate (5 cm x 12 cm) : fixed mesh, final temperature field, experimental last front, and calculated successive fronts .

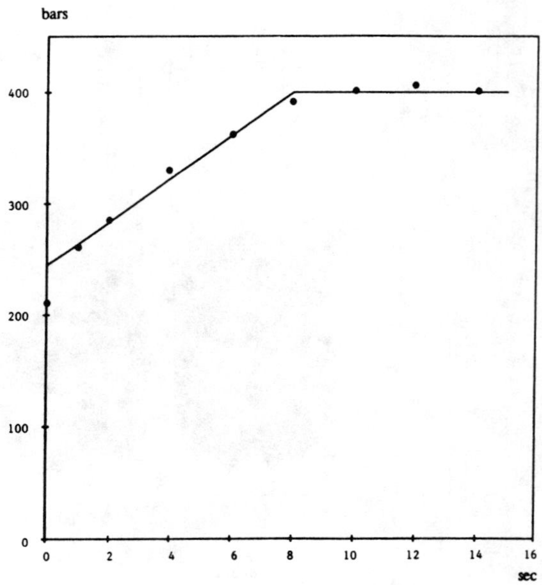

FIGURE 11. Incomplete filling of a polyethylene plate : evolution of the imposed pressure at the sprue (• : experimental values).

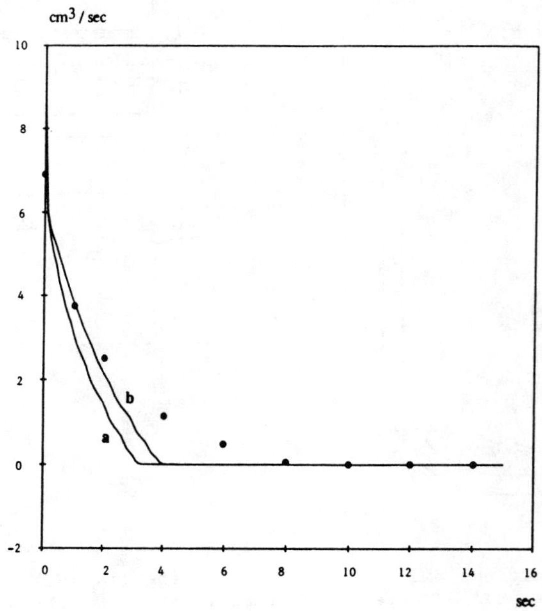

FIGURE 12. Incomplete filling of a polyethylene plate : comparison of experimental versus calculated flow rates
(• : experimental values ; a : thermal thickness = 5 mm ; b : 8 mm).

ACKNOWLEDGMENT

This work was supported by the E.E.C., in the frame of the BRITE Project RI1B-0087-F(CD). The authors wish to acknowledge the French Company La Télémécanique for performing the experiments and providing the material data.

REFERENCES

Bender, C.M., and Orszag, S.A., *Advanced Mathematical Methods for Scientists and Engineers*, McGraw Hill, New York, 1978.

Couniot, A., Crochet, M.J., and Dupret, F., 1990, in preparation.

Couniot, A., Dheur, L., Hansen, O., and Dupret, F., A Finite Element Method for Simulating Injection Molding of Thermoplastics, *Proc. 3rd Int. Conf. on Numerical Methods in Industrial Forming Processes* (NUMIFORM), Fort Collins, Colorado, pp.235-241, 1989.

Dheur, L., Vanderschuren, L., and Dupret, F., 1990, in preparation.

Dupret, F., and Vanderschuren, L., Asymptotical Behavior of the Equations Governing the Filling Stage in the Injection Molding Process, 1990, in preparation.

Dupret, F., and Vanderschuren, L., Calculation of the Temperature Field in Injection Molding, *AIChE Jl.*, vol.34, n°12, p.1959-1972, 1988.

Hieber, C.A., *Injection and Compression Molding Fundamentals*, ed. A.I. Isayev, Marcel Dekker, New York, 1987.

Hieber, C.A., and Shen, S.F., A Finite Element/Finite Difference Simulation of the Injection-Molding Filling Process, *J. Non-Newt. Fluid Mech.*, vol.7, p.1, 1980.

Lord, H.A., and Williams, G., Mold-Filling Studies for the Injection Molding of Thermoplastic Materials : Part II : The Transient Flow of Plastic Materials in the Cavities of Injection-Molding Dies, *Polym. Eng. Sci.*, vol.15, n°8, p.569, 1975.

Schlichting, H., *Boundary-Layer Theory*, McGraw Hill, New York, 1968.

Shen, S.F., Polymer-Flow Modelling and Simulation of the Injection-Molding Process, *Proc.2nd Int. Conf. on Num. Meth. in Industrial Forming Proc.*, Gothenburg, Sweden, p.51, 1986.

Vanderschuren, L,.and Dupret, F., Three-Dimensional Calculation of the Temperature Field during Injection Molding, *Proc.2nd Int. Conf. on Num. Meth. in Industrial Forming Proc.*, Gothenburg, Sweden, pp.199-204, 1986.

Williams, G., and Lord, H.A., Mold-Filling Studies for the Injection Molding of Thermoplastic Materials : Part I : The Flow of Plastic Materials in Hot- and Cold-Walled Circular Channels, *Polym. Eng. Sci.*, vol.15, n°8, p.553, 1975.

Session Summary: Heat Transfer Problems in Polymer Processing

Yasuo Kurosaki, Tokyo Institute of Technology

We have many kind of heat transfer problems in polymer processings such as polymerization and reactive processing, manufacturing of polymeric composites, plastics and rubber extrusion, polymer blending, mixing and compounding, and injection molding process. In this seminar, the numerical simulation and the visualization in injection molding process were presented by Professor Dupret and me, respectively. Both of them were related to the mechanisms of residual stress generation in the injection-molded products.

I believe that one of the final objectives of these investigations is to design and to create high quality products. Many researchers so far have been done to achieve this purpose; relations between the defects appears on the product and its molding conditions have been studied by means of a "trial and error" approach. These studies, however, have not revealed the phenomena occurring in the molding die sufficiently, and the information about the heat transfer problems in injection molding process is limited. In this sense, the goal of our studies seems to be still far.

When the detailed mechanism of injection molding process is once made clear, we must be able to actively control the process to obtain high quality products in the near future. For example, the sinks appearing on the surface of polymer products is usually undesirable defect; they often are inevitable. If the sinks can be controlled by the injection molding conditions, it must be interesting for manufacturers of polymer plastic products.

The recent experimental study in our laboratory have clarified the fact that the location of sink could be actively controlled by the cooling rate on the mold walls. This means that the sinks can be confined only in the region where the sink does not lower the quality of the product, e.g., the back-surface of the molded products, the rim of a plastic lens, etc. This is sure to be feasible to design the molding cavity. Our result should be published elsewhere.

In order to obtain better understandings of the heat transfer problems in polymer processing, it is important to know what occurs in manufacturing of the polymer products. I am sure that there still remain numerous heat transfer problems worth challenging in polymer processing.

Session Summary: Key Issues in Injection Molding

F. Dupret, Université Catholique de Louvain, Belgium

In injection molding, major objectives concern both experiments and development of numerical methods.

Experiments are necessary for characterizing the polymers, for understanding the physics, and for validating the numerical tools. In particular, the following material data should be obtained : (i) the dependence of viscosity versus shear rate, temperature and, maybe, pressure; (ii) the dependence of specific mass versus pressure and temperature; (iii) any thermal property, such as conductivity or heat capacity of the polymer; (iv) the appropriate viscoelastic models and material functions characterizing the fluid behavior during the filling stage; (v) the relaxation functions characterizing the viscoelasticity of the solidifying polymer during the cooling down stage; (vi) any data related to the physics of solidification.

Among phenomena which remain not completely understood, one should mention the formation and behavior of welding lines, the thermal and viscous contact between wall and polymer, kinematical effects such as jet emission from gates, and the general physics of polymers. Basic experiments are pressure measurements, determination of successive fronts and welding lines during the filling stage, kinematical analysis of fountain flow, stress measurements during the filling, packing and cooling stages, and analysis of the residual stresses.

On the other hand, numerical methods should be investigated in order to use the full predictive capability of modern computers and so to provide an alternative to the experience based tuning of industrial molds. Developping injection molding simulation programs is difficult, since computing time must remain acceptable, which means that strong numerical methods must be used with a constant view to stability and precision concerns. Moreover, several problems of numerical geometry (related to the ability of handling complex shapes) render the programming task very intricate. In order to achieve this complex set of objectives, strongly established simplified models are required. In particular, the Hele-Shaw approximation should always be used, together with the correct associated simplified model of fountain flow. Other modeling problems are to measure the level of flow stresses and to take into account in an appropriate manner the gate and runner system, the wall thermal influence, and the presence of a solidified layer.

Numerical tools should be built with the ability of being extended to a wide range of applications. Important issues are the molding of semi-cristalline polymers, of thermosets (taking chemical reactions into account), and of composite materials (with prediction of fiber orientation).

PAPERMAKING

New Drying and Dewatering Processes in Papermaking

JEFFREY D. LINDSAY
The Institute of Paper Science and Technology
Atlanta, Georgia 30318, USA

ABSTRACT

Current trends and advances in water removal technology for the paper industry are discussed. Most such innovations incorporate heat transfer mechanisms. Recent research results for two new water removal processes are presented. These new processes, impulse drying and displacement dewatering, have been proposed as possible means of saving energy while improving paper properties. While displacement dewatering has not yet lived up to its promise, impulse drying appears to have significant commercial potential. An improved understanding of phase-change heat transfer in paper is being sought to bring the impulse drying concept to fruition. Broad recommendations for future research are given.

INTRODUCTION

Water removal from a wet fibrous web is a critical operation in papermaking. Over 300,000 liters of water typically enter the paper machine for each ton of paper produced. Most of this water is removed by drainage and mechanical pressing. Only the final 0.5% or less of this water must be removed by evaporation, but doing so uses most of the energy needed for papermaking and requires a vast amount of equipment. There is definitely an economic incentive for improved water removal techniques. The motivation to adopt a new technology may not be strong, however, if only energy savings are offered. New (and thus potentially risky) water removal technology should also offer benefits in paper properties if it is to be accepted by the industry. Improved paper properties can mean new markets, or can allow a papermaker to meet specifications by using an alternative source of fibers.

In the critical areas of pressing and drying, a number of recent innovations have been developed which may result in energy savings or property improvements, or both. Many of these advances rely on some form of heat transfer. In this paper we will review a number of these advances, and will then focus on recent research involving two novel and related water removal strategies, impulse drying and displacement dewatering, for which energy savings and improvements in paper properties have been claimed. Phase-change heat transfer in impulse drying will receive special emphasis.

REVIEW OF ADVANCES IN MODERN DRYING AND DEWATERING

Innovations in Pressing Operations

Wet pressing is a time-honored process in which water is pressed out of a wet sheet by passing the sheet and an absorbent felt through a nip between two rolling cylinders. Since water not removed in the press section requires more expensive evaporative drying, efficient wet pressing is always desired.

In recent years, many purely mechanical advances have been made to improve water removal in wet pressing. Improved felt designs, double-felted nips, advanced roll covers, drilled rolls, grooved rolls, and innovative nip schemes are all examples. The development of the long or extended nip press (1) is one of the most important innovations, allowing a long, controlled pressure pulse to be applied to the wet sheet and a felt. Many of the most promising innovations, however, rely on heat transfer to improve or modify pressing operations. A number of these processes will be treated here.

Hot pressing. Hot pressing refers to a widespread practice of heating the web before it enters a press nip. Higher temperatures have several beneficial effects: reduced water viscosity means less resistance to water flow from the sheet to the felt; lower water surface tension means reduced rewet from capillary suction (bringing water from the felt back to sheet upon exiting the nip); and a softened fiber network means less resistance to compression (2).

Hot pressing today is achieved through steam showers before the nip of a press (3). Cross-directional control of steam delivery is now possible, allowing improved control over sheet moisture profiles. Sheet temperatures range from 60°C to 95°C. As a rule of thumb, a 10°C increase in sheet temperature tends to give a 1% increase in the solids content after pressing, which can mean about a 4% decrease in energy demand in the dryer section. Sheet properties are often enhanced, with various operations reporting improvements in moisture profiles and gains in either sheet bulk or density (not on the same machine, obviously). Higher machine speeds and the ability to use less expensive furnishes are also reported (4).

Efficient heat transfer with a steam shower is obtained by using steam near the saturation temperature, for superheated steam must first cool before it can deliver its heat of vaporization to the sheet. Heat transfer is also improved removing air in the condensing zone. This can be done in several ways, such as putting the steam shower close to the sheet surface, having a high-velocity steam curtain at the leading edge of the shower, or using a vacuum to remove the air (4).

The Tem-Sec press. The Tem-Sec or Direct Action Hot Press, a recent and now commercialized innovation, is a variation of wet pressing in which a large, heated, central roll acts as a common element for two or more press nips. Each nip is somewhat longer than conventional nips because of the size of the central roll (1.5-3.0 m in diameter). The central roll is internally heated with low-pressure steam, and each of the felts for the two nips may be heated, resulting in double-sided heating of the paper. Between the nips, the sheet is wrapped next to the hot central roll.

Like hot pressing, the Tem-Sec press enhances water removal by reducing water viscosity and by increasing the compressibility of the sheet (5). Good heat transfer is obtained by the two-sided heating and the close contact with the heated roll. Exiting dryness levels of 52-58% in linerboard have been reported. Slightly improved sheet strength is also claimed.

Impulse drying. Impulse drying is a novel water removal process which was proposed by Wahren (6) and subsequently developed at The Institute of Paper Science and Technology (IPST). At a superficial level, impulse drying can be described as a variation of wet pressing, with one roll heated to 250-375°C (see Figure 1). This is a much higher temperature than that used in Tem-Sec pressing or other existing thermally-assisted pressing processes. As a wet sheet passes through such a nip, intense heat transfer occurs, with peak fluxes ranging from

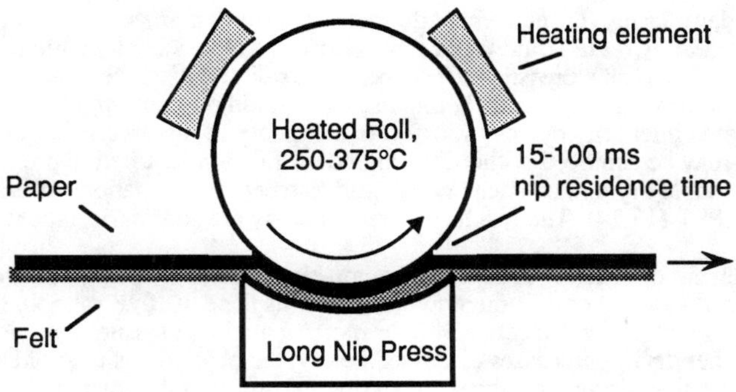

FIGURE 1. The impulse drying concept.

2 to 8 MW/m^2, rapidly tailing off to fluxes around 0.2-1.0 MW/m^2 during the 15 to 100 ms event. Through the interaction of heat transfer and other mechanisms, a large quantity of liquid water is expelled from the sheet; some water is also vaporized. The process has the potential to give higher dryness levels than wet pressing while using less energy than conventional cylinder drying (7).

Impulse drying not only offers the potential for energy and capital savings over traditional dewatering and drying methods, but can give significantly improved paper properties as well (7-9). The high surface temperatures in impulse drying enhance surface fiber conformability and interfiber bonding, resulting in increased tensile strength and surface smoothness. Researchers at The Pulp and Paper Research Institute of Canada (10) have examined impulse drying in newsprint, and have recently reported that in addition to the potential for substantial energy savings, impulse drying gives sheets with improved smoothness and higher breaking lengths and burst strength, although tear index and brightness may suffer slight losses. Strength improvement has been one of the major causes for industrial interest in impulse drying, as it would allow a required strength to be achieved with a lower weight sheet or with alternative raw materials (11).

A vapor-liquid displacement mechanism has been proposed as part of the physics of impulse drying. The high temperatures at the surface of the sheet correspond to vapor pressures which exceed the hydraulic pressures generated in much or all of a press nip. As a result, phase-change heat transfer will occur in the nip, leading to a pressurized vapor zone which can increase or sustain the hydraulic pressure gradient which drives liquid water out of the sheet into the felt. For example, Burton (8,9) conducted dynamic *in-situ* measurements of sheet density, vapor pressure, and temperature during simulated impulse drying. He concluded that a steam layer forming in the sheet helps to drive liquid water into the felt during impulse drying. A hot pressing effect also plays a role, as heat transfer warms the water and reduces its viscosity, making removal easier. Increased compression of the heated fiber webs will also contribute to water removal.

The commercialization of impulse drying has been delayed by the problem of delamination (12). In some fibrous webs, the vapor pressures generated can exceed the sheet strength as the sheet exits the nip, resulting in catastrophic sheet failure. The phenomenon of delamination must be understood and controlled before a commercial implementation of impulse drying is possible.

The physics of impulse drying have been the subject of some controversy recently, particularly in light of the troublesome delamination problem. Recent research results toward understanding impulse drying physics and the problem of delamination, with promising means for delamination reduction, will be discussed below.

Displacement dewatering. Displacement dewatering denotes a process in which a pressurized gas phase is used to increase liquid water removal from a mechanically compressed sheet. Water is removed by both a pressing effect and a vapor-liquid displacement mechanism. This process is similar to what may occur in impulse drying, although in impulse drying the gas phase is generated internally by heat transfer. In the theory of displacement dewatering, high dryness levels may be achieved without the high densification required in pressing alone. The potential to decouple dryness and density has lead to recent investigations of displacement dewatering at IPST (13,14). The results of a recent study are summarized below.

If true displacement dewatering is achieved, relatively little vapor would be required to pass through the sheet – ideally only enough to uniformly displace the free water in the interconnected pores of a compressed sheet (1 MPa or more of applied pressure may be required). Unlike some other drying processes discussed below, the objective is to avoid significant vapor flow rather than drying the sheet by passing gas through. In practice, however, the gas in a displacement dewatering process is likely to break through some pores in the sheet and also remove water by entrainment and evaporation. This undesired effect is a common instability in gas-liquid displacement in porous media, and is known as "viscous fingering" (15).

Several approaches can be taken to reduce viscous fingering and thus improve the efficiency of a displacement process. Heat transfer in particular can play a role in decreasing instabilities (16). As a hot viscous finger penetrates into cooler liquid, the gas begins to cool and contract, thus decreasing the growth rate of the viscous finger. The most promising strategy, however, may be to use a condensible vapor as the displacing phase, as when steam displaces water. As fingers of steam penetrate into the sheet and contact cooler liquid, they will condense. The viscous fingers become self-sealing to a degree. The combined effects of heat transfer and condensation are believed to make superheated steam a good candidate for displacement dewatering.

Several technologies have been developed in the past which share some features with the displacement dewatering concept. Use of a gas phase to remove water by entrainment, as well as by evaporation, was patented by Holden in 1966 (17) and extended by Kawka and co-workers (18,19) over a number of years. In some devices, gas passes through a sheet that is under low mechanical pressure applied by porous belts, wires, or felts (20). Kawka's work focuses on blowing air through paper, especially absorbent papers and boards, and does not really deal with displacement. In general, the blow-through process, with light mechanical compression, is severely limited in the dryness that can be achieved in short times, and exposure times on the order of a second may be required (21).

A related concept of gas-liquid displacement in dewatering was addressed in a patent awarded to Gottwald, et al., in 1967 (22). Their proposed device was a heated drum at 120-250°C, wrapped with a wet web held in place by a porous belt under enough tension to cause at least 0.03 MPa (5 psi) of pressure on the web. They claim that vapor generated at the drum-web interface would drive liquid water into the porous belt, reducing the evaporative load on subsequent dryers. The proposed physics seem questionable, as a saturated liquid layer is not likely to exist under these conditions, but the possibility of *in-situ* steam-liquid displacement clearly was envisioned. This process can be viewed as a precursor to impulse drying.

Innovations in the Drying of Paper

Here we briefly review the current state of evaporative drying in paper, dealing with conventional cylinder drying after a discussion of several innovations from recent years.

Press drying. In the 1970s, Setterholm et al. (23,24) developed a method of drying paper while under mechanical compression which resulted in improved physical properties. The high strength developed in high yield furnishes was especially noteworthy. The heating and restraint times were large, measured in seconds rather than milliseconds. The apparatus for press drying typically includes two heated platens; wire screens may be added above or below the paper to provide a path for vapor escape. Platen temperatures from 150 to over 200°C have been used, with restraining pressures of 0.4 to 4 MPa. The drying method is purely evaporative; the effect of pressure is to increase the heat transfer and elevate the internal temperatures.

Several others have continued exploring the potential of this new process. Recent results again emphasis the high strength properties which develop, especially wet strength or water resistance, when pulps with high lignin content are used (25,26). The flow of polymeric materials, lignin and hemicellulose, at elevated temperatures accounts for much of the property improvement observed in this process (27). Hemicellulose flow improves bonding between fibers, and while the presence of lignin decreases dry strength, its interfiber flow at high temperatures can cover and protect previously formed bonds, leading to greatly improved water resistance in the paper (28). These mechanisms are probably similar to those that occur in the upper layers of a sheet during impulse drying.

The high residence times required for this process severely limit its application, but pilot machines have been developed which attempt to provide the necessary restraint over a long period of time (29). Press-drying is most effective when the incoming sheet dryness is between 50 and 70%, so some pre- and post-drying with a conventional dryer section may be required. In the future, press drying concepts may be especially fruitful in the drying of paper made from hardwood, where a naturally high lignin content leads to high strength.

Thermal vacuum drying. Lehtinen (30-32) has developed a novel evaporative drying method in which vaporized water from a sheet is removed by condensation on a cool surface. In the initial form of this process, a sheet is placed in a sealed enclosure in contact with a hot platen. Between the sheet and the chilled platen is a highly porous water receiver such as a felt or open matt. Air is removed by vacuum prior to the beginning of the process. Drying can occur at low temperatures due to the partial vacuum in the system, meaning that inexpensive waste heat can be used for drying. The vacuum also helps to compress the sheet and enhance conductive heat transfer from the hot platen. The process is said to yield drying rates much greater than in conventional cylinder drying. More recent forms of this process, better suited for commercial implementation, put the sheet between two continuous temperature-controlled belts. The ability to control temperature and the mechanical pressure on the belts offers increased control over paper properties. Low speed may impede commercialization.

Through drying. Through drying is a well developed technology in which heated air is passed through a highly porous sheet under minimal compression to evaporate water (33-36). Because the hot air contacts wet fibers across a large surface area inside the sheet, heat transfer is very efficient. A good discussion of the physics of this evaporative technique is given by Rohrer and Gardiner (37). Tissue and toweling are prime grades for through drying, although various filter grades, roofing felts, wiper grades, and many wet-laid nonwovens can be used (35). Both cylindrical and flat bed through dryers are made.

High velocity gas impingement in tandem with through drying has been investigated and patented by Burgess et al. at The Pulp and Paper Research Institute of Canada (38,39). In this "Papridryer" system, vacuum pressure inside a drilled suction roll pulls hot air through the sheet, decreasing boundary layer heat transfer resistance and causing internal heat transfer in the sheet. Randall (40) has investigated a similar concept using a honeycomb shell for the porous roll; the technique is aimed at grades with too much flow resistance for standard through-drying. It employs roll vacuum pressures up to about 0.012 MPa or 1.7 psi, and dwell times of 0.6 to 3.0 seconds. In the future, through drying will continue to be an important method for drying lightweight and highly porous grades.

Improved impingement processes. In addition to the Papridryer process mentioned above, impingement heat transfer plays several roles in the paper industry. The Yankee dryer for tissue grades represents a well developed impingement system where convective heat transfer is used to remove large amounts of water from highly porous tissue webs in a brief time. Impingement is also used in the drying of coatings, often in series with infrared heating. Flotation systems, with impinging streams of air supporting the sheet, are common (41).

One recent advance in impingement heat transfer with great potential for the paper industry will be noted here. Page and Seyed-Yagoobi (42) have developed a new impingement concept, Radial Jet Reattachment or RJR, in which the heated jet is deflected away from the heat transfer surface by a radial nozzle. The deflected jet reattaches to the surface, creating an intense recirculation zone inside the reattachment region. The intense turbulence and the disruption of the boundary layer result in substantially improved heat transfer rates. Perhaps even more importantly, the net force exerted on the surface can be positive, negative, or zero, depending on the angle of the deflected jet. Thus intense impingement heat transfer can be achieved without damaging the paper or without disrupting a wet coating. With negative forces possible, it should function well in a flotation dryer.

Dielectric heating. Radio frequency and microwave heating of paper has been explored as a means of paper drying. These forms of dielectric heating offer a great advantage over conventional drying methods in that the whole volume of the paper is heated rather than just the surface. Unlike conductive, convective, and typical radiative heating, drying is not dependent on surface heat transfer, and temperature and vapor pressure gradients do not dominate the process. The high costs of these techniques may limit their practical potential in the paper industry to profile correction, in which conventional drying is supplemented by occasional local dielectric heating to establish a uniform moisture profile in the cross-direction of a moving paper web. Given the importance of moisture profile control, however, dielectric heating is a technology which is likely to play an increasingly important role in the paper industry. The traditional way of achieving uniform moisture profiles has been to overdry the paper to 3 or 4% moisture, when drying to 5-10% moisture is all that is needed. Overdrying requires a reduction in machine speed and higher energy costs. Experience in Europe has indicated that higher machine speeds are possible with radio frequency drying for profile control, although accurate figures are not available (43).

Superheated steam drying. David et al. (44) have directly applied superheated steam to paper to cause evaporative drying. The authors suggest that energy savings may be possible because none of the latent heat of vaporization of water need be lost, since the outgoing steam can be used in subsequent processes. This technology is still in the developmental stage, but may offer some important benefits in sheet properties. Using steam at 350-450°C, strength improvements similar to those of press drying were found, without the loss in bulk that comes from pressing. In the absence of oxygen, cellulose degradation at high temperatures (a potential problem in air drying) was suppressed. Brightness properties were not reduced by the high temperatures. Superheated steam drying requires further development, but may prove to be of commercial importance in the future.

Developments in conventional cylinder drying. Evaporative drying of paper on hot cylinders is one of the most studied and best understood water removal processes in the paper industry. In spite of decades of experience with cylinder drying, recent advances often involve simple procedures which may improve performance and energy efficiency. For example, simply improving the testing diagnostics for dryer sections will help productivity and allow better heat transfer strategies to be implemented. Cross-directional measurements of dryer variables such as surface temperature and pocket humidity represent a potentially important advance in this area (45). Furthermore, while it is known that noncondensibles in a dryer will significantly reduce heat transfer and dryer efficiency, the industry-wide practice of using continuous bleeds to remove noncondensibles now appears to be an unnecessary and wasteful practice, especially if proper purging of the cylinders is done before start-up (46). Simply shutting off unneeded bleeds could thus result in large energy savings.

Several other modern trends, all fairly simple, can lead to improved performance. Operating at higher steam pressures is a general trend, but the marginal gains in heat transfer are limited. Controlling the condensed liquid layer inside the dryer is especially important. Most efficient operation occurs when the condensate forms a puddle at the bottom of the dryer, but at higher speeds rimming occurs, resulting in a continuous condensate layer inside the drum which impedes heat transfer (47). Solutions include the addition of "spoiler bars" in the dryer to break up the condensate layer, and improved siphoning techniques to minimize the volume of condensate (an example of a possible advance in siphon design can be found in [48]).

Recent design improvements allow better mass transfer from the sheet by eliminating open draws of paper and the associated high humidity pockets (49). A number of other studies have shown significant improvements in performance are possible by modifying the gas flows associated with the dryer section (50-52).

NEW RESEARCH RELATED TO IMPULSE DRYING

Investigations of the Fundamental Physics of Impulse Drying

A number of recent studies at The Institute of Paper Science and Technology have sought to clarify the physics of impulse drying. A combination of both experimental and computational tools has been applied.

Flash x-ray visualization. In an effort to visualize the steam-water interface in impulse drying, Zavaglia and Lindsay (53) have used flash x-ray radiography to track the motion of x-ray-absorbing silver nitrate solution added to the upper layers of a linerboard sheet. Impulse drying and wet pressing events were simulated with a falling-weight press-nip device. During the pressing event, a 30-nanosecond burst of x-rays was sent horizontally through the sheet and the felt, passing then to x-ray sensitive film. The location of the silver nitrate solution in the z-direction could be observed in the radiograph.

Figure 2 shows a computer-enhanced radiograph taken of an impulse drying event. The light rectangle is the paper on top of a felt, bounded by metal surfaces above and below. Silver nitrate solution which was on the top of the sheet at the beginning of the event is now in the middle of the sheet, moving towards the felt. The radiograph was taken late in the event, during the sheet expansion phase. When the same experiment was done with a cool upper platen, the resulting wet pressing event did not remove the silver nitrate from the surface, but smeared it in the direction of the felt. The presence of a region virtually free of silver nitrate in the upper layers of the sheet implies that vapor has formed and displaced the liquid.

The peak mechanical pressure in this case was several times greater than what occurs in typical press nips, probably resulting in a more intense impulse drying process later in the nip. More recent radiographs (53) taken under more typical pressing conditions again show a low density region developing in the upper layers of the sheet in impulse drying, but the extent of the low density zone is not as great. It is believed that the observed low density region represents a dry zone; a larger two-phase zone containing both silver nitrate solution and steam may not be distinguishable from a saturated zone without further improvements in the experimental method. In any case, the existence of a vapor zone and the motion of fluid through a sheet can be observed with the flash x-ray technique.

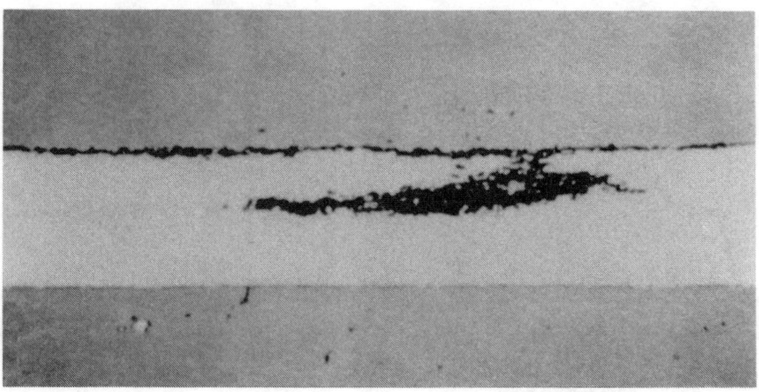

FIGURE 2. Computer-enhanced flash x-ray radiograph of an impulse drying event in linerboard on a felt. The dark region in the paper-felt zone is a drop of silver nitrate solution.

Evaporation in impulse drying. Measurements of water added to the felt by impulse drying and by wet pressing show that about 90% of the extra water removed by impulse drying in 250 g/m^2 wet sheets is in liquid form (the possibility of steam breaking through the sheet and condensing in the felt is easily ruled out by an examination of felt surface temperature in related measurements) (54). This is consistent with earlier energy balance and mass balance measurements made by Lavery (55,56). Enhanced *liquid* water removal is the key to potential energy savings in impulse drying.

Rewet reduction. An investigation of liquid water removal by impulse drying as a function of applied mechanical pressure provided indirect evidence that mechanisms other than displacement alone contribute to higher water removal (54). As a result, it was hypothesized that part of the enhanced water removal in impulse drying is due to the elimination of rewet (rewet is the transfer of water back into the sheet from the felt at the end of the nip by capillary or suction forces). With a pressurized vapor zone in the sheet, the normal process of rewet may be greatly reduced or reversed in impulse drying. This possible contribution to liquid water removal merits further investigation.

Temperature history. Measurement of local temperature histories within a sheet provides useful information. Burton (8,9) reported a measurement of internal sheet temperatures at three different layers inside a linerboard sheet during a simulated impulse drying event. In spite of uncertainties in the data, the temperature history in each layer is consistent with a displacement model of impulse drying, and could be interpreted as evidence that a distinct steam-liquid interface was moving through the sheet.

More extensive measurements of internal temperature propagation have recently been made at IPST (57, see also 54). Extremely fine thermocouples were sandwiched between thin, wet sheets of bleached kraft paper, each sheet having a basis weight of 50 g/m^2. During impulse drying with an MTS electrohydraulic press simulator, the temperature at each layer (including the felt-sheet interface) could be tracked in time. Sample results are given in Figure 3. Figure 3a shows the temperature at the felt-paper interface beneath a single 50 g/m^2 sheet. Three thermocouples at different locations on that interface were used, two of which gave similar results. Nonuniformities or a thermocouple problem may account for the third curve. The upper two curves show traits found in several of the measurements: a steep rise to a plateau above the ambient boiling temperature, followed by a rapid temperature rise which then levels off. Measurements at three transverse locations are shown in Figure 3b. Here three sheets have been stacked, and single thermocouples have been placed between the sheets. The upper curve from the thermocouple closest to the surface does not show an intermediate plateau. The second curve, showing data from the interface between the middle and bottom sheets, does show an S-shaped rise followed by a nearly flat region. The thermocouple at the sheet-felt interface shows only a gradual temperature rise.

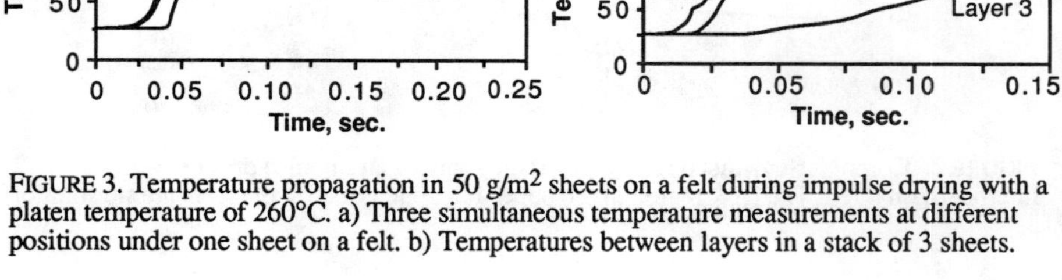

FIGURE 3. Temperature propagation in 50 g/m^2 sheets on a felt during impulse drying with a platen temperature of 260°C. a) Three simultaneous temperature measurements at different positions under one sheet on a felt. b) Temperatures between layers in a stack of 3 sheets.

A vapor-liquid displacement mechanism can account for the above results. The existence of plateau regions above 100°C is strong evidence for a two-phase zone where the vapor and liquid are in equilibrium at an elevated pressure. The two-phase zone at any point may only be temporary, and as it moves away, a dry zone with higher temperatures follows. In some cases, the two-phase zone is very thin (or nonexistent), so a sharp steam-water interface may be a good description of the process. Regions of slow temperature rise show the effect of transient conduction heating through a saturated liquid zone. In short, the data are consistent with the proposed displacement mechanism of impulse drying, and provide new evidence that extended two-phase zones may be formed during impulse drying. (The interpretation of these recent data has been aided by an examination of the numerical results from this study.)

Numerical modeling. In order to overcome the remaining roadblocks to industrial implementation of impulse drying, our physical understanding must be advanced. The difficulty of directly observing transport processes inside the nip, combined with uncertainties in interpreting experimental data, suggest that new tools are needed to supplement experimental studies. Numerical modeling is such a tool. While modeling cannot replace observation, the combined application of modeling and observation can lead to insights not available with either approach alone.

Recently, Lindsay (54) applied computational techniques to examine heat transfer in an idealized impulse drying process. The transient behavior of dry, two-phase and saturated zones in a one-dimensional, rigid porous medium was examined with a finite-difference moving-boundary model, MIPPS (Moving Interface Problems in Porous Systems). The problem included two internal moving boundaries each with changing temperature, pressure, and fluid velocity. Flow and heat transfer in the porous medium, including some capillary pressure effects, were accounted for. Temperature-dependent properties were used.

An early version of the model, MIPPS-I, attempted to simulate a heat-pipe mechanism of capillary wicking by using a wicking rate as an adjustable parameter (58). Only dry and saturated zones were allowed to exist, with wicking bringing fluid from the saturated zone back near the surface through a small subset of pores. A wicking rate of about 0.2-0.5 kg/m^2s then tended to give results consistent with measured data (primarily heat flux data). In MIPPS-II, the awkward wicking rate concept was eliminated, and a two-phase equilibrium zone was allowed to form (54,59). Specifically, bound or trapped water in pores remained immobile as vapor advanced until the liquid had been evaporated by heat transfer. The predictions of this presumably more correct model gave results similar to those of MIPPS-I. Sample results are shown in Figure 4. Predicted temperature profile results and the associated heat transfer mechanisms are qualitatively depicted in Figure 5.

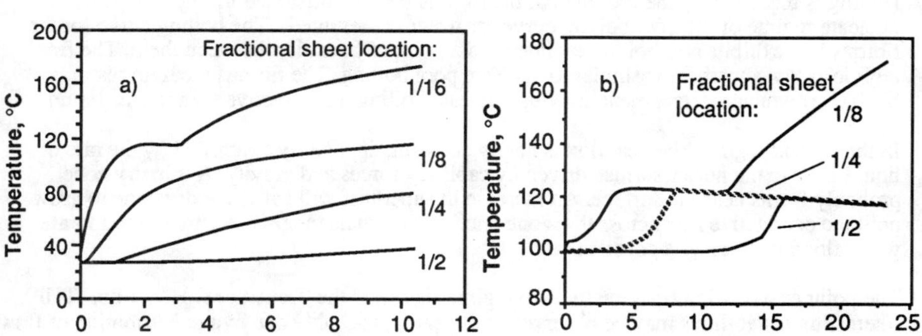

FIGURE 4. Two sample MIPPS-II predictions of local temperature histories in different paper sheets using different but realistic impulse drying conditions.

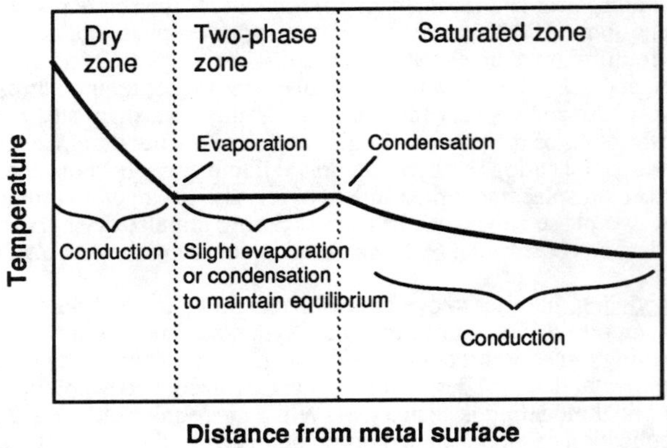

FIGURE 5. Temperature profile in a sheet and associated heat transfer processes during impulse drying (based on numerical predictions). Sketch is not drawn to scale.

Predicted local rates of evaporation and condensation have nearly equal magnitudes, giving a net evaporation rate in the nip close to zero. The process of evaporation and condensation is analogous to a heat pipe and gives heat fluxes beyond that of conduction alone. Simultaneous evaporation and condensation in the sheet is not an assumption but is a direct consequence of thermodynamics and conservation of energy.

Following recent modifications, MIPPS has been used as a design tool to choose materials and layer thicknesses in composite platen materials which provide some control over the heat flux delivered to the paper during impulse drying. The importance of controlling the thermal behavior of the hot surface is discussed below.

Boiling in fibrous media. In an attempt to overcome the lack of fundamental information about phase-change heat transfer in fibrous media (as opposed to the coarse granular materials for which boiling in porous media has been studied), Rudemiller and Lindsay (60) have studied boiling in thick ceramic fiber matts. Details of the experimental equipment are given in (61).

Boiling is modified by the presence of the fibrous bed, as illustrated in Figure 6 (only the nucleate regime of the pool boiling curve for water is presented). The boiling curve for the fibrous bed exhibits two boiling regimes and a point of transition between them. The first regime at low superheat is similar to nucleate pool boiling. The fibrous medium restricts bubble growth and detachment, making nucleate boiling less effective than in free liquid.

In the second regime, the heat flux is nearly constant, apparently controlled by the rate of liquid flow to the heater surface driven by capillary forces and gravity. A thin dry zone probably forms near the surface. An increase in superheat will cause the dry zone to grow until the conductive flux across that zone once again equals the flux required to evaporate water flowing down at a constant rate.

The point of transition between the two regimes is called the Transitional Heat Flux (THF). Interesting instabilities may be observed at this point, probably due to a redistribution of flow pathways for liquid and vapor. Macbeth has discussed some related phenomena in boiling in other types of porous media (62).

These steady-state experiments cover conditions far removed from impulse drying, which is a vastly more complicated process. However, the study represents a first step in characterizing phase-change processes in fibrous media. Further work is underway in this area.

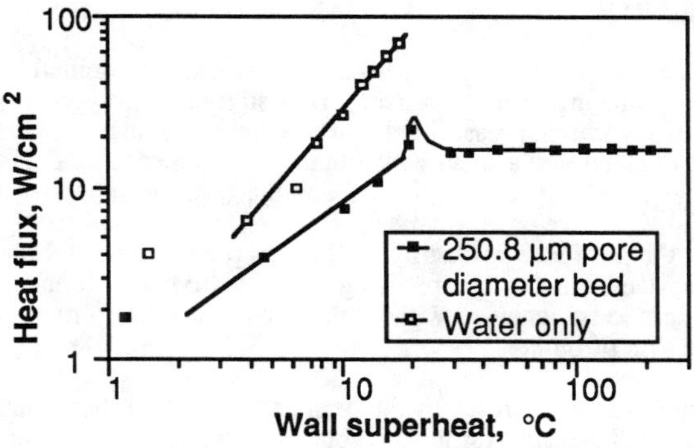

FIGURE 6. Typical measured boiling curves for pure water and water in a fibrous bed.

New Strategies for Delamination Control

As mentioned above, the problem of delamination has been the major stumbling block for the impulse drying process. In some furnishes, the internal vapor pressure is great enough to blow apart the sheet as it leaves the mechanical restraint of the nip. Wet sheets are more likely to delaminate than dry ones. Orloff (63) found that for a given sheet, delamination is a strong function of the operating conditions, especially nip residence time, platen temperature, and applied pressure. Conditions that improve water removal tend to increase the severity or likelihood of delamination. Burkhead et al. (64) also found that delamination was sensitive to felt moisture, with about 10-20% felt moisture giving better performance than either dry felts or wetter felts. They suggested that improvements in felt design might extend the operating window of delamination-free impulse drying. At that time, however, the prospects for delamination control looked bleak. Papermakers would not be satisfied with narrow operating windows for a new process. If it could not work well over a wide spectrum of conditions and for a variety of furnish types, the process would probably not be commercialized.

Important new advances in delamination control have now come from the extensive research of Orloff at IPST (65). Orloff has sought to control delamination by modifying the heat flux delivered to the sheet. His objective was to deliver a high heat flux early in the nip to generate the impulse drying effect, and then to significantly lower the heat flux near the end of the nip to reduce the internal vapor pressure. Simply lowering the temperature of a steel platen does prevent delamination, but also greatly reduces water removal. Initial work with porous platens did prevent delamination, but water removal was poor because steam vented into the platens, resulting in excessively low internal vapor pressures and preventing displacement. Orloff then developed novel composite platens with an external layer of impermeable, low-thermal diffusivity ceramic to modify the delivered heat flux. He found that the ceramic coated platens could give higher water removal in impulse drying without delamination than was possible with steel platens. (In one test series, for example, a moisture ratio change of 1.5 was possible without delamination using ceramic platens, while for the steel platens, delamination began at a moisture ratio change of about 1.1.) The use of ceramic coated rolls appears promising for effective impulse drying without delamination.

Current research in delamination control is concerned with optimizing the thermal properties of plasma-sprayed ceramic coated rolls by modifying the porosity of the ceramic. While measurements of surface heat flux in metal and ceramic coated platens have been made, experimental uncertainties have required improvements in the technique. These improvements are being achieved. Heat flux measurements will be combined with numerical predictions from MIPPS to guide design of future composite roll surfaces.

RECENT RESEARCH IN DISPLACEMENT DEWATERING

The displacement dewatering process has now been critically examined by Lindsay (14), following initially promising results from an earlier study by Sprague (13). The concept was tested with an electrohydraulic press, fitted with special heads that allowed gas (superheated steam or air) to be injected into a sheet while undergoing a mechanical pressing event. Finely drilled bronze plates acted as pressing platens, and the holes in the upper plate served as channels for gas flow. A wire or fabric mesh between the sheet and the upper platen distributed the load and the gas more uniformly. A felt was placed under the sheet. The concept was to apply pressurized gas during a pressing event to drive additional water into the felt, allowing high dryness to be achieved without the need to operate at high mechanical pressures which cause loss of bulk.

Dewatering, as expected, was limited by the permeability of the sheet, making long gas application times necessary. Lindsay examined displacement dewatering events ranging from 70 to 350 ms, with gas pressures up to 0.6 MPa (90 psi) and low applied mechanical pressures of 1.2-2.4 MPa (180-340 psi). For both superheated steam and cool air, displacement dewatering gave greatly improved dewatering compared to wet pressing under the same low-pressure press conditions. Steam was also much more effective than air, with solids levels up to 60% being possible in sheets initially around 20% solids, but much of the effect was due to evaporative drying after substantial heat transfer to the paper.

The most critical issue, bulk control, can be examined in several ways. In comparing densification in displacement dewatering to wet pressing run with the same low mechanical pressures and long nip times, displacement dewatering appears promising. The dryness-density curves show that displacement dewatering is superior in preserving bulk. However, when wet pressing is done under more conventional conditions (high-pressure, low-dwell time), the density-dryness results show that normal wet pressing usually gives less densification (higher bulk) than displacement dewatering. A set of data showing these features is presented in Figure 7. It appears that a creep effect occurs during long nip times, even at low mechanical pressure, and the resultant loss of springback leaves a denser sheet than expected. Steam displacement, while effective at water removal, densifies the sheet even more as it heats and softens fibers, lowering their compressive resistance.

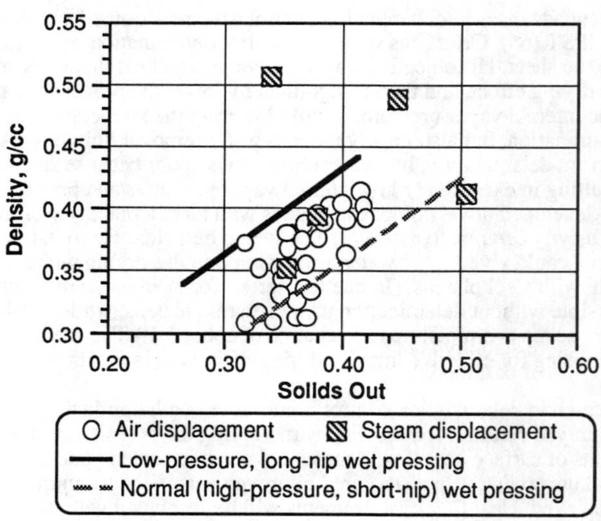

FIGURE 7. Density development in displacement dewatering of 100 and 150 gsm linerboard handsheets. Results from normal and low-pressure wet pressing are also shown.

The promise of improved bulk control thus has not been realized in the experiments done to date. However, it may be that a combination of higher vapor pressure and lower residence time may eliminate the apparent creep effect and maintain bulk in the sheet. The next stage of experimental work will seek to find operating conditions and pulp types where displacement is effective in preserving bulk and removing water.

FUTURE RESEARCH NEEDS

Many recent advances in water removal in paper incorporate heat transfer. Depending on the application, the benefits of heat transfer may include reduction of water viscosity or surface tension, sheet softening, induced flow of polymeric materials for better wet strength, improved fiber bonding, vapor generation to cause displacement, enhanced stability of a vapor-liquid displacement front, or simply vaporization of water. In every case, however, there is a need for more fundamental information about heat transfer in paper. Fundamental mechanisms are poorly understood, partially because of the complexity of paper and the difficulty of conducting measurements in a material that is both thin and heterogeneous.

Novel measurement techniques may open new doors in elucidating transport processes in paper. Flash x-ray visualization has proven helpful in studying impulse drying, and is now being applied to regular pressing operations to study in-plane shear (66). Infrared thermography is another visualization method which is advancing heat transfer research in paper (42). Numerous other measurement techniques may be or are becoming of use, including x-ray tomography, advanced microscopy methods, NMR and other forms of spectroscopy.

Numerical models of commercial pressing and drying processes were not reviewed in this paper, but represent a rich avenue for future progress. A major frustration to those developing such models, however, is the lack of data on transport parameters in paper (67). For example, a drying model may require information about the relationship between capillary pressure and saturation in paper, but this relationship may change significantly at elevated temperatures or under light mechanical constraint. The necessary data are unavailable, but could be measured. Even more basic thermodynamic properties, such as the heat of vaporization of water, change significantly when water is in the presence of small pores. Current research on the thermodynamics of vicinal water in cellulose (68) may prove of value here. A database of data on effective diffusion coefficients, contact resistance parameters, and other mass and heat transfer properties would be of great value to many researchers in addition to those developing models. Some progress is being made (69,70), but much more work is needed.

With regard to the research focused on in this paper, both impulse drying and displacement dewatering require further research in many areas. Plasma-sprayed ceramic coatings on rolls will soon be tested in a pilot impulse drying machine and may prove successful in controlling delamination, but more effective means may be discovered once the physics of impulse drying are more fully understood. Gamma radiation methods will soon be applied at IPST to study details of saturation profiles during boiling in a fibrous medium, and further studies are planned to give more insight into the complex phase-change heat transfer process that occurs in impulse drying. Bench-scale tests of displacement dewatering concepts need to be supported by more fundamental studies of vapor-liquid interaction in porous media. Since the issue of displacement stability is important to both impulse drying and displacement dewatering, studies in paper or a model fibrous medium may be helpful in guiding future thinking.

The importance of paper properties must be kept in mind as novel processes are developed. Heat transfer, already of value in enhancing some paper properties, may be useful in creating new new products rather than just new processes. For example, high-temperature processes like impulse drying may catalyze development of novel composite materials featuring paper and thermoplastics or thermosetting resins. Special high-lignin furnishes with polymer additives may be used to achieve new, inexpensive materials that compete with traditional plastics. Creative minds are needed to pursue the possibilities.

SUMMARY

Many recent innovations in water removal in paper involve heat transfer operations. Recent research results from two of these processes, impulse drying and displacement dewatering, have been presented. Impulse drying has received a considerable amount of attention and is approaching commercialization as new means for delamination control are being found. This unique application of intense phase-change heat transfer offers energy savings as well as property improvements. Displacement dewatering, a cousin of impulse drying, is still early in the developmental stage, and faces significant barriers in achieving the goal of good dewatering while maintaining bulk. With further research into the fundamentals of transport processes in paper, new advances in water removal and property development are expected.

LITERATURE CITED

1. Wicks, L., Continued Development and Experience with the Extended Nip Press, *Tappi J.*, vol. 66, no. 4, pp. 61-64, 1983.

2. Back, E. L., Temperature Effects on Dewatering, *Appita*, vol. 41, no. 3, pp. 217-223, 1988.

3. Batty, R. C., Daunais, R., and Pye, I. T., Hot Pressing of Newsprint, *J. Pulp Paper Sci.*, vol. 9, no. 4, pp. TR106-TR111, 1983.

4. Cutshall, K., Bell, N., and Hudspeth, D., "Hot Pressing," *Proc. of TAPPI 1988 Practical Aspects of Pressing and Drying Seminar*, Atlanta, Georgia, March 14-18, 1988.

5. Walker, K., Advances in Hot Pressing Technology, *Tappi J.*, vol. 73, no. 8, pp. 99-101, 1990.

6. Wahren, D., Methods and Apparatus for the Rapid Consolidation of Moist Paper Webs, U. S. Patent 4 324 613, 1982.

7. Lavery, H. P., High Intensity Drying Processes - Impulse Drying, Second Annual Report for U.S. DOE under Contract No. DE-FG02-85CE40738, The Institute of Paper Science and Technology, Atlanta, Georgia, Feb. 1988.

8. Burton, S. W., An Investigation of Z-direction Density Profile Development During Impulse Drying, Ph.D. Thesis, The Institute of Paper Chemistry, Appleton, Wisconsin, 1986.

9. Burton, S. W., and Sprague, C. H., The Instantaneous Measurement of Density Profile Development During Web Consolidation, *J. Pulp Paper Sci.*, vol. 13, no. 5, pp. J145-J149, 1987.

10. Sparkes, D. G., and Poirier, D., Impulse Drying at Paprican, *Pulp Paper Can.*, vol. 91, no. 4, pp. T147-T150 (1990).

11. Ellis, R. L., Impulse Drying Shows Promise in Reducing Energy Costs, *American Papermaker*, vol. 53, no. 10, pp. 35-36, 1990.

12. Crouse, J. W., Woo, Y. D. and Sprague, C. H., Delamination — A Stumbling Block to Implementing Impulse Drying Technology for Linerboard, *Tappi J.*, vol. 72, no. 10, pp. 211-215, 1989.

13. Sprague, C. H., New Concepts in Wet Pressing, Final report prepared for the US DOE under contract AC02-84CE40685, U.S. Dept. of Energy, Office of Industrial Programs, Washington, D.C., March 1986.

14. Lindsay, J. D., Experimental Evaluation of Displacement Dewatering, AIChE Annual Meeting, Chicago, Illinois, Nov. 11-16, 1990.

15. G. M. Homsy, Viscous Fingering in Porous Media, *Ann. Rev. Fluid Mechanics*, vol. 19, pp. 271-311, 1987.

16. C. A. Miller, Stability of Moving Surfaces in Fluid Systems with Heat and Mass Transport, *AIChE J.*, vol. 19, no. 5, pp. 909-915, 1973.

17. Holden, G. R., Apparatus for Dewatering of Fibrous Webs in Papermaking and Similar Machines, U. S. Patent 3 284 285, Nov. 8, 1966.

18. Kawka, W., Theoretical and Experimental Analysis of the Blow-Through Process of Fibrous Web Dewatering, trans., *Przeglad Papier.*, vol. 39, no. 11/12, pp. 403-407, 1983.

19. Kawka, W., and Ingielewicz, H., A New Technology for Producing Porous Papers Using Through Dryers, trans., *Przeglad Papier.*, vol. 30, no. 1, pp. 10-18, 1974.

20. Kawka, W., Ingielewicz, H., and Marek, I., Study of Equipment for Intense Dewatering and Drying of Porous Paper Products,trans., *Przeglad Papier.*, vol. 34, no. 3, pp. 82-87, March 1978.

21. Kawka, W., Optimum Consolidation of Absorbent Fibrous Webs on Paper Machines with Air Blow-Through Presses, trans., *Przeglad Papier.*, vol. 41, no. 5, pp.163-169, 1985.

22. Gottwald, B. C., Halsey, B. S., Williams, R. C., and Haigh, J. M., Continuous Process of Drying Uncoated Fibrous Webs, U.S. Patent 3 354 035, Nov. 21, 1967.

23. Setterholm, V. C., Benson, R. E., Wichman, J. R., and Auchter, R. J., Z-direction Restraint: A New Approach to Papermaking, USDA Forest Service, Research Paper FPL 256, Forest Products Laboratory, Madison, Wisconsin, 1975.

24. Setterholm, V. C., Press Drying Overview, *Proc. TAPPI Annual Meeting*, New York, 1979.

25. Back, E. L., The Present State of Press Drying. Part 1: The Fundamentals of Press-Drying, in *The Role of Fundamental Research in Papermaking*, ed. J. Brander, Proc. of the Seventh Fundamental Research Symposium, Cambridge, England, Sept. 1981, Mechanical Eng. Publ. Ltd., London, 1981.

26. Back, E. L., A Review of Press Drying, in *Advances in Drying*, ed. A. S. Mujumdar, Hemisphere, Washington, D. C., 1987.

27. Byrd, V. L., Flow and Adhesion of Hemicellulose and Lignin During Press Drying, *Proc. TAPPI Annual Meeting*, New York, 1979.

28. Byrd, V. L., Press-drying Saves Energy and Enhances Properties, *Paper Technology*, vol. 30, no. 5, pp. 24-28, 1989.

29. Swenson, R., The Present State of Press-Drying of Paper - Part 2: The Application of Press-drying to Paper-making, in *The Role of Fundamental Research in*

Papermaking, ed. J. Brander, Proc. of the Seventh Fundamental Research Symposium, Cambridge, England, Sept. 1981, Mechanical Eng. Publ. Ltd., London, 1981.

30. Lehtinen, J., A Versatile Web Drying and Consolidation Method Under Development, Paperi ja Puu, vol. 70, no. 4, 1988.

31. Lehtinen, J., Press Consolidation - The Condensing Belt Systems, *Proc. PIRA Intl. Conf. on New Technology in Web Consolidation and Drying*, Sussex, England, 1986.

32. Lehtinen, J., Die Entwicklung des Condebelt-Prozesses, *Das Papier*, vol. 42, no. 10A, pp. 154-159, 1988.

33. Walser, R., and Swenson, R. S., Air Through-Drying of Paper, *Tappi J.*, vol. 51, no. 4, pp.184-190, 1968.

34. Holik, H., Zur Durchstromtrocknung von Papier, *Das Papier*, vol. 26, no. 4, pp.153-161, 1972.

35. Villalobos, J. A., Paper Machine Hoods, Pocket Ventilation, Heat Recovery and Thru-dryers, *Proc. TAPPI 1988 Practical Aspects of Pressing and Drying Seminar*, Atlanta, Georgia, March 1988.

36. Bryand, E. T., Energy-conserving Drying Methods, *Paper Tech. and Ind.*, vol. 16, no. 1, pp. 24-29, 1975.

37. Rohrer, J. W. and Gardiner, F. J., Through-drying: Heat Transfer Mechanism and Machine System Response, *Tappi J.*, vol. 59, no. 4, pp. 82-87, 1976.

38. Burgess, B. W., and Chapman, S. M., Turbulent Drying Process, U.S. Patent 3 418 723, 1968.

39. Burgess, B. W., Koller, W., and Pye, E., The Papridryer Process Part II - Mill Trials, *Pulp Paper Mag. Can.*, vol. 73, no. 11, pp. 73-81, 1972.

40. Randall, K. R., Using High-velocity Impingement Air to Improve Through Drying Performance on Semi-Permeable Webs, in *Drying 84*, ed. A. S. Mujumdar, pp. 254-263, Hemisphere, Wash., D. C., 1984.

41. Krizek, F., Impinging Jet Systems to Support and Dry Paper, *Drying Tech.*, vol. 4, no. 2, pp. 271-294, 1986.

42. Page, R. H., and Seyed-Yagoobi, J., A New Concept for Air or Vapor Impingement Drying, *Tappi J.*, vol. 73, no. 9, pp. 229-234, 1990.

43. P. L. Jones, High Frequency Dielectric Heating in Paper Making, *Drying Tech.*, vol. 4, no. 2, pp. 217-244, 1986.

44. David, M., Mujumdar, A. S., Crotogino, R. H. and Douglas, W. J. M., Effect of Superheated Steam Drying on Paper Properties, *Proc. 74th Annual Meeting*, pp. B233-B237, Canadian Pulp and Paper Association, Montreal, Canada, Jan. 1988.

45. Rosenberger, G. W., and Reese, J. R., Profile Testing Improves Dryer Section Diagnostics, *Tappi J.*, vol. 73, no. 8, pp. 89-94, 1990.

46. Perrault, R. D., Troubleshooting Design and Operation of the Dryer Section, *Proc. TAPPI 1988 Practical Aspects of Pressing and Drying Seminar*, Atlanta, Georgia, March 14-18, 1988.

47. Han, S. T., Drying of Paper, *Tappi J.*, vol. 53, no. 6, pp. 1034-1046, 1970.

48. Anon., *Paper Age*, vol. 106, no. 7, p. 23, 1990.

49. Palazzolo, S., No-draw Drying, *Tappi J.*, vol. 73, no. 9, pp.225-228, 1990.

50. Reese, J. R., High-humidity Hoods Conserve Energy and Improve Runnability, *Tappi J.*, vol. 72, no. 4, pp. 87-90, 1990.

51. Fagerholm, L., Aerodynamical Properties of Dryer Fabrics for High Speed Machines, *Proc. 1990 Tappi Engineering Conference*, Seattle, Washington, Sept. 24-27, 1990.

52. Hill, K. C., and Reese, J. R., Dryer Section Rebuild Requirements, *Proc. 1989 Tappi Engineering Conf.*, Atlanta, Georgia, Sept. 1989.

53. Zavaglia, J., and Lindsay, J. D., Flash X-ray Visualization of Multiphase Flow in Impulse Drying, *Tappi J.*, vol. 72, no. 9, pp. 79-85, 1989.

54. Lindsay, J. D., The Physics of Impulse Drying: New Insights from Numerical Modelling, in *Fundamentals of Papermaking*, Vol. 2, ed.C. F. Baker and V. W. Punton, Trans. of the Ninth Fundamental Research Symposium, Cambridge, England, Sept. 16-22, 1989, Mech. Eng. Publ. Ltd., London, 1989.

55. Lavery, H., Project 3470 Status Report to the Engineering Project Advisory Committee, The Institute of Paper Science and Technology, Atlanta, Georgia, Oct. 22, 1987.

56. Lavery, H. P., New Mechanisms for Water Removal From Paper Through Impulse Drying, AIChE Summer Meeting, Minneapolis, Minnesota, Aug. 16-20, 1987.

57. Sprague, C. H., unpublished data, The Institute of Paper Science and Technology, Atlanta, Georgia, 1988.

58. Lindsay, J. D. and Sprague, C. H., MIPPS: A Numerical Moving Boundary Model for Impulse Drying, *J. Pulp Paper Sci.*, vol. 15, no. 4, pp. J135-J141, July 1989.

59. J. D. Lindsay, Advances in the Numerical Modeling of Impulse Drying, *Proc. 1989 TAPPI Eng. Conf.*, Atlanta, Georgia, Sept. 1989.

60. Rudemiller, G. R. and Lindsay, J. D., An Investigation of Boiling Heat Transfer in Fibrous Porous Media, in *Heat Transfer 1990*, ed. G. Hestroni, Proc. of the Ninth International Heat Transfer Conference, Jerusalem, Israel, July 1990.

61. Rudemiller, G. R. and Lindsay, J. D., Apparatus for Investigating Boiling Phenomena in a Fibrous Porous Medium, Int. Commun. Heat Mass Transfer, vol. 16, no. 6, pp. 785-794, 1989.

62. Macbeth, R. V., Boiling on Surfaces Overlayed with a Porous Deposit: Heat Transfer Rates Obtainable by Capillary Action, United Kingdom Atomic Energy Authority Report AEEW-R711, June 1971.

63. Orloff, D. I., High Intensity Drying Processes – Impulse Drying, Report 4 prepared for the U.S. DOE, Office of Industrial Programs, under contract DE-FG02-

85CE40738, The Institute of Paper Science and Technology, Atlanta, Georgia, May 1990.

64. J. R. Burkhead, J. D. Lindsay, J. R. Burns, and D. I. Orloff, The Effect of Felt Saturation and Other Felt Properties on Delamination in Impulse Drying, *Proc. 1989 TAPPI Eng. Conf.*, Atlanta, Sept. 1989.

65. Orloff, D. I., High-Intensity Drying Processes – Impulse Drying, Annual report prepared for the U.S. DOE, Office of Industrial Programs, under contract No. DE-FG02-85CE40738, The Institute of Paper Science and Technology, Atlanta, Georgia, Sept. 1990.

66. Aidun, C. K., personal communication, The Institute of Paper Science and Technology, Atlanta, Georgia, 1990.

67. Seyed-Yagoobi, J., personal communication, Mech. Eng. Dept., Texas A&M University, 1990.

68. Etzler, F. M., personal communication, The Institute of Paper Science and Technology, Atlanta, Georgia, 1990.

69. Kerekes, R. J., A Simple Method for Determining the Thermal Conductivity and Contact Resistance of Paper, *Tappi J.*, vol. 63, no. 9, pp. 137-140, 1980.

70. Fletcher, L. S., and Seyed-Yagoobi, J., personal communication, Texas Drying Research Consortium, Mech. Eng. Dept., Texas A&M Univ., 1990.

Drying Mechanism and Its Numerical Simulation in Wet and Thin Porous Media

HIROAKI KUNO, YOUNOSUKE HOSHI, and YASURO TAKAHASHI
Hiroshima Research and Development Center
Mitsubishi Heavy Industries, Ltd.
5007 Itozaki-cho, Mihara 729-03, Japan

MAKIO IWABUCHI
Advanced Technology Research Center
Mitsubishi Heavy Industries, Ltd.
8-1 Sachiura 1-chome, Kanazawa-ku, Yokohama 236, Japan

1. INTRODUCTION

In the recent paper industry, the greatest interests in the technological innovation are to increase the machine speed and to produce paper of good quality without non-uniform shrinkage or curl. The small energy-consumption for the production process will be the third requirement of importance in the near future.

The drying section of the wet web of paper, in which the moisture is reduced approximately from 60% to 8% within a very short term, is one of the most important processes. And increasing the machine speed requires to improve the drying rate of the wet web of paper. The amount of non-uniform shrinkage or curl may be constrained by controlling the temperature and moisture profile during the drying process of the web. Accordingly, it seems to be unavoidable for the appropriate drying process design satisfying the above requirements to get the adequate knowledge on the heat and mass transfer mechanism of the drying process.

It is, however, so difficult to clarify the process of moisture removal through the paper sheet by the direct measuring that mathematical simulation of the drying process would be the only practical method for investigating the moisture and temperature change within thin and porous media such as a paper sheet. As for the numerical simulation on the overall drying characteristics along the machine direction, the useful results have been given by a number of researchers as described in Kirk's liturature review.[(1)] Though the uniform physical condition along the thickness direction (z-direction) was assumed in the most of their simulations, the another kind of simulation which describes more exactly the moisture and temperature change in the z-direction is required to investigate the mechanism of shrinkage or curl of paper. The later was developed by several researchers.[(2), (3), (4)] But verifying their sumulation on moisture profile in the z-direction is still insufficient because of the difficulty of measuring it to date.

In this consequence, the authors have conducted the development of the numerical simulation for the drying process of paper as well as the experiment including the measurement of moisture profiles in the z-direction by newly developed technique as described in the followings.

2. ANALYTICAL METHOD AND NUMERCIAL SIMULATION

2.1 Paper drying process

Paper is made from a suspension of fibers in water, with small quantities of nonfibrous additives such as fillers, size, retention aids and colours etc. The concentration of the fibrous suspension is usually between 0.5 and 1.5%.[1] Figure 1 shows the layout of a paper machine. The majority of the water is drained out in the wire section called " former " which consists of foils, suction boxes and plastic wires. Then the wet web passes through the press section and the moisture content is reduced to a value of about 58%. The web then passes into the dryer section which consists of a series of cast iron cylinders, usually of 1500 or 1800mm (5 or 6 ft) in diameter, which are heated by steam. The paper in contact with the cylinder is covered by " dry felts " which tend to restrain shrinkage and prevent cockling. There may be some 30 or even 60 cylinders. The web moisture content is reduced to about 8% for several seconds in the dryer section.

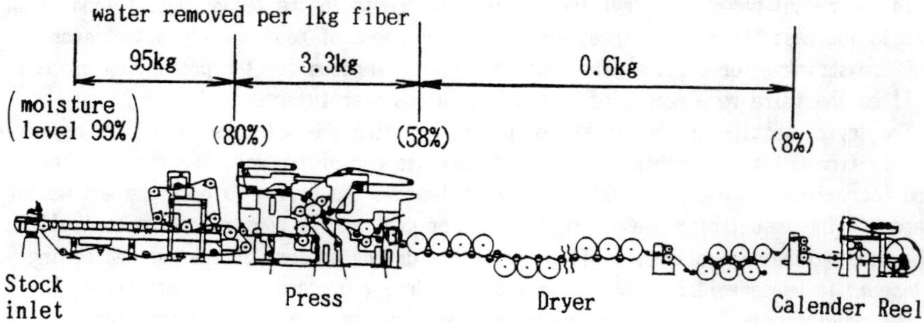

FIGURE 1. Layout of a paper machine

2.2 Analytical model

Paper consists of fine fibers which form a porous network-structure.[5] Figure 2 shows a photomicrograph of cross-section of a sliced paper sheet. The typical dimensions of fibers are 2-5mm long and 0.01-0.03mm in diameter. A large amount of water is contained between fibers and also within central pores of fibrous cells in the early drying stage. At the moisture level exceeding about 20% by weight fraction, the liquid evaporates on the surface of the fibers and the liquid water between fibers is moved by the capillary effect. When the liquid water carried by the capillary force ceases to exist and the vapor pressure difference between cell wall surfaces and cell center pores becomes large to some extent, vaporized water in cell itself diffuses to the outside through a very narrow path of cell wall as free molecular flow.

Figure 3 presents the phenomena relating heat and mass transfer within drying paper as mentioned above.

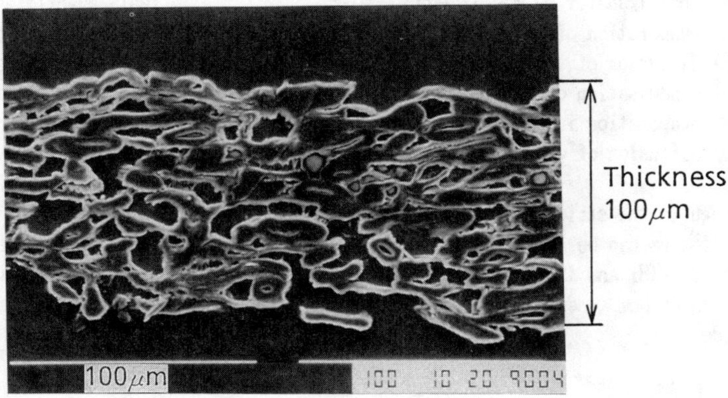

FIGURE 2. Photomicrograph of cross-section of a sliced paper sheet(60g/m ^{2})

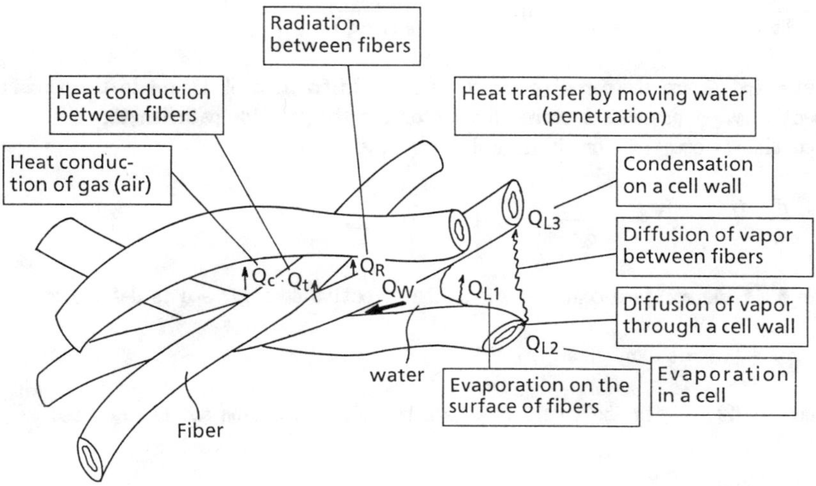

FIGURE 3. Heat and mass transfer in wet porous media

They are :

(1) heat conduction between fibers
(2) heat conduction in air between fibers
(3) radiation between fibers
(4) heat transfer by moving liquid water
(5) evaporation of liquid water on the surface of fibers
(6) diffusion of vapor between fibers
(7) condensation on cell wall surface
(8) evaporation in cells
(9) diffusion of vapor through cell wall

Though in the strict model all of the above phenomena must be taken into account, some of them can be neglected according to the drying condition. In the present analysis items (3), (8) and (9) are neglected and the properties in porous media itself are regarded as homogeneity, which means the configuration of individual fibers is not considered.

2.3 Basic equations

In order to clarify the moisture removal in porous media, one dimensional basic equations along the z-direction for gas phase and solid phase including liquid portion must be solved.

When the evaporated vapor flows through the porous media with air, the continuity equation for gas phase is given by

$$\frac{\partial \rho_g}{\partial t} + \frac{1}{\eta} \frac{\partial}{\partial z} (\rho_g U_g) = \frac{\dot{m} S_v}{\varepsilon (1-\zeta)} \tag{1}$$

where ε and ζ are the porosity and the volumetric fraction of liquid to the porosity respectively, η presents a correction factor for the gas flow path length.

A continuity equation for the liquid phase is given by

$$\frac{\partial}{\partial t} (\rho_e \phi) = \frac{\partial}{\partial z} (\rho_L U_L) - \dot{m} S_v \tag{2}$$

where ϕ is the moisture content. ρ_e is the effective density, and is defined as

$$\rho_e = (1-\varepsilon) \rho_F + \varepsilon \zeta \rho_L \tag{3}$$

In equation(2), U_L is the liquid flow rate by capillary action and is expressed by

$$U_L = D_L \frac{\partial \phi}{\partial z} \tag{4}$$

where D_L is the capillary migration coefficient.

The propotion of vapor to the gas phase is also presented by a following conservation equation of vapor.

$$\frac{\partial}{\partial t}(\rho_g m_w) + \frac{1}{\eta}\frac{\partial}{\partial z}(\rho_g U_g m_w)$$

$$= \frac{1}{\eta^2}\frac{\partial}{\partial z}\left(\frac{\mu_g}{Sc_g}\frac{\partial m_w}{\partial z}\right) + \frac{\dot{m}Sv}{\varepsilon(1-\zeta)} \quad (5)$$

where m_w is the mass fraction of vapor.

The momentum equation for the gas phase is

$$-\frac{1}{\eta}\frac{\partial P}{\partial z} - f = 0 \quad (6)$$

where f is the flow loss through porous media and is defined as

$$f = \lambda \frac{1}{de}\frac{1}{2}\rho_g U_g^2 \quad (7)$$

where de is the equivalent diameter of gas flow channel within porous media, and is assumed to be in proportion to the cube root of gas volume.

$$de = deo\ \{\varepsilon\ (1-\zeta)\}^{1/3} \quad (8)$$

In the liquid and the solid phases, an energy equation is given by

$$\frac{\partial}{\partial t}(\rho_e C_{pe} T_s) + \frac{\partial}{\partial z}(\rho_L C_{pL} U_L T_s)$$

$$= \frac{\partial}{\partial z}\left\{\frac{\mu_L}{Pr_L}\frac{\partial}{\partial z}(C_{pe}T_s)\right\} + Q_H - \dot{m}\ Sv\ r_H \quad (9)$$

where Q_H is the inner heat generation. The heat transfer between gas and solid in micro porous media is so large that gas temperature Tg in the control volume is assumed to be nearly equal to the solid temperature Ts.
In the above equations, the evaporation or the condensation rate $\dot{m}$ is given by

$$\dot{m} = \frac{2\sigma}{2-\sigma}\frac{1}{\sqrt{2\pi}}\frac{P_s - P_w}{\left(\frac{R_g}{M_w} T_g\right)^{1/2}} \quad (10)$$

where σ is the evaporation or the condensation accomodation factor, and M_W is the molecular weight of water.

2.4 Simulation

Equations for the heat and mass transfer in porous media were converted into difference equations which could be numerically solved by the Newton-Raphson's method.

As a calculation model, a hand-made paper sheet which was used for our experiment was chosen here. It was heated from one side by a hot surface plate. Its geometry and boundary conditions were as follows, a thickness of 800μm, a hot plate surface temperature of 383 K, an initial moisture content of 70% and mass transfer coefficient of 0.011m/s on the open surface. It is also assumed that an outflow of vapor to the hot surface does not exist.

The results of the simulations are shown in Figures 4,5 and 6. In this simulation, temperature and moisture profiles are affected by the equivalent diameter of flow channels d_{eo} and the capillary migration coefficient D_L. In figure 4, moisture begins to decrease slowly in the No.1 layer of hot plate side. But the temperature of the hot plate side layer begins to rise rapidly and approaches the saturated temperature.

Figure 5 shows results in the case where D_L is very small. The moisture content of the hot plate side layer decreases more rapidly than that of center or outer layers, so the moisture gradient becomes large.

If d_{eo} becomes small as in Figure 6, the temperature of the No.8 layer near the hot air side rises slowly than that of layers close to the hot plate.

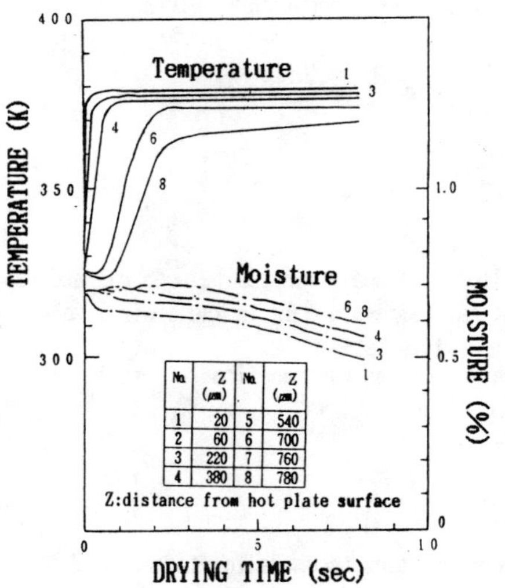

FIGURE 4. Temperature and Moisture Profiles by simulation
: d_{eo} = 10μm and D_L = 200×10^{-9}(m^2/s)

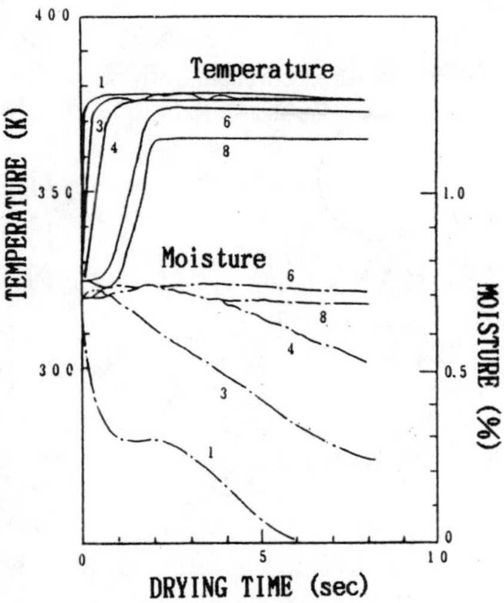

FIGURE 5. Temprature and Moisture Profiles by simulation
d_{eo} = 10μm , D_L = 20×10^{-9} (m^2/s)

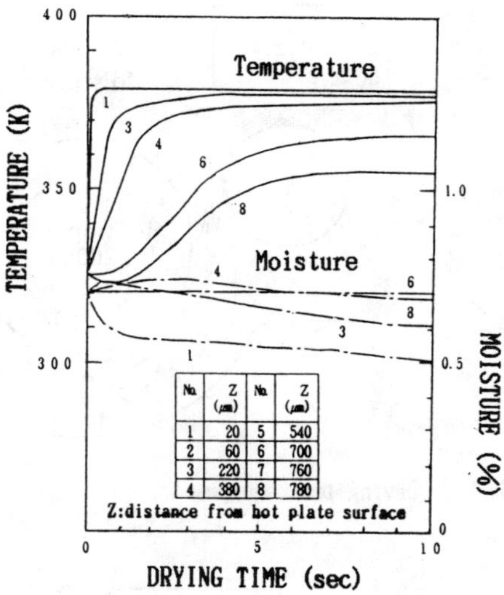

FIGURE 6. Temprature and Moisture Profiles by simulation
d_{eo} = 5μm , D_L = 20×10^{-9} (m^2/s)

3. EXPERIMENT

3.1 Experimental procedure

The drying test equipment shown in Figure 7 was used to measure the moisture and temperature profile in porous media during drying. The hot roll reciprocatively rotates and continuously heats the wet sheet of paper. The conditioned hot air flows outside the roll, and the wet sheet are held between the felt and the heating roll, the felt permiability is 5000cc/min · cm² at 0.5inch H_2O

A test specimen of wet sheet was formed by 6 sub-layers of the thin hand-made sheet of about 60g/m ² (B.D.) and fine thormocouples and electrodes were inserted between them.

Figure 8 shows the principle used in the measuring the moisture profiles. The impedance of thin wet porous media tends to be in proportion to the moisture. [6] This characteristic was used for measuring it.

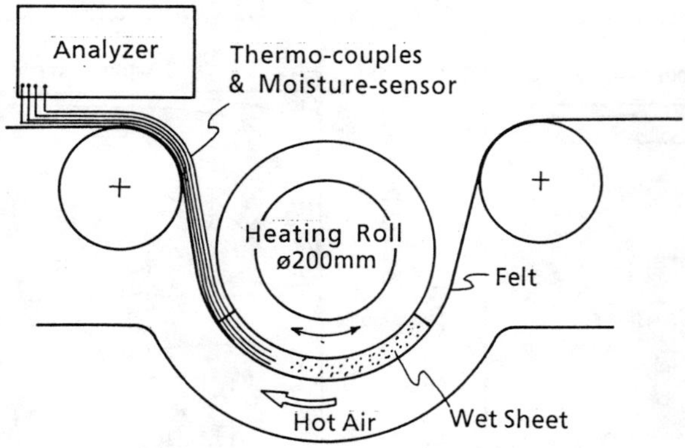

FIGURE 7. Drying test equipment

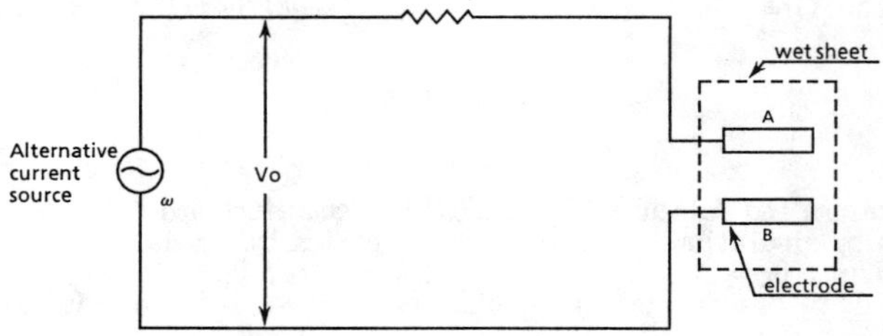

FIGURE 8. Principle of a measurement method of the moisture profile

3.2 Test Results

A wet sheet consisting of 6 sub-layers was pressed under the load of $3.5\times10^5 N/m^2$ for 5 minutes so that the contact resistance between each layer becomes negligible.

Results of the measurement are shown in Figures 9 and 10. The temperature of each thermocouple began to rise rapidly in the early drying stage. When the temperature in a layer approached to the saturated condition, the temperature rise became moderate and then nearly constant. When the amount of moisture in the heating roll side sub-layer reduced below about 25%, the temperature of the layer began to rise again. The temperature re-rise gradually became late toward the felt side. The saturation temperature seems to decrease as drying continues. The reason is not yet clear, but the partial vapor pressure change in drying process may be regarded as one of the reasonable explanations.

The moisture profile is shown in Figure 10. As the paper begins to dry, the moisture in the hot side layer decreases more rapidly than that in any of outside layers.

The vapor generated in paper sub-layers passes through the porous media and is released from the felt side. To check the reliability of measurement, we compared the data from the impedance method with those obtained from the direct weighing.

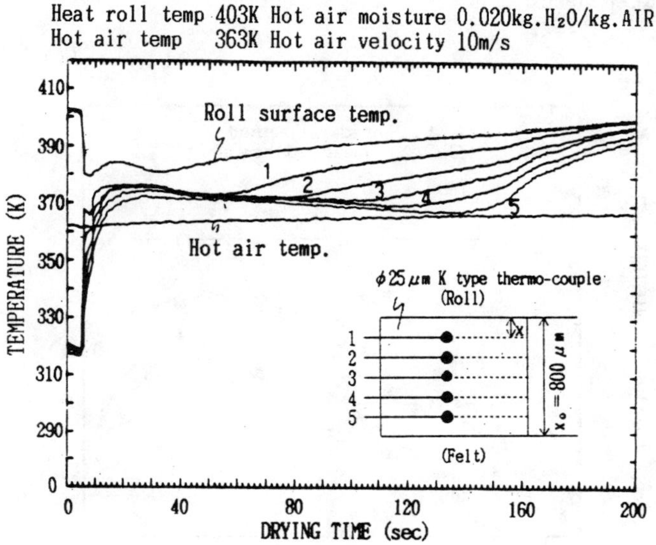

FIGURE 9. Temperature Profile during drying process of hand-made sheet

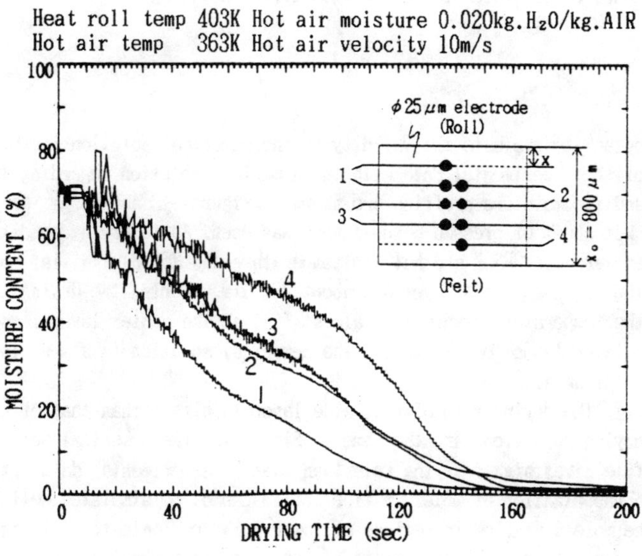

FIGURE 10. Moisture Profile during drying Process of hand-made sheet

Figure 11 shows this comparison. These data were measured at the roll temperature of 400 K and the air temperature of 300 K. As shown in the results the data from the impedance method coincided with those by the direct weighing in the case of 40 seconds after the paper started to dry.

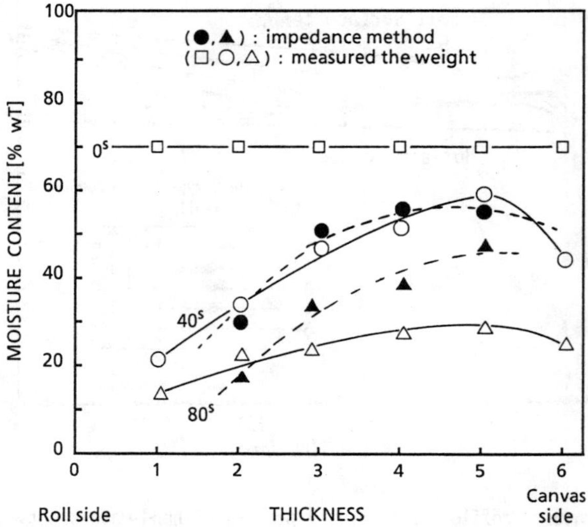

FIGURE11. Comparison of impedance method with direct weighing

4. DISCUSSION

It is important to evaluate the validity of the numerical solution in this paper. Comparisons of the experimental data with the numerical solution regarding to the temperature and the moisture profiles are shown in Figures 12 and 13. As previously described, a wet sheet of pressed 6 sub-layers was used. As to drying conditions, roll surface temperature was 383 K and hot air temperature was 353 K. Initial moisture was about 70%. The thickness of a layer was about 200 μm. During the drying time from 10 to 30 sec., the temperature gradient is almost flat at the center layers, and that of the felt side layer tends to be lower. The numerical solution is almost in accordance with experimental data.

In Figure 13, the drying rate of roll side layer is higher than that of a center layer. The drying conditions are the same in Figure 12. The numerical solution of the moisture profile gives higher drying rate than that of experimental data. It is considered that the vaporization of water in fiber cells cannot be simulated well because of neglecting the molecular flow in regions of lower moisture content. This suggests that the homogeneous model adopted here cannot always describe the actual phenomena.

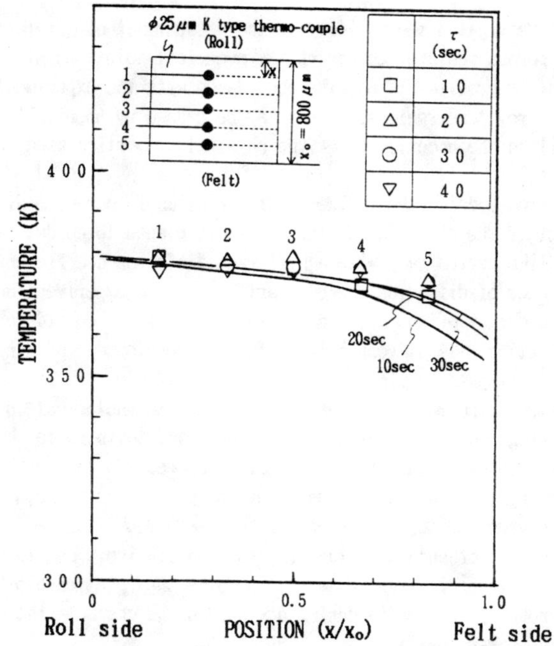

FIGURE12. Comparison of temperature profiles from the experiment and the simulation

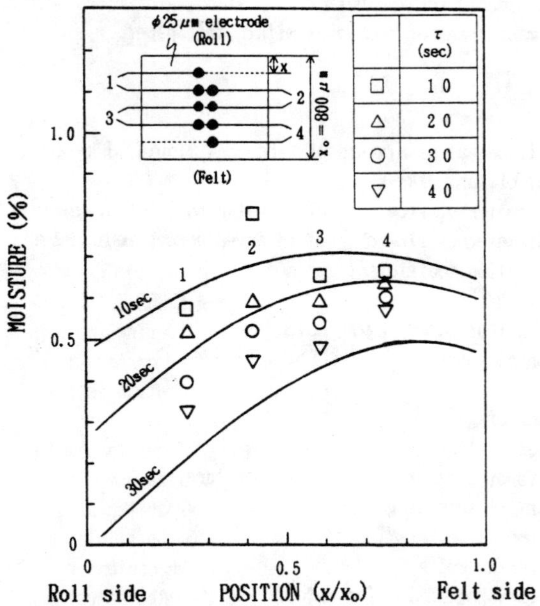

FIGURE 13. Comparison of moisture profiles from the experiment and the simulation

5. CONCLUSION

The authors have investigated the validity of the present method simulating the drying process in thin porous media by comparing the numerical solutions with the measured temperature and moisture profiles. But the accordance with the experimental data was unsatisfactory in low moisture regions. This may be caused by unsuitability of the present model as well as the uncertainty of the physical properties used for the above calculations.

For improvement of the model, the influence of diffusion in porous media must be investigated. To clear the effect of diffusion, it is planed to conduct an experiment on drying in a gas, like herium gas, with a different diffusion coefficient from that of air. If the influence of diffusion can be clarified by the experiment and be taken into the model, the accuracy of the numerical simulation will be considerably improved. Furthermore the simulation will be very helpful for the development of a new paper drying process such as the impulse drying. (7)

In the paper industry, the most important goals in the technological innovation are to realize manufacturing process for high quality paper and to increase the productivity under the condition of saving energy and saving space. To insure the quality, it has been proposed to dry a wet web under restraint for preventing it from shrinking and to dry it alternatively on the top and reverse side to avoid curling. In addition, to evaluate the effectiveness of such a drying process quantitatively and to find the optimum operating conditions, it is necessary to acquire the relationship between the moisture transfer through porous media during drying and the characteristics of the paper structure or the paper quality.

The author and co-workers have aleady undertaken the investigation corresponding to the above requirements. In this investigation, it should be emphasized that the numerical simulation is unavoidable for the better understanding of micro-scopic phenomena such as the heat and mass transfer mechanism within thin paper.

Nomenclature

C_{Pe} : average specific heat as mixture of fiber and liquid, J/kg·K
C_{pg} : specific heat of gas, J/kg·K
d_e : equivalent diameter of flow channel in wet porous media, μm
d_{eo} : equivalent diameter of flow channel in dryed porous media, μm
D_L : capillary migration coefficient, m^2/s
f : friction loss, N/m^3
M_w : molecular weight of water, kg/kg-mol
m_w : mass fraction of vapor
P : Pressure, Pa
Pr_g : Prandtl number of gas
Pr_L : Prandtl number of liquid
P_s : equilibrium vapor pressure to solid temperature, Pa
P_w : partial vapor pressure of gas phase, Pa
Q_H : inner heat generation, W/m^3
R_g : gas constant, J/kg-mol K
r_H : latent heat of vaporization for water, J/kg

Sc_g : Schmidt number of gas
S_v : Specific surface area for vaporization, m^2/m^3
T_g : gas temperature, K
T_s : solid temperature, K
t : time, sec
U_g : gas velocity, m/s
U_L : liquid flow rate by capillary action, m/s
V_0 : voltage, V
ε : porosity
ζ : volumetric fraction of liquid to the porosity
η : correction factor for a gas flow path length
λ : friction factor
μ_g : viscosity of gas, Pas
μ_L : viscosity of liquid, Pas
ρ_e : effective density of solid and liquid, kg/mä
ρ_F : density of fiber, kg/m^3
ρ_g : density of gas , kg/m^3
ρ_L : density of liquid, kg/m^3
σ : evaporation (or condensation) accomodation factor
ϕ : moisture content

References

(1) L.A.Kirk " A Literature Review of Computer Simulation of Paper Drying " ADVANCES IN DRYING volume 3, ed. Arun S. Mujumdar, pp.1-37, Hemisphere Publishing Co., 1984.

(2) F.T. Hartley and R.J.Richards " Hot Surface drying of paper-the development of a diffusion model " Tappi vol.57, No.3, pp.157-160, 1974.

(3) S.T.Han and J.F.Matters " Vapor Transport in Fiber Mat During Drying " Tappi vol.19, No.1, pp.1-4, 1966.

(4) Peter F.Lee and Jeffrey A Hinds " Optimizing Dryer Perfomance : Modeling Heat and Mass Transfer within A Moist Sheet of Paper or Board " Tappi Engineering Confference Procceedings, pp.117-124, 1981.

(5) Mikael Rigdahl and Holger Hollmark " Network Mechanics " Paper Structure and Properties, ed. J.Anthony Bristow, Petter Kolseth, Marcel Dekker, Inc.

(6) M. Rueff, R. Charuel " Drying paper : Measurement of moisture Content ", the Fifth International Symposium on Drying, 1986.

(7) J.D.Lindsay " The physics of Impulse Drying : New Insights from Numerical Modelling " FUNDAMENTALS OF PAPERMAKING volume 2, ed. C.F. Baker, pp.679-729, mechanical Engineering Publications Limited, 1989.

Sc_g : Schmidt number of gas
S_V : Specific surface area for vaporization, m^2/m^3
T_g : gas temperature, K
T_s : solid temperature, K
t : time, sec
U_g : gas velocity, m/s
U_L : liquid flow rate by capillary action, m/s
V_0 : voltage, V
ε : porosity
ζ : volumetric fraction of liquid to the porosity
η : correction factor for a gas flow path length
λ : friction factor
μ_g : viscosity of gas, Pas
μ_L : viscosity of liquid, Pas
ρ_e : effective density of solid and liquid, kg/mä
ρ_F : density of fiber, kg/m^3
ρ_g : density of gas , kg/m^3
ρ_L : density of liquid, kg/m^3
σ : evaporation (or condensation) accomodation factor
ϕ : moisture content

References

(1) L.A.Kirk " A Literature Review of Computer Simulation of Paper Drying " ADVANCES IN DRYING volume 3, ed. Arun S. Mujumdar, pp.1-37, Hemisphere Publishing Co., 1984.

(2) F.T. Hartley and R.J.Richards " Hot Surface drying of paper-the development of a diffusion model " Tappi vol.57, No.3, pp.157-160, 1974.

(3) S.T.Han and J.F.Matters " Vapor Transport in Fiber Mat During Drying " Tappi vol.19, No.1, pp.1-4, 1966.

(4) Peter F.Lee and Jeffrey A Hinds " Optimizing Dryer Perfomance : Modeling Heat and Mass Transfer within A Moist Sheet of Paper or Board " Tappi Engineering Confference Procceedings, pp.117-124, 1981.

(5) Mikael Rigdahl and Holger Hollmark " Network Mechanics " Paper Structure and Properties, ed. J.Anthony Bristow, Petter Kolseth, Marcel Dekker, Inc.

(6) M. Rueff, R. Charuel " Drying paper : Measurement of moisture Content ", the Fifth International Symposium on Drying, 1986.

(7) J.D.Lindsay " The physics of Impulse Drying : New Insights from Numerical Modelling " FUNDAMENTALS OF PAPERMAKING volume 2, ed. C.F. Baker, pp.679-729, mechanical Engineering Publications Limited, 1989.

Summary of General Discussion of Session on Papermaking

S. Motakef, Massachusetts Institute of Technology
H. Ozoe, University of Kyushu

Two papers were presented in this session. The following is a summary of research needs in paper drying as presented by Professor Lindsay.

In the general area of paper drying, three specific target areas were identified. First, increasing the scope of data base on thermophysical properties of paper as related to heat and mass transfer. In this context, measurement of liquid diffusion coefficients and correlating it with the internal structure of the paper appear necessary. The second focus area identified is the need for development of non-intrusive, on-line, real-time techniques for measurement of moisture content across the paper thickness. The third area is development of drying techniques which would provide for more control over paper properties. The subtopics introduced here were (a) dewatering without densification; (b) considerations related to tensile strength of the wet paper; and (c) increasing the quality of papers made from recycled fiber.

One of the promising areas in paper-drying is impulse drying. The issues that need to be identified in this technique are:

- delamination control;
- fundamental transport processes;
- use of novel materials, such as coatings and composite materials, for the dryers;
- identification of the optimum time-history of heat input into the paper;
- investigations into boiling in fibrous media.

PHYSICAL PROPERTIES

Numerical Evaluation of the Physical Properties of Magnetic Fluids Suitable for Heat Transfer Control

TOSHIO AIHARA and JOO-KYUN KIM
Institute of Fluid Science
Tohoku University
2-1-1 Katahira, Aoba-ku, Sendai 980, Japan

ABSTRACT

Recently, magnetic fluids of several kinds have been developed and applied to various engineering uses because of their unique functional characteristics. In the present paper, existing studies on the physical properties, functional characteristics, flow and heat transfer of the magnetic fluids are briefly reviewed. Then, applicability of the magnetic fluids to the active control of heat transfer is discussed through a numerical analysis of forced convective heat transfer in a circular tube with a uniform wall temperature in the presence of a transversely applied magnetic field, using a comparatively universal expression of fluid magnetization.

INTRODUCTION

Until recently, there were no liquids which could be magnetized to be comparable with the magnetization of ferromagnetic substances. In 1965, Dr. Pappel[1] developed a magnetic fluid which is a colloidal solution of very fine ferromagnetic particles of about 10 nm diameter stably suspended in a nonconductive liquid. Thereafter, magnetic fluids of various kinds have been developed and their application to engineering uses attempted. Presently, magnetic fluids are in wide use in seals, bearings, magnetostatic support, jet printers, separation of nonmagnetic particles, flow control and drag reduction, dampers, actuators, sensors, transducers, and medical application.

Although several studies on heat transfer of magnetic fluids in a magnetic field have been carried out, we are particularly interested in the active control of heat transfer by the magnetic field. In the conventional "thermal control" technique, temperature control of a device is performed indirectly by adjusting the flow rate or bulk temperature of the working fluid; therefore, quick control of the thermal phenomenon is usually difficult. Recently, we proposed the concept of "heat-transfer control", which means the active and direct control of heat transfer, which in a sense corresponds to the boundary-layer control in fluid mechanics. It is well known that a magnetic fluid has various functional characteristics. Hence, we may expect magnetic fluids to have possibilities of application to heat-transfer control.

In the present report, the possibilities are examined through a numerical evaluation of the physical properties of magnetic fluids. Namely, existing studies on the physical properties of magnetic fluids, functional characteristics, flow and heat transfer are first reviewed. Then, our original research on a numerical evaluation of the physical properties of magnetic fluids suitable for heat-transfer control is described. In the latter, we considered the forced convective heat transfer of a magnetic fluid in a circular tube with a uniform wall temperature in the presence of a transversely applied magnetic field. The effects of the temperature-dependent physical properties of magnetic fluids on the heat transfer characteristics are discussed for various kinds of magnetization.

MAGNETIC FLUIDS

A typical magnetic fluid [2] consists of a stable colloidal suspension of monodomain ferromagnetic particles in a nonconductive base fluid, as shown in Figure 1. A surfactant, or dispersant, covering the particles prevents aggregation of the particles; Brownian motion and Neel rotation prevent particle sedimentation in gravitational or magnetic fields. In general, 1 m^3 of a magnetic fluid contains about 10^{23} ferromagnetic particles. There are several bibliographies of papers, books, and patents by Zahn and Shenton [3], Charles and Rosensweig [4], and Kamiyama and Rosensweig [5]. At present, magnetic fluids are prepared mainly by the size reduction method [6] and chemical precipitation method [7].

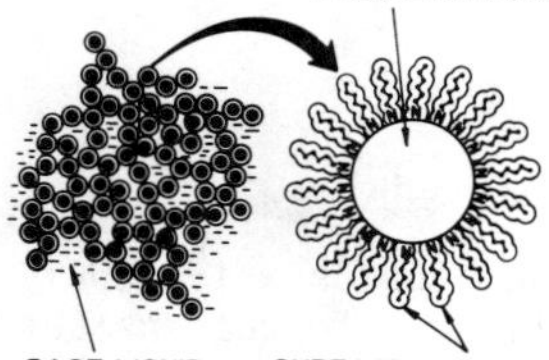

FIGURE 1. Concept of a magnetic fluid [2].

TABLE 1. Nominal properties of ferrofluids at 298 K [6]

Carrier fluid	Magnetic saturation ($A \cdot m^{-1}$)	Density ($kg \cdot m^{-3}$)	Viscosity ($N \cdot s \cdot m^{-2}$)	Pour point (K)	Boiling point (K)	Surface tension ($mN \cdot m^{-1}$)	Thermal conductivity ($W \cdot m^{-1} \cdot K^{-1}$)	Specific heat ($kJ \cdot m^{-3} \cdot K^{-1}$)	Thermal expansion coefficient ($m^3 \cdot m^{-3} \cdot K^{-1}$)
Hydrocarbon	15,900	1050	0.003	278	350	28	0.15	1715	9.0×10^{-4}
Fluorocarbon	31,800	1250	0.006	281	350	28	0.19	1840	8.6×10^{-4}
Ester	7,960	2050	2.50	239	456	18	0.20	1966	10.6×10^{-4}
	15,900	1150	0.014	217	422	26	0.31	3724	8.1×10^{-4}
Water	15,900	1180	0.007	273	299	26	0.59	4184	5.2×10^{-4}

PHYSICAL PROPERTIES OF MAGNETIC FLUIDS

The physical properties of commercially available magnetic fluids are listed by Rosensweig [6]; some of them are included in Table 1.

Magnetization Ferromagnetic particles in a magnetic fluid are carriers of magnetic moment. Very fine particles ensure their monodomain character. The greater the field magnitude, the more the particles are aligned with the field direction. Finally, the particles are completely aligned at high field strengths, and the magnetization is saturated.

With the assumption of no magnetic interaction between particles, the following approximate expression for the magnetization M of a magnetic fluid is obtained from Langevin's theory [8], which gives the superparamagnetic relationships:

$$\left.\begin{aligned} M &= \phi M_d(\coth \xi - 1/\xi) \\ \xi &= \mu_0 m H/kT. \end{aligned}\right\} \quad (1)$$

where μ_0 is the magnetic permeability of a vacuum, m the magnetic moment of a particle, H the magnetic field intensity, k the Boltzmann constant, T the absolute temperature, ϕ the volume fraction of solids in the magnetic fluid, and M_d the spontaneous magnetization of the ferromagnetic material.

In Figure 2, Langevin's classical theory is compared with the experimental results [9]. With increasing magnetic field intensity, Langevin function ($\coth \xi - 1/\xi$) approaches unity.

The spontaneous magnetization M_d decreases with increasing temperature T; then, at the Curie point T_C, the ferromagnetic is transformed into the paramagnetic. The behavior of decrease in the magnetization is shown in Figure 3 [9]. Figure 4 shows the temperature dependence of the saturation magnetization M_s of ferromagnetic fluids[9]. At lower temperatures, Mn-Zn ferrite particles have stronger temperature dependence than magnetite particles.

Viscosity Magnetic fluids have a specific mechanism of internal friction which appears in the case of relative rotation of ferromagnetic particles in the magnetic field.

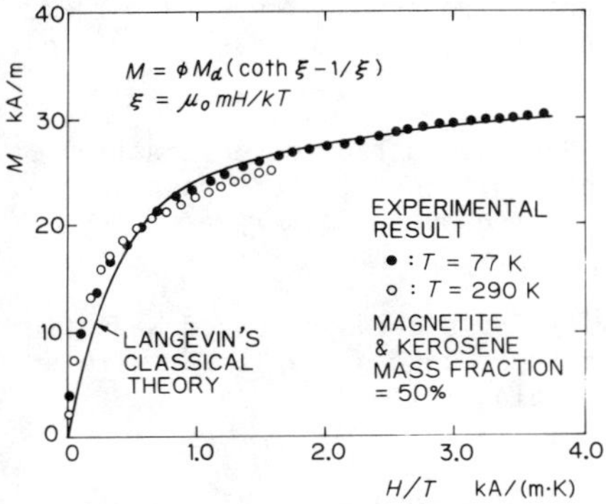

FIGURE 2. Magnetization of a magnetic fluid [9].

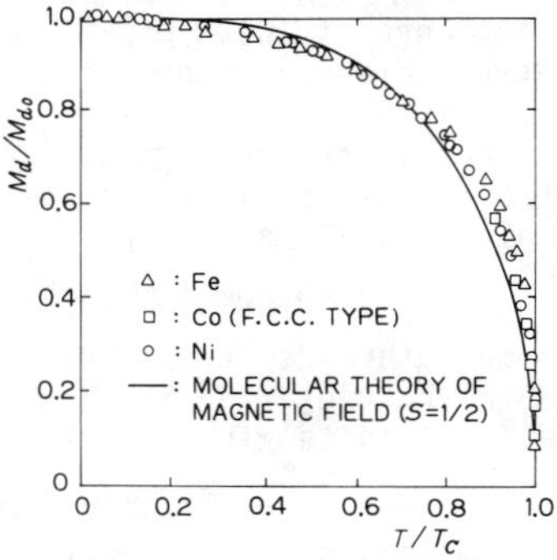

FIGURE 3. Spontaneous magnetization vs. temperature [9].

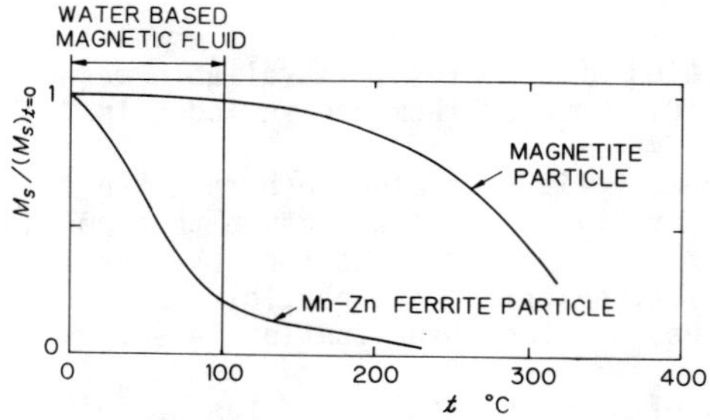

FIGURE 4. Temperature dependence of saturation magnetization of ferromagnetic fluids [9].

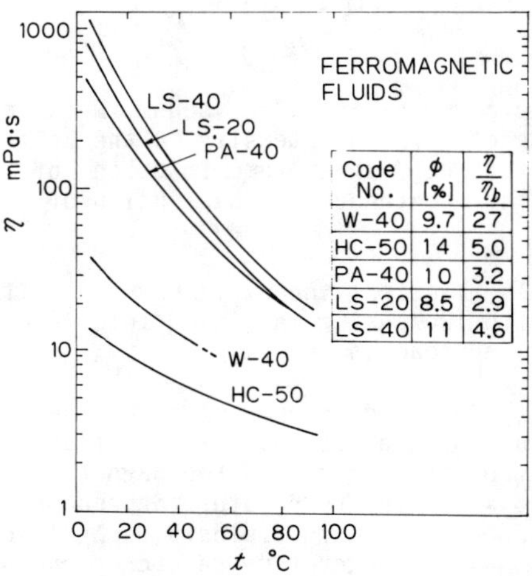

FIGURE 5. Temperature dependence of viscosity [2].

The viscosity of a magnetic fluid depends on the shear rate $\dot{\gamma}$, the magnetization M, and the magnetic field intensity H. According to Rosensweig et al.[10], the viscosity shows the minimum constant value when $10^{-4} < \dot{\gamma}\eta_b/MH$ and the maximum constant value when $0 < \dot{\gamma}\eta_b/MH < 10^{-6}$. In the range of $10^{-6} < \dot{\gamma}\eta_b/MH < 10^{-4}$, the viscosity is dependent on the magnetic field and shear rate.

The viscosity of the magnetic fluid also depends on the volume fraction ϕ of the ferromagnetic particles. The measured viscosity of suspensions [11] with noninteracting particles is expressed approximately as

$$\eta = \eta_b exp[(2.54 + 2.7\phi^2)/(1 - 0.609\phi)]. \qquad (2)$$

The temperature dependence of the viscosity for a kerosene-based magnetic fluid in the range 233 K < T < 303 K at a shear rate $\dot{\gamma}$ = 1312 s^{-1} with no magnetic field is given by the Andrade formula [12],

$$\eta = Bexp(E/RT) \qquad (3)$$

where E = 15.8 X 10^3 J/mol, B = 5.9 X 10^{-3} MPa•s, and R = 8.31 J/(mol•K). Figure 5 shows a typical temperature dependence of the viscosity of ferromagnetic fluids [2].

Thermal Conductivity Tareef [13] obtained the following expression for the thermal conductivity of a solid-liquid mixture:

$$\lambda = \lambda_b\left[\frac{2\lambda_b + \lambda_p - 2\phi(\lambda_b - \lambda_p)}{2\lambda_b + \lambda_p - \phi(\lambda_b - \lambda_p)}\right] \tag{4}$$

where λ_b is the thermal conductivity of the base liquid including the surfactant, and λ_p the thermal conductivity of ferromagnetic particles, respectively.

Density and Specific Heat[8] The density of a magnetic fluid can be written as

$$\rho = \rho_p\phi + (1-\phi)\rho_b. \tag{5}$$

A similar expression holds for the specific heat of the colloid:

$$c_p = (c_p)_p\phi + (1-\phi)(c_p)_b. \tag{6}$$

Although the density and specific heat to be obtained by the mixing rule [8,12] are more reasonable than those obtained by Eqs.(5) and (6), the colloid properties are not appreciably affected by the existence of surfactants because of their low concentration.

SURFACE INSTABILITY

It is possible to control remotely the shape of the free-surface/interface of a magnetic fluid by utilizing its surface instability. The instability can be attributed mainly to the magnetic body force and magnetic pressure jump on the surface.

Figure 6 shows a typical example of the spike formation caused by the normal-field instability. When a small disturbance occurs in an initially flat layer of magnetic fluid, it causes a concentration of magnetic flux at the peaks of disturbance. While an increase in the magnetic flux enhances the magnetization at the peaks, a decrease in the flux in the depression reduces the magnetization. The insufficient pressure in the depression is balanced by the gravitational hydrostatic pressure drop between the top and the depression. The surface becomes wavy, and the form of the surface becomes stationary due to the equilibrium of

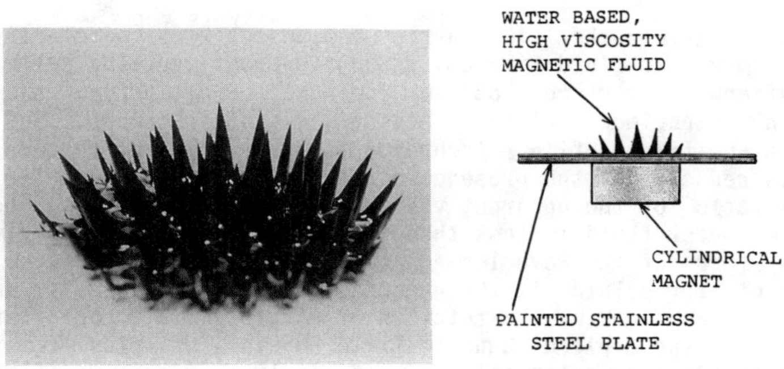

FIGURE 6. Formation of spikes (S. Kamiyama, 1990).

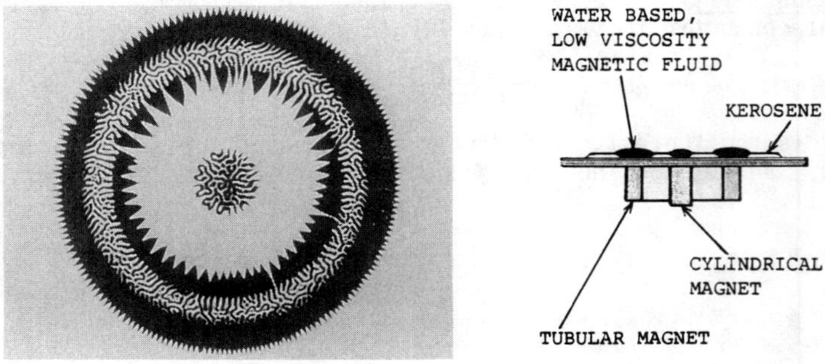

FIGURE 7. Formation of Labyrinth ring (S. Kamiyama, 1990).

magnetic, gravitational and capillary forces. Figure 7 shows a labyrinthine instability.

Detailed discussions of the surface instability have been described in books[2,6,8,9].

EXISTING STUDIES OF MAGNETIC FLUID FLOW

Recently, many researchers have studied the flow behavior of magnetic fluids. The results of studies are included in some books[2,6,8,9]. In the following, a brief review of some studies is presented.

Professor Kamiyama and his group have performed extensive studies of the flow behavior of magnetic fluids. Through the experiment on pipe flow of water- and kerosene-based magnetic fluids in a nonuniform transverse magnetic field[14], they clarified that the resistance coefficient increases with magnetic field intensity in a laminar flow regime and shows little magnetic field effect in a turbulent flow regime. Two-dimensional flows of magnetic fluid between two parallel plates in transverse and longitudinal magnetic fields were experimentally investigated by Sawada et al.[15]; the increasing rate of the resistance coefficient was examined with respect to the magnetic field length and Reynolds number.

Kamiyama and his coworker [16] carried out an analysis to clarify the effects of a magnetic fluid film coated on a current-carrying cylinder on the free-stream structure and hydrodynamic drag force, and also conducted an experiment on the effects of Reynolds number and film thickness on the film profile. According to their results, the drag of a cylinder is reduced by the presence of a magnetic fluid coating only when the ratio of the apparent viscosity of magnetic fluid to the viscosity of outer fluid is less than 1/4. Shukla and Kumar derived a generalized form of the Reynolds equation governing the pressure in a thin film of ferrofluid in the presence of a transversely applied magnetic field, considering the rotation of particles[17], and clarified the effects of the applied magnetic field and the Brownian relaxation time on the bearing characteristics of infinitely long squeeze film and a slider bearing.

An analysis of the one-dimensional two-phase flow of a magnetic fluid in a diverging duct was also carried out by Kamiyama and Inoue [18] for a low-quality case, taking into account the slip and pulsation of gas bubbles. Kamiyama and Ishikawa[19] studied experimentally the effect of magnetic field on the two-phase flow characteristics of a temperature-sensitive magnetic fluid.

Kamiyama et al. [20] also performed an analysis of the effects of steady and unsteady magnetic fields on pulsating pipe-flow characteristics of magnetic fluid in the laminar regime. Yamane et al. [21] studied experimentally the laminar flow of a water-based ferromagnetic fluid with an alternating, travelling magnetic field. The pressure difference was found to be induced in the direction opposite to the field.

Kumar and Chandra [22] studied theoretically the flow of a magnetic fluid through a channel in the presence of a transversely applied magnetic field, treating it as a two-phase fluid with internal rotation of magnetic particles.

EXISTING STUDIES ON HEAT TRANSFER OF MAGNETIC FLUIDS

Comprehensive surveys of literature on heat transfer of liquid metals in the presence of magnetic fields were given by Romig [23], Viskanta [24], and Davidson et al.[25]. However, there are only a few studies on the heat transfer of magnetic fluids.

Although a magnetic fluid is a colloidal suspension of very fine magnetic particles in a Newtonian fluid, most of the existing theoretical studies on magnetic fluids have been made on the basis of the continuum fluid and isotropic transport property models. However, the role of these phenomena is negligibly small, and there are no serious problems in considering convective heat transfer[24].

Neuringer and Rosensweig [26] found that vorticity may be generated by thermomagnetic interaction even in the absence of viscosity and that this leads to the development of augmented Bernoulli relationships. They also obtained an analytical solution for the problem of source flow with heat addition in order to display the thermomagnetic and magneto-mechanical effects attendant upon simultaneous heat addition and fluid motion in the presence of a magnetic field.

Most of the research on convective heat transfer of a magnetic fluid is restricted to the problem of thermal instability by free convection. When a temperature distribution exists in a magnetic fluid, there arises thermomagnetic convection due to the temperature dependence of its magnetization. However, even in the case of an isothermal fluid, free convection occurs owing to magnetophoretic transport of particles in the presence of a nonuniform magnetic field. This phenomenon was studied by Blūms[27]; he clarified that the nonuniformity of magnetization due to magnetodiffusion of colloidal particles is large enough to produce magnetic convection.

Stiles and Kagen [28] carried out a theoretical investigation of the magnetic Benard convection in a horizontal layer of a magnetic fluid, confined by infinite diamagnetic metal plates in a vertical magnetic field. A related theoretical analysis has been carried out by Gotoh and Yamada[29], considering convective instability in a ferrofluid confined by ferromagnetic plates of infinite magnetic permeability. They pointed out that a strong external magnetic field increases the viscous dissipation, and the temperature dependence of viscosity cannot be neglected when the plate spacing is small and the base liquid is aqueous. Okanaga et al.[30] carried out a numerical investigation of unsteady free convection of magnetic fluids in a cubic enclosure with and without a magnetic field and around a circular pipe. The average Nusselt number is increased by applying a magnetic field, and the steady state is reached in a shorter time than in the case of no magnetic field.

Aihara and his coworkers [31] conducted a numerical study of forced convective heat transfer of a temperature-sensitive magnetic fluid in a circular tube with a uniform temperature. The temperature dependence of viscosity and the magnetic field substantially affect the developing velocity profile, but little affect the developing temperature profile, as shown in Figure 8. The development of a thermal boundary layer is slow owing to the large Prandtl number of the fluid.

Berkovsky et al.[32] examined experimentally that flow and heat transfer for a circular cylinder, a channel, and tubes are affected by magnetic fluid coatings. The possibility of the control of hydrodynamics and heat transfer in magnetic fluid flows with a free surface was investigated by Pogirnitskaya et al.[33]. An increase of the magnetic fluid layer thickness resulted in the enhancement of heat transfer from the cylinder, and the maximum enhancement achieved was 60%.

The boiling heat transfer of magnetic fluid was obeserved experimentally by Nakatsuka et al.[34] using a horizontal platinum wire in the presence of parallel and vertical magnetic fields. The test fluid was a isoparaffine-based magnetic fluid including Mn-Zn ferrite particles, and the boiling temperature of the base fluid is about 180°C. The maximum enhancement of heat transfer was about 180%.

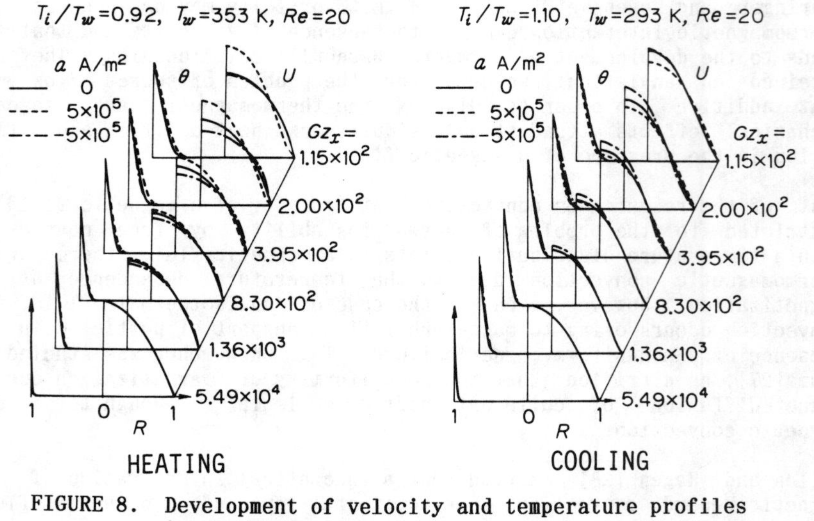

FIGURE 8. Development of velocity and temperature profiles in a tube [31].

Mathematical Formulation and Numerical Procedure   The physical model to be considered is shown in Figure 9. A magnetic fluid of temperature T_i flows into a circular tube with a uniform wall temperature T_w. The circular tube is located in a transverse magnetic field H whose intensity changes linearly along the axial direction of the tube, but is constant in the radial direction. The gradient of the magnetic field intensity is a(=dH/dx).

For simplification, some assumptions are introduced as follows:
(1) The flow is incompressible, steady, axisymmetric, and laminar.
(2) The magnetic fluid is a continuum and it has the isotropic transport properties.
(3) The magnetization of the fluid is a function of the magnetic field intensity and its temperature.
(4) The influences of gravity, viscous dissipation, internal heat generation, magnetocaloric effect of the fluid are neglected.
(5) The inlet velocity profile is fully developed.

Hence, the governing equations are expressed as follows:

Mass conservation:

$$\frac{\partial}{\partial x}(\rho u) + \frac{1}{r}\frac{\partial}{\partial r}(\rho r v) = 0 \tag{7}$$

Axial momentum conservation:

$$\rho u\frac{\partial u}{\partial x} + \rho v\frac{\partial u}{\partial r} = -\frac{\partial p}{\partial x} + \frac{\partial}{\partial x}(2\eta\frac{\partial u}{\partial x}) + \frac{1}{r}\frac{\partial}{\partial r}[r\eta(\frac{\partial u}{\partial r} + \frac{\partial v}{\partial x})] + \mu_0 M\frac{\partial H}{\partial x} \tag{8}$$

Radial momentum conservation:

$$\rho u\frac{\partial v}{\partial x} + \rho v\frac{\partial v}{\partial r} = -\frac{\partial p}{\partial r} + \frac{\partial}{\partial x}[\eta(\frac{\partial v}{\partial r} + \frac{\partial u}{\partial x})] + \frac{1}{r}\frac{\partial}{\partial r}(2r\eta\frac{\partial v}{\partial r}) - \frac{2\eta v}{r^2} \tag{9}$$

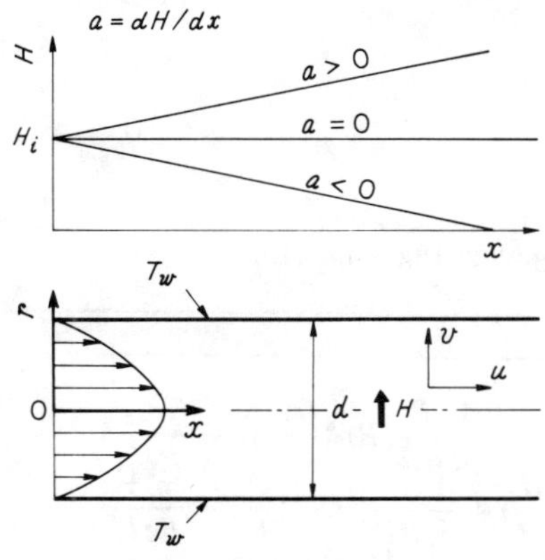

FIGURE 9. Physical model.

TABLE 2. Dimensionless variables

X	R	U	V	P
$\frac{x}{d}$	$\frac{r}{d}$	$\frac{u}{\overline{u}_i}$	$\frac{v}{\overline{u}_i}$	$\frac{p}{\rho_i \overline{u}_i^2}$

θ	M^*
$\frac{T - T_i}{T_i - T_w}$	$\frac{M}{M_i}$

ρ^*	η^*	c_p^*	λ^*
$\frac{\rho}{\rho_i}$	$\frac{\eta}{\eta_i}$	$\frac{c_p}{c_{pi}}$	$\frac{\lambda}{\lambda_i}$

Energy conservation:

$$\rho c_p u\frac{\partial T}{\partial x}+\rho c_p v\frac{\partial T}{\partial r}=\frac{\partial}{\partial x}(\lambda\frac{\partial T}{\partial x})+\frac{1}{r}\frac{\partial}{\partial r}(r\lambda\frac{\partial T}{\partial r}). \tag{10}$$

Applying the dimensionless variables listed in Table 2, we can represent the non-dimensional conservation equations in a general form as

$$\frac{1}{R}(\frac{\partial}{\partial X}(\rho^* RU\varphi - R\Gamma\frac{\partial\varphi}{\partial X})+\frac{\partial}{\partial R}(\rho^* RV\varphi - R\Gamma\frac{\partial\varphi}{\partial R}))=S^* \tag{11}$$

where φ stands for the dimensionless variables U, V and θ. The values of Γ and S* for each dependent variable are listed in Table 3. The dimensionless physical properties c_p*, ρ*, λ* and η* are the functions of the fluid temperature T, and M* is a function of fluid temperature T and magnetic field intensity H. The magnetic Grashof number Gr_m is defined as

$$Gr_m=\frac{\mu_0 M_i a d^3\rho_i}{\eta_i^2} \tag{12}$$

The term of $Gr_m M^*/Re^2$ in Table 3 expresses the body force due to the magnetic field. The local magnetic flux intensity H is represented by $H = H_i + ax$.

The boundary conditions for the present analysis are as follows:

$$\left.\begin{array}{l} U=1-R^2,\ V=0,\ P=0,\ \theta=0\ \ for\ \ X=0,\ \ 0\le R\le 0.5 \\ \partial U/\partial R=0,\ V=0,\ \partial\theta/\partial R=0\ \ for\ \ X>0,\ \ R=0 \\ U=0,\ V=0,\ \theta=1\ \ for\ \ X>0,\ \ R=0.5 \\ \partial U/\partial X=0,\ V=0,\ \partial\theta/\partial X=0\ \ for\ \ X=2000,\ \ 0\le R<0.5. \end{array}\right\} \tag{13}$$

The bulk temperature T_m is defined as

$$T_m=\int\rho c_p uT dA/\int\rho c_p u dA \tag{14}$$

and the local heat transfer coefficient h_x and local Nusselt number Nu_x are defined as

$$h_x=\frac{\lambda_w}{T_w-T_m}(\frac{\partial T}{\partial r})_w=\frac{\lambda_w}{d}\frac{1}{1-\theta_m}(\frac{\partial\theta}{\partial R})_w \tag{15}$$

$$Nu_x=\frac{h_x d}{\lambda_w}=\frac{1}{1-\theta_m}(\frac{\partial\theta}{\partial R})_w \tag{16}$$

The Graetz number Gz_x is defined as

$$Gz_x=PrRe\frac{d}{x}. \tag{17}$$

TABLE 3. Variables in general form for governing equations

φ	Γ	S^*
1	0	0
U	$\frac{\eta^*}{Re}$	$-\frac{\partial P}{\partial X}+\frac{1}{Re}[\frac{\partial}{\partial X}(\eta^*\frac{\partial U}{\partial X})+\frac{1}{R}\frac{\partial}{\partial R}(R\eta^*\frac{\partial V}{\partial X})]+\frac{Gr_m}{Re^2}M^*$
V	$\frac{\eta^*}{Re}$	$-\frac{\partial P}{\partial R}+\frac{1}{Re}[\frac{\partial}{\partial X}(\eta^*\frac{\partial U}{\partial R})+\frac{1}{R}\frac{\partial}{\partial R}(R\eta^*\frac{\partial V}{\partial R})-\frac{\eta^* V}{ReR^2}$
θ	$\frac{\lambda^*}{Pec_p^*}$	$-\frac{1}{Pe}[\frac{\partial\theta}{\partial X}\frac{\partial}{\partial X}(\frac{1}{c_p^*})+\frac{\partial\theta}{\partial R}\frac{\partial}{\partial R}(\frac{1}{c_p^*})]$

The numerical analysis was made basically by Patankar's control volume method[35]. Namely, the governing equations were discritized by the power law scheme; the numerical calculation of the velocity field was carried out by the SIMPLE method and line-by-line iteration technique, and that of the temperature field by the point-by-point iteration technique. A non-uniform staggered grid system, 51 X 41, was adopted. The details of the numerical method are similar to those in the authors' previous study [36].

Results and Discussion First, the applicability of a commercial magnetic fluid to heat-transfer control is examined. As an example, a kerosene-based magnetic fluid with 50% mass concentration of fine Mn-Zn ferrite particles is considered.

Figure 10 shows the temperature dependence of all physical properties of the magnetic fluid in the dimensionless form rendered by the values at T = 320 K. In the numerical calculation, the values of the physical properties were evaluated using the approximate correlations, Equations (18)-(22). These were derived from existing data [2,12,19,37].

$$\eta = 1.12 \times 10^{-5} exp(2860/T) \quad [Pa \cdot s] \tag{18}$$

$$c_p = 187 + 4T \quad [J/(kg \cdot K)] \tag{19}$$

$$\rho = 1447 - 0.1T \quad [kg/m^3] \tag{20}$$

$$\lambda = 0.2319 - 1.9 \times 10^{-4} T \quad [W/(m \cdot K)] \tag{21}$$

$$\begin{aligned} M = & -6.13 \times 10^{-4} T^3 - 2.596 \times 10^{-1} T^2 \\ & + 1.484 \times 10^2 T + 1.157 \times 10^5 \quad [A/m] \ (at \ \ H = 446 \ \ kA/m) \end{aligned} \tag{22}$$

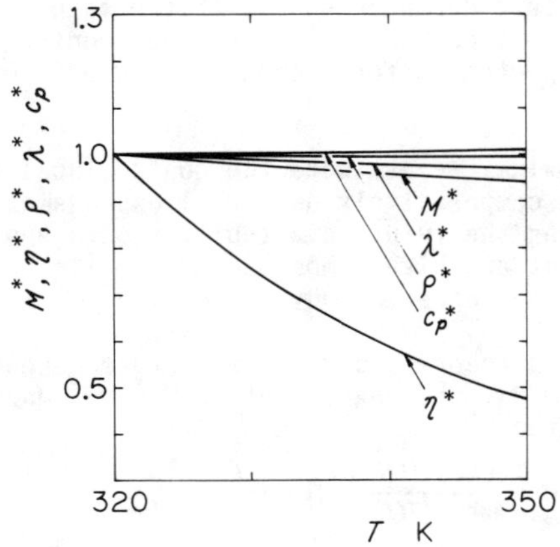

FIGURE 10. Temperature dependence of physical properties of a kerosene based magnetic fluid with 50% mass concenteration of Mn-Zn ferrite particles.

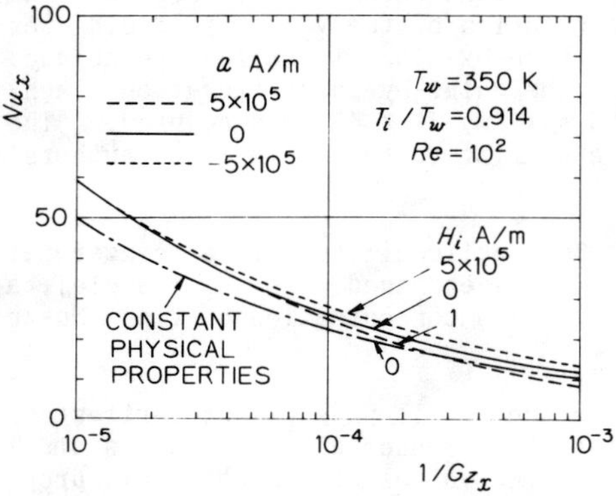

FIGURE 11. Local Nusselt numbers for commercial magnetic fluid.

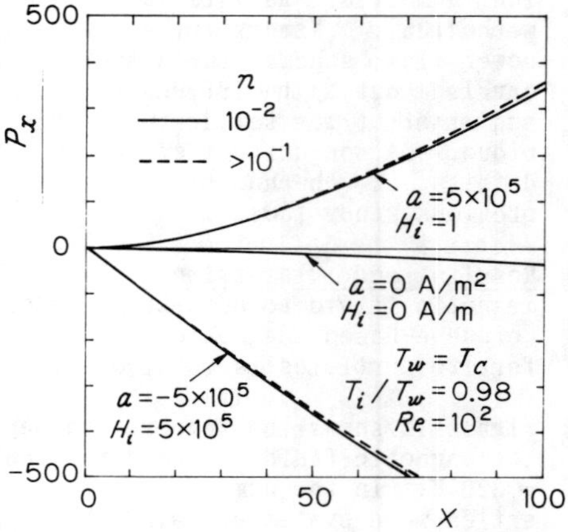

FIGURE 12. Pressure distribution.

Figure 11 shows the numerical results of the local Nusselt number Nu_x of a heated tube. The value of the Nusselt number is greater for the case of negative gradient of the magnetic field intensity (a<0) than that for no magnetic field (a=0); the Nusselt number for the case of positive gradient (a>0) is lower than that for no magnetic field. This tendency is closely related to the velocity profiles shown in Figure 8.

In the case of a commercial magnetic fluid, the magnetic force is not so great as to increase the velocity gradient at the wall and to control the convective heat transfer remarkably. These numerical results include the effects of the temperature dependence of both the magnetic properties and the thermophysical properties. Namely, the great controllability of heat transfer can not be expected as for a channel flow of a currently available magnetic fluid.

Therefore in the following, we estimate the controllability of convective heat transfer with a comparatively universal expression of the fluid magnetization, assuming the future availability of a new magnetic fluid of the low Curie point and high-temperature base liquid, in order to clarify the effect of the magnetic body force alone.

According to Langevin's theory and the temperature dependence of spontaneous magnetization M_d, the magnetization M of a magnetic fluid is assumed to be represented by

$$M \cong \frac{3.26 \times 10^4 (H/T)}{334 + (H/T)}[1 - (\frac{T}{T_C})^n] \tag{23}$$

where T_C is the temperature of Curie point.

The numerical results obtained are shown Figures 12 through 16. The effect of the magnetic force on pressure distribution in the heated tubes is shown in Figure 12. In the figure, the result for no magnetic force (H=0) is also plotted for comparison. The pressure distribution is more affected by magnetic force than by viscous force. In general, the difference in pressure distribution caused by heating or cooling is small; this is mainly owing to the very slow development of the thermal boundary layer.

Figure 13 is a plot of the local Nusselt number in the case of the wall temperature T_w being equal to the Curie point temperature T_C, the temperature ratio T_i/T_w = 0.5, and the positive gradient of the magnetic field intensity. With increasing the temperature-dependence index n of the saturation magnetization, the convective heat transfer can be suppressed; however, there is no difference in the Nusselt number for n greater than 10^{-1}.

Although the influence of the index n is rather weak in the case of T_i/T_w = 0.5, the heat-transfer controllability becomes good for the temperature ratio T_i/T_w, as shown in Figure 14.

Figure 15 shows the case of the fluid inlet temperature T_i equal to the Curie point. In this case, the heat-transfer controllability is also good; with increasing index n, the convective heat transfer is enhanced.

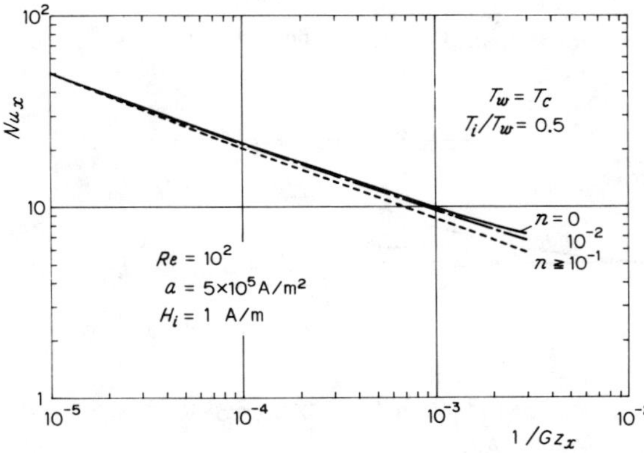

FIGURE 13. Local Nusselt number for heating at T_i/T_w = 0.5.

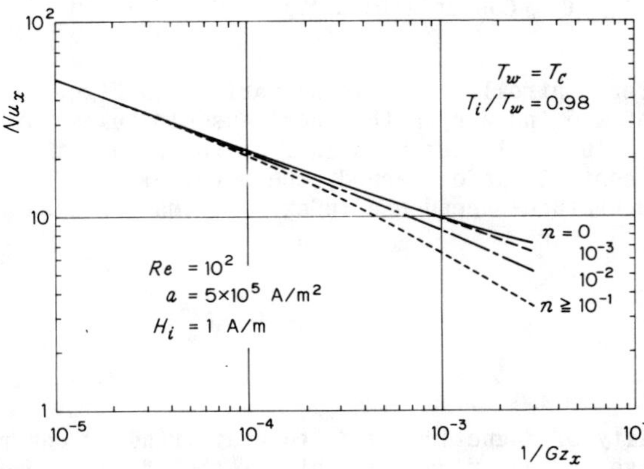

FIGURE 14. Local Nusselt number for heating at T_i/T_w = 0.98.

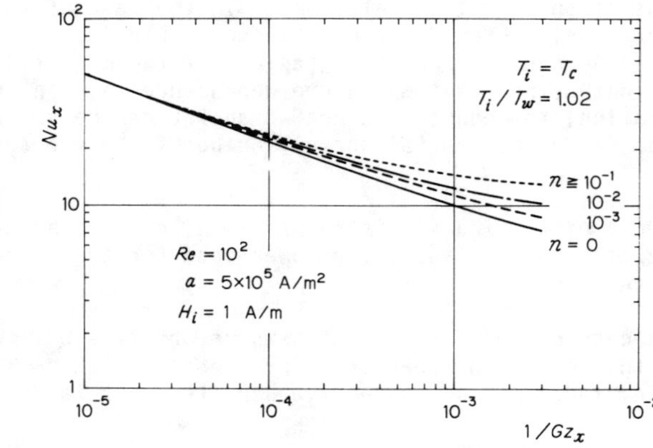

FIGURE 15. Local Nusselt number for cooling at T_i/T_w = 1.02.

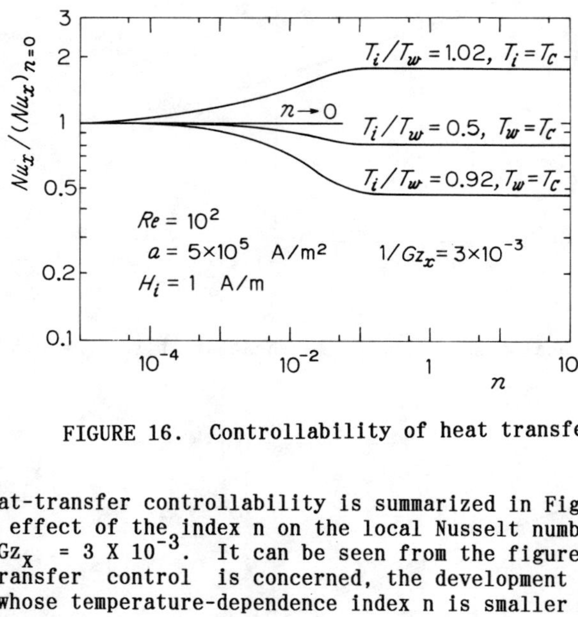

FIGURE 16. Controllability of heat transfer.

The heat-transfer controllability is summarized in Figure 16 in the form of the effect of the index n on the local Nusselt number at the location of $1/Gz_x = 3 \times 10^{-3}$. It can be seen from the figure that, as far as heat-transfer control is concerned, the development of a new magnetic fluid whose temperature-dependence index n is smaller than 10^{-1} would be worthless.

CONCLUDING REMARKS

The applicability of magnetic fluids to heat-transfer control has been examined through a numerical analysis of the forced convective heat transfer in a circular tube with a uniform wall temperature in the presence of a transversely applied magnetic field.

The possibility of heat-transfer control was found to be comparatively low with a currently available magnetic fluid. Hence, the controllability of heat transfer with a magnetic fluid having the magnetization of arbitrary temperature-dependence has been estimated by assuming constant thermophysical properties, in order to clarify the effect of the magnetic body force.

When the temperature ratio T_i/T_w approaches unity and the temperature-dependence index n is increased, the heat transfer controllability becomes good. As far as the heat-transfer control is concerned, developing of a new magnetic fluid with an index $n<10^{-1}$ is worthless. Anyway, the possibility of heat-transfer control is not so great in the case of utilizing only the magnetic body force as we expected. It is necessary to examine the magnetocaloric effect as well.

Using the magnetic fluid film coated on a solid surface and the spatial and temporal phase-change of a magnetic fluid as a controllable turbulence promotor may be another promising tool for the active control of heat transfer.

ACKNOWLEDGMENT

The authors wish to express their gratitute to Professor S. Kamiyama of the Institute of Fluid Science, Tohoku University for his encouragement in doing this work, and to Mr. M. Hongoh, Mr. T. Shimoyama, and Mr. K. Okuyama of the same Institute for their help in preparing the manuscript.

NOMENCLATURE

- a = gradient of the magnetic field intensity
- c_p = specific heat of magnetic fluid
- d = diameter of a circular tube
- Gr_m = magnetic Grashof number, Equation (12)
- Gz_x = Graetz number, Equation (17)
- h_x = local heat transfer coefficient, Equation (15)
- H = magnetic field intensity
- k = Boltzmann constant
- m = magnetic moment of a particle
- M = magnetization
- M_d = spontaneous magnetization of ferromagnetic material
- M_{d0} = spontaneous magnetization at 0 K
- M_s = saturation magnetization of magnetic fluid
- Nu_x = local Nusselt number, Equation (16)
- p = pressure
- P = dimensionless pressure, Table 2
- Pr = Prandtl number, $\eta_i c_p/\lambda_i$
- Re = Reynolds number, $\bar{u}_i \rho_i d/\eta_i$
- S = quantum number of electron spin
- t = Celsius temperature

T = absolute temperature
T_m = bulk temperature
T_C = absolute temperature at the Curie point
u, v = velocity components in axial- and radial-direction
$\bar{u}_i$ = average velocity at the inlet of a circular tube
U, V = dimensionless velocity components in axial- and radial-direction
x, r = axial and radial coordinates
X, R = dimensionless axial and radial coordinates, Table 2

Greek symbols

$\dot{\gamma}$ = shear rate
η = absolute viscosity of magnetic fluid
θ = dimensionless temperature, Table 2
λ = thermal conductivity of magnetic fluid
λ_p = thermal conductivity of ferromagnetic particle
μ_0 = magnetic permeability of a vacuum
ρ = density of magnetic fluid
ϕ = volume fraction of ferromagnetic particles

Subscripts and superscript

b = value of base liquid
i = value at inlet
p = value of ferromagnetic particles
w = value at solid wall
* = normalized by fluid property at the inlet temperature

REFERENCES

1. Pappell, S.S., Low Viscosity Magnetic Fluid Obtained by the Collodial Suspension of Magnetic Particles, U.S.Patent 3215572, 1965.

2. Kamiyama, S., Magnetic Fluids; Production, Its Characteristics and Application, Ohyou Gijyutsu Shuppan, Tokyo, 1989. (in Japanese)

3. Zahn, M. and Shenton, K.E., Magnetic Fluids Bibliography, IEEE Trans. Magn., vol.MAG-16, no. 2, pp. 387-388, 1980.

4. Charles,S.W. and Rosensweig,R.E., Introduction to the Magnetic Fluids Bibliography, J. of Mag. Mag. Mat., vol.39, pp. 191-192, 1983.

5. Kamiyama,S. and Rosensweig,R.E., Introduction to the Magnetic Fluids Bibliography, J. of Mag. Mag. Mat., vol.65, pp. 401-412, 1987.

6. Rosensweig,R.E., Ferrohydrodynamics, Cambridge Univ. Press, Cambridge, 1985.

7. Khalahala,S.E. and Reimers,G.W., Magnetic Liquids, U.S.Patent 3843540, 1974.

8. Bashtovoy, V.G., Berkovsky, B.M. and Vislovich, A.N., Introduction to Thermomechanics of Magnetic Fluids, Hemisphere Pub., N.Y., pp. 24-28,

1988.

9. Kamiyama, S., Fundamentals of Magnetic Fluids, Sangyo Tosho, Tokyo, 1989. (in Japanese)

10. Rosensweig, R.E., Kaiser, R. and Miskolczy,G., Viscosity of Magnetic Fluid in a Magnetic Field, J. of Coll. Interface Sci., vol.29, pp. 680-686, 1969.

11. Vand, V. J., Viscosity of Solution and Suspensions. I. Theory, J. of Phys. and Coll. Chem., vol.52, no. 2, pp. 277-299, 1948.

12. Matusevich, N. P., Orlov, L. P., Samoilov, V. B. and Fertman, V. E., Production and Properties of Magnetic Fluids, Heat Transfer-Soviet Research, vol.19, no. 3, pp. 25-66, 1987.

13. Popplewell,J. and Al-Qenaie,A., Thermal Conductivity Measurements on Ferrofluids Containing Metallic Cobalt and Iron Particles, J. of Mag. Mag. Mat., vol.65, pp. 215-218, 1987.

14. Kamiyama,S., Koike,K. and Oyama,T., Pipe Flow Resistance of Magnetic Fluids in a Nonuniform Transverse Magnetic Field, J. of Mag. Mag. Mat. vol.39, pp. 23-26, 1983.

15. Sawada,T., Tanahashi,T. and Ando,T., Two-Dimensional Flow of Magnetic Fluid Between Two Parallel Plates, J. of Mag. Mag. Mat., vol.65, pp. 327-329, 1987.

16. Kamiyama, S. and Satoh, A., Steady Flow Around a Circular Cylinder Coated with a Magnetic Fluid Film, Trans. Japan Soc. Mech. Engrs., vol.53, pp. 2395-2402, 1987. (in Japanese)

17. Shukla, J.B. and Kumar, D., A Theory for Ferromagnetic Lubricatin, J. of Mag. Mag. Mat., vol.65, pp. 375-378, 1987.

18. Kamiyama, S. and Inoue, S., A Study on the Two-Phase Flow Characteristics of a Magnetic Fluid, Trans. Japan Soc. Mech. Engrs., vol.54, pp.80-86, 1988. (in Japanese)

19. Kamiyama, S. and Ishikawa, H., Two-Phase Flow Characteristics of a Temperature-Sensitive Magnetic Fluid, Trans. Japan Soc. Mech. Engrs., vol.55, pp. 741-747, 1989. (in Japanese)

20. Kamiyama, S., Koike, K. and Ikeda, Y., Pulsating Pipe-Flow Characteristics of Magnetic Fluid, Trans. Japan Soc. Mech. Engrs., vol.54, no.508, pp.3331-3336, 1988. (in Japanese)

21. Yamane, R., Ishibashi,Y. and Inagaki,A., Ferromagnetic Fluid Flows in a Rectangular Channel with Travelling Magnetic Fields, J. of Mag. Mag. Mat., vol.65, pp. 324-326, 1987.

22. Kumar,D. and Chandra,P., Flow of a Magnetic Fluid in a Channel, J. of Mag. Mag. Mat., vol.73, pp. 361-366, 1988.

23. Romig, M.F., The Influence of Electric and Magnetic Fields on Heat Transfer to Electrically Conducting Fluids, in Advances in Heat Transfer, ed. Irvine,Jr.T.F. and Hartnett,J.P., vol.1, pp. 267-354, Academic, N.Y.,

1964.

24. Viskanta, R., Electric and Magnetic Fields, in Handbook of Heat Transfer Fundamentals, ed. Rohsenow,W.M., Hartnett,J.P. and Ganic,E.N., pp. 10/1-10/45, McGraw-Hill, N.Y., 1985,

25. Davidson, J.H., Kulacki,F.A. and Dunn,P.F., Convective Heat Transfer with Electric and Magnetic Fields, in Handbook of Single-Phase Convective Heat Transfer, ed. Kakaç,S., Shah,R.K. and Aung,A., pp. 9/1-9/27, John Wiley & Sons, N.Y., 1987.

26. Neuringer, J.L. and Rosensweig,R.E., Ferrohydrodynamics, Physics of Fluids, vol.7, no. 12. pp. 1927-1937, 1964.

27. Blūms,E., Free Convection in an Isothermic Magnetic Fluid Caused by Magnetophoretic Transport of Particles in the presence of a Non-Uniform Magnetic Field, J. of Mag. Mag. Mat., vol.65, pp. 343-346, 1987.

28. Stiles, P. J. and Kagan, M., Thermoconvective Instability of a Horizontal Layer of Ferrofluid in a Strong Vertical Magnetic Field, Abstracts of 5th Int. Conf. Magnetic Fluids, Riga, pp.181-182,1989.

29. Gotoh, K. and Yamada,M., Thermal Convection in a Horizontal Layer of Magnetic Fluid, J. of Phys. Soc. Japan, vol.51, no.9, pp.3042-3048, 1982.

30. Okanaga, H., Saitoh,A., Sawada,A. and Tanahashi,T., A Simulation on Free-Convection Control of a Magnetic Fluid, Proc. Union, Conf. of Magnetic Fluid in Japan, pp. 65-68, Tokyo, 1989. (in Japanese)

31. Maruyama, S., Aihara, T. and Kishimoto, H., Heat Transfer of Temperature-Sensitive Magnetic Fluid in a Circular Tube with a Uniform Temperature, Proc. of 9th Int. Heat Transfer Conf., Paper No. 20-MSC-15, pp. 329-334, Jerusalem, 1990.

32. Berkovsky, B.M., Bashtovoy, V.G. and Krakov, M.S., Hydrodynamics and Heat Hransfer when Affected by Magnetofluid Coatings, Abstracts of 5th Int. Conf. Magnetic Fluids, Riga , pp. 179-180, 1989.

33. Pogirnitskaya, S.G., Reks, A.G., Chernobay, V.A. and Taits,E.M., Heat Transfer Enhancement in Magnetofluid Flows with a Free Surface, Abstracts of 5th Int. Conf. Magnetic Fluids, Riga, pp. 191-192, 1989.

34. Nakatsuka, K., Hama, Y. and Takahashi, J., Heat Transfer and in Temperature-Sensitive Magnetic Fluid, Abstracts of 5th Int. Conf. Magnetic Fluids, Riga, pp. 189-190, 1989.

35. Patankar, S.V., Numerical Heat Transfer and Fluid Flow, McGraw-Hill, New York, 1980.

36. Aihara,T., Kim,J.-K. and Maruyam,S., Effects of Temperature-Dependent Fluid Properties on Heat Transfer Due to an Axisymmetric Impinging Gas Jet Normal to a Flat Surface, Wärme-und Stoffübertragung, vol.25, pp.145-153, 1990.

37. Sasaki,H, Masuda,H., Hukuya, H. and Mizuta, I., Thermal Conductivity of Magnetic Fluids Measuring by Unsteady State Hot-Wire Method, Mem. Inst. Fluid Sci, Tohoku Univ., vol.1, pp.93-110, 1990. (in Japanese)

Convection and Thermophysical Properties of Molten Semiconductors

KOICHI KAKIMOTO, SHIN NAKAMURA, and TAKETOSHI HIBIYA
Fundamental Research Laboratories
NEC Corporation
34, Miyukigaoka, Tsukuba 305, Japan

ABSTRACT

Flow visualization of molten silicon convection was performed using X-ray radiography with a tracer particle. The molten silicon flow is unsteady because of its low Prandtl number. Reliable thermophysical data of molten semiconductors data are required for numerical simulation of crystal growth. Experimental determination of the thermophysical properties including thermal conductivity measurement of molten InSb under microgravity is reviewed. Future of research on melt flow and thermophysical properties is also forecasted.

1. INTRODUCTION

High quality semiconductor wafers are required for ultralarge-scale integrated circuits (ULSIs) and opto-electronics devices. Almost all bulk single crystals of silicon and GaAs are fabricated using a melt-growth technique, such as Czochralski (CZ), Bridman and floating-zone (FZ) methods, where transfer of heat of crystallization determines growth rate. There is a comparatively small number of works which characterize molten-state semiconductors, although there has been a huge amount of work concerning the solid state physics of semiconductors.

One of the important subjects for crystal growth of semiconductors is the effort to understand and control convection of molten semiconductors. In a CZ furnace, heat exchange by conduction, convection and radiation coexist. In particular the convection problem becomes very complex because a molten semiconductor has a low Prandtl number like molten metals. Therefore, the flow becomes periodic and unstable. Since the temperature field near the solid-liquid interface is strongly coupled with fluid flow and is affected by its oscillation, the growth rate of the crystals and their distribution coefficients are not kept constant. Therefore, distribution of impurities and oxygen, which controls electronic characteristics, are not spatially homogeneous but irregular. This is one of the reasons for growth striation formation(1).

Characterization of thermophysical properties is the other research field where not only thermophysical data requisite to computer simulation is obtained but where also a clue to the comprehension of the physics of molten semiconductors is found(2).

In this paper, a newly developed experimental technique for flow visualization of molten silicon during crystal growth and its scientific results are discussed. Measurement of thermophysical properties, including thermal conductivity measurements of molten semiconductors under microgravity, are also presented. The problems which should be overcome for both of the above mentioned experimental approaches were attributed to the fact that molten semiconductors have a high melting temperature, are corrosive, opaque in visible light, electrically conductive, and so on. These characteristics are common among low Prandtl number melts, such as molten metals.

2. *IN-SITU* OBSERVATION OF MOLTEN SILICON FLOW DURING CRYSTAL GROWTH

2.1 Conventional Approach to Melt Flow

In the past decade computer simulations have been performed for flow in molten semiconductors which are characterized as the low Prandtl number melts to compensate for experimental difficulties. In 1983 Crochet et al.[(3)] performed a challenging work on time-dependent flow-simulation for the CZ case. After that, several works on time-dependent flow-simulation in a CZ furnace were reported, although thermal-boundary conditions were imposed on the crucible wall, the melt surface and the solid-liquid interface position[(4-7)]; the method was a so-called "local model". It is not possible to judge whether the calculated result can really reproduce the flow in the CZ furnace, because the flow is extremely sensitive to thermal-boundary conditions[(8)].

In order to solve the above problems, a global model, which can relax the imposed thermal boundary conditions, such as crucible temperature, heat flux on the free surface of the melt and solid-liquid interface position, has been introduced. This model includes numerical calculations of the heat exchange by radiation between boundaries of the skeleton and external walls with fixed temperatures, although flow was assumed to be steady and axisymmetric[(9,10)].

It is widely accepted that homogeneity of impurity distribution is strongly coupled with temperature field at the solid-liquid interface and flow field. In order to understand such phenomena by numerical simulation, an extremely refined mesh is needed, because boundary-layer thickness for impurity concentration is very thin. This is due to the small value of the diffusion coefficient of impurities in the melt.

As mentioned above, a lot of progressive numerical simulations have been performed. However, there are as of yet few experiments on fluid flow of molten semiconductors and molten metals, while some measurements of temperature oscillation at a fixed point in the fluid crucible have been performed[(11)].

2.2 X-Ray Radiography System for Flow Visualization

We have performed direct observation of molten silicon flow by X-ray radiography with a tungsten particle tracer[(12)]. For this purpose we developed a new system which consists of a Czochralski furnace, the X-ray source (tungsten target: 160 kV, 3 mA), the X-ray detector (CsI) and the data processing system, as shown in **Fig. 1**. A specially designed tracer particle was also prepared, as shown in **Fig. 2**. A tungsten particle was used, in consideration of X-ray absorbance and stability at high temperatures. The diameter of the tungsten particle was determined according to the spot size of the X-ray source,

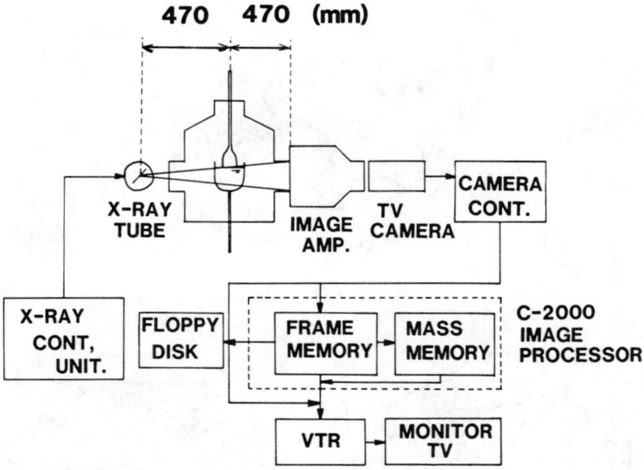

FIGURE 1. *In-situ* observation system for molten silicon flow using X-ray radiography technique(12).

spatial resolution of the X-ray detector and the density of the molten silicon. In order to prohibit corrosion and to adjust its density, the particle was coated with silica glass and then with carbon. Carbon coating assures wettability of the tracer with molten silicon. Total diameter of the tracer was 2.5 mm. Flow was observed with and without a growing crystal from a 75 mm diameter crucible.

2.3 Flow Velocity

The flow pattern of a silicon melt in a CZ furnace is sensitive to flow domain as shown in the literatures(12,13). The flow becomes oscillatory and time-dependent for deep melts. This is also the case for high rotation rates of a crystal and/or crucible. Otherwise, the flow becomes axisymmetric and almost steady as presented in the literature. The typical pattern of the time-dependent flow is shown in **Fig. 3**; (a) total flow, (b) up-flow and

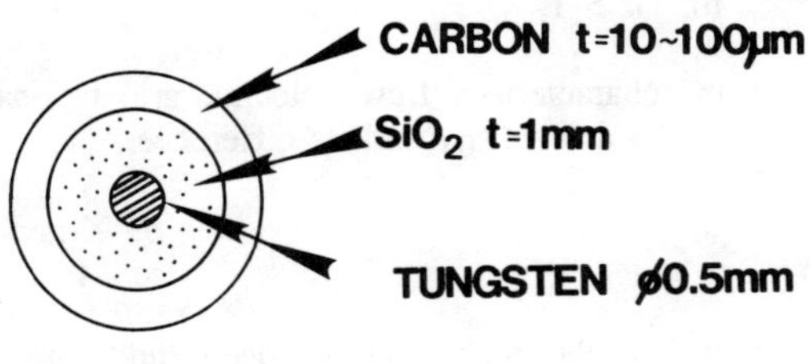

FIGURE 2. Cross section of a tracer for molten silicon(12).

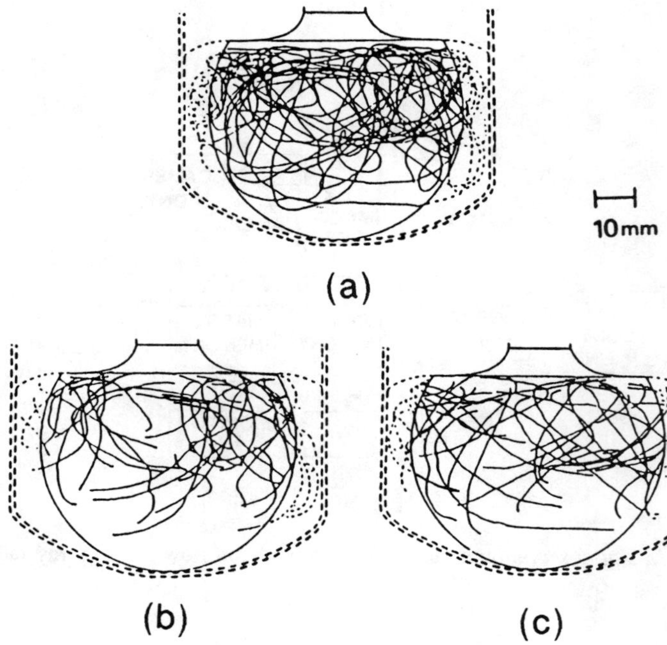

FIGURE 3. Particle paths of one specific tracer for 5 min: (a) total flow, (b) up-flow and (c) down-flow[8].

(c) down-flow. If axisymmetric and steady flow exists in a crucible, down-flow should be concentrated in the center of the crucible. However, the stream lines of the down-flow are distributed over the whole region of the melt. Consequently, the experimentally obtained flow pattern is neither axisymmetric nor steady, but time-dependent. Characteristic flow velocity obtained by the experiment is several cm/sec. The characteristic flow velocity obtained from the analytical solution by Ostrach is as follows[14]:

$$U = O[\, g\, L\, \Delta\rho/\rho \,]^{1/2} \quad \text{for Gr} > 1, \tag{1}$$

where U, g and L are the characteristic flow velocity, gravitational acceleration and characteristic length. ρ and Δρ are density and its difference.

2.4 Flow Instability

As mentioned in Section 2.3, the melt flow of deep- and large-diameter domains becomes unsteady and oscillatory. **Figure 4** shows a flow-instability map for a shallow

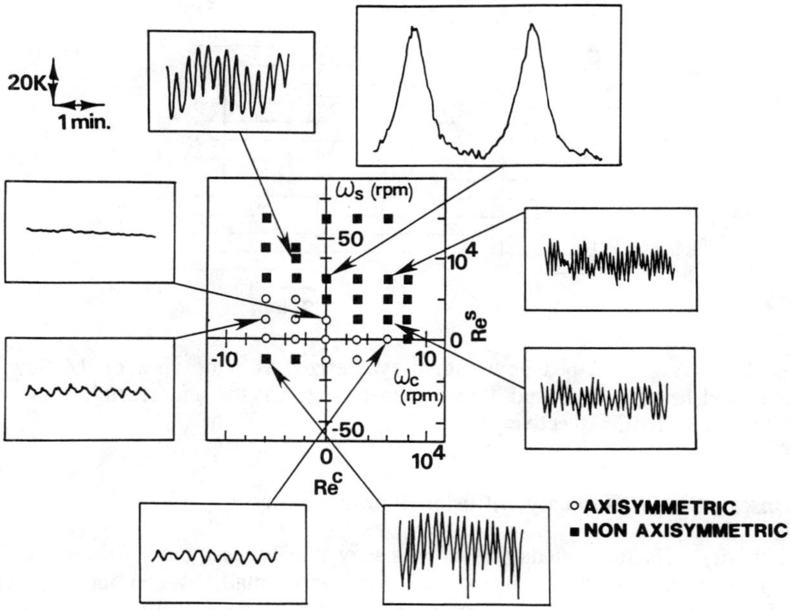

FIGURE 4. Map for flow instability and temperature oscillation for the shallow melt with aspect ratio (melt radius/depth) of 1.2[(13)].

melt[(13)]. Diameter and the height of the melt were 75 mm and 31 mm, respectively. The ordinate and the abscissa represent crystal and crucible rotation rates; the Reynolds numbers caused by both the crystal and crucible rotations are superimposed on the axes. Temperature oscillation at a fixed point in the melt in a real-time scale is also shown for the corresponding crucible and crystal rotation conditions. The open circles represent the flow identified as almost steady and axisymmetric, while the non-steady and non-axisymmetric flows are shown by the closed squares. It is clear that the crystal-growth condition with a larger amplitude of temperature fluctuation is identical to the non-axisymmetric flow condition and that the condition with smaller amplitude of temperature fluctuation is identical to the axisymmetric flow condition. This strongly suggests that the iso-rotational condition for the crystal and the causes flow instability.

2.5 Three-Dimensional Observation

Three-dimensional flow visualization for molten silicon is now available[(15)]. Two sets of X-ray radiography systems, which were aligned 90° to each other, were attached to the crystal puller. By identifying the tracer position on the monitor screen for each observation system, it was pinpointed in an X-Y-Z ordinate system. **Figure 5** shows the top view of a particle path. For a melt with a diameter of 75 mm and height of 40 mm, the particle path showed a gear-shaped pattern, and azimuthal flow direction was the same as that of the crucible rotation (crystal and crucible rotation rates were + 1 and - 1 rpm, respectively). Through three-dimensional observation, more detailed information on molten silicon flow will be collected.

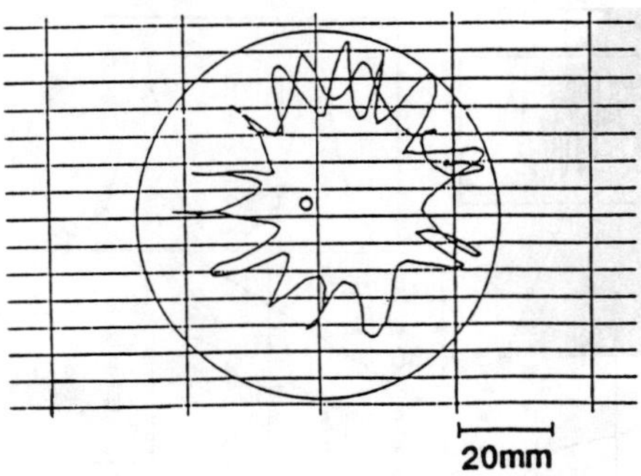

FIGURE 5. A gear-shaped flow pattern synthesized as a top view of the flow: melt diameter and height are 75 and 40 mm,respectively. Crystal and crucible rotation rates were +1 and -1 rpm, respectively[15].

2.6 Observation of Impurity Diffusion Boundary Layer

The impurity diffusion boundary layer was very thin because the diffusion coefficients of impurities in the semiconductor melts are very small, as mentioned previously. Therefore, it was very difficult to observe the layer near the walls, such as the solid-liquid interface, melt surface and melt-crucible interface; *in-situ* observations of the layer involved several difficulties. However, observations of the impurity-diffusion boundary layer were performed using the same X-ray radiography system as that used for convection observation[16].

The difference in absorption coefficients between molten silicon and impurity was used to visualize concentrated impurity atoms in the diffusion boundary layer; indium was used as an impurity atoms in the present experiment. Transmitted X-ray intensity is governed by the absorption equation as shown in eq. (2),

$$I = I_o \exp (- \Sigma \alpha_i c_i l), \qquad (2)$$

where I and I_o are transmitted and incident X-ray intensities. The α_i, c_i and l are the absorption coefficient, concentration of material component of "i" (for example, silicon and impurity), and transmission length of the X-ray, respectively. If the absorption coefficient of the impurity atom is different from that of silicon, a contrast in the impurity-concentration region can be recognized in a transmitted image. **Figure 6** shows the transmitted X-ray image taken during the growth of In-doped silicon[16]. A black belt caused by absorption attributed to concentrated doped-indium can be identified just beneath the solid-liquid interface. The layer was as thin as 0.3 mm.

3. THERMOPHYSICAL PROPERTIES

3.1 Conventional Data for Thermophysical Properties

For computer simulation of molten semiconductor flow, several thermophysical data are required, such as specific heat capacity, density, volume expansion coefficient, surface

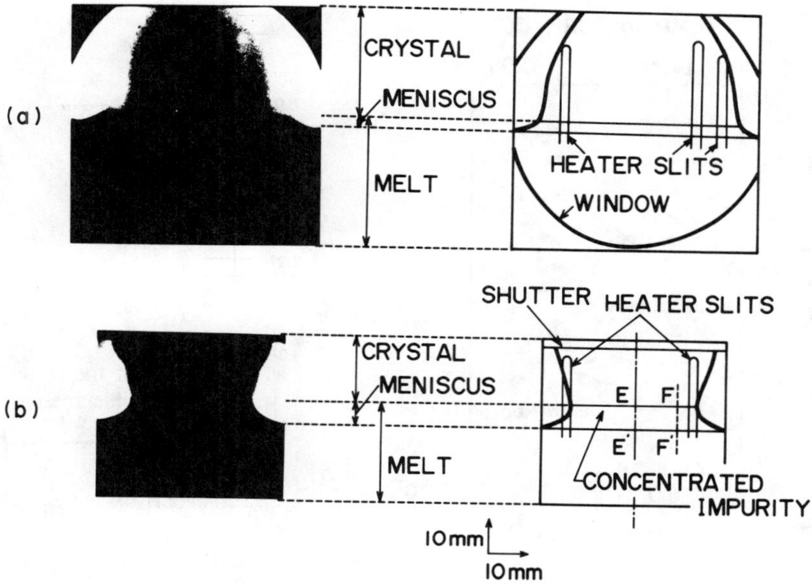

FIGURE 6. Transmitted X-ray images which show concentrated impurities just beneath the solid-liquid interface: without doped impurities (a) and with impurities (b)[16].

tension, viscosity, thermal conductivity, emissivity, and so on. Research on thermophysical properties has been carried out continuously, and mainly by Russian scientists[2]. However, particularly for thermal conductivity, amount of reliable data is quite limited because of the experimental difficulty of measurement. This is attributed to high melting temperatures, corrosivity, convection in melts, electrical conductivity, high vapor pressure, etc.

3.2 Viscosity

Viscosity of molten GaAs was measured using the oscillating cup method as a function of temperature and composition, as shown in **Figs. 7a and b**[17,18]. As shown in **Fig. 7a**, the viscosity values were smaller than those reported by Glazov et al.[2], but showed similar temperature dependence above 1300°C. However, a remarkable change in the viscosity was found between the melting temperature (1238°C) and 1300°C. **Figure 7b** shows composition dependence of viscosity for the Ga-As system, where viscosity has the maximum value at the stoichiometric composition.

Viscosity measurement during direct synthesis of GaSb made it possible to analyze reaction kinetics of GaSb from metallic Ga and Sb[19]. Viscosity change during the synthesis reaction measured at various temperature is shown in **Fig. 8**. Through this measurement, activation energy for the reaction was determined to be 15.5 kcal/mol.

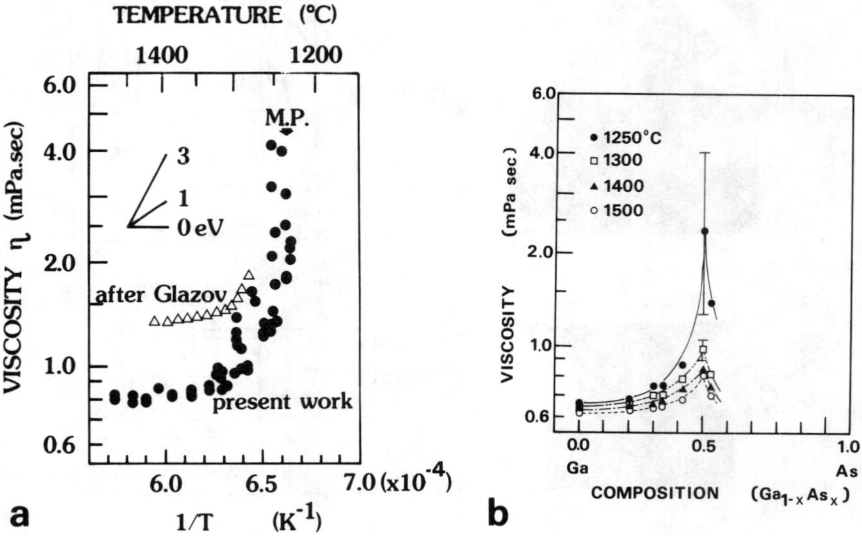

FIGURE 7. Viscosity of molten GaAs as a function of inverse temperature (a)[17] and of composition (b)[18].

3.3 Thermal Conductivity⁺

Thermal conductivity is one of the most difficult thermophysical properties to measure accurately. Therefore, thermal conductivities only for Si, Ge, GaSb and InSb of the molten state were experimentally determined among group IV and III-V compound semiconductors. Moreover, the data reported from different research groups did not agree. Buoyancy convection is one of the reasons measurement is difficult.

The thermal conductivity of molten InSb was measured using the transient hot wire method under microgravity on December 6, 1989[20]. A specially prepared thermal

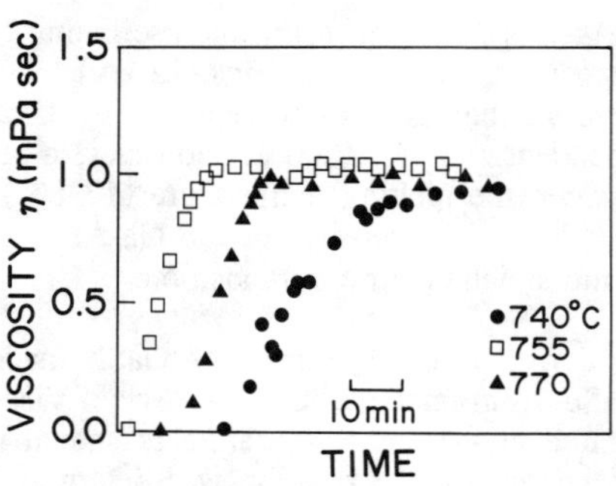

FIGURE 8. Viscosity change of GaSb as a function of reaction time[19].

conductivity measurement facility (TCMF)[21] was developed and flown with the TEXUS-24 sounding rocket. We have also developed a special ceramic sensor using a green sheet technique, where the sensing wire is fabricated on an alumina substrate and coated with a thin insulation layer, for applying the transient hot wire method to the measurement on board the rocket. The working equation for the measurement is derived by Takegoshi et al. as follows[22]:

$$\lambda_l + \lambda_s = (q \,/\, 2\pi) \,/\, (d\,(\Delta T) \,/\, d(\ln t)), \tag{3}$$

where λ_l and λ_s are thermal conductivities of a liquid sample and a substrate, respectively; q is input electric power; ΔT is the temperature increase of the sensing wire; and t is measurement time.

Figure 9a shows a temperature increase of the sensing wire immersed in the molten InSb both under microgravity and on earth at 557°C. **Figure 9b shows** apparent thermal conductivities calculated from **Fig. 9a** using eq.(3); the apparent conductivities correspond to a sum of thermal conductivities of molten InSb and an alumina substrate. As shown in **Fig. 9a**, the deviation of temperature increase from a linear relationship observed on the earth is attributed to buoyancy convection due to wire heating. This convection causes an increase in apparent thermal conductivity values measured on the earth, while the measured values under microgravity were almost constant. From the experiments, thermal conductivities of molten InSb were determined to be 15.8 W/m-K at 557°C and 18.2 W/m-K at 617°C[20].

3.4 Density and Volume Expansion Coefficient of Molten Silicon

Volume expansion coefficient β is related to density ρ as follows:[1]

$$\beta = (1/\rho) \,/\, (d\rho/dT), \tag{4}$$

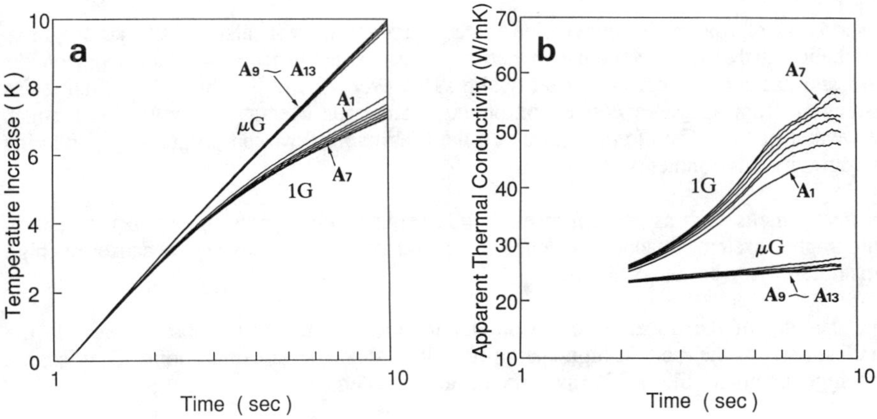

FIGURE 9. Results of thermal conductivity measurement under microgravity and on earth at 557°C. Temperature increase of the sensing wire as a function of logarithmic time (a), and apparent thermal conductivities as a function of logarithmic time (b)[20].

where T is temperature. A typical volume expansion coefficient value is $10^{-4}K^{-1}$ for molten state and $10^{-5}K^{-1}$ for solid state[2]. The volume expansion coefficient value for molten state is larger than that for solid state because atoms in the molten state feel unharmonicity of the potential well rather than atoms in the solid state. Measured values of density for molten silicon using the different methods agree with each other; calculated thermal expansion coefficients were 1.44 - 1.45 x $10^{-4}K^{-1}$[2,23].

The volume expansion coefficient is one of the most important parameters to describe flow velocity, as shown in eq. (1). The characteristic flow velocity which is obtained from eq. (1) is about 20 mm/s. This is almost identical to the experimentally obtained value[12].

4. FUTURE OF RESEARCH ON MELT FLOW AND THERMOPHYSICAL PROPERTIES

4.1 Melt Convection

In the future, experiments on flow visualization of molten semiconductors will utilize new technologies. Observation of flow in large-diameter crucibles, which are utilized in commercial crystal production will be required because flow velocity and flow pattern in a large-diameter melt are supposed to be a little bit different from those in small-diameter melts. For this purpose, a synchrotron orbital radiation (SOR) system which can supply a high-intensity X-ray source is needed.

Flow observation of a melt in a magnetic field will be a challenging task, since application of a magnetic field during crystal growth has a dampening effect on flow velocity; thus an improvement in crystal quality is expected[24]. However, it is not yet clear how convection can be suppressed in the magnetic field. Special care should be taken in using an electromagnetic shield for an image intensifier, because it is sensitive to magnetic stray fields.

Observation of Marangoni convection under microgravity will also be an exciting area. It is believed that a Marangoni convection exists on the surface of a melt even on the earth and has some effect on crystal quality. However, it is rather difficult to distinguish Marangoni from buoyancy convection on earth, since the Marangoni convection is much the weaker of two. Therefore, research on the Marangoni flow can progress by utilizing microgravity environments.

For heavy melts such as gallium arsenide (GaAs) and indium phosphide (InP), a neutron radiography system attached to nuclear reactor will be necessary because of high transmittance to the heavy melts.

The diameter of the tracer needs to be smaller in order to clearly observe flow in the smallest parts of the melt. A high-spatial-resolution detector and small point X-ray source with high intensity, like an X-ray lens are also needed.

4.2 Numerical Simulation

For future trends in numerical simulation, dynamic system simulation which is able to be performed by time-integration of the Lagrange derivative of crystals and melts in order to perform defect generation in growing crystals must be developed. Time-dependent simulation for the melt flow should also be performed so as to calculate striation formation by combining the finite element method or the finite difference method with the spectral method. Finally, microscopic calculations like molecular dynamics (MD) together with macro heat transfer model at the solid-liquid interface should be adopted. Numerical simulation research has not used the condition of single crystals for heat and mass transfer calculation. If it becomes possible to perform the above operation, "ideal and perfect" crystals can be created on a computer.

4.3 Thermophysical Properties

At present there is little reliable thermophysical data about molten semiconductors, although a build-up of the data base is urgent for the crystal growth research community. An increase in experiment opportunities of crystal growth under microgravity requires thermophysical data in order to design experiments and to simulate them by computer. Measurement of thermophysical properties is also a subject of the space experiments. The TEMPUS, German space experiment facility for containerless processing using electromagnetic positioning and inductive heating in an ultra-high-vacuum, will offer new opportunities to measure thermophysical properties of molten materials, such as specific heat capacity, surface tension and viscosity[(25)]. Since the contamination of melts through the crucible wall and the lack of appropriate crucible materials often cause difficulty for thermophysical property measurement, the development of the containerless processing technique under microgravity will encourage the build-up of a data base of thermophysical properties.

Microgravity environments attained for 5 to 10 seconds through free fall can also supply thermal conductivity data, since the transient hot wire method[(20)] only requires a 5 - 10 second microgravity period.

[+] Measurement of thermal conductivity of molten InSb under microgravity was carried out as a research activity of Space Technology Corporation, Japan.

REFERENCES

1. Rosenberger, F. E., *Fundamentals of Crystal Growth I*, pp.452-478, Springer Verlag, Berlin, Heidelberg, New York, 1979.

2. Glazov, V. M., Chizhevskaya, S. N., and Glagoleva, N. N., *Liquid Semiconductors*, Plenum Press, New York, 1969.

3. Crochet, M. J., Wouters, P. J., Geyling, F. T., and Jordan, A. S., Finite-Element Simulation of Czochralski Bulk Flow, *Journal of Crystal Growth*, vol.65, nos. 1/3, pp.153-165, 1983.

4. Mihelčić, M., and Wingerath, K., and Pirron, Chr., Three-Dimensional Simulation of the Czochralski Bulk Flow, *Journal of Crystal Growth*, vol.69, nos. 2/3, pp.473-488, 1984.

5. Munakata, T, and Tanasawa, I., Computer Simulation and Animation of Oscillatory Flow in the Melt during the Czochralski Growth of Single Crystal, *Proc. Int. Symp. Supercomputers for Mechanical Engineering, Tokyo,* pp.50-58,

1988.

6. Sawada, I., Okazawa, K., and Matsumiya, T., Numerical Simulation of Oscillatory Flow in Silicon Melt during CZ Crystal Growth, 9th Int. Conf. Crystal Growth, Sendai, Paper 22-VA-07, 1989.

7. Battaro, A., and Zebib, A., Three Dimensional Thermal Convection in Czochralski Melt, *Journal of Crystal Growth*, vol.97, no. 1, pp.50-58, 1987.

8. Kakimoto, K., Eguchi, M., Watanabe, M., and Hibiya, T., Natural and Forced Convection of Molten Silicon during Czochralski Single Crystal Growth, *Journal of Crystal Growth*, vol.94, no. 2, pp.412-420, 1989.

9. Dupret, F., Ryckmans, Y., Wouters, P., and Crochet, M. J., Numerical Calculation of the Global Heat Transfer in a Czochralski Furnace, *Journal of Crystal Growth*, vol.79, no. 1/3, pp.84-91, 1986.

10. Atherton, L. J., Derby, J. J., and Brown, R. A., Radiative Heat Exchange in Czochralski Crystal Growth, *Journal of Crystal Growth*, vol.84, no. 1, pp.57-78, 1987.

11. Yamashita, K., Aoki, T., Kitagawa, K., Kawata, Y., Fujino, N., and Shiraiwa, T., Fundamental Aspects of Molten Si during the Czochralski Crystal Growth, 9th Int. Conf. Crystal Growth, Sendai, Paper 23a-C-07, 1989.

12. Kakimoto, K., Eguchi, M., Watanabe, H., and Hibiya, T., Direct Observation by X-Ray Radiography of Convection of Molten Silicon in the Czochralski Growth Method, *Journal of Crystal Growth*, vol.88, no. 3, pp.365-370, 1988.

13. Kakimoto, K., Eguchi, M., Watanabe, H., and Hibiya, T., Molten Silicon in the Czochralski Configuration, *Journal of Crystal Growth*, vol.102, nos. 1/2, pp.16-20, 1990.

14. Ostrach, S., Convective Effect in Crystal Grown from Melt, *Journal of Crystal Growth*, vol.55, no 3, pp.614-646, 1981.

15. Watanabe, M., Eguchi, M., Kakimoto, K., and Hibiya, T., 3-Dimensional Observation of Molten Si Convection by double-Beam X-Ray Radiography, 51st Autumn Meeting, Jpn. Soc. Appl. Phys., Morioka, Paper 28p-V-4/I, 1990.

16. Kakimoto, K., Eguchi, M., Watanabe, H., and Hibiya, T., In-situ Observation of Impurity Diffusion Boundary Layer in Silicon Czochralski Growth, *Journal of Crystal Growth*, vol.99, nos. 1/4, pp.665-669, 1990.

17. Kakimoto, K, and Hibiya, T., Temperature Dependence of Viscosity of Molten GaAs by an Oscillating Cup Method, *Applied Physics Letters*, vol.50, no. 18, pp.1249-1250, 1987.

18. Kakimoto, K and Hibiya, T., Composition Dependence of Viscosity for Molten $Ga_{1-x}As_x$ ($0.0 \leq x \leq 0.53$), *Applied Physics Letters*, vol.52, no. 19, pp.1576-1577, 1988.

19. Kakimoto, K., and Hibiya, T., Kinetic Study by viscosity Measurement on

Direct Synthesis of Gallium Antimonide, *Journal of Applied Physics*, vol. 66, no. 9, pp.4181-4183, 1989.

20. Nakamura, S., Hibiya, T., Yamamoto, F., and Yokota, T., Measurement of Thermal Conductivity for Molten Semiconductors using Transient Hot Wire Method under Microgravity, 27th Natl. Heat Transfer Symp. Japan, Nagoya, Paper E-2203, 1990.

21. Yamamoto, F., Nakamura, S., Hibiya, T., Yokota, T., Grothe, D., Harms, H. and Kyr, P., Developing a Measuring System for Thermal Conductivity in Molten Semiconductors on Board the TEXUS Rocket, *Proc. CSME Mechanical Engineering Forum, Toronto*, vol.4, pp.1-6, 1990.

22. Takegoshi, E., Imura, S., Hirasawa, Y., and Takenaka, T., A Method of Measuring the Thermal Conductivity of Solid Materials by Transient Hot Wire Method of Comparison, *Bulletin of the JSME, vol.25, no. 201, pp.395-402, 1982.*

23. Lucas, L.-D., Volume Spécifique de Métaux et Alliages Liquides à Hautes Températures, *Memoires Scientifiques Rev. Metallurg.*, vol.61, no 1, pp.1-24, 1964.

24. Hoshi, K., Suzuki, T., Okubo, Y. and Isawa, N., CZ silicon Crystal Grown in Transverse Magnetic Fields, Electrochemical Society Spring Meeting, St. Louis, Paper 324, 1980.

25. Feuerbacher, B. and Herlach, D. M., Containerless Processing of Metals in Space, 40 th Cong. Int. Astronautical Federation, Malaga, Spain, 1989.

Measurement Techniques of Thermophysical Properties of High Temperature Melts

AKIRA NAGASHIMA
Department of Mechanical Engineering
Keio University
3-14-1 Hiyoshi, Kohoku, Yokohama 223, Japan

ABSTRACT

In numerical simulations and theoretical predictions of thermo–fluid dynamic behavior of high temperature melts, thermophysical property data are needed as key informations. However, experimental difficulties at high temperatures in addition to strong corrosion and high electrical conductivity of the sample liquid prevent precise determination of such needed properties as the viscosity, the thermal conductivity and the surface tension. Present status of availability and quality of these data are briefly explained. Activities of the author's group are described on developing new measurement techniques, producing new data sets and establishing reference data for high temperature melts.

1. TECHNOLOGICAL IMPORTANCE OF HIGH TEMPERATURE MELTS

Technological importance of high temperature melts is increasingly visible in various applications of material production at high temperature. For production of high quality single crystals, thermophysical properties of the melt have to be known. For controlling heat transfer and thermal behavior in fusion reactor and other high temperature energy storage and conversion devices, informations on properties of high temperature molten salts are requisite. Molten carbonate salts are intended to be used in high temperature fuel cells and their delicate control is possible only with knowledge of thermophysical properties. The properties needed for these purposes are the viscosity, the thermal conductivity, the surface tension, the expansion coefficient, the melting and solidifying temperatures and so on.

In addition to conventional needs such as calculations of flow resistance and conduction heat losses, the most important usage of thermophysical property informations is to enable precise computer simulations of various technological processes such as single crystal formation, fusion reactor system and high temperature fuel cell system. So-called high temperature melts(HTM) are materials which are solid at room temperature and turn to liquid only at high temperatures. In the past, molten metals and molten salts for metallurgy were the main target. However, in recent years they also include molten semiconductors, high temperature molten salts, molten glasses, molten polymers, molten volcanic lava, and so on [1][2].

Among various thermophysical properties, the viscosity, the thermal conductivity and the surface tension of high– temperature melts are the most needed and, on the other hand, the most poorly established properties. Viscosity is used, for example, in flow simulation calculations. Thermal conductivity is needed in in heat–transfer calculations. Surface tension is needed in the study of surface behavior or droplet and in the prediction of Marangoni convection.

The difficulties of experimental determination of thermophysical properties are different for each different kind of high temperature melts. As listed in Table 1, the most serious difficulties are strong corrosion and strong electrical conductivity of the sample. These are especially serious for liquid metals and molten salts and cause such troubles as chemical reaction of the sample with a container wall, rapid degradation of the sample, electrical disturbance to sensors, sensitive impurity effect, unpredictable supercooling and non–Newtonian behavior near the melting temperature.

Table 1. Experimental difficulties of high temperature melts.

	Corrosion	Electr. conduct.	High melt. temp.	High vap. pres.	Purifi- cation	Non- Newton- ean
Molten salts	xx	xx	x	-	x	-
Liquid metals	xx	xx	x	-	xx	-
Molten semi- conductor materials	xx	x	xx	x	xx	x
Molten glasses	-	-	x	-	-	x
Molten polymers	-	-	-	-	-	x
Lava etc.	-	-	xx	-	-	x

2. AVAILABILITY AND QUALITY OF THERMOPHYSICAL PROPERTY DATA FOR HIGH TEMPERATURE MELTS

Since extreme difficulty exists in the experimental determination of thermophysical properties of high temperature melts, available data are limited and their reliability is often questionable. For example, even for the thermal conductivity of molten NaCl, one of the most studied melts at high temperatures, only four sets of data had been available covering narrow temperature range above its melting point and recently only one set of data by the present authors group was added. And these sets of data show systematic deviation of more than 400% from each other. This situation is not only for NaCl but also for most of other salts. This fact suggests the necessity of critical evaluation of existing data in addition to new experimental study.

The most studied materials among high temperature melts are molten salts which include pure salts and mixtures. Extensive data compilations have been done at the Molten Salts Data Center of Rensselaer Polytechnical Institute in USA largely due to the effort of Prof.G.J. Janz.

A number of experts cooperated with him to establish standard reference data on various properties including the viscosity for two selected materials, namely KNO_3 and NaCl. The accumulated data and their background are published in a series of reports mostly in the Journal of Physical and Chemical Reference Data [3]. Although these reports cover extensive range of properties, substances and temperature, they also show scarcity of reliable data on important properties and materials. The most urgent needs of further studies are on the thermal conductivity and the surface tension. Even for such better–known properties as the viscosity, re–evaluation of recommended data is needed in order to catch up rapid progress of measurement technique in recent years. In 1983, the viscosity of NaCl recommended by the Molten Salts Data Center was shifted up to 30 % from their recommendation of 1968.

3. THERMAL DIFFUSIVITY MEASUREMENT BY FORCED RAYLEIGH SCATTERING METHOD

Among currently available methods to measure the thermal conductivity of fluids, the transient hot–wire method is considered as the most accurate method. The method is especially advantageous for compatibility to two major troubles at high temperatures, namely thermal convection and radiation, However, the strong electrical conductivity of the sample prohibits application of the method to high temperature melts.

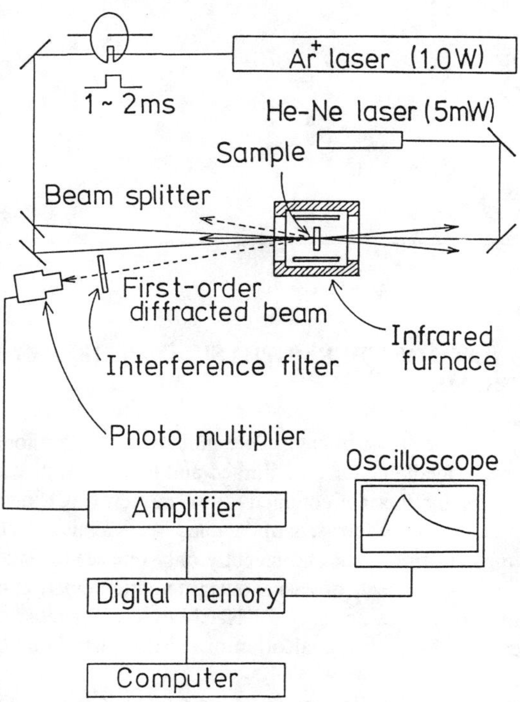

Fig.1
Schematic diagram of the forced Rayleigh scattering apparatus

The present author's group developed an optical method named the forced Rayleigh scattering method, which is the only method applicable to high temperature molten salts[4]. The principle of the method is as follows. A laser beam for heating, Fig.1, is splitted into two beams and they are set to cross in the thin (about 1 mm) layer of the molten salt. The fringe pattern of two beams creates a periodic pattern of temperature and the density, and then it creates a corresponding pattern of the refractive index. If another laser beam having a different wave length is set to pass through that heating–beam spot, the periodic pattern of the refractive index acts as an optical grating which creates diffracted beams of the first, the second, ... and the nth order. If the heating is terminated in short time, the created temperature distribution dies out with time due to heat conduction in the fluid layer. This means that the thermal diffusivity of the sample can be calculated from the decay time constant of the diffracted beam intensity shown in Fig.2.

This method has distinct advantages for overcoming two most serious error sources in high temperature fluid measurement, namely the convection error and the radiation error. One run of the measurement completes within 1 ms which is enough short to make effect due to convection negligibly small. Number of fringes in a heated spot of about 2 mm is 50 or more, and the actual temperature rise of the sample is less than 0.1 C. Therefore, although the temperature gradient is enough large, the temperature rise is extremely small. This means, the radiation error in this method is at least one order smaller than that in other conventional methods.

Figure 3 shows an example of comparison of the thermal conductivity of NaCl. The forced Rayleigh scattering method

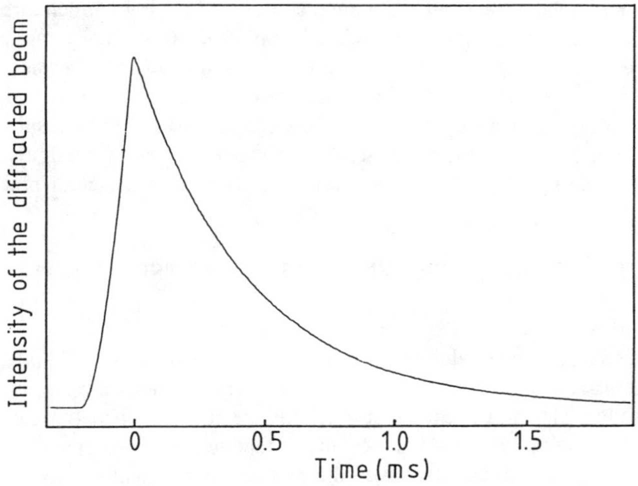

Fig.2
Typical recording of decay curve

significantly extended the measured range of temperature. Results by the forced Rayleigh scattering method are considerably lower by 400 % than most of previous data. Also, the former is the only set of data which show negative temperature dependence. This large difference is common among most of high temperature molten salts. The most probable reasons are convection and radiation errors in older measurements.

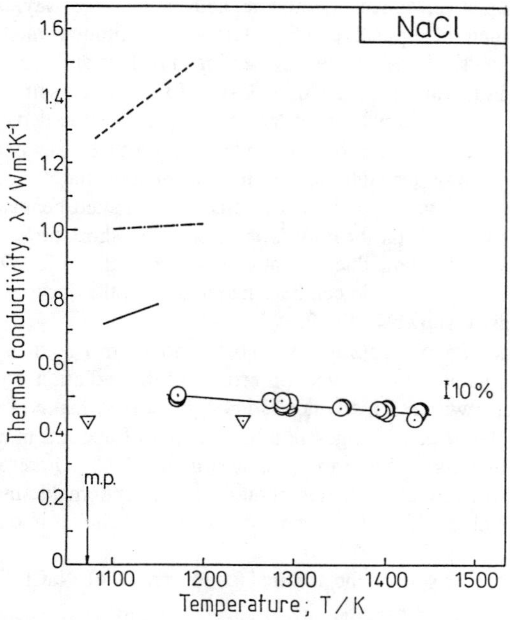

Fig.3
Thermal conductivity of molten NaCl – Comparison of available data

At high temperatures, no reference standards are available for liquid thermal conductivity. Therefore a new method can not be checked with a standard data and it is extremely difficult to judge which one among significantly different data sets is correct. The present author proposes that every researcher measures a common substance in every temperature range with his new method. Figure 4 explains the principle. Applicable temperature range of each apparatus was expanded step by step by improving insulation coating of transient hot–wire cell (THW). The highest temperature was attained by the forced Rayleigh scattering method (FRS). Toluene was measured by all of five apparatuses and the results were compared. Similarly "common" substances to be measured are proposed in the figure. Thus even without any reference standard, we can at least compare data by different new methods.

4. THERMAL CONDUCTIVITY MEASUREMENT BY THE TRANSIENT HOT–WIRE METHOD

Although at present the most accurate method to measure the thermal conductivity of fluids is the transient hot–wire method, it is not applicable for high temperature melts because of its high electrical conductivity. The most significant merit of the method is that it can eliminates the error due to convection, and this is especially attractive for measurement of high temperature fluids. The authors group challenged the problem in two ways. Firstly, a fine quartz capillary (outer diameter: about 60 – 90 um) which contained liquid mercury was used as the heating element [6]. The thin quartz wall was used as an insulating layer for highly conductive liquid. This new probe was used for mixtures of KNO_3 and $NaNO_3$.

As shown in Fig.5, the results gave answer to the long– lasted question, namely which of linear dependence on concentration given by McDonald and coworkers [7] or non– linear dependence by Bloom and coworkers [8] was correct. Secondly, a platinum wire was covered with thin layer of ceramics by ion–plating technique [9]. A typical results by this insulated probe are shown in Fig.6. This shows that the new data are the lowest among

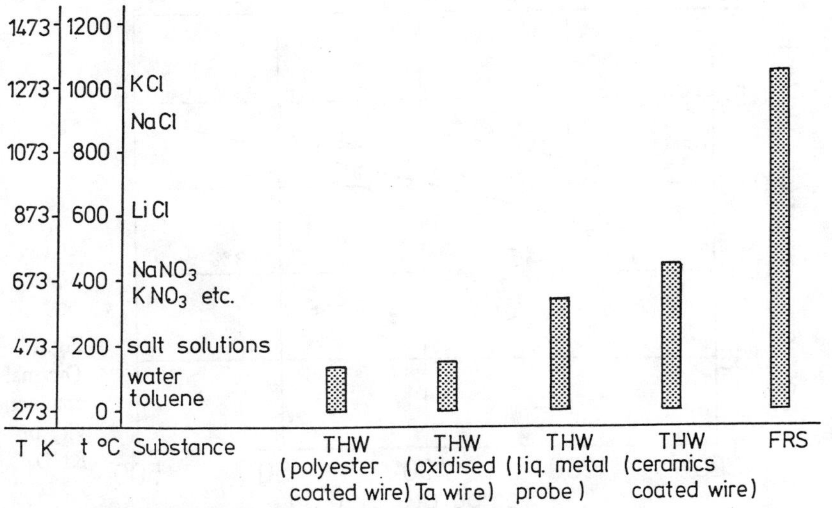

Fig.4 Step–up comparison by common substances

available data. The temperature range has been significantly expanded. Furthermore, negative temperature dependence was confirmed in accordance with results by the forced Rayleigh scattering method of Fig.3.

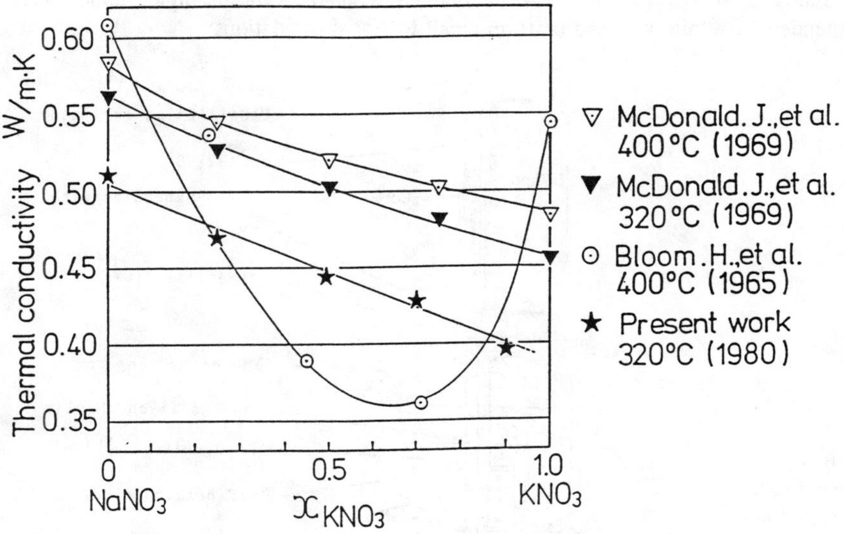

Fig.5 Composition dependence of the thermal conductivity of KNO_3–$NaNO_3$ mixtures

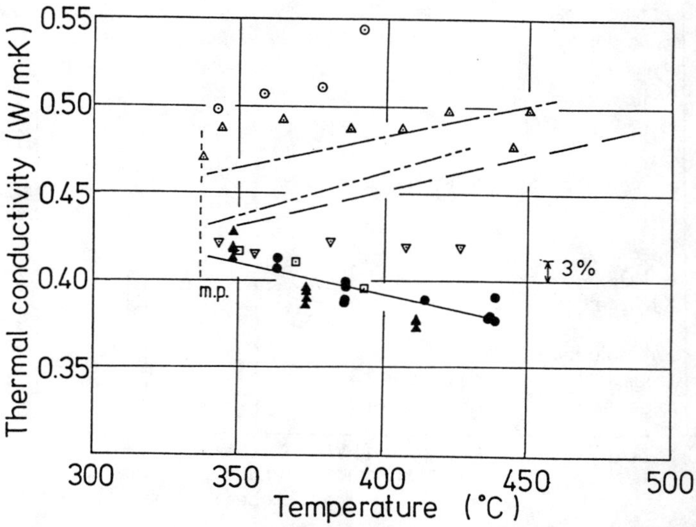

Fig.6 Thermal conductivity of KNO_3

5. VISCOSITY MEASUREMENT BY THE OSCILLATING–CUP METHOD

Viscosity is a key property to study the structure of liquids and also the most important transport property for engineering applications.

The present author's group performed a series of measurements on the viscosity of molten salts and liquid metals. The oscillating–cup method was used. It is the absolute method based on rigorous theoretical working equation and considered as one of the few most precise methods applicable to high temperature melts. A typical setting–up of the apparatus is shown in Fig.7. A cylindrical container with sample liquid inside is suspended by a thin wire and perform small torsional oscillations.

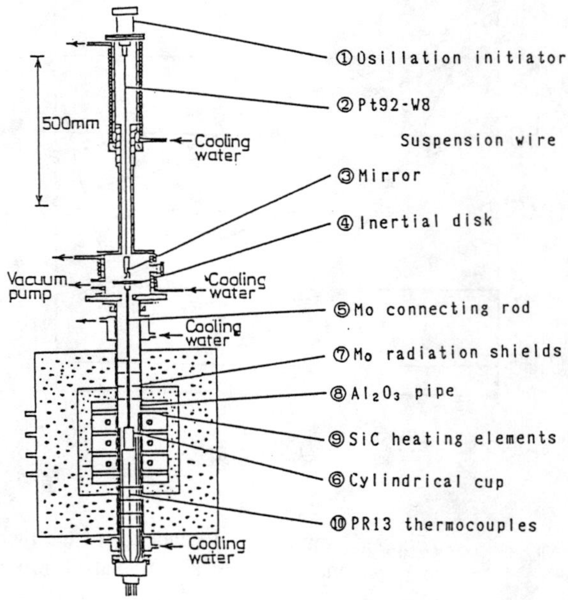

Fig.7 Main part of oscillating cup viscometer

The viscosity of the sample is calculated from damping characteristics of the oscillation due to viscous drag inside the cup. The advantage of the method is that the liquid is confined in the small cup and this is especially important for highly corrosive liquids at high temperatures. One example of the measured results is shown in Fig.8 for NaCl [10]. The figure also shows the present status of other available data. Here only the data published after 1975 are shown. As mentioned earlier, the older recommendation of the Molten Salts Data Center and all of data published before 1974 are much larger and show different temperature dependences. In Fig.8, agreement among these recent data is reasonably good. The mutual

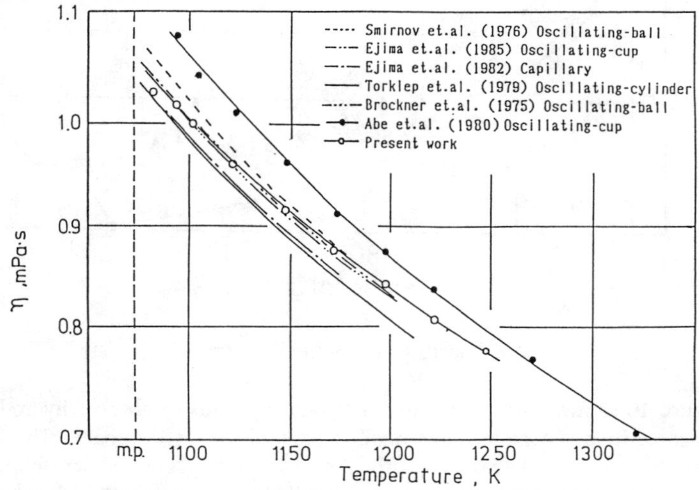

Fig.8 Viscosity of NaCl (after 1975)

discrepancy of about 7 % is one of the smallest for liquids at this high temperature.

For liquid metals, the situation is poorer. In many cases, agreement between experimental data is poor and the reason of disagreement between different authors is unknown. Figure 9 shows the viscosity of molten lithium by the present authors group and by other investigators. Disagreement of up to 50 % is observed for this industrially important material. The present author succeeded to detect the probable reasons of these disagreements for lithium [11]. There are two reasons; calculations using inadequate working equation and degradation of the sample due to chemical reaction with molybdenum container used by some of researchers. This was rare case in which the origin of disagreement was detected. If the origin is not known, correction or recalculation is not possible for previous data and also the confidence level of new measurement can not be established.

6. SURFACE TENSION MEASUREMENT BY SURFACE–WAVE METHOD

The surface tension is very sensitive to physical as well as chemical interactions, and therefore, its measurement is extremely difficult at high temperatures. The present authors group used the pendant–drop method for ordinary liquids but recently started to develop two kinds of surface–wave methods for high temperature melts. One of the methods is to observe macroscopic hydrodynamic waves and the other is to detect microscopic waves due to molecular fluctuations, both at the surface of liquids in the container.

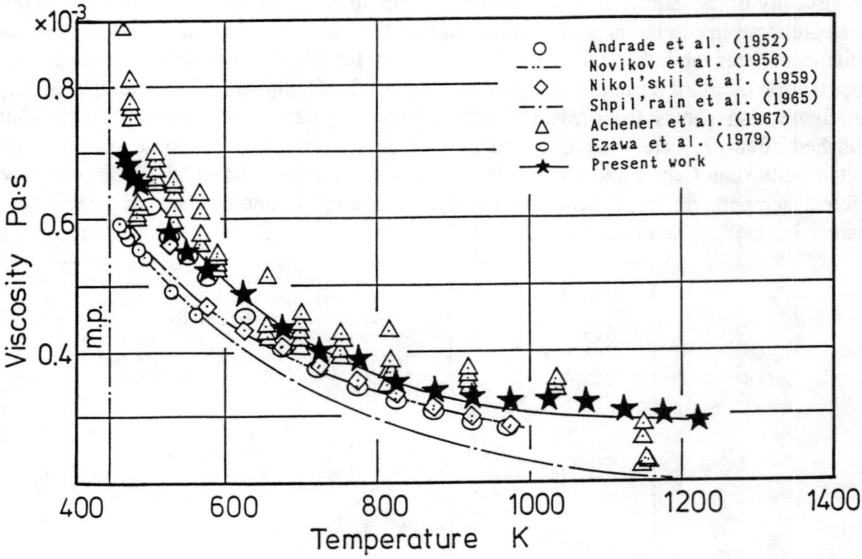

Fig.9 Viscosity of molten lithium

Figure 10 shows schematic diagram of one apparatus to observe hydrodynamic waves. The sample fluid is contained in a small container in a closed furnace. The standing waves is produced by a flat plate directly driven by a speaker system. Other diving methods such as oscillation of the total system on a controlled bench or remote driving of

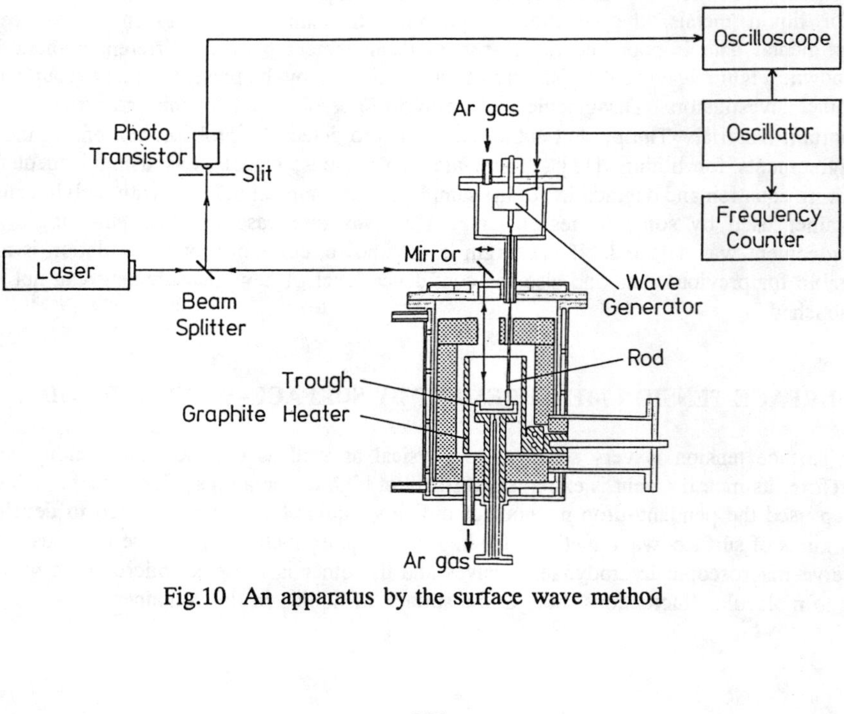

Fig.10 An apparatus by the surface wave method

a sinker are also possible. The characteristics of waves are detected by an optical system with a He–Ne laser. The surface tension is calculated by means of hydrodynamic equation. The advantage of the method is that the surface of the sample liquid is in contact only with vapor of the sample. This is important for highly reactive liquids. At the present stage, the study is concentrated on finding optimum dimensions of the driving system, refining the optical sensing system and excluding disturbances from outside vibrations.

7. FUTURE CHALLENGES

Considering strong industrial demands for thermophysical properties of high temperature melts, the present situation of availability and also quality of informations is chaotic. The most urgent task at he present stage is to develop new measurement techniques which are applicable at temperatures above, al least, 1000 K for materials with chemical reactivity and high electrical conductivity. New methods are needed especially for the thermal conductivity and the surface tension for which conventional methods are not applicable. Recently, an European group made a critical survey of measuring techniques of the thermal conductivity for molten materials at 600 – 1800 K [12]. The present author had chances to exchange comments with them in the course of their preparation of this material and therefore had completely agreed with their conclusion. Figure 11 is taken from this reference and shows predicted gradings of classical methods (a) and non–classical methods (b). Among classical methods, only the modified transient hot–wire method (THW)

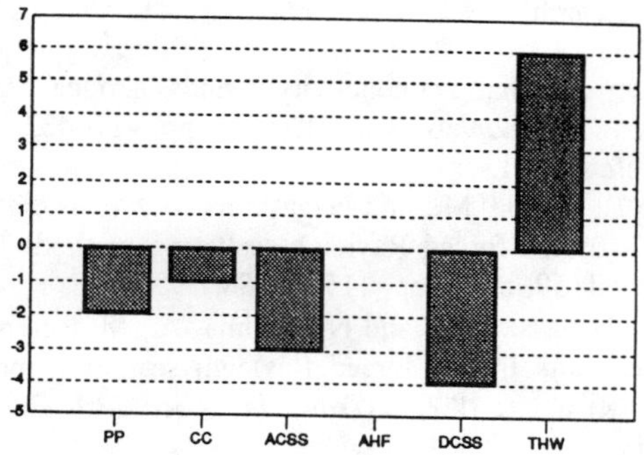

Fig.11 Evaluation of methods for thermal conductivity of high temperature liquids (by Mardolcar et al.)

(a) Classical methods

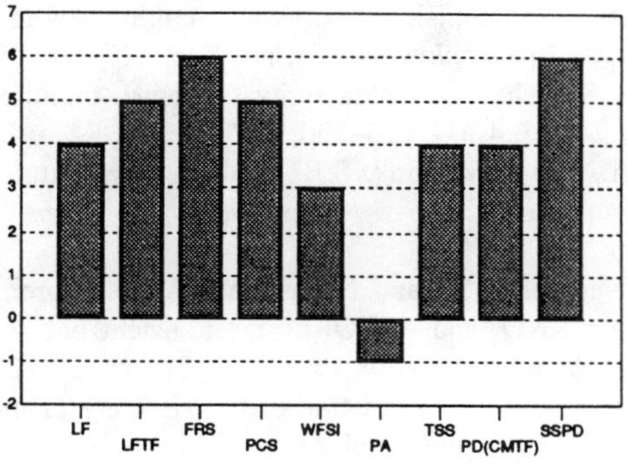

(b) Non–classical methods

described in the section 4 of the present report has given positive credits and all other conventional methods have evaluated negatively. Among non–classical methods, the forced Rayleigh scattering method (FRS) described in the section 3 of the present report and the simple source photo deflecting technique (SSPD) have given the highest credits. However, in the conclusion they stated that all the methods studied have advantages and disadvantages and none of them seem to be totally satisfying.

We need new breakthroughs in the measurement techniques which are expected to be;

(1) transient methods which are capable of subsecond or, hopefully, millisecond measurements,

(2) non–contact methods, probably, optical sensing method with levitation technique, and

(3) methods applicable for highly corrosive or high vapor–pressure fluids.

These are same also for the surface tension.

The task needed is the critical evaluation of existing data. Many of data listed in past databooks for high temperature melts are not reliable and exclusion of unreliable data is urgently needed. And also selection and recommendation selected data will help the improvement of measuring techniques and prediction methods.

The creation of an international body for collecting, evaluating and disseminating of thermophysical property informations is also one of our important tasks.

REFERENCES

1. Nagashima, A., Thermophysical properties research on fluids at the technology frontier: Status of current research, *Appl. Mech. Rev.*, vol. 41, no 3, pp.113–128, 1988.
2. Nagashima, A., Viscosity, thermal conductivity, and surface tension of high–temperature melts, *Int. J. Thermophys.*, vol. 11, no. 2, pp. 417–432, 1990.
3. See REFERENCES in the reference 1.
4. Nagasaka, Y., Hatakeyama, T., Okuda, M., and Nagashima, A., Measurement of the thermal diffusivity of liquids by the forced Rayleigh scattering method: Theory and experiment, *Rev. Sci. Instr.*, vol. 59, no. 7, pp. 1156–1168, 1988.
5. Nakazawa, Y., Akabori, M., Nagasaka, Y., and Nagashima, A., Measurement of the thermal diffusivity of molten salts by the forced Rayleigh scattering method – Measurement of alkali chlorides above 1000 C, *Trans. Jap. Soc. Mech. Eng.*, vol.56B, no. 525, pp. 1467– , 1990.
6. Omotani, T., Nagasaka, Y., and Nagashima, A., Thermal conductivity measurement of KNO_3–$NaNO_3$ mixtures using a transient hot–wire emthod with a liquid metal–in–capillary probe, *Int. J. Thermophys.*, vol. 3, no. 1, pp. 17– , 1982.
7. McDonald, J., and Davis, H. T., Thermal conductivity of binary mixtures of alkali nitrates, *J. Phys. Chem.*, vol. 74, no. 4, pp. 725–730, 1970.
8. Bloom, H., Doroszkowski, A., and Ticklebank, S.B., Molten Salt Mixtures, IX. The thermal conductivities of molten nitrate systems, *Aust. J. Chem.*, vol. 18, pp.1171–1176, 1965.
9. Kitade, S., Kobayashi, Y., Nagasaka, Y., and Nagashima, A., Measurement of the thermal conductivity of molten KNO_3 and $NaNO_3$ by the transient hot–wire method using ceramic–coated probes, *High Temp.–High Pres.*, vol. 21, pp. 219–224, 1989.
10. Ito, T., Kojima, N., Nagashima, A., Redetermination of the viscosity of molten NaCl at elevated temperatures, *Int. J. Thermophys.*, vol.10, no. 4, pp. 819–831, 1989.

11. Ito, Y., Minami, K., and Nagashima, A., Viscosity of liquid lithium by an oscillating cup viscometer in the temperature range 464–923 K, *Int. J. Thermophys.*, vol.10, no.1, pp. 173–182, 1989.
12. Mardolcar, U. V., and Niet de Castro, C. A., Measurement of High Temperature Thermal Conductivity of Liquids, Univ. Lisboa, 1990.

Session Summary: Comments on the Need for a Material Property Data Base

F. P. Incropera, Purdue University
A. Nagashima, Keio University

Throghout the seminar, many of the presenters pointed to the existence of uncertain material properties and to the need for establishing a reliable material property data base. In view of the many categories of materials currently in various stages of development and application, generation of this data base represents an enormous task. It was indicated, for example, that in laser material processing, assessments of material response required knowledge of the spectral and temperature dependence of the absorption coefficient, as well as the temperature dependence of thermophysical properties. In processes which involve transformation from a melt to a solid, such as crystal grouth, the casting of metals, and the molding of polymers, process simulation requires a complete range of fluid properties such as surface tension and expansion coefficient, as well as thermal conductivity, specific heat and viscosity. Such properties depend strongly on temperature and, in the case of multicomponent systems, on composition. In multicomponent systems there is an additional need to characterize morphological conditions in the mushy region and to establish relations between dendritic microstructures and exchange coefficients for momentum, energy, or species transfer between solid and liquid phases.

In pointing to the need for an improved material property data base, the presenters were motivated, either implicitly or explicitly, by the desire to mathematically simulate manufacturing processes involving fabrication or application of the material. However, another presenter remarked that the role of such models should be to infer important trends and to obtain an improved understanding of process physics. If that premise is accepted, the need to obtain accurate material properties is diminished. It is only when one wishes to obtain a reliable quantitative description of process details that the need for a reliable material property data base becomes significant.

We would argue that it is important to work towards developing numerically reliable prediction tools for material processes. Such tools are vital to establishing the influence of process variables, to identifying optimal operating conditions, and to implementation of process controls. Moreover, the explosive growth in computing power (CPU speeds and core memory sizes of approximately 20 GFLOPS and 2000 MB in 1990) is making it possible to simulate all pertinent physical features of a process at microscale, as well as macroscale, levels. What is clearly needed to gain the full benefits of such modeling capabilities is the availability of a reliable and comprehensive material property data base. In fact, it is clear that, at this point in time, the development of modeling and computational capabilities is far outstripping efforts to obtain material properties required for effectively implementing these capabilities. There is a clear and growing need for research activity whichd focuses on the properties of modern engineering materials. Measurement techniques must be refined; comprehensive data for important classes of materials must be acquired and critically assessed; and computerized data bases must be established to facilitate ready retrieval.

The session on physical properties consisted of three papers. One paper introduced a technologically new material consisting of a magnetic fluid for which property data are needed. A second paper described the measurement of thermophysical properties under microgravity conditions, and the third paper assessed the status of thermophysical property data of high temperature melts for which measuring techniques are not yet established. It was symbolic the session on thermophysical properties concluded the seminar.

Histrically, the need to obtain reliable thermophysical properties for a material has been identified after applications and related mathematical simulations have been proposed. However, this time delay in obtaining basic data is an impediment to the rapid development of new technologies. New initiatives must be undertaken to identify critical needs for material properties and measurement methods and to develop reliable property data bases for important classes of materials.

Index